Advances in Environmental Remote Sensing—Volume I

Advances in Environmental Remote Sensing—Volume I

Topic Editors

Zhengqiang Li
Zhongwei Huang
Chi Li
Kai Qin
Han Wang
Tianhe Wang
Jie Luo

MDPI

Basel • Beijing • Wuhan • Barcelona • Belgrade • Novi Sad • Cluj • Manchester

Topic Editors
Zhengqiang Li
Aerospace Information Research Institute, Chinese Academy of Sciences
Beijing
China

Zhongwei Huang
Lanzhou University
Lanzhou
China

Chi Li
Washington University in St. Louis
St. Louis
USA

Kai Qin
China University of Mining and Technology
Xuzhou
China

Han Wang
China University of Mining and Technology
Xuzhou
China

Tianhe Wang
Lanzhou University
Lanzhou
China

Jie Luo
Chinese Academy of Sciences
Beijing
China

Editorial Office
MDPI AG
Grosspeteranlage 5
4052 Basel, Switzerland

This is a reprint of the Topic, published open access by the journals *Atmosphere* (ISSN 2073-4433), *Geosciences* (ISSN 2076-3263), *International Journal of Environmental Research and Public Health* (ISSN 1660-4601) and *Remote Sensing* (ISSN 2072-4292), freely accessible at: https://www.mdpi.com/topics/environment_remote_sensing.

For citation purposes, cite each article independently as indicated on the article page online and as indicated below:

Lastname, A.A.; Lastname, B.B. Article Title. *Journal Name* **Year**, *Volume Number*, Page Range.

Volume I
ISBN 978-3-7258-1477-0 (Hbk)
ISBN 978-3-7258-1478-7 (PDF)
https://doi.org/10.3390/books978-3-7258-1478-7

Set
ISBN 978-3-7258-1475-6 (Hbk)
ISBN 978-3-7258-1476-3 (PDF)

Contents

About the Editors

Zhengqiang Li

Zhengqiang Li is a Research Fellow at the Aerospace Information Research Institute of the Chinese Academy of Sciences (CAS). He is the Director of the State Environment Protection Key Laboratory of Satellite Remote Sensing and the Dean of the School of Remote Sensing Science and Technology of Aerospace Information Technology University. Zhengqiang Li received his Ph.D in Optics from CAS in 2004 and worked at the University of Lille in France from 2004 to 2009. He engaged in research covering optical remote sensing, polarized satellite monitoring and the atmospheric environment, and participated in more than 30 research and development projects. He has published over 340 papers, more than 30 patents and 5 monographs, was selected as the winner of the National Outstanding Youth Fund of China and was in the Top 2% of Scientists Worldwide by Stanford University. Zhengqiang Li also serves as the President of the Atmospheric Environment Remote Sensing Society (AERSS) and as a national representative in the international organization of the Space for Climate Observatory (SCO).

Zhongwei Huang

Prof. Huang Zhongwei focuses on lidar remote sensing of the atmosphere, bioaerosol detection and its climate effects. He developed international first-class advanced laser atmospheric detection technologies integrating laser remote sensing, depolarization, Raman and fluorescence. The first lidar atmospheric remote sensing research center with a complete research system in Northwest China was established at Lanzhou University. He proposed a new polarization lidar retrieval method using aerosol microphysical properties, which significantly improved the accuracy of lidar identification of cloud and aerosol types. Furthermore, he proposed and organized a large-scale field observation experiment (DuBi) focusing on dust-bioaerosols in China, significantly bridging the gap in atmospheric microorganism research in certain regions of Northern China. He has conducted extensive and long-term studies pertaining to lidar network observations of aerosol–cloud interactions in Northwest China and has comprehensively analyzed the evolutionary characteristics of aerosol optical properties along dust transport paths. He first found a fast, long-range transport path of dust aerosols from East Asia to the Arctic region, which provided a novel perspective for studying the climate effect of dust aerosols, particularly on the Arctic climate. Based on active remote sensing techniques, he studied the physicochemical properties of dust originating from East Asia after its long-range transport to oceanic regions and observed the real-time evolution of dust aerosols to form water clouds for the first time. By exploring the principles of environmental adaptability, transportation and the dissemination of bioaerosols, he discovered the influence of ambient temperature and humidity on the spread of COVID-19, thereby making remarkable contributions to the advancement of atmospheric remote sensing research in China.

Chi Li

Chi Li is currently a staff scientist in the Department of Energy, Environmental and Chemical Engineering, Washington University, in St. Louis. He obtained a Ph.D in Atmospheric Science from Dalhousie University in 2018. He is broadly interested in the abundances, impacts and processes of various atmospheric compositions that affect the quality of air and the changing climate. His work involves using various observation platforms and modeling tools to unveil how atmospheric composition forms, changes and affects planetary health. Li has published over 40 journal articles and has an H-index of 21 (Web of Science). He has received several prestigious awards, including the Killam Predoctoral Scholarship and the William Leiper Memorial Scholarship. Li was chosen to participate in the 15th Atmospheric Chemistry Colloquium for Emerging Senior Scientists.

Kai Qin

Kai Qin, Ph.D., is a Professor and the Head of the Department of Remote Sensing and GIS at China University of Mining and Technology. He was a Visiting Scientist with the German Aerospace Center in 2016/2017 and a Visiting Professor with Yonsei University in 2022. His research interests include improving satellite retrievals of aerosol and trace gases, assessing air quality using satellite measurements, and estimating methane emissions. He has published around 90 peer-reviewed papers. He organized a series of conferences on the remote sensing of atmospheric environments. Kai Qin serves as Secretary-General of the Atmospheric Environment Remote Sensing Society, Secretary-General of the Commission on Digital Energy of the Chinese National Committee of the International Digital Earth Society (ISDE), a Board Member of *Remote Sensing*, an Editorial board member of *Journal of Remote Sensing* and more.

Han Wang

Han Wang is a master's supervisor, an associate professor at the School of Environment and Spatial Informatics of China University of Mining and Technology, a member of the International Atmospheric Remote Sensing Society (AERSS) and a member of the Remote Sensing and Environmental Science Alliance of Universities in Jiangsu, Hong Kong, and Macao. He graduated from the University of Chinese Academy of Sciences in 2015 with a Ph.D. in Optics. His main research focuses on aerosol inversion based on polarized remote sensing data, analyses of the polarized reflection characteristics of ground surfaces and related works. He refined the spectral relationship between the polarized reflectance of land surfaces and proposed an aerosol adaptive polarization inversion method (adaptive land–atmospheric decoupling algorithm, ALAD). Currently, he has published more than 20 related papers.

Tianhe Wang

Tianhe Wang is a professor at the College of Atmospheric Sciences, Lanzhou University, China. His research focuses on aerosol–cloud remote sensing, dust transport mechanisms, and aerosol–cloud–climate interactions. He has presided over five National Natural Science Foundations of China (NSFCs) and participated in a number of national-level projects, including the Second Tibetan Plateau Scientific Expedition and Research Program (STEP). He has published more than 60 papers in professional academic journals such as *Remote Sensing of Environment*. He has won first prize of the Natural Science of Gansu Province award (3/5, 2012) and second prize of the National Natural Science award (3/5, 2013). Some of his results were selected as research highlights in *Science of the Total Environment* (2021). He is the deputy secretary-general of the Atmospheric Environmental Remote Sensing Society (AERSS) and a reviewer of more than 20 international academic journals.

Jie Luo

Jie Luo obtained his Ph.D from the State Key Laboratory of Fire Science at the University of Science and Technology of China. He previously served as an assistant researcher at the Aerospace Information Research Institute of the Chinese Academy of Sciences and a senior research specialist at the Zhejiang Lab. Currently, he is a professor at Hangzhou Dianzi University, specializing in research on scattering modeling and the climate effects of absorbing aerosols, ocean satellite remote sensing, forest fire detection and climate effects. He has published over 20 academic papers in prestigious journals such as *Atmospheric Chemistry and Physics*, *Journal of Geophysical Research—Atmospheres*, and *Optics Express*. He has led the National Natural Science Foundation of China Youth Project and participated in national key research and development projects of the Ministry of Science and Technology, the National Natural Science Foundation of China and others. From 2019 to 2020, he visited Washington University in St. Louis, USA, for exchange and collaboration. He has served as a Guest Editor for journals like *Remote Sensing* and *Frontiers in Environmental Science* and he is a reviewer for multiple international academic journals including *Optics Express* and *Atmospheric Measurement Techniques*.

International Journal of
Environmental Research and Public Health

MDPI

Article

Construction and Optimization of an Ecological Network in the Yellow River Source Region Based on MSPA and MCR Modelling

Jia Liu [1,2], Jianjun Chen [1,3,*], Yanping Yang [1], Haotian You [1,3] and Xiaowen Han [1,3]

1 College of Geomatics and Geoinformation, Guilin University of Technology, Guilin 541004, China
2 Shenzhen Data Management Center of Planning and Natural Resources (Shenzhen Geospatial Information Center), Shenzhen 518040, China
3 Guangxi Key Laboratory of Spatial Information and Geomatics, Guilin University of Technology, Guilin 541004, China
* Correspondence: chenjj@glut.edu.cn

Abstract: The source region of the Yellow River (SRYR) is an important water conservation and farming area in China. Under the dual influence of the natural environment and external pressure, ecological patches in the region are becoming increasingly fragmented, and landscape connectivity is continuously declining, which directly affect the landscape patch pattern and SRYR sustainable development. In the SRYR, morphological spatial pattern analysis (MSPA) and landscape index methods were used to extract ecologically important sources. Based on the minimum cumulative resistance model (MCR), Linkage Mapper was used to generate a potential corridor, and then potential stepped stone patches were identified and extracted by the gravity model and betweenness centrality to build an optimal SRYR ecological network. The distribution of patches in the core area of the SRYR was fragmented, accounting for 80.53% of the total grassland area. The 10 ecological sources based on the landscape connectivity index and 15 important corridors identified based on the MCR model were mainly distributed in the central and eastern regions of the SRYR. Through betweenness centrality, 10 stepped stone patches were added, and 45 planned ecological corridors were obtained to optimize the SRYR ecological network and enhance east and west connectivity. Our research results can provide an important reference for the protection of the SRYR ecosystem, and have important guiding significance and practical value for ecological network construction in ecologically fragmented areas.

Keywords: source region of the Yellow River (SRYR); landscape connectivity; morphological spatial pattern analysis (MSPA); minimum cumulative resistance model (MCR); ecological network

Citation: Liu, J.; Chen, J.; Yang, Y.; You, H.; Han, X. Construction and Optimization of an Ecological Network in the Yellow River Source Region Based on MSPA and MCR Modelling. *Int. J. Environ. Res. Public Health* **2023**, *20*, 3724. https://doi.org/10.3390/ijerph20043724

Academic Editor: Deyong Yu

Received: 5 January 2023
Revised: 12 February 2023
Accepted: 17 February 2023
Published: 20 February 2023

1. Introduction

Building an ecological civilization is a "millennium plan" for sustainable development in China, where the ecological environment has always been a key concern [1]. With rapid social and economic development, the ecological environment is still deteriorating globally, large ecological patches that maintain ecosystem stability are gradually fragmenting the landscape, and patch connectivity is being reduced, thereby greatly hindering species migration and material energy spread, which are serious threats to ecosystem structure and function [2]. The SRYR is an important water conservation and livestock farming base in China, and includes numerous ecological patches. The President of China, Xi Jinping, once emphasized at a symposium on ecological protection and high-quality development of the Yellow River Basin: "The Yellow River Basin is an important ecological barrier and an important economic zone in China" [3]. In recent years, soil erosion, water and soil loss, human activities, rodent browsing, and other phenomena have plagued the SRYR. Consequently, the SRYR ecosystem structure has lost its inherent balance, and suffered functional decline and weakened recovery ability. Therefore, it is urgent to construct and optimize the SRYR ecological network, to scientifically and effectively promote patch connectivity. Additionally, ecological network construction and optimization are highly significant for maintaining ecological security [4], optimizing ecological patterns [5], and improving ecosystem quality [6].

Based on landscape ecology, island biogeography, and population theory, ecological networks comprehensively analyze the distribution and connection of ecological patches in space [7]. Since the 1990s, ecological network research has involved all ecosystem aspects, including energy flow, material cycles, information transmission, and ecological network structure and composition [8]. For example, Marc et al. [9] measured local, regional, and inter-sample network diversity (α-, γ-, and β-diversity) to describe how ecological interactions change over space and time. Isadora et al. [10] developed a spatial model that identifies and prioritizes riparian corridors to improve landscape connectivity. Ecological network construction simplifies ecological patches in a region into ecological nodes to build an ecological corridor and ecological network. Presently, ecological network construction methods include models such as graph theory [11], landscape suitability [12], minimum consumption distance [13], current theory [14], and thermodynamic law [15]. Commonly used software include ConeforSensinode, Circulitscape, Guidos, Zzonation, and Marxan. Among them, the commonly used ecological network construction method is the least cumulative resistance model (MCR). The MCR model construction is mainly about source selection and resistance surface construction.

For source selection, considering the impact of habitat quality and human activities, Gao et al. [16] extracted ecological sources based on ecosystem service function and ecological sensitivity to construct ecological resistance surface. Yu et al. [17] selected Dengkou County, a typical ecologically fragile area, as an ecological source area and improved ecological network stability by optimizing the spatial layout of ecological nodes. However, most current studies selected scenic parks or large nature reserves with good habitat patches as ecological sources, although this approach is somewhat subjective. In recent years, a morphological spatial pattern analysis (MSPA) method focusing on structural connections has gradually been integrated into ecological network construction and analysis. Based on Ritters research, Vogt et al. [18] combined the convolution algorithm with the mathematical morphological mapping algorithm proposed by Soille [19], and proposed a new method for a landscape connectivity analysis based on the principles of expansion, corrosion, and open–close operation, i.e., morphological spatial pattern analysis. This algorithm can divide the binary image into seven non-overlapping categories (namely, core area, bridge area, loop, branch, edge area, pore, and island patch). Then, the landscape types that are important to maintain patch connectivity are identified, which increases the scientific rigor of the selection of ecological sources and ecological corridors. For example, using the methods of morphological spatial pattern analysis (MSPA) and landscape connectivity, Xiao et al. [20] combined the graphic theory and quantitative analysis to evaluate the spatio-temporal pattern and network connectivity changes of ecological networks in Zhengzhou. Yi et al. [14], based on a morphological spatial pattern analysis and circuit theory, focused on the importance of human activities in tropical southwest China to the optimization of the Asian elephant ecological network.

Construction of the resistance surface represents the degree of interference encountered when the target species moves between patches, and it will seriously influence the ecological corridor and ecological network research outputs [21]. Presently, scholars constructed the resistance surface based on various methods, such as expert scoring, entropy weighting, landscape development intensity index, and biological behavior resistance estimation. Based on the TOPSIS model of entropy weight, Li [22] constructed an evaluation model of the eco-geological environmental carrying capacity. Li et al. [23] took the Sichuan-Yunnan ecological barrier as a typical national complex ecological barrier area, and proposed to construct a sustainable Sichuan-Yunnan ecological barrier based on the cycle theory and future land growth changes. Some scholars modified the resistance surface according to the actual situation, to scientifically judge and simulate the potential ecological corridor.

Yu Gao [24] proposed a landscape resistance surface construction method based on a habitat quality assessment, and compared it with a resistance surface constructed using the entropy coefficient and expert scoring methods, and found it more suitable for ecological network research in the scattered Changzhou landscape. However, due to differences in land nutrients and environmental elevations, there may also be differences between the same land use types. Currently, most studies are based on professional knowledge and overall rating of some land use types to construct the landscape resistance surface, which leads to heavy dependence of the landscape resistance surface on the grade coefficient. The MCR model can solve this problem well. Moreover, in general, the combination of MSPA and MCR has been applied to ecological networks in urban landscapes in the central and eastern regions of China, but it has rarely been used in the field of natural landscapes and biological protection in the northwest inland areas.

Although MSPA can identify patches that are important for maintaining landscape connectivity, it still requires assistance from the overall connectivity index (IIC), possibility connectivity index (PC), and equivalent connection proposed by Pascual-Hortal et al. [25]. In addition to patch abundance and spatial arrangement, these indices combine the dispersal specificity of plant habitats. Wu et al. [26] took the Guangdong-Hong Kong-Macao Greater Bay Area as an example, and found that the overall ecological connectivity of ecological networks at all scales showed a gradual upward trend, and the overall connectivity index (IIC) and the possible connectivity index (PC) gradually increased with the increase in the maximum dispersal distance of species. Javier Babi Almenar [27] integrated a landscape index analysis, including the overall connectivity index (IIC), probable connectivity index (PC), and equivalent connectivity index (EC) to show that from 1999 to 2007, habitat fragmentation and loss increased ecological connectivity in Luxembourg, Western Europe. Although the MCR model can judge and simulate the potential ecological corridor by constructing the regional cumulative resistance surface, for corridor relative importance, it is necessary to analyze the interaction strength between patches through a gravity model [28]. This method mainly combines the network structure index and gravity model to obtain the important patch node rank classification and potential corridor suitability analysis through quantitative calculation, to make the research results more consistent with ecological principles. L. Thiault [29] evaluated an ecological network of marine-protected areas established on Moorea and French Polynesia through a progressive BACIPS method. For the ecological network evaluation index, scholars also measured the ecological service value based on the probability of occurrence of a certain species in ecological patches [30]. However, this method lacks consideration of the spatial relationship between landscape ecological elements and is therefore unsuitable.

In summary, this paper took the SRYR alpine grassland as the research object, and based on the MSPA method, identified and extracted the core area landscape type with the best ecological function in the research area. According to the overall connectivity (IIC), probable connectivity (PC), and patch importance (dPC) in the landscape index, core area patches were quantitatively evaluated, to select the ecological sources. The least-LCPs method was used to generate the ecological corridor through the MCR model, and patch interaction intensity was determined based on the gravity model. Then, through betweenness centrality, we identified patches with a better mediating effect as stepped stones to identify potential corridors, and to build the SRYR ecological network. Our research results can provide a basis for the construction and planning of the SRYR ecological network, and also have a certain reference value for ecosystem protection in other regions.

2. Materials and Methods

2.1. Study Area

The SRYR, as one of the sources of three rivers, is an important water conservation area on the Qinghai-Tibet plateau [31]. It is located at N 33°56′~35°51′ and E 95°55′~98°40′, with an altitude of 4200 m to 5266 m and a total area of 12.54×104 km^2. The SRYR is extremely rich in grassland resources, covering about 80% of the source area, and is one of the most important livestock farming bases on the Qinghai-Tibet plateau [32]. Ecosystem stability in this region could guarantee the ecological security of China and even East Asia.

2.2. Data Sources

The land use data in this study were obtained from the 30 m spatial resolution land cover data in 2020 from the Resources and Environmental Data Center, Chinese Academy of Sciences (https://www.resdc.cn/ (accessed on 5 December 2021)) (Figure 1). The land use types in the SRYR include grassland, construction land, cultivated land, shrub, wetland, forest land, ice and snow, water area, and bare land. The vegetation cover data (NDVI spatial distribution map) were obtained from 2020 Landsat OLI downloaded from the National Aeronautics and Space Administration (NASA) (https://www.nasa.gov/ (accessed on 2 January 2021)). And the data were preprocessed by ENV5.1 released by Exelis Visual Information Solutions in Colorado, USA, and ArcGIS 10.8 released by the Environmental Systems Institute in RedLands, California, USA. Elevation data were obtained from Geospatial Data Cloud (https://www.gscloud.cn/ (accessed on 15 March 2022)). The road data were obtained from the National Tibetan Plateau Scientific Data Center (http://data.tpdc.ac.cn/ (accessed on 20 March 2022)). The Arctic 1: 1 million road data set (2014) is tailored, and the data contain two types, namely, 5 main roads (RMR3) and 321 branch roads (ROR3).

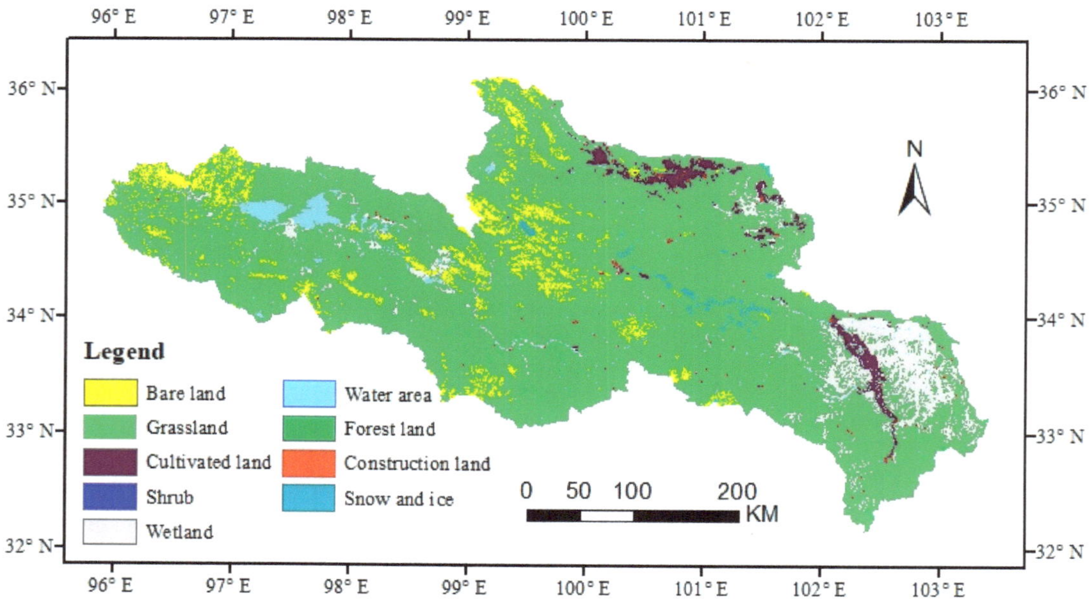

Figure 1. Spatial distribution map of land use types in the SRYR.

2.3. Research Methods

2.3.1. Ecological Source Extraction Based on the MSPA Method

- MSPA method

The ecological network is composed of a "source" and an ecological corridor connected to the source. Generally, the "source" is selected as a patch with a larger area in the landscape, and the MSPA method is used to identify landscape connectivity from the pixel level. Important ecological patches (such as source areas, corridors) can more accurately classify the spatial pattern of raster images in a functional structure, thereby increasing the scientific nature of ecological source areas and ecological corridor selection [33]. Firstly, based on the land use-type data of the SRYR (Figure 1), the grassland of the nine types was set as the foreground of the MSPA, and the other types were set as the background of the MSPA. At the same time, the data were converted into binary raster files in TIFF format. Secondly, a landscape pattern analysis of raster data was conducted with the eight-neighborhood analysis method using Guidos Toolbox analysis software to obtain seven landscape types with different landscape functions (Table 1) [19]. Finally, according to the MSPA classification of landscape types, the core area which plays an important role in maintaining the connectivity of the regional landscape was determined as the basis for selecting ecological source patches.

Table 1. Definition of MSPA landscape types and corresponding ecological representations.

Landscape Type	Definition	Ecological Elements and Characterization
Core area	Foreground pixels with background pixels larger than the set parameters	Ecological source patches, high vegetation coverage, can provide a larger habitat for species, and has important significance for biodiversity protection
Connecting bridge	Linear pixels connecting at least 2 core areas	It has the characteristics of an ecological corridor, which is conducive to species migration and connection of the domestic landscape. The greater its number, the better the connectivity between the patches
Marginal zone	Refers to the boundary between the core area and the external background pixels, which is linear	Located on the edge of the core area, it can reduce the impact of the external environment and human interference
Feeder	Linear pixels connecting the boundary (edges and pores) or corridors (circles and bridges) on one side only, and foreground pixels on the other side	Represents the most marginal area where the green landscape pixels communicate material energy
Ring road	Linear pixels connected to the same core area	Shortcut for material energy exchange within the core area
Isolated island	Small and isolated area	Less organic matter exchange and flow with the outside world, mostly small green spaces in cities or rural areas
Gap	Transition area between core area and non-green landscape patches	With edge effect, it can play a role in the peripheral edge of the area that hinders species movement in the core area

- Landscape connectivity index

To maintain regional ecosystem stability and protect biodiversity, a landscape connectivity index was introduced. The landscape connectivity assessment evaluates species migration between patches, material energy exchange, and the biological movement ability of information flow [34]. Among the many landscape connectivity evaluation methods, graph theory can simultaneously quantify structural and functional characteristics [35]. Presently, based on a graph theory-based connectivity evaluation, researchers often use three landscape indices (overall connectivity (IIC), possible connectivity (PC), and patch importance (dPC)) to measure important landscape pattern and function indicators, which can better reflect the connectivity level between core patches in the area.

$$IIC = \frac{\sum_{i=1}^{n}\sum_{j=1}^{n}\frac{a_i a_j}{1+nl_{ij}}}{A_L^2} \tag{1}$$

$$PC = \frac{\sum_{i=1}^{n}\sum_{j=1}^{n} a_i \cdot a_j \cdot p_{ij}^*}{A_L^2} \tag{2}$$

$$dPC = \frac{PC - PC_{remove}}{PC} \times 100\% \tag{3}$$

In Equations (1)–(3), n is the total number of patches in the area; a_i and a_j are the areas of patches i and j, respectively; nl_{ij} is the connection between patch i and patch j, which is patch i. The maximum product of all path probabilities between blocks j, A_L is the total area of the landscape in the study area. IIC represents the connectivity index value ($0 \leq IIC \leq 1$). If $IIC = 0$, there is no connection between ecological patches; if $IIC = 1$, the entire landscape is a habitat patch. PC represents the possible patch connectivity index in the study area landscape; after PC_{remove} removes patch i from the landscape, the connectivity index value of the landscape ($0 \leq PC \leq 1$), the greater the PC value, the greater the patch connectivity. In this study, Conefor Sensinode 2.6 software was selected, and the connection distance threshold was set at 5000 m with a connection probability of 0.5, and EdgeWidth was 1. The landscape connection degree of the core patch obtained after the MSPA in the SRYR was evaluated by IIC, PC, and dPC landscape indexes. Moreover, the 10 core patches with the highest dPC value were used as ecological sources for the development and reproduction of biological species.

2.3.2. Ecological Resistance Surface Construction

Biological species migration from one ecological source to another requires overcoming different resistances to carry out the material exchange, energy flow, and gene exchange [36]. Since roads strongly impact ecological patches, by dividing originally large ecological patches and leading to ecosystem disorder in the region, there were two road factors in the resistance indicator selection. Based on the MSPA and landscape connectivity evaluations, we selected six resistance factors, including elevation, aspect, land use type, vegetation coverage, distance from main roads, and distance from branch roads in combination with the principles of quantification and select ability. The impact of each resistance factor on the ecological source area was divided into five resistance scores, and the corresponding weight of each resistance factor (Tables 2 and 3) was determined according to the SPSS principal component analysis method, so that the "comprehensive weighted index sum method" was used to build the minimum cumulative resistance surface under ArcGIS support.

Table 2. SPSS principal component analysis weight determination process.

Ecological Resistance		First Principal Component	Second Principal Component
Number of roads	Distance from branch road	0.93	0.27
	Distance from main road	0.89	0.44
	Slope	0.82	0.57
	Land use type	0.77	0.63
	Vegetation coverage	0.68	0.44
	Elevation	0.01	0.98
Characteristic root of principal component		3.39	2.13
Coefficients in linear combinations	Distance from branch road	0.50	0.19
	Distance from main road	0.48	0.30
	Slope	0.45	0.39
	Land use type	0.42	0.43
	Vegetation coverage	0.37	0.30
	Elevation	0.01	0.67
Variance of principal components		56.50	35.57
Coefficients in the integrated scoring model	Distance from branch road	0.23	
	Distance from main road	0.26	
	Slope	0.21	
	Land use type	0.06	
	Vegetation coverage	0.08	
	Elevation	0.16	

Table 3. Rating and weighting of resistance factors.

Resistance Factor	Grading Index	Resistance Value	Weight
Elevation	0–3500	1	0.16
	3500–4000	2	
	4000–4500	3	
	4500–5000	4	
	>5000	5	
Slope	South	1	0.08
	Southwest, Southeast	2	
	East, West	3	
	Northwest, Northeast	4	
	North	5	
Land use type	Cultivated land, shrub, grassland, forest land	1	0.06
	Water area	2	
	Wetland	3	
	Construction land	4	
	Ice and snow	5	
Vegetation coverage	80–100%	1	0.21
	60–80%	2	
	40–60%	3	
	20–40%	4	
	<20%	5	
Distance from main road (m)	1300–2000	1	0.26
	1000–1300	2	
	700–1000	3	
	400–700	4	
	<400	5	
Distance from branch road (m)	400–700	1	0.23
	300–400	2	
	200–300	3	
	100–200	4	
	<100	5	

2.3.3. Ecological Network Construction Based on the MCR Model

- Ecological corridor extraction based on the MCR model

The basic principle of the minimum cumulative resistance model is the "source-sink" theory. By calculating the minimum cumulative resistance distance between the source and the target to determine the best path for species migration and diffusion, it can avoid external interference to a minimum. It reflects the possibility and tendency for material energy and biological species movement among ecological patches in the landscape [29]. The simplicity of its construction, the extension of the elements, and the wide range of applications, have led to its wide use. The minimum cumulative resistance model (*MCR*) was modified by multiple experts to obtain the following formula:

$$MCR = f \sum_{i=n}^{i=m} \left(D_{ij} \times R_i\right)_{min} \tag{4}$$

In Equation (4), D_{ij} represents the spatial distance from the source point i to the space unit j, and R_i represents the resistance coefficient of the space unit i.

In this study, based on the source and resistance surface obtained by the previous method, we used the Linkage Mapper tool, based on the principle of minimum path, to automatically draw the corridor of ecological patches and determine the priority protection level, to establish the network and map connection, and gradually analyze the landscape composition of its potential corridor network.

- Patch interaction based on a gravity model

In this study, the interaction matrix among the eight ecological foci was constructed using a gravity model, and the interaction intensity between the patches was quantitatively evaluated, to scientifically combine interaction intensity with the actual research area situation. The situation was combined to construct an ecological network map in line with the SRYR. The gravity model formula is as follows:

$$G_{ij} = \frac{N_i N_j}{D_{ij}^2} = \frac{\left[\frac{1}{P_i} \times \ln(S_i)\right]\left[\frac{1}{P_j} \times \ln(S_j)\right]}{\left(\frac{L_{ij}}{L_{max}}\right)^2} = \frac{L_{max}^2 \ln(S_i S_j)}{L_{ij}^2 P_i P_j} \quad (5)$$

where G_{ij} is the interaction strength between patch i and patch j; N_i and N_j are the weight coefficients of patch i and patch j, respectively; and D_{ij} is the standardized resistance value of the potential corridor between patch i and patch j. P_i is the overall resistance value of patch i; S_i is the area of patch i; L_{ij} is the cumulative resistance value of the potential corridor between patch i and patch j; and L_{max} is the maximum resistance value of all corridors in the research area.

- Selection of stepped stones based on betweenness centrality

Betweenness centrality is a concept proposed by American sociologist Professor Linton C. Freeman [37]. It refers to the ratio of the shortest path that passes through a certain point and connects the two points to the total number of shortest path lines between the two points in the network, and it is a main indicator to measure the importance of nodes in the graph. In this study, the betweenness centrality module in the Matrix Green analysis tool was used to calculate Green space patches with good intermediary function in the ecological network, and 10 Green space patches were identified as stepped stones, according to their scores, to construct the planned ecological network.

$$G_i^B = \frac{1}{(N-1)(N-2)} \sum_{j=1;k=1;j\neq k\neq i}^{N} \frac{n_{ik}(i)}{n_{ik}} \quad (6)$$

where N is the number of nodes in the network; n_{ik} is the number of shortest paths between nodes j and k; $n_{ik}(i)$ is the number of shortest paths between two nodes j and k passing through node i. In the ecological network, the higher the betweenness centrality nodes, the more obvious the role as a hub in the network, which can be used as an important stepped stone.

3. Results

3.1. Landscape Pattern Analysis Based on the MSPA Method

The SRYR core area of the landscape type was 99,560.85 km^2, accounting for 80.53% of the total area of grassland, and was distributed mainly in the northeast of the study area. However, the distribution of the core areas in the west was more fragmented. The edge area and perforation were mainly distributed between the core area and the background, with a relatively large area of 1725.99 km^2 and 2283.26 km^2, respectively. The three landscape types of loop, branch, and bridge were mainly distributed in the western region, with an area of 396.59 km^2, 356.52 km^2, and 213.66 km^2, respectively. The landscape area of islet was the smallest at 108.83 km^2, accounting for only 0.09% of the total area (Figure 2 and Table 4).

3.2. Research Area Landscape Connectivity Evaluation

Ten core areas with high dPC values were selected as ecological sources, and the results showed that there was a big difference in dPC values among different ecological sources. The number of ecological sources in the west was far less than that in the east, and the northern and southern regions lacked the distribution of ecological sources. Patch 8 had the largest dPC value of 542.80 km^2, dPC was 83.17, and dIIC was 82.65 (Table 5). It was located in the east of the SRYR and mainly distributed with cultivated land patches. Secondly, the dPC values of patches 9, 4, 7, 10, 6, and 5 distributed in the east decreased successively, mainly distributed in wetland patches. Patches 1, 2, and 3 located in the west of the SRYR had low dPC values, and their dIIC and dPC values were all less than 1. They were mainly regional small, fragmented patches, and mainly distributed with bare patches.

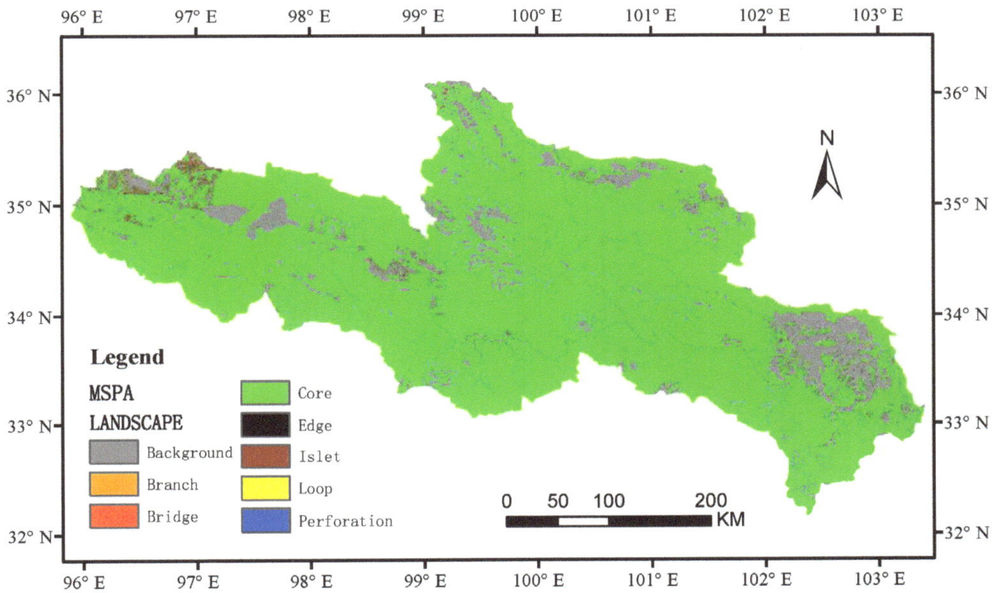

Figure 2. Spatial distribution of MSPA analysis of land cover in the SRYR.

Table 4. MSPA classification of land cover in source region of Yellow River.

Landscape Type	Area (km^2)	Total Area of Grassland Landscape (%)	Total Area (%)
Core	99,560.85	95.14%	80.53%
Bridge	356.52	0.34%	0.29%
Edge	1725.99	1.65%	1.40%
Branch	213.66	0.2%	0.17%
Loop	396.59	0.38%	0.32%
Islet	108.83	0.1%	0.09%
Perforation	2283.2568	2.18%	1.85%

Table 5. Core area ranking based on landscape connectivity.

Serial Number	Patch Number	dPC	dIIC	Area/km^2
1	42,005	83.16607	82.64966	542.80
2	38,907	14.47701	9.222232	60.57
3	42,452	11.74287	11.68335	198.64
4	41,491	5.056864	0.201331	28.24
5	42,872	2.993351	2.024486	34.42
6	40,841	1.177503	0.147397	24.17
7	40,157	0.671658	0.115662	21.41
8	32,221	0.529326	0.595473	48.58
9	15,241	0.489112	0.558735	47.05
10	29,832	0.439163	0.492475	44.18

3.3. Ecological Network Construction Based on the MCR Model

The minimum cumulative resistance of the ecological network in the SRYR decreased from west to east. The northwestern region had the highest resistance, with a resistance value of 4.50, mainly distributed in ice land, bare land, and the water area. The southeast had the least resistance, with a resistance value of 0, mainly distributed in cultivated land and wetland (Figure 3). The cost distance values of the 10 ecological sources expanded from the source region to the source region boundary of the Yellow River and gradually increased, with the maximum consumption distance value of 619,734 and the minimum value of 0. In total, 45 potential ecological corridors were identified based on the MCR model. At the same time, 15 important corridors were obtained by the gravity model, which were mainly distributed in the east and less in the west (Figure 4).

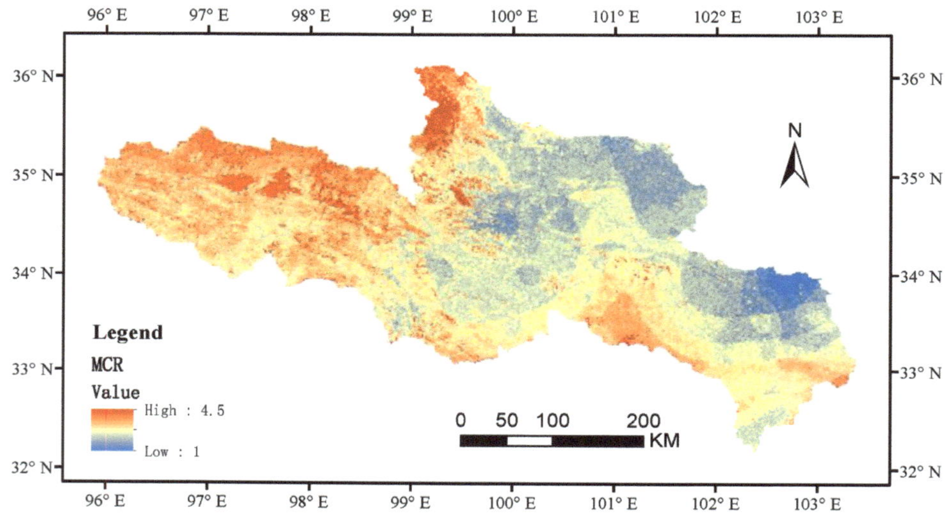

Figure 3. Minimum cumulative resistance surface in the SRYR.

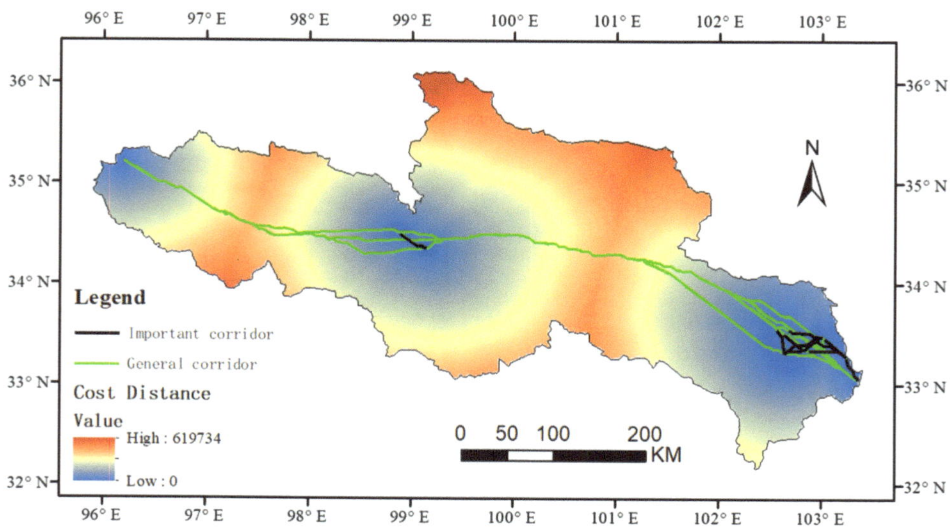

Figure 4. Corridor grade spatial distribution in the SRYR.

The interaction matrix between patches in the ecological source region obtained from the gravity model shows that the strength of the interaction between patches in the western region was greater than that in the eastern region. The interaction intensity between patches 5 and 6 was the highest, with a value of 371,577.6 (Table 6). The two patches covered adjacent wetland patches, and the landscape connectivity was the strongest. Secondly, the intensity of interactions between patches 6, 7, and 8 was higher, and the values were 173,416.3 and 155,876.7, respectively. The interaction intensity between patch 1 with a large amount of bare land in the west and other ecological source patches was relatively small, indicating that the landscape connectivity between the bare land patch and the eastern ecological source patch was poor.

Table 6. Patch interaction matrix based on the gravity model.

Patch Number	1	2	3	4	5	6	7	8	9	10
1	0.00	323.93	252.76	77.68	49.61	65.76	62.34	70.09	46.86	46.56
2	0.00	0.00	26,297.31	287.90	167.93	224.31	212.86	241.46	148.76	140.45
3	0.00	0.00	0.00	314.99	180.83	241.86	229.56	260.80	158.47	148.40
4	0.00	0.00	0.00	0.00	15,168.88	25,857.77	31,132.04	40,773.64	5665.64	3260.00
5	0.00	0.00	0.00	0.00	0.00	371,577.62	50,788.60	20,412.32	19,742.59	5238.65
6	0.00	0.00	0.00	0.00	0.00	0.00	155,876.68	173,416.31	22,879.16	7387.04
7	0.00	0.00	0.00	0.00	0.00	0.00	0.00	113,138.58	15,454.03	6547.94
8	0.00	0.00	0.00	0.00	0.00	0.00	0.00	0.00	0.00	4693.51
9	0.00	0.00	0.00	0.00	0.00	0.00	0.00	0.00	0.00	19,887.67
10	0.00	0.00	0.00	0.00	0.00	0.00	0.00	0.00	0.00	0.00

3.4. Ecological Network Construction and Optimization

According to the overall ecological network constructed, the distribution of ecological sources and ecological corridors in the SRYR is unbalanced on the whole, and the landscape connectivity between the eastern and western regions is poor. Therefore, in order to maintain the balance of the ecological network system, we optimized the ecological network of the SRYR by adding "stepped stone" patches. Among the patches in the core area, 10 green patches with higher scores of betweenness centrality were selected as stepped stone patches (Table 7). Further, one hundred ninety planned ecological corridors were constructed by combining ten ecological sources and selected stepped stone patches, among which eight were important corridors. Finally, the optimized land cover ecological network planning map of the SRYR was obtained (Figure 5).

Table 7. 10 "Stepped stone" patches based on betweenness centrality.

Serial Number	Patch Number	BC	Area/km^2
1	1211	1.54	6.67
2	3082	19.50	84.67
3	19,730	21.78	94.54
4	21,979	29.97	130.12
5	25,172	8.18	35.52
6	26,247	19.69	85.46
7	26,760	12.70	55.15
8	34,533	1.70	7.36
9	37,005	30.41	132.03
10	40,752	6.29	27.33

In addition, the network closure index (α index), network connectivity index (β index), and network connectivity rate index (γ index) in the network analysis method [38] were used to calculate the ecological network quality of the study area before and after planning. It was found that each index was higher than the value before planning (Table 8). The results showed that the planned ecological network significantly improved the connectivity level of ecological patches in the study area and increased the effectiveness of network connectivity.

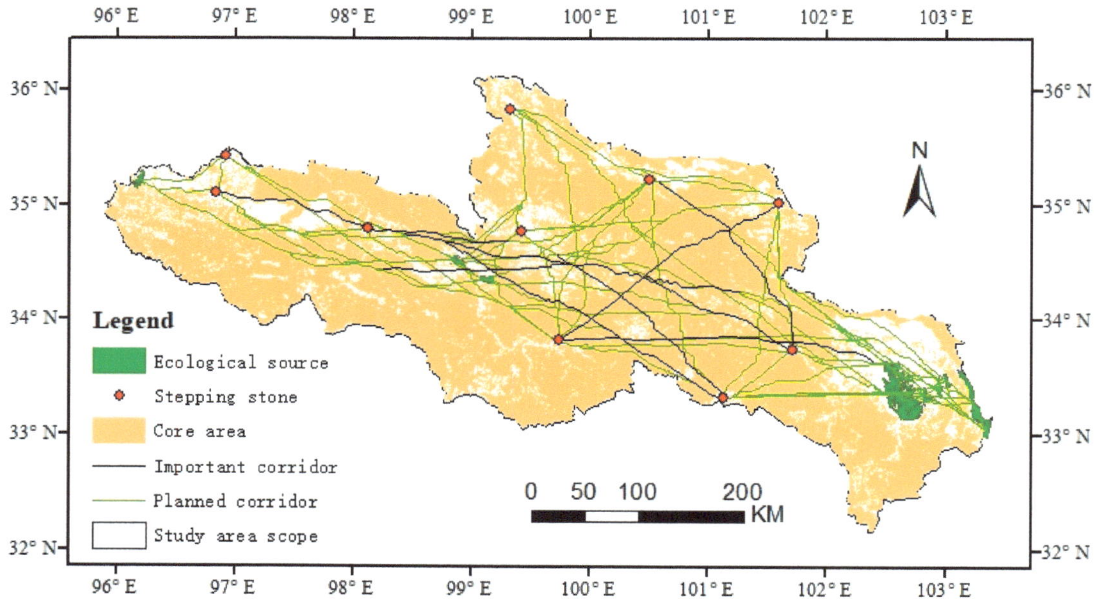

Figure 5. Land cover ecological network planning map of the SRYR.

Table 8. Ecological network quality before and after planning based on network structure indices.

Network Structure Index	Before Planning	After Planning
Network closure index (α index)	0.73	1.72
Line point rate index (β index)	2.00	4.22
Network connectivity index (γ index)	53.33	2723.33

4. Discussion

4.1. Landscape Pattern Analysis Based on the MSPA Method

According to the results of the landscape pattern analysis based on the MSPA method, the patches in the northeastern core area of the study area were mostly large patches with good spatial connectivity, while the patches in the west core areas of the study area were relatively fragmented, which hindered the material exchange of biological species to a certain extent. The edge area was 1725.99 km^2, accounting for 1.65% of the total grassland area, and the perforation area accounted for 2.18% of the total grassland area. The area of the edge area and perforation area was only smaller than that of the core area, indicating that the grassland landscape in the study area had a better edge effect, which could reduce the interference brought by external factors. As a way of animal migration within the patch, the loop was conducive to species migration within the same patch, accounting for 0.38% of the total grassland area. As a structural corridor for material exchange and energy flow in the process of interspecies migration in the ecological network, bridges accounted for 0.34% of the total area of grassland. Branch represented the interrupted ecological corridors in the ecological network, and had certain connectivity in the study area, accounting for 0.2% of the total grassland area. As an isolated grassland patch, the islet patch could be used as a stepped stone for organisms. Its area was small, accounting for 0.1% of the total grassland area.

When performing a landscape MSPA analysis, setting the research scale and edge width has a greater impact on the results [39]. When setting the study scale, increasing the size of the image grid will result in the disappearance of small landscape elements or their conversion to the non-core MSPA category [40]. Setting the edge width represents the size range in which the patch produces edge effects. The edge effect is an important ecological process in nature reserve function design, which is closely related to species habitat protection, community dynamics, ecological restoration, and so on [41]. In this study, we set the edge width to 1 by default. However, the edge effect is specific and complex, and its width varies with different landscape areas, landscape types, and patch shapes. Therefore, the width of the edge effect set in this study may not be suitable for some species. When setting the influence range of the edge effect, it is necessary to consider the protection object and the shape and suitability of the study area landscape [42]. Wickham [43] analyzed the green infrastructure in various states in the USA based on MSPA to explore the effects of edge effects and neighborhood rules on the spatial and temporal pattern of green infrastructure. Jonathan Phillips [44], for the North Carolina coastal plain, identified three edge effect types, and found that their effects might be related to the unique geomorphologic control along the boundary, and within boundary resistance differences. Therefore, the scale effect and edge effect of MSPA should be further compared and analyzed, so as to explore the influence of the edge effect on the construction of an ecological network.

4.2. Ecological Network Construction Based on MCR Method

4.2.1. Landscape Connectivity Analysis

According to the results of the landscape connectivity evaluation in the study area, 10 ecological sources were selected in the study area according to the value of patch importance (dPC), and the larger the value of dPC, the better the patch connectivity. On the whole, the distribution of ecological source areas was extremely uneven. The ecological source areas were mainly located in the east, which had good natural conditions of a high vegetation coverage rate, providing a large ecological source area for the SRYR, which was more suitable for species migration and material and energy exchange, and more conducive to species survival to a certain extent. However, in the western region, vegetation coverage was low, a large number of alpine grasslands were distributed, and ecological patches with good ecological functions were lacking, so there was no distribution of the ecological source, resulting in poor overall connectivity and serious east-west faults in the study area. Patch 8 in the east, as the largest patch of dPC, indicated that the large area of swamp had an impact on the overall ecological network connectivity level of patches, and had more ecological functions than other patches in the study area. This is more conducive to the protection of biodiversity and a greater degree of rich species diversity. Therefore, in the future conservation of ecological diversity, priority should be given to the protection of large ecological patches. The patches in other ecological sources were mainly regional, small, fragmented patches with relatively small dPC value and poor landscape connectivity. Meanwhile, it is necessary to construct a foot patch in the western and central regions to strengthen the connectivity between the landscapes in the study area, maintain the balance of the ecosystem and the value of ecological services, construct an ecological network in the study area, and focus on protecting the patches with poor connectivity, so as to improve the habitat suitability and landscape connectivity.

When landscape indices dPC and dIIC were used to calculate landscape patch connectivity, the connection distance threshold must be set. If the distance between patches is greater than the threshold, the patches are considered disconnected. Moreover, setting the connection distance threshold requires considering species diffusion distance, as it often differs between species. The threshold distance used in this paper was 5000 m, and the connected probability was 0.5. Devi [45] proposed a graph theory method to determine the optimal threshold distance for forest patches in a potential connectivity alternative for tropical deciduous forests in the Eastern Ghats of India (optimum threshold 250 m). Louise Geldenhuys [46] reported good within-patch connectivity when the grassland community spreading distance in Mpumalanga was between 50 and 1000 m, and 99.6% of the total habitat area was connected with a single patch at a 1000 m threshold distance. Therefore, setting the connection distance threshold has an important effect on quantifying the structural and functional connectivity of plaques [47]. Landscape connectivity will gradually improve with the increase in diffusion distance, and the feasibility of the results needs to be further verified in subsequent studies. Ultimately, landscape connectivity can be used as an important indicator to measure landscape pattern and function, and provide an important basis for ecological protection. Catia Matos [48], using the graph theory method and incorporating the landscape connectivity model, studied the widespread but rapidly diminishing pond amphibian Triturus cristatus, and the results were critical for predicting the impacts on its migration and dispersal. Santiago Saura [49] confirmed that graphical structure and habitat availability metrics can better analyze regional landscape connectivity for various forest habitats in Lleida (northeastern Spain). This method can be adapted to map different levels of ecological and spatial details, while still maintaining a coherent framework for identifying key elements in the landscape network.

4.2.2. Ecological Network Construction

According to the results of ecological network construction based on the MCR model, the connectivity of the central and eastern parts of the study area was better, and the ecological corridors were more dense, which is conducive to species migration between patches. The intensity of the interaction between patches indicates the importance of patch connectivity and the importance of the corridor between patches. The patches of ecological source areas in the eastern region had greater interaction intensity, closer distance, and a relatively large area, and the exchange and propagation of species energy were simpler and more extensive. Therefore, in landscape planning, it is necessary to strictly control and protect this type of ecological corridor, ensure connectivity between patches, and avoid damage due to natural disasters or external disturbances. The interaction intensity between patch 1 and other patches in the west was the smallest, indicating that patch 1 and other patches had greater landscape resistance, which hindered the migration of species. In future ecosystem planning, the corridor connection between the two patches should be increased to improve the habitat suitability of the corridor. However, from the overall view of the ecological network constructed, the eastern and western parts of the study area were not connected by corridors, and the ecological network was not perfect. Therefore, in order to maintain the balance of the ecological network system, it is particularly important to optimize the ecological network of the SRYR by planning and designing new ecological sources and ecological corridors in the study area.

As an important species source and habitat, core patch is an important functional node in constructing the ecological network. Therefore, on the basis of protecting the core patches, we should reasonably plan the "stepped stone" patches to build a bridge of material and energy exchange between the eastern and western regions, and enhance the overall connectivity of the ecological network in the SRYR. Increasing the number of "stepped stones" and decreasing the distance between "stepped stones" can effectively improve the survival rate of species during migration [50]. In this study, 145 additional planned ecological corridors were obtained by adding stepped stone patches, thus constructing the overall ecological network planning diagram of the SRYR. Among them, important corridors were mainly concentrated in the middle of the SRYR, connecting the main ecological sources and having great significance for biodiversity protection, so they are the key areas to be protected in the ecological planning. The planning corridor was optimized on the original general corridor to better communicate between the eastern and western regions and optimize the overall structure of the ecological network. The planned ecological network significantly improved the connectivity level of ecological patches in the study area and increased the effectiveness of the network connection.

4.3. Limitations and Future Research Directions

In this study, based on the principles of landscape ecology, MSPA and MCR methods were used to construct and optimize the ecological network in the SRYR, which provided an important indication for ecosystem protection. However, there are still some limitations. Firstly, when constructing the minimum cumulative resistance surface, only elevation, vegetation coverage, road, and land use type were selected as resistance factors, without considering the impact of objective factors such as human factors on the ecological source area. Secondly, in the analysis of landscape connectivity, there was a lack of research on the impact of scale effects such as edge width and distance threshold in the research results. Thirdly, due to the large basin area of the SRYR, complex landforms are formed under the influence of internal precipitation, glaciers, evapotranspiration, and wind. Therefore, the applicability of the research results still needs to be further discussed. Finally, in future studies, the time scale should be expanded to study the changes of the land cover ecological network in the SRYR in a long time series.

5. Conclusions

In this paper, the SRYR is taken as the research area. From the perspective of ecological landscape connectivity, the ecological source region is identified based on the MSPA method, and the land cover ecological network of the SRYR is constructed and optimized by combining the MCR model. The conclusions are as follows.

(1) The core landscape area of the SRYR was 99,560.85 km^2, accounting for 80.53% of the total grassland area, which was mainly distributed in the northeast region, with relatively large marginal areas and void patches. Ring roads, branch lines, and connecting bridges were mainly distributed in the western region. The island had the smallest landscape area.

(2) The dPC values of different ecological sources in the SRYR were significantly different. The number of ecological sources in the west was much less than that in the east, and the northern and southern regions lack the distribution of ecological sources. The patches with large dPC values were located in the east, mainly distributed in cultivated land and wetland, while the dPC value was less than 1 in the western ecological source area. The patches were mainly regional, small, fragmented patches, mainly distributed in bare land.

(3) The minimum cumulative resistance of the ecological network in the SRYR decreased from west to east, and the northwestern region showed the highest resistance, with a value of 4.5. The eastern part had the lowest resistance, with a resistance value of 1.16. Meanwhile, 45 potential ecological corridors were identified based on the MCR model, among which the important corridors were mainly distributed in the eastern part.

(4) In total, 190 planned ecological corridors were obtained by combining 10 core area patches with increased betweenness centrality, which optimized the ecological network of the SRYR. The optimized network structure index was much higher than before.

The results show that the ecological network based on the MCR model is poor, and the eastern and western parts lack connectivity. The optimized ecological network effectively improves the connectivity of the whole ecological patches in the SRYR, and promotes the material exchange and energy flow among the core regions. The results of this study provide an important basis for the sustainable development of the SRYR, and provide a reference for the research and protection of fragile ecosystems. However, the influences of the spatial scale, time scale, and resistance factor on the research results still need to be further discussed.

Author Contributions: Conceptualization, J.L.; methodology, J.L. and J.C.; validation, J.C. and Y.Y.; formal analysis, J.L.; resources, J.C., Y.Y., H.Y. and X.H.; data curation, J.L.; writing—original draft preparation, J.L.; writing—review and editing, J.L. and J.C.; visualization, J.L. supervision, J.C., Y.Y., H.Y. and X.H.; funding acquisition, J.C., Y.Y., H.Y. and X.H. All authors have read and agreed to the published version of the manuscript.

Funding: This study was supported by grants from the Guangxi Science and Technology Base and Talent Project (GuikeAD19245032), the National Natural Science Foundation of China (41801030, 41861016), the Guangxi Key Laboratory of Spatial Information and Geomatics (19-050-11-22), and the Research Foundation of Guilin University of Technology (GUTQDJJ2017069).

Institutional Review Board Statement: Not applicable.

Informed Consent Statement: Not applicable.

Data Availability Statement: Acknowledgement for the data support from the "National Tibetan Plateau Scientific Data Center" (http://data.tpdc.ac.cn/ (accessed on 20 March 2022)).

Acknowledgments: The authors sincerely thank the editors and anonymous reviewers for their helpful and constructive comments which have greatly improved the manuscript.

Conflicts of Interest: The authors declare no conflict of interest.

References

1. Yan, J.; Bocharnikov, V. Knowledge and Understanding of Ecological Civilization: A Chinese Perspective. *BRICS J. Econ.* **2022**, *3*, 231–247. [CrossRef]
2. Spanowicz, A.G.; Jaeger, J.A.G. Measuring landscape connectivity: On the importance of within-patch connectivity. *Landsc. Ecol.* **2019**, *34*, 2261–2278. [CrossRef]
3. Xi, J. Speech at the symposium on ecological protection and quality development in the Yellow River basin. *Water Resour. Dev. Manag.* **2019**, *No.46*, 1–4. [CrossRef]
4. Lan, Y.; Jianjun, C.; Yang, Y.; Ling, M.; You, H.; Han, X. Landscape Pattern and Ecological Risk Assessment in Guilin Based on Land Use Change. *Int. J. Environ. Res. Public Health* **2023**, *20*, 2045. [CrossRef] [PubMed]
5. Yang, Y.; Jianjun, C.; Lan, Y.; Zhou, G.; You, H.; Han, X.; Wang, Y.; Shi, X. Landscape Pattern and Ecological Risk Assessment in Guangxi Based on Land Use Change. *Int. J. Environ. Res. Public Health* **2022**, *19*, 1595. [CrossRef]
6. Richter, B.; Behnisch, M. Integrated evaluation framework for environmental planning in the context of compact green cities. *Ecol. Indic.* **2019**, *96*, 38–53. [CrossRef]
7. Sanclements, M.; Record, S.; Rose, K.; Donnelly, A.; Chong, S.; Duffy, K.; Hallmark, A.; Heffernan, J.; Liu, J.; Mitchell, J.; et al. People, infrastructure, and data: A pathway to an inclusive and diverse ecological network of networks. *Ecosphere* **2022**, *13*, e4262. [CrossRef]
8. Zhang, X.; Liu, K.; Wang, S.; Wu, T.; Li, X.; Wang, J.; Wang, D.; Zhu, H.; Tan, C.; Ji, Y. Spatiotemporal evolution of ecological vulnerability in the Yellow River Basin under ecological restoration initiatives. *Ecol. Indic.* **2022**, *135*, 108586. [CrossRef]
9. Ohlmann, M.; Miele, V.; Dray, S.; Chalmandrier, L.; O'Connor, L.; Thuiller, W. Diversity indices for ecological networks: A unifying framework using Hill numbers. *Ecol. Lett.* **2019**, *22*, 737–747. [CrossRef]
10. Salviano, I.; Gardon, F.; Santos, R. Ecological corridors and landscape planning: A model to select priority areas for connectivity maintenance. *Landsc. Ecol.* **2021**, *36*, 3311–3328. [CrossRef]
11. Hashemi, R.; Darabi, H. The Review of Ecological Network Indicators in Graph Theory Context: 2014–2021. *Int. J. Environ. Res.* **2022**, *16*, 24. [CrossRef]
12. Magliocca, N.; Summers, D.; Curtin, K.; McSweeney, K.; Price, A. Shifting landscape suitability for cocaine trafficking through Central America in response to counterdrug interdiction. *Landsc. Urban Plan.* **2022**, *221*, 104359. [CrossRef]
13. Lu, C.; Ji, W.; Muchen, H.; Ma, T.; Mao, J. Evaluation of efficiency and resilience of agricultural water resources system in the Yellow River Basin, China. *Agric. Water Manag.* **2022**, *266*, 107605. [CrossRef]
14. An, Y.; Liu, S.; Sun, Y.; Shi, F.; Beazley, R. Construction and optimization of an ecological network based on morphological spatial pattern analysis and circuit theory. *Landsc. Ecol.* **2021**, *36*. [CrossRef]
15. Hou, J.; He, X.; Ni, X. Hydration mechanism and thermodynamic simulation of ecological ternary cements containing phosphogypsum. Mater. *Today Commun.* **2022**, *33*, 104621. [CrossRef]
16. Gao, M.; Hu, Y.; Bai, Y. Construction of ecological security pattern in national land space from the perspective of the community of life in mountain, water, forest, field, lake and grass: A case study in Guangxi Hechi, China. *Ecol. Indic.* **2022**, *139*, 108867. [CrossRef]
17. Yu, Q.; Yue, D.; Wang, Y.; Kai, S.; Fang, M.; Ma, H.; Zhang, Q.; Huang, Y. Optimization of ecological node layout and stability analysis of ecological network in desert oasis: A typical case study of ecological fragile zone located at Deng Kou County(Inner Mongolia). *Ecol. Indic.* **2018**, *84*, 304–318. [CrossRef]
18. Vogt, P.; Ferrari, J.R.; Lookingbill, T.R.; Gardner, R.H.; Riitters, K.H.; Ostapowicz, K. Mapping functional connectivity. *Ecol. Indic.* **2009**, *9*, 64–71. [CrossRef]
19. Soille, P.; Vogt, P. Morphological segmentation of binary patterns. *Pattern Recognit. Lett.* **2009**, *30*, 456–459. [CrossRef]
20. Zhe-tao, X.; Ying, Z.; Li-jun, H. The evolution of spatial and temporal patterns of Zhengzhou ecological network based on MSPA. *Arab. J. Geosci.* **2021**, *14*, 1107. [CrossRef]
21. Wang, Q.; Liu, S.; Liu, Y.; Wang, F.; Liu, H.; Yu, L. Effects of urban agglomeration and expansion on landscape connectivity in the river valley region, Qinghai-Tibet Plateau. *Glob. Ecol. Conserv.* **2022**, *34*, e2004. [CrossRef]
22. Li, X. TOPSIS Model with Entropy Weight for Eco Geological Environmental Carrying Capacity Assessment. *Microprocess. Microsyst.* **2021**, *82*, 103805. [CrossRef]
23. Chen, L.; Wu, Y.; Gao, B.; Zheng, K.; Wu, Y.; Wang, M. Construction of ecological security pattern of national ecological barriers for ecosystem health maintenance. *Ecol. Indic.* **2023**, *146*, 109801. [CrossRef]

24. Gao, Y.; Ma, L.; Liu, J.; Zhuang, Z.; Huang, Q.; Li, M. Constructing Ecological Networks Based on Habitat Quality Assessment: A Case Study of Changzhou, China. *Sci. Rep.* **2017**, *7*, 46073. [CrossRef]
25. Pascual-Hortal, L.; Saura, S. Comparison and development of new graph-based landscape connectivity indices: Towards the priorization of habitat patches and corridors for conservation. *Landsc. Ecol.* **2006**, *21*, 959–967. [CrossRef]
26. Wu, J.; Zhang, S.; Wen, H.; Fan, X. Research on Multi-Scale Ecological Network Connectivity—Taking the Guangdong–Hong Kong–Macao Greater Bay Area as a Case Study. *Int. J. Environ. Res. Public Health* **2022**, *19*, 15268. [CrossRef] [PubMed]
27. Babí Almenar, J.; Bolowich, A.; Elliot, T.; Geneletti, D.; Sonnemann, G.; Rugani, B. Assessing habitat loss, fragmentation and ecological connectivity in Luxembourg to support spatial planning. *Landsc. Urban Plan.* **2019**, *189*, 335–351. [CrossRef]
28. Kong, F.; Yin, H.; Nakagoshi, N.; Zong, Y. Urban green space network development for biodiversity conservation: Identification based on graph theory and gravity modeling. *Landsc. Urban Plan.* **2010**, *95*, 16–27. [CrossRef]
29. Thiault, L.; Kernaléguen, L.; Osenberg, C.; Lison De Loma, T.; Chancerelle, Y.; Siu, G.; Claudet, J. Ecological evaluation of a marine protected area network: A progressive-change BACIPS approach. *Ecosphere* **2019**, *10*, e025076. [CrossRef]
30. Boitani, L.; Falcucci, A.; Maiorano, L.; Rondinini, C. Ecological Networks as Conceptual Frameworks or Operational Tools in Conservation. *Conserv. Biol. J. Soc. Conserv. Biol.* **2008**, *21*, 1414–1422. [CrossRef]
31. Zhao, X.Q.; Ma, Y.S.; Zhou, H.K. *Restoration and Sustainable Management of Degraded Ecosystems in the Three River Source Areas*; Science Press: Beijing, China, 2011.
32. Wang, G.X.; Li, Y.S.; Wang, Y.B. *Landsurface Processes and Environmental Changes River Headwater Regions of QINGHAI-TIBET Plateau*; Science Press: Beijing, China, 2010.
33. Gandini, M.L.; Lara, B.D.; Moreno, L.B.; Cañibano, M.A.; Gandini, P.A. Landscape dynamics of Paspalum quadrifarium grasslands analyzed by Morphological Spatial Pattern Analysis (MSPA). *PeerJ Inc.* **2018**, 1–24. [CrossRef]
34. Pelorosso, R.; Gobattoni, F.; Geri, F.; Leone, A. PANDORA 3.0 plugin: A new biodiversity ecosystem service assessment tool for urban green infrastructure connectivity planning. *Ecosyst. Serv.* **2017**, *26*, 476–482. [CrossRef]
35. Cook, E. Landscape Structure Indices for Assessing Urban Ecological Networks. *Landsc. Urban Plan.* **2002**, *58*, 269–280. [CrossRef]
36. Marrotte, R.; Gonzalez, A.; Millien, V. Landscape resistance and habitat combine to provide an optimal model of genetic structure and connectivity at the range margin of a small mammal. *Mol. Ecol.* **2014**, *23*, 3983–3998. [CrossRef]
37. Freeman, L. Centrality in Social Networks' Conceptual Clarification. *Soc. Netw.* **1979**, *1*, 215–239. [CrossRef]
38. Zhang, Y.; Hu, W.; Min, M.; Zhao, K.; Zhang, S.; Liu, T. Optimization of ecological connectivity and construction of supply-demand network in Wuhan Metropolitan Area, China. *Ecol. Indic.* **2023**, *146*, 109799. [CrossRef]
39. Jianjun, C.; Chen, Z.; Huang, R.; You, H.; Han, X.; Yue, T.; Zhou, G. The Effects of Spatial Resolution and Resampling on the Classification Accuracy of Wetland Vegetation Species and Ground Objects: A Study Based on High Spatial Resolution UAV Images. *Drones* **2023**, *7*, 61. [CrossRef]
40. Ostapowicz, K.; Vogt, P.; Riitters, K.; Kozak, J.; Estreguil, C. Impact of scale on morphological spatial pattern of forest. *Landsc. Ecol.* **2008**, *23*, 1107–1117. [CrossRef]
41. La Puma, I.; Lathrop, R.G.; Keuler, N. A large-scale fire suppression edge-effect on forest composition in the New Jersey Pinelands. *Landsc. Ecol.* **2013**, *28*, 1815–1827. [CrossRef]
42. Lahti, D. The "edge effect on nest predation" hypothesis after twenty years. *Biol. Conserv.* **2001**, *99*, 365–374. [CrossRef]
43. Wickham, J.; Riitters, K.; Vogt, P. A national assessment of green infrastructure and change for the conterminous United States using morphological image processing. *Landsc. Urban Plan.* **2010**, *94*, 186–195. [CrossRef]
44. Phillips, J. Edge effects in geomorphology. *Phys. Geogr.* **2013**, *20*, 53–66. [CrossRef]
45. Devi, B.; Murthy, M.; Debnath, B.; Jha, C. Forest patch connectivity diagnostics and prioritization using graph theory. *Ecol. Model.* **2013**, *251*, 279–287. [CrossRef]
46. Geldenhuys, L.; Rouget, M.; Lötter, M.C. Landscape connectivity of the grassland biome in Mpumalanga, South Africa. *Austral Ecol.* **2014**, *40*, 67–76. [CrossRef]
47. Ng, C.; Xie, Y.; Yu, X. Integrating landscape connectivity into the evaluation of ecosystem services for biodiversity conservation and its implications for landscape planning. *Appl. Geogr.* **2013**, *42*, 1–12. [CrossRef]
48. Matos, C.; Petrovan, S.; Wheeler, P.; Ward, A. Landscape connectivity and spatial prioritization in an urbanising world: A network analysis approach for a threatened amphibian. *Biol. Conserv.* **2019**, *237*, 238–247. [CrossRef]
49. Saura, S.; Rubio, L. A Common Currency for the Different Ways in Which Patches and Links can Contribute to Habitat Availability and Connectivity in the Landscape. *Ecography* **2010**, *33*, 523–537. [CrossRef]
50. Luo, Y.; Wu, J.; Wang, X.; Peng, J. Using stepping-stone theory to evaluate the maintenance of landscape connectivity under China's ecological control line policy. *J. Clean Prod.* **2021**, *296*, 126356. [CrossRef]

International Journal of
Environmental Research and Public Health

MDPI

Article

Remote Sensing Application in Mountainous Environments: A Bibliographic Analysis

Simbarashe Jombo [1,2,*], Mohamed A. M. Abd Elbasit [1,2], Anesu D. Gumbo [3] and Nthaduleni S. Nethengwe [3]

1 Department of Physical and Earth Sciences, Sol Plaatje University, Kimberley 8300, South Africa
2 Risk, Vulnerability and Science Centre, Sol Plaatje University, Kimberley 8300, South Africa
3 Department of Geography and Environmental Sciences, University of Venda, Thohoyandou 0950, South Africa
* Correspondence: simbarashejombo@gmail.com

Abstract: Advancement in remote sensing platforms, sensors, and technology has significantly improved the assessment of hard-to-access areas, such as mountains. Despite these improvements, Africa lags in terms of research work published. This is of great concern as the continent needs more research to achieve sustainable development. Therefore, this study applied a bibliometric analysis of the annual production of publications on the application of remote sensing methods in mountainous environments. In total, 3849 original articles between 1973 and 2021 were used, and the results indicate a steady growth in publications from 2004 (n = 26) to 2021 (n = 504). Considering the source journals, *Remote Sensing* was the top-ranked, with 453 total publications. The University of the Chinese Academy of Sciences was the highest-ranking affiliation, with 217 articles, and China produced the highest number of publications (n = 217). Keywords used between 1973 and 1997, such as "Canada", "alps", and "GIS", metamorphosed into "remote sensing" between 1998 and 2021. This metamorphosis indicates a change in the areas of interest and an increase in the application of remote sensing methods. Most studies were conducted in the Global North countries, and a few were published in low-impact journals within the African continent. This study can help researchers and scholars better understand the progress and intellectual structure of the field and future research directions in the application of remote sensing methods in mountainous environments.

Keywords: data scarcity; publishing equity; mountain; remote sensing; sustainable development; Africa; bibliometric analysis

Citation: Jombo, S.; Abd Elbasit, M.A.M.; Gumbo, A.D.; Nethengwe, N.S. Remote Sensing Application in Mountainous Environments: A Bibliographic Analysis. *Int. J. Environ. Res. Public Health* **2023**, *20*, 3538. https://doi.org/10.3390/ijerph20043538

Academic Editor: Paul B. Tchounwou

Received: 5 January 2023
Revised: 9 February 2023
Accepted: 15 February 2023
Published: 17 February 2023

1. Introduction

Mountain formation has been attributed to plate tectonics, in which pieces of the Earth's crust smash against each other [1]. The three main types of mountains mainly found in Africa are volcanic, fold, and block [2]. As there is no proper definition of a mountain in the literature, the general understanding is that it is distinctively elevated land compared to the surrounding areas, with steep sides and exposed bedrock [3]. Due to their physical characteristics, mountains provide goods and ecosystem services that humans and animals use [4]. Mountains are one of the most vital ecosystems for the world population because they offer clean water and energy that supports biodiversity [4,5]. The mountain elevation provides for cooler climates and its rugged terrain discourages human intrusion [6]. These characteristics promote the existence of species that are endemic to the area [7].

Many rivers originate from mountains, making them known as "water towers", supporting more than 2 billion lives worldwide [8]. A good example is the Maluti mountains in Lesotho, which are associated with high precipitation and cloud cover, reduced evaporation, and the provision of fresh potable water to the citizens of the kingdom [9]. Land use and land cover (LULC) and climate changes make mountain ecosystems vulnerable to decay while limiting their provision of ecosystem goods and services [5]. These vulnerabilities have affected the ecological stability of mountains and the livelihoods of communities that access socioeconomic benefits from the mountains [10]. This is especially true in Africa, where most of the population is rural, residing in headwater catchments that are dominated by rugged and mountainous terrain [11]. There exists a beneficial relationship between man and the environment in these areas that can inform management to plan and provide people with equitable and sustainable ecological services [12].

The natural environment offers resources for communities to utilize, and communities, in turn, maintain these environments [13]. Human populations in rural areas depend on freshwater resources of sufficient quantity and quality to sustain their dominant agricultural activities for their livelihoods [14]. The Pungwe River basin in the Zimbabwean part is a good example of a basin that provides livelihood opportunities to inhabitants in the area [11]. The catchment is generally mountainous [15], having the second-highest mountain and the fourth-tallest waterfall in Africa [15]. The area produces enough water resources to cater for the commercial and subsistence farming activities in the area, provides a good environment for fish populations (the most common source of animal protein in rural areas), and promotes ecotourism, which provides jobs for the local communities [16]. Degradation of the mountainous environment that supports this system directly affects the livelihoods of communities that lag behind the successes of sustainable development in Africa [16].

With current projections in climate change trends pointing towards a gloomy future, especially for Africa, it is necessary to fully understand the impact of such change [9,17]. Africa, based on its adaptive capacity, has a limited ability to cope with climate change [2]. The livelihoods of communities in rural Africa are closely related to the availability and use of natural resources, which results in LULC changes that threaten human well-being in the region [18]. Climate change will increase the frequency and magnitude of extreme weather events (e.g., heatwaves, droughts, floods, and hailstorms) [19]. As temperatures increase, some animal species endemic to the cool climates provided by mountainous environments will disappear, and there will be crop damage and failure in these environments [5]. Understanding climate and LULC changes will allow decision-makers to come up with adaptation strategies to combat these alterations [17,20]. However, mountainous regions are data-scarce, a result of their inaccessibility that limits in situ data capturing [21]. However, with modern science, data can be collected and evaluated ex situ to understand changes that a place is undergoing [22]. Remote sensing has become one of the most widely used ex situ data collection and analysis approaches for environmental assessments [10,20,23]. The analysis approach is an alternative source that is quick, easy to use, and intrinsically spatialized [22,24].

Remote sensing is a method of collecting information about objects by analyzing data collected by sensors that are not in contact with the objects of interest [25,26]. The data can be used to map, model, and monitor mountain ecosystem patterns [27]. Moreover, the data can provide comprehensive and cost-effective geospatial information acquired at varying spatial scales, temporal frequencies, and spectral properties [21–23]. The application of remote sensing has been successfully used in mountainous environments to classify habitats [28], estimate daily land surface temperature [24], water extraction [29], detect vegetation cover [30], detect fire events [27], evaluate snowpack simulations [31], monitor ecosystem services [5], detect changes in mountain glaciers [32], and mapping mountain forests [33] among many functions. Remote sensing application has advanced where it is used in combination with geographic information systems (GIS) systems in evaluating glacier and permafrost dangers in mountains [33].

Remote sensing methods using optical and radar technology are increasingly crucial for understanding environmental dynamics in mountainous areas [34]. These methods include the use of unmanned aerial vehicles (UAVs), very high spatial and temporal resolution data, and geographical information systems (GIS) data—such as digital elevation models (DEMs)—in modelling, mapping, and monitoring changes in mountainous regions [35,36]. Persistent clouds, frequent and heavy snowfall, and output data accessibility are particularly problematic for optical satellite sensors [37]. These problems can be reduced by using radar sensors that can penetrate clouds and measure mountain deformation rates [38]. With the massive growth in the amount of free and open-access data available, artificial intelligence (AI) and cloud computing are starting to enhance the processing of these new datasets [37]. In mountainous areas, remote sensing has advanced, it has moved beyond simply analyzing images from a single satellite sensor to combining data from several satellite sensors and examining their long-term spatiotemporal properties [39]. Knowledge of the application and use of remote sensing in mountainous environments can help to increase understanding of environmental dynamics [40]. This understanding is useful for decision-makers, natural resource management officers, and other stakeholders in making decisions for conserving and management of resources in mountainous regions.

There has been a rise in research in mountainous areas using passive optical data with high spectral and temporal resolution. By capturing multiple bands and high spectral resolution, the remote sensing data assist in distinguishing features in mountainous regions [41]. Remote sensing advancements have a significant impact on the methods used in monitoring mountainous environments. The use and type of remote sensing data vary from polar to equatorial regions. For example, more observations from the polar regions are seen on the Moderate-Resolution Imaging Spectroradiometer (MODIS) aboard the polar orbit satellite Terra as compared to equatorial regions [42]. Remote sensing methods are convenient in glacial mapping. Satellite imageries, such as IKONOS and Quickbird, have been used to monitor glacial surfaces in three dimensions due to their capability to acquire stereoscopic images, from which elevation data can be extracted [43]. However, their use is constrained at broad spatial scales by their high costs, small swath sizes, and lengthy revisit intervals. Remote sensing is applied to predict future water resources, and glacial hazards and study earth crust movement in mountainous regions [43]. Several indices, such as the Normalized Difference Snow Index (NDSI) and Normalized Difference Vegetation Index (NDVI), are used to separate snow and ice from dark areas such as rocks and monitor vegetation changes [44]. The application of remote sensing methodologies in mountainous environments consists of several processing and analytical techniques. These techniques include image pre-processing, which is important in correcting systematic and non-systematic errors present in remotely sensed images [21,45]. Image post-processing involves the extraction of information from the pre-processed images such as the classification of mountainous environments using either pixel-based or object-based classification methods [21]. Knowledge of the vulnerability of mountain regions to LULC and climate changes and the dependence of communities residing in these regions emphasizes the need for the application of remote sensing strategies in mountain environments [21]. Although mountains are important and fragile, research on mountain environments is still scarce. Existing studies in mountainous environments have been limited and focus on monitoring shifting cultivation [46], measuring, modelling, and monitoring ecosystem services [17], estimation and mapping of soil properties [47], and vulnerability assessments [4]. Understanding possible scenarios of trends in both anthropogenic and natural changes can aid in creating adaptation strategies that are informed by science [32]. This gives a sense of the direction of adaptation that can be followed. Africa is resource-limited and, therefore, requires sound and cost-effective scientific evidence that informs decision-making.

It is against this background that this study applied a bibliometric analysis approach to understand the distribution of relevant literature that applied remote sensing techniques in mountainous environments. This analysis is important because it helps researchers provide an integrated understanding of progress, gaps, directions, and targets for future research studies. Bibliometric analysis popularity is attributed to the advancement, availability, and accessibility of bibliometric software such as VOSviewer 1.6.19 and scientific databases such as Web of Science (WoS) and Scopus [48]. Bibliometric analysis for this study will show the point of view of Africa in terms of research carried out in mountainous areas versus the whole world. The study summarized annual production, source journals, affiliations, collaborations, and country scientific production. Moreover, the study outlined common research topics, co-occurrence networks, and thematic evolution of keywords in publications focusing on the application of remote sensing methods in mountainous environments. The data source, materials, and description of the R statistical software (Version 4.2.2) and packages were used in this study.

2. Materials and Methods

2.1. Bibliographic Database

The data used in this study included authors, keywords, citations, source journals, and countries of publications obtained from the Web of Science and Scopus databases. The WoS belongs to Clarivate Analytics and is one of the oldest databases, with more than 1.5 billion references dating back to 1900 [35]. Scopus database has over 17 million researchers profiled, 81 million curated documents, 80,000 institution profiles, and 7000 publishers [36]. For this study, the search for the article was guided by the terms: "mountain", "mountainous", and "remote sensing" published between 1 January 1973 and 31 December 2021. A total of 3343 original articles were downloaded from the WoS database and 660 from the Scopus database (Figure 1). The downloaded articles were merged, 154 duplicate articles were removed, and 3849 articles were retained.

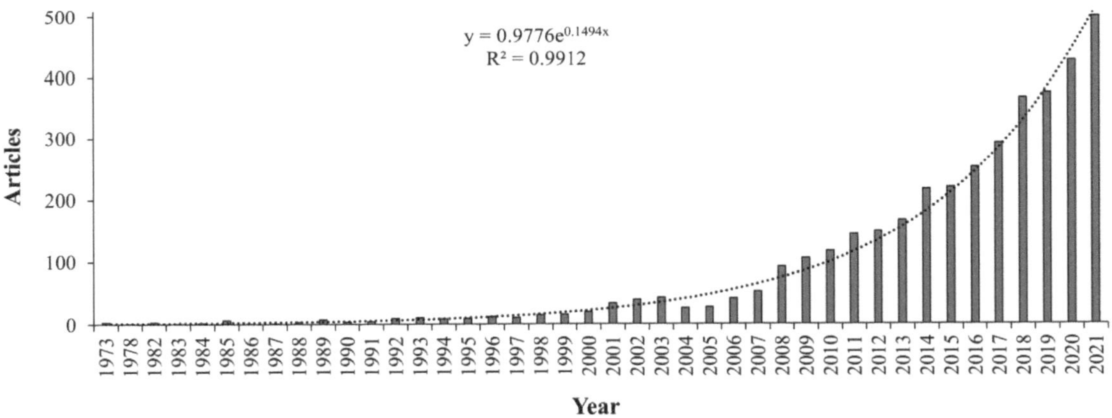

Figure 1. The total number of articles published between January 1973 and December 2021 on the application of remote sensing methods in mountainous environments.

2.2. R Statistical Application

The bibliometric package in R statistical software analyzed the data generated from the databases. The bibliometrix R-package is written in the R language, which is freely available for generating bibliometric maps using effective statistical algorithms [49]. The data transported into R were translated into a bibliographic data frame and structured for duplication, and the duplicated records were presented as a single document. The bibliometrix package analyzed data by creating a bibliographic coupling, collaboration, co-citation, and co-occurrence network. Spelling errors in articles and associations were checked and corrected before visualizing the author's names, keywords plus, and keywords.

3. Results

3.1. Publication Time Series Analysis

From 1973 to 2021, a total of 3849 articles were published focusing on remote sensing and mountain studies. Figure 1 shows how the articles were distributed for the selected period of study. There was minimal production of articles between 1973 and 1989. A steady increase in articles was noticed between 1990 and 2007. An exponential increase was recorded for the period 2008 to 2021. In terms of publishing journals, the top three were *Remote Sensing, Remote Sensing of Environment*, and *the International Journal of Remote Sensing* (Table 1). A total of 14 of the top main-source journals were from European countries, 4 were from the USA, 1 from China and 1 from India (Table 1). Table 1 also shows that the top 20 main source journals had 8 journal names including the words "remote sensing".

Table 1. Top 20 main source journals, country, and the total number of publications.

Rank	Journal Name	Country	Number
1	*Remote Sensing*	Switzerland	415
2	*Remote Sensing of Environment*	USA	222
3	*International Journal of Remote Sensing*	UK	181
4	*Journal of Mountain Science*	China	154
5	*IEEE Transactions on Geoscience and Remote Sensing*	USA	66
6	*Geomorphology*	Netherlands	58
7	*Journal of Glaciology*	UK	50
8	*Mountain Research and Development*	Switzerland	50
9	*IEEE Journal of Selected Topics in Applied Earth Observations and Remote Sensing*	USA	46
10	*Arabian Journal of Geosciences*	Germany	43
11	*ISPRS Journal of Photogrammetry and Remote Sensing*	Netherlands	40
12	*International Journal of Applied Earth Observation and Geoinformation*	Netherlands	39
13	*Canadian Journal of Remote Sensing*	UK	38
14	*Forest Ecology and Management*	Netherlands	37
15	*Cryosphere*	Germany	34
16	*Environmental Earth Sciences*	Germany	34
17	*Hydrological Processes*	UK	33
18	*Forests*	Switzerland	32
19	*Journal of Geophysical Research Atmospheres*	USA	32
20	*Journal of the Indian Society of Remote Sensing*	India	31

3.2. Affiliations, Collaborations, Country Scientific Production, and Top-Cited Articles

From these results, 4480 institutions contributed to the analyzed publications on the application of remote sensing techniques in mountainous environments. The University of Chinese Academy of Sciences in China was the highest-ranked affiliation, with 217 articles (Table 2). China and USA were the top-ranked nations, with 9 affiliations that published research on the application of remote sensing methods in mountainous environments (Table 2). Authors affiliated with Chinese institutions produced the highest number of publications on the topic. China had a total of 1167 articles, with an intra-country or single country publication (SCP) collaboration index of 863 and 304 for the inter-country or multiple country publication (MCP) collaboration index, as shown in Figure 2. This indicates that most of the corresponding authors in the published articles on the application of remote sensing methods in mountainous environments were from China. The USA was the second-ranked country, with 762 publications—554 for SCP and 208 for MCP (Figure 2). Most countries had higher SCP compared to MCP values (Figure 2). The global distribution of publications is shown in Figure 3 with most publications produced in China (n = 4659) followed by the USA (n = 3969) and Germany (n = 1088). A few publications were produced in the African continent.

The global distribution of publications is shown in Figure 3 and the darker the color, the more publications have been produced. Most publications were produced in China (n = 4659), followed by the USA (n = 3969) and Germany (n = 1088). A few publications were produced in the African continent, and there was no single publication in Mauritius (Figure 3).

Table 2. The rank of the top 20 affiliations, country, and the number of articles published.

Rank	Affiliations	Country	Articles
1	University of Chinese Academy of Sciences	China	217
2	Beijing Normal University	China	209
3	The Institute of Mountain Hazards and Environment	China	183
4	Institute of Remote Sensing and Digital Earth	China	175
5	Chinese Academy of Sciences	China	136
6	University of Colorado Boulder	USA	120
7	Institute of Geographic Sciences and Natural Resources	China	109
8	Jet Propulsion Laboratory	USA	94
9	Northwest Institute of Eco-Environment and Resources	China	91
10	University of Maryland	USA	90
11	University of Idaho	USA	89
12	United States Forest Service	USA	88
13	University of Zurich	Switzerland	86
14	Colorado State University	USA	84
15	The University of Arizona	USA	84
16	The University of Oklahoma	USA	83
17	SETI Institute	USA	76
18	University of Marburg	Germany	76
19	Institute of Tibetan Plateau Research	China	75
20	Lanzhou University	China	73

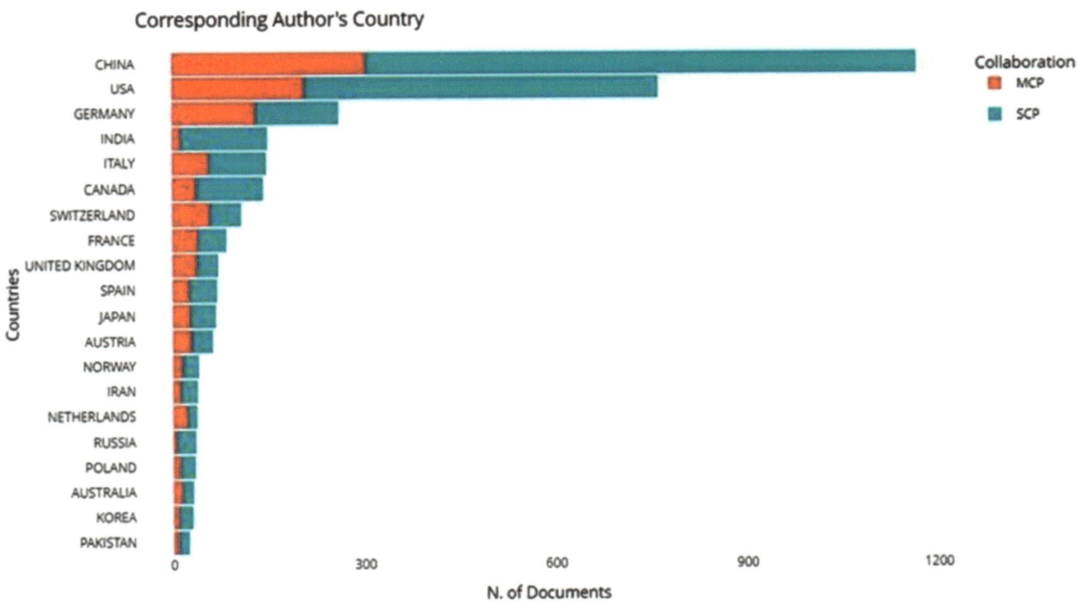

Figure 2. Top 20 countries publishing articles on the application of remote sensing methods in mountainous environments. The country collaborations are represented as inter-country (MCP) and intra-country (SCP) collaboration indices.

The study revealed that the top five most cited publications focusing on the application of remote sensing methods in mountainous environments were from the *Journal of Glaciology*, *Scientific Bulletin*, *Applied Geography*, and *Remote Sensing of Environment* (Table 3). The top 20 publications cited were written between 2002 and 2019, with a highest total citations (TC) of 587 and citations per year (TCpY) of 65.22 for a publication with Pfeffer as the first author (Table 3). The TC and TCpY for the top 20 cited articles ranged from 87 to 587 and from 7.2 to 73.8, respectively. A total of 14 corresponding authors were affiliated with Chinese institutions; 5 were from the USA, and 1 was from Germany. The *Remote Sensing of Environment* had seven publications, followed by the *Journal of Glaciology* ($n = 2$), and the rest had a single publication in the top 20 cited articles focused on the application of remote sensing in mountainous environments.

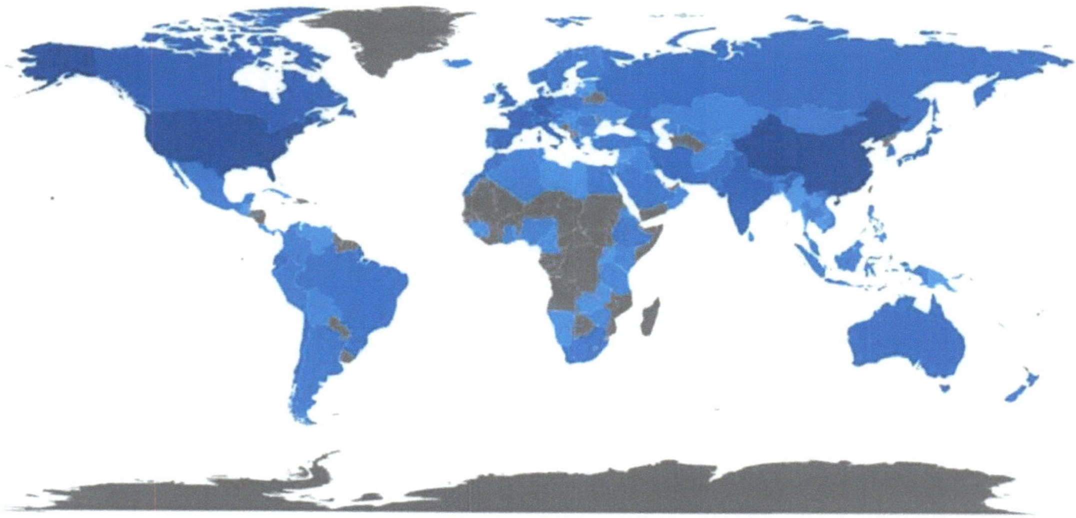

Figure 3. The global distribution of publications in this study.

Table 3. Top 20 articles cited, corresponding author's name, year, title, source, total citations (TC), and total citations per year (TCpY)on the application of remote sensing methods in mountainous environments.

Rank	First Author's Name and Year	Title	Source	TC	TCpY
1	Pfeffer et al. [50]	The Randolph Glacier Inventory: A Globally Complete Inventory of Glaciers	*Journal of Glaciology*	587	65.2
2	Guo et al. [51]	The Second Chinese Glacier Inventory Data Methods and Results	*Journal of Glaciology*		
3	Gong et al. [52]	Stable Classification with Limited Sample Transferring a 30 m Resolution Sample Set Collected in 2015 to Mapping 10 m Resolution Global Land Cover in 2017	*Science Bulletin*	295	73.8
4	Su et al. [53]	Characterizing Landscape Pattern and Ecosystem Service Value Changes for Urbanization Impacts at an Ecoregional Scale	*Applied Geography*	243	22.1
5	Zhu et al. [54]	A Flexible Spatiotemporal Method for Fusing Satellite Images with Different Resolutions	*Remote Sensing of Environment*	235	33.6
6	Xiao et al. [55]	Characterization of Forest Types in NorthEastern China using Multitemporal SPOT4 Vegetation Sensor Data	*Remote Sensing of Environment*	218	10.4
7	Li et al. [56]	Eco-environmental Vulnerability Evaluation in Mountainous Region using Remote Sensing and GIS: A Case Study in the Upper Reaches of Minjiang River China	*Ecological Modelling*	190	11.2

Table 3. *Cont.*

Rank	First Author's Name and Year	Title	Source	TC	TCpY
8	Huang et al. [57]	Mapping Major Land Cover Dynamics in Beijing Using All Landsat Images in Google Earth Engine	*Remote Sensing of Environment*	177	29.5
9	Zhang et al. [58]	A 2010 Update of National Land use cover Database of China at 1:100,000 Scale Using Medium Spatial Resolution Satellite Images	*Remote Sensing of Environment*	172	19.1
10	Wulfmeyer et al. [59]	The Convective and Orographically-induced Precipitation Study (COPS): The Scientific Strategy, The Field Phase, and Research Highlights	*Quarterly Journal of the Royal Meteorological Society*	148	12.3
11	Chen et al. [60]	A Mangrove Forest Map of China In 2015 Analysis of Time Series Landsat 78 and Sentinel1A Imagery in Google Earth Engine Cloud Computing Platform	*ISPRS Journal of Photogrammetry and Remote Sensing*	139	23.2
12	Muno et al. [61]	A Catalog of Xray Point Sources from Two Megaseconds of Chandra Observations of the Galactic Center	*Astrophysical Journal Supplement Series*	134	9.8
13	Nie et al. [62]	A Regional-scale Assessment of Himalayan Glacial Lake Changes Using Satellite Observations From 1990 to 2015	*Remote Sensing of Environment*	121	20.2
14	Ma et al. [63]	Response of Hydrological Processes to Landcover and Climate Changes in Kejie Watershed Southwest China	*Hydrological Processes*	114	8.1
15	Li and Sheng [64]	An Automated Scheme for Glacial Lake Dynamics Mapping using Landsat Imagery and Digital Elevation Models: A Case Study in the Himalayas	*International Journal of Remote Sensing*	113	10.3
16	Chen et al. [65]	Forested Landslide Detection Using Lidar Data and the Random Forest Algorithm: A Case Study of the Three Gorges China	*Remote Sensing of Environment*	109	12.1
17	Yin et al. [66]	An Assessment of the Biases of Satellite Rainfall Estimates over the Tibetan Plateau and Correction Methods Based on Topographic Analysis	*Journal of Hydrometeorology*	108	7.2
18	Zhang et al. [67]	Regional Differences of Lake Evolution Across China During 1960s–2015 and its Natural and Anthropogenic Causes	*Remote Sensing of Environment*	107	26.8
19	Jiapaer et al. [68]	Vegetation Dynamics and Responses to Recent Climate Change in Xinjiang using Leaf Area Index as an Indicator	*Ecological Indicators*	100	12.5
20	Yao et al. [69]	Spatiotemporal Pattern of Gross Primary Productivity and Its Covariation with Climate in China Over the Last Thirty Years	*Global Change Biology*	87	17.4

3.3. *Remote Sensing Data Used in the Top 20 Articles Cited*

The highest number of studies (n = 12) used freely available Landsat satellite images in their studies (Table 4). LiDAR and radar data were used in two studies whilst only a single study used hyperspectral data (Table 4).

Table 4. Overview of the total number of studies, sensor type, and cost of the remote sensing data used in the top 20 articles cited in the application of remote sensing in mountainous environments.

Data Type	Number of Studies	Sensor	Acquisition Cost
Landsat	12	Multispectral	Free
Sentinel	2	Multispectral	Free
MODIS	2	Multispectral	Free
Meteosat Second Generation-8 (MSG-8), LiDAR and radar	1	Multispectral, LiDAR, and Radar	Free and High
SPOT	1	Multispectral	Free
Tropical Rainfall Measuring Mission (TRMM)	1	Radar	Free
LiDAR	1	LiDAR	High
Advanced CCD Imaging Spectrometer (ACIS)	1	Hyperspectral	Free

3.4. Word Cloud, Co-Occurrence Network, and Thematic Evolution

The word cloud shown in Figure 4 gives information on the most used keyword in the published articles. The size of the keyword implies the number of occurrences in the publication. Remote sensing was the word most used, with an occurrence of 504 times, followed by climate change (n = 304) and vegetation (n = 283). Mountain was the 20th-most used word, with an appearance of 115. The closeness of keywords to each other implies their interrelation during the time under investigation.

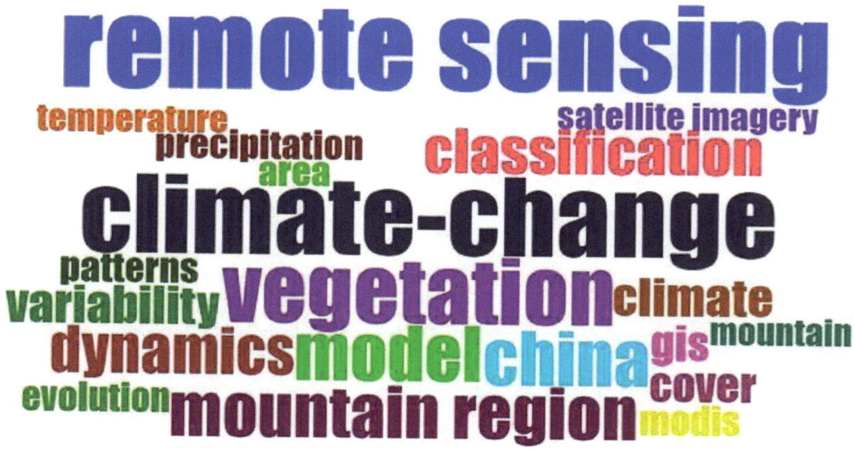

Figure 4. Word cloud showing the top 20 keywords commonly used in studies focused on the application of remote sensing methods in mountainous environments.

The research theme on the application of remote sensing methods in mountainous environments was categorized into three colored groups. The highest number of keywords were in the blue cluster (n = 23), followed by both the red and green clusters (n = 13) as illustrated in Figure 5. The blue cluster has keywords including "climate change", "model", "variability", "climate", and "cover", while the red cluster had keywords including "remote sensing", "mountain region", "China", "satellite imagery", and "modis". The keyword most used was "remote detection", followed by "climate change", which were in the red and blue groups, respectively (Figure 5). There was a great connection between "remote sensing", "China", and "mountain region" in the red group (Figure 5).

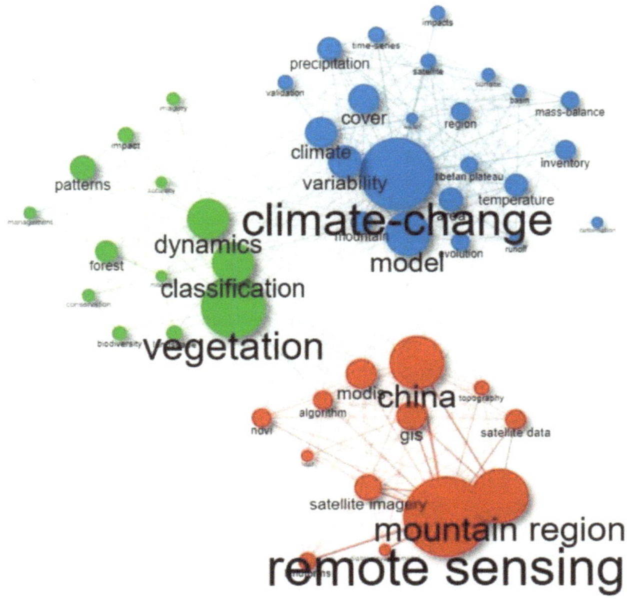

Figure 5. Co-occurrence network for the keywords used in this study. Each node represents a keyword, the size of the node shows the number of occurrences of the keyword, and the thickness of the line shows the degree of connection.

Thematic evolution of keywords shows that five keywords ("model", "Canada", "GIS", "mountainous terrain", and "satellites") metamorphosed between 1973 and 2017 into "remote sensing" between 1998 and 2021 (Figure 6). The keywords "alps", "California", and "model" metamorphosed into "model" between 1998 and 2021. The keywords "Canada", "California", and "alps" also metamorphosed into "climate change" during the 1998–2021 period. The keyword "satellites" changed to "remote sensing", and "vegetation" during the 1998–2021 period (Figure 6).

Figure 6. Thematic evolution of keywords from 1973 to 2021 for the publications used in this study.

4. Discussion

4.1. Bibliographic Analysis

The results obtained from a thorough search of the application of remote sensing methods in mountainous environments between 1973 and 2021 showed significant growth in publications. This is a clear indication that more researchers are interested in applying remote sensing methods in mountainous environments. Gathering information about the Earth using remote sensing methods has seen significant evolution since the 1800s [70]. At the time, what appears primitive today was the technology of the time, utilizing pigeons, kites, and hot air balloons to gather information about the earth [71]. Technological advancements in the 20th century saw the invention of airplanes [20]. Concurrently, photography was also developing enabling the capturing of aerial photographs [72]. The need to accurately map how the land surface looks were enhanced by the invention of satellite technology in the 1970s [70]. Since then, the images have been enhanced, and with an increase in technology and understanding systems, satellite imagery has helped scientists understand the environment, detect changes, and predict likely future scenarios [27,45,57]. Mountainous region studies have manipulated the remotely sensed data and have brought about a better understanding of these once hard-to-reach areas [5,10,28]. This evolution in remote sensing from simple tools to highly sophisticated satellite images and, currently, drone-based images producing ultra-high-resolution data [10,35] is part of the reason why the study saw the trend that was followed in Figure 1. In recent times, the use of airborne and spaceborne sensors is replaced by the use of UAVs and small unmanned aerial vehicles (sUAVS) that collect high-quality aerial image data that can help in managing mountainous regions [73]. More articles can now be produced compared to previous decades, when science was still being developed. It is expected that this trajectory will be followed in the future as systems are understood, procedures are refined, and the use of technology is enhanced.

The distribution of research that uses remote sensing in mountainous areas shows that developed nations are the main contributors (Figure 2). China, Europe, and the USA have significant work that utilizes remote sensing work compared to other regions. In Africa, there is a distinct disconnect in the production of this work, as shown in Figure 2. This can be explained by several factors that are inherent to the African situation. Socio-economic and political vulnerabilities have crippled Africa's progress to contribute to the body of knowledge. During the 1970s, most African countries were battling colonial inequalities, and those who had gained independence were often plunged into civil unrest because of power dynamics [74]. This rendered the continent to contribute late to scientific research. Access is another factor that has resulted in Africa not producing much research on remote sensing applications in mountainous areas [75]. Most technologies are developed and maintained by developed nations such that there are limited native scientists conversant with the procedures and application of remote sensing [76].

Data scarcity has been reported to be a major barrier to environmental assessments in the region [2,9,45]. Remotely sensed data would become the most obvious to use when understanding systems in these data-scarce regions. However, copyrights and ownership of remotely sensed data limit these regions from acquiring quality data to conduct research [77]. This puts Africa in a lagging position as the continent depends on free datasets, which usually have a coarse resolution [78,79]. As images are rendered obsolete in developed nations, they are made available to developing nations for free [80]. Though new to these countries, the information will not have temporal significance, resulting in research based on these datasets not qualifying for publication in reputable journals with a wide readership [81]. Limitations in research publications of this nature in Africa can also be attributed to language barriers [82]. The journals studied are English-based, which limits contributions from French-speaking nations. This is elaborated in Figure 2 without research from the central and west African regions. However, it is different for countries such as South Africa and Zimbabwe because of their well-established tertiary education system [83], the use of English as a language, and government initiatives that promote scientific research [84].

4.2. Associations and Production

The top five affiliations where the publications in this study were produced are from China. Chinese institutions published more studies because Chinese researchers use advanced remote sensing applications to monitor mountainous environments and almost two-thirds of the country is high in elevation [23]. This is supported by Li, Pei, Zhao, Xiao, Sang, and Zhang [20], who highlighted that remote sensing methods are being applied by Chinese institutions, including government agencies, research organizations, and universities, in monitoring mountainous environments.

Most publications written by intra-country authors were also in China, with an SCP collaboration index of 863. From the perspective of the spatial distribution of publications in the world, China, the USA, and Germany were the countries with the most publications (Figure 3). This highlights that China has invested so much in science and development; hence, they have the most research studies that focus on the applications of remote sensing methods in mountainous environments. A total of 8 out of 10 countries with high SCP and MCP values are from the Global North, consisting of wealthy and technologically advanced countries. This is in line with Wang, Zhao, and Wang [19], who highlight that most developed papers are from the Global North because they are wealthy and their governments assist them in promoting science and development through research and publication.

The most cited author is Pfeffer et al. [50], with an article published in the *Journal of Glaciology* that has a TC of 587 (Table 3). They were followed by Guo, Liu, Xu, Wu, Shangguan, Yao, Wei, Bao, Yu, Liu, and Jiang [51] who published in the *Scientific Bulletin* journal with a TC of 295 and a TCpY of 73.8 (Table 3). In these studies, remote sensing methods were applied in glacier mountains, and this shows their potential to the global audience, hence high TC and TCpY rates. Other papers applied remote sensing methods in several areas, including urbanization, forestry, land cover changes, glacial lake changes, climate change, and rainfall estimations in mountainous environments [53,55,57,62,63,66]. Weiss and Walsh [21] highlighted that remote sensing applications in mountainous environments continue to be common because of the ongoing and improving utility of imageries in solving real-world problems and providing solutions that can support sustainable mountain management.

4.3. Cloud of Words, Co-Occurrence Associations, and Thematic Progress

Remote sensing is the commonly used keyword, followed by "climate change". The keyword "remote sensing" is close to "satellite imagery", "temperature", and "classification" (Figure 4). This shows that studies have been growing over the years, focusing on the application of remote sensing methods using satellite imageries by employing classification techniques. This is in line with Praticò et al. [85], who highlighted that there have been developments in the application of remote sensing methods in the classification of data using satellite imageries in mountainous environments. China was the only country on the word cloud that showed that most studies were conducted in this mountainous nation. Zhao, Bian, and Li [35] highlight that since 2002, the application of remote sensing methods and digital elevation techniques has gradually increased over the years in China for various reasons, including the availability of research funds and technological advances.

The keywords used in the studies are connected, with "remote sensing", "climate change", and "vegetation" as the most common words (Figure 5). These keywords are connected because of the increase in remote sensing studies focusing on topics of great interest, including climate change and vegetation studies. This is in agreement with Xu et al. [86], who highlighted an increase in remote sensing research studies focusing on vegetation ecosystem response to changing climate in the 21st centuryThis also explains the metamorphosis of the keywords: "alps", "California", "model", "Canada", "GIS", "mountainous terrain", "remote sensing", and "satellites" for the period 1973–1997 into "remote sensing", "model", "climate change", and "vegetation" for the 1998–2021 period. Other keywords that were in the co-occurrence network include "classification", "mountain region", "gis", "modis", and "forest", which are some of the areas and data sets that are of great concern in recent studies.

4.4. Importance of the Study

In recent years, with the development and advancement of satellite imagery, sensors, and techniques, the application of remote sensing methods has been broadly used in the monitoring and management of mountainous environments [4]. The research studies in mountainous environments using remote sensing methods provide knowledge and information that can improve the level of monitoring and formulation of policies that assist in the management and promotion of sustainable mountain management [35]. However, with all these advances, African research in this regard still lags. Several factors affect the ability of the continent to produce internationally recognized papers. The lag in terms of technology and economics renders the research of developing countries "old news" for publishers. However, what is important is not the level of sophistication of methods employed in the African context but rather an understanding of processes and how they impact their livelihoods. If the information on Africa is published in journals with a wide readership, the dissemination of information will increase. This will show gaps in science and promote sustainable development efforts in the region. With literature e.g., [4,9,17,19,45] pointing out that Africa will be greatly impacted by the effects of climate change, it only becomes the next best idea to improve and promote research coming out of the continent.

4.5. Limitations

Despite the success of this study in showing the literature that has been published over the years, some limitations were encountered. The study excluded some publications from government agencies, nature conservancies, intergovernmental organizations, and other institutions that have information on the application of remote sensing methods in mountainous environments. Expanding the search across other databases, such as Google Scholar, will help improve the analysis results. The other limitation is that the results were all based on publications authored in the English language. Due to the high volume of publications in China, there might be other studies written in Chinese characters that were not used. This limitation can be addressed by including various languages while searching the research databases.

5. Conclusions

This study analyzed global research and publication trends for remote sensing applications in mountainous environments from 1973 to 2021. The study comprised, among others, annual production, source journals, affiliations, collaborations, countries, citations, and keywords. The results indicated steady growth in the number of publications since 2004 and the main source journal was *Remote Sensing*, with a total of 415 publications. Considering the affiliations, the University of Chinese Academy of Sciences was top-ranked, with 217 articles, and China produced the highest number of publications (n = 4659). This provides China with a leading position to strengthen other countries on the application of remote sensing in mountainous environments through more collaborations, funding and the provision of technological knowledge. The top-cited article had authors from various countries in the Global North, with a TC of 587 and a TCyP of 65.2. The countries in the Global North can assist other countries in the Global South, especially developing countries in Africa, with resources and funding to develop research using remote sensing methods in mountainous environments. The interconnections and evolution of the keywords over the years hint at the relatedness of remote sensing, climate, and vegetation for future studies. This study offers knowledge of development trends and hotspots that can help in future research that focuses on the application of remote sensing methods in mountainous environments. Research from all countries should strengthen collaborations and exchange of ideas to increase the number of studies focused on topical issues that need remote sensing applications in mountainous environments.

Author Contributions: Conceptualization, S.J.; methodology, S.J.; software, S.J.; writing—original draft preparation, S.J.; A.D.G. and M.A.M.A.E.; writing—review and editing, S.J.; visualization, S.J.; supervision, M.A.M.A.E. and N.S.N. All authors have read and agreed to the published version of the manuscript.

Funding: This research received no external funding.

Institutional Review Board Statement: Not applicable.

Informed Consent Statement: Not applicable.

Data Availability Statement: Not applicable.

Conflicts of Interest: The authors declare no conflict of interest.

References

1. Wilson, J.; Burke, K. Two types of mountain building. *Nature* **1972**, *239*, 448–449. [CrossRef]
2. McCarthy, T. *The Story of Earth & Life: A Southern African Perspective on a 4.6-Billion-Year Journey*; Penguin Random House South Africa: Cape Town, South Africa, 2013.
3. Fisher, P.; Wood, J.O. What is a Mountain? Or The Englishman who went up a Boolean Geographical Concept but Realised it was Fuzzy. *Geography* **1998**, *83*, 247–256.
4. Thakur, S.; Negi, V.S.; Dhyani, R.; Satish, K.V.; Bhatt, I.D. Vulnerability assessments of mountain forest ecosystems: A global synthesis. *Trees For. People* **2021**, *6*, 100156. [CrossRef]
5. Carvalho-Santos, C.; Monteiro, A.; Arenas-Castro, S.; Greifeneder, F.; Marcos, B.; Portela, A.; Honrado, J. Ecosystem Services in a Protected Mountain Range of Portugal: Satellite-Based Products for State and Trend Analysis. *Remote Sens.* **2018**, *10*, 1573. [CrossRef]
6. Wang, Y.; Berardi, U.; Akbari, H. Comparing the effects of urban heat island mitigation strategies for Toronto, Canada. *Energy Build.* **2016**, *114*, 2–19. [CrossRef]
7. Rahbek, C.; Borregaard, M.K.; Colwell, R.K.; Dalsgaard, B.; Holt, B.G.; Morueta-Holme, N.; Nogues-Bravo, D.; Whittaker, R.J.; Fjeldså, J. Humboldt's enigma: What causes global patterns of mountain biodiversity? *Science* **2019**, *365*, 1108–1113. [CrossRef]
8. Elmore, A.C.; Alexiev, N.; Craig, V. Understanding the World's Water Towers through High-Mountain Expeditions and Scientific Discovery. *One Earth* **2020**, *3*, 561–563. [CrossRef]
9. Grab, S.W.; Knight, J.; Holmes, P.; Boardman, J. Southern African montane environments. In *Southern African Landscapes and Environmental Change*; Earthscan: Oxford, UK, 2018; pp. 153–180.
10. Duan, P.; Wang, Y.; Yin, P. Remote Sensing Applications in Monitoring of Protected Areas: A Bibliometric Analysis. *Remote Sens.* **2020**, *12*, 772. [CrossRef]
11. Gumbo, A.D.; Kapangaziwiri, E.; Chikoore, H.; Pienaar, H.; Mathivha, F. Assessing water resources availability in headwater sub-catchments of Pungwe River Basin in a changing climate. *J. Hydrol. Reg. Stud.* **2021**, *35*, 100827. [CrossRef]
12. Katrandzhiev, K.; Gocheva, K.; Bratanova-Doncheva, S. Whole System Data Integration for Condition Assessments of Climate Change Impacts: An Example in High-Mountain Ecosystems in Rila (Bulgaria). *Diversity* **2022**, *14*, 240. [CrossRef]
13. Bernués, A.; Rodríguez-Ortega, T.; Ripoll-Bosch, R.; Alfnes, F. Socio-Cultural and Economic Valuation of Ecosystem Services Provided by Mediterranean Mountain Agroecosystems. *PLoS ONE* **2014**, *9*, e102479. [CrossRef] [PubMed]
14. Pointet, T. The United Nations World Water Development Report 2022 on groundwater, a synthesis. *LHB* **2022**, *108*, 2090867. [CrossRef]
15. Swedish International Development Cooperation Agency. *Development of the Pungwe River Basin Joint Integrated Water Resources Management Strategy*; Swedish International Development Cooperation Agency: Stockholm, Sweden, 2004.
16. Swedish International Development Cooperation Agency. *Training Material in Integrated Water Resources Management for River Basin Organisations Case Study: Pungwe River Basin in Mozambique*; Swedish International Development Cooperation Agency: Stockholm, Sweden, 2008.
17. Fisher, B.; Turner, R.K.; Burgess, N.D.; Swetnam, R.D.; Green, J.; Green, R.E.; Kajembe, G.; Kulindwa, K.; Lewis, S.L.; Marchant, R.; et al. Measuring, modeling and mapping ecosystem services in the Eastern Arc Mountains of Tanzania. *Prog. Phys. Geogr. Earth Environ.* **2011**, *35*, 595–611. [CrossRef]
18. Mekonen, A.A.; Berlie, A.B. Rural households' livelihood vulnerability to climate variability and extremes: A livelihood zone-based approach in the Northeastern Highlands of Ethiopia. *Ecol. Process.* **2021**, *10*, 55. [CrossRef]
19. Wang, Z.; Zhao, Y.; Wang, B. A bibliometric analysis of climate change adaptation based on massive research literature data. *J. Clean. Prod.* **2018**, *199*, 1072–1082. [CrossRef]

20. Li, J.; Pei, Y.; Zhao, S.; Xiao, R.; Sang, X.; Zhang, C. A Review of Remote Sensing for Environmental Monitoring in China. *Remote Sens.* **2020**, *12*, 1130. [CrossRef]
21. Weiss, D.J.; Walsh, S.J. Remote Sensing of Mountain Environments. *Geogr. Compass* **2009**, *3*, 1–21. [CrossRef]
22. Jombo, S.; Adam, E.; Tesfamichael, S. Classification of urban tree species using LiDAR data and WorldView-2 satellite imagery in a heterogeneous environment. *Geocarto Int.* **2022**, 1–24. [CrossRef]
23. Wang, L.; Zhang, G.; Wang, Z.; Liu, J.; Shang, J.; Liang, L. Bibliometric Analysis of Remote Sensing Research Trend in Crop Growth Monitoring: A Case Study in China. *Remote Sens.* **2019**, *11*, 809. [CrossRef]
24. Neteler, M. Estimating Daily Land Surface Temperatures in Mountainous Environments by Reconstructed MODIS LST Data. *Remote Sens.* **2010**, *2*, 333–351. [CrossRef]
25. Avery, T.E.; Berlin, G.L. *Fundamentals of Remote Sensing and Airphoto Interpretation*; Macmillan: New York, NY, USA, 1992.
26. DeFries, R. Remote Sensing and Image Processing. In *Encyclopedia of Biodiversity*; Elsevier: Amsterdam, The Netherlands, 2013; pp. 389–399.
27. Fornacca, D.; Ren, G.; Xiao, W. Performance of Three MODIS Fire Products (MCD45A1, MCD64A1, MCD14ML), and ESA Fire_CCI in a Mountainous Area of Northwest Yunnan, China, Characterized by Frequent Small Fires. *Remote Sens.* **2017**, *9*, 1131. [CrossRef]
28. Varela, R.A.D.; Rego, P.R.; Iglesias, S.C.; Sobrino, C.M. Automatic habitat classification methods based on satellite images: A practical assessment in the NW Iberia coastal mountains. *Environ. Monit. Assess.* **2008**, *144*, 229–250. [CrossRef] [PubMed]
29. Kaplan, G.; Avdan, U. Water extraction technique in mountainous areas from satellite images. *J. Appl. Remote Sens.* **2017**, *11*, 046002. [CrossRef]
30. Kim, M.-K.; Daigle, J.J. Detecting vegetation cover change on the summit of Cadillac Mountain using multi-temporal remote sensing datasets: 1979, 2001, and 2007. *Environ. Monit. Assess.* **2011**, *180*, 63–75. [CrossRef]
31. Revuelto, J.; Lecourt, G.; Lafaysse, M.; Zin, I.; Charrois, L.; Vionnet, V.; Dumont, M.; Rabatel, A.; Six, D.; Condom, T.; et al. Multi-Criteria Evaluation of Snowpack Simulations in Complex Alpine Terrain Using Satellite and In Situ Observations. *Remote Sens.* **2018**, *10*, 1171. [CrossRef]
32. Podgórski, J.; Kinnard, C.; Pętlicki, M.; Urrutia, R. Performance Assessment of TanDEM-X DEM for Mountain Glacier Elevation Change Detection. *Remote Sens.* **2019**, *11*, 187. [CrossRef]
33. Vega Isuhuaylas, L.; Hirata, Y.; Ventura Santos, L.; Serrudo Torobeo, N. Natural Forest Mapping in the Andes (Peru): A Comparison of the Performance of Machine-Learning Algorithms. *Remote Sens.* **2018**, *10*, 782. [CrossRef]
34. Chen, J.; Zhang, J.; Wu, T.; Hao, J.; Wu, X.; Ma, X.; Zhu, X.; Lou, P.; Zhang, L. Activity and Kinematics of Two Adjacent Freeze–Thaw-Related Landslides Revealed by Multisource Remote Sensing of Qilian Mountain. *Remote Sens.* **2022**, *14*, 5059. [CrossRef]
35. Zhao, W.; Bian, J.; Li, A. Research Center for Digital Mountain and Remote Sensing Application, Institute of Mountain Hazards and Environment. *Mt. Res. Dev.* **2019**, *39*, P1–P4. [CrossRef]
36. Mulders, M.A. Advances in the application of remote sensing and GIS for surveying mountainous land. *Int. J. Appl. Earth Obs. Geoinf.* **2001**, *3*, 3–10. [CrossRef]
37. Taylor, L.S.; Quincey, D.J.; Smith, M.W.; Baumhoer, C.A.; McMillan, M.; Mansell, D.T. Remote sensing of the mountain cryosphere: Current capabilities and future opportunities for research. *Prog. Phys. Geogr. Earth Environ.* **2021**, *45*, 931–964. [CrossRef]
38. Arenson, L.U.; Kääb, A.; O'Sullivan, A. Detection and analysis of ground deformation in permafrost environments. *Permafr. Periglac. Process.* **2016**, *27*, 339–351. [CrossRef]
39. Lam, N.S.-N.; Cai, H.; Zou, L. Editorial for the Special Issue: "Human-Environment Interactions Research Using Remote Sensing". *Remote Sens.* **2022**, *14*, 2720. [CrossRef]
40. Rosas-Chavoya, M.; López-Serrano, P.M.; Hernández-Díaz, J.C.; Wehenkel, C.; Vega-Nieva, D.J. Analysis of Near-Surface Temperature Lapse Rates in Mountain Ecosystems of Northern Mexico Using Landsat-8 Satellite Images and ECOSTRESS. *Remote Sens.* **2021**, *14*, 162. [CrossRef]
41. Morley, P.J.; Donoghue, D.N.M.; Chen, J.-C.; Jump, A.S. Integrating remote sensing and demography for more efficient and effective assessment of changing mountain forest distribution. *Ecol. Inform.* **2018**, *43*, 106–115. [CrossRef]
42. Jiao, Z.-H.; Mu, X. Global validation of clear-sky models for retrieving land-surface downward longwave radiation from MODIS data. *Remote Sens. Environ.* **2022**, *271*, 112903. [CrossRef]
43. Racoviteanu, A.E.; Williams, M.W.; Barry, R.G. Optical remote sensing of glacier characteristics: A review with focus on the Himalaya. *Sensors* **2008**, *8*, 3355–3383. [CrossRef]
44. Stow, D.A.; Hope, A.; McGuire, D.; Verbyla, D.; Gamon, J.; Huemmrich, F.; Houston, S.; Racine, C.; Sturm, M.; Tape, K.; et al. Remote sensing of vegetation and land-cover change in Arctic Tundra Ecosystems. *Remote Sens. Environ.* **2004**, *89*, 281–308. [CrossRef]
45. Jombo, S.; Adam, E.; Odindi, J. Quantification of landscape transformation due to the Fast Track Land Reform Programme (FTLRP) in Zimbabwe using remotely sensed data. *Land Use Policy* **2017**, *68*, 287–294. [CrossRef]
46. Takeda, S.; Suzuki, R.; Thein, H.M. Three-Year Monitoring of Shifting Cultivation Fields in a Karen Area of the Bago Mountains, Myanmar. Nihon Shinrin Gakkai Taikai Koen Yoshishu. Available online: https://www.burmalibrary.org/docs20/Shinya-TAKEDA-red.pdf (accessed on 16 December 2022).

47. Mammadov, E.; Nowosad, J.; Glaesser, C. Estimation and mapping of surface soil properties in the Caucasus Mountains, Azerbaijan using high-resolution remote sensing data. *Geoderma Reg.* **2021**, *26*, e00411. [CrossRef]
48. Donthu, N.; Kumar, S.; Mukherjee, D.; Pandey, N.; Lim, W.M. How to conduct a bibliometric analysis: An overview and guidelines. *J. Bus. Res.* **2021**, *133*, 285–296. [CrossRef]
49. Aria, M.; Cuccurullo, C. bibliometrix: An R-tool for comprehensive science mapping analysis. *J. Informetr.* **2017**, *11*, 959–975. [CrossRef]
50. Pfeffer, W.T.; Arendt, A.A.; Bliss, A.; Bolch, T.; Cogley, J.G.; Gardner, A.S.; Hagen, J.-O.; Hock, R.; Kaser, G.; Kienholz, C.; et al. The Randolph Glacier Inventory: A globally complete inventory of glaciers. *J. Glaciol.* **2014**, *60*, 537–552. [CrossRef]
51. Guo, W.; Liu, S.; Xu, J.; Wu, L.; Shangguan, D.; Yao, X.; Wei, J.; Bao, W.; Yu, P.; Liu, Q.; et al. The second Chinese glacier inventory: Data, methods and results. *J. Glaciol.* **2015**, *61*, 357–372. [CrossRef]
52. Gong, P.; Liu, H.; Zhang, M.; Li, C.; Wang, J.; Huang, H.; Clinton, N.; Ji, L.; Li, W.; Bai, Y.; et al. Stable classification with limited sample: Transferring a 30-m resolution sample set collected in 2015 to mapping 10-m resolution global land cover in 2017. *Sci. Bull.* **2019**, *64*, 370–373. [CrossRef] [PubMed]
53. Su, S.; Xiao, R.; Jiang, Z.; Zhang, Y. Characterizing landscape pattern and ecosystem service value changes for urbanization impacts at an eco-regional scale. *Appl. Geogr.* **2012**, *34*, 295–305. [CrossRef]
54. Zhu, X.; Helmer, E.H.; Gao, F.; Liu, D.; Chen, J.; Lefsky, M.A. A flexible spatiotemporal method for fusing satellite images with different resolutions. *Remote Sens. Environ.* **2016**, *172*, 165–177. [CrossRef]
55. Xiao, X.; Boles, S.; Liu, J.; Zhuang, D.; Liu, M. Characterization of forest types in Northeastern China, using multi-temporal SPOT-4 VEGETATION sensor data. *Remote Sens. Environ.* **2002**, *82*, 335–348. [CrossRef]
56. Li, A.; Wang, A.; Liang, S.; Zhou, W. Eco-environmental vulnerability evaluation in mountainous region using remote sensing and GIS—A case study in the upper reaches of Minjiang River, China. *Ecol. Model.* **2006**, *192*, 175–187. [CrossRef]
57. Huang, H.; Chen, Y.; Clinton, N.; Wang, J.; Wang, X.; Liu, C.; Gong, P.; Yang, J.; Bai, Y.; Zheng, Y.; et al. Mapping major land cover dynamics in Beijing using all Landsat images in Google Earth Engine. *Remote Sens. Environ.* **2017**, *202*, 166–176. [CrossRef]
58. Zhang, Z.; Wang, X.; Zhao, X.; Liu, B.; Yi, L.; Zuo, L.; Wen, Q.; Liu, F.; Xu, J.; Hu, S. A 2010 update of National Land Use/Cover Database of China at 1:100000 scale using medium spatial resolution satellite images. *Remote Sens. Environ.* **2014**, *149*, 142–154. [CrossRef]
59. Wulfmeyer, V.; Behrendt, A.; Kottmeier, C.; Corsmeier, U.; Barthlott, C.; Craig, G.C.; Hagen, M.; Althausen, D.; Aoshima, F.; Arpagaus, M.; et al. The Convective and Orographically-induced Precipitation Study (COPS): The scientific strategy, the field phase, and research highlights. *Q. J. R. Meteorol. Soc.* **2011**, *137*, 3–30. [CrossRef]
60. Chen, B.; Xiao, X.; Li, X.; Pan, L.; Doughty, R.; Ma, J.; Dong, J.; Qin, Y.; Zhao, B.; Wu, Z.; et al. A mangrove forest map of China in 2015: Analysis of time series Landsat 7/8 and Sentinel-1A imagery in Google Earth Engine cloud computing platform. *ISPRS J. Photogramm. Remote Sens.* **2017**, *131*, 104–120. [CrossRef]
61. Muno, M.; Bauer, F.E.; Baganoff, F.; Bandyopadhyay, R.; Bower, G.; Brandt, W.; Broos, P.; Cotera, A.; Eikenberry, S.; Garmire, G. A catalog of X-ray point sources from two megaseconds of Chandra observations of the Galactic Center. *Astrophys. J. Suppl. Ser.* **2009**, *181*, 110. [CrossRef]
62. Nie, Y.; Sheng, Y.; Liu, Q.; Liu, L.; Liu, S.; Zhang, Y.; Song, C. A regional-scale assessment of Himalayan glacial lake changes using satellite observations from 1990 to 2015. *Remote Sens. Environ.* **2017**, *189*, 1–13. [CrossRef]
63. Ma, X.; Xu, J.; Luo, Y.; Prasad Aggarwal, S.; Li, J. Response of hydrological processes to land-cover and climate changes in Kejie watershed, south-west China. *Hydrol. Process. Int. J.* **2009**, *23*, 1179–1191. [CrossRef]
64. Li, J.; Sheng, Y. An automated scheme for glacial lake dynamics mapping using Landsat imagery and digital elevation models: A case study in the Himalayas. *Int. J. Remote Sens.* **2012**, *33*, 5194–5213. [CrossRef]
65. Chen, W.; Li, X.; Wang, Y.; Chen, G.; Liu, S. Forested landslide detection using LiDAR data and the random forest algorithm: A case study of the Three Gorges, China. *Remote Sens. Environ.* **2014**, *152*, 291–301. [CrossRef]
66. Yin, Z.-Y.; Zhang, X.; Liu, X.; Colella, M.; Chen, X. An Assessment of the Biases of Satellite Rainfall Estimates over the Tibetan Plateau and Correction Methods Based on Topographic Analysis. *J. Hydrometeorol.* **2008**, *9*, 301–326. [CrossRef]
67. Zhang, G.; Yao, T.; Chen, W.; Zheng, G.; Shum, C.K.; Yang, K.; Piao, S.; Sheng, Y.; Yi, S.; Li, J.; et al. Regional differences of lake evolution across China during 1960s–2015 and its natural and anthropogenic causes. *Remote Sens. Environ.* **2019**, *221*, 386–404. [CrossRef]
68. Jiapaer, G.; Liang, S.; Yi, Q.; Liu, J. Vegetation dynamics and responses to recent climate change in Xinjiang using leaf area index as an indicator. *Ecol. Indic.* **2015**, *58*, 64–76. [CrossRef]
69. Yao, Y.; Wang, X.; Li, Y.; Wang, T.; Shen, M.; Du, M.; He, H.; Li, Y.; Luo, W.; Ma, M. Spatiotemporal pattern of gross primary productivity and its covariation with climate in China over the last thirty years. *Glob. Change Biol.* **2018**, *24*, 184–196. [CrossRef] [PubMed]
70. Patino, J.E.; Duque, J.C. A review of regional science applications of satellite remote sensing in urban settings. *Comput. Environ. Urban Syst.* **2013**, *37*, 1–17. [CrossRef]
71. de Sherbinin, A.; Balk, D.; Yager, K.; Jaiteh, M.; Pozzi, F.; Giri, C.; Wannebo, A. *A CIESIN Thematic Guide to Social Science Applications of Remote Sensing*; Center for International Earth Science Information Network (CIESIN) of Columbia University: New York, NY, USA, 2002.

72. Anderson, K.; Griffiths, D.; DeBell, L.; Hancock, S.; Duffy, J.P.; Shutler, J.D.; Reinhardt, W.; Griffiths, A. A grassroots remote sensing toolkit using live coding, smartphones, kites and lightweight drones. *PLoS ONE* **2016**, *11*, e0151564. [CrossRef] [PubMed]
73. Burchfield, D.R.; Petersen, S.L.; Kitchen, S.G.; Jensen, R.R. sUAS-based remote sensing in mountainous areas: Benefits, challenges, and best practices. *Pap. Appl. Geogr.* **2020**, *6*, 72–83. [CrossRef]
74. Mama, A. Sheroes and Villains: Conceptualizing Colonial and Contemporary Violence Against Women in Africa 1. In *Feminist Genealogies, Colonial Legacies, Democratic Futures*; Routledge: Oxford, UK, 2013; pp. 46–62.
75. Clark, V.R.; Mukwada, G.; Hansen, M.; Adelabu, S.; Magaiza, G.; le Roux, A.; Bredenhand, E.; Otomo, P.V.; Steenhuisen, S.-L.; Franke, A. The Afromontane Research Unit: Driving Connections and Capacity Building for the Sustainable Development of Southern African Mountains. *Mt. Res. Dev.* **2021**, *41*, P1. [CrossRef]
76. Chasmer, L.E.; Ryerson, R.A.; Coburn, C.A. Educating the next generation of remote sensing specialists: Skills and industry needs in a changing world. *Can. J. Remote Sens.* **2022**, *48*, 55–70. [CrossRef]
77. Boulila, W.; Khlifi, M.K.; Ammar, A.; Koubaa, A.; Benjdira, B.; Farah, I.R. A Hybrid Privacy-Preserving Deep Learning Approach for Object Classification in Very High-Resolution Satellite Images. *Remote Sens.* **2022**, *14*, 4631. [CrossRef]
78. Wilkie, D.S. Remote Sensing Imagery for Resource Inventories in Central Africa: The Importance of Detailed Field Data. *Hum. Ecol.* **1994**, *22*, 379–403. [CrossRef]
79. Mngadi, M.; Odindi, J.; Mutanga, O.; Sibanda, M. Quantitative remote sensing of forest ecosystem services in sub-Saharan Africa's urban landscapes: A review. *Environ. Monit. Assess.* **2022**, *194*, 242. [CrossRef]
80. Varshney, K.R.; Chen, G.H.; Abelson, B.; Nowocin, K.; Sakhrani, V.; Xu, L.; Spatocco, B.L. Targeting villages for rural development using satellite image analysis. *Big Data* **2015**, *3*, 41–53. [CrossRef] [PubMed]
81. Tijssen, R. Africa's contribution to the worldwide research literature: New analytical perspectives, trends, and performance indicators. *Scientometrics* **2007**, *71*, 303–327. [CrossRef]
82. Yevide, A.S.; Wu, B.; Khan, A.S.; Zeng, Y.; Liu, J. Bibliometric analysis of ecosystem monitoring-related research in Africa: Implications for ecological stewardship and scientific collaboration. *Int. J. Sustain. Dev. World Ecol.* **2016**, *23*, 412–422. [CrossRef]
83. Mohamedbhai, G. Massification in higher education institutions in Africa: Causes, consequences and responses. *Int. J. Afr. High. Educ.* **2014**, *1*, 59–83. [CrossRef]
84. Deslandes, C.; Kaufmann, L.M.; Anderson, J.R. A systematic and integrative review of qualitative research exploring experiences of acculturation and education among African-born migrants. *Int. J. Intercult. Relat.* **2022**, *86*, 240–257. [CrossRef]
85. Praticò, S.; Solano, F.; Di Fazio, S.; Modica, G. Machine Learning Classification of Mediterranean Forest Habitats in Google Earth Engine Based on Seasonal Sentinel-2 Time-Series and Input Image Composition Optimisation. *Remote Sens.* **2021**, *13*, 586. [CrossRef]
86. Xu, W.; Liu, H.; Zhang, Q.; Liu, P. Response of vegetation ecosystem to climate change based on remote sensing and information entropy: A case study in the arid inland river basin of China. *Environ. Earth Sci.* **2021**, *80*, 132. [CrossRef]

International Journal of
Environmental Research and Public Health

MDPI

Article

Effects of Household Resource Utilization Behaviors on Giant Panda Habitat under the Background of Aging: Evidence from Sichuan Province

Zhenjiang Song [1,2], Baoshu Wu [3,†], Yue Huang [2], Shubin Zhu [1,2], Lan Gao [4,*,†] and Yi Li [4,*,†]

1 Institute of New Rural Development, Jiangxi Agricultural University, Nanchang 330045, China
2 College of Economics and Management, Jiangxi Agricultural University, Nanchang 330045, China
3 School of Business Administration, Jiangxi University of Finance and Economics, Nanchang 330032, China
4 College of Economics and Management, South China Agricultural University, Guangzhou 510642, China
* Correspondence: gaolan@scau.edu.cn (L.G.); xiaoyi1983524@126.com (Y.L.);
Tel.: +86-137-2484-9061 (L.G.); +86-139-2228-0524 (Y.L.)
† These authors contributed equally to this work (Co-first author: qgwubaoshu@163.com (B.W.); Tel.: +86-138-2508-2092 (B.W.)).

Abstract: The Giant Panda (Ailuropoda melanoleuca) is a flagship species for endangered wildlife conservation and is a specific relic species in China. Its habitat conservation has received widespread attention around the world. Since 2010, the phenomenon of an aging labor force gradually appeared within the Giant Panda Nature Reserve and its surrounding communities. Under the new labor force structure, households' resource utilization behavior has had different characteristics, which has led an evolution in giant panda habitats. This study is based on a questionnaire and geographic data. It reveals the internal mechanisms of households' resource utilization behavior impacting giant panda habitat patterns under the ongoing trend of labor force aging. The study shows that labor force aging has promoted rising ecological niche widths and falling ecological niche overlaps. These could drive a growth in giant panda habitat globally. From a spatial perspective, nature reserves with lower comprehensive ecological niche widths and higher ecological niche overlaps face greater conflict between conservation and development. However, the phenomenon of labor force aging mitigates these ecological conflicts to a certain extent.

Keywords: aging labor force; resource utilization behavior; giant panda habitat; ecological niche; ecological niche overlap

Citation: Song, Z.; Wu, B.; Huang, Y.; Zhu, S.; Gao, L.; Li, Y. Effects of Household Resource Utilization Behaviors on Giant Panda Habitat under the Background of Aging: Evidence from Sichuan Province. *Int. J. Environ. Res. Public Health* **2022**, *19*, 15417. https://doi.org/10.3390/ijerph192215417

Academic Editors: Paul B. Tchounwou and Lingxin Chen

Received: 26 September 2022
Accepted: 17 November 2022
Published: 21 November 2022

1. Introduction

At present, the conflict between giant panda conservation and development can no longer be simply classified as a scramble for living space between a growing population and the endangered giant panda. The changes in household labor force have complicated this problem [1,2]. Following reforms that promote the opening up of Chinese society, the separation barrier between rural and urban structures has been increasingly deconstructed [3,4]. The households of the central and western regions have continued to migrate to the southeastern coast to improve their livelihoods [5,6]. After more than 40 years of rural migrant workers migrating into cities, the aging of the labor force in the central and western regions has become significant [7,8]. Agricultural laborers are now predominantly middle-aged and elderly people [9,10]. However, the aging problem is more significant when combined with livelihood stress and institutional constraints in villages around Chinese nature reserves, which are located in high mountain valleys and cold alpine region [10,11]. Therefore, the trade-off between conservation and development has been transformed into a trade-off between small-scale family farming and the expansion of giant panda habitats. The extent of this trade-off has been gradually reducing with the weakening of labor behavior. Therefore, the Chinese government has proposed an improvement of the weak links between these areas and is promoting human-giant panda coexistence [12]. At present, the habitat conservation of giant panda and its umbrella animals has entered a new stage: a harmonious development period of humanistic coordination and ecological conservation.

Many previous studies have explored the effect of household resource utilization behavior on biodiversity under the context of agricultural labor force aging. At present, the interference factors of giant panda habitat include deforestation, bamboo and its shoots cutting, fire utilization, hunting, road, herb gathering, grazing, cultivated land utilization and firewood cutting. However, the cutting of forest and bamboo have been restricted by the cuts target quota. Fire utilization and hunting have been expressly prohibited. The roads among the nature reserves have been abandoned. A series of precious Chinese herbal medicine grow in alpine meadows outside the giant panda habitat, while the proportion of herb gathering was lower. Meanwhile, grazing was usually found in alpine meadows far from the giant panda habitat and the behavior of bamboo shoots cutting was strictly limited to March and April, due to which interference was relatively small. However, cultivated land utilization and firewood cutting were closely related to the livelihood of households, which were common in Sichuan giant panda habitats. Therefore, this paper took these two resource utilization behaviors as typical representatives. Regarding the change in household resource utilization behavior, cultivated land in the edge of the forest has experienced frequent geological disasters (such as debris flow and landslides) [10,11,13], serious soil erosion [14,15] and cultivated land fragmentation [16]. These disasters have intensified the abandonment of cultivated land utilization. In addition, a large area of cultivated land presides in high mountain valleys, which is hard and steep, meaning that the aging labor force are unable to cultivate it [17,18]. The utilization rate of cultivated land has decreased with the increasingly aging labor force [19,20]. Fuelwood is cut from brushwood in high mountain valleys. However, as the labor capacity of the aging population is restricted [10,21], hydropower stations provide free electricity for households in the villages around the nature reserves and have replaced fuelwood [11,22]. Meanwhile, fuelwood utilization has been decreasing with the aging of the agricultural labor force [23,24]. Some reforms surrounding resource utilization have disappeared, such as the expanded Ecological Public-Welfare Forests [25,26], a larger Giant Panda National Park which covered important nature reserves and surrounding communities [27,28] and intensified regulations of planting behavior and firewood cutting. Under the diverse pressures of regulations [29,30], weakened labor capacity [10] and natural condition constraints [31,32], households have gradually accepted livelihood transformation and energy structure revolution.

At present, there are some research directions surrounding the effect of human activities on the giant panda habitat, such as the ecological research method and economic research method. The ecological research method sets an observation point to compare the differences between infrared image probability of giant panda [33]. It also uses the method of catch-rate-per-day and one clamp to explore the change of habitat at multiple time quantum, using capture rate as a research basis [34]. It also used FRAGSTATS to analyze the landscape pattern to explore the evolution of habitats [35]. The economic research method uses the casual model, which analyzes the effect of an economic activity (or disturbance activity) on giant panda habitats [36–38]. In addition, the geographical research method is also applicable. Some scholars have integrated the concepts of ecology and have taken natural geographic units as evaluation units. This could evaluate the habitat of giant pandas using a static mathematical analysis method, such as the multi-index comprehensive evaluation method, grey relational analysis or the entropy method [39]. However, a static evaluation model is unable to reflect the trajectory characteristics of giant panda habitat changes. Therefore, an analytical framework of dynamically changing spatio-temporal process should be created for the evaluation of giant panda habitats. As such, this research adopts the ecological niche model to reveal the trajectory of changes in giant panda habitats, under the context of human activity transformation, through the combination of "state" and "potential". In the model, the ecological niche overlap accurately reflects the quantified competitive relations. The model is able to analyze the change in giant panda habitat under the context of current cultural and environmental changes in a multi-dimensional dynamic view. Therefore, the research aims to build an ecological niche model with macro-micro data interaction to explore the changes in giant panda habitat under the context of agricultural labor force aging in Sichuan giant panda nature reserves, using nine nature reserves as examples. This study could be used as a reference for the further enhancement of the habitat level of the giant panda.

2. Research Design

2.1. Interaction between Household Resource Utilization Behavior and Habitat Pattern

2.1.1. Interactive Mechanisms between Household Resource Utilization Behavior and Habitat Pattern

The interactive mechanisms between household resource utilization behavior and habitat pattern occur between individual decision clustering and group decisions, and between re-clustering and overall decisions. In a rural household survey, households were considered as independent individuals, and a group consisted of households with similar decision-making behaviors. Meanwhile, a village constituted a decision-making whole. However, the clustering method was nonadditive. This was because households did not have the main direction when they made these decisions. Furthermore, these short-term changes were accumulated across time and space. Therefore, the concept of mapping was introduced into the clustering process. In the macro scale, a village was regarded as the interference source. Therefore, individuals could cluster into a whole, and villages were used as evaluation units. From the point of view of space, the research could reveal the impact of household resource utilization behavior on panda habitats. Furthermore, it allowed for the exploration of the relationship between the resource utilization behavior of the microcosmic body (household) and the mapping result of their interaction (macro-habitat landscape pattern) through multi-scale transformation (shown in Figure 1) [40].

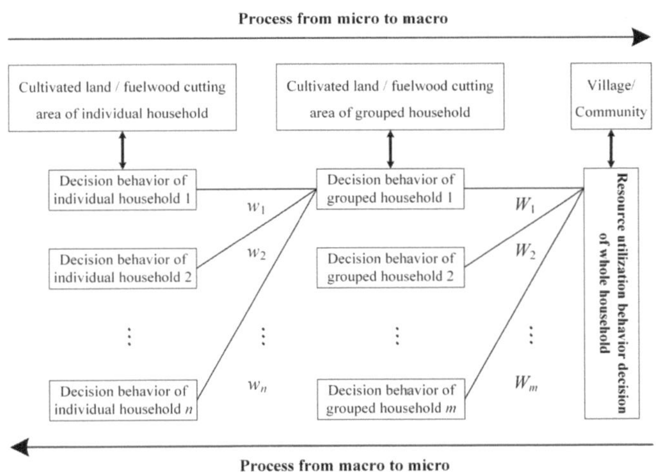

Figure 1. Spatial manifestation and transformation mechanism of households' resource utilization behavior decision.

2.1.2. Construction of Individual Household Resource Utilization Decision Model

- Construction of Belief Model

Belief referred to the cognition of the current system state and the estimates of the future state of agents; this definition of belief was based on a study by Wooldridge [40,41]. Consequently, the beliefs of households could be understood as the cognition of the current resource utilization behavior under the background of nature reserve regulation institution and the estimates of the future state under the context of aging labor force dominance.

$$Belief_{t+1} = f(P_t, Belief_t), \tag{1}$$

In the equation, $Belief_t$ was the cognition of the current resource utilization behavior under the background of nature reserve regulation institution. P_t was the household perception of different influencing factors in resource utilization activities.

- Construction of Household Resource Utilization Decision Model

Aspiration was defined as the expected achievement state under the background of nature reserve regulation institution [40]. Therefore, households' decisions could be calculated through the following equation:

$$Decision_{t+1} = f(Belief_t, Aspiration_t), \tag{2}$$

In the equation, $Decision_t$ was the decision of a household at the time t. $Aspiration_t$ represented the importance of household resource utilization patterns under the $Aspiration_t$. The importance of the kth resource utilization pattern by the household and j expressed the decision of time t with $Belief_t$ and $Aspiration_t$ of the household at the time t. The model controlled for the following conditions:

o The household resource utilization decisions should be mainly considered over the past five years.
o There was no significant difference in the technical level of resource utilization among households.
o The physical geographical environment of the households were similar.
o There was no technological progress in the field of household resource utilization during the study period.
o The external institutional environment remained unchanged during the study period.
o There was no significant difference in labor literacy among households.
o The price of resource market was stable during the study period.
o The data of household resource utilization behavior came from the survey of the research group.

Therefore, a household resource utilization decision model was constructed as follows:

$$Decision_{jkt} = UA_{jkt} \times I_{jkt} / \sum_{k=1}^{n} \left(UA_{jkt} \times I_{jkt} \right), \tag{3}$$

In the equation, $Decision_t$ was the importance of the kth resource utilization pattern by the household j at the time t. UA_{jkt} was the area/amount/intensity of the kth resource utilization pattern by the household j at the time t. I_{jkt} was the revenue of the kth resource utilization pattern by the household j at the time t. In the paper, k = 2. Therefore, the importance of the resource utilization methods was not ranked.

2.1.3. Construction of Final Household Resource Utilization Decision Model

The final household resource utilization decision depended on ecological value recognition, social pressure, environmental constraints and labor literacy, under the context of aging labor force. Therefore, the decision equation for the following period could be constructed as follows:

$$Decision_{jk,\,t+1} = f(Decision_{ikt} \mid I = 1, 2, 3, 4), \tag{4}$$

In the equation, $Decision_{u,\,t+1}$ was the decision of the household for the next period. $Decision_{it}$ was the ecological value identity, social pressure, environmental constraints and labor literacy acting on household decision, respectively, at the time t.

In study area, there were few conversions of cropland and forest land. Land rent was relatively stable. Two rounds of reforestation had been completed. There was no government-led reforestation during the study period. The reduction in cultivated land area was the result of individual household's decision-making behaviors. The research objects were the resource (cultivated land and fuelwood) utilization decision of households alone. This was also the case for part-time households (operating eco-tourism), part-time households (out-migration for work) and non-farm households, under the context of aging labor force dominance. Therefore, the final resource utilization decisions of individual households were mainly influenced by ecological value identity, social pressure, environmental constraints and labor literacy. These were shown as choices between different resource utilization pattern. Therefore, the above equation could be further transformed:

$$Decision_{jk,t+1} = \sum_{x=1}^{4} w_x Decision_{jkt}, \quad (5)$$

In the equation, w was the weight coefficient of the different factors on the effect of household decision behavior. $w_x = a_x / \sum_{x=1}^{4} a_x$ and $w_1 + w_2 + w_3 + w_4 = 1$ [40,41].

2.2. Ecological Niche Evaluation Model

2.2.1. Comprehensive Evaluation Index System of Ecological Niche

Based on the theory of "multi-dimensional super volume", and the characteristics of the giant panda habitat level, this paper has developed a comprehensive evaluation index system with 4 dimensions and 14 measurement indexes (shown in Table 1).

Table 1. Comprehensive evaluation index system of ecological niches at the habitat level of giant panda.

Target	Dimension	Measurement Index	Explanation of Indexes
Ecological Niche Width of Giant Panda Habitat	X_1: Resource Utilization Strength level dimension	x_1: Cultivated land utilization decision of community	Obtained by MAS model calculation, $x_1 \in [0, 1]$
		x_2: Fuelwood utilization decision of community	Obtained by MAS model calculation, $x_2 \in [0, 1]$
	X_2: Resource Utilization Support force level dimension	x_3: Proportion of available cultivated land in community	$AACL/TACL$
		x_4: Proportion of available fuelwood land/miscellaneous shrub forest in community	$AAFL(AMSSF)/TAFL$
		x_5: Proportion of practitioners	$LFQH/TPR$
	X_3: Resource Utilization Constraint level dimension	x_6: Perception on intensity change of planting industry regulation	Obtained by MAS model calculation, $x_6 \in [1, 5]$
		x_7: Perception on intensity change of firewood cutting regulation	Obtained by MAS model calculation, $x_7 \in [1, 5]$
		x_8: Changing rate of cultivated land area over the last five years	Obtained by MAS model calculation, $x_8 \in [0, 1]$
		x_9: Changing rate of fuelwood cutting distance over the last five years	Obtained by MAS model calculation, $x_9 \in [0, 1]$
		x_{10}: Changing rate of fuelwood cutting labor-hour over the last five years	Obtained by MAS model calculation, $x_{10} \in [0, 1]$
	X_4: Giant Panda Sustainable Habitat Level Dimension	x_{11}: Changing rate of giant panda habitat	$(AGPH4th - AGPH3rd)/AGPH3rd$
		x_{12}: Changing rate of wild giant panda population	$(PGPH4th - PGPH3rd)/PGPH3rd$
		x_{13}: Area percentage of staple food bamboo	$ASFB/ANR$
		x_{14}: Owned percentage of staple food bamboo in giant panda population	$ASFB/PGPH4th$

Note: (1) Area of available cultivated land was shortened to *AACL*. (2) Area of available fuelwood land/miscellaneous shrub forest was shortened to *AAFL/AMSSF*. (3) Area of giant panda habitat during the 3rd survey report on giant pandas was shortened to *AGPH*3rd. (4) Area of giant panda habitat during the 4th survey report on giant pandas was shortened to *AGPH*4th. (5) Area of staple food bamboo was shortened to *ASFB*. (6) Area of nature reserve was shortened to *ANR*. (7) Total area of cultivated land was shortened to *TACL*. (8) Total area of forestland was shortened to *TAFL*. (9) Labor force quantity of household was shortened to *LFQH*. (10) Total population of region was shortened to *TPR*. (11) Population of giant panda habitat during the 3rd survey on giant pandas was shortened to *PGPH*3rd. (12) Population of giant panda habitat during the 4th survey on giant pandas was shortened to *PGPH*4th.

2.2.2. Calculating Method of Ecological Niche Width

- Data Standardization

The purpose of the data standardization was to combine the units and dimensions of each index, which were nondimensionalized. The paper adopted the standard deviation method for standardization. The equation was as follows:

$$X_{ij} = \left| \frac{x_{ij} - \overline{x_j}}{\delta_j} \right|, \tag{6}$$

In the equation, x_{ij}, x_{ij}, $\overline{x_j}$, and δ_j were the standardized value, original value, arithmetic mean and standard deviation of the jth index in ith community, respectively.

- Calculation of index Weights

The index weights were calculated by the following equation:

$$V = \frac{\delta_j}{\overline{x_j}}, \tag{7}$$

$$W_j = \frac{V}{\sum\limits_{j=1}^{n} V_j}, \tag{8}$$

In the equation, V was the variable coefficient of each index. W_j was the weight of the jth index.

- Calculation of Ecological Niche Width

The calculating equation of ecological niche width was as follows:

$$N_i = \frac{W_{ik}(S_i + A_i P_i)}{\sum\limits_{j=1}^{n} (S_j + A_j P_j)}, \tag{9}$$

In the equation, $i, j = 1, 2, \ldots, n$. N_i was the relative ecological niche in the community i. S_i and S_j were the state of communities i and j, respectively. P_i and P_j were the potentials of community i and j, respectively. A_i and A_j were the conversion coefficients of dimension. $S_i + A_i P_i$ indicated the absolute ecological niche of community i. W_{ik} was the k-item weight of the community i. The value range of the ecological niche width was between 0 and 1. When the ecological niche width was closer to 1, this indicated a higher ecological level [42].

The comprehensive ecological niche width could obtain the relative ecological niche width through an arithmetic method. The equation was as follows:

$$M_i = \sum_{j=1}^{n} N_{ij} w_j, \tag{10}$$

In the equation, M_i was the comprehensive ecological niche in the community i. N_{ij} was the relative ecological niche of the j-dimension in the community i. w_j was the weight of the j-dimension [42].

- Calculation of Ecological Niche Overlap

The ecological niche overlap reflected the relationship of ecological competition.

$$O_{jk} = \frac{\sum\limits_{i=1}^{n} P_{ij} P_{ik}}{\sqrt{\sum\limits_{i=1}^{n} P_{ij}^2 \sum\limits_{i=1}^{n} P_{ik}^2}}, \tag{11}$$

In the equation, O_{jk} was the ecological niche overlap value of species k to species j. n was the total resource status. P_{ij} and P_{ik} were the percentage ownerships of resources i in the total resources utilized by species j and species k, respectively [42].

3. Study Areas and Data Sources

3.1. Study Areas

The study areas were the Sichuan Giant Panda Nature Reserves and their surrounding communities. According to the sampling prescription, four national natural reserves and five provincial nature reserves were sampled in this study. These were distributed across 35 communities (villages), 17 towns, 9 counties and 6 cities in Sichuan Province (shown in Figure 2). The area of giant panda habitat in the study area was 302,940.03 hm^2, which accounted for 29.39% of total giant panda habitat area in Sichuan Province. The population of wild giant panda was 304, which accounted for 34.23% of the total wild giant panda population. Much of the youth labor force has migrated outside of the area and the aging of the labor force is significant. In addition, cultivated land has encroached on the edge of forestland. Furthermore, households still rely on fuelwood. These factors have disturbed the habitat of giant panda. Meanwhile, the researchers consulted the administrative staff of the Forestry and Grassland Administration of Sichuan Province, each giant panda nature reserve administration and each giant panda nature reserve station. The nature reserves in the samples were found across the entire scope of Giant Panda National Park. Therefore, the samples were assumed to be representative in revealing current problems faced by nature reserves.

3.2. Data Sources

3.2.1. Survey Data and Study Sample

At present, the Wawushan Giant Panda Nature Reserve is primarily located in lots that do not belong to any individual and is a stated-owned forest. Some communities were located on the edge of the nature reserve. There were less than 10 communities with significant relevance to the nature reserves and 3–4 adjoining communities. In this paper, we adopted a stratified sampling method. Four communities were sampled in each nature reserve. Two communities were located inside the nature reserves. The community presiding inside the Tangjiahe Giant Panda Nature Reserve was formed by merging two villages. Meanwhile, two communities were located outside the nature reserves. In each community, 15–18 households were randomly sampled as respondents.

The data were obtained from household questionnaires during July 2018, October 2018, January 2019 and May 2019. A total of 557 questionnaires were distributed, and 538 valid questionnaires were obtained after the rejection of invalid questionnaires. The samples were in the experimental areas of the giant panda nature reserve and their surrounding communities, within a 5 km geographic buffer zone outside the nature reserves. Some communities were included in the scope of the Giant Panda National Park. At present, the phenomenon of labor force aging is significant in the study areas, and the resource utilization behavior of households have been constrained (as shown in Table 2).

3.2.2. Survey Data on Giant Panda

The area of giant panda habitat, the population of wild giant panda, the area of nature reserve and the area of staple food bamboo were obtained from *The 3rd National Survey Report on Giant Panda in China* [43] and *The Pandas of Sichuan: The 4th survey report on giant panda in Sichuan Province* [33].

3.2.3. Land Use Data

The land use data included in this paper were sourced from public Landset TM/ETM+ /OLI images on Geospatial Data Cloud (http://www.gscloud.cn (accessed on 10 September 2022)), which was released by the Computer Network Information Center, Chinese Academy of Sciences. The bands, strip numbers and row numbers of two period images were used. The images were taken between June and September. This defined period made it easy to identify the information in the earth surface. The images were processed through steps such as band compositions, geometric correction, image mosaic, geometry clipmaps, supervised classification and the unification of coordinate systems. Following his, the land use patterns were classified as cultivated land, forestland, grassland, shrubland, wetland, water bodies, tundra, artificial surfaces, bareland and permanent snow and ice. These were based on the *Current Land Use Classification* (GB/T 21010-2017).

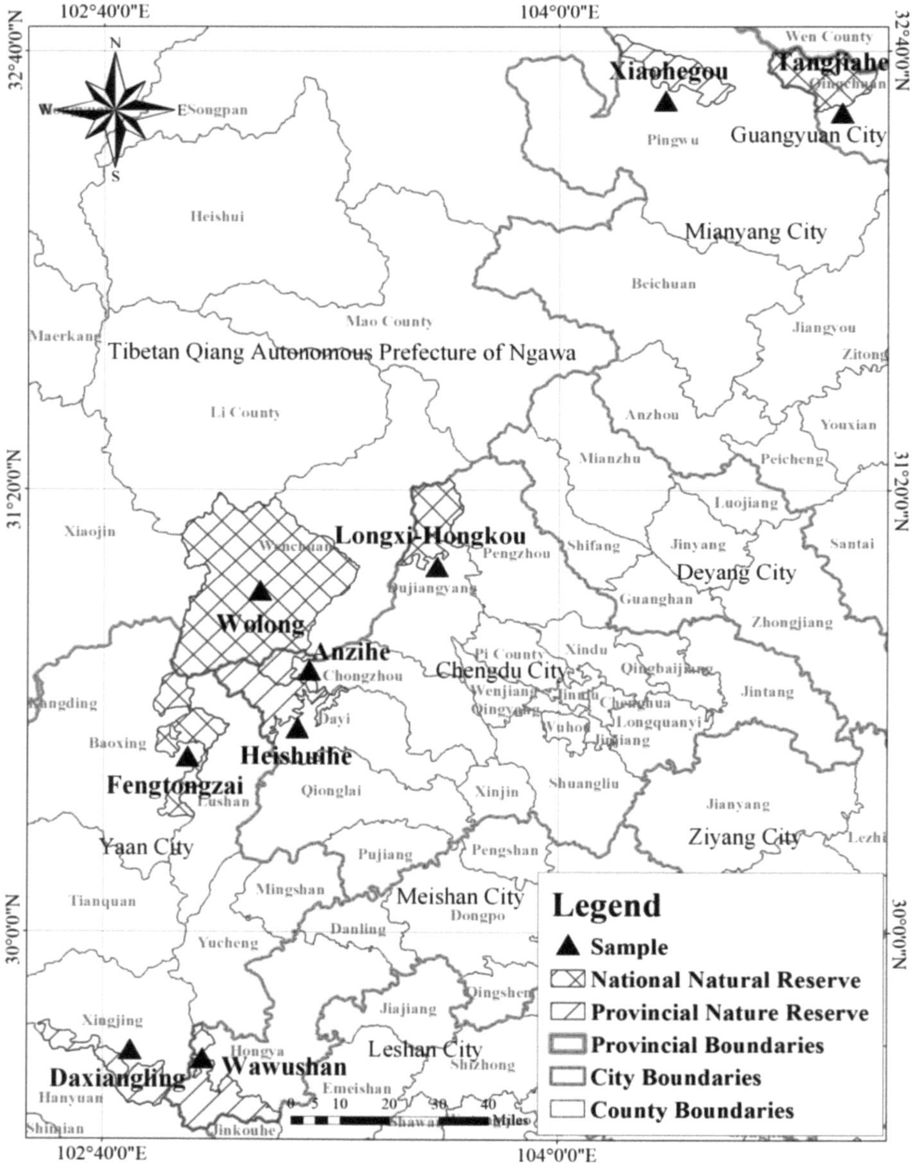

Figure 2. Study area.

Table 2. Characteristic of study areas.

Variables	Explanation of Variables	Mean	Standard Deviation	Minimum	Maximum
Labor force aging	Quantity of labor force aging/Total quantity of labor force	0.289	0.327	0	1
Proportion of aging labor force in the field of planting	Quantity of aging labor force in the field of planting/Total quantity of labor force	0.478	0.287	0	1
Proportion of aging labor force in the field of fuelwood utilization	Quantity of aging labor force in the field of fuelwood utilization/Total quantity of labor force	0.455	0.281	0	1
Educational level of aging labor force	1 = Elementary school and below; 2 = Junior high school; 3 = Senior high school and above	1.271	0.887	1	3
Health level of aging labor force	1 = Serious disease and disability; 2 = Chronic disease; 3 = Health	2.385	1.439	1	3
Type of household's livelihoods	1 = Totally dependent on agriculture; 2 = Part-time household; 3= Non-agricultural household	2.017	0.620	1	3
Soil quality	1 = Poor; 2 = General; 3 = Well	1.704	1.120	1	3
Degree of cultivated land fragmentation	(Quantity of cultivated land plots − 1)/(Total area of cultivated land/Area of minimum cultivated land plot in the study area)	0.733	1.507	0	1
Slope of Fuelwood cutting destination	1 = Gentle; 2 = Steeper; 3 = Extremely steep	2.617	1.241	1	3
Restriction of fuelwood cutting	1 = Existence; 0 = Inexistence	0.684	0.474	0	1
Mean distance of fuelwood cutting destination	Kilometer	4.891	4.247	0	16

Note: (1) From the perspective of households' resources utilization, the proportion of aging labor force in the field of planting and fuelwood utilization were 47.80% and 45.50%. Therefore, aging labor force has become the main body of resource utilization activities. (2) From the perspective of labor force literacy, lower educational level and sub-health status was common among aging labor forces, whose mean values were 1.271 and 2.385. (3) From the perspective of resource utilization constraints, the trend towards non-agriculture (2.017) has been obvious in the study area. Meanwhile, less fertile soil (1.704) and fragmented cultivated land (0.733) could restrict the planting behavior of households. Extremely steep topography (2.617), strongly restriction of fuelwood cutting (0.684), and far distance of fuelwood cutting destination (4.891) could restrict the fuelwood utilization behavior of households.

3.2.4. Geographic Information Data

A series of basic vector data came from the Basic Geographic Database of China (http://www.webmap.cn/commres.do?method=result100W (accessed on 10 September 2022)) and the 4th survey on giant pandas in Sichuan Province. This data had a 1:1 million scale and the 4th survey on giant panda in Sichuan Province [41], which included roads, rivers, the points of human activity and the points of giant panda activity trace. The points of cultivated land utilization were obtained from the map of cultivated land in LUCC. The points of firewood utilization came from the survey data. A 30 m Digital Elevation Model (DEM) was obtained from the Geospatial Data Cloud, which was produced by the Chinese Academy of Sciences Computer Network Information Center (http://www.gscloud.cn/ (accessed on 10 September 2022)). Slope and slope direction were obtained based on the DEM.

4. Pattern of Giant Panda Habitat under the Different Age Structure of the Labor Force

4.1. Width of Comprehensive Ecological Niche

In this section, the measurement indexes of state were the values during 2013, and the measurement indexes of potential were the increment of each index during the period of 2013 to 2018. The dimensional transformation coefficient was one year ago. According to the methods discussed above, the ecological niche value of the giant panda habitat could gain. This included the old-age group (the aging labor force ≥55 years ago) dominating the natural resource utilization behavior of households, which was shortened to OAG and the young adult group (the young adult labor force <55 years ago) dominated the natural resource utilization behavior of households, which was shortened to YAG (shown in Figure 3).

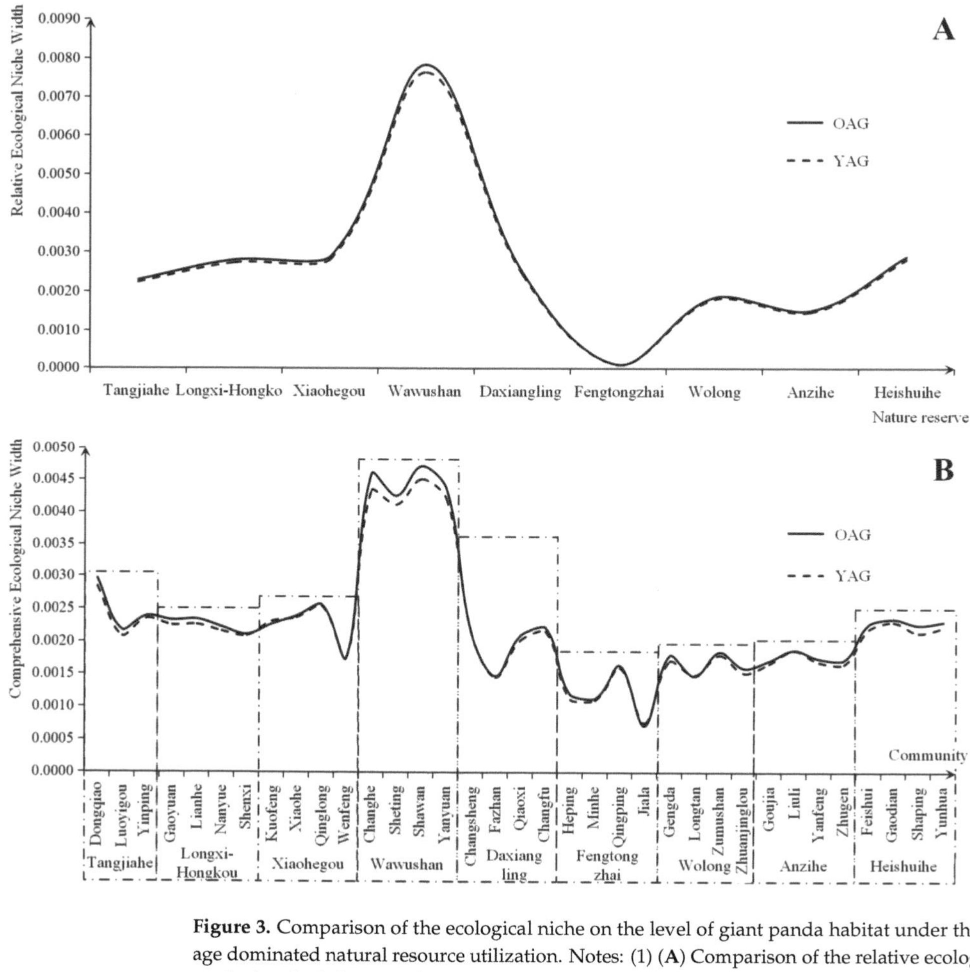

Figure 3. Comparison of the ecological niche on the level of giant panda habitat under the different age dominated natural resource utilization. Notes: (1) (**A**) Comparison of the relative ecological niche on the level of giant panda sustainable habitat. (**B**) Comparison of the comprehensive ecological niche on the level of giant panda sustainable habitat. (2) In the above two figures, the ecological niche widths of the OAG were generally higher than the YAG. The differences between the ecological niche widths of the OAG and YAG were smaller in the nature reserves with better giant panda habitat. (3) The level of giant panda habitat in the Daxiangling Mountain System Reserve was higher than others. The nature reserves in the Min Mountains were second. However, the giant panda habitat in the nature reserves in Qionglai Mountains faced certain risks.

Regarding the relative ecological niche of the level of giant panda sustainable habitat, although the level of OAG was generally higher than YAG, the difference between their ecological niche widths was small (as shown in Figure 3). As shown in the results in Figure 3A, the aging phenomena created higher levels of sustainable habitat for giant pandas. For example, the level of giant panda sustainable habitat in the Wawushan Nature Reserve was in the front of the region and displayed significant differences between the differentiation of giant panda sustainable habitat due to the labor force age structure (the ecological niche width in the OAG was 0.0078 and the ecological niche width in the YAG was 0.0076). The Xiaohegou Nature Reserve and the Heishuihe Nature Reserve displayed similar results. There were two possible reasons for this phenomenon. Firstly, labor force aging has prompted older households to shift toward less labor-intensive work, which is less disruptive to the ecological environment, such as beekeeping and white tea cultivation. These industries couple regional natural resource utilization and conservation objectives. Secondly, the aging labor force in the region either lived alone in the village, had moved to a city or lived with their children. Some young adults had left the area for employment elsewhere, meaning that the permanent population were children, women and the elderly. The impact of the size of the permanent population on divided households created miniaturization. This reduced the demand of energy in households and weakened the labor force resources, which drove households to prefer easy and cheap clean energy (such as Hydropower, and marsh gas) to adapt to the changing household structure. Therefore, the aging labor force in the region reduced the intensity of fuelwood cutting to adapt to the aging of the labor force. However, the relative ecological niche widths of the level of giant panda sustainable habitats in the Fengtongzhai Nature Reserve were lower and age had no significant effect on the sustainable habitat differentiation of the giant panda (The Ecological Niche Widths of OAG and YAG were 0.0001). There were three main reasons for this. Firstly, the industrial cluster of ecotourism was dominated by nuclear and main households, and the aging trend was not significant in the communities built after the Wenchuan Earthquake, such as David Town in Baoxing County. Secondly, the reconstructed communities were usually equipped with infrastructure for clean energy, such as marsh gas and solar energy. However, some business operations still relied on wood burning, such as for bonfire parties and heating by brazier. Therefore, households' livelihoods still relied on firewood. Thirdly, Tibetans grew field corn to feed yak without intensive cultivation, which did not rely highly on the literacy of the labor force. Therefore, the intensity of planting behavior in the region did not fluctuate significantly with the change in the labor force structure. Therefore, the relative ecological niche widths of the level of giant panda sustainable habitat was at a relatively low level.

The widths of OAGs were generally higher than YAGs and labor force aging was more significant in the relatively higher comprehensive ecological niche widths of giant panda nature reserves (as shown in Figure 3B). According to the results, the level of the comprehensive ecological niche widths in the Wawushan Nature Reserve were the highest in the region (The ecological niche widths of the OAG in Changhe Village, Sheting Village, Shawan Village and Yanyuan Village were 0.0046, 0.0043, 0.0047 and 0.0044, respectively. The ecological niche widths of the YAG in these villages were 0.0043, 0.0041, 0.0045 and 0.0042, respectively). However, there was a large gap in the ecological niche widths between the OAG and YAG. This was attributed to the exclusivity of natural resource utilization, which only allowed these resources to be used within a given time and quantity of labor force. However, the aging labor force was weaker than the young labor force in terms of resource acquisition. This was because of their weak physiological function and ecological identity. Hence, the ecological niche widths of the OAG were higher than for the YAG. The Tangjiahe Nature Reserve, Longxi-Hongkou Nature Reserve, Xiaohegou Nature Reserve and Heishuihe Nature Reserve came second. The comprehensive ecological niche widths in Daxiangling Nature Reserve (The Ecological Niche Widths of OAG in Changsheng Village, Fazhan Village, Qiaoxi Village and Changfu Village had results of 0.0022, 0.0015, 0.0021 and 0.0022, respectively. The ecological niche widths of the YAG in these villages were 0.0022, 0.0015, 0.0020 and 0.0022, respectively). The results for the ecological niche width for the OAG in the Fengtongzhai Nature Reserve in Heping Village, Minhe Village, Qingping Village and Jiala Village were 0.0012, 0.0011, 0.0016 and 0.0007, respectively. The ecological niche widths of the YAG in these villages were 0.0011, 0.0011, 0.0016 and 0.0008, respectively. These were the most unstable. This was because Fazhan Village was close to the boundary of the nature reserve and it was also located in the southern gate of the Giant Panda National Park, leading to the development of ecotourism and overstretched natural resources. Meanwhile, ecotourism was also a leading industry in Heping Village and Qingping Village, which had a similar problem to Fazhan Village. Furthermore, Jiala Village was similar to Jiarong Tibetan village. Grazing, bonfire burning, and corn cultivation were widespread in Jiala Village, where households relied on natural resources to earn a living. Therefore, its ecological niche width was lower. In addition, the comprehensive ecological niche widths of the OAG in the Fengtongzhai Nature Reserve were slightly higher than the YAG. It also showed that the aging labor forces were weakened, which had a lesser effect on ecology. Therefore, the phenomenon of labor force aging had a regulatory effect on resource utilization decisions.

In view of the spatial pattern, nature reserves with a significantly aging labor force usually had high comprehensive ecological niche widths. The nature reserves with the best habitat were found in the Wawushan Nature Reserve in the Daxiangling Mountains. The nature reserves with relatively low comprehensive ecological niche widths were found in the Qionglai Mountains (as shown in Figure 4). Therefore, the spatial pattern of comprehensive ecological niche widths on the level of giant panda habitat further verifies the previous research results, which showed that the nature reserves and their closely related communities with a significantly aging labor force usually had relatively high ecological niche widths. Labor force aging further drove households to reduce the intensity of their resource utilization, while indirectly supporting the conservation of giant panda habitat.

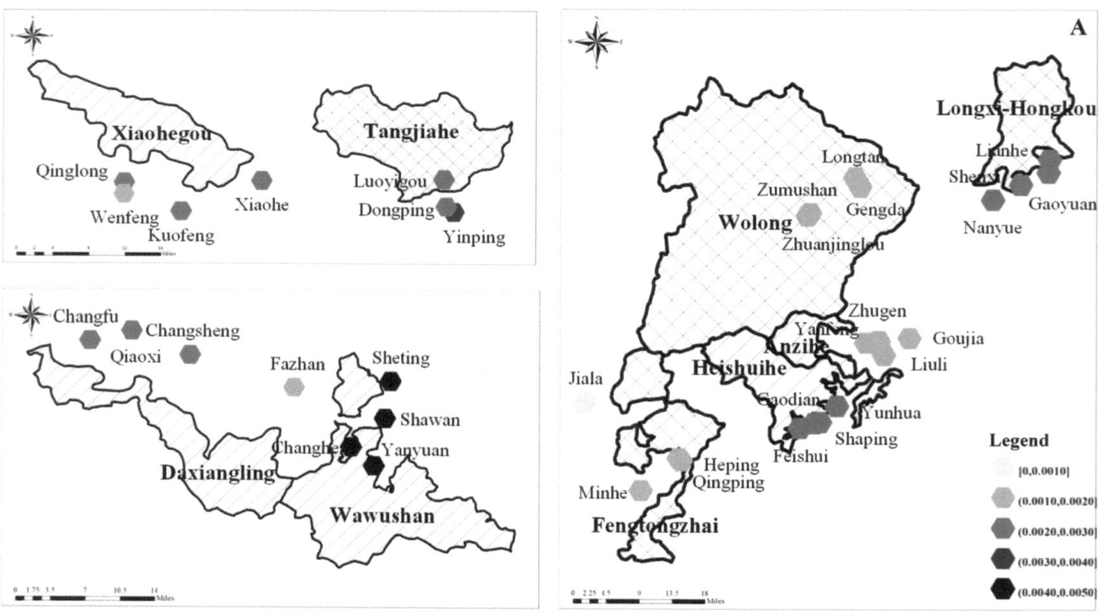

Figure 4. *Cont.*

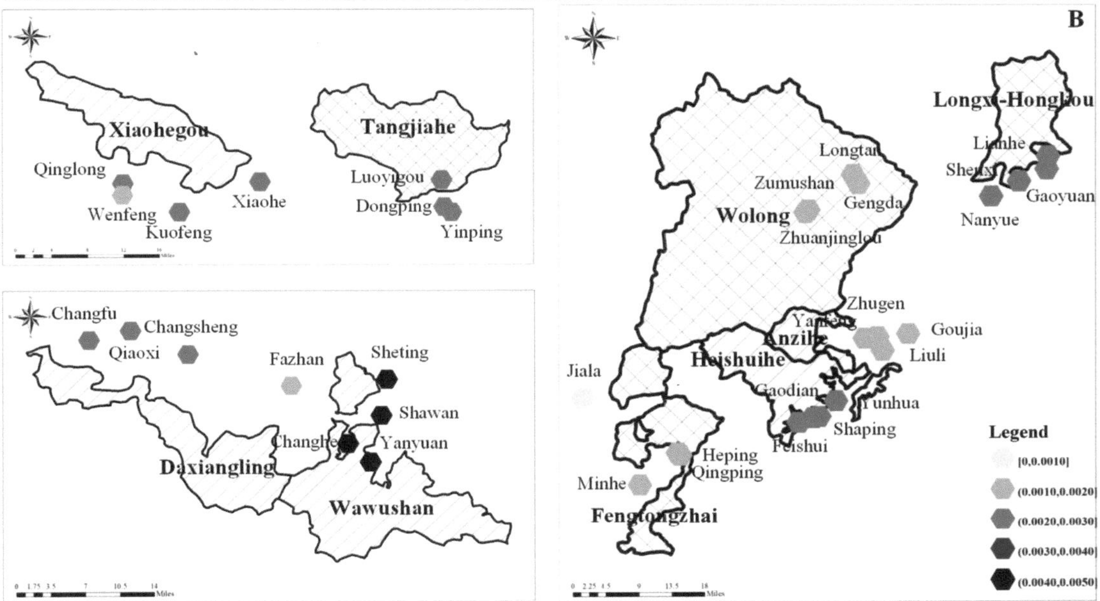

Figure 4. Comprehensive Ecological Niche on the level of giant panda habitat under the different age dominated natural resource utilization. Notes: (**A**) Comprehensive Ecological Niche of OAG. (**B**) Comprehensive Ecological Niche of YAG. Map Content Approval Number: CS (2019) 282.

4.2. Ecological Niche Overlap

The purpose of measuring the ecological niche overlap was to explore the competitive relationship between the communities that were close to the nature reserves (the ecological niche overlap indexes are shown in Figure 5).

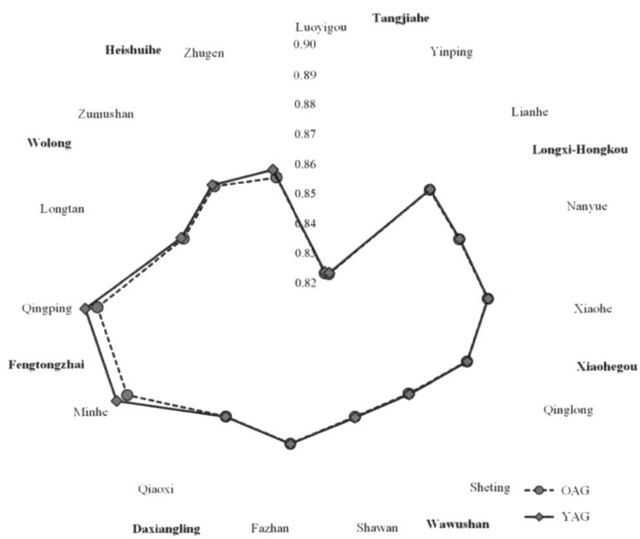

Figure 5. Ecological Niche Overlap Index between a part of communities and their neighboring communities.

The ecological niche overlap commonly existed in the OAG and the YAG. The ecological niche overlap of the YAG was slightly higher than the OAG. The values in some villages were higher than 0.800, such as in Luoyigou Village and Yinping Village in the Tangjiahe Nature Reserve. The ecological niche overlaps of Changfu Village in the Daxiangling Nature Reserve were over 0.9000. Some problems were common in these communities. Firstly, some communities were located inside the nature reserves, where there were experimental zones at an elevation of above 1 km. In these communities, households relied on the resources inside the nature reserves. Their resource utilization behavior always disturbed giant panda habitat, such as in Luoyigou Village in the Tangjiahe Nature Reserve, Lianhe community in the Longxi-Hongkou Nature Reserve and Wolong Special Administrative Region in the Wolong Nature Reserve. Secondly, some communities depended on nature reserves to operate ecotourism, which were far away from the nature reserves. However, they also had a relatively weak regulation to intensify the intensity of household resource utilization behavior where giant panda habitats would be disturbed. Thirdly, some communities relied on hydropower before the small hydropower station was shut down in 2017. These communities were Sheting Village, Shawa Village and Yanyuan Village in the Wawushan Nature Reserve. However, regional fuelwood utilization behavior significantly expanded after 2017, which caused the intensified human activities to disturb the habitat of giant pandas (as shown in Figure 5).

As shown in Figure 5, the ecological niche overlaps of the YAG were higher than the OAG in the Tangjiahe Nature Reserve. This indicates that the weakened labor force had relatively lower disturbances on giant panda habitats. The aging labor force reduced the intensity of resource utilization to restore ecology. This had a negative effect on the growth of the ecotourism economy. Therefore, the OAG had lower ecological niche overlaps. In the Wawushan Nature Reserve, the aging labor forces in Changhe Village were predominantly engaged in agroforestry to keep bees and plant white tea for the improvement of giant panda habitat level. Therefore, these aging labor forces were more sympathetic to environmentally friendly production behavior and the ecological niche overlaps of the OAG were lower than the YAG. In the Daxiangling Nature Reserve, households relied on the forestry industry in Changsheng Village. They mainly engaged in the cultivation and conservation of bamboo, willow cedar and yellow cypress. The main labor force was the OAG. Households actively supported ecological protection to improve their livelihoods and to reduce the disturbance of production activities on the ecological environment. They aimed to achieve conservation development. Therefore, it is clear that households made rational choices for sustainable livelihoods under the context of labor force aging, who relied on eco-industries.

Positive differences in ecological niche overlaps on the level of giant panda habitat were mainly found in most of regional communities under the context of labor force aging. Negative differences were mainly found in the communities which were far away from the nature reserves and had relatively frequent human activities and a relatively complex population structure, such as Wenfeng Village and Kuofeng Village in the Xiaohegou Nature Reserve, Feishui Village in the Heishuihe Nature Reserve, Jiala Village in the Fengtongzhai Nature Reserve, and Changfu Village in the Daxiangling Nature Reserve. A positive difference in the ecological niche overlap was the ratio of ecological niche overlaps of the OAG to ecological niche overlaps of the YAG, of which the data range was (0, 1). The data range of the negative differences was $(1, +\infty)$. This also showed that there were spatial differences in the effects of the aging labor force on the competing relationships between giant panda habitat levels and ecology. This further shows that the communities who lived under a special geographical environment faced resource utilization constraints and labor force weakening. This weakened the intensity of household resource utilization behavior to reduce the disturbance of giant panda habitat. Therefore, it could indirectly weaken the ecological conflicts and improve the number of giant panda sustainable habitats (as shown in Figure 6).

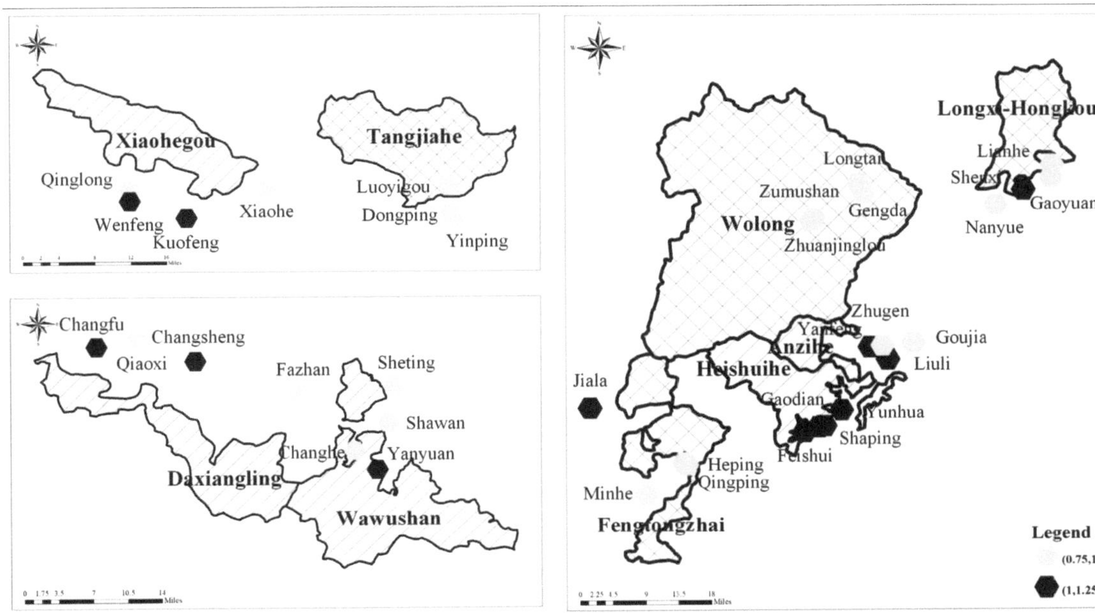

Figure 6. Comparison of ecological niche overlaps on the level of giant panda habitat under the labor force aging. Note: Comparison of ecological niche overlaps on the level of giant panda habitat = ecological niche overlap of OAG/Ecological Niche Overlap of YAG. Map Content Approval Number: CS (2019) 282.

4.3. Analysis of General Tendency

This paper used the trend analysis tool, ArcGIS10.1, to identify the 3D Map with comprehensive ecological niche widths and ecological niche overlaps (as shown in Figure 7). In general, the spatial distribution of the U-shaped comprehensive ecological niche widths were similar to the inverted U-shaped ecological niche overlaps. The lowest value of ecological niche width corresponded to the highest value of ecological niche overlap. The data showed that the extreme values were in the Fengtongzhai Nature Reserve and the Wolong Nature Reserve. These were caused by human activities blocking the internal structure of the nature reserves. The data indicated that human activity destroyed the internal structure of the Fengtongzhai Nature Reserve and the Wolong Nature Reserve, which caused the low values of ecological niche widths and the high values of ecological niche overlaps. The communities distributed inside the Wolong Nature Reserve were stripe-shaped, along the central main road, which caused the isolation of giant panda habitat in the core zone and buffer zone. To develop the regional economy, some villages were defined as being outside of the Fengtongzhai Nature Reserve boundary to create a calabash-shaped nature reserve. However, the habitats on both sides of the nature reserve were isolated and faced increased disturbance from human activities. As shown in Figure 7, the differences between the ecological niche widths of the OAG and YAG were primarily in the northern region, such as in the Tangjiahe Nature Reserve, which had a relatively high level of community co-management to drive the effective transformation of the household resource utilization structure. The differences in the ecological niche overlaps were mainly in the southern region, such as in the Wawushan Nature Reserve. The livelihoods of households depended on diverse resources. Additionally, the research further confirmed that labor force aging could weaken the resource utilization behavior of households and promote sustainable habitats for giant pandas.

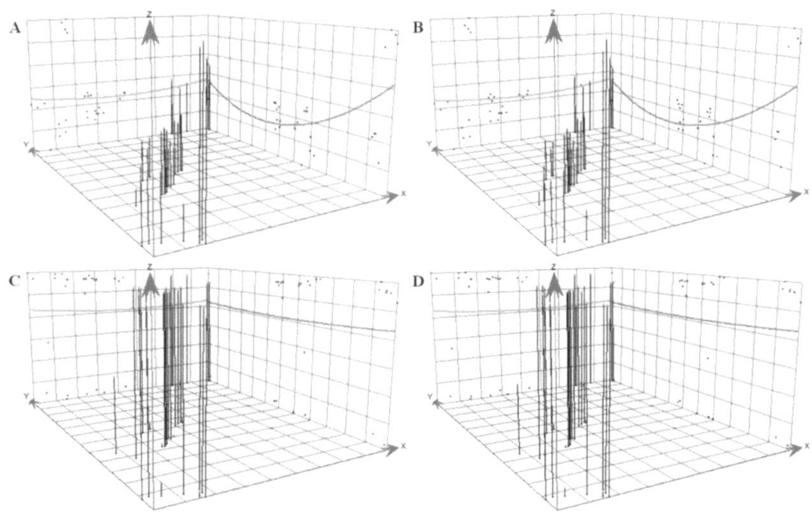

Figure 7. Comparison between ecological niche widths and overlaps on the level of giant panda habitat under the different age dominated natural resource utilization. Note: (**A**) Trend of ecological niche widths in OAG. (**B**) Trend of ecological niche widths in YAG. (**C**) Trend of ecological niche overlaps in OAG. (**D**) Trend of ecological niche overlaps in YAG.

5. Conclusions and Discussion

This article analyzed the evolution of giant panda habitats in nature reserves in the Sichuan Province, from a humanistic perspective, through the household resource utilization decision model and ecological niche model, which were able to identify challenges to giant panda habitat conservation. The results showed that labor force aging has contributed to an increase in the comprehensive ecological niche widths and a decrease in the comprehensive ecological niche widths overlaps, which also somewhat alleviated ecological conflict. Furthermore, it improved the sustainable habitat levels of the giant panda.

From a global perspective, the different age-dominated resource utilization behaviors created disparate intensities of resource utilization. Household resource utilization activities created less disturbance and competitive characteristics of giant panda habitats in significant aging regions. Rising ecological niche widths and falling ecological niche overlaps appeared in these regions.

From a spatial perspective, the nature reserves in the middle region had lower comprehensive ecological niche widths and higher ecological niche overlaps with an effect on terrain structure, population distribution, boundary of nature reserves and community layout. This was the case in areas such as the Wolong Nature Reserve and the Fengtongzhai Nature Reserve. These nature reserves faced a higher conflict between conservation and development. However, the phenomenon of labor force aging could somewhat alleviate the ecological conflict, which primarily manifested in disparate ecological niche widths and ecological niche overlaps under the context of labor force weakening.

This paper aimed to evaluate the differentiation of giant panda habitats through ecological niche theory under the context of labor force aging. This is a preliminary study. In a follow-up study, we will further explore the effect of household natural resource utilization on the differentiation of giant panda habitats under different labor force structures to reveal the complex problems facing giant panda habitat conservation. The further study will be the foundation for alleviating the conflict between conservation and development.

Author Contributions: Z.S. wrote the paper with contributions from all co-authors; Z.S., B.W., L.G. and Y.L. conceived and designed the research. Z.S., Y.H. and S.Z. contributed to the data processing. All authors have read and agreed to the published version of the manuscript.

Funding: This research was funded by Ministry of Education, Humanities and Social Sciences Project (grant number: 22YJCZH150), Projects Supported By Jiangxi Provincial Social Science Foundation (grant number: 21GL47), Humanities and Social Science Project of Jiangxi Province (grant number: 21Y02742), Special fund project of graduate student innovation of Jiangxi Province (grant number: YC2022-B113), Humanities and Social Science Project of Jiangxi Province (grant number: JD18072) and National Natural Science Foundation of China (grant number: 71761147003).

Institutional Review Board Statement: Not applicable.

Informed Consent Statement: Not applicable.

Data Availability Statement: Survey data were obtained from household questionnaires during July 2018, October 2018, January 2019 and May 2019. Survey data on giant panda were obtained from *The 3rd National Survey Report on Giant Panda in China* and *The Pandas of Sichuan: The 4th survey report on giant panda in Sichuan Province*. The land use data of this paper was sourced from public Landset TM/ETM+/OLI images on Geospatial Data Cloud (http://www.gscloud.cn (accessed on 10 September 2022)), which was released by Computer Network Information Center, Chinese Academy of Sciences. A series of basic vector data came from the Basic Geographic Database of China (http://www.webmap.cn/commres.do?method=result100W (accessed on 10 September 2022)) and the 4th survey on giant pandas in Sichuan Province.

Conflicts of Interest: The authors declare no conflict of interest.

References

1. Xu, J.; Wang, Q.; Kong, M. Livelihood changes matter for the sustainability of ecological restoration: A case analysis of the Grain for Green Program in China's largest Giant Panda Reserve. *Ecol. Evol.* **2018**, *8*, 3842–3850. [CrossRef] [PubMed]
2. Kang, D.; Li, J. Giant panda protection: Challenges and hopes. *Environ. Sci. Pollut. Res.* **2019**, *26*, 18001–18002. [CrossRef] [PubMed]
3. Li, Y.; Gu, S.; Gao, P.; Liu, S.; Liu, W.; Hong, Y.; Fan, G.; Hong, Y. Special articles for commemorating the 70th anniversary of the founding of new China. *Econ. Res. J.* **2019**, *54*, 4–23.
4. He, X.; Yin, Z. The small-scale peasant economy and path selection of agricultural modernization: A review of the agricultural modernization radicalism. *China Rev. Political Econ.* **2015**, *6*, 45–65.
5. Jiang, C.; Li, J.; Wang, Y.; Zhao, W. The characteristics, problems and future choices of households' income increase in China in recent years. *J. Nanjing Agric. Univ. (Soc. Sci. Ed.)* **2021**, *21*, 1–21. [CrossRef]
6. Xiang, D.; Mei, Y. The China's experience in green poverty reduction: Policy evolution and practice Model. *J. Nanjing Agric. Univ. (Soc. Sci. Ed.)* **2021**, *21*, 43–53. [CrossRef]
7. Cheng, Y.; Gao, S.; Li, S.; Zhang, Y.; Rosenberg, M. Understanding the spatial disparities and vulnerability of population aging in China. *Soc. Sci. Electron. Publ.* **2019**, *6*, 73–89. [CrossRef]
8. Liu, J.; Liu, Q. The characteristics, influence and coping strategies of the new form of aging society in China: Interpretation based on the data of the Seventh Census. *Popul. Econ.* **2021**, *42*, 13–24.
9. Chen, J.; Han, Q. Effect of agricultural productive services on farmland abandonment. *J. South China Agric. Univ. (Soc. Sci. Ed.)* **2021**, *20*, 23–34. [CrossRef]
10. Song, Z.; Gao, L. Effect of feminization of labor force on households' firewood utilization behavior in nature reserves: Evidence from Sichuan Province. *J. Agro-For. Econ. Manag.* **2021**, *20*, 356–365. [CrossRef]
11. Shen, J.; Song, Z.; Duan, W.; Zhang, Y. Exploring local challenges and adaptation strategies in the establishment of National Parks in giant panda habitats. *Glob. Ecol. Conserv.* **2021**, *30*, e01764. [CrossRef]
12. General Office of the CPC Central Committee (GOCPCCC); General Office of the State Council of the People's Republic of China (GOSCPRC). *Guidance on System of Natural Protected Areas with National Parks as the Main Body*; GOCPCCC: Beijing, China; GOSCPRC: Beijing, China, 2019.
13. Li, H.; Shi, X. Cherry growers' adaptation to climate change and meteorological disasters based on Theory of Planned Behavior. *Clim. Change Res.* **2019**, *15*, 301–312. [CrossRef]
14. Chen, X.; Xin, G.; Wei, C. An analysis of farmland abandonment in poor mountainous areas and its influencing factors: A case study of four villages in two A case study of four villages in two twonships of Youyang County, Chongqing. *J. Southwest Univ. (Nat. Sci. Ed.)* **2016**, *38*, 166–174. [CrossRef]
15. Yang, J.; Song, Z.; Li, Z. Assessment of cultivated land ecological safety based on PSR Method: The case study of main grain production area of the Lower Yangtze Region. *Res. Soil Water Conserv.* **2017**, *24*, 301–307, 313. [CrossRef]
16. Wang, M.; Liu, Y.; Kuai, H.; Zhou, X. The effects of land fragmentation and land quality on the technical efficiency of grain production based on 354 rice planters on the Jianghan Plain. *Resour. Sci.* **2017**, *39*, 1488–1496. [CrossRef]

17. Han, S.; Xie, Y.; Hao, M.; Pu, L. Research on the difference of investment behavior of different farm household type. *Res. Soil Water Conserv.* **2005**, *12*, 83–85. [CrossRef]
18. Li, M.; Zhao, L. Agricultural labor force aging phenomenon and the effect on agricultural production: Evidence from Liaoning Province. *Issues Agric. Econ.* **2009**, *10*, 12–19. [CrossRef]
19. Yang, J.; Yang, G.; Hu, X. Impact of Agricultural Labor Aging on Farmland Use Efficiency of Rural Households:An Empirical study from Regions of Differing Economic Development Levels. *Resour. Sci.* **2011**, *33*, 1691–1698.
20. Qiao, Z.; Huo, X. Impact of Agricultural Labor Aging on Farmland Use Efficiency. *J. South China Agric. Univ. (Soc. Sci. Ed.)* **2017**, *16*, 61–73. [CrossRef]
21. Wang, P.; Zhang, N.; Zhang, J. The Impact of Livelihood Capital on Rural Residents' Energy Consumption Structure: Analysis Based on the Survey Data of Rural Households in Shaanxi. *J. Stat. Inf.* **2020**, *35*, 121–128.
22. Li, M.; Gong, Q.; Pan, J. Evaluation and Strategic Transformation of China's Small Hydropower under the Goal of Carbon Neutrality. *J. Beijing Univ. Technol. (Soc. Sci. Ed.)* **2022**, *22*, 86–104. [CrossRef]
23. Yuan, L.; Zhang, J.; He, K.; Lai, X.; Wang, X. Livelihood capital, diversified non-agricultural livelihood strategies and farmers'willingness to consume on clean energy: Take bio-natural gas as an example. *Resour. Environ. Yangtze Basin* **2022**, *31*, 244–257.
24. Liu, Z.; Yao, J. Household energy consumption in China: An application of the LA-AIDS Model. *J. Soc. Sci. Hunan Norm. Univ.* **2020**, *49*, 78–85. [CrossRef]
25. Liu, C. The Reforms of Collective Forestland Tenures Transfer in China: Retrospectives, key issues and path selection. *Reform* **2020**, *36*, 133–147.
26. Zhu, L.; Ma, B.; Wen, Y. Research on problems and perfecting countermeasures for new round Reform of Collective Forest Property Right System. *J. Northwest AF Univ. (Soc. Sci. Ed.)* **2017**, *17*, 143–151. [CrossRef]
27. Wang, J. An integrated management rights system for National Parks in China: The institutional design of the National Parks Law. *Jinan J. (Philos. Soc. Sci.)* **2020**, *42*, 10–23.
28. Zhong, L.; Xiao, L. Chinese National Park System Pilot: Establishing path and research issues. *Resour. Sci.* **2017**, *39*, 1–10. [CrossRef]
29. Wang, W.; Chen, L. The phasic features and development strategies of China's energy industry over 40 years of Reform and Opening-up. *Reform* **2018**, *9*, 55–65.
30. Sun, W.; Hu, W.; Yan, M.; Lv, C. Influencing factors of firewood consumption of rural households in restricted development zones:A case study of the Nujiang Prefecture in Yunnan Province. *Geogr. Res.* **2014**, *33*, 1694–1705. [CrossRef]
31. Qin, Q.; Liu, J.; Ma, B.; Tan, H.; Xu, Y.; Wen, Y. Analysis on the utilization mode and influencing factors of natural resources in Sichuan giant panda reserve and its surrounding communities. *Issues For. Econ.* **2020**, *40*, 345–352. [CrossRef]
32. Tian, C.; Li, M.; Li, Y.; Liu, Q. Regional characteristics of carbon emission reducing effect of rural household biogas project and strategies for future development of the project: A case study of Sichuan Province, China. *J. Ecol. Rural. Environ.* **2016**, *32*, 432–438. [CrossRef]
33. Sichuan Provincial Department of Forestry and Grassland (SPDFG). *The Pandas of Sichuan: The 4th Survey Report on Giant Panda in Sichuan Province*; Sichuan Science and Technology Publishing House: Chengdu, China, 2015.
34. Fu, Y.; Gengqiu, Z.; Tu, F.; Miao, N.; Fei, L.; Zhao, C. Composition and diversity of small mammals in China Panda Valley, Dujiangyan, Sichuan Province. *Sichuan J. Zool.* **2015**, *34*, 570–573. [CrossRef]
35. Zhang, S.; Ran, J.; Tang, M.; Du, B.; Yang, Q.; Liu, S. Landscape pattern analysis of red panda habitat in Liangshan Mountains. *Acta Ecol. Sin.* **2008**, *28*, 4787–4795.
36. Li, J.; Shen, G. *The Habitat of Giant Pandas*; Higher Education Press: Beijing, China, 2012.
37. Duan, W.; Zhao, Z.; Liu, M.; Wen, Y. Research on natural resources dependence of surrounding rural households. *J. Agrotech. Econ.* **2016**, *251*, 93–102. [CrossRef]
38. Wu, J.; Ren, Y.; Wen, Y. Analysis on wild edible herbs collection of rural households surrounding nature reserves based on tourist' requirement. *For. Econ.* **2018**, *40*, 67–74. [CrossRef]
39. Zhong, Y.; Ye, C. Study on the spatial differentiation and types of agricultural modernization based on Niche Theory in Poyang Lake Region. *Jiangsu Agric. Sci.* **2016**, *44*, 572–578. [CrossRef]
40. Wooldridge, M. Intelligent Agents: The Key Concepts. In *Multi-Agent Systems and Applications II*; Springer: Berlin, Germany, 2002.
41. Chen, H.; Liang, X.; Xi, J. *Research on Land Utilization Change and Household Response in Eco-Fragile Region*; Science Press: Beijing, China, 2011.
42. Huang, Y. Mathematical methods in niche theory study. *Chin. J. Appl. Ecol.* **1994**, *5*, 331–337. [CrossRef]
43. National Forestry and Grassland Administration of the People's Republic of China. *The 3rd National Survey Report on Giant Panda in China*; Science Press: Beijing, China, 2006.

International Journal of *Environmental Research and Public Health*

MDPI

Article

CO Detection System Based on TDLAS Using a 4.625 μm Interband Cascaded Laser

Kun Li [1], Boyang Wang [1], Mingyao Yuan [1], Zhixiong Yang [1,2,*], Chunchao Yu [1] and Weijian Zheng [1]

1 Kunming Institute of Physics, Kunming 650223, China
2 Yunnan Normal University, Kunming 650029, China
* Correspondence: 13888240701@139.com

Abstract: During industrial operations and in confined places, carbon monoxide (CO) may collect in harmful proportions if ventilation is insufficient or appliances are not properly maintained. When the concentration of CO is too high, it might result in suffocation, coma, or even death. The detection of tiny concentrations of CO plays an important role in safe production. Due to the selective absorption of specific wavelengths of light by gas molecules, lasers have a wide range of applications in the field of gas detection. In this paper, a tunable diode laser absorption spectroscopy (TDLAS) system for CO detection was constructed using an interband cascaded laser (ICL) with a central wavelength of 4.625 μm. The modulated signal generated by the FPGA module was output to the laser controller to modulate the laser. The signal received by the detector was input to the FPGA module. After lock-in amplification, the second harmonic signal of high frequency modulation was output. Several concentrations of CO that were dispersed via static gas distribution were identified. A CO detection system with an open optical path was constructed, and the detection distance was about 8 m. The minimum detectable concentration is around 10.32 ppmm. The concentration of CO in the open optical path was 510.6 ppmm, according to the calibration of the detected concentration. The remote detection system based on TDLAS using an ICL can be used to monitor CO in the open optical path.

Keywords: gas detection; carbon monoxide; remote measurement

Citation: Li, K.; Wang, B.; Yuan, M.; Yang, Z.; Yu, C.; Zheng, W. CO Detection System Based on TDLAS Using a 4.625 μm Interband Cascaded Laser. *Int. J. Environ. Res. Public Health* **2022**, *20*, 12828. https://doi.org/10.3390/ijerph191912828

Academic Editor: Paul B. Tchounwou

Received: 15 September 2022
Accepted: 4 October 2022
Published: 7 October 2022

1. Introduction

Incomplete combustion of carbon fuel produces carbon monoxide (CO) during industrial processes such as coking and steelmaking, as well as in enclosed spaces such as mines and submarines. Suffocation, unconsciousness, or even death can occur when CO concentrations are exceeded [1]. The temperature of the coal seam has a direct impact on the amount of CO that is produced during coal mining. It commonly serves as an indicator gas for a coal seam fire warning [2]. The main methods of detecting CO are electrochemical test, catalytic combustion, and laser detection [3]. The electrochemical approach relies on the electrical signal generated by the chemical reaction of the target gas and the sensor electrode; in addition, it is easily affected by other gases. Through the process of catalytic combustion, heat is produced. With the increasing temperature, the detection electrode wire's resistance increases. It is typically used to detect flammable gases. Laser detection is based on the selective absorption of light by gas molecules, which has the advantages of fast detection and low detectable concentration.

In order to determine the concentration of the detected gas, laser direct absorption detection technique measures the attenuation of a certain laser over a set length of gas [4–8]. The tunable diode laser absorption spectroscopy (TDLAS) system uses a wavelength-tunable semiconductor laser to obtain gas absorption spectra within the tuning range. The modulated laser is incident on the detector after passing through the gas, and the concentration of the target gas is proportional to the harmonic signal intensity of the high-frequency modulation signal. Usually, the second harmonic signal extracted by the lock-in amplifier is used as the detection signal. The concentration of the gas to be measured is obtained through the calibration of the standard concentration gas [9–13]. Some of the disadvantages associated with the laser direct absorption detection technique are a low signal-to-noise ratio, large amount of data calculation, etc. Many of them can be improved with TDLAS use. Semiconductor diode lasers found immediate application in the 1970s as much needed tunable sources for high-resolution laser spectroscopy commonly referred to as TDLAS. In reference [14], available semiconductor lasers for spectroscopy in the near- and mid-infrared spectral region have been reviewed together with the main features of TDLAS. A theoretical description of the wavelength-modulation spectrometry technique is given by Kluczynski P et al. [15]. Chen et al. [16] introduced an instrument that takes advantage of a mid-infrared quantum cascaded laser (QCL) operating at 4.8 μm and a mercury cadmium telluride (HgCdTe) mid-infrared detector. Low detection sensitivity down to 50 nmol/mol level in 4 s acquisition time was achieved using a multipass cell with a 76 m absorption path length. The simultaneous atmospheric pressure measurement of the trace gases methane (CH_4) and CO using an open-path sensor based on TDLAS has been described. The detection limit of 0.58 parts per million by volume (ppmv) for CO and 0.4 ppmv for CH_4 was accomplished at 1 s averaging time by using a distribution feedback (DFB) laser operating at 2.33 μm [17]. Shao et al. [18] adopted a DFB laser to constantly monitor CO and CH_4 in an atmosphere based on TDLAS. The absorption signals of the sample gases were improved using a multipass absorption cell with a 72 m optical path length.

The laser detection of CO is typically performed in a multipass absorption cell. Gas detection in an open environment is typically necessary in practical applications. The telemetry detection of CO in an open optical path has received relatively little attention. CO has a substantially higher absorption coefficient at 4.6 μm than it does at 2.3 μm. The interband cascaded laser is currently the dominating light source for gas detection in the infrared spectrum. It benefits from room temperature functioning, a small linewidth, and an adjustable wavelength. In this paper, a TDLAS telemetry system is constructed using an interband cascade laser with a center wavelength of 4.625 μm. To accurately identify low concentrations of CO, the gas concentration is measured in accordance with the concentration calibration. The detection device can be employed in industrial parks, coal mining sites, and restricted places for fixed-point monitoring of low-concentration CO.

2. TDLAS Detection Principle

The light intensity is attenuated by the gas being measured in TDLAS, which uses a laser to emit a specific wavelength of light that is absorbed. The gas concentration directly relates to the degree of light intensity attenuation. It adheres to the Lambert-Beer law [19].

$$I[v(t)] = I_0[v(t)]\exp\{-\alpha[v(t)]cL\}, \tag{1}$$

where I_0 is the intensity of incident laser, and I is the intensity of transmitted laser α is the absorption coefficient of the gas, and V is the frequency of the laser. c is the gas concentration, and L is the light path length. The direct absorption technology detected can reversely calculate the concentration of the gas by the light intensity after attenuation.

When detecting trace gases, wavelength modulation technology can significantly reduce noise, boost the signal-to-noise ratio, and lower the detection threshold. The instantaneous frequency of laser emission at this time is

$$v(t) = v_0 + \sigma_v \cos(2\pi ft), \tag{2}$$

where v_0 is the selected laser center frequency, σ_v is the modulation signal amplitude, and f is the modulation frequency.

The Fourier series expansion of Equation (2) can be obtained

$$I(v_0, t) = \sum_{n=0}^{\infty} A_n(v_0)\cos(n2\pi ft), \tag{3}$$

The attenuation signal after gas absorption is received by detector and the second harmonic signal is obtained after phase-locked amplification [20]

$$A_2(v_0) = \frac{I_0 cL}{4}{\sigma_v}^2 \frac{d^2\sigma(v)}{dv^2}\bigg|_{v=v_0}, \tag{4}$$

Therefore, the concentration of the gas to be measured can be inverted by the direct current component I_0 and the second harmonic amplitude

$$c = \frac{A_2(v_0)}{I_0 KL}, \tag{5}$$

where K is the calibration constant. Equation (5) shows that the gas concentration and second harmonic amplitude have a linear relationship. The determined gas concentration can be derived by inversion through standard gas calibration.

The selection of spectral lines is critical for gas measurement using TDLAS technology. The absorption spectra of CO and NO were shown in Figure 1. According to the HITRAN2012 spectral library, the pressure was 1 atm and the temperature was 296 K. In order to reduce the interference of N_2O on CO detection, a laser source with a central wavelength of 4.625 μm was selected.

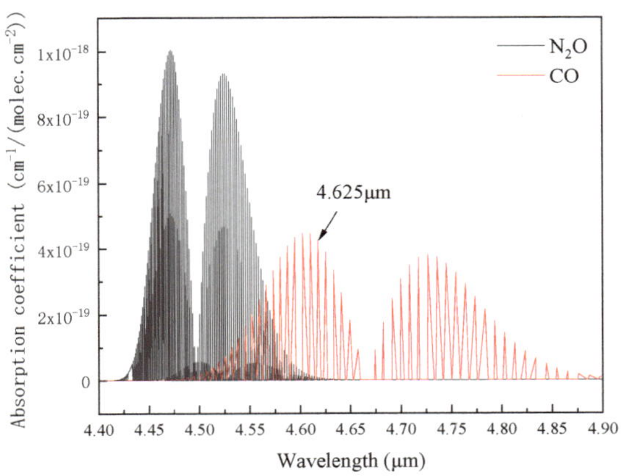

Figure 1. Absorption spectra of CO and N_2O.

3. Experimental Equipment

The CO telemetry system based on TDLAS was shown in Figure 2. The light source was a quantum cascade laser with a central wavelength of 4.625 μm. At a constant temperature, the laser source was tuned around 15 nm by current, with a linewidth of less than 2.14×10^{-4} nm. When the working temperature was 20 °C and the input current of the laser was 72 mA, the output laser wavelength was 4.625 μm and the output power was 7.9 mW. The diameter of the laser was 3 mm, and the divergence angle was 35° × 55°. After being collimated by the collimation mechanism, the spot size at 3 m was 4.5 mm. About 30% energy was lost after the collimator. The light reflected from the corner reflector traveled via the focusing lens before concentrating on the detector. The receiver was a HgCdTe detector with a working cut-off wavelength of 6 μm. The optical area was 1×1 mm^2, the acceptance angle was 36°, and the time constant was 50 ns. The optical axis distance between the laser and detector was around 45 mm, and the diameter of the corner mirror was 50.8 mm. When the corner mirror was 8 m away from the laser source, the incident light and reflected light were at the edge of the corner mirror. A sawtooth wave with a frequency of 5 Hz and a sine wave with a frequency of 31.4 kHz were both produced by the field programmable gate array (FPGA) module. The modulation signal was loaded into the laser controller to modulate the laser source. The signal received by the detector was input to the FPGA module, and the second harmonic signal was output after the phase-locking operation. The second harmonic signal was gathered by a data collection card and shown on the computer software.

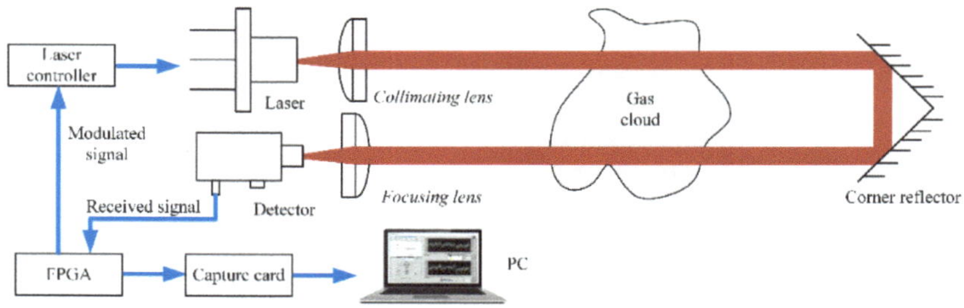

Figure 2. Gas remote detection system based on TDLAS.

To determine the gas concentration, the gas detection system based on TDLAS requires calibration using standard concentration gas. One easy way to set up standard gas concentrations is through static gas distribution. Calculations can be performed to determine the gas concentration when a specific amount of gas is supplied to a container with a known volume. A gas chamber with an inner diameter of 150 mm and a length of 1 m was put in the optical path. An inelastic gas bag was used to contain the CO gas after it was discharged from the high-pressure gas cylinder at the same pressure as the atmosphere. After the air chamber was evacuated, the CO gas was retrieved from the air bag and injected into it. After that, nitrogen was added to the gas chamber to bring the pressure there closer to that of the atmosphere.

4. Results and Discussion

The calibration concentration was determined by detecting CO in the gas chamber using an electrochemical gas detector. The electrochemical gas detector had a detection range of 1–1000 ppm and a detection accuracy of 1 ppm. As shown in Figure 3, CO was fed into the gas chamber in amounts of 0.2 mL, 0.5 mL, 1 mL, 2 mL, and 4 mL. The measured concentrations were 12 ppm, 27 ppm, 52 ppm, 112 ppm, and 217 ppm, respectively. The actual concentration of the gas was rather close to the theoretically calculated concentration, and the error may have resulted from a volume error in the injected gas. Through concentration calibration, the concentration of CO in the open optical path can be directly obtained.

Different CO concentrations in the gas chamber were measured using the CO detection system based on TDLAS. In Figure 4, the detection results were displayed. The amplitude of the second harmonic signal (ASH) grew linearly with the gas concentration, confirming the concept. Calculations revealed that the ratio of ASH to the gas concentration was 4.431×10^{-3}/ppm. The lower detection limit of CO in the gas chamber was approximately 12 ppmm. (1 ppmm means that the concentration of 1 ppm gas was distributed in the length of 1 m.) CO gas at a concentration of 240 ppb could be detected if a multipass absorption cell with an optical path length of 50 m was utilized. The CO gas concentration corresponding to the amplitude of various second harmonics of the gas detection system might be determined using the ASH of CO gas with varied concentrations in the gas chamber as the standard value. The signal-to-noise ratio can be improved, low-frequency background noise can be suppressed, and the limit of detection concentration can be lowered compared with the laser direct absorption detection technique. It can detect CO with low concentration; thus, gas leakage can be found at an early stage to keep people safe.

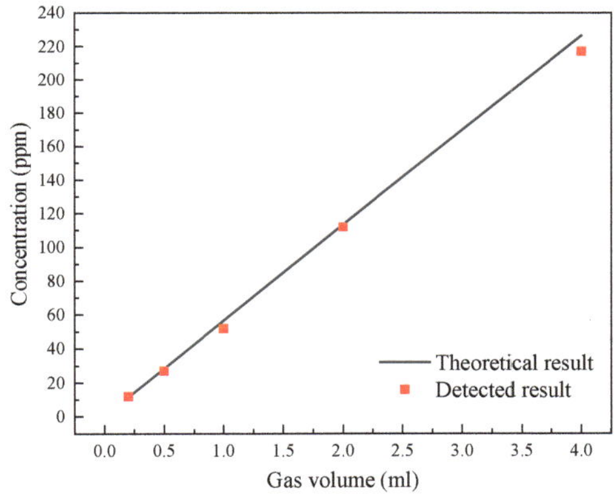

Figure 3. CO concentration calibrated with an electrochemical gas detector as a function of CO gas volume added into the gas chamber.

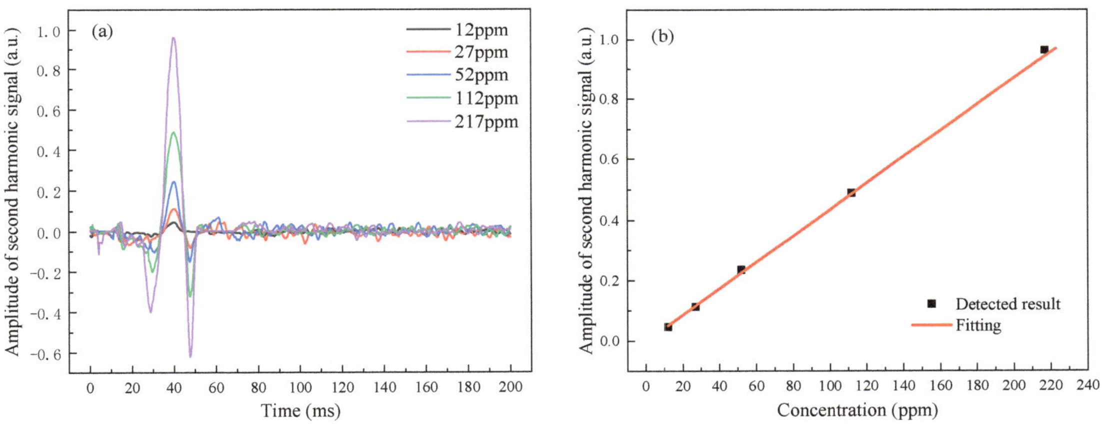

Figure 4. Detected result of CO with different concentrations in the gas chamber based on TDLAS: (**a**) waveforms of the second harmonic signal; and (**b**) amplitudes of the second harmonic signal.

The scanning period of the laser modulation signal was 200 ms. By continually collecting the detection signal for 50 times, it was possible to achieve the stability of the gas detection result within 10 s. Figure 5 displayed the stability of the ASH for gases at various concentrations. The relative standard deviations of CO detection signals with concentrations of 12 ppm, 27 ppm, 52 ppm, 112 ppm, and 217 ppm were 0.019, 0.029, 0.024, 0.02, and 0.017, respectively. The measurement results of the CO detection system based on TDLAS were stable because the low-frequency background noise could be suppressed. The detection result is stable and reliable; therefore, this system can work for a long time to detect CO gas in the environment.

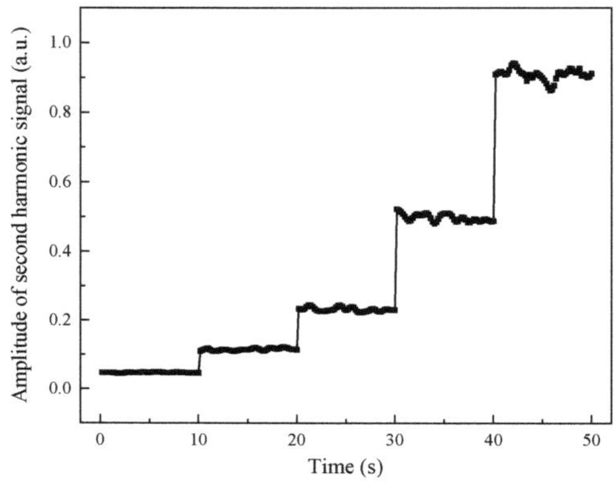

Figure 5. The fluctuation of ASH with different concentrations of CO in the gas chamber.

The silicon slices used for the gas cell windows had a thickness of 5 mm, and their transmittance at 4.625 μm was approximately 96.3%. Through the gas cell, about 14% of the laser's power was lost. The ASH was proportional to the gas concentration and the laser emission power. According to the calibration of the standard concentration gas in the gas chamber, the ratio of the ASH in the open optical path to the gas concentration was 5.152×10^{-3}/ppm (4.431×10^{-3}/ppm/(1–14%)), and the minimum detectable concentration was about 10.32 ppmm (12 ppmm×(1–14%)). CO gas was sprayed in the optical path directly, and the measured second harmonic signal was shown in Figure 6a. The ASH was 2.638, and the concentration of CO gas in the optical path was approximately 510.6 ppmm. The concentration of CO in the optical path within two minutes was shown in Figure 6b. The detection result increased significantly right away after the gas was sprayed, and the concentration fluctuation during the gas diffusion was not significant. The CO content in the optical path immediately decreased as soon as the ventilation fan was turned on. The concentration of CO was much lower after one minute of ventilation, although there was still some CO gas present in comparison to before degassing. The diffusion of CO gas in the optical path was generally uniform because of the air movement. This system can detect the concentration of CO in the environment and give an alarm when the concentration exceeds the standard value. The detectable concentration is low, so that the danger can be detected in time. In confined spaces, timely ventilation can quickly reduce CO in the environment, thereby reducing the impact on health.

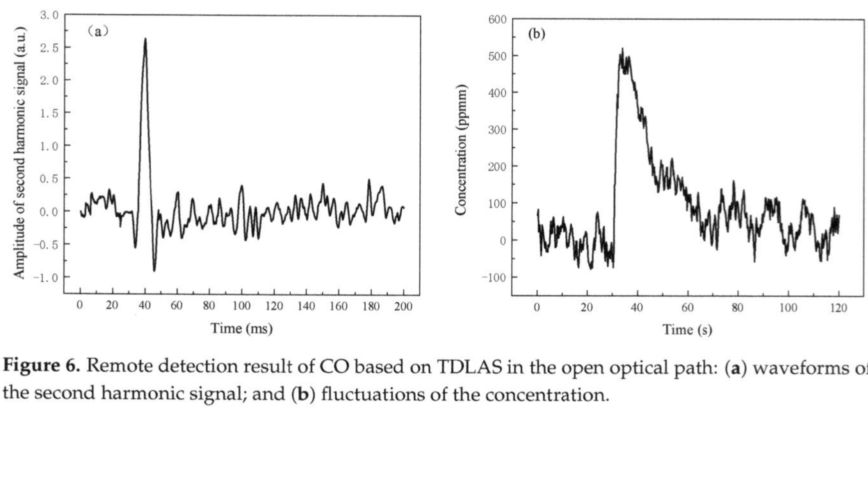

Figure 6. Remote detection result of CO based on TDLAS in the open optical path: (**a**) waveforms of the second harmonic signal; and (**b**) fluctuations of the concentration.

5. Conclusions

The laser detection of CO usually uses a long optical path gas chamber to improve the detection sensitivity, and there are few studies on the telemetry system. A CO gas telemetry system based on TDLAS was constructed using a 4.625 μm mid-infrared ICL with a high CO absorption coefficient in order to increase the detection sensitivity. The standard concentration of CO gas was prepared by the static gas distribution method. The standard concentration gas was calibrated in the gas chamber. The results of the gas detection were stable and reliable. The detection distance for CO gas in the open optical was roughly 8 m. The lower limit of detection concentration was approximately 10.32 ppmm, according to the calibration of standard concentration gas in the gas chamber. The gas detection distance can be extended by increasing the diameter of the corner mirror and the laser intensity.

Author Contributions: Methodology, C.Y. and W.Z.; software, B.W.; data curation, K.L. and M.Y.; writing—original draft preparation, K.L.; writing—review and editing, K.L.; supervision, C.Y.; project administration, W.Z.; funding acquisition, Z.Y. All authors have read and agreed to the published version of the manuscript.

Funding: This research was funded by National Defense Innovation Special Zone projects, grant number 19H86318ZD01100302.

Institutional Review Board Statement: Not applicable.

Informed Consent Statement: Not applicable.

Data Availability Statement: Not applicable.

Acknowledgments: The authors would like to thank the support of Kunming Institute of Physics in this work.

Conflicts of Interest: The authors declare no conflict of interest.

References

1. Gu, L.S.; Pei, Y.J.; Wang, L.N.; Guo, H.L. TDLAS carbon monoxide concentration detection system. *Ship Sci. Tech.* **2019**, *41*, 134–137.
2. Ma, L. Application of TDLAS Technology in the Field of Coal Mine Gas Disaster Detection. *S. Coal Mines* **2015**, *46*, 87–89.
3. Liang, Y.T.; Tian, F.C.; Feng, W.B.; Shao, Z.L.; Meng, X.N.; Chen, C.F. Research progress of coal mine gas detection technology in China. *J. CHN. Coal Soc.* **2021**, *46*, 1701–1714.
4. Zheng, W.X.; Zheng, C.T.; Yao, D.; Yang, S.; Dang, P.P.; Wang, Y.D. Development of a Mid-Infrared Interband Cascade Laser Methane Sensor. *Acta Opt. Sinica* **2018**, *38*, 0328013. [CrossRef]
5. Lv, M.; Wang, Y.D.; Chen, C. Development of Mid-Infrared Trace-CO Detector with Long-Path Differential Optical Absorption Spectroscopy (LP-DOAS). *Spectrosc. Spect. Ana.* **2017**, *37*, 2278–2282.
6. Roy, A.; Chakraborty, A.L. Quantum cascade laser-based in situ measurement of atmospheric CO and CO_2 in Gandhinagar using 1f and 2f wavelength modulation spectroscopy. In Proceedings of the Workshop on Recent Advances in Photonics, Guwahati, India, 13–14 December 2019.
7. Zhang, H.L.; Wu, T.; He, X.D. Progress of Measurement of Infrared Absorption Spectroscopy Based on QCL. *Spectrosc. Spect. Ana.* **2019**, *9*, 2751–2757.
8. Chen, X.; Yang, C.G.; Hu, M.; Shen, J.K.; Niu, E.C.; Xu, Z.Y.; Fan, X.L.; Wei, M.; Yao, L.; He, Y.B.; et al. Highly-sensitive NO, NO_2, and NH_3 measurements with an open-multipass cell based on mid-infrared wavelength modulation spectroscopy. *Chin. Phy. B.* **2018**, *4*, 040701. [CrossRef]
9. Xie, Y.; Chang, J.; Chen, X.; Sun, J.; Zhang, Q.; Wang, F.; Zhang, Z.; Feng, Y. A DFB-LD Internal Temperature Fluctuation Analysis in a TDLAS System for Gas Detection. *IEEE Photonics J.* **2019**, *3*, 6801708. [CrossRef]
10. Feng, Y.; Chang, J.; Chen, X.; Zhang, Q.; Wang, Z.; Sun, J.; Zhang, Z. Application of TDM and FDM methods in TDLAS basedmulti-gas detection. *Opt. Quant. Electron.* **2021**, *53*, 195. [CrossRef]
11. Wang, C.; Jiang, M.; Huang, Q.; Zhou, W.; Jiang, Y.; Tian, S. Detection and calibration method for decomposition of sulfur hexafluoride based on TDLAS. In Proceedings of the Applied Optics and Photonics China, Beijing, China, 30 November–2 December 2020.
12. Wang, Y.; Wei, Y.; Liu, T.; Sun, T.; Kenneth, T.V. TDLAS Detection of Propane/Butane Gas Mixture by Using Reference Gas Absorption Cells and Partial Least Square Approach. *IEEE Sens. J.* **2018**, *18*, 8587–8596. [CrossRef]
13. Chen, H.; Ju, Y.; Han, L. Research on the Relationship Between Modulation Depth and Center of High Order Harmonic in TDLAS Wavelength Modulation Method. *Spectrosc. Spect. Ana.* **2021**, *41*, 3676–3681.

14. Werle, P.; Slemr, F.; Maurer, K.; Kormann, R.; Muecke, R.; Jaenker, B. Near- and mid-infrared laser-optical sensors for gas analysis. *Opt. Lasers Eng.* **2002**, *37*, 101–114. [CrossRef]
15. Kluczynski, P.; Axner, O. Theoretical description based on Fourier analysis of wavelength-modulation spectrometry in terms of analytical and background signals. *Appl. Opt.* **1999**, *38*, 5803–5815. [CrossRef]
16. Chen, C.; Wang, B.; Li, C.G.; Li, J.; Wang, Y.D. A Trace Gas Sensor Using Mid-Infrared Quantum Cascaded Laser at 4.8μm to Detect Carbon Monoxide. *Spectrosc. Spect. Ana.* **2014**, *34*, 838–842.
17. Dang, J.; Kong, L.; Zheng, C.; Wang, Y.; Sun, Y.; Yu, H. An open-path sensor for simultaneous atmospheric pressure detection of CO and CH_4 around 2.33μm. *Opt. Lasers Eng.* **2019**, *123*, 1–7. [CrossRef]
18. Shao, L.; Fang, B.; Zheng, F.; Qiu, X.; He, Q.; Wei, J.; Li, C.; Zhao, W. Simultaneous detection of atmospheric CO and CH_4 based on TDLAS using a single 2.3μm DFB laser. *Spectrochim. Acta A* **2019**, *222*, 117118. [CrossRef] [PubMed]
19. Tang, F.; Liu, S.G.; Lv, Q.S.; Li, X.T.; He, S.K.; Zeng, X.Z.; Yue, Y.Q. Experimental study on the detection of trace H_2S and H_2O in high-voltage combination appliances based on TDLAS technology. In Proceedings of the Advanced Sensor Systems and Applications IX, Hangzhou, China, 20 November 2019.
20. Werle, P. A review of recent advances in semiconductor laser based gas monitors. *Spectrochim. Acta A* **1998**, *54*, 197–236. [CrossRef]

International Journal of
Environmental Research and Public Health

MDPI

Article

Study on Spatiotemporal Variation Pattern of Vegetation Coverage on Qinghai–Tibet Plateau and the Analysis of Its Climate Driving Factors

Xiaoyu Deng [1], Liangxu Wu [2], Chengjin He [1] and Huaiyong Shao [1,*]

[1] College of Earth Sciences, Chengdu University of Technology, Chengdu 610059, China; dxy@cdut.edu.cn (X.D.); 2020020014@stu.cdut.edu.cn (C.H.)
[2] College of Global Change and Earth System Science, Beijing Normal University, Beijing 100875, China; 202121490037@mail.bnu.edu.cn
* Correspondence: shaohuaiyong@cdut.edu.cn

Abstract: As one of the most sensitive areas to global environmental change, especially global climate change, the Qinghai–Tibet Plateau is an ideal area for studying global climate change and ecosystems. There are few studies on the analysis of the vegetation's driving factors on the Qinghai–Tibet Plateau based on large-scale and high-resolution data due to the incompetence of satellite sensors. In order to study the long-term vegetation spatiotemporal pattern and its driving factors, this study used the enhanced spatial and temporal adaptive reflectance fusion model (ESTARFM) to improve the spatial resolution of the GIMMS NDVI3g (8 km) data of the Qinghai–Tibet Plateau in 1990 and 1995 based on the MODIS NDVI (500 m) data. The research on the spatiotemporal pattern and driving factors of vegetation on the Qinghai–Tibet Plateau from 1990 to 2015 was carried out afterward, with combined data including topographic factors, annual average temperature, and annual precipitation. The results showed that there was a strong correlation between the actual MODIS NDVI image and the fused GIMMS NDVI3g image, which means that the accuracy of the fused GIMMS NDVI3g image is reliable and can provide basic data for the accurate evaluation of the spatial and temporal patterns of vegetation on the Qinghai–Tibet Plateau. From 1990 to 2015, the overall vegetation coverage of the Qinghai–Tibet Plateau showed a degrading trend at a rate of −0.41%, and the degradation trend of vegetation coverage was the weakest when the slope was ≥25°. Due to the influence of the policy of returning farmland to forests, the overall degradation trend has gradually weakened. The significant changes in vegetation in 2010 can be attributed to the difference in the spatial distribution of climatic factors such as temperature and precipitation. The area with reduced vegetation in the west was larger than the area with increased vegetation in the east. The effects of temperature and precipitation on the distribution, direction, and degradation level of vegetation coverage were varied by the areal differentiation in different zones.

Keywords: vegetation coverage; ESTARFM; data fusion; climate change; topography

Citation: Deng, X.; Wu, L.; He, C.; Shao, H. Study on Spatiotemporal Variation Pattern of Vegetation Coverage on Qinghai–Tibet Plateau and the Analysis of Its Climate Driving Factors. *Int. J. Environ. Res. Public Health* **2022**, *20*, 8836. https://doi.org/10.3390/ijerph19148836

Academic Editor: Paul B. Tchounwou

Received: 10 May 2022
Accepted: 16 July 2022
Published: 21 July 2022

1. Introduction

Vegetation plays an irreplaceable role in the global mass–energy exchange, and it is interactive with climate change by affecting water, the carbon cycle, and soil composition [1–3]. Some studies have linked the vegetation change on the Qinghai–Tibet Plateau to global warming [4]. The Qinghai–Tibet Plateau, known as " the third pole of the earth", is located in a high-altitude and high-cold environment [5]. Its fragile ecosystem and complex geographic environment have resulted in a drastic change in the vegetation dynamics in the region and made it more sensitive to the change in global climate and ecosystems. Thus, it is considered to be an ideal research area for vegetation cover dynamics and driving force analysis research [6]. For example, Zhou et al. found that the vegetation on the Qinghai–Tibet Plateau exhibited an overall green profile from 1982 to 2012, using GIMMS NDVI3g data [7]. Xu et al. found that the vegetation coverage of the

Qinghai–Tibet Plateau increased from 1982 to 2000 by extracting vegetation coverage derived information from the leaf area index satellite images [8]. Zhang et al. compared the results of vegetation trend changes based on vegetation indices from multiple data sources and found that different data sources may cause variance in trend changes [9]. Therefore, the dynamic monitoring of vegetation coverage on the Qinghai–Tibet Plateau and the analysis of driving factors are of great significance to the evaluation of construction in the Qinghai–Tibet Plateau ecological environment and environmental protection projects.

It should be noted that the current remote sensing data sources usually cannot have both a long time series and a high spatial resolution. For example, predecessors studied the vegetation changes in Northeast China from 1982 to 2009 [10], on the Mongolian Plateau from 1982 to 2006 [11], in Xinjiang from 1982 to 2013 [12], and on the Qinghai–Tibet Plateau from 1982 to 2015 [13] using GIMMS NDVI3g data, which is one of the longest time-series NDVI data sets [14]. However, the spatial resolution of this data was low, which was suitable for large-scale research, and had a weak ability to describe changes in regional small-scale vegetation coverage. The data collected by satellite sensors, such as the Earth Observation System (SPOT)-VEGETATION, MODIS, and ENVISAT Medium Resolution Imaging Spectrometer (MERIS), have a high time resolution. The effects of clouds can be reduced or even eliminated by synthesizing images within a certain period [14]. Thus, the quality of remote sensing data is improved, which expands the usage of the regional vegetation trend analysis on a global scale [15–17]. However, the demand for improving the spatial resolution of the data remains unsolved. Although in comparison with GIMMS NDVI3g, MODIS NDVI products have higher accuracy and can better express the time-series changes of vegetation. Many scholars have analyzed the changes in vegetation cover and its influencing factors on the Qinghai–Tibet Plateau with different temporal and spatial resolutions based on MODIS data [13,18,19]. It is inappropriate to be applied in studies of long time series due to its time coverage (July 2001 to present). Therefore, there is a need to improve the resolution of early remote sensing data for long-term vegetation change monitoring research.

There are great differences in vegetation types and distribution among different regions, so it is difficult to evaluate the regional vegetation growth status based on a single data set. It is urgent to reconstruct multi-source spatiotemporal data. At present, the spatiotemporal fusion methods based on remote sensing data were roughly divided into two types: on the one hand, the transformation model is based on systematic error, the main idea of which was that the reflectivity of different objects in different data sources has strong consistency, and the deviation was mainly caused by the system error of the sensor itself, so it was necessary to establish a certain regular conversion relationship for the data of different sensors. Li et al. established the data mapping relationship between pixels through pixel-by-pixel regression analysis and analyzed the response of vegetation dynamic changes on the Qinghai–Tibet Plateau to hydrothermal conditions in the past 40 years [20]. Zhang et al. obtained long-term remote sensing data by establishing a linear regression equation between GIMMS NDVI3g and MODIS data [21]. On the other hand, the principle of the data fusion method based on the reconstruction model was to realize the fusion of temporal and spatial information between sensors of different data sources by establishing the relationship between different data sources.

Predecessors have done a lot of research on the fusion of remote sensing data in time and space. For example, the STARFM (spatial and temporal reflection fusion model) proposed by Gao et al. [22] can fuse Landsat and MODIS surface reflectance to generate synthetic reflectance data with high temporal and spatial resolution. Using the STARFM model, Semmens et al. effectively monitored crop water use and soil water status at the farmland scale [23], while Singh fully extracted the phenology of the crop during the 8-year growing season [24]. Gevaert and Javier Garcia-Haro evaluated STARFM and believe that it is practically more effective in constructing time profiles with more high-resolution images [25]. Bhandari et al. [26] and Schmidt et al. [27] evaluated the ability of STARFM to synthesize long-term time-series data and obtained the phenological characteristics of different vegetation for research. Zhu et al. improved the STARFM algorithm and proposed the enhanced spatial and temporal adaptive reflectance fusion model (ESTARFM) to reflect the temporal and spatial changes of the land surface more accurately [28]. Moreover, in areas with complex terrain, ESTARFM is frequently used to generate synthetic remote sensing images with high spatial and temporal resolution due to its extraordinary fusion effect [29,30]. This method not only considered the temporal variation characteristics of pixel reflectivity but also considered the spatial reflectivity variation characteristics of pixel reflectivity, which effectively improved the accuracy of fusion. However, very few studies have focused on the synthesis of the GIMMS NDVI3g and MODIS NDVI on the Qinghai–Tibet Plateau spatiotemporally based on ESTARFM and further analyzed the spatiotemporal pattern of vegetation from 1990 to the present. For this reason, based on the ESTARFM, the present article developed spatiotemporal fusion of the Qinghai–Tibet Plateau, using GIMMS NDVI3g and MODIS NDVI to monitor the evolution of vegetation.

Therefore, this study used GIMMS NDVI3g and MODIS NDVI data to analyze the temporal and spatial patterns and driving factors of vegetation cover on the Qinghai–Tibet Plateau. Its main purposes were as follows. Firstly, we used the ESTARFM model to establish the functional relationship between the two data sources to improve the spatial resolution of the GIMMS NDVI3g data. Secondly, we quantified the vegetation coverage of the Qinghai–Tibet Plateau. Finally, we analyzed long-term vegetation change characteristics and their influencing factors on the Qinghai–Tibet Plateau.

2. Materials and Methods

2.1. Study Area

Known as the "roof of the world", the Qinghai–Tibet Plateau (26°00′ N–39°47′ N, 73°19′ E–104°47′ E) is located in the mid-latitude alpine area with an average altitude of higher than 4000 m, an area of about 2.57 million km^2, and solar radiation of 0.5861–0.7954 MJ cm^{-1} a^{-1} and is the largest area in China with the highest solar radiation value. This region has a complex terrain, numerous towering mountains, such as the Tanggula Mountains, Kunlun Mountains, and the Himalayas. The connection of the major mountains causes a huge terrain gap in the territory, showing an obvious altitudinal zonality. The Qinghai–Tibet Plateau is therefore characterized by high altitudes, forming natural conditions significantly different from other mid-latitude warm temperate zones and subtropical regions [31]. The weather is cold all year round with the average annual temperature in the hinterland below 0 °C. Moreover, the uneven distribution of precipitation, such as in the Himalayan mountains, greatly blocks the northward movement of the warm and wet air in the southern ocean, resulting in the decline of annual precipitation from 2000 mm to below 50 mm in the region and a climate type change from the humid southeast to the arid northwest [5]. The climate difference between different regions is obvious, the dry and wet were distinct, the northwest region is cold and arid, and the southeast region is warmer, hotter, and more humid. The Qinghai–Tibet Plateau has a long snow cover in winter and covers a large area, which is the source of the Yellow River, the Yangtze River, the Yarlung Zangbo River, and other important rivers. The Qinghai–Tibet Plateau contains rich herbaceous resources, and the main vegetation types are mountain forests, mountain shrubs, alpine meadows, alpine grasslands, alpine deserts, etc. [32]. Due to the harsh climate and high altitude, human activities are scarce. The local natural vegetation is well preserved. Such a unique geographical location and climate characteristics make the Tibetan Plateau a "sensor" of global climate change, where the vegetation is highly sensitive to climate change. As an eco-environmental fragile area, this region serves as an important ecological barrier in China. The vegetation changes pose a significant impact on the hydrological characteristics, carbon cycle, and surface stability of this area's surface [33] as well as the climate and ecology of China or even the world. Therefore, it is of great significance to explore the spatial and temporal distribution characteristics of vegetation cover on the Qinghai–Tibet Plateau and analyze the main driving factors of regional vegetation changes.

2.2. Data Sources and Pre-Processing

2.2.1. NDVI Data

As one of the sensitive indicators of vegetation changes, NDVI is widely applied in large-scale and regional vegetation monitoring research [33–35]. The MODIS NDVI used in this study was acquired from the MODIS 13Q1 Vegetation Index data set product (Geospatial Data Cloud Platform of the Computer Network Information Center of the Chinese Academy of Sciences,http://www.gscloud.cn/ (accessed on 10 January 2022)), with a spatial resolution of 500 m × 500 m and the temporal resolution of 30 days. To smooth the period of NDVI, the self-adapted filtering operation Savitzky–Golay was used to reduce the influences caused by noise such as outliers. GIMMS NDVI3g data set was downloaded from the ECOCAST website (https://ecocast.arc.nasa.gov/data/pub/gimms/3g.v1/ (accessed on 12 January 2022)), with the spatial resolution of 8 km × 8 km and the temporal resolution of 15 days.

These two kinds of NDVI original data sets have discrepancies in spatial and temporal resolutions. In order to enhance the fusion of the two data sets, reduce the influence of cloud and other atmospheric effects on the data, and improve the data accuracy our study adopted the maximum synthesis method to synthesize the MODIS NDVI data sets in 2000, 2005, 2010, and 2015 and GIMMS NDVI data sets in 1990, 1995, 2000, 2005, 2010 and 2015, respectively [36]. In addition, to meet the data requirements of the ESTARFM, the spatial resolution of the GIMMS NDVI data sets was configured to 500 m × 500 m after resampling, using the nearest neighbor pixel method.

2.2.2. Climatic and Auxiliary Data

The meteorological data set includes the average annual temperature and annual precipitation in 1990, 1995, 2000, 2005, 2010, and 2015, with a spatial resolution of 1 km × 1 km and the terrain data spatial resolution of 500 m × 500 m. The climate data set, topographic data set, and Qinghai–Tibet Plateau partitioned data were obtained from the Resource and Environmental Science Data Center of the Chinese Academy of Sciences (http://www.resdc.cn (accessed on 20 January 2022)). The vegetation type data came from the National Tibetan Plateau Data Center (https://www.tpdc.ac.cn (accessed on 3 February 2022)). The climate data set were adjusted with Australia's ANUSPLIN interpolation software to interpolate climatic factors such as temperature and precipitation. Specific principles are provided in the reference [37].

To unify the data standards, the spatial resolution of the data set resampled by the nearest-neighbor pixel method is 500 × 500 m. All data were converted into a unified projected coordinate system WGS_1984_Zone_48N and cropped out in the study area.

2.3. Methods

2.3.1. ESTARFM Spatiotemporal Fusion Algorithm

Obtained the ESTARFM algorithm by improving the STARFM algorithm [28]. This algorithm can generate fused data with high time and space resolution by combining the time or space advantages of the two types of data sources. Based on the original model, the weighting method was adjusted by the ESTARFM fusion model according to the heterogeneity of the pixels, thus the prediction results by setting conversion coefficients were improved (Algorithm 1). Greater heterogeneity led to higher prediction accuracy, which preserved more details of spatial features [4]. In this study, based on MODIS NDVI with a spatial resolution of 500 m × 500 m, the ESTARFM algorithm was used to obtain the fused GIMMS NDVI3g data from 1990 and 1995 in the Tibetan Plateau, respectively, with a spatial resolution of 500 m × 500 m. As shown in Figure 1:

Algorithm 1: Pseudocode of the ESTARFM

Input: two fine-resolution images at t_m and t_n, three coarse-resolution images at t_m , t_n and t_p
Output: fine-resolution image at t_p
If $F(x,y,t_k,B) = a \times C(x,y,t_k,B) + b:$
$T_k = \frac{1/\left|\sum_{j=1}^{w}\sum_{l=1}^{w} C(x_j,y_l,t_k,B) - \sum_{i=1}^{w}\sum_{l=1}^{w} C(x_j,y_l,t_p,B)\right|}{\sum_{k=m,n}\left(1/\left|\sum_{j=1}^{w}\sum_{l=1}^{w} C(x_j,y_l,t_k,B) - \sum_{i=1}^{w}\sum_{l=1}^{w} C(x_j,y_l,t_p,B)\right|\right)}, (k = m,n).$
Then
$F(x_{w/2}, y_{w/2}, t_p, B) = T_m \times F_m(x_{w/2}, y_{w/2}, t_p, B) + T_n \times F_n(x_{w/2}, y_{w/2}, t_p, B)$
Check convergence
$|F(x_i, y_i, t_k, B) - F(x_{w/2}, y_{w/2}, t_k, B)| \leq \delta(B) \times 2/m$
Compute average absolute difference (AAD) and average absolute (AD)

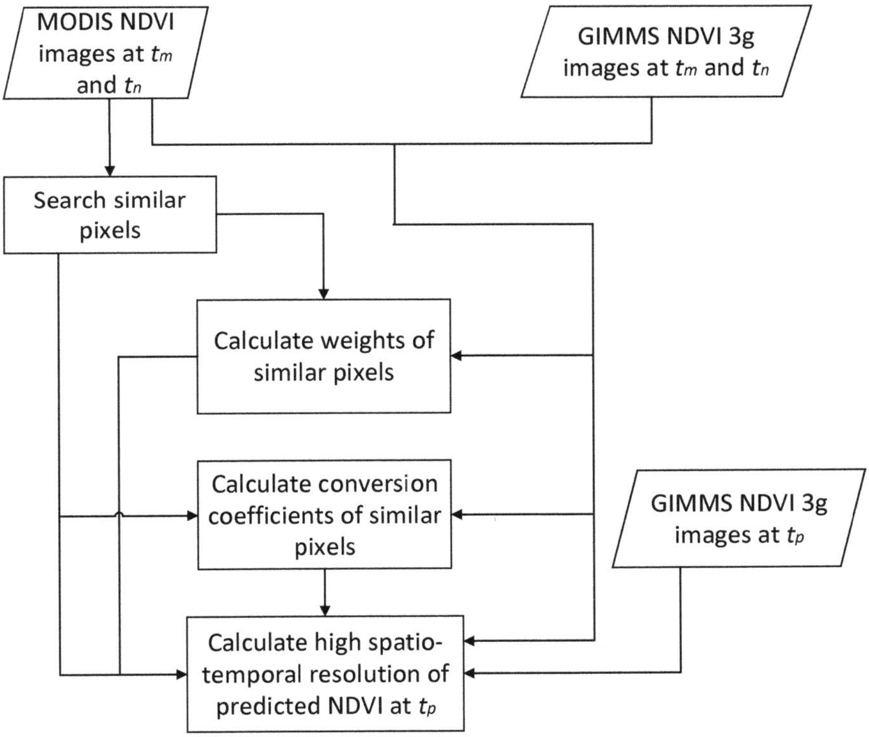

Figure 1. Schematic explanation of the enhanced spatial and temporal adaptive reflectance fusion model (ESTARFM) for fusing GIMMS NDVI3g and MODIS NDVI.

After preprocessing such as registration and cropping of the two images, it is assumed that the systematic deviation between MODIS NDVI and GIMMS NDVI was solely attributed to the discrepancy in NDVI values, and there was no significant difference between the images in the two periods, as shown in Formula (1).

$$M(x,y,t_p,B) = M(x,y,t_0,B) + a \times (G(x,y,t_p,B) - G(x,y,t_0,B)) \quad (1)$$

In which M represents the MODIS NDVI data, G represents the GIMMS NDVI3g data, (x,y) represents the pixel location, B represents the image band, t_0 and t_p represent the time of data acquisition, and a is the conversion coefficient, depending on the system error of the sensor.

Complexity and uncertainty of the actual surface conditions caused deviations in the actual pixel information prediction, therefore it was assumed that the change of the

NDVI value of the mixed pixel is a typical epitome of NDVI amongst different land cover types, and the change of the mixed pixel NDVI over time is the weighted sum of the NDVI variations of the pixels from different land cover types. The proportion of each component in different land cover types remains unchanged, as shown in Formula (2).

$$M(x,y,t_p,B) = M(x,y,t_0,B) + v(x,y) \times (G(x,y,t_p,B) - G(x,y,t_0,B)) \tag{2}$$

$v(x,y)$ represents the correlation coefficient corresponding to the i-th similar pixel after the decomposition of mixed pixels. Formula (2) reveals that the change in the NDVI value of a pixel is the most similar to its neighbors. Images were fused according to the following steps. Using the moving window set by the adjacent pixels, similar pixels can be found. Whilst using the mutual relationship among the pixels, the value of the center pixel of the image can be obtained, as shown in Formula (3).

$$M(x_{w/2},y_{w/2},t_p,B) = M(x_{w/2},y_{w/2},t_0,B) + \sum_{i=1}^{N} W_i \times v_i \times (G(x_i,y_i,t_p,B) - G(x_i,y_i,t_0,B)) \tag{3}$$

In which N represents the number of similar pixels in the center prediction pixel, (x_i,y_i), W, and v_i represent the position, the weight, and the conversion coefficient of the i-th similar pixel, respectively.

The MODIS NDVI image at a time t_p is obtained by fusing the MODIS NDVI at time m in the first period and the GIMMS NDVI3g at a time t_p, which is recorded as $M_m\ (x_{w/2},y_{w/2},t_p,B)$. The observation data at a time n in the second period and the MODIS NDVI data at the time t_p were fused and denoted as $M_n\ (x_{w/2},y_{w/2},t_p,B)$. The MODIS NDVI data at the time t_p tended to be more accurate after weighting the two results. The weights are calculated by the changes between GIMMS at the time t_m and time t_n, and the GIMMS at time t_p, respectively, as shown in Formula (4).

$$T_k = \frac{1/\left|\sum_{j=1}^{W}\sum_{i=1}^{W} G(x_i,y_j,t_k,B) - \sum_{j=1}^{W}\sum_{i=1}^{W} G(x_i,y_j,t_p,B)\right|}{\sum_{k=m,n}\left(1/\left|\sum_{j=1}^{W}\sum_{i=1}^{W} G(x_i,y_j,t_k,B) - \sum_{j=1}^{W}\sum_{i=1}^{W} G(x_i,y_j,t_p,B)\right|\right)}, (k=m,n) \tag{4}$$

The value of the center pixel can be obtained by Formula (5):

$$M(x_{w/2},y_{w/2},t_p,B) = T_m \times M_m\ (x_{w/2},y_{w/2},t_p,B) + T_n \times M_n\ (x_{w/2},y_{w/2},t_p,B) \tag{5}$$

2.3.2. The Calculation of Vegetation Coverage

Vegetation coverage, as an important indicator for studying global climate change and vegetation growth status, has received extensive attention from scholars worldwide [33,35,38]. The calculation method is shown in Formula (6):

$$F = (NDVI - NDVI_n)/(NDVI_i - NDVI_n) \tag{6}$$

where F represents the vegetation coverage; $NDVI$ is the normalized vegetation index of the pixel; $NDVI_i$ and $NDVI_n$ are, respectively, the vegetation index of vegetation-covered land and bare soil land. $NDVI$ values with 95% confidence $(NDVI_i)$ and 5% confidence $(NDVI_n)$ in the study area were chosen.

2.3.3. Linear Regression Analysis

The simple linear regression model was used to assist the calculation of various characteristics and laws of vegetation coverage in single-pixel [39].

$$A = \frac{m \times \sum_{i=1}^{m}(i \times F_i) - (\sum_{i=1}^{m} i) \times (\sum_{i=1}^{m} F_i)}{(m \times \sum_{i=1}^{m} i^2 - \sum_{i=1}^{m} i^2)} \tag{7}$$

where A represents the change rate of vegetation coverage over time in the research period; m represents the total research years; and F_i represents the vegetation coverage of the i-th year. Thus, the variation range of vegetation coverage can be calculated as follows:

$$B = A \times (n-1) \tag{8}$$

If $\boldsymbol{A}$ value is negative, the vegetation coverage of pixels from 1990 to 2015 shows A downward trend; otherwise, it shows an upward trend. If A value is zero, it shows no obvious change in trend; the larger the absolute value of A is, the more obvious the change is.

2.3.4. Hurst Index

According to the Hurst index, based on the R/S analysis method, the continuity of vegetation coverage over time can be calculated effectively, as long as the value is in the range of [0, 1]. It has received extensive attention in hydrology [40,41], geography [42], climatology [43], and so on. The calculation procedure is demonstrated as follows:

The mean of vegetation coverage over time is determined by Formula (9), with the time set $t = 1, 2, \ldots, n$.

$$\overline{f}(\tau) = \frac{1}{\tau}\sum_{t=1}^{\tau} f(t), \tau = 1, 2, \ldots, n \tag{9}$$

The cumulative deviation of time t:

$$X(t,\tau) = \sum_{t=1}^{\tau}\left(f(t) - \overline{f}(t)\right), 1 \le t \le \tau \tag{10}$$

The range of vegetation coverage over time set:

$$R(\tau) = \max_{1\le t\le\tau} X(t,\tau) - \min_{1\le t\le\tau} X(t,\tau),\ \tau = 1, 2, \ldots, n \tag{11}$$

Construct the standard deviation sequence:

$$S(\tau) = \sqrt{\frac{1}{\tau}\sum_{t=1}^{\tau}\left(f(t) - \overline{f}(t)\right)^2},\ \tau = 1, 2, \ldots, n \tag{12}$$

Hurst index:

$$R(\tau)/S(\tau) = (c\tau)^H \tag{13}$$

In the formulas above, H is the Hurst index. Time set t is not continuous with $H = 0.5$. Time set t has a distinctive feature of continuity with $H > 0.5$. The continuous trend becomes more and more obvious as H is approaching 1. The time set has a distinctive feature of anti-continuity when $H < 0.5$. The closer H is to 0, the more obvious the anti-continuous trend is.

2.3.5. Partial Correlation Analysis

The complexity and the correlation which exist between elements of geographic systems render the relation between driving factors become the key to driving analysis. Using partial correlation analysis, the correlation between vegetation coverage and a single, climate-driving factor can be determined by ruling out the influences of other climate driving factors [44].

$$R_{ab,c} = \frac{R_{ab} - R_{ac}R_{bc}}{\sqrt{(1 - R_{ac}{}^2) + (1 - R_{bc}{}^2)}} \tag{14}$$

In the formula, $R_{ab,c}$ is the partial correlation index between a and b, based on the control variable c; R_{ab}, R_{bc}, and R_{ac} are, respectively, the simple correlation coefficients between a and b, b and c, and a and c, in which a, b, and c are, respectively, the vegetation coverage index, mean annual temperature, and annual precipitation. Two variables are positively correlated, when $R_{ab,c} > 0$, and vice versa. Greater absolute value $R_{ab,c}$ stands for a greater correlation between the two variables.

The calculation of the simple correlation coefficient is as follows:

$$R_{xy} = \frac{\sum_{i=1}^{n}((x_i - \overline{x})(y_i - \overline{y}))}{\sqrt{\sum_{i=1}^{n}(x_i - \overline{x})^2 \sum_{i=1}^{n}(y_i - \overline{y})^2}} \tag{15}$$

where, R_{xy} is the simple correlation coefficient between x and y, with the range of $[-1, 1]$; x_i is the value of the variable x in the i-th year; y_i is the value of a variable y in the i-th year; $\overline{x}$ is the mean of a variable x during the study period; $\overline{y}$ is the mean of a variable y during the study period.

3. Results

3.1. Data Fusion

We summarized and plotted the values of pixels in the fused GIMMS NDVI3g image and the real MODIS NDVI image in 2000, 2005, 2010, and 2015, as shown in Figure 2. The fused GIMMS NDVI3g data and the real MODIS NDVI data were in good agreement. The scattered points of the pixels were evenly distributed along the trend line of $y = x$, indicating a decent prediction effect.

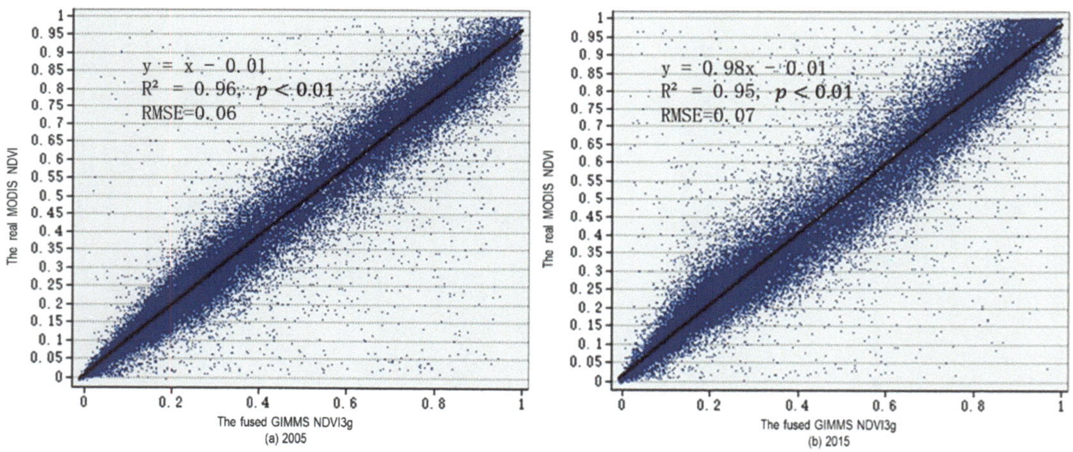

Figure 2. Scatter plot of pixel values corresponding to the fused GIMMS NDVI3g and MODIS NDVI.

By comparing the details of GIMMS NDVI3g data, fused GIMMS NDVI3g data, and real MODIS NDVI data in 2000, 2005, 2010, and 2015 (Figure 3), higher pixel resolutions were observed with the fused GIMMS NDVI3g data and real MODIS NDVI data, where the texture of the ground features was clear and the details of some characteristic areas can be identified. In addition, the detailed features of the regional vegetation coverage can be well reproduced.

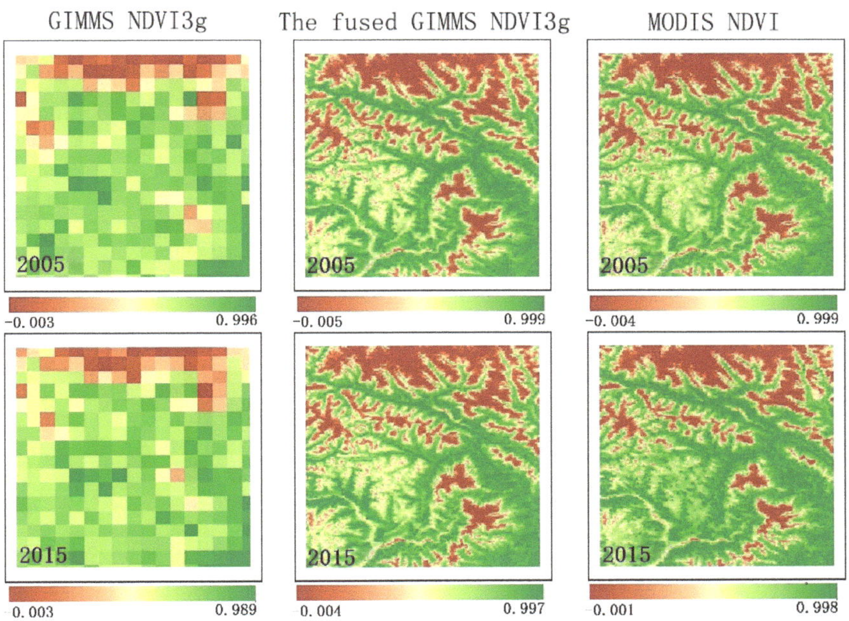

Figure 3. Detail comparison of fused GIMMS NDVI3g and real MODIS NDVI data.

3.2. Analysis of Vegetation Coverage Characteristics

The Qinghai–Tibet Plateau is a vast territory with complex and diverse climate types. In order to better analyze the vegetation distribution and change characteristics of the Qinghai–Tibet Plateau, the study area was divided into 18 zones according to the Qinghai–Tibet Plateau zoning data and vegetation coverage type data. (Table 1). Notably, the VIIBiib, VIIIBi, VIIICi, VIIICii, and VIIIBii zones had more obvious vegetation degradation trends, whereas the VIAiic zone had a more obvious vegetation improvement trend. The absolute values of *A* in all the zones were above 0.8%, implying that the vegetation changes in the Qinghai–Tibet Plateau had extreme distribution characteristics. In areas with low average vegetation coverage, such as desert zones and sub-zones, the vegetation tends to degenerate, while vegetation trends improve as the average vegetation coverage becomes higher.

Table 1. Characteristics of different zones on the Qinghai–Tibet Plateau.

Zones	Codes	Average Vegetation Coverage		*A* Value(%) of 1990–2015
		1990	2015	
Warm temperate shrubbery, semi-shrubbery, and bare land zone	VIIBiib	0.17	0.13	−0.87
Southern temperate sylvosteppe zone	VIAiia	0.84	0.86	0.45
Southern temperate desert steppe sub-zone	VIAiic	0.75	0.8	1.03
Alpine steppe zone	VIIIBi	0.39	0.36	−0.84
Alpine shrub and meadow zone	VIIIAi	0.89	0.9	0.19
Alpine meadow zone	VIIIAii	0.68	0.72	0.58
Cold temperate coniferous forest zone in subtropical mountain	IVBiii	0.8	0.81	0.2
Warm temperate shrub, semi-shrub and desert sub−zone	VIIBiia	0.3	0.29	−0.38
Alpine desert zone	VIIICi	0.21	0.18	−0.88
Temperate desert zone	VIIICii	0.3	0.25	−1.08
Temperate steppe zone	VIIIBii	0.53	0.49	−0.92
Mid-subtropical evergreen broad-leaved forest zone	IVBi	0.99	0.98	−0.24
North tropical seasonal rain forest, semi evergreen season	VBi	0.87	0.87	0.1
Northern mid-subtropical evergreen broad-leaved forest sub-zone	IVAiia	0.95	0.94	−0.35
Northern subtropical mixed evergreen and deciduous broad-leaved zone	IVAi	0.98	0.98	−0.07
Warm temperate deciduous oak forest zone	IIIii	0.98	0.98	0.03
Temperate shrub and semi-fruticous desert zone	VIIBib	0.55	0.56	−0.13
Temperate semi-shrub and fruticous desert zone	VIIBi	0.19	0.18	−0.23

In 2000, the State Council issued the "Several Opinions on Doing a Good Job in Returning Cultivated Land to Forest and Grassland" [45]. Returning farmland to forest and grassland policy was implemented nationwide in China, which generally promoted vegetation growth. In order to explore the change in vegetation on the Qinghai–Tibet Plateau since the promulgation of this policy, the profile of vegetation coverage and Hurst index of the Qinghai–Tibet Plateau around 2000 and 1990–2015 were calculated in this study, as shown in Figure 4. From 1990 to 1995, 37.46% of the area was subject to vegetation degradation. The land with A values less than -0.5%, between -0.5% and 0%, and between 0% and 0.5% accounted for 9.93%, 27.53%, and 54.42% of the total area, respectively, while the land with A values over 0.5% accounted for 8.12% of the total area. From 2000 to 2015, 52.47% of the area suffered from vegetation degradation, with an average growth trend (A) of 0.02%. The land with A values less than -5%, between -5% and 0%, and between 0% and 5% accounted for 0.23%, 52.24%, and 47.37% of the total area, respectively, while the land with A values over 5% accounted for 0.15% of the total area. From 1990 to 2015, the area with vegetation degradation increased to 73.49% with an average growth trend (A) of -0.041%. The land with A values less than -5%, between -5% and 0%, between 0% and 5%, and over 5% accounted for 0.10%, 73.39%, 26.42%, and 0.08% of the total area, respectively. According to the distribution of vegetation temporal profile from 1990 to 2015 (Figure 4c), it is clear that the vegetation changes on the Qinghai–Tibet Plateau were spatially heterogeneous and specifically manifest as degradation in the west and rise in the east. Furthermore, the degradation of vegetation was prevalent in the study area. Areas with a Hurst index less than 0.5, over 0.5, and equal to 0.5 accounted for 9.16%, 80.59%, and 10.25% of the total area, respectively. The spatial distribution of the Hurst index indicated that the change of vegetation characteristics in most parts of the Qinghai–Tibet Plateau was continuous, whereas the continuity was not evident in the western Sichuan and the northern Yunnan regions. The comprehensive vegetation change trend and Hurst index showed that 22.14% of the Tibetan Plateau continued to improve from 1990 to 2015 ($A > 0$ and $H > 0.5$).

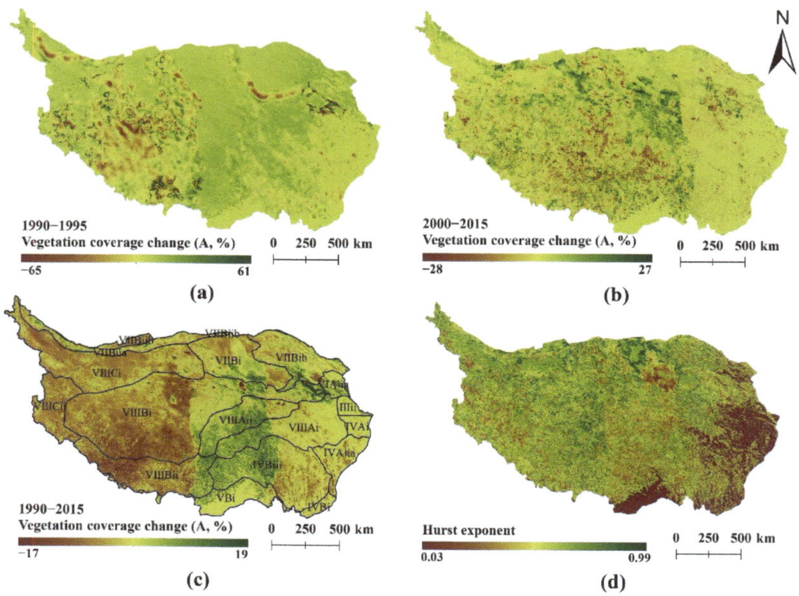

Figure 4. The vegetation change trend and Hurst index on the Qinghai–Tibet Plateau, (**a**) is the vegetation change trend from 1990 to 1995, (**b**) is the vegetation change trend from 2000 to 2015, (**c**) is the vegetation change trend from 1990 to 2015, (**d**) is the Hurst index from 1990 to 2015.

3.3. Impact of the Topographic on Vegetation Coverage

The terrain of the Qinghai–Tibet Plateau is complex and changeable, with ascending attitude from the southeast toward the northwest. Regarding previous studies, this paper extracted two types of terrain factors, namely slope, and aspect, to explore their driving mechanism for vegetation change [46]. The slope was divided into 0–3, 3–8, 8–15, 15–25, and greater than 25 degrees by the natural breakpoint method. At the same time, the slope aspect was calculated by the elevation data, and the aspect data was divided into three categories: flat slopes with a result of −1; sunny slopes were generally south, southwest, west, and northwest and the value range was between 157.5°and 337.5°; shady slopes were generally northeast, east, north, and southeast and the value range was 337.5°–360° and 0°–157.5°. In the aspect of the spatial scale, when the slope was over 15°or located on the slope surface, the vegetation cover was in good condition. In terms of the time scale, the vegetation coverage driven by terrain factors was developing toward continuous degradation ($A < 0$ and Hurst > 0.5). The degradation trend appeared to be less significant with steeper slopes. It should be highlighted that when the terrain slope was between 0°and 3°, the vegetation coverage in 1990 and 2015 was 0.38 and 0.36, respectively. In contrast, with the A value of -0.58% (Hurst > 0.6) and the terrain slope of over 25°, the vegetation coverage in 1990 and 2015 increased to 0.71 and 0.70, respectively. In 2015, the vegetation coverage was 0.71 and 0.70, respectively. Moreover, as the *A* value came to −0.08% (Hurst > 0.6), the impact of this aspect on the vegetation coverage was smaller than that of the slope (Table 2).

Table 2. Vegetation coverage and its variation characteristics on the Qinghai–Tibet Plateau under different terrain factors.

Topographic Factors		Average Vegetation Coverage 1990	Average Vegetation Coverage 2015	*A* Value (%) from 1990 to 2015	Hurst Index from 1990 to 2015
Slope (°)	0–3	0.38	0.36	−0.58	0.67
	3–8	0.5	0.5	−0.51	0.67
	8–15	0.58	0.57	−0.37	0.65
	12–25	0.63	0.62	−0.23	0.65
	≥25	0.71	0.7	−0.08	0.64
Aspect	Flat	0.35	0.34	−0.45	0.65
	Sunny slope	0.52	0.5	−0.4	0.66
	Shady slope	0.53	0.51	−0.44	0.66

3.4. Link between Climate Change and Vegetation Coverage

Based on the partial correlation analysis (Figure 5), it could be found that the average annual temperature and annual precipitation in different areas of the Qinghai–Tibet Plateau from 1990 to 2015 had significant differences due to the interference of vegetation coverage. In general, their changes drove the decrease in vegetation coverage, with annual mean temperature having a greater impact on vegetation coverage than annual precipitation. Rt > 0, that is, 9.59% of the total area was driven by the change of average annual temperature promoting vegetation coverage. A lot of the 47.67% of the land had Rp > 0, where the change of annual precipitation promotes vegetation coverage, while the change in annual precipitation was negatively correlated with the vegetation coverage in the rest of the area.

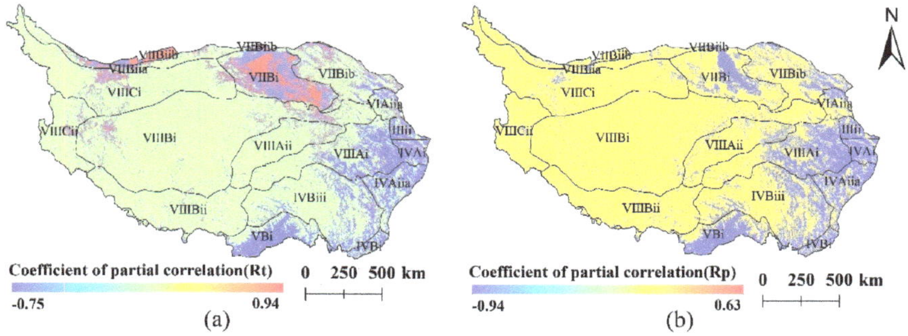

Figure 5. The coefficient of partial correlation between vegetation coverage and climate factors, (**a**) is the coefficient of partial correlation between vegetation coverage and average annual temperature, and (**b**) is the coefficient of partial correlation between vegetation coverage and annual precipitation.

In the alpine region of the Qinghai–Tibet Plateau, vegetation coverage was positively correlated with annual mean air temperature and negatively correlated with annual precipitation in the alpine shrub and meadow zone (VIIIAi) and in the alpine desert zone (VIIICi). The annual precipitation and the average annual temperature, respectively, had a negative and positive correlation with the vegetation coverage. In the temperate desert region, the average annual temperature and annual precipitation had significant differences in the driving degree of vegetation coverage changes, which showed that the correlation degree between vegetation coverage and the annual air temperature was greater than that of annual precipitation in the temperate shrub/semi-shrub desert sub-region (VIIBib) and alpine desert region (VIIICi).

The degree of correlation is greater than the annual precipitation. The average annual temperature had a stronger correlation with the vegetation coverage compared to the annual precipitation.

The changes in vegetation coverage, temperature, and precipitation with time are shown in Figure 6. In general, global warming caused by the increase of annual average temperature had a positive effect on the alpine vegetation change on the Qinghai–Tibet Plateau but a negative effect on the vegetation change in the temperate desert regions. The increase in annual precipitation had a significant negative effect on alpine vegetation and temperate desert vegetation on the Tibetan Plateau, presumably due to the hysteresis, with a certain hysteresis effect. The annual vegetation coverage changes in the alpine shrub and meadow zone (VIIIAi) and alpine desert zone (VIIICi) were more sensitive to the average annual temperature than to the annual precipitation, due to the location of the alpine region. Vegetation coverage had a significant positive correlation with the average annual temperature and a weakly negative correlation with the annual precipitation. In addition, hysteresis was one of the driving factors influencing the annual precipitation (Figure 6a–d). Although both belong to the temperate desert region, the temperate shrub/semi-fruticous desert zone (VIIBib) and the temperate semi-shrub/fruticous desert zone (VIIBi) showed that the vegetation coverage was affected by the average annual temperature due to different water and heat conditions. There were differences in the driving degree and driving direction of annual precipitation. In temperate shrub and semi-fruticous desert zone (VIIBib), the average annual temperature had a positive effect on vegetation coverage, while the annual precipitation had a negative driving effect. Furthermore, the impact of annual precipitation before 2010 was more obvious. The combined effect of annual average temperature and annual precipitation was responsible for the improvement of vegetation coverage after 2010. In the temperate semi-shrub and fruticous desert zone (VIIBi), the average annual temperature and annual precipitation, with similar growth curves, both showed a negative effect on vegetation coverage (Figure 6e–h).

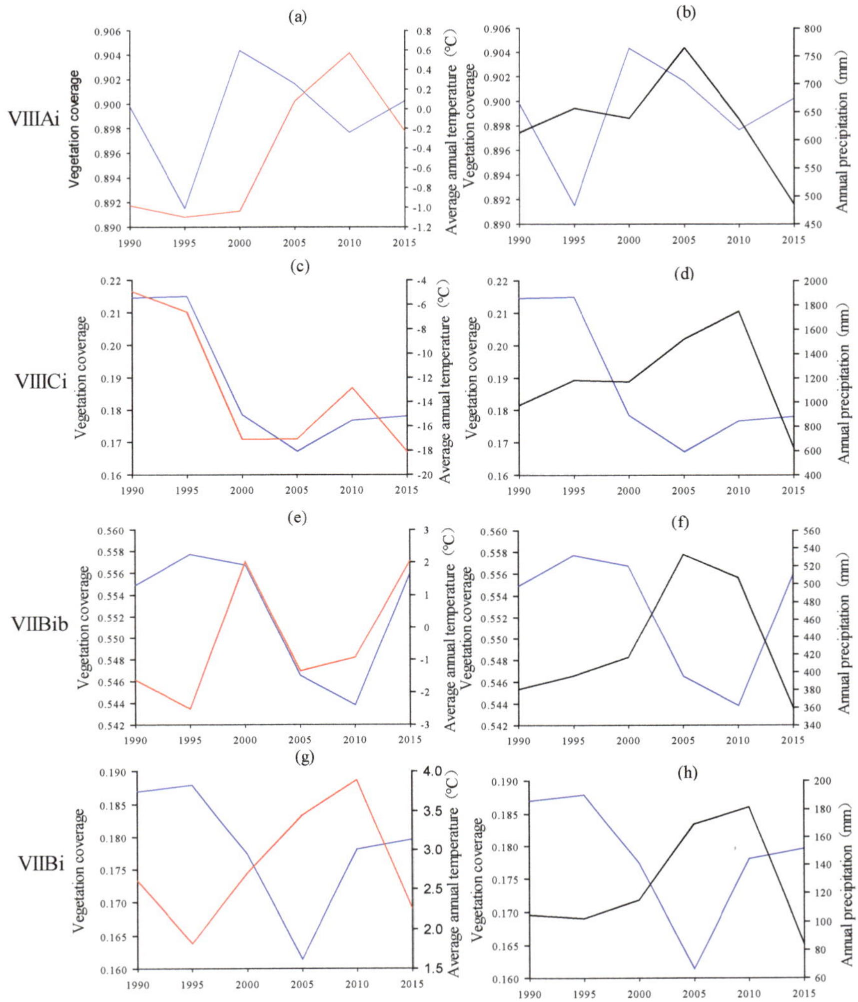

Figure 6. The change curves of vegetation coverage, average annual temperature, and annual precipitation in VIIIAi (**a**,**b**), VIIICi (**c**,**d**), VIIBib (**e**,**f**), VIIBi (**g**,**h**). The change curves of vegetation coverage, average annual temperature, and annual precipitation are, respectively, blue, red, and black.

4. Discussion

Biological and environmental factors were two main causes for the change of NDVI, and the latter had a stronger influence [47]. Therefore, it is reasonable for the ESTARFM model to assume that the change characteristics in a small range are similar [28]. The adjustment of weighting and conversion coefficient can improve the results to a certain extent and retain more spatial details. The high spatiotemporal-resolution image that fused in this study, based on the ESTARFM, had preeminent spatial details and a strong correlation with the original MODIS data ($R^2 \geq 0.95$, $p < 0.01$), which facilitated the study of the characteristics of vegetation changes on the Qinghai–Tibet Plateau. This result demonstrated the feasibility of the ESTARFM model to study the vegetation change characteristics of the Tibetan Plateau, and further proved the scientific nature of the ESTARFM model to integrate medium- and high-resolution images.

From 1990 to 2015, the vegetation coverage of different slopes and slope directions on the Qinghai–Tibet Plateau showed a trend of degradation. When the slope was more than 25°, the vegetation coverage had the weakest degradation trend ($A = -0.08\%$, Hurst = 0.64), and the slope with the strongest degradation trend was between 0°and 3° ($A = -0.58\%$, Hurst = 0.67). According to the analysis, the policy implemented by the State Council in 2000, that all the farmland with a slope of >25°should be returned to forests and grasslands in the Qinghai–Tibet Plateau area, was responsible for the improved vegetation coverage. The areas with a slope between 0°and 3° were mainly intermountain basins and low mountains, where the degradation of grassland was significant and widespread, and the ecological environment was poor. The data suggest that restoration measures, such as returning farmland, further protection, and restoration need to be carried out in order to achieve significant results. The vegetation cover and its degradation degree on the shady slope was slightly higher than that of the sunny slope (Table 2), which further proves that the vegetation cover change on a shady slope was more sensitive than that of the sunny slope [46].

The total change percentage for vegetation coverage was −0.41% from 1990 to 2015 and 0.02% from 2000 to 2015. The results showed that the conversion of cropland to forest and grassland in China since 2000 improved the vegetation coverage in this region to a certain extent, which further proved that the vegetation coverage in the Tibetan Plateau had decreased [48].

From 1990 to 2015, the vegetation coverage on the Tibetan Plateau increased in the east while decreasing in the west with the larger land area (Figure 4c). In 2010, the vegetation coverage of the Qinghai–Tibet Plateau changed significantly as well as the average annual temperature and the annual precipitation (Figure 6). Previous studies and analyses illustrated that the spatiotemporal distribution of major climatic factors is one of the main reasons for their complexity [49,50]. The analysis of this study showed that the distribution, trend, and direction of vegetation coverage were affected differently by temperature and precipitation due to regional differences in different zones. The northeast and northwest of the Qinghai–Tibet Plateau were mainly deserts, where vegetation coverage had a strong and positive correlation with temperature but a negative correlation with precipitation. The data indicated that vegetation coverage was sensitive to changes in water and heat conditions. In this study, it was found that the precipitation change in the northeast was the main reason for vegetation degradation (Figure 6), which was consistent with previous research results [5,51]. In addition, a stronger positive correlation between temperature and vegetation coverage in the alpine desert area of the Qinghai–Tibet Plateau was observed, and the change of hysteresis in vegetation coverage caused by precipitation is more significant compared to temperature. On the southeastern Qinghai–Tibet Plateau, temperature and precipitation had similar change trends, and they both negatively affected vegetation coverage.

The area was mainly located in temperate grassland and subtropical broad-leaved forest, with good water and heat conditions. Therefore, in terms of climate impact, it was believed that the effects of temperature and climate on vegetation growth in the area had reached equilibrium or saturation. If significant climate change occurs, the growth of vegetation will be inhibited. Combining our results with those of previous studies [5], it was concluded that the uncertainty of other factors was an important reason for the de-concentration of partial correlation between vegetation coverage and temperature and precipitation in the northwest part of the Qinghai–Tibet Plateau (Figure 5). Therefore, further studies are needed to determine the driving factors of vegetation change in this area.

The ESTARFM model required two pairs of input data, and the accuracy of the input data was high; due to the high altitude of the Qinghai–Tibet Plateau, the cloud cover was more serious, which affected the accuracy of data fusion. The spatial resolution of the data fusion in this study was 500 m. Therefore, follow-up research can calculate the vegetation coverage before fusion and then remove and process outliers. The influencing factors of vegetation change are complex, and future research should consider this more comprehensively and use higher spatial resolution data to study the vegetation conditions of the Qinghai–Tibet Plateau.

5. Conclusions

The fused NDVI results based on the ESTARFM model were used to study vegetation changes on the Tibetan Plateau with complex terrain. From 1990 to 2015, the overall vegetation coverage of the Qinghai–Tibet Plateau was degraded by 0.41%. However, under the policy of returning farmland to forest and grassland, the overall degradation trend degree and the degradation trend degree of slope $\geq 25^{\circ}$ were weakened. In a local scale, the area of vegetation reduced in the west was greater than that of vegetation increased in the east. The regional variances in different zones led to various influences on temperature and precipitation, in terms of distribution, change degree, and change in the direction of vegetation coverage. The main causes of the significant changes in 2010 were attributed to the spatial distribution of climate factors such as temperature and precipitation. This study evaluated the vegetation coverage of the Qinghai–Tibet Plateau and revealed its key characteristics for terrain and some climatic factors. The knowledge generated from this study provides an important scientific basis for local government decision-making and economic development.

The data used in this study has a long time interval, and some details may be neglected. Two main climatic factors, average annual temperature and annual precipitation, were selected for this study. Given the complexity of the geographic system and human activities, other types of climatic factors should be considered as well. Thus, the specific response mechanism needs to be further studied.

Author Contributions: Conceptualization, X.D., L.W. and H.S.; formal analysis, X.D. and L.W.; funding acquisition, H.S.; investigation, X.D., L.W. and C.H.; methodology, X.D., L.W. and H.S.; validation, X.D., L.W. and C.H.; writing—original draft, X.D. and L.W.; writing—review and editing, L.W. and C.H. All authors have read and agreed to the published version of the manuscript.

Funding: This research was funded by the National Natural Science Foundation of China (Grant No. 41401659).

Institutional Review Board Statement: Not applicable.

Informed Consent Statement: Not applicable.

Data Availability Statement: The data presented in this study are available contained within the article.

Acknowledgments: The authors thank the anonymous reviewers for providing invaluable comments on the original manuscript.

Conflicts of Interest: The authors declare no conflict of interest.

References

1. Newbold, T.; Hudson, L.N.; Hill, S.; Contu, S.; Lysenko, I.; Al, E. Global effects of land use on local terrestrial biodiversity. *Nature* **2015**, *520*, 45–50. [CrossRef] [PubMed]
2. Pielke, R.A., Sr. Atmospheric science. Land use and climate change. *Science* **2005**, *310*, 1625-6. [CrossRef] [PubMed]
3. Sterling, S.M.; Ducharne, A.; Polcher, J. The impact of global land-cover change on the terrestrial water cycle. *Nat. Clim. Chang.* **2013**, *3*, 385–390. [CrossRef]
4. Chen, M.L.; Cun-Jun, L.I.; Guan, Y.L.; Zhou, J.P.; Wang, D.Y.; Luo, Z.Q. Generation and application of high temporal and spatial resolution images of regional farmland based on ESTARFM model. *Acta Agron. Sin.* **2019**, *45*, 1099–1110. [CrossRef]
5. Li, L.H.; Zhang, Y.L.; Liu, L.S.; Wu, J.S.; Wang, Z.F.; Li, S.C.; Zhang, H.M.; Zu, J.X.; Ding, M.J.; Paudel, B. Spatiotemporal Patterns of Vegetation Greenness Change and Associated Climatic and Anthropogenic Drivers on the Tibetan Plateau during 2000–2015. *Remote Sens.* **2018**, *10*, 1525. [CrossRef]

6. Piao, S.L.; Fang, J.Y.; He, J.S. Variations in vegetation net primary production in the Qinghai-Xizang Plateau, China, from 1982 to 1999. *Clim. Chang.* **2006**, *74*, 253–267. [CrossRef]
7. Zhou, Y.K.; Fan, J.F.; Wang, X.Y. Assessment of varying changes of vegetation and the response to climatic factors using GIMMS NDVI3g on the Tibetan Plateau. *PLoS ONE* **2020**, *15*, e0234848. [CrossRef]
8. Xu, X.; Chen, H.; Levy, J.K. Spatiotemporal vegetation cover variations in the Qinghai-Tibet Plateau under global climate change. *Chin. Sci. Bull.* **2008**, *53*, 915–922. [CrossRef]
9. Zhang, Y.; Song, C.; Band, L.E.; Sun, G.; Li, J. Reanalysis of global terrestrial vegetation trends from MODIS products: Browning or greening? *Remote Sens. Environ.* **2017**, *191*, 145–155. [CrossRef]
10. Mao, D.H.; Wang, Z.M.; Luo, L.; Yang, G. Correlation Analysis between NDVI and Climate in Northeast China based on AVHRR and GIMMS Data Sources. *Remote Sens. Technol. Appl.* **2012**, *27*, 81–89.
11. Bao, G.; Qin, Z.H.; Bao, Y.H.; Zhou, Y. Spatial-temporal Changes of Vegetation Cover in Mongolian Plateau during 1982–2006. *J. Desert Res.* **2013**, *33*, 918–927. [CrossRef]
12. Liu, Y.; Li, C.Z.; Liu, Z.H.; Deng, X.Y. Assessment of spatio-temporal variations in vegetation cover in Xinjiang from 1982 to 2013 based on GIMMS-NDVI. *Acta Ecol. Sin.* **2016**, *36*, 6198–6208. [CrossRef]
13. Ding, J.; Liu, X.Y.; Guo, Y.C.; Ren, H.R. Study on Vegetation Change in the Qinghai-Tibet Plateau from 1980 to 2015. *Ecol. Environ. Sci.* **2021**, *30*, 288–296. [CrossRef]
14. Stellmes, M.; Udelhoven, T.; Röder, A.; Sonnenschein, R.; Hill, J. Dryland observation at local and regional scale—Comparison of Landsat TM/ETM+ and NOAA AVHRR time series. *Remote Sens. Environ.* **2010**, *114*, 2111–2125. [CrossRef]
15. Hui, G.; Yue, H.; Li, X.; Bao, A.; Yang, S.; Meng, F. Dynamic Changes of Fractional Vegetation Cover along the Mainstream of the Tarim River. *J. Desert Res.* **2016**, *36*, 1472. [CrossRef]
16. Bo, C.; Bo, Z.; Ma, B.; Wang, G.; Min, T.; Zhang, Y.; Jia, Y. Spatial and Temporal Variations of NDVI in Gansu, China from 2000 to 2014. *J. Desert Res.* **2018**, *38*, 418. [CrossRef]
17. Sarmah, S.; Jia, G.; Zhang, A.; Singha, M. Assessing seasonal trends and variability of vegetation growth from NDVI3g, MODIS NDVI and EVI over South Asia. *Remote Sens. Lett.* **2018**, *9*, 1195–1204. [CrossRef]
18. Wang, H.W.; Qi, Y.; Huang, C.L.; Li, X.Y.; Deng, X.H.; Zhan, J.L. Analysis of vegetation changes and dominant factors on the Qinghai-Tibet Plateau, China. *Sci. Cold Arid Reg.* **2019**, *11*, 150–158. [CrossRef]
19. Duan, H.C.; Xue, X.; Wang, T.; Kang, W.P.; Liao, J.; Liu, S.L. Spatial and Temporal Differences in Alpine Meadow, Alpine Steppe and All Vegetation of the Qinghai-Tibetan Plateau and Their Responses to Climate Change. *Remote Sens.* **2021**, *13*, 669. [CrossRef]
20. Li, H.; Zhang, C.G.; Wang, S.Z.; Ma, W.D.; Liu, F.G.; Chen, Q.; Zhou, Q.; Xia, X.S.; Niu, B.C. Response of vegetation dynamics to hydrothermal conditions on the Qinghai—Tibet Plateau in the last 40 years. *Acta Ecol. Sin.* **2022**, 1–14. [CrossRef]
21. Zhang, Y.; Zhang, C.B.; Wang, Z.Q.; Yang, Y.; Li, J.L. Spatiotemporal dynamics of grassland coverage in response to climate change from 1982 to 2012 in the Three Rivers Source Region, China. *Pratacultural Sci.* **2017**, *34*, 1977–1990. [CrossRef]
22. Feng, G.; Masek, J.G.; Schwaller, M.R.; Hall, F.F. On the Blending of the Landsat and MODIS Surface Reflectance: Predicting Daily Landsat Surface Reflectance. *IEEE Trans. Geosci. Remote Sens.* **2006**, *44*, 2207–2218. [CrossRef]
23. Semmens, K.A.; Anderson, M.C.; Kustas, W.P.; Gao, F.; Vélez, M. Monitoring daily evapotranspiration over two California vineyards using Landsat 8 in a multi-sensor data fusion approach. *Remote Sens. Environ.* **2016**, *185*, 155–170. [CrossRef]
24. Singh, D. Evaluation of long-term NDVI time series derived from Landsat data through blending with MODIS data. *Atmosfera* **2011**, *25*, 43–63. [CrossRef]
25. Gevaert, C.M.; García-Haro, F. A comparison of STARFM and an unmixing-based algorithm for Landsat and MODIS data fusion. *Remote Sens. Environ.* **2015**, *156*, 34–44. [CrossRef]
26. Santosh, B.; Stuart, P.; Tony, G. Preparing Landsat Image Time Series (LITS) for Monitoring Changes in Vegetation Phenology in Queensland, Australia. *Remote Sens.* **2012**, *4*, 1856–1886. [CrossRef]
27. Schmidt, M.; Udelhoven, T.; Gill, T.; Roeder, A. Long term data fusion for a dense time series analysis with MODIS and Landsat imagery in an Australian Savanna. *J. Appl. Remote Sens.* **2012**, *6*, 2240–2246. [CrossRef]
28. Zhu, X.L.; Chen, J.; Gao, F.; Chen, X.H.; Masek, J.G. An enhanced spatial and temporal adaptive reflectance fusion model for complex heterogeneous regions. *Remote Sens. Environ.* **2010**, *114*, 2610–2623. [CrossRef]
29. Hao, G.B.; Wu, B.; Zhang, L.F.; Fu, D.J.; Li, Y. Temporal and Spatial Variation Analysis of the Area of Siling Co Lake in Tibet Based on ESTARFM(1976–2014). *J. Geo-Inf. Sci.* **2016**, *18*, 14. [CrossRef]
30. Kim, K.; Ursula, G.; Rasmus, F.; Claudia, K. An ESTARFM Fusion Framework for the Generation of Large-Scale Time Series in Cloud-Prone and Heterogeneous Landscapes. *Remote Sens.* **2016**, *8*, 425. [CrossRef]
31. Yue, B.O.; Xiaolan, L.I.; Wang, C. Seasonal characteristics of the interannual variations centre of the Tibetan Plateau snow cover. *J. Glaciol. Geocryol.* **2014**, *36*, 1353–1362. [CrossRef]
32. Huang, W.J.; Zeng, T.Y.; Huang, X.D. Spatiotemporal dynamics of alpine grassland phenology on the Tibetan Plateau. *Pratacultural Sci.* **2019**, *36*, 919, 1032–1043. [CrossRef]
33. Li, J.; Peng, S.; Zhi, L. Detecting and attributing vegetation changes on China's Loess Plateau. *Agric. For. Meteorol.* **2017**, *247*, 260–270. [CrossRef]
34. Piao, S.; Wang, X.; Ciais, P.; Zhu, B.; Liu, J. Changes in satellite-derived vegetation growth trend in temperate and boreal Eurasia from 1982 to 2006. *Glob. Chang. Biol.* **2011**, *17*, 3228–3239. [CrossRef]

35. Zheng, K.; Wei, J.Z.; Hua, C.; Zhang, X.L.; Huang, F.Q.; Li, F.M.; Ye, J.S. Impacts of climate change and human activities on grassland vegetation variation in the Chinese Loess Plateau. *Sci. Total Environ.* **2019**, *660*, 236–244. [CrossRef]
36. Zhang, L.C.; Hu, L.J.; Zheng, Z.J.; Li, S.; Hou, X.G. Comparison of snow line elevation extraction method based on remote sensing image: A case of north slope of Tianshan Mountains. *Arid Land Geogr.* **2015**, *38*, 788–796. [CrossRef]
37. Hutchinson, M.F. The Application of Thin Plate Smoothing Splines to Continent-Wide Data Assimilation. *Bur. Meteorol. Res. Rep.* **1991**, *27*, 104–113.
38. Jong, R.D.; Bruin, S.D.; Wit, A.D.; Schaepman, M.E.; Dent, D.L. Analysis of monotonic greening and browning trends from global NDVI time-series. *Remote Sens. Environ.* **2011**, *115*, 692–702. [CrossRef]
39. Zhao, L.; Dai, A.; Dong, B. Changes in global vegetation activity and its driving factors during 1982–2013. *Agric. For. Meteorol.* **2018**, *249*, 198–209. [CrossRef]
40. Chen, C.D.; Niu, A.J. Application of R/S Method in the Tendency Prediction of Reservoir Water Inflow. *J. Water Resour. Water Eng.* **2010**. [CrossRef]
41. Koirala, S.R.; Gentry, R.W. Hurst Analysis of Hydrologic and Water Quality Signals. In Proceedings of the World Environmental & Water Resources Congress 2010, Providence, RI, USA, 16–20 May 2010; pp. 4537–4546.
42. Kong, B.; Wang, E.; Li, Z. The effect of high temperature environment on rock properties—An example of electromagnetic radiation characterization. *Environ. Sci. Pollut. Res.* **2018**, *25*, 29104–29114. [CrossRef] [PubMed]
43. Koutsoyiannis, D. Climate change, the Hurst phenomenon, and hydrological statistics. *Int. Assoc. Sci. Hydrol. Bull.* **2003**, *48*, 3–24. [CrossRef]
44. Liu, Y.; Wang, Y.; Peng, J.; Wei, H.; Zhang, X. Land ecological suitability assessment for forest coupled with the resilience perspective: A case study in Wangqing county, Jilin province, China. *Acta Geogr. Sin.* **2015**, *70*, 476–487. [CrossRef]
45. The State Council. Several Opinions on Doing a Good Job in Returning Cultivated Land to Forest and Grassland. *For. China* **2000**, *10*, 3–4.
46. Xu, G.C.; Zhang, J.X.; Li, P.; Li, Z.B.; Lu, K.X.; Wang, X.K.; Wang, F.C.; Cheng, Y.T.; Wang, B. Vegetation restoration projects and their influence on runoff and sediment in China. *Ecol. Indic.* **2018**, *95*, 233–241. [CrossRef]
47. Rao, Y.H.; Zhu, X.L.; Chen, J.; Wang, J.M. An Improved Method for Producing High Spatial-Resolution NDVI Time Series Datasets with Multi-Temporal MODIS NDVI Data and Landsat TM/ETM+ Images. *Remote Sens.* **2015**, *7*, 7865–7891. [CrossRef]
48. Zhang, Y.L.; Qi, W.; Zhou, C.P.; Ding, M.J.; Liu, L.S.; Gao, J.G.; Bai, W.Q.; Wang, Z.F.; Zheng, D. Spatial and temporal variability in the net primary production of alpine grassland on the Tibetan Plateau since 1982. *J. Geogr. Sci.* **2014**, *24*, 269–287. [CrossRef]
49. Lehnert, L.W.; Wesche, K.; Trachte, K.; Reudenbach, C.; Bendix, J. Climate variability rather than overstocking causes recent large scale cover changes of Tibetan pastures. *Sci. Rep.* **2016**, *6*, 24367. [CrossRef]
50. Cong, N.; Shen, M.; Yang, W.; Yang, Z.; Zhang, G.; Piao, S. Varying responses of vegetation activity to climate changes on the Tibetan Plateau grassland. *Int. J. Biometeorol.* **2017**, *61*, 1433–1444. [CrossRef]
51. Pang, G.; Wang, X.; Yang, M. Using the NDVI to identify variations in, and responses of, vegetation to climate change on the Tibetan Plateau from 1982 to 2012. *Quat. Int.* **2016**, *444*, 87–96. [CrossRef]

International Journal of *Environmental Research and Public Health*

MDPI

Article

Spatiotemporal Variation and Driving Forces Analysis of Eco-System Service Values: A Case Study of Sichuan Province, China

Chengjin He [1], Huaiyong Shao [1,*] and Wei Xian [2]

[1] College of Earth Sciences, Chengdu University of Technology, Chengdu 610059, China; 2020020014@stu.cdut.edu.cn

[2] College of Resources and Environment, Chengdu University of Information Technology, Chengdu 610225, China; xianwei@cuit.edu.cn

* Correspondence: shaohuaiyong@cdut.edu.cn

Abstract: Sichuan Province is an important ecological barrier in the upper reaches of the Yangtze River. Therefore, it is critical to investigate the temporal and spatial changes, as well as the driving factors, of ecosystem service values (ESVs) in Sichuan Province. This paper used land use data from 2000, 2005, 2010, 2015, and 2020 to quantify the spatiotemporal changes in the ESVs in Sichuan Province. Correlation coefficients and bivariate spatial autocorrelation methods were used to analyze the trade-offs and synergies of ESVs in the city (autonomous prefecture) and grid scales. At the same time, we used a Geographical Detector model (GDM) to explore the synergies between nine factors and ESVs. The results revealed that: (1) In Sichuan Province, the ESVs increased by 0.77% from 729.26 $\times$ 10^9 CNY in 2000 to 741.69 $\times$ 10^9 CNY in 2020 (unit: CNY = Chinese Yuan). Furthermore, ecosystem services had a dynamic degree of 0.13%. Among them, the ESVs of forestland were the highest, accounting for about 60.59% of the total value. Among the individual ecosystem services, only food production, environmental purification, and soil conservation decreased in value, while the values of other ecosystem services increased. (2) The ESVs increased with elevation, showing a spatial distribution pattern of first rising and then decreasing. The high-value areas of ESVs per unit area were primarily distributed in the forestland of the transition area between the basin and plateau; The low-value areas were distributed in the northwest, or the urban areas with frequent human activities in the Sichuan Basin. (3) The tradeoffs and synergies between multi-scale ecosystems showed that ecosystem services were synergies-dominated. As the scale of research increased, the tradeoffs between ecosystems gradually transformed into synergies. (4) The main driving factors for the spatial differentiation of ESVs in Sichuan Province were average annual precipitation, average annual temperature, and gross domestic product (GDP); the interaction between normalized difference vegetation index (NDVI) and GDP had the strongest driving effect on ESVs, generally up to 30%. As a result, the distribution of ESVs in Sichuan Province was influenced by both the natural environment and the social economy. The present study not only identified the temporal and spatial variation characteristics and driving factors of ESVs in Sichuan Province, but also provided a reference for the establishment of land use planning and ecological environmental protection mechanisms in this region.

Keywords: ecological service values; spatial-temporal variation; tradeoffs and synergies; multi-scale; driving forces

Citation: He, C.; Shao, H.; Xian, W. Spatiotemporal Variation and Driving Forces Analysis of Eco-System Service Values: A Case Study of Sichuan Province, China. *Int. J. Environ. Res. Public Health* **2022**, *20*, 8595. https://doi.org/10.3390/ijerph19148595

Academic Editor: Paul B. Tchounwou

Received: 25 April 2022
Accepted: 12 July 2022
Published: 14 July 2022

1. Introduction

Ecosystem services refer to the products and benefits which humans obtain from ecosystems. The provision of such services can occur directly or indirectly, depending on the structure, processes, and functions of ecosystems [1,2]; ecosystem services are essential for maintaining life on Earth and the ecosystem integrity [3,4]. The Millennium Ecosystem

Assessment (MEA) promulgated by the United Nations in 2005 divided ecosystem services into four categories: provisioning services, regulating services, supporting services and cultural services, and quantifying the importance of ecosystems to human well-being is one of its main objectives [5]. Ecosystem service values (ESVs) are a monetary quantification of ecosystem services. In general, scientific evaluation of ESVs is conducive to improve people's awareness of biodiversity conservation, optimize land use structure, and provide a reference for regional ecological security management and sustainable development [6,7].

There are generally two ways to quantify the ESVs [8]: one is based on the unit price of ecological products, using the shadow engineering method, market price method, carbon tax method, and other methods to calculate the ESVs [9–12]. This method has high data requirements, complex calculations, and thus a unified and versatile evaluation standard is difficult to achieve. The other is in concert with the economic value of the unit area of the ecosystem, multiplying the value coefficient of the corresponding land use type area to obtain the ESVs [13], this method was proposed by Costanza in 1997, and applied for the assessment of ecosystem services all over the world [3]. However, the method is susceptible to subjective factors and insensitive to the temporal and spatial changes in the properties and quality of ecosystems [14,15]. To realize the dynamic change of ecosystem service values, a dynamic equivalent factor combined with remote sensing was proposed [8]. At present, the ecosystem adjustment coefficient is generally determined by incorporating vegetation coverage [16,17], net primary productivity [18–21], and normalized vegetation index [22,23], and the calculated ESVs have qualified spatial-temporal resolution and high degree of credibility.

There are various degrees of trade-offs and synergies among ecosystem services due to their complex and dynamic interactions [24]. The enhancement, in term of provision level, of one ecosystem service at the expense of the provision of other ecosystem services is referred to as a trade-off, whereas synergy is the simultaneous increase or decrease of two ecosystem services [25]. With the rapid growth of the global economic population and the growing shortage of resources, the study on ecosystem service trade-offs and synergies not only is of great significance to global environmental changes and improvement of the regional ecological environment but also provides a theoretical basis for the rational development and utilization of resources [26,27]. Therefore, exploring the complex interactions behind ecosystem services has become a popular topic among the scientific communities in the past few years [28]. To date, tradeoffs and synergies among ecosystem services have been analyzed at a global scale [29,30], national scale [31], watershed scale [32,33], and landscape scale [34,35]. However, ecosystem service trade-offs and synergies are dependent on spatiotemporal scales, and the synergistic relationship of ecosystem service tradeoffs at the regional scale is not able to represent the ecosystem service relationship on a small scale [36,37]. The tradeoffs and synergies of ecosystem services vary over time and space. Moreno-Llorca et al. analyzed the relationship between four ecological services in the Sierra Nevada Mountains of Spain from three nested spatial scales of biosphere reserves, watersheds, and grid cell levels [38]. Yang et al. investigated the trade-offs and synergies among ecosystem services in the Yellow River Basin and its eight subbasins. Their findings revealed definite secondary basin differences and regional regularities, implying that tradeoffs and synergies were scale-dependent [39].

Most of the aforementioned studies had carried out detailed research on the evaluation of ESVs and their temporal and spatial changes. Although temporal and spatial changes in ESVs are important, potential factors that affect changes in ESVs still need to be considered. Understanding the ESVs and their driving forces helps to achieve the goal of sustainable development of various ecosystems and the harmonious coexistence of human society and natural ecosystems [40]. At present, some scholars have discussed the relationship between land use change and ecosystem services [1,41,42], and also analyzed the impact of ecological restoration policies on changes in ecosystem services [43–45]. However, previous studies have shown that changes in ESVs are the result of a combination of multiple driving factors. Therefore, a comprehensive analysis of the impact of natural factors [46–49], socio-economic factors [49–51], and political factors [44,52] on the ESVs is helpful to understand the ecological environment protection and formation mechanism.

Sichuan Province is located on the eastern edge of the Qinghai-Tibet Plateau. The region has large undulating terrain, complex geological structure, frequent natural disasters, and sensitive and fragile ecosystems. Affected by human factors, the ecological environment has been seriously degraded. In addition, it has been successively included in the "Returning Farmland to Forest Project", "Returning Grazing to Grassland", the Qinghai-Tibet Plateau Region (Sichuan Province) Ecological Construction and Environmental Protection Planning, and other ecological projects and planning to protect and construct the ecological environment. Therefore, it is very important to fully understand the temporal and spatial variation characteristics of ESVs in Sichuan Province and their influencing factors for regional ecological environment protection and sustainable development.

Previous studies have analyzed the relationship between land use change and ESVs on the one hand [53–55], and the relationship between regional ESVs and driving factors on the other hand [56,57]. Therefore, the value coefficient and evaluation model were modified according to the actual situation in the study area. Based on the land use data in 2000, 2005, 2010, 2015, and 2020, the temporal and spatial variation characteristics and influencing factors of ESVs were analyzed. The main objectives of this study were: (1) to quantify ESVs and reveal the spatial distribution characteristics of ESVs; (2) to identify trade-offs and synergies between the values of individual ecosystem services through correlation analysis; (3) to use bivariate spatial autocorrelation analysis at different scales to reveal the spatial heterogeneity of trade-offs and synergies among the six groups of ecosystem services; (4) to quantify the degree of impact of driving factors on ecosystem service value.

Study Area

Our study area is located at the intersection of the Qinghai-Tibet Plateau and the Middle-Lower Yangtze plain. Sichuan Province (26°03′~34°19′ N, 97°21′~108°12′ E) covers an area of484,000 km^2. The landform of Sichuan Province varies greatly from east to west, the terrain is complex and diverse, and the terrain is high in the west and low in the east (Figure 1). The western part is plateau and mountainous, and the altitude is mostly above 3 km. The eastern part is a basin and a hill, and the altitude is mostly between 0.5 and 2 km. Sichuan Province has three major climates: The subtropical humid and semi-humid climate in the Sichuan Basin is, respectively, divided into four distinct seasons, with the same period of rain and heat, the average annual temperature is 16~18 °C, with 1000–1200 mm of precipitation. The subtropical semi-humid climate in the mountains of southwest Sichuan is not clearly distinguished between the four seasons, the annual average temperature is 12~20 °C, with 900–1200 mm of precipitation. The alpine plateau in northwest Sichuan has an alpine climate with a great difference in altitude and significant temperature changes. The average annual temperature is 4~12 °C, with 500–900 mm of precipitation.

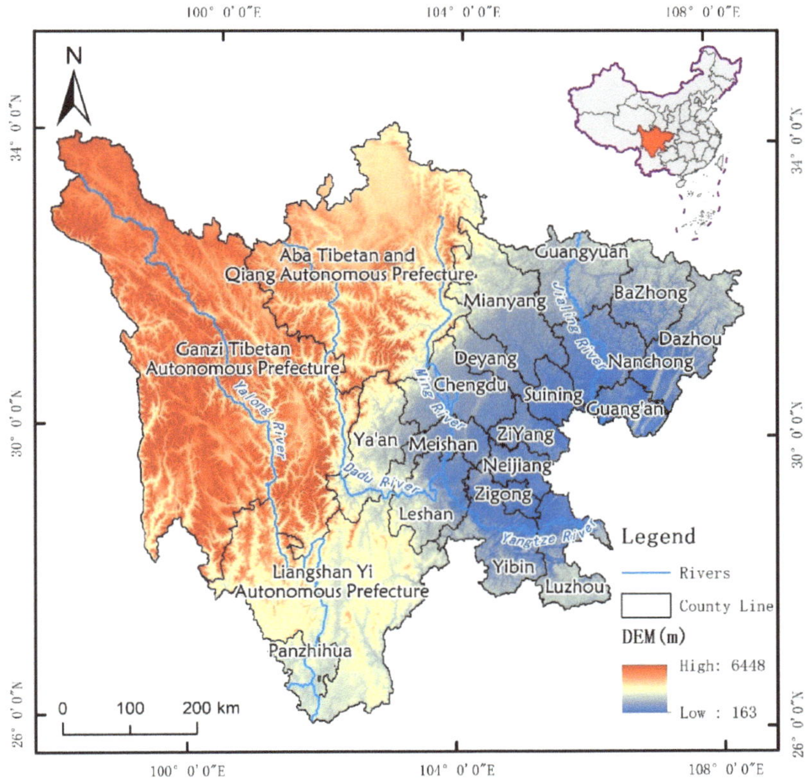

Figure 1. Location of Sichuan Province.

2. Materials and Methods

2.1. Factor Selection and Setting

We referred to the research of Xie et al. on the equivalent scale of ecological services per unit area of terrestrial ecosystems in China [58] and combine the natural and socioeconomic conditions of the study area to classify ecosystem services into four categories: provisioning, regulating, supporting and cultural services, and further subdivided into 9 services (Table 1). According to previous studies, it was determined that the farmland equivalent factor of Sichuan Province is 1.35 times that of the national farmland [59]. The average value of the coniferous forest and shrub forest was selected for forestland. The equivalent factors of grassland and wetland were set according to the research results of Zheng [60]. The setting of the equivalent factor of bare land and construction land is based on the research results of Li [18].

Table 1. Ecological service value per unit area.

Ecosystem Services	Farmland	Forestland	Grassland	Water	Bare Land	Construction Land	Wetland
Provisioning services							
Food production	1.35	0.205	0.1	0.1	0.01	0.01	0.51
Raw material production	0.135	0.475	0.14	0.01	0	0	0.5

Table 1. *Cont.*

Ecosystem Services	Farmland	Forestland	Grassland	Water	Bare Land	Construction Land	Wetland
Regulating services							
Gas regulation	0.675	1.555	0.51	0	0	0	1.9
Climate regulation	1.2015	4.65	1.34	0.46	0	0	3.6
Hydrological adjusting	0.81	3.345	0.98	20.38	0.03	−7.51	24.23
Environmental purification	2.214	1.385	0.44	18.18	0.01	−2.46	3.6
Supporting services							
Soil conservation	1.971	1.89	0.62	0.01	0.02	0.02	2.31
Biodiversity	0.9585	1.725	0.56	2.49	0.34	0.34	7.87
Cultural services							
Aesthetic landscape	0.0135	0.755	0.25	4.34	0.01	0.01	4.73

2.2. Data Sources and Processing

The land use data of five periods (2000, 2005, 2010, 2015, and 2020) were obtained from the 30 m resolution annual China land cover dataset (CLCD) [61], Then it was divided into 7 categories: farmland, forestland, grassland, water, bare land, construction land, and wetland. Based on the MODIS dataset in the Google Earth Engine platform. The Modified Normalized Difference Water Index (MNDWI) and the Remote Sensing Ecological Index (RSEI) were calculated in the study area. Regardless of price fluctuations, according to the website of the State Bureau of Grain and Material Reserves (http://www.lswz.gov.cn (accessed on 12 January 2022)) and the Sichuan Statistical Yearbook, the average price of grain in 2010 was 1.87 (unit: CNY/kg). The study area was divided into 1 km grids, and the land use data of each grid and 21 prefecture-level administrative districts were extracted based on the land use data of five periods.

Considering the availability of data and the fact that changes in ESVs were influenced by a variety of factors such as the natural environment and the social economy, this paper identified the following nine driving factors: Elevation and slope were derived from Geospatial Data Cloud (http://www.gscloud.cn (accessed on 20 January 2022)). The annual average temperature, annual average precipitation, and soil organic carbon content were obtained from the National Tibetan Plateau Data Center (http://data.tpdc.an.cn (accessed on 15 February 2022)). Population density, gross domestic product (GDP), normalized difference vegetation index (NDVI), and river data were from the Resource and Environmental Science and Data Center (https://www.resdc.cn/ (accessed on 10 March 2022)). Finally, we analyzed the spatial differentiation characteristics between them and ESVs.

2.3. Ecosystem Service Values

2.3.1. Ecosystem Service Assessment Model

Ecological service values are dynamic value that changes over time and varies with the type, size, and quality of regional ecosystems. Considering the impact of temporal and spatial changes in ecological quality on ESVs, RSEI and MNDWI were chosen to correct the ESVs of each pixel at each moment. The formula is as follows:

$$ESV(Si, j, t_h) = \sum_{f=1}^{q} VC_{if} \times A_{Si} \times R(Si, j, t_h) \quad (1)$$

In Formula (1), when the i-th pixel is the j-th land use type, $ESV(Si, j, t_h)$ represents the ESV at the research moment t_h. The value coefficient of the f-th ecosystem service function of the j-th land use type is denoted by VC_{if} (CNY/ha). A_{Si} represents the pixel area (ha). $R(Si, j, t_h)$ is the ecological quality correction coefficient at the research moment t_h when the i-th pixel is of the j-th land use type.

$$VC_{if} = E_{if} \times C_{crop} \tag{2}$$

where E_{if} is the equivalent coefficient of the f-th ecosystem service function of the j-th land use type, representing the weight coefficient of each ecosystem service value. The standard equivalence coefficient C_{crop} (CNY/ha) is based on the ecological service equivalence table per unit area of China's ecosystems, combined with the social and economic development, the economic value of the natural ecosystem is 1/7 of the food production service value provided by the existing unit area of cultivated land without human input. The economic value of an equivalent factor of ecological service value in Sichuan Province is calculated to be 1403.56 CNY/ha.

$$R(Si, j, t_h) = \frac{e(Si, j, t_h)}{\sum_{i=1}^{n} e(Si, j, t_h)/n} \tag{3}$$

where $e(Si, j, t_h)$ is the ecological condition index of the j-th land use type of the i-th pixel when the study year is t_h. $\sum_{i=1}^{n} e(Si, j, t_h)/n$ is the average value of the ecological condition index of all pixels of the same land type at the same time.

RSEI is to reflect the impact of changes in external factors such as human activities, climate change, and environmental state changes on the environment. In addition to quantitatively evaluating the ecological quality of the area, RSEI can also visualize the ecological environment of the study area, and support the analysis, prediction, and assessment of temporal and spatial changes in the ecological environment quality of the study area [62]. MNDWI can quickly extract water body information [63]. Li et al. corrected the equivalent factor pixel by pixel through RSEI and MNDWI, which can effectively display the temporal and spatial changes of ESVs in each pixel [64]. In order to better distinguish the ecological status between pixels, this paper introduced RSEI and MNDWI to construct $e(Si, j, t_h)$.

$$e(Si, j, t_h) = RSEI + MNDWI \tag{4}$$

The formula for calculating MNDWI is as follows (5):

$$MNDWI = \frac{\rho_{green} - \rho_{mir}}{\rho_{green} + \rho_{mir}} \tag{5}$$

RESI is defined as a function of greenness, wetness, heat, and dryness components, where greenness uses the normalized vegetation index (NDVI) to describe the growth and change of regional vegetation; The land surface temperature (LST) obtained by thermal infrared remote sensing inversion represents heat; The land surface moisture (LSM) is represented by the wetness component obtained by the tasseled-cap transformation of the multispectral image; Normalized differential build-up and bare soil index (NDBSI) composed of the index-based built-up index (IBI) and bare soil index (SI) was selected to indicate dryness.

$$RSEI = f(Greeness, Wetness, Heat, Dryness) \tag{6}$$

Then, the four indicators such as NDVI, LSM, LST, and NDBSI are normalized to be between 0 and 1; Secondly, perform principal component analysis on the multi-band images synthesized by the four indicators, using the first principal component (PC1) as the starting remote sensing ecological index $RSEI_0$; Finally, the RSEI obtained by normalizing $RSEI_0$ ranges from 0 to 1. The larger the RSEI value, the better the ecological condition [65,66].

2.3.2. Dynamic Degree of Ecological Service Values

The dynamic changes in regional ecological service values were analyzed using the dynamic degree of ESVs, following the formula below:

$$K_{esv} = \frac{ESV_b - ESV_a}{ESV_a} \times \frac{1}{T} \times 100 \tag{7}$$

where K_{esv} reflects the intensity of ecosystem service values changes with time, T represents the period, ESV_a is the initial ecological service value in a period, and ESV_b is the value of terminated ecological service within a period.

2.4. Methods of Analysis

2.4.1. Correlation Analysis

Correlation analyses were conducted to determine whether there were synergies or trade-offs between these ecosystem services, according to Formula (8). The higher the value, the stronger the correlation between the two.

$$V_{xy} = \frac{\sum_{i=1}^{n}(x_i - \overline{x})(y_i - \overline{y})}{\sqrt{\sum_{i=1}^{n}(x_i - \overline{x})^2}\sqrt{\sum_{i=1}^{n}(y - \overline{y})^2}} \tag{8}$$

where V_{xy} represents the correlation coefficient between the two ecosystems. n represent the total number of ecosystem services. The value of ecosystem services is represented by x_i and y_i, with $\overline{x}$ and $\overline{y}$ being the averages of the corresponding ecosystem service value.

2.4.2. Bivariate Spatial Autocorrelation Analysis

Spatial autocorrelation includes global autocorrelation and local autocorrelation and is mainly used to describe whether the spatial distribution between variables is clustered. In order to describe the correlation between multiple variables, Anselin et al. proposed a bivariate spatial autocorrelation based on the Moran index to reveal the correlation characteristics of the spatial distribution of different elements [67]. This method was introduced into ecosystem services, and GeoDa software was used to calculate the Moran index to evaluate the correlation between ecosystem services, and the local indicators of spatial association (LISA) were used to measure whether the ecosystem services have agglomeration.

2.4.3. Geographic Detector Model

The geographic detector model includes four types of detectors that are used as a new statistical method to investigate the spatial and temporal differentiation characteristics of things and their driving factors. Its central idea is based on the assumption that if an independent variable has a significant influence on a dependent variable, the independent variable's spatial distribution and the dependent variable's spatial distribution should be consistent [68]. The dominant factors and their interactions in the spatial differentiation of ESVs in Sichuan Province were analyzed using factor detection and interaction detection in this paper. The following is the formula:

$$q = 1 - \frac{1}{N\sigma^2}\sum_{h=1}^{L} N_h\sigma_h^2 \tag{9}$$

In the Formula (9), q represents the explanatory power of the influencing factors on the spatial differentiation characteristics of ecosystem service value, and its value range is [0, 1]. The greater the value, the greater the interpretive ability of the independent variable X to the dependent variable Y. On the contrary, it is smaller; L is the number of categories of variable Y or driving factor X; N and σ^2 represent the total number of samples in the study area and the discrete variance of the entire area, respectively; N_h and σ_h^2 represent the number of samples and the dispersion variance in the h area.

3. Results

3.1. Spatial-Temporal Changes in Landuse

The main land use types in the study area were forest land, followed by grassland and cropland (Figure 2). From 2000 to 2020, among all land use types, forest land increased significantly by 2.06% (Table 2). In the past 20 years, due to urban development, the situation of cropland occupation was more significant, the cropland area had decreased by 8377 km^2, but the area of impervious land had increased by 1926 km^2. With the development of animal husbandry, overgrazing led to a significant reduction in the grassland area in the study area, by 5763 km^2.

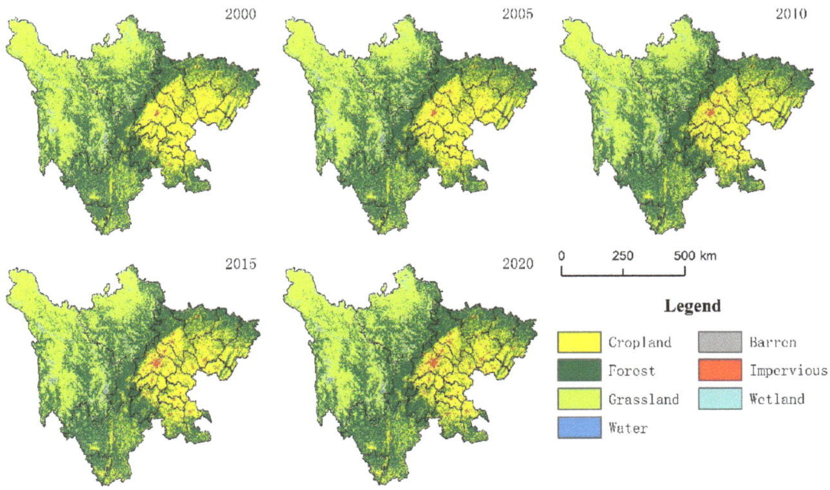

Figure 2. Spatial distribution of land use from 2000 to 2020.

Table 2. Land use area in Sichuan Province from 2000 to 2020.

Types	Areas/(km^2)					2000–2020
	2000	2005	2010	2015	2020	
Cropland	120,296	118,517	117,373	116,143	111,919	−1.73%
Forest	190,412	192,560	194,622	195,452	200,371	2.06%
Grassland	163,397	162,129	160,408	159,285	157,634	−1.19%
Water	4033	4674	4957	4875	4234	0.04%
Barren	3280	3424	3181	3974	4677	0.29%
Impervious	1907	2313	2993	3833	4453	0.53%
Wetland	418	126	209	181	455	0.01%

3.2. Spatial and Temporal Changes in Ecological Services

3.2.1. Characteristics of Temporal Development

According to Formulas (1)–(6), the ESVs of seven land use types in Sichuan Province from 2000 to 2020 were calculated. The calculation results of grid data in the study area were counted and shown in Tables 3 and 4. The ESVs and dynamic degree of all cities (autonomous prefectures) were calculated by Formulas (1)–(7), and the results were shown in Figure 3.

Table 3. Changes in the value of ecological services in Sichuan Province from 2000 to 2020.

Ecosystem Services	ESV (10^9 CNY)					The Average Annual Increasing Rate	K_{esv}
	2000	2005	2010	2015	2020	2000–2020	2000–2020
Food production	30.64	30.34	30.17	29.94	29.27	−0.89%	−0.76%
Raw material production	18.21	18.28	18.36	18.37	18.61	0.43%	0.36%
Gas regulation	64.73	64.87	65.11	65.08	65.71	0.30%	0.25%
Climate regulation	175.69	176.44	177.33	177.44	179.72	0.46%	0.38%
Hydrological adjusting	136.41	137.45	138.43	137.31	137.36	0.14%	0.12%
Environmental purification	94.26	95.39	95.86	95.05	92.88	−0.29%	−0.24%
Soil conservation	98.12	97.99	98.1	97.88	97.97	−0.03%	−0.03%
Biodiversity	77.21	77.32	77.74	77.7	78.34	0.29%	0.24%
Aesthetic landscape	28.86	29.24	29.62	29.6	29.85	0.68%	0.56%
Total value	724.13	727.32	730.72	728.38	729.71	0.15%	0.13%

Table 4. Ecological service value and proportion of land use in Sichuan Province from 2000 to 2020.

Types	ESV/(10^9 CNY)					Proportion/%				
	2000	2005	2010	2015	2020	2000	2005	2010	2015	2020
Cropland	157.42	155.09	153.6	151.98	146.45	21.59	21.14	20.79	20.57	19.75
Forest	427.05	431.86	436.49	438.35	449.39	58.56	58.87	59.08	59.34	60.59
Grassland	113.23	112.35	111.15	110.37	109.23	15.53	15.32	15.05	14.94	14.73
Water	25.92	30.05	31.88	31.34	27.21	3.55	4.1	4.32	4.24	3.67
Barren	0.19	0.2	0.19	0.23	0.28	0.03	0.03	0.03	0.03	0.04
Impervious	−2.57	−3.11	−4.03	−5.16	−5.99	0.35	0.42	0.55	0.7	0.81
Wetland	2.89	0.87	1.44	1.25	3.15	0.4	0.12	0.2	0.17	0.42
Total	729.26	733.54	738.77	738.7	741.69	100	100	100	100	100

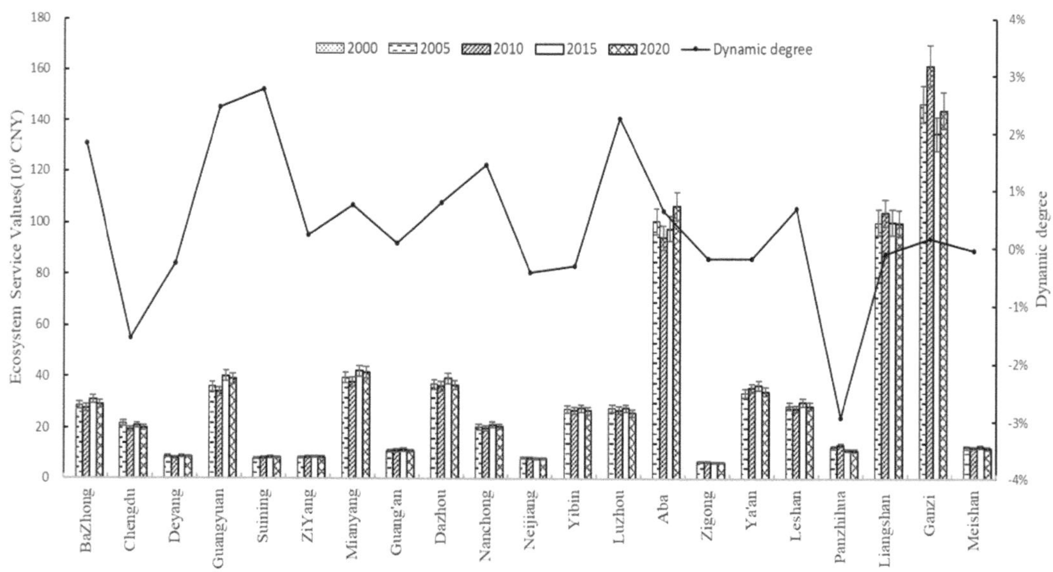

Figure 3. The value and dynamics of ecological waiters in various cities (autonomous regions) in Sichuan Province from 2000 to 2020.

According to Table 4, total ESVs increased from 724.13 × 10^9 CNY in 2000 to 729.71 × 10^9 CNY in 2020, with an average annual growth rate of 0.15%, a dynamic degree of 0.13%, and an increased rate of 0.77%. Among them, the most prominent ecological service was the regulating service, accounting for 65.19% of the total ESVs, followed by the supporting service, the provisioning service, and the cultural service.

In the provisioning services, the ESVs of food production and raw material production were low, showing four stages: gradual decline, rapid decline, slow rise, and rapid rise. In the regulating services, the average annual growth rate (0.46%) and the growth rate (2.3%) of climate regulation were the largest, with an increase of nearly 4.0 × 10^9 CNY in 20 years, and the minimum growth rate from 2010 to 2015 was only 0.06%. The ESVs of hydrological adjusting and environmental purification showed an upward trend, with a decline rate of 0.13 and 0.14% from 2010 to 2015, so the ESVs reached their peak in 2010. In the supporting services, soil conservation and biodiversity accounted for a similar proportion, but the two trends were opposite, with increases of −0.16 and 1.46%. In terms of cultural services, the ESVs of aesthetic landscapes were relatively low, showing a steady upward trend with an increase of 3.42%. In summary, the four ecosystem services in the study area are mainly regulating services and supporting services. In contrast, provisioning services and cultural services accounted for a smaller proportion but were more variable during the study period. The ESVs of climate regulation, hydrological adjusting, and soil conservation were higher in terms of individual ecological services.

Over the past 20 years, only ESVs in food production, environmental purification, and soil conservation have declined, while other services have increased.

The results in Table 4 showed that among the seven land types in the study area, forest land had the highest ESVs, followed by farmland, grassland, water, wetland, and bare land, and the smallest was construction land. From 2000 to 2020, except for farmland, grassland, and construction land, the ESVs of all other land types increased, but the degree of ESVs change was different for each type of land. Forestland, farmland, and grassland were the main contributors to ESVs in Sichuan Province, and the contribution rate of forest land remains above 58%. The range of ESVs changes in water and wetland was more obvious. The ESVs of water reduced by 14.65% between 2010 and 2020. The ESVs provided by water reached a high of 31.83 × 10^9 CNY in 2010. Wetland ESVs increased, and the growth rate reached the maximum (1.51%) from 2015 to 2020 and increased to 0.26 × 10^9 CNY. The proportion of grassland ESVs has decreased by 0.8% over the last 20 years. Bare land accounted for a small proportion of 0.03%, and ESVs fluctuations were small.

Figure 3 reflected the changes in ESVs in cities (autonomous prefectures). Among them, three autonomous prefectures (Ganzi Tibetan Autonomous Prefecture, Liangshan Yi Autonomous Prefecture, and Aba Tibetan Autonomous Prefecture) had relatively high ESVs. In the past 20 years, the ESVs of each region have increased or decreased to varying degrees. Overall, the most dynamic cities were Panzhihua, Suining, Guangyuan, and Luzhou, while Meishan was the least dynamic (−0.04%). From 2000 to 2005, the ESVs of Ziyang, Liangshan, Mianyang, Chengdu, Aba, Ya'an, and Panzhihua decreased by 0.02, 0.06, 0.22, 0.28, 0.38, 0.46 and 3.54%, other regions had increased, and the largest growth rate was Luzhou (3.41%). From 2005 to 2010, the ESVs of 14 cities (autonomous prefectures) in Sichuan Province declined, among which Chengdu had the largest decline (1.82%), and among the remaining areas, the ESVs of the Ganzi had the largest increase of 1.69%; From 2010 to 2015, except for Neijiang, Zigong, Liangshan, Panzhihua and Ganzi, the overall ecology has improved. From 2015 to 2020, the ESVs of Aba and Ganzi increased by 1.47 and 1.14%, respectively, and the ESVs of the rest of the regions showed a downward trend. Overall, the cities (autonomous prefectures) with the largest ESVs fluctuations during the study period were Garze, Luzhou, Panzhihua, and Guangyuan.

3.2.2. Spatial Distribution Characteristics

According to Formulas (1)–(6), the ESVs of the 1-km grid in Sichuan Province from 2000 to 2020 were calculated. GIS software was used for spatial mapping, and the ESVs were divided into 8 levels by the natural breakpoint method. Finally, the ESVs of each city (autonomous prefecture) in Sichuan Province during the study period were obtained. The spatial distribution of ESVs was shown in Figure 4.

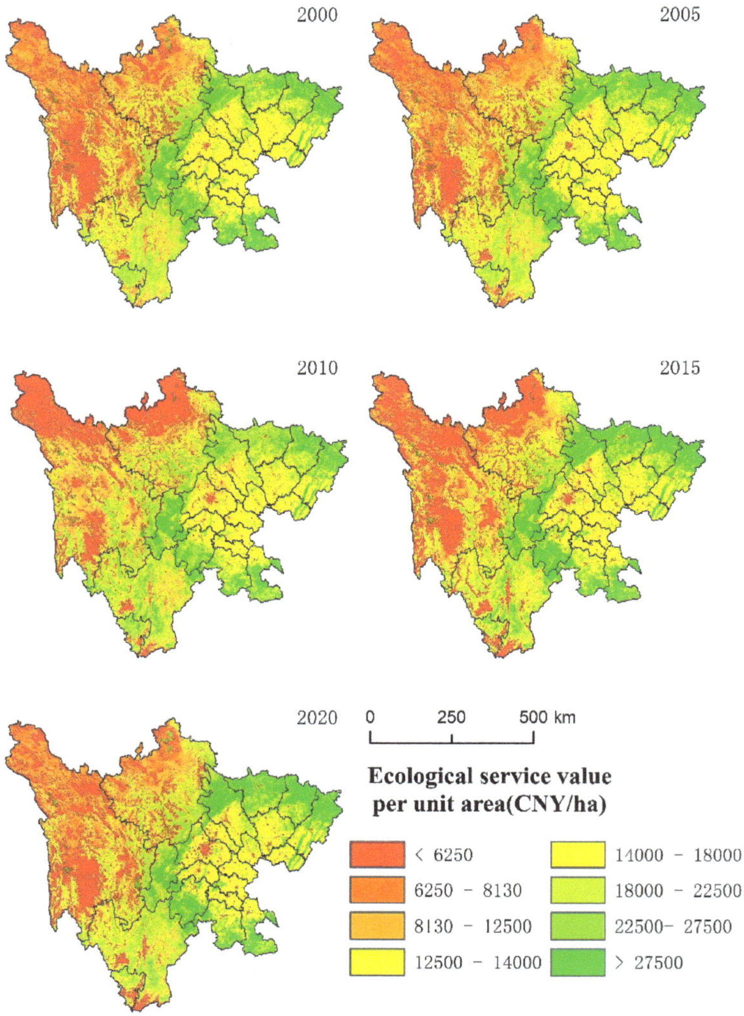

Figure 4. Spatial distribution of ESVs in Sichuan Province from 2000 to 2020.

In 2000, the high-value areas of the unit area ESVs were mainly distributed in the transition zone between the Sichuan Basin and the Western Sichuan Plateau, and the Zoige Plateau in the northern part of Aba Tibetan and Qiang Autonomous Prefecture. The low-value areas were mainly distributed in the northwest area and the central urban area of Chengdu. The unit area ESVs of the eastern cities were generally high, ranging from 12,500–18,000 CNY/ha. In 2005, the unit area ESVs of Aba Tibetan and Qiang Autonomous Prefecture increased, but there was no significant change in other areas. The ESVs per unit area increased in all cities in 2010, particularly in eastern Sichuan Province. The ESVs per unit area decreased between 2015 and 2020. Compared with the initial stage of the study, the unit area ESVs in the central region showed a downward trend, from 11,400 CNY/ha to 12,500 CNY/ha. Among them, the lowest value of Chengdu and its surrounding cities exceeded 6250 CNY/ha, and the area was constantly expanding. Overall, most of the high-value areas were located in forestland, and gradually decreased to both sides with the change of altitude. The low-value areas were mostly grassland and bare land in the northwest. Compared with the initial period, the unit area ESVs showed an upward trend, and the high-value area expanded at the end of the study.

3.3. Tradeoffs and Synergies Analysis

3.3.1. Correlation Analysis of Ecosystem Services

We analyzed the correlation between individual ecosystem services by using Formula (8), and the results were shown in Table 5.

Table 5. Correlation of ecosystem services in Sichuan Province.

Correlation	Provisioning Services		Regulating Services				Supporting Services		Cultural Services
	Food Production	**Raw Material Production**	**Gas Regulation**	**Climate Regulation**	**Hydrological Adjusting**	**Environmental Purification**	**Soil Conservation**	**Biodiversity**	**Aesthetic Landscape**
Food production	1								
Raw material production	−0.268	1							
Gas regulation	−0.134	0.990	1						
Climate regulation	−0.290	0.999	0.986	1					
Hydrological adjusting	−0.179	0.480	0.461	0.491	1				
Environmental purification	0.283	0.034	0.063	0.046	0.790	1			
Soil conservation	0.527	0.678	0.772	0.660	0.271	0.228	1		
Biodiversity	−0.020	0.862	0.881	0.857	0.744	0.436	0.738	1	
Aesthetic landscape	−0.357	0.518	0.474	0.535	0.947	0.704	0.169	0.744	1

Note: Correlation is significant at the 0.01 level (2-tailed).

Overall, various ecosystem services were positively correlated at the 0.01 significance level, accounting for 83%, and synergy was the dominant relationship of ecosystem services in Sichuan Province. In the provisioning services, food production was negatively correlated with raw material production, gas regulation, climate regulation, hydrological adjusting, biodiversity, and aesthetic landscape, while positively correlated with other ecosystem services. Among them, the positive correlation between environmental purification and raw material production was weak. In the regulation services, except for the weak correlation among gas regulation, climate regulation, and environmental purification, other ecosystem services showed a significant positive correlation. In the supporting services, there was a weak negative correlation between biodiversity and food production, and environmental purification and biodiversity were positively correlated with the other ecological services, of which biodiversity was closely related to various other services. In terms of cultural services, a positive correlation was found between aesthetic landscapes and various ecosystem services, except food production.

3.3.2. Analysis of Multi-Scale Tradeoffs and Synergies in Ecosystem Services

To further understand the relationship between ecosystem services in Sichuan Province, we analyzed the bivariate spatial autocorrelation between six pairs of ecosystem services at the city (autonomous prefecture) and 5 km grid scales based on GeoDa software. When Moran's > 0, it means a positive correlation, which is a synergistic effect; when Moran's < 0, it means a negative correlation, which is a tradeoff effect.

The results in Table 6 showed that the global autocorrelation indices between the six pairs of ecosystem services on the two scales were positive and passed the 5% significance level test. The results showed that there are synergistic effects among the four ecosystem services. In Figures 5 and 6, the characteristics of high-high aggregation and low-low aggregation indicated that the two ecosystem services exhibited synergistic effects, and high-low aggregation and low-high aggregation represented the trade-off effect between the two. At the city (autonomous region) scale, there were trade-off effects between provisioning services and cultural services, regulating services and cultural services, provisioning services, and supporting services in Luzhou. However, there was a trade-off effect for a large number of regions at the grid-scale. Based on the two scales, the correlation between ecosystem services was concluded as follows: as the research scale became larger, the scope of synergistic effects gradually expanded, and the trade-off effect was gradually transformed into a synergistic effect.

Table 6. Global spatial autocorrelation of four ecosystem services in Sichuan Province.

Ecosystem Services	Scale	Provisioning Services and Regulating Services	Provisioning Services and Supporting Services	Provisioning Services and Cultural Services
Moran's I	City-scale	0.435	0.482	0.362
	Grid-scale	0.377	0.669	0.01
Ecosystem Services	**Scale**	**Regulating Services and Supporting Services**	**Regulating Services and Cultural Services**	**Supporting Services and Cultural Services**
Moran's I	City-scale	0.432	0.348	0.36
	Grid-scale	0.709	0.605	0.453

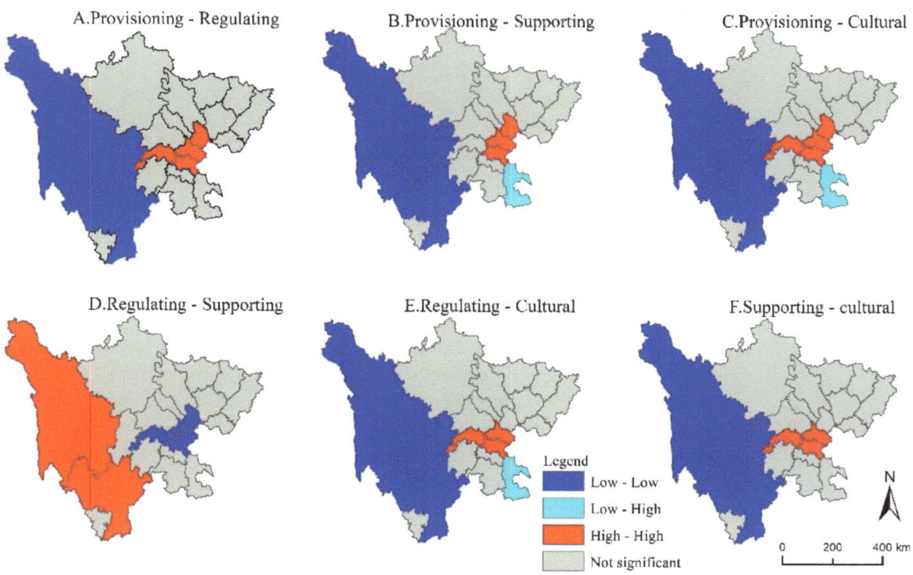

Figure 5. LISA cluster map of four ecosystem services in Sichuan Province at the city-scale.

At the city (autonomous prefecture) scale, six pairs of ecosystem service functions were distributed in the western and central parts of the study area. However, there were tradeoffs between provisioning and supporting services, provisioning and cultural services, and regulating and cultural services, respectively, in Luzhou. The specific performance was as follows: In addition to regulating and supporting services, the ecosystem services of

Ganzi Tibetan Autonomous Prefecture and Liangshan Yi Autonomous Prefecture showed low-low clustering. High-high clusters were distributed in different cities but mainly concentrated in the central part of Sichuan Province. For example, the high-high clusters of provisioning and regulating services, and provisioning and cultural services were mainly distributed in Meishan, Neijiang, Ziyang, and Suining (Figure 5A,C); the high-high clusters of regulating and cultural services, and supporting and culture were mainly distributed in Meishan, Neijiang, and Ziyang (Figure 5E,F); the high-high clusters of provisioning and supporting services were distributed in Nejiang, Ziyang and Suining (Figure 5B); while the high-high clusters of regulating and supporting services were mainly distributed in the western Ganzi Tibetan Autonomous Prefecture and Liangshan Yi Autonomous Prefecture in the western part of the study area (Figure 5D).

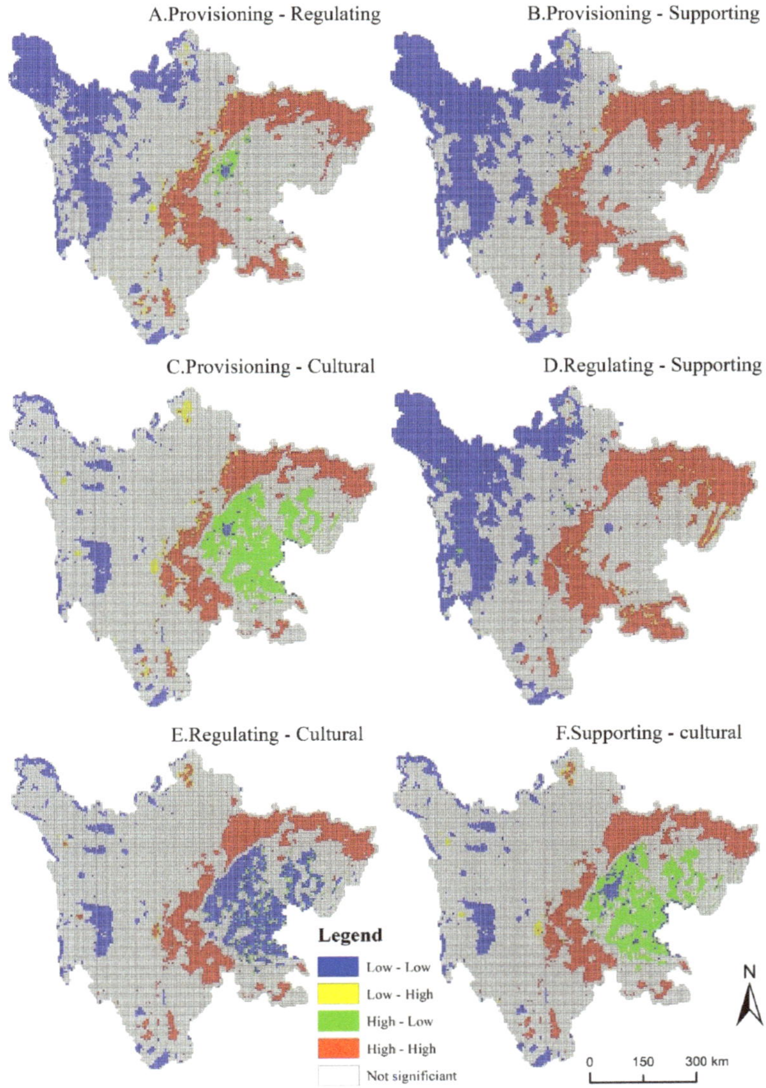

Figure 6. LISA cluster map of 4vecosystem services in Sichuan Province at grid-scale.

At the grid-scale, the spatial distribution characteristics of the tradeoffs and synergies among the six pairs of ecosystem services were significantly different. Synergies were dominant, and tradeoffs existed in a few grids. The results in Figure 6 showed that the synergies among ecosystem services were distributed in blocks, while the distribution of tradeoffs was more scattered. The low-high clusters were mainly distributed around the high-high clusters, and the high-low clusters were distributed in a ring around the low-low clusters. Compared with the city (autonomous prefecture) scale, the synergistic effect at the grid-scale was distributed in the northwest of the study area, or the transition zone between the Sichuan Basin and the Western Sichuan Plateau. The results were as follows: the synergies of provisioning and regulating services, provisioning, and supporting services, and regulating and supporting services were distributed in the northwest of the study area, the area around the Sichuan Basin, and a small part of Chengdu (Figure 6A,B,D). The high-high clustering areas of the six pairs of ecosystem services were mainly distributed in forestland. The tradeoff effects of provisioning and regulating services, provisioning, and cultural services, and regulating and cultural services were more significant. Among them, high-low clusters were distributed around low-low clusters, mainly concentrated in low-altitude areas (Figure 6A,C,F).

3.4. Driving Force Analysis

3.4.1. Single Factor Detection of Ecosystem Service Value

The geographic detector model was implemented based on the GD package in R. To make the calculation simple and combined with the actual situation of the study area, we selected nine influencing factors related to the natural environment and social economy to study their driving force on ESV. Among them, X1, X2, X3, X4, X5, X6, X7, X8, and X9 represent elevation, slope, annual average precipitation, annual average temperature, NDVI, and distance from the river, population density, GDP, and soil organic carbon content, respectively. The geographic detector required the independent variable X to be a discretized variable, so it is necessary to discretize the driving factor data. The GD package provides 6 discretization methods, and the optimal method and optimal classification of the data discretization are determined by algorithms, so as to obtain the most explanatory q-value. The results of the factor detector detection are shown in Table 7.

Table 7. Detection results of driving factors for spatial differentiation of ESVs in Sichuan Province.

Driving Factors	2000	2005	2010	2015	2020
X1	0.247301	0.261944	0.155072	0.207354	0.216714
X2	0.012226	0.014617	0.024748	0.016525	0.016403
X3	0.157459	0.157646	0.162215	0.20978	0.156716
X4	0.253969	0.2528	0.166463	0.20753	0.219138
X5	0.105141	0.142849	0.061232	0.201623	0.154678
X6	0.004143	0.003967	0.002635	0.002828	0.002635
X7	0.113036	0.130435	0.06769	0.089033	0.093796
X8	0.223148	0.242428	0.188704	0.240205	0.191877
X9	0.106206	0.106207	0.077228	0.088671	0.087548

Among them, factors such as temperature, elevation, and GDP had a relatively large contribution rate and were the main driving factors; while precipitation, population density, soil organic carbon, etc. had relatively small contribution rates to the spatial differentiation of ESVs and were secondary driving factors. From the perspective of the impact of driving factors on ESVs, temperature changes within a certain range promote the growth of vegetation, and ESVs increased accordingly. The spatial differentiation of ESVs was influenced by elevation. Lower elevations were more conducive to agricultural development and urban expansion, and ecological land was significantly destroyed, whereas higher elevations had fewer human activities, but the ecological environment for vegetation growth was harsh, and ESVs were also relatively low; GDP can reflect the strength of human activities, and areas with higher GDP value had lower ESVs and vice versa.

There are some differences in the explanatory power of each factor for ESVs across years, but it is generally consistent. Due to the proposal of the western development strategy at the end of the 20th century, the industrial scale of Sichuan Province continued to expand from 2000 to 2010, and the level of environmental pollution was relatively high, so the explanatory power of most driving factors to ESVs was weakened. After the "Eleventh Five-Year Plan", Sichuan Province has stepped up efforts to protect the ecological environment, and the contradiction between man and nature has been gradually eased. Therefore, after 2010, the ability of each driving factor to explain ESVs in the study area has steadily increased.

3.4.2. Interaction Factor Detection of Ecosystem Service Values

The interaction detector is used to test the interaction between the two influencing factors, that is, whether the two factors will increase or decrease the explanatory power of the ESVs when they act together. The interaction of nine influencing factors on ESVs was obtained with the help of the interaction detector module. The interactive detection results were shown in Figure 7. All driving forces interacted to improve the spatial distribution and differentiation of ESV, and the effects were not independent. This demonstrated that the interaction of multiple factors affected ecosystem services in Sichuan Province from 2000 to 2020. The strongest interaction was between NDVI and GDP, which typically reached 30%. The interaction between the slope and the distance from the river was the weakest, accounting for less than 4% of the total; it can be found that the interaction of elevation, GDP, and other factors had a greater impact on the distribution of ESVs in Sichuan Province. The distribution of land use was influenced by elevation. The intensity of human activities was closely related to the level of GDP value. As a result, the interaction between the natural environment and the social economy influenced the distribution of ESVs in Sichuan Province.

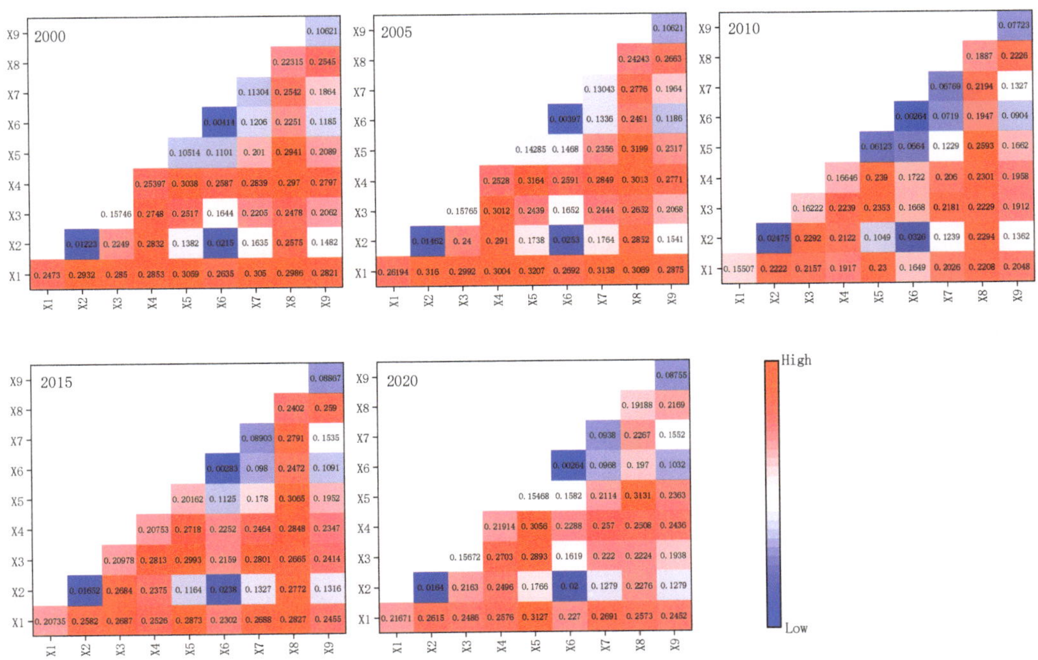

Figure 7. Interaction-driven results of ESVs in Sichuan Province.

4. Discussion

4.1. Spatial-Temporal Variation of Ecosystem Services

This paper quantified the ESVs in Sichuan Province in 2000, 2005, 2010, 2015, and 2020 based on land use data. The rational planning and utilization of land resources were of great significance to the ecological environment protection and sustainable development of Sichuan Province. In this paper, the RSEI and MNDWI indices were introduced to correct the ESVs of the study area, and the results showed that the total ecosystem service values in Sichuan Province had improved. The calculation results of ESVs deviated from previous studies because this paper took into account differences in ESVs between different pixels of the same land type, but the changing trend was essentially the same [53]. The results showed that the spatial heterogeneity of ecosystem services is closely related to the spatial distribution of land use [69–71]. In recent 20 years, the area of forestland and impervious land in Sichuan province had increased, while the area of cropland had decreased, which was in line with the trend of increasing forest resources and expanding building area in China, and was related to the project of "returning farmland to the forest". Forestland and grassland cover about 70% of the total area of Sichuan province, leading to the change of regional ESVs. However, the stable distribution of land use types did not significantly change the spatial pattern of ecosystem services.

In response to the current situation of the reduction of grassland and cultivated land and the increase of forest land and construction land in Sichuan Province, the western plateau region should vigorously implement grassland protection systems such as grazing prohibition and fallow, the balance between grass and livestock, and grassland ecological compensation [72]. The eastern region should take advantage of its unique location in the core of the Chengdu-Chongqing economic circle, and actively integrate into the construction of the "Belt and Road" and the Yangtze River Economic Belt, focusing on environmental improvement, and in the process of improvement Repair the environment, protect the ecology during development, comprehensively improve the ecological environment, and achieve green development [73].

4.2. Scale Effects of Tradeoffs and Synergies

The tradeoffs and synergies of ecosystem services are spatially heterogeneous and temporally dynamic and change over time and space. The correlation coefficient and Moran index can reveal the trade-offs and synergies of ecosystem services on the temporal and spatial scales. The correlation coefficients of the nine ecosystem services quantified tradeoffs and synergies over time, with fast feedback on raw material production and biodiversity services. However, the cycle of environmental purification was long, and there was a lag in other ecosystem services. The results showed that the correlation coefficient between gas regulation, climate regulation, and raw material production was close to 1, which indicated that vegetation in the study area had a regulating effect on gas and climate, and could also promote the production of raw materials. They had a strong synergistic effect on mutual promotion. On the contrary, the tradeoff between food production and other services was weak, indicating a conflict between food production and environmental protection, reflecting the competition between cultivated land and other land uses [74].

A bivariate spatial autocorrelation method was used at the city (autonomous prefecture) and grid scales to quantify the spatial synergy and tradeoff effects among six pairs of ecosystem services. The results showed that the Moran's I of the six pairs of ecosystems on both scales were all positive, indicating that the relationship between ecosystem services in Sichuan Province was mainly determined by synergistic effects (Table 6). At the same time, three conclusions can be drawn from the results of the binary space autocorrelation analysis: Firstly, the Moran's I obtained at the grid-scale was generally larger than the Moran's I calculated at the city (autonomous region) scale. Secondly, the tradeoff effects among the six pairs of ecosystems at the grid-scale were distributed around the synergistic effect. However, there was not necessarily a trade-off effect between ecosystem services at the city scale, indicating that the tradeoffs and synergies of the same ecosystem services at different regions and scales were also different [19]. Finally, the tradeoffs and synergies among ecosystems at the city scale were mainly distributed in Garze Tibetan Autonomous Prefecture, Liangshan Yi Autonomous Prefecture, and some eastern regions. On the grid-scale, tradeoffs and synergies were only distributed in the northwestern part of the study area, as well as in the transition area between the basin and the western Sichuan Plateau (Figures 5 and 6).

4.3. Driving Factor Analysis

Exploring the relationship between ESVs and driving factors provides a basis for ecosystem service management and decision-making. This study quantitatively analyzed the relationship between ESVs and driving factors in the study area and identified the interaction between factors. Ecosystem services in Sichuan Province were the result of the interaction between natural and human factors. However, due to the large proportion of forest land and grassland in the study area, our research focused on the influence of natural factors. The results showed that the annual average temperature was the main driving force, which was consistent with the existing research on the driving force of ecosystem services [48,75–77]. The second most important driving factor was the elevation. The spatial heterogeneity of elevation leads to changes in the regional ecological environment, thereby changing the type of land use, which in turn affected the value of regional ecosystem services [46,78].

The spatial differentiation of ESVs in the study area was caused by the interaction of multiple factors. Only analyzing the impact of a single driving factor on ESVs was not able to reveal the contribution of the synergistic effect of driving factors on ESVs. The results of this study showed that the contribution of the interaction of GDP and other drivers to ESVs was generally higher than that of GDP alone to ESVs. Therefore, it is crucial to understand the impact of the interaction between drivers on ESVs. For example, Pan et al. found the synergy of human activities, landscape pattern changes, and natural factors led to the spatial differentiation of ESVs in the study area [79]; Fang et al. explored the impact of natural and anthropogenic factors on the ecosystem service values of the Yangtze and Yellow River basins using a geographically weighted regression model and a geographic detector model, and the results showed that the combined effect of driving factors was much higher than the individual effect [80].

4.4. Uncertainties and Further Work Outlook

The interaction mechanisms of ecosystem services are complex, and their evaluation relies on the assessment of similar biological communities, but large ecosystems contain diverse communities and habitat types. Therefore, it is impossible to accurately quantify ESVs. In this study, the secondary land use types of forest land in Sichuan Province were combined, and the equivalent factor was set as the average value of coniferous forest and shrub forest, and the subcategories of land use types were not evaluated in detail.

Although the equivalence factors were revised according to the actual situation of the study area, RSEI was sensitive to phenological changes. Therefore, different image acquisition times will affect the calculation accuracy of RSEI. Considering the high spatial correlation of RSEI, Zhu et al. calculated the remote sensing ecological index based on a moving window, reducing the impact of long-distance features on specific research blocks [81]. Furthermore, this study analyzed the impact of driving factors on the ESVs; however, because the selection of driving factors focused on natural factors, the analysis' results were biased. Human disturbance factors [82–84], natural factors [47,85], socioeconomic factors [51,57], and policy factors [52] also affect the correlation between ecosystem services. Therefore, the spatial aggregation distribution of ecosystem service trade-offs and synergies will be different. Quantifying and modeling tradeoffs and synergies at multiple scales is an important part of ecosystem services research. Therefore, it is necessary to model the tradeoffs and synergies between climate change, land use change, human activity impacts, policy changes, and ecosystem services in future research, as well as forecast future spatiotemporal changes in regional ecosystem services.

5. Conclusions

Based on the evaluation model of ecosystem services from 2000 to 2020, nine kinds of ESVs in cities (autonomous prefectures) and 1 km grids were obtained in Sichuan Province. The results revealed that the total amount of ecosystem services increased by 5.58×10^9 CNY, indicating that ecosystem services have improved. The spatial heterogeneity of ESVs was significant, and the ESVs showed a spatial pattern of first increase and then decrease with the increase of altitude. However, due to the stability of the ecosystem structure, the ecosystem pattern in Sichuan Province has not changed significantly in the past 20 years. The results of correlation analysis showed that the synergistic effect of ecosystem services dominated, and only food production and other services showed a weaker tradeoff effect. Bivariate spatial autocorrelation analysis showed that the four main services had different degrees of synergy at different scales, and the tradeoff effect of ecosystems was more significant at small scales. In addition, the contribution of drivers to ESVs was quantified using a geographic detector model, and it was found that the combined effects of drivers were much higher than their individual effects. Therefore, the relationship between driving factors and ESVs should be fully considered in the construction of ecological civilization in the future. In particular, when formulating development policies, relevant departments need to find a balance between development and protection to achieve coordinated development of ecosystem services at different levels, and to ensure ecosystem stability while steadily increasing ESVs.

Author Contributions: Writing—original draft preparation, C.H.; investigation, H.S.; visualization, C.H.; software, C.H.; validation, C.H.; writing—review and editing, C.H. and H.S.; project administration, W.X.; supervision, W.X. and H.S. All authors have read and agreed to the published version of the manuscript.

Funding: This research was funded by the National Natural Science Foundation of China (Grant No. 41401659) and the Sichuan Province Natural Resources Science and Technology Project (Grant No. AZ20210002).

Institutional Review Board Statement: Not applicable.

Informed Consent Statement: Not applicable.

Data Availability Statement: The data presented in this study are available contained within the article.

Acknowledgments: The author thanks anonymous reviewers for providing invaluable comments on the original manuscript.

Conflicts of Interest: The authors declare no conflict of interest.

References

1. Tolessa, T.; Senbeta, F.; Kidane, M. The impact of land use/land cover change on ecosystem services in the central highlands of Ethiopia. *Ecosyst. Serv.* **2017**, *23*, 47–54. [CrossRef]
2. Brownson, K.; Guinessey, E.; Carranza, M.; Esquivel, M.; Hesselbach, H.; Madrid Ramirez, L.; Villa, L. Community-Based Payments for Ecosystem Services (CB-PES): Implications of community involvement for program outcomes. *Ecosyst. Serv.* **2019**, *39*, 100974. [CrossRef]
3. Costanza, R.; d Arge, R.; de Groot, R.; Farber, S.; Grasso, M.; Hannon, B.; Limburg, K.; Naeem, S.; O'Neill, R.V.; Paruelo, J.; et al. The value of the world's ecosystem services and natural capital. *Nature* **1997**, *387*, 253–260. [CrossRef]
4. Crossman, N.D.; Burkhard, B.; Nedkov, S.; Willemen, L.; Petz, K.; Palomo, I.; Drakou, E.G.; Martin-Lopez, B.; McPhearson, T.; Boyanova, K.; et al. A blueprint for mapping and modelling ecosystem services. *Ecosyst. Serv.* **2013**, *4*, 4–14. [CrossRef]
5. Assessment, M.E. Ecosystems and human well-being: Synthesis. *Phys. Teach.* **2005**, *34*, 534.
6. Dai, E.F.; Wang, X.L.; Zhu, J.J.; Zhao, D.S. Methods, tools and research framework of ecosystem service trade-offs. *Geogr. Res.* **2016**, *35*, 1005–1016. [CrossRef]
7. Konstantinova, E.; Brūnia, L.; Perēvica, A.; Ivitere, M. Assessment of Ecosystems Services for Sustainable Development and Land Use Management. *Soc. Integr. Educ. Proc. Int. Sci. Conf.* **2017**, *4*, 257. [CrossRef]
8. Xie, G.D.; Zhang, C.X.; Zhang, C.S.; Xiao, Y.; Lu, C.X. The value of ecosystem services in China. *Resour. Sci.* **2015**, *37*, 1740–1746.

9. He, H.; Pan, Y.Z.; Zhu, W.Q.; Liu, X.L.; Zhang, Q.; Zhu, X.F. Measurement of terrestrial ecosystem service value in China. *Chin. J. Appl. Ecol.* **2005**, *16*, 1122–1127. [CrossRef]
10. Richmond, A.; Kaufmann, R.K.; Myneni, R.B. Valuing ecosystem services: A shadow price for net primary production. *Ecol. Econ.* **2007**, *64*, 454–462. [CrossRef]
11. Gashaw, T.; Tulu, T.; Argaw, M.; Worqlul, A.W.; Tolessa, T.; Kindu, M. Estimating the impacts of land use/land cover changes on Ecosystem Service Values: The case of the Andassa watershed in the Upper Blue Nile basin of Ethiopia. *Ecosyst. Serv.* **2018**, *31*, 219–228. [CrossRef]
12. Cetin, N.I.; Bourget, G.; Tezer, A. Travel-cost method for assessing the monetary value of recreational services in the Ömerli Catchment. *Ecol. Econ.* **2021**, *190*, 107192. [CrossRef]
13. Constanza, R.; de Groot, R.; Sutton, P.; van der Ploeg, S.; Anderson, S.J.; Kubiszewski, I.; Farber, S.; Turner, K.R. Changes in the global value of ecosystem services. *Glob. Environ. Change* **2014**, *26*, 152–158. [CrossRef]
14. Xie, G.D.; Zhang, C.X.; Zhen, L.; Zhang, L.M. Dynamic changes in the value of China's ecosystem services. *Ecosyst. Serv.* **2017**, *26*, 146–154. [CrossRef]
15. Wang, Y.; Pan, J.H. Building ecological security patterns based on ecosystem services value reconstruction in an arid inland basin: A case study in Ganzhou District, NW China. *J. Clean. Prod.* **2019**, *241*, 118337. [CrossRef]
16. Lei, J.C.; Wang, S.; Wang, J.M.; Wu, S.Q.; You, X.B.; Wu, J.; Cui, P.; Ding, H. Effects of land use change on ecosystem services value of Xunwu County. *Acta Ecol. Sin.* **2019**, *39*, 3089–3099. [CrossRef]
17. Xu, L.F.; Xu, X.G.; Luo, T.; Zhu, G.R.; Ma, Z.W. Services based on land use: A case study of Bohai Rim. *Geogr. Res.* **2012**, *31*, 1775–1784. [CrossRef]
18. Li, X.S.; Zhu, Y.M.; Zhao, L.; Tian, J.J.; Li, J. Ecosystem services value change in Qinglong County from dynamically adjusted value coefficients. *Chin. J. Eco Agric.* **2015**, *23*, 373–381. [CrossRef]
19. Liu, H.; Wu, J.; Chen, X.L. Study on spatial-temporal change andtrade-off/synergy relationships of ecosystem services in the Danjiangkou water source area. *Acta Ecol. Sin.* **2018**, *38*, 4609–4624. [CrossRef]
20. Sutton, P.C.; Anderson, S.J.; Costanza, R.; Kubiszewski, I. The ecological economics of land degradation: Impacts on ecosystem service values. *Ecol. Econ.* **2016**, *129*, 182–192. [CrossRef]
21. Xie, G.D.; Zhang, C.X.; Zhang, L.M.; Chen, W.H.; Li, S.M. Improvement of the Evaluation Method for Ecosystem Service Value Based on Per Unit Area. *J. Nat. Resour.* **2015**, *30*, 1243–1254. [CrossRef]
22. Roces-Díaz, J.V.; Díaz-Varela, R.A.; Álvarez-Álvarez, P.; Recondo, C.; Díaz-Varela, E.R. A multiscale analysis of ecosystem services supply in the NW Iberian Peninsula from a functional perspective. *Ecol. Indic.* **2015**, *50*, 24–34. [CrossRef]
23. Fu, B.L.; Li, Y.; Wang, Y.Q.; Zhang, B.; Yin, S.B.; Zhu, H.L.; Xing, Z.F. Evaluation of ecosystem service value of riparian zone using land use data from 1986 to 2012. *Ecol. Indic.* **2016**, *69*, 873–881. [CrossRef]
24. Lee, H.; Lautenbach, S. A quantitative review of relationships between ecosystem services. *Ecol. Indic.* **2016**, *66*, 340–351. [CrossRef]
25. Bennett, E.M.; Peterson, G.D.; Gordon, L.J. Understanding relationships among multiple ecosystem services. *Ecol. Lett.* **2009**, *12*, 1394–1404. [CrossRef]
26. Howe, C.; Suich, H.; Vira, B.; Mace, G.M. Creating win-wins from trade-offs? Ecosystem services for human well-being: A meta-analysis of ecosystem service trade-offs and synergies in the real world. *Glob. Environ. Change* **2014**, *28*, 263–275. [CrossRef]
27. Maes, J.; Paracchini, M.L.; Zulian, G.; Dunbar, M.B.; Alkemade, R. Synergies and trade-offs between ecosystem service supply, biodiversity, and habitat conservation status in Europe. *Biol. Conserv.* **2012**, *155*, 1–12. [CrossRef]
28. Feng, Y.; Cao, Y.G.; Li, S.P.; Wang, S.F.; Liu, S.H.; Bai, Z.K. Trade-offs and synergies of ecosystem services: Development history and research characteristics. *J. Agric. Resour. Environ.* **2022**, *39*, 11–25. [CrossRef]
29. Petz, K.; Alkemade, R.; Bakkenes, M.; Schulp, C.J.E.; van der Velde, M.; Leemans, R. Mapping and modelling trade-offs and synergies between grazing intensity and ecosystem services in rangelands using global-scale datasets and models. *Glob. Environ. Change* **2014**, *29*, 223–234. [CrossRef]
30. Paiola, A.; Assandri, G.; Brambilla, M.; Zottini, M.; Pedrini, P.; Nascimbene, J. Exploring the potential of vineyards for biodiversity conservation and delivery of biodiversity-mediated ecosystem services: A global-scale systematic review. *Sci. Total Environ.* **2020**, *706*, 135839. [CrossRef]
31. Juerges, N.; Arts, B.; Masiero, M.; Hoogstra-Klein, M.; Borges, J.G.; Brodrechtova, Y.; Brukas, V.; Canadas, M.J.; Carvalho, P.O.; Corradini, G.; et al. Power analysis as a tool to analyse trade-offs between ecosystem services in forest management: A case study from nine European countries. *Ecosyst. Serv.* **2021**, *49*, 101290. [CrossRef]
32. Swallow, B.M.; Sang, J.K.; Nyabenge Bundotich, D.K.; Duraiappah, A.K.; Yatich, T.B. Tradeoffs, synergies and traps among ecosystem services in the Lake Victoria basin of East Africa. *Environ. Sci. Policy* **2009**, *12*, 504–519. [CrossRef]
33. Bai, Y.; Zheng, H.; Ouyang, Z.Y.; Zhuang, C.W.; Jiang, B. Modeling hydrological ecosystem services and tradeoffs: A case study in Baiyangdian watershed, China. *Environ. Earth Sci.* **2013**, *70*, 709–718. [CrossRef]
34. García, A.M.; Santé, I.; Loureiro, X.; Miranda, D. Green infrastructure spatial planning considering ecosystem services assessment and trade-off analysis. Application at landscape scale in Galicia region (NW Spain). *Ecosyst. Serv.* **2020**, *43*, 101115. [CrossRef]
35. Castro, A.J.; Verburg, P.H.; Martín-López, B.; Garcia-Llorente, M.; Cabello, J.; Vaughn, C.C.; López, E. Ecosystem service trade-offs from supply to social demand: A landscape-scale spatial analysis. *Landsc. Urban Plan.* **2014**, *132*, 102–110. [CrossRef]
36. Meng, H.B.; Zhou, Q.G.; Li, M.H.; Zhou, L.; Liu, X.W.; Peng, C.H. Study of the Spatio-temporal Changes in Ecosystem Services and Trade-offs / Synergies Relationship in the Three Gorges Reservoir Area. *J. Ecol. Rural. Environ.* **2021**, *37*, 566–575. [CrossRef]

37. Raudsepp-Hearne, C.; Peterson, G.D.; Bennett, E.M. Ecosystem service bundles for analyzing tradeoffs in diverse landscapes. *Proc. Natl. Acad. Sci. USA* **2010**, *107*, 5242–5247. [CrossRef]
38. Moreno-Llorca, R.; Vaz, A.S.; Herrero, J.; Millares, A.; Bonet-Garcia, F.J.; Alcaraz-Segura, D. Multi-scale evolution of ecosystem services' supply in Sierra Nevada (Spain): An assessment over the last half-century. *Ecosyst. Serv.* **2020**, *46*, 101204. [CrossRef]
39. Yang, J.; Xie, B.P.; Zhang, D.G. Spatial-temporal heterogeneity of ecosystem services trade-off synergy in the Yellow River Basin. *J. Desert Res.* **2021**, *41*, 78–87. [CrossRef]
40. Garmendia, M.; Borja, Á.; Breton, F.; Butenschön, M.; Marín, A.; Miller, P.I.; Morisseau, F.; Xu, W. Challenges and difficulties in assessing the environmental status under the requirements of the Ecosystem Approach in North African countries, illustrated by eutrophication assessment. *Environ. Monit. Assess.* **2015**, *187*, 289. [CrossRef]
41. Wang, X.Z.; Wu, J.Z.; Liu, Y.L.; Hai, X.Y.; Shanguan, Z.P.; Deng, L. Driving factors of ecosystem services and their spatiotemporal change assessment based on land use types in the Loess Plateau. *J. Environ. Manag.* **2022**, *311*, 114835. [CrossRef] [PubMed]
42. Talukdar, S.; Singha, P.; Mahato, S.; Praveen, B.; Rahman, A. Dynamics of ecosystem services (ESs) in response to land use land cover (LU/LC) changes in the lower Gangetic plain of India. *Ecol. Indic.* **2020**, *112*, 106121. [CrossRef]
43. He, J.H.; Pan, Z.Z.; Liu, D.F.; Guo, X.L. Exploring the regional differences of ecosystem health and its driving factors in China. *Sci. Total Environ.* **2019**, *673*, 553–564. [CrossRef]
44. Chen, Y.J.; Yu, Z.R.; Li, X.D.; Li, P.Y. How agricultural multiple ecosystem services respond to socioeconomic factors in Mengyin County, China. *Sci. Total Environ.* **2018**, *630*, 1003–1015. [CrossRef] [PubMed]
45. Zhang, H.Y.; Fan, J.W.; Cao, W.; Zhong, H.P.; Harris, W.; Gong, G.L.; Zhang, Y.X. Changes in multiple ecosystem services between 2000 and 2013 and their driving factors in the Grazing Withdrawal Program, China. *Ecol. Eng.* **2018**, *116*, 67–79. [CrossRef]
46. Du, S.H.; Wang, Q.; Guo, L. Spatially varying relationships between land-cover change and driving factors at multiple sampling scales. *J. Environ. Manag.* **2014**, *137*, 101–110. [CrossRef]
47. Runting, R.K.; Bryan, B.A.; Dee, L.E.; Maseyk, F.J.F.; Mandle, L.; Hamel, P.; Wilson, K.A.; Yetka, K.; Possingham, H.P.; Rhodes, J.R. Incorporating climate change into ecosystem service assessments and decisions: A review. *Glob. Change Biol.* **2017**, *23*, 28–41. [CrossRef]
48. Sannigrahi, S.; Zhang, Q.; Pilla, F.; Joshi, P.K.; Basu, B.; Keesstra, S.; Roy, P.S.; Wang, Y.; Sutton, P.C.; Chakraborti, S.; et al. Responses of ecosystem services to natural and anthropogenic forcings: A spatial regression based assessment in the world's largest mangrove ecosystem. *Sci. Total Environ.* **2020**, *715*, 137004. [CrossRef]
49. Rana, R.B.; Garforth, C.; Sthapit, B.; Jarvis, D. Influence of socio-economic and cultural factors in rice varietal diversity management on-farm in Nepal. *Agric. Hum. Values* **2007**, *24*, 461–472. [CrossRef]
50. Beilin, R.; Lindborg, R.; Stenseke, M.; Pereira, H.M.; Llausàs, A.; Slätmo, E.; Cerqueira, Y.; Navarro, L.; Rodrigues, P.; Reichelt, N.; et al. Analysing how drivers of agricultural land abandonment affect biodiversity and cultural landscapes using case studies from Scandinavia, Iberia and Oceania. *Land Use Policy* **2014**, *36*, 60–72. [CrossRef]
51. Chen, S.; Li, G.; Xu, Z.G.; Zhuo, Y.F.; Wu, C.F.; Ye, Y.M. Combined Impact of Socioeconomic Forces and Policy Implications: Spatial-Temporal Dynamics of the Ecosystem Services Value in Yangtze River Delta, China. *Sustainability* **2019**, *11*, 2622. [CrossRef]
52. Zhao, M.M.; Zhou, L.H.; Chen, Y.; Zhang, J.S.; Guo, X.L.; Wang, R. The Influence of Ecological Policies on Changes of Land Use and Ecosystem Service Value in Hangjinqi, Inner Mongolia, China. *J. Desert Res.* **2016**, *36*, 842–850. [CrossRef]
53. Fan, M.; Chen, L. Spatial characteristics of land uses and ecological compensations based on payment for ecosystem services model from 2000 to 2015 in Sichuan Province, China. *Ecol. Inform.* **2019**, *50*, 162–183. [CrossRef]
54. Fan, M.; Xiao, Y.T. Impacts of the grain for Green Program on the spatial pattern of land uses and ecosystem services in mountainous settlements in southwest China. *Glob. Ecol. Conserv.* **2020**, *21*, e00806. [CrossRef]
55. Peng, W.F.; Zhou, J.M.; Yang, C.J.; Zhao, J.F.; Luo, H.L. Research on Ecosystem Service Values based on Land Use Change in Sichuan Province. *Resour. Environ. Yangtze Basin* **2014**, *23*, 1053–1062. [CrossRef]
56. Dai, X.A.; Johnson, B.A.; Luo, P.L.; Yang, K.; Dong, L.X.; Wang, Q.; Liu, C.; Li, N.W.; Lu, H.; Ma, L.; et al. Estimation of Urban Ecosystem Services Value: A Case Study of Chengdu, Southwestern China. *Remote Sens.* **2021**, *13*, 207. [CrossRef]
57. Han, R.; Sun, S.Q.; Guo, L.; Chen, Y.Z. Evolution of Ecosystem Service Value and Analysis of Driving Forces in the East Region of Sichuan Province, China. *J. Ecol. Rural. Environ.* **2019**, *35*, 1136–1143. [CrossRef]
58. Xie, G.D.; Lu, C.X.; Leng, Y.F.; Zheng, D.; Li, S.C. Ecological assets valuation of the Tibetan Plateau. *J. Nat. Resour.* **2003**, *18*, 189–196. [CrossRef]
59. Xie, G.D.; Xiao, Y.; Zhen, L.; Lu, C.X. Study on ecosystem services value of food production in China. *Chin. J. Eco-Agric.* **2005**, *13*, 10–13.
60. Zheng, D.F.; Wang, Y.H.; Hao, S.; Xu, W.J.; Lv, L.T.; Yu, S. Spatial-temporal variation and tradeoffs/synergies analysis on multiple ecosystem services: A case study in the Three-River Headwaters region of China. *Ecol. Indic.* **2020**, *116*, 106494. [CrossRef]
61. Yang, J.; Huang, X. The 30 m annual land cover dataset and its dynamics in China from 1990 to 2019. *Earth Syst. Sci. Data* **2021**, *13*, 3907–3925. [CrossRef]
62. Zhou, M.; Liu, Y.; Meng, L.; Jiao, K.; Li, Q.; Zou, X. Evaluation of Ecological Environment Quality of Dongjiang River Headwaters Based on Remote Sensing Ecological Index During 2000 2019. *Bull. Soil Water Conserv.* **2021**, *41*, 231–240. [CrossRef]
63. Xu, H. A Study on Information Extraction of Water Body with the Modified Normalized Difference Water Index (MNDWI). *J. Remote Sens.* **2005**, *9*, 589–595. [CrossRef]

64. Li, J.; Liang, J.X.; Wu, Y.; Yin, S.Q.; Yang, Z.L.; Hu, Z.Q. Quantitative evaluation of ecological cumulative effect in mining area using a pixel-based time series model of ecosystem service value. *Ecol. Indic.* **2021**, *120*, 106873. [CrossRef]
65. Hu, X.S.; Xu, H.Q. A new remote sensing index for assessing the spatial heterogeneity in urban ecological quality: A case from Fuzhou City, China. *Ecol. Indic.* **2018**, *89*, 11–21. [CrossRef]
66. Xu, H.; Ding, F.; Wen, X. Urban Expansion and Heat Island Dynamics in the Quanzhou Region, China. *IEEE J. Sel. Top. Appl. Earth Obs. Remote Sens.* **2009**, *2*, 74–79. [CrossRef]
67. Anselin, L.; Syabri, I.; Smirnov, O. *Visualizing Multivariate Spatial Correlation with Dynamically Linked Windows*; University of California: Santa Barbara, CA, USA, 2002.
68. Wang, J.F.; Xu, C.D. Geodetector: Principle and prospective. *Acta Geogr. Sin.* **2017**, *72*, 116–134. [CrossRef]
69. Liu, Y.B.; Hou, X.Y.; Li, X.W.; Song, B.Y.; Wang, C. Assessing and predicting changes in ecosystem service values based on land use/cover change in the Bohai Rim coastal zone. *Ecol. Indic.* **2020**, *111*, 106004. [CrossRef]
70. Martinez-Harms, M.J.; Bryan, B.A.; Figueroa, E.; Pliscoff, P.; Runting, R.K.; Wilson, K.A. Scenarios for land use and ecosystem services under global change. *Ecosyst. Serv.* **2017**, *25*, 56–68. [CrossRef]
71. Sannigrahi, S.; Bhatt, S.; Rahmat, S.; Paul, S.K.; Sen, S. Estimating global ecosystem service values and its response to land surface dynamics during 1995–2015. *J. Environ. Manag.* **2018**, *223*, 115–131. [CrossRef]
72. Zheng, X.; Zhang, J.; Cao, S. Net value of grassland ecosystem services in mainland China. *Land Use Policy* **2018**, *79*, 94–101. [CrossRef]
73. Liu, Y.; Yang, R.; Sun, M.; Zhang, L.; Liu, Q. Regional sustainable development strategy based on the coordination between ecology and economy: A case study of Sichuan Province, China. *Ecol. Indic.* **2022**, *134*, 108445. [CrossRef]
74. Yang, Q.Q.; Xu, G.L.; Li, A.J.; Liu, Y.T.; Hu, C.S. Evaluation and trade-off of ecosystem services in the Qingyijiang River Basin. *Acta Ecol. Sin.* **2021**, *41*, 9315–9327. [CrossRef]
75. Jiang, C.; Li, D.Q.; Wang, D.W.; Zhang, L.B. Quantification and assessment of changes in ecosystem service in the Three-River Headwaters Region, China as a result of climate variability and land cover change. *Ecol. Indic.* **2016**, *66*, 199–211. [CrossRef]
76. Terrado, M.; Acuña, V.; Ennaanay, D.; Tallis, H.; Sabater, S. Impact of climate extremes on hydrological ecosystem services in a heavily humanized Mediterranean basin. *Ecol. Indic.* **2014**, *37*, 199–209. [CrossRef]
77. Wang, X.F.; Chu, B.Y.; Feng, X.M.; Li, Y.H.; Fu, B.J.; Liu, S.R.; Jin, J.M. Spatiotemporal variation and driving factors of water yield services on the Qingzang Plateau. *Geogr. Sustain.* **2021**, *2*, 31–39. [CrossRef]
78. Dewan, A.M.; Yamaguchi, Y. Land use and land cover change in Greater Dhaka, Bangladesh: Using remote sensing to promote sustainable urbanization. *Appl. Geogr.* **2009**, *29*, 390–401. [CrossRef]
79. Pan, N.H.; Guan, Q.Y.; Wang, Q.Z.; Sun, Y.F.; Li, H.C.; Ma, Y.R. Spatial Differentiation and Driving Mechanisms in Ecosystem Service Value of Arid Region: A case study in the middle and lower reaches of Shule River Basin, NW China. *J. Clean. Prod.* **2021**, *319*, 128718. [CrossRef]
80. Fang, L.L.; Wang, L.C.; Chen, W.X.; Sun, J.; Cao, Q.; Wang, S.Q.; Wang, L.Z. Identifying the impacts of natural and human factors on ecosystem service in the Yangtze and Yellow River Basins. *J. Clean. Prod.* **2021**, *314*, 127995. [CrossRef]
81. Zhu, D.Y.; Chen, T.; Wang, Z.W.; Niu, R.Q. Detecting ecological spatial-temporal changes by Remote Sensing Ecological Index with local adaptability. *J. Environ. Manag.* **2021**, *299*, 113655. [CrossRef]
82. Huang, M.Y.; Yue, W.Z.; Fang, B.; Feng, S.R. Scale response characteristics and geographic exploration mechanism of spatial differentiation of ecosystem service values in Dabie Mountain area, central China from 1970 to 2015. *Acta Geogr. Sin.* **2019**, *74*, 1904–1920. [CrossRef]
83. Divinsky, I.; Becker, N.; Bar, P. Ecosystem service tradeoff between grazing intensity and other services—A case study in Karei-Deshe experimental cattle range in northern Israel. *Ecosyst. Serv.* **2017**, *24*, 16–27. [CrossRef]
84. Benra, F.; Nahuelhual, L.; Gaglio, M.; Gissi, E.; Aguayo, M.; Jullian, C.; Bonn, A. Ecosystem services tradeoffs arising from non-native tree plantation expansion in southern Chile. *Landsc. Urban Plan.* **2019**, *190*, 103589. [CrossRef]
85. Wang, R.S.; Pan, H.Y.; Liu, Y.H.; Tang, Y.P.; Zhang, Z.F.; Ma, H.J. Evolution equivalent and driving force of ecosystem in Leshan City service value based on dynamic. *Acta Ecol. Sin.* **2022**, *42*, 76–90. [CrossRef]

International Journal of
Environmental Research and Public Health

MDPI

Article

Spatial-Temporal Evolution and Driving Forces of Drying Trends on the Qinghai-Tibet Plateau Based on Geomorphological Division

Yi Liu [1], Zhongyun Ni [1,2,3,*], Yinbing Zhao [1,2,3,4], Guoli Zhou [1], Yuhao Luo [1], Shuai Li [1], Dong Wang [2] and Shaowen Zhang [2]

1 College of Tourism and Urban-Rural Planning, Chengdu University of Technology, Chengdu 610059, China; nica6611@gmail.com (Y.L.); zhaoyinbing06@cdut.cn (Y.Z.); zhouguoli@stu.cdut.edu.cn (G.Z.); 2020050872@stu.cdut.edu.cn (Y.L.); shuailee.cn@gmail.com (S.L.)
2 College of Earth Sciences, Chengdu University of Technology, Chengdu 610059, China; wdong6985@163.com (D.W.); 2021020918@stu.cdut.edu.cn (S.Z.)
3 School of Geography, Archaeology & Irish Studies, National University of Ireland, Galway (NUIG), H91 CF50 Galway, Ireland
4 Human Geography Research Center of Qinghai Tibet Plateau and Its Eastern Margin, Chengdu 610059, China
* Correspondence: nizhongyun2012@mail.cdut.edu.cn or theoneyun@gmail.com

Abstract: The Qinghai–Tibet Plateau (QTP) is a sensor of global climate change and regional human activities, and drought monitoring will help to achieve its ecological protection and sustainable development. In order to effectively control the geospatial scale effect, we divided the study area into eight geomorphological sub-regions, and calculated the Temperature-Vegetation Drought Index (TVDI) of each geomorphological sub-region based on MODIS Normalized Difference Vegetation Index (NDVI) and Land Surface Temperature (LST) data, and synthesized the TVDI of the whole region. We employed partial and multiple correlation analyses to identify the relationship between TVDI and temperature and precipitation. The random forest model was further used to study the driving mechanism of TVDI in each geomorphological division. The results of the study were as follows: (1) From 2000 to 2019, the QTP showed a drought trend, with the most significant drought trend in the central region. The spatial pattern of TVDI changes of QTP was consistent with the gradient changes of precipitation and temperature, both showing a gradual trend from southeast to northwest. (2) There was a risk of drought in the four seasons of the QTP, and the seasonal variation of TVDI was significant, which was characterized by being relatively dry in spring and summer and relatively humid in autumn and winter. (3) Drought in the QTP was mainly driven by natural factors, supplemented by human factors. The driving effect of temperature and precipitation factors on TVDI was stable and significant, which mainly determined the spatial distribution and variation of TVDI of the QTP. Geomorphological factors led to regional intensification and local differentiation effects of drought, especially in high mountains, flat slopes, sunny slopes and other places, which had a more significant impact on TVDI. Human activities had local point-like and linear impacts, and grass-land and cultivated land that were closely related to the relatively high impacts on TVDI of human grazing and farming activities. In view of the spatial-temporal patterns of change in TVDI in the study area, it is important to strengthen the monitoring and early warning of changes in natural factors, optimize the spatial distribution of human activities, and scientifically promote ecological protection and restoration.

Keywords: climate change; Temperature-Vegetation Drought Index (TVDI); random forest; geomorphological division; Sanjiangyuan

Citation: Liu, Y.; Ni, Z.; Zhao, Y.; Zhou, G.; Luo, Y.; Li, S.; Wang, D.; Zhang, S. Spatial-Temporal Evolution and Driving Forces of Drying Trends on the Qinghai-Tibet Plateau Based on Geomorphological Division. *Int. J. Environ. Res. Public Health* **2022**, *20*, 7909. https://doi.org/10.3390/ijerph19137909

Academic Editor: Nir Krakauer

Received: 19 May 2022
Accepted: 25 June 2022
Published: 28 June 2022

1. Introduction

Drought represents a natural disaster bearing a wide range of impacts that often occur at broad spatial scales and can last for a long time. Not only does drought seriously affect the growth of vegetation, the water cycle, and human life, but also destroys the stability of the ecological environment and restricts the development of the social economy. However, drought is difficult to monitor in time since the occurrence of drought is a slow and dynamic process. This process begins with the reduction of precipitation, followed by the loss of soil moisture and the rise of surface temperature, which eventually leads to vegetation stress and a reduction in productivity [1]. In recent years, many researchers have proposed drought indices or improved existing indexes, such as the Vegetation Condition Index (VCI) [2–4], Normalized Difference Water Index (NDWI) [5], Vegetation Health Index (VHI) [6], and Drought Frequency Index (DFI) [7], which have been used to monitor the onset, duration, and intensity of drought and to explore the regional applicability of drought indicators. Among them, the most commonly used drought indices were the Standardized Precipitation Index (SPI) [8] and the Palmer Drought Severity Index (PDSI) [9]. SPI is based on the cumulative probability of precipitation at different time scales and was developed to reflect the impact of changes in water resources on groundwater and agriculture during the study period, thereby revealing the degree of drought in the study area [10], which is often used to capture the drought situation at multiple scales [11]. PDSI is a comprehensive index that uses precipitation and air temperature data to evaluate drought status, and is widely employed to study agricultural drought [12]. However, SPI only uses precipitation data and ignores other drought variables, and thus cannot fully capture the degree of drought in the study area [13]. PDSI has a fixed time scale and cannot be used to evaluate the drought at long time scales [14]. The index may lag behind emerging drought conditions and may not capture drought conditions in mountainous areas with complex climate [15].

Therefore, more comprehensive indicators need to be considered. Using the correlation between the surface temperature and a vegetation index [16], Carlson et al. [17] proposed the Different Water Index (DWI), which uses the Soil–Vegetation–Atmosphere-Transfer (SVAT) model to estimate and describe the trapezoid feature of the characteristic space composed of vegetation cover and surface soil water content. Sandholt et al. [18] proposed a simplified surface drought index, Temperature-Vegetation Drought Index (TVDI), based on temperature, radiation balance, and other factors. TVDI is an index which can monitor the degree of drought in a region by establishing a model based on the relationship between a vegetation index and a surface temperature index. This method only needs to use remote sensing images to monitor the severity of drought, which is beneficial to study the spatial and temporal characteristics of regional drought. Compared with the Standardized Precipitation Index (SPI) and the Standardized Precipitation Evapotranspiration Index (SPEI), TVDI is relatively simpler and has higher accuracy [19]. In addition, a conspicuous feature of TVDI is that it correlates negatively with surface soil moisture [20,21]. Thus, this method can be used to detect changes in soil moisture [22] and drought conditions [23,24]. Chen et al. [25] demonstrated that TVDI can effectively monitor soil water dynamics during the growing season in the Huang-Huai-Hai Plain. Chen et al. [26] combined the TVDI and the VHI method to monitor drought in Central America and assessed the affected agricultural areas.

TVDI is commonly used to study areas across the entire range of surface moisture content, from dry to wet, and from bare soil areas with no vegetation cover to areas completely covered with vegetation [18]. However, affected by factors such as changes in vegetation cover and complex and variable terrain, the edges of the dry and wet fitted by TVDI are unstable, and TVDI cannot clearly express the spatial edge in a theoretical sense [27]. At a large scale, changes in geomorphological types provide precipitation, air temperature, evapotranspiration, and other factors with spatial heterogeneity, and then control the differences in regional vegetation spatial distribution and vegetation cover. Finally, these changes in turn have an impact on the regional climate and the ecological environment. Therefore, the uncertainty of soil moisture estimation by TVDI will increase with soil surface heterogeneity [28], and the TVDI method is only suitable for areas with little topographic variability [29]. The differences in geomorphological divisions not only reflect the changes of regional geological tectonic movement characteristics, but also indicate the differences in ecogeographic systems, such as climate, water resource allocation, biology, and soil properties on a regional scale. Geomorphological division may effectively address heterogeneity-related issues that arise when using the TVDI method in dry and wet edges, and can effectively reduce the error and improve the calculation accuracy.

As an important ecological security barrier in China, the Qinghai–Tibet Plateau (QTP) plays an important role in water conservation, biodiversity protection, soil and water conservation, and is a 'sensor' of climate change in Asia and the Northern Hemisphere. The QTP has a very significant warming effect in the background of global warming [30]. Elevated temperatures have increased evapotranspiration and drought probability, which has seriously restricted the development of local agriculture and the animal husbandry economy. Moreover, factors such as vegetation cover, the intensity of human activities, and the distribution of plateau lakes and glaciers have affected drought severity in the QTP. Therefore, methods to effectively monitor the spatial-temporal variability of drought in such a large area as the QTP have become the focus of attention.

However, previous studies have shown significant differences in the spatial and seasonal characteristics of drought in the QTP. For example, Wang et al. [31] pointed out that the central QTP has become wetter using SPI and SPEI to study the drought in China from 1961 to 2012. Bin et al. [32] demonstrated that the southeastern QTP is an area prone to drought, and is a high-risk area by using the SPI method and monthly precipitation data from 616 meteorological stations. Wang et al. [33] believed that drought duration, drought severity, and drought frequency decreased in the central QTP from 1950 to 2006. Kai et al. [34] pointed out that there was a significant wet trend in the eastern QTP from 1961 to 2012 using SPEI, SPI, and the Reconnaissance Drought Index (RDI) method. Feng et al. [35] believed that in the future this region would become wetter in spring, more arid in winter, and more arid in summer by using the SPEI-TH (Standardized Precipitation Evapotranspiration Index-Thornthwaite) method after analyzing the meteorological data of stations in the QTP from 1970 to 2017. Wang et al. [36] believed that the degree of drought in the QTP was relieved by wet conditions in spring, slightly wet conditions in summer and autumn, and more arid conditions in winter from 1994 to 2013. Overall, in terms of spatial characteristics, drought severity in the central and southeastern QTP were found to be changing differently, while the drought status in other regions was not clear. In terms of seasonal characteristics, the drought severity in spring, summer, and autumn differed, and drought occurred in winter for both regions.

In terms of the driving factors of drought in the QTP, previous studies focused on the discussion of factors, such as precipitation, temperature, elevation, slope, and aspect, the distribution of lakes on the plateau, human activities, and other factors. For example, precipitation is often used to discuss changes in the degree of drought in the region due to the significant correlation between precipitation and TVDI [37]. Temperature affects regional heat distribution and is one of the important factors affecting drought changes [38]. As for geomorphological factors, elevation produces vertical differences in regional precipitation and temperature [39], and the slope and aspect produce regional differences in precipitation and air temperature [40]. The QTP has a wide area of lakes, glaciers, and permafrost, and thus the melting of glaciers and the freezing and thawing of permafrost affect the seasonal changes of drought in the region [41]. Human activities are also an important external driving force affecting changes in drought [42]. For example, the expansion of urban construction decreases vegetation cover, while water resource utilization for domestic, industrial and mining uses affect groundwater levels and recharge. Due to the different intensities of human activities, differences in land use will lead to differences in drought severity [43]. However, in an area as large as the QTP, changes in drought must be the result of the comprehensive action of many factors, and such research has received little attention.

Random forest is an algorithm that integrates multiple decision trees through the idea of integrated learning [44]. This algorithm is one of the most commonly used and most powerful supervised learning algorithms. It does not require dimensionality reduction and feature selection. It can better process high-dimensional data and solve regression and classification problems. Moreover, it can be trained quickly and implemented easily [45]. The random forest algorithm can obtain estimates that are easier to explain and understand, such as IncMSE and IncNodePurity. IncMSE, which is the increase in mean squared error, indicates the increase of the error estimated by the random forest model relative to the original error after the variable is randomly selected [46]. IncNodePurity, which is the increase in node purity, indicates the degree of influence for a variable on each decision tree node [47]. Both indices can judge the importance of the influencing factor, and the larger the index value, the more important the variable. Moreover, after controlling the interaction between exploratory variables, the individual marginal effect of each factor can also be visualized [48]. However, at present, few researchers have systematically compared the differences and similarities of the relative importance of drought factors in the QTP using the random forest algorithm.

Therefore, the spatial-temporal differences and driving forces of the arid state in the QTP still warrant attention. This paper attempts to: (1) analyze the feasibility of using the TVDI method to study drought in the QTP, (2) clarify the spatial-temporal variation of drought in the QTP, (3) evaluate the main driving forces affecting the QTP and comprehend the specific range of various factors that affect drought. Based on the previous rules of geomorphological division [49], the workflow of our study was to: (1) divide the QTP into eight geomorphological sub-regions; (2) use the TVDI method to calculate the annual TVDI and seasonal TVDI of eight regions from 2000 to 2019; (3) analyze the spatial-temporal variation characteristics of annual and seasonal drought in the QTP, and summarize the variation of TVDI in the study area; (4) use the partial correlation and the multiple correlation analyses to investigate the relationship between air temperature, precipitation, and TVDI; (5) select eight factors in four categories including meteorology, landform, land use and land cover (LULC), and human activities to construct a random forest model, explore the main factors driving the drying trends in each geomorphological sub-division, and rank the contribution of each driving factor. The results of this paper provide a theoretical basis for the division of drought risk in the QTP, the prevention of drought hazards and the sustainable development of regional economy.

2. Study Area

The QTP is the highest plateau in the world and the largest plateau in China. It has an important impact on the formation of geographical elements, such as climate and rivers in China. The latitude and longitude of the QTP ranges from 25°59′56″ N to 39°47′35″ N and 73°30′22″ E to 104°39′50″ E (Figure 1), covering six provinces, including the Tibet Autonomous Region and Qinghai Province, the southern parts of the Xinjiang Autonomous Region, the western parts of Sichuan Province, and small portions of Yunnan Province and Gansu Province.

Referring to the previous results of geomorphological division of the QTP [49], we divided the study area into 8 geomorphic areas (Appendix B, Table A2). The topography of the QTP is complex, with high elevation mountains and many basins, canyons, and lakes. The eight factors are shown in Figure 2. The elevation of the QTP is higher in the west and lower in the east. The alpine areas on its edge bear large undulations and low interior. Affected by its own geographical location, the QTP has formed a unique plateau climate zone. The precipitation in QTP decreases from southeast to northwest (Figure 2b). The temperature has a clear latitudinal zonality, decreasing from south to north (Figure 2c). The land cover types in QTP are dominated by grassland (Figure 2h), which is widely distributed in all parts of the QTP and mainly has low to medium canopy cover. Other land cover types include Gobi, bare land, sandy land, saline–alkali land, and other areas with vegetation cover below 5%, which are distributed in the northern part of the QTP. Forests are mainly distributed in the southeast of the QTP due to the influence of precipitation and temperature. Cultivated, urban, and rural built-up lands are mainly distributed in low-lying, warm valleys with abundant water sources.

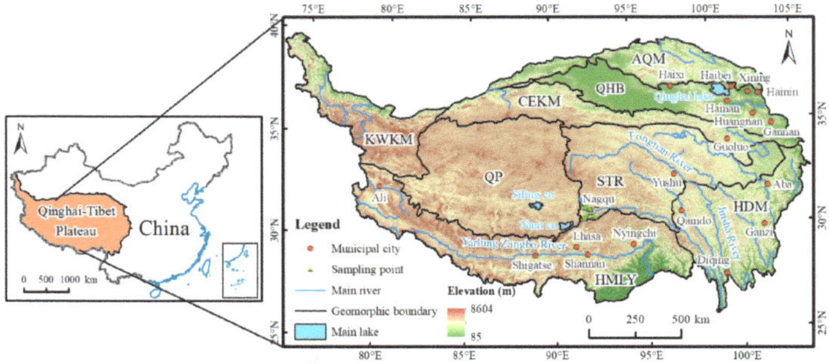

Figure 1. Map of the Qinghai–Tibet Plateau (QTP). AQM is the high valley area of the Altun–Qilian Mountains. QHB are the high mountain areas of Qaidam-Yellow River-Huangshui River Basin. KWKM are the high mountain areas of Karakorum and Western Kunlun Mountains. CEKM are the high mountain areas of the central and eastern Kunlun Mountains. QP are the lake and basin areas of the Qiangtang Plateau. STR are the mountains sources of the Yangtze River, Yellow River, and Lancang River (Three Rivers or Sanjiangyuan) and the valley bottom of the upper reaches of Three Rivers. HMLY are the high mountain areas of the Himalayan Mountains. HDM are the high mountain and valley areas of Hengduan Mountains. (Appendix A of Table A1 for acronyms).

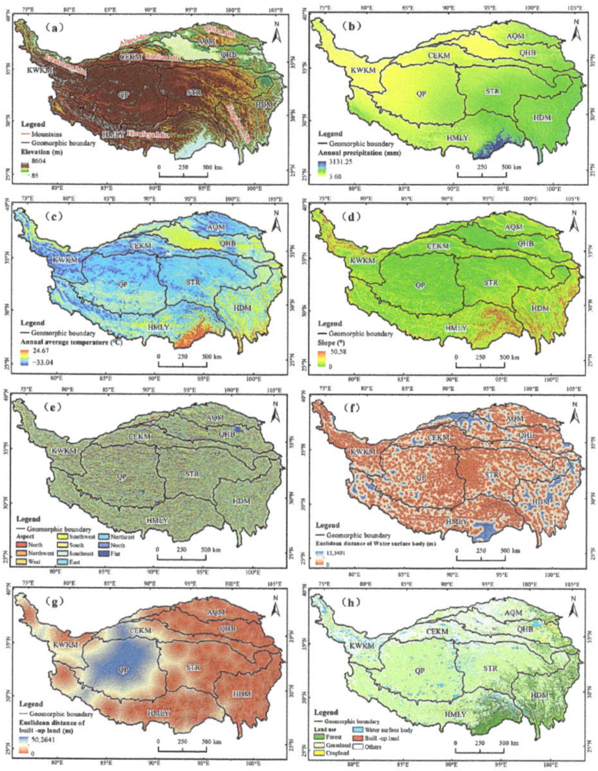

Figure 2. Maps of the eight factors. (**a**) Elevation; (**b**) Annual precipitation; (**c**) Annual average temperature; (**d**) Slope; (**e**) Aspect; (**f**) Euclidean distance of water surface body; (**g**) Euclidean distance of built-up land; (**h**) Land use and Land cover.

3. Materials and Methods

3.1. Materials

3.1.1. MODIS Data

The relationship between NDVI (Normalized Difference Vegetation Index) and surface temperature can be used to calculate TVDI [50] to study the spatial-temporal drought evolution of the QTP. We used NDVI and surface temperature data from 2000 to 2019 from NASA (https://search.earthdata.nasa.gov/, accessed on 1 March 2020). We selected the MOD13A3 product (MOD/Terra monthly vegetation index data product with spatial resolution of 1 km) for NDVI data, and the MOD11A2 product (MOD/Terra 8-day land surface temperature data product with spatial resolution of 1 km) for surface temperature data. The data projection coordinate system was WGS_1984_UTM_Zone_45N. After batch splicing, band extraction, resampling, and projection, the data were processed using the maximum value synthesis method to obtain annual and quarterly NDVI and daytime surface temperature of the eight geomorphological divisions of the QTP. Using the NDVI and surface temperature data, we calculated TVDI. We used Savizky–Golay filtering to eliminate poor quality observations, and the maximum synthesis method was used to eliminate deviation in the data caused by atmospheric interference and cloud cover.

3.1.2. Soil Moisture Data

There is a negative correlation between soil moisture and TVDI, and the relationship between soil moisture at a depth of 10 cm and TVDI is more significant [25]. Therefore, we used the measured 10 cm soil moisture data to verify whether the TVDI method could capture the drought conditions in the QTP. The soil moisture data were obtained from the National Qinghai–Tibet Plateau scientific data center (http://www.tpdc.ac.cn/zh-hans/, accessed on 6 February 2021). Due to the lack of soil moisture measurements at some sampling points, we selected data collected in summer from 2010 to 2016. After we eliminated the missing sampling point data in this period, we calculated the mean value.

3.1.3. Driving Factors

Drought severity in the region is not only affected by the natural environment, but also human activities. Precipitation and temperature are important factors that affect drought variation, and these two factors determine the spatial distribution of drought grade and drought frequency in the region [51]. Geomorphic factors lead to regional differentiation of drought, among which elevation plays a critical role in the spatial-temporal differentiation of drought spread in plateau mountains [52]. Human activities such as the increase in domestic water supplies and the expansion of urban areas can accelerate the spread of drought [53]. Changes in land cover will affect the supply of atmospheric water, thus affecting the severity of regional drought [54]. Lakes, glaciers and other water areas will also affect the seasonal variation of regional drought [41]. So, we adopted the random forest method and selected six driving factors, including precipitation, temperature, elevation, slope, aspect, the Euclidean distance of water surface land, the Euclidean distance of built-up land, and LULC to explore the main driving factors of drought severity change in various geomorphological divisions of the QTP. The source of data for the driving factors is shown in Table 1, and the coordinate system of the data was WGS_1984_UTM_Zone_45N.

The slope and aspect of the geomorphological factors were extracted from the DEM data, and the Euclidean distance of water surface body was used to represent the degree and scope of influence of the water surface body in the QTP, and the human activity factor was described by the Euclidean distance of built-up land. The land use types were combined according to the attribute table, and finally the data of six main land use types, including forest, grassland, cultivated land, water surface body, built-up land were obtained.

Table 1. Data and data sources.

Data Class	Data	Data Sources	Spatial Resolution	Time of Data
Climatic factor	Precipitation Temperature	National Qinghai Tibet Plateau scientific data center (http://www.tpdc.ac.cn/zh-hans/, accessed on 7 April 2021)	1 km	2000–2017 2000–2017
Geomorphological factor	Elevation Slope Aspect	China Geological Survey (https://www.cgs.gov.cn/, accessed on 21 April 2021)	1 km	2015
Accessibility factor	Water surface body Built-up land	National Qinghai Tibet Plateau scientific data center (http://www.tpdc.ac.cn/zh-hans/, accessed on 21 April 2021)	1 km	2015
Land use factor	LULC	National Qinghai Tibet Plateau scientific data center (http://www.tpdc.ac.cn/zh-hans/, accessed on 21 April 2021)	1 km	2015

3.2. Method

3.2.1. TVDI Calculation

TVDI is an index based on the correlation between a vegetation index and a surface temperature index, and serves to reflect drought severity in a region [18]. After constructing the *NDVI-LST* feature space with the same interval of *NDVI* values combined with the maximum composite method and the minimum composite method, the TVDI value can be calculated using Formula (1).

$$TVDI = \frac{LST - LST_{min}}{LST_{max} - LST_{min}} \tag{1}$$

$$LST_{min} = a_1 + b_1 \times NDVI \tag{2}$$

$$LST_{max} = a_2 + b_2 \times NDVI \tag{3}$$

In Formula (1), LST represents the surface temperature of any pixel in the study area. In Formula (2), LST_{min} is the fitted wet-edge function, representing the minimum surface temperature. In Formula (3), LST_{max} is the fitted dry-edge function, representing the maximum surface temperature [18]. a_1 and b_1 are the linear fitting coefficients of the wet-edge function, while a_2 and b_2 are the linear fitting coefficients of the dry-edge function. The value of TVDI is between 0 to 1, and the higher the value, the drier the area. In areas with obvious geomorphic differences and large scale areas, there may be errors in using the TVDI method to study drought. Therefore, this paper adopts the method of geomorphological divisions to reduce the uncertainty of the results.

3.2.2. Correlation Analysis

Partial correlation analysis was used to analyze the relationship between TVDI and precipitation and temperature [55]. The partial correlation coefficient between temperature, precipitation, and TVDI was obtained using Formula (4).

$$R_{xy,z} = \frac{Rxy - RxzRyz}{\sqrt{(1 - {R_{xz}}^2)}\sqrt{(1 - {R_{yz}}^2)}} \tag{4}$$

In Formula (4), $R_{xy,\ z}$ means that the dependent variable z is fixed, and the partial correlation coefficient between the independent variables x and y. Rxy, Rxz, and Ryz are the Pearson correlation coefficients of variables x and y, variables x and z, and variables

y and z, respectively. If $R_{xy,z} > 0$, it means a positive correlation; if $R_{xy,z} < 0$, it means a negative correlation. The larger the coefficient, the higher the correlation. The significance test was performed using a t-test.

Similarly, in order to study the relationship between TVDI and precipitation and temperature in the QTP, we applied the multiple correlation analysis method. The calculation formula is shown in Formula (5):

$$R_{xy,z} = \sqrt{1 - \left(1 - R_{xy}^2\right)\left(1 - R_{xy,z}^2\right)} \tag{5}$$

In Formula (5), $R_{xy,z}$ is the multiple correlation coefficient between the dependent variable x and the independent variables y and z. The larger the correlation coefficient, the closer the linear correlation between the elements or variables. The correlation was tested by F-test.

The partial correlation coefficient between TVDI and annual precipitation and average annual temperature from 2000 to 2017 was calculated by Formula (5), then the t-test of the partial correlation coefficient and the F-test of the multiple correlation were performed on the calculation results. The statistics that meet the classification conditions were divided into four categories, as shown in Table 2.

Table 2. Criteria for TVDI-driven zoning (It was reprinted with the permission from Ref. [56] Copyright 2022, Chen).

Types of TVDI Changes	**Zoning Criteria**		
	$r_{TVDI\ P,T}$	$r_{TVDI\ T,P}$	$R_{TVDI,\ TP}$
Precipitation driven	$t \geq t_{0.05}$		$F \geq F_{0.05}$
Temperature driven		$t \geq t_{0.05}$	$F \geq F_{0.05}$
Temperature and precipitation driven	$t \leq t_{0.05}$	$t \leq t_{0.05}$	$F \geq F_{0.05}$
Other drive types			$F \leq F_{0.05}$

In Table 2, $r_{TVDI\ P,T}$ is the partial correlation coefficient between TVDI and annual precipitation, $r_{TVDI\ T,P}$ is the partial correlation coefficient between TVDI and annual average temperature, $R_{TVDI,\ TP}$ represents the multiple correlation coefficient between TVDI and temperature and precipitation, t and F are the statistical values of the t-test and F-test, respectively, and $t_{0.05}$ and $F_{0.05}$ are the 0.05 significance levels of the t-test and the F-test, respectively.

3.2.3. Linear Trend Analysis

Linear trend analysis is a method of predicting the change trend of a variable by performing linear regression analysis on a variable that changes over time. The linear trend analysis method can be used to analyze the change of each pixel. We applied this method to analyze the changes of TVDI in the QTP in the recent 20 years. Formula (6) was used to calculate the linear trend rate [57].

$$TVDI\_Slope = \frac{n \times \sum_{i=1}^{n}(i \times TVDI_i) - \sum_{i=1}^{n} i \sum_{i=1}^{n} TVDI_i}{n \times \sum_{i=1}^{n} i^2 - \left(\sum_{i=1}^{n} i\right)^2} \tag{6}$$

In Formula (6), i is the year of 2001, 2002, 2003 ... 2019, n is the length of the time series; $TVDI_i$ is the TVDI value of the year i, $TVDI\_Slope$ is the regression value of TVDI pixels. $TVDI\_Slope > 0$ indicates that TVDI in the time series is increasing, while $TVDI\_Slope < 0$ means that the TVDI is decreasing.

To test the consistency of the trend of TVDI with time, we used a t-test to judge the level of significance of the change. The p value was calculated using the obtained $TVDI\_Slope$ of change and the results of the t-test, and the two significance judgment criteria were a

95% confidence level and a 99% confidence level. According to Table 3, the trend in TVDI was divided into six levels.

Table 3. Classification criteria of TVDI trend.

Grading Criteria		TVDI Trend
$TVDI\_Slope < 0$	$p > t_{0.01}$	Extremely significant decrease
	$p \geq t_{0.05}$ and $p \leq t_{0.01}$	Significant decrease
	$p < t_{0.05}$	Non-significant decrease
$TVDI\_Slope > 0$	$p < t_{0.05}$	Non-significant increase
	$p \geq t_{0.05}$ and $p \leq t_{0.01}$	Significant increase
	$p > t_{0.01}$	Extremely significant increase

3.2.4. Random Forest Algorithm

We selected eight factors: annual precipitation, annual average temperature, elevation, slope, aspect, Euclidean distance of water surface body, Euclidean distance of built-up land, and LULC. These eight factors were resampled to a spatial resolution of 4 km × 4 km in ArcGIS 10.3 software. The degree of influence of these eight factors on TVDI was analyzed using the Random Forest regression (RF) model, which was implemented by calling the "Random Forest" package in R studio 3.6 software (R Studio, Boston, MA, USA). Each group of samples was divided into two subsets, 70% of the training set and 30% of the test set. On the basis of determining the optimal parameters, the fitting calculation was performed to obtain the importance ranking (IncNodePurity) and local relative importance of the driving factors [48]. The larger the IncNodePurity value, the higher the relative importance of the factors and the deeper the influence on the regional TVDI. The analysis of relative importance of the driving factors can reveal the relationship between a certain range of driving factors and TVDI, where the higher the local relative importance value, the higher the contribution to TVDI.

3.2.5. Verification of TVDI

To verify the accuracy of the TVDI method for monitoring drought, the common method is to perform correlation analysis between measured soil moisture data and TVDI. Due to the lack of soil moisture data in some years and some sampling points, we selected summer soil moisture data from 2010 to 2015 to verify the accuracy of TVDI method and compare the advantages and disadvantages between the two TVDI data with and without geomorphological divisions. In Figures 3 and 4, the y-axis represents the measured soil moisture content at 10 cm, and the x-axis is the TVDI value extracted from the location of the sampling point. It can be seen from Figures 3 and 4 that there is a clear negative correlation between soil moisture and TVDI in the summer from 2010 to 2015. With increased soil moisture, TVDI shows an obvious negative trend. We found TVDI to have a significant linear relationship ($p < 0.01$) with soil moisture in each year. Moreover, in comparing Figures 3 and 4, we found that the correlation between TVDI data retrieved by geomorphological division and soil moisture data was higher than that between TVDI data retrieved directly without geomorphological division and soil moisture data, and the R^2 value of the former were all over 0.41. Therefore, the TVDI data retrieved by the geomorphological division method can better reflect the soil moisture status, and also show that the TVDI method is suitable for studying the drought status of the QTP.

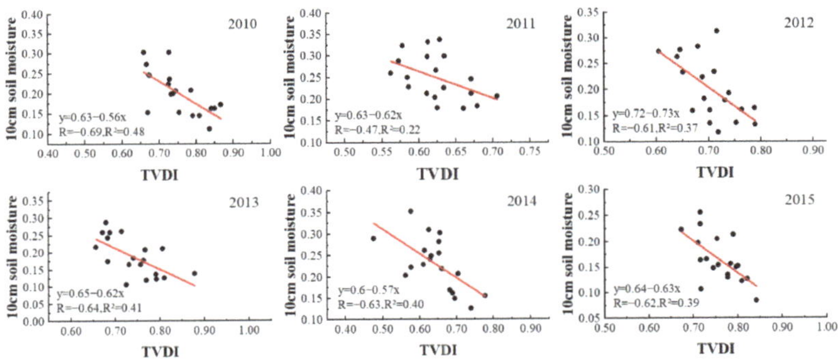

Figure 3. Correlation between TVDI retrieved directly without geomorphological division and 10 cm soil moisture in summer from 2010 to 2015.

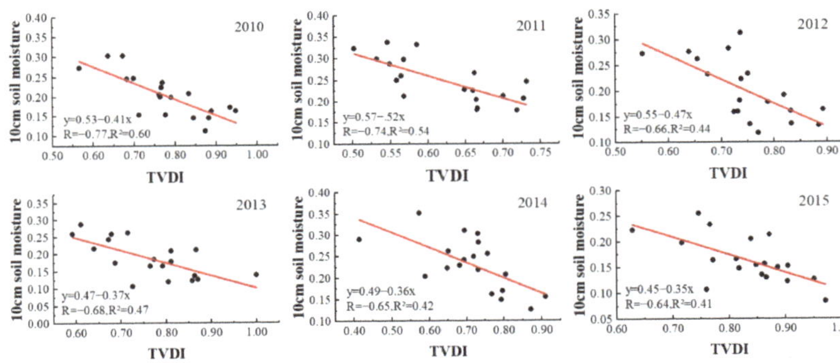

Figure 4. Correlation between TVDI retrieved by geomorphological division and 10 cm soil moisture in summer from 2010 to 2015.

4. Results

4.1. Characteristic of NDVI-LST Feature Space

Figure 5 shows the feature spaces of the eight geomorphological divisions that we constructed, which basically satisfied the triangular or trapezoidal space theory. The maximum and minimum LST values corresponding to NDVI were extracted through the TVDI VTCL plug-in of the ENVI 5.3 software, and the NDVI-LST space was constructed. When constructing the NDVI-LST feature space of the QP, the range of NDVI value from 0 to 1 was selected. It should be noted that when constructing the NDVI-LST feature space of the other seven regions, we excluded values of NDVI < 0.2 from the analysis. After determining the NDVI range of each district, we set the sample size in the plug-in to 1 to obtain the annual and quarterly NDVI-LST feature spaces from 2000 to 2019. Because the NDVI-LST feature space of the eight geomorphological divisions in 2001 had the best fit (the average R^2 is 0.69), we took the characteristic space of 2001 as an example. Figure 4 shows that there was an obvious linear relationship between the maximum and minimum surface temperature and NDVI. With increased NDVI, the maximum surface temperature decreased gradually, while the minimum surface temperature increased slowly. It can be seen from Appendix B Table A3 that the average R^2 of the HDM was 0.35, with the low value of R^2, poor relative fitting, and low accuracy. Other regions have high R^2 values, a good fit, and high accuracy, among which the QHB, QP, and HMLY were the best (the average value of R^2 is 0.75).

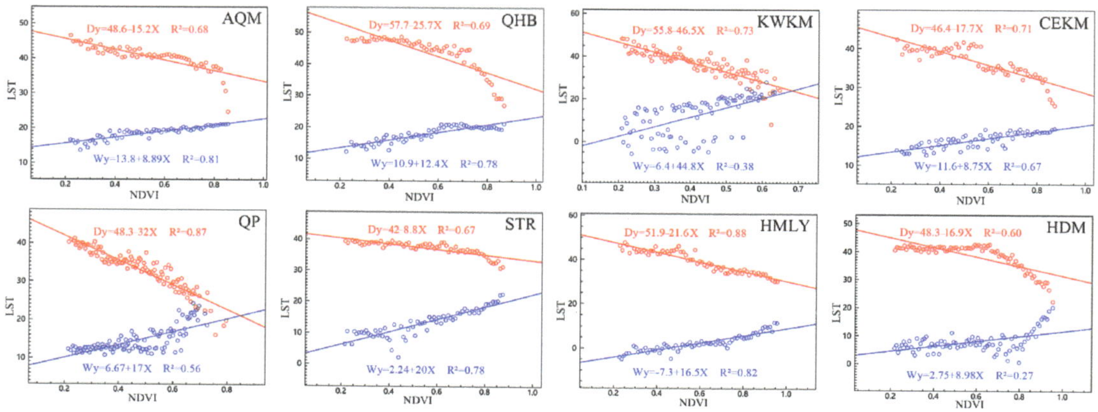

Figure 5. NDVI-LST spatial characteristics of eight geomorphological divisions of the QTP (taking 2001 as an example)(The horizontal axis represents NDVI. The vertical axis represents LST. The blue line is the fitted wet edge equation. The red line is the fitted dry edge equation. The sample points are represented by small hollow circles).

4.2. *Temporal and Spatial Variation Characteristics of TVDI*

The dry-edge and wet-edge equations were used to calculate the TDVI value of each pixel at different times, and the TVDI value was applied as the classification index of drought in the QTP [58]. TVDI was divided into five grades (Table 4) [21,24,59]: extremely wet ($0 \leq$ TVDI ≤ 0.2), wet ($0.2 <$ TVDI ≤ 0.4), normal ($0.4 <$ TVDI ≤ 0.6), dry ($0.6 <$ TVDI ≤ 0.8), and extremely dry ($0.8 <$ TVDI ≤ 1.0). When the TVDI was between [0, 0.6], the drought type of the region was no drought; when the TVDI was between (0.6, 0.8], the drought type of the region was drought; and when the range was between (0.8, 1.0], the drought type of the region was severe drought.

Table 4. TVDI classification criteria of drought in the QTP.

Classification of TVDI	**[0, 0.2]**	**(0.2, 0.4]**	**(0.4, 0.6]**	**(0.6, 0.8]**	**(0.8, 1.0]**
Drought rank **Drought types**	Extremely wet No drought	Wet No drought	Normal No drought	Dry Drought	Extremely dry Severe drought

4.2.1. Spatial Variation Characteristics of TVDI

Figure 6 shows that there were obvious regional differences in the spatial distribution of the mean value of TVDI in the QTP from 2000 to 2019, which was mainly characterized by drought in the southwest and north, and wet conditions in the northeast and southeast. Overall, the average annual value of TVDI in the QTP was 0.56, which was normal. However, the regional drought situation was not consistent. For example, for QHB the average annual TVDI was 0.66, which we classified as being in drought. The no-drought areas were distributed in the eastern region of the STR, HMLY, HDM, the Qilian Mountains, and the Kunlun Mountains. Most of these areas were forest and grassland, accounting for 56.9% of the area of the QTP. The drought areas were mainly distributed in the southwest QP, in the west HMLY and most of the KWKM, and grassland was the main land use type, accounting for 33.4% of the area of the QTP. The severe drought areas accounted for 9.7% of the QTP, and were mainly distributed in the Qaidam Basin, the western part of AQM, and the river valleys of HMLY.

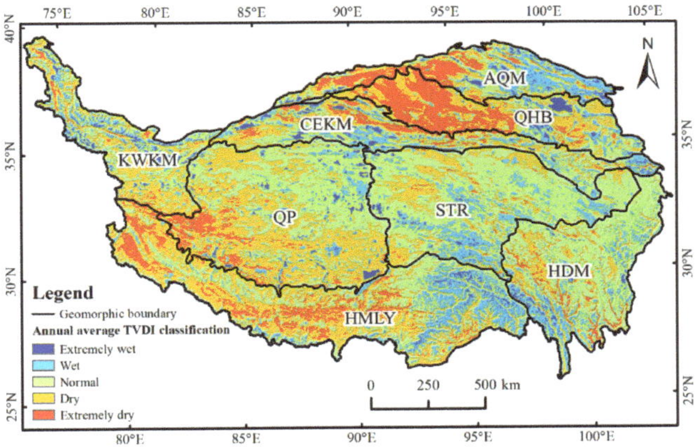

Figure 6. Annual average TVDI classification of the QTP.

The QTP was drier in spring and summer and humid in autumn and winter (Figure 7). Spring was the driest season of the year. The southern part of the QTP was humid, and the southwestern and northern parts were dry. Most of the QHB, the southern part of the QP, and the central and western parts of the HMLY were relatively dry. In summer, the QTP was generally wet in the south and dry in the north, in which drought severity in the southwest was reduced and drought severity in the north increased. In terms of spatial distribution, the QHB, the Altun Mountain, the west of HMLY, and the southwest of QP were relatively arid. In autumn, extremely wet areas in the QTP decreased, while wet and normal areas increased. Dry areas were concentrated in the north and southwest, such as the QHB, QP, and the west of HMLY. In winter, the central part of the QTP was wet, the western part was dry, and the wet areas increased. Winter was the wettest season of the year with the smallest arid area. Only the south of QP and the mid of HMLY were in a dry state. In short, the AQM and QP were prone to spring drought, the HMLY was prone to autumn drought, the QHB was prone to spring drought and summer drought, and the central QHB was in a state of drought for a long time, and other areas had no obvious seasonal drought.

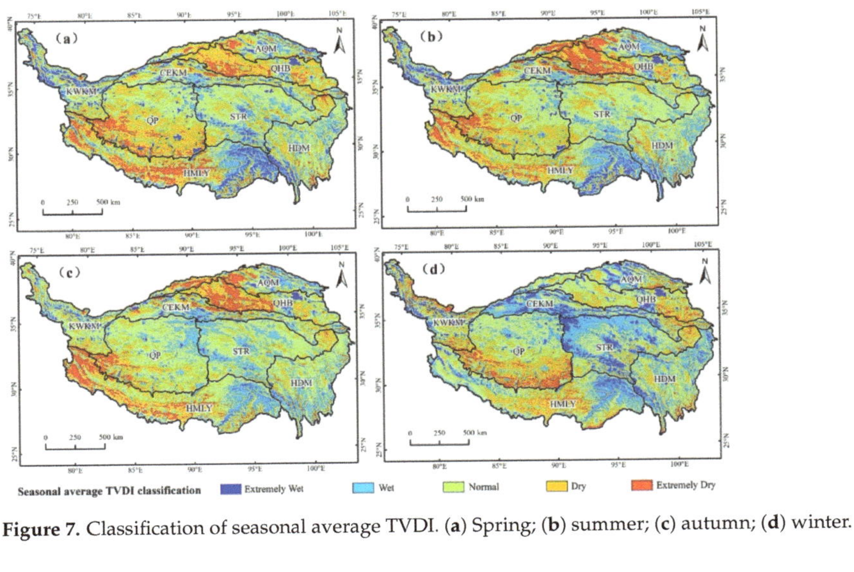

Figure 7. Classification of seasonal average TVDI. (**a**) Spring; (**b**) summer; (**c**) autumn; (**d**) winter.

4.2.2. Temporal Variation Characteristics of TVDI

The QTP has exhibited an obvious trend in drought severity over the past 20 years, and the central region has become the center of increased drought (Figure 8). The area with aggravated drought accounted for 52.9% of the area of the QTP, the area with a significant increase in TVDI accounted for 5.8% of the QTP, and the area with a significant and extremely substantial increase accounted for 2.3% of the QTP. These two types of drought variation are mainly distributed in the southeast of QP, the southwest of STR, the southwest of AQM, the most of KWKM, and the south of CEKM.

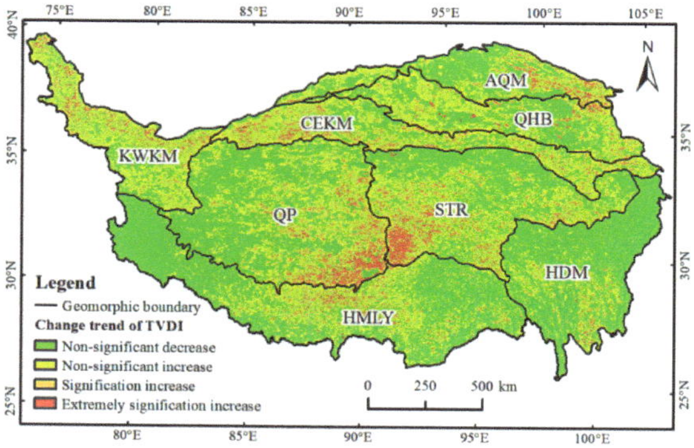

Figure 8. TVDI trends in QTP.

The land use types in these areas were mainly unused land and grassland, and the regional characteristics were high elevation and low vegetation cover. The areas with non-significant increase in TVDI accounted for 44.8% of the total area, mainly distributed in the KWKM, CEKM, QP, STR, and other areas. The areas with non-significant decrease in TVDI were mainly distributed in the northwest of HMLY, HDM, and QP. Most of these areas were grassland and forest, and accounted for 47.1% of the whole area. In general, there were obvious differences in drought severity trends in the QTP, with significant drying trends in the eastern and southeastern parts of QP and the western and southwestern parts of STR. There was a risk of drought in the Kunlun Mountains and its north, and the risk of drought in the HDM and HMLY regions decreased.

The QTP had a drying trend in all seasons except autumn (Figure 9). Spring was the season with the largest drying trend. The AQM, CEKM, STR, and HMLY were the centers of increasing drought in spring. The land use types in these areas are mainly grassland and unused land, accounting for 61.7% of the total area. The areas with a drying trend in summer were mainly distributed in the mid of HMLY, the southeast of QP, and the southwest of STR, which accounted for 60% of the entire region. Autumn was the season with the smallest area with a drying trend, which accounted for 48.9% of the total area of the region without a drying trend. There were three areas with strong drying trends in winter, QP, QHB, and AQM, and the areas with drying trends accounted for 58.2% of the total area. Overall, there was a significant drying trend in spring, the drought risk in the northern QTP weakened in summer, but the drought risk in the southern part remained very high. In autumn, most of the areas were the wettest and had a weaker trend in drought. In winter, the risk of drought in the northern QTP increased significantly.

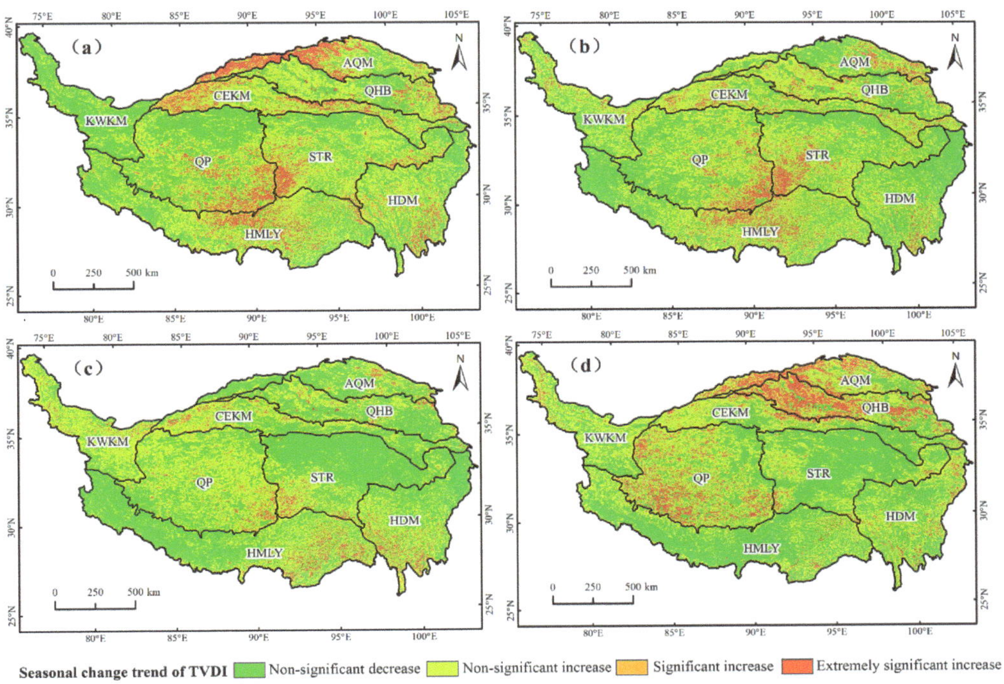

Figure 9. Seasonal TVDI trends of the QTP. (**a**) Spring; (**b**) summer; (**c**) autumn; (**d**) winter.

4.2.3. Spatial and Temporal Variation Characteristics of TVDI Based on Cluster Analysis

According to the distribution and variation characteristics of TVDI in various regions of the QTP, the method of cluster analysis was used to deeply explore the law of TVDI variation in the QTP. Based on annual average TVDI, $TVDI\_Slope$ of the annual average TVDI, and the trend classification of the annual average TVDI, we selected nine indicators in each district, maximum and minimum annual average TVDI (TVDI-max and TVDI-min), mean annual average TVDI (TVDI-mean), the standard deviation of annual mean TVDI (TVDI-SD), the mean value of $TVDI\_Slope$ (slope-mean), the area proportion of areas with non-significant decrease (TVDI-t3), the area proportion of areas with non-significant increase (TVDI-t4), the area proportion of significantly increased areas (TVDI-t5), and the area proportion of areas with extremely significant increase (TVDI-t6) (Table 5). We then used the Pearson correlation to conduct a case-by-case systematic cluster analysis (Figure 9).

Table 5. Indicators of cluster analysis in each geomorphological division.

Geomorphological Division	TVDI-Max	TVDI-Min	TVDI-Mean	TVDI-SD	Slope-Mean	TVDI-t3	TVDI-t4	TVDI-t5	TVDI-t6
AQM	1.00	0.00	0.55	0.24	3.75	47.34%	44.11%	6.19%	2.37%
QHB	0.97	0.01	0.66	0.20	1.04	42.64%	51.48%	4.07%	1.82%
KWKM	1.00	0.00	0.55	0.18	1.09	20.64%	69.28%	8.21%	1.87%
CEKM	1.00	0.00	0.54	0.21	1.54	25.59%	62.20%	9.96%	2.25%
QP	0.99	0.00	0.59	0.15	5.94	46.76%	42.58%	6.96%	3.70%
STR	0.94	0.00	0.50	0.10	9.69	41.10%	46.45%	8.04%	4.41%
HMLY	1.00	0.03	0.59	0.18	−3.43	58.71%	37.09%	3.41%	0.79%
HDM	1.00	0.00	0.52	0.15	−1.83	73.79%	24.25%	1.57%	0.40%

The TVDI of the QTP demonstrated obvious gradients in its characteristics. It can be found from Figure 10 that the QTP are divided into three categories: HMLY and HDM are one category, QP, STR, and AQM are one category, and the last category includes KWKM,

CEKM, and QHB. The HMLY and HDM areas are located in the southern part of the QTP, and their annual average TVDI values were 0.59 and 0.52, respectively, and the drought severity was at the non-drought level. The standard deviation of their annual TVDI was not much different, the mean value of slope was less than 0, and the areas with non-significant decrease had the largest ratio of area to the total area. They were characterized by a low degree of drought and a trend towards wetter conditions.

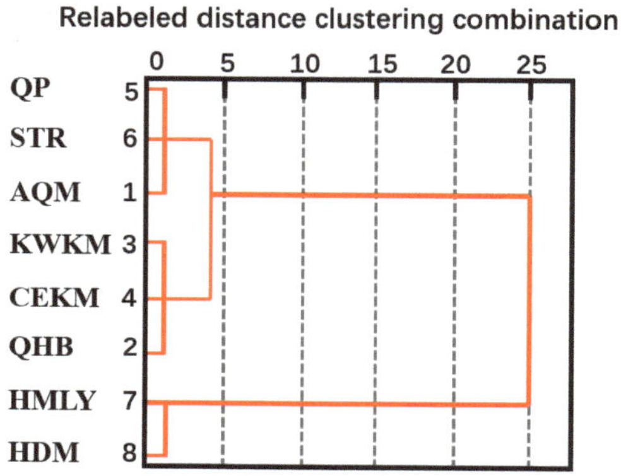

Figure 10. Pedigree of systematic cluster analysis.

The QP and STR are in the central QTP, and the AQM is in the northern part of the QTP. The average TVDI of the three areas was between 0.5 and 0.6, and the drought category was normal. The average slope was large, and the proportion of the area with non-significant decrease to the total area was between 41.10% and 47.34%. These three areas were characterized by a lower degree of drought and increased regional drought risk.

The KWKM, CEKM, and QHB are in the northern-central QTP. The average value of QHB was relatively high, and the drought severity was drought. The standard deviation of the annual average TVDI of the three places ranged from 0.18 to 0.21, and the average value of slope was between 1.04 and 1.54. Among the three regions, the difference in results was small. The ratio of the non-significant decrease area to the total area decreased in the three regions, and the ratio of the non-significant increase area to the total area was the highest, which indicated that the three regions were characterized by a higher degree of drought and had a significant risk of drought in the region. In summary, the drought severity and drought risk in the QTP increased from the southeast to the northwest, which indicated that the spatial variation pattern of TVDI was roughly consistent with the spatial variation of precipitation and temperature in the southeast–northwest gradient.

4.3. *Characteristics of TVDI Drivers*

4.3.1. Climate Driven Characteristics of TVDI

Figure 11 shows that the temperature, precipitation, and TVDI of the QTP had different trends. Annual precipitation had a negative trend, average annual temperature had a positive trend, and the average annual TVDI had a positive trend. The changes of these three indicated that the study area had a tendency of drought. From 2000 to 2017, the annual precipitation in the QTP had a negative trend, with a trend of −0.39 mm/year. The annual precipitation in the QTP decreased from the southeast to the northwest. Among them, the HMLY and HDM had the most precipitation, and the Qaidam Basin, AQM, and the western QP had the least precipitation. From 2000 to 2017, the average annual temperature of the QTP has fluctuated upward at a rate of 0.02 °C/year. The spatial distribution of the average annual temperature for many years was uneven. The areas with lower temperature were distributed in the Kunlun Mountains, the northern QP, and STR, and the areas with higher temperature are distributed in HMLY, the Qaidam Basin, and HDM. From 2000 to 2019, the TVDI had an insignificant upward trend, with a cyclical change of roughly six years.

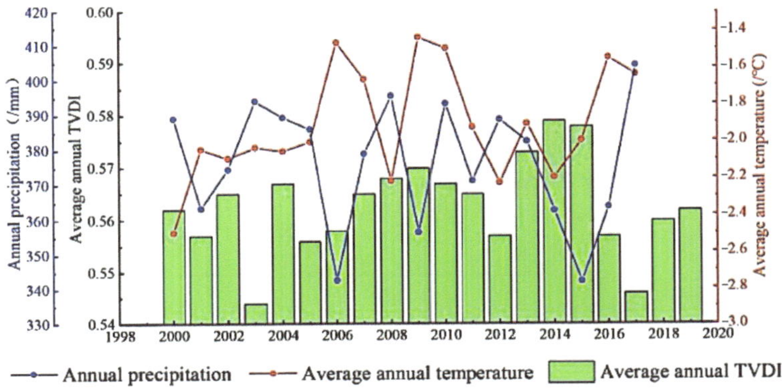

Figure 11. Changes in annual precipitation, average annual temperature and average annual TVDI of the QTP.

There are regional differences in the effects of precipitation and temperature on the TVDI of the QTP. The influence of meteorological factors on TVDI was quantitatively analyzed using the partial correlation coefficient of TVDI, annual precipitation, and annual average temperature obtained by pixel-by-pixel calculation. Figure 12a shows that the partial correlation coefficients between TVDI and annual precipitation in the QTP ranged from −0.97 to 0.93. The areas where TVDI were positively correlated with annual precipitation accounted for 53.56% of the total area of the whole region, which were distributed in the HMLY, STR, and QHB. The areas with a negative correlation accounted for 46.44% of the total area, which were mostly distributed in the QP and HDM. According to the spatial characteristics of partial correlation between TVDI and average annual temperature in the QTP, Figure 12b shows that the partial correlation coefficient between TVDI and temperature ranged from −0.89 to 0.98. Among them, the areas with a positive correlation accounted for 64.06% of the entire region, and most of them were distributed in the QP, STR, and the east wing of the HMLY, KWKM and CEKM. The areas with a negative correlation accounted for 35.94% of the area and were mainly distributed in the west wing of the HMLY and HDM.

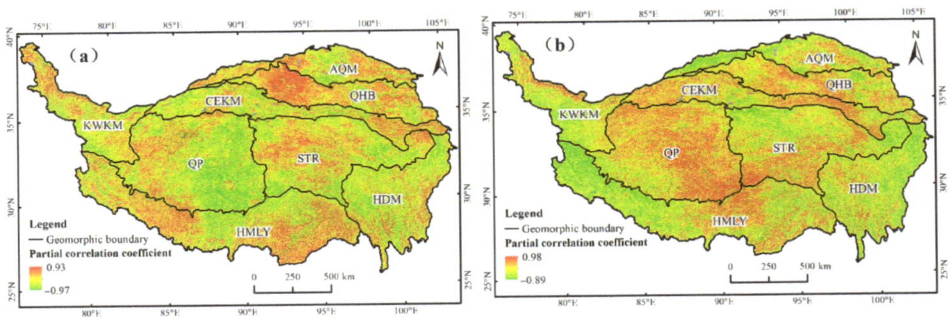

Figure 12. The distribution of partial correlation coefficients between TVDI and annual precipitation (**a**) and annual average temperature (**b**).

After calculating the *t*-test of the partial correlation coefficient and the F-test of the multiple correlation coefficient, according to Table 1, the driving force of the climatic factor of the QTP was obtained (Figure 13). The factors affecting the change in drought severity in the QTP were complex. The drought in most areas was affected by other factors. The areas driven by a single factor of temperature or precipitation were distributed in clusters, and the areas dominated by temperature and precipitation were scattered. The distribution of the areas driven by precipitation was concentrated, accounted for 1.32% of the total area, and were mainly located in the Qaidam Basin and scattered in the STR and HMLY. The land use types in these areas were mainly unused land and the vegetation coverage was low. The areas driven by temperature accounted for 3.65% of the total area of the region, which were distributed in all regions of the QTP and concentrated in areas with higher elevation, such as QP. The areas driven by precipitation and temperature account for 5.95% of the total area, which were widely distributed in various regions of the QTP, mostly in HMLY, HDM, and AQM.

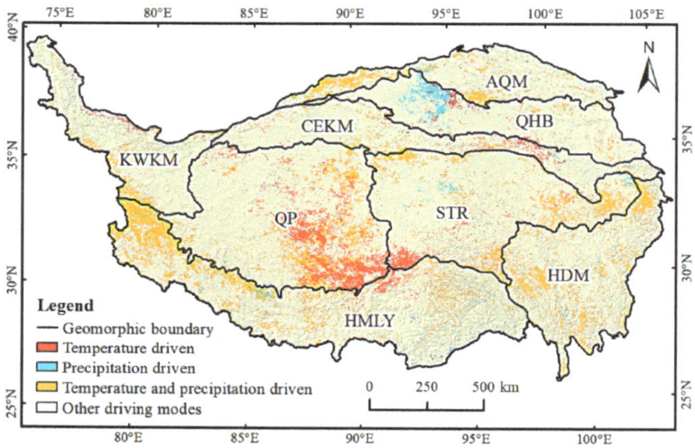

Figure 13. Climatic factor-driven zones.

4.3.2. Driver Characteristics of TVDI

Figure 14 shows that the evolution of drought in the QTP was affected by various factors, but different topography, vegetation types, and other factors led to differences in the main driving factors of each geomorphological division. The drought situation in the AQM was mainly driven by precipitation, and secondly affected by temperature. In the KWKM area, the driving effect of the Euclidean distance factor to the water surface body was more significant. The driving factors of QP drought were mainly temperature and Euclidean distance of water surface body, and other factors had less influence. The drought severity in the HMLY was driven by precipitation, followed by temperature and elevation. The drought conditions of QHB, CEKM, STR, and HDM were comprehensively affected by precipitation and temperature, and other factors had relatively little effect. In general, the influence of climatic factors (temperature and precipitation) on the drought conditions of the QTP was relatively stable, geomorphological factors (slope, aspect and elevation) played an important role in the drought conditions in each region, and accessibility factors (Euclidean distance of water surface body and Euclidean distance of built-up land) and land use factors had relatively little influence.

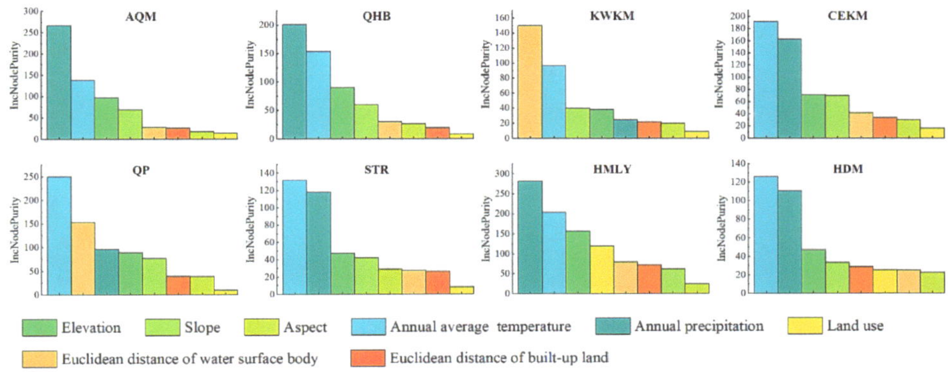

Figure 14. Ranking of the importance of factors in different geomorphological divisions of the QTP.

According to the random forest model, we obtained a local dependence map of the driving factors in each geomorphological division (Figure 15), to further explore the relationship between each driving factor and TVDI. We found that the overall trends of the relative importance of driving factors in each region were similar, but there were obvious differences in the sensitivity intervals of the driving factors.

First, the relative importance of annual average temperature increased between −20 °C and 20 °C, and maintained a high level from 0 °C to 16 °C. The relative importance of HDM peaked at 16 °C, and the relative importance of other regions peaked from 0 °C to 5 °C.

Second, the relative importance of annual precipitation fluctuated and decreased from 0 mm to 1500 mm, and the impact on regional drought was more profound when the precipitation was less than 500 mm. The trend of the relative importance of the QHB and CEKM showed a double peak, and the trend of the relative importance of other regions had a single peak.

Third, the relative importance of elevation had a fluctuating and negative trend overall, but there were obvious differences in the sensitive range of elevation in each area. The relative importance of HMLY elevation had two peaks, about 0 m and 4700 m, respectively, and the relative importance of elevation in other areas ranged from 3000 m to 5200 m.

Fourth, the relative importance of the slope changed similarly in different areas, with an upward trend from 0° to 5°, a downward trend from 5° to 30°, and remained stable above 20°.

Fifth, the relative importance of the aspect had a clear peak, rising first and then falling between 0° and 360°, and had a high level from 175° to 200° (sunny slope).

Sixth, the relative importance of the Euclidean distance of built-up land did not change significantly in the HMLY, KWKM, and QHB, and the relative importance of other regions had a downward trend. The Euclidean distance of built-up land was the most sensitive from 0 m to 1000 m. Seventh, the relative importance of the Euclidean distance of water surface body was on the rise, at a high level from 1600 m to 3000 m, and remained stable above 3000 m. Finally, the relative importance of land use types varied significantly among different types, with grassland and cultivated land generally being the highest, built-up land and other land use being higher, and the forest and water surface body being the lowest.

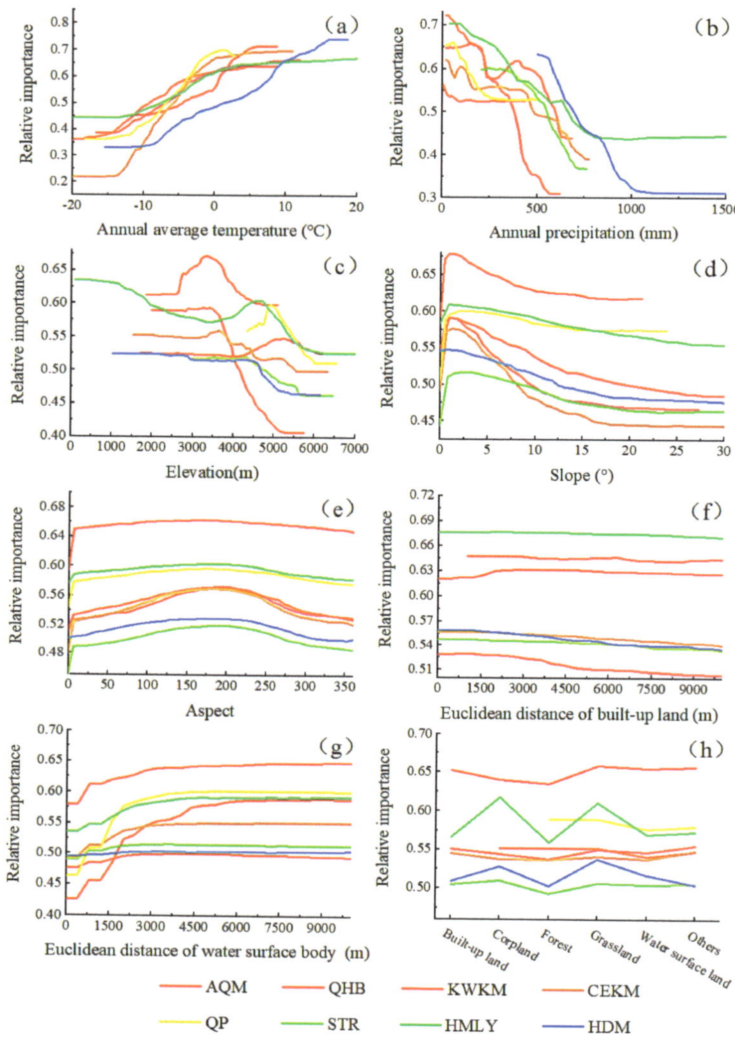

Figure 15. Relative importance of driving factors in different geomorphic regions. (**a**) Annual average temperature; (**b**) Annual precipitation; (**c**) Elevation; (**d**) Slope; (**e**) Aspect; (**f**) Euclidean distance of built-up land; (**g**) Euclidean distance of water surface body; (**h**) Land use and Land cover.

5. Discussion

5.1. Analysis of NDVI-LST Feature Space

Considering the characteristics of LST and NDVI data in the study area, determining a reasonable range of NDVI for regression fitting and obtaining the appropriate dry- and wet-edge equation was the key for calculating TVDI. When the vegetation coverage is less than 15%, NDVI is a poor indicator of vegetation cover in the region [60]. Moreover, when determining the NDVI range in the feature space, the corresponding LST minimum value in the low-value area of NDVI has a tail decline phenomenon [61]. The above two phenomena may lead to the deviation of the simulated dry and wet edge from the theoretical boundary. Therefore, when determining the NDVI range of the feature space in our study, to avoid regional differences and the occurrence of tail subsidence, except for the QP, we did not use

NDVI values < 0.2 in the other seven regions, which ensured that the dry and wet edges of each region conformed to the theoretical boundary.

Combined with Figure 4 and Appendix B Table A3, it can be concluded that the feature spaces in different phases and different regions were triangular. Thus, as NDVI increased, the maximum and minimum values of LST gradually approached and converged at one point. The overall characteristics of the data in the study area were consistent, so the NDVI-LST relationship could be used to accurately simulate the drought state [62]. The fitting of the dry and wet edges in the eight geomorphological regions were all good, which indicated that the NDVI-LST feature space of each region can accurately reflect the LST distribution and NDVI characteristics. Among them, the dry- and wet-edge coefficients (R^2) of the QHB, QP, and HMLY were the most stable and had the best fit, while the dry- and wet-edge coefficients (R^2) of the HDM fluctuated greatly and the fit was relatively poor, which may be related to deep-cut landforms and higher vegetation coverage [23].

5.2. Analysis of Spatial and Temporal Variation Trend of TVDI

The overall spatial distribution of the QTP was drought in the southwest and north, and wet in the northeast and south, which was consistent with previous research results [31]. The QTP was drier in spring and summer, and wetter in autumn and winter. In spring and summer, the western part of the QTP was relatively dry, which was caused by the vegetation in the growing season, a large demand for water, high temperature, high evapotranspiration, and less precipitation in the region. In autumn, with the decrease of temperature and water evaporation, the water required for vegetation growth decreases, and snowfall occurs in high-elevation areas, which replenishes water resources, increasing the soil water content in the central part of the QTP, thus alleviating the drought. In winter, under the combined influence of land-source water vapor and ocean-source water vapor, the central region of the QTP is obviously humid [63]. In addition, it is also affected by factors such as the decrease of temperature, vegetation dormancy, the expansion of snow cover area and the increase of snow depth [64].

In the past 20 years, the degree of drought in more than half of the QTP has increased significantly, and the central QTP has become the center of aggravated drought. Other studies have demonstrated similar findings [65,66]. The decrease in precipitation and increase in evapotranspiration in the QTP decreased soil moisture and caused a drying trend [67]. The QTP demonstrates a drying trend in other seasons except autumn, which was also concluded by Feng et al. [35] using the SPEI method. As the seasons changed, the drought risk varied in different regions, such as AQM. The AQM was prone to spring drought and combined with the drying trend in the region, the risk of drought was further increased.

Affected by the climatic gradient, the drought pattern of the QTP generally had a southeast–northwest gradient. According to the statistical characteristics of TVDI in eight geomorphological areas, the QTP could be divided into three categories by cluster analysis. The first category included the HMLY and HDM, which had a low degree of drought and a wetting trend. The second category included the QP, STR, and AQM, which had a low degree of drought and a drying trend. The third category included the KWKM, CEKM, and QHB, which had a higher degree of drought and an increased risk of drought. The drought severity and variation characteristics of these three regions were different, which was not only related to differences in precipitation, temperature, elevation, land use type, and vegetation cover, but also affected by significant geomorphological effects. As the Yangtze River, Yellow River, Yarlung Zangbo River, and other rivers flow through the HMLY and HDM areas, there was abundant precipitation, sufficient water resources and high soil humidity. The QP, STR, and AQM are located in inland areas, with high elevation, low temperature, and less precipitation. However, lakes, glaciers, and other water bodies have large areas, so water resources are more abundant and the regional drought is low. The KWKM, CEKM, and QHB are located deep inland and receive little precipitation. The land use type is dominated by grassland and unused land and drought conditions are susceptible to climate change.

5.3. Analysis of Driving Force of TVDI Change

The change of TVDI in the QTP was mainly affected by climate factors, but there were slight differences in the driving factors of TVDI in each geomorphological division. The AQM and HMLY were dominated by precipitation, the QP was dominated by temperature, and the QHB, CEKM, STR, and HDM were affected by temperature and precipitation. The driving factor of TVDI in the KWKM was the Euclidean distance of water surface body, which was due to the wide distribution of glaciers and lakes affected by monsoon [68]. The leading factors of TVDI in all regions of the QTP echo the results of climate drivers by partial correlation analysis, which also indirectly showed the reliability of the random forest method. In addition, the random forest results of TVDI data obtained by direct inversion without geomorphological division also showed that the QTP was dominated by climate factors, followed by geomorphological characteristics, which was consistent with the results we obtained.

5.3.1. Meteorological Factors

There were various factors affecting drought in the QTP, among which precipitation and temperature played a significant role [69,70]. Precipitation affects water resources, and temperature affects evapotranspiration, which interact and jointly affect drought in the QTP. In recent years, the decrease of precipitation and the increase of temperature in the QTP have changed the soil moisture [71], which has affected the growth of vegetation, weakened the ability of soil to retain water, and led to increased drought in the QTP. Among them, increased temperature has increased snow melt and evapotranspiration in the study area [41], which is an important reason for the spatial-temporal change of drought in the QTP [72]. Combined with analysis of relative importance, there were large spatial differences in the relationship between TVDI and climate factors in the QTP. When the annual average temperature was greater than 0 °C, the relationship between the TVDI and the annual average temperature of the QTP was the most sensitive. When the annual precipitation was less than 500 m, this factor had the most important impact on the change of TVDI. The areas with high temperature and low precipitation are in the western parts of HMLY, the southern parts of QP, and the northwestern parts of KWKM and the QHB. Due to the combined influence of precipitation and temperature, these regions are prone to drought and extreme drought.

Climate change largely affects the duration of drought [73]. A warming climate will increase the risk of drought [74], especially the frequency and duration of drought in arid areas [75]. The reduction of regional precipitation will lead to stronger and more frequent drying periods and the intensification of evaporation induced by global warming, and this situation will increase the probability of drought in the region [76]. From the climate driven zone map, it was found that drought in the central QTP was dominated by temperature, and drought in the northern basin was driven by precipitation, of which temperature was more driven than precipitation. The Qaidam Basin is located deep within the interior of the continent and is surrounded by high mountains, which affects the moist air flow into the basin, resulting in less precipitation in the region. The TVDI in QHB was dominated by precipitation and had a positive correlation with the annual precipitation. With decreased precipitation, the drought in this area increased. The drought in QP was dominated by temperature, and there was a positive correlation between TVDI and temperature in this area. Affected by the temperature rise of the QTP in the recent 20 years, QP has become the central area with an enhanced drying trend.

5.3.2. Geomorphological Factors

Geomorphological characteristics are important factors that affect the construction of regional basic factors such as mountains, climate, vegetation, roads, and human activities. The uplift of the QTP has not only formed a unique plateau climate [77], but also had a profound impact on the composition of modern atmospheric circulation patterns, the establishment of the Asian monsoon system [78], regional and global climate change, the development and evolution of desert loess, plateau ice and snow, lake and river water systems and ecosystems, and the origin and evolution of human beings [79]. Among them, the thermal effect of the QTP has an impact on the Asian monsoon and precipitation variability [80], resulting in regional differentiation of precipitation and heat generation [81], which directly or indirectly cause drought, water shortage, and deserts. The distribution of water and heat in the region affected by elevation results in a gradient of drought conditions in the QTP [82], especially in HMLY. The HMLY Valley area is blocked by the southern high mountains to the water vapor from the North Indian Ocean, resulting in a large amount of precipitation falling on the windward slope, and the area has been in a dry state for a long time. At elevations between 3000 m and 5200 m, the impact on TVDI is most significant, with rainfall and temperature changing positively in the high-altitude range [35].

Slope affects soil properties, soil nutrient content, and vegetation types to a certain extent. With the increase of slope, precipitation infiltration is difficult, and the loss rate of soil water intensifies, which further affects the likelihood of drought in the region [83]. The smaller the slope, the better the soil moisture retention. Slopes less than 5° were the most sensitive to TVDI in the study area. Such areas are mainly distributed in the central and northern QTP, with high elevation and local precipitation less than 500 mm, and slope had a low degree of influence on TVDI.

Aspect affects the amount of solar radiation, water content, and evapotranspiration received by the slope, resulting in aspect-related differences in drought conditions [40]. When located on a sunny slope, TVDI was most sensitive to aspect. Compared with the shady slope, the sunny slopes have longer sunshine hours, receive more solar radiation, and better develop vegetation, which increases the regional evapotranspiration intensity, reduces soil water content, and causes drought differentiation. Compared with factors such as temperature, precipitation, and elevation, the aspect had less influence on the TVDI, which indicated that the humidity change caused by the aspect had a relatively small impact on the regional drought change.

5.3.3. Accessibility Factors

With economic development and the expansion of urban land, the impact of human activities on regional drought has increased [84–87] and the impact of human activities on drought is greater than the duration of drought [88]. There are two aspects to the impact of human activities on drought. First, human activities directly act on surface water and groundwater resources, consume and change the occurrence and regeneration conditions of water resources, and cause or exacerbate drought [89]. Second, due to urban and rural construction, traffic route development, deforestation, wasteland removal, and other activities that change the land surface, the soil moisture and roughness are reduced and the surface reflectivity of the underlying surface is increased, thereby changing the production and confluence law and evaporation composition, which ultimately leads to localized reductions in precipitation and increased drought. In terms of the accessibility impact of cities and towns, the closer to cities and towns, the more intense human activities become and the greater their effects on landform and surface cover. Therefore, the impact of the Euclidean distance of built-up land demonstrated significant distance attenuation, and its sensitive range was mostly between 0 m to 1000 m. Due to the relatively small proportion of built-up land in the QTP, it had little impact on the overall changes in drought on the QTP.

There are many frozen soils, glaciers, and lakes in the QTP, especially in KWKM and QP. Soil moisture in the areas surrounding rivers and water bodies gradually declines as one moves further away from the water bodies, which causes a wet to arid trend as the distance from water bodies increases. Under the background of global warming and the imbalance of water towers in Asia, the glaciers in the QTP are in a state of continuous melting, their solid water is melting rapidly, and their liquid water is increasing [90]. This affects the hydrological process in the study area and relieves the drought pressure in some areas [91]. In our study, the Euclidean distance to water surface body was used to analyze the impact of glaciers, lakes, and other waters on TVDI. We found that the Euclidean distance of water surface body had the characteristics of distance attenuation on the impact of regional drought. When it was greater than 1600 m, the relationship between TVDI and this factor was the most sensitive. In addition, the formation and melting of glaciers and permafrost produce obvious seasonal differences in the drought situation of the QTP. In summer, the QTP had high temperature, high evaporation, and a large demand for water by vegetation, but its drought degree was less affected by glacier and frozen soil melting than in spring [92]. In winter, the soil of the QTP gradually freezes, the regional evapotranspiration decreases, and the soil water holding capacity increases, making winter the wettest season of the year. In the short term, glacier melting will alleviate drought in some areas, but long-term glacier melting will reduce the ability of arid areas to cope with drought and increase the risk of drought.

5.3.4. Land Use Type Factor

Differences in vegetation cover and surface roughness of different land use types lead to differences in their impact on TVDI [93]. The results of local dependence analysis showed that grassland and cultivated land had the greatest impact on TVDI, followed by urban and rural built-up land and other land, and that the forest and water bodies had the lowest impact. The crops in the QTP are mainly highland barley, and the sowing and growing season spans from March to September. The cultivated land is generally adjacent to built-up land and is greatly affected by human activities [94]. The ecosystem stability of cultivated land and built-up land is relatively weak, and the soil water holding capacity is weak, which affects the change of TVDI [95]. Forest is mainly distributed in the southeastern QTP, which has rich water resources, good soil coverage, less evaporation, strong soil water holding capacity, and non-significant impact on TVDI [96]. The water area includes lakes, permanent glaciers, reservoirs, ponds, beaches, and other water bodies, which are humid and have low drought risk. Since the water changes to the surrounding area, the soil moisture gradually decreases with a trend from wet to drought. Grassland and unused land are the most widely distributed in the study area and are easily affected by climate fluctuations, have relatively low vegetation coverage, poor stability, and strong ground evaporation, and thus have a more significant impact on TVDI.

5.4. Applications and Limitations

As an important ecological security barrier in China, the QTP plays a role in water conservation, biodiversity protection, and soil and water conservation. It is a 'sensor' of climate change in Asia and the Northern Hemisphere. Drought affects regional vegetation status, the water cycle, and human life, and can thereby destroy the stability of the ecological environment and restrict social and economic development. Using geomorphological division, we applied the TVDI method to identify the temporal and spatial variation of drought in the QTP from 2000 to 2019 in an effort to provide a scientific basis for the prevention and control of regional drought and promote sustainable development of the region. Under the background of global warming, we found that the QTP had a tendency of drought, and the central QTP has become the hotspot of increased drought severity. In response to the potential challenges of drought risk, the government should actively adjust and optimize drought prevention and control measures. In areas with low drought risk, such as HDM and HMLY, the government should strengthen the ecological protection of cultivated land and important areas such as rivers, lakes, and surrounding areas, control the intensity of human reclamation activities, and closely monitor changes in the hydrological conditions of grasses. In areas with perennial droughts, such as QHB, KWKM, and CEKM, where the drying trend was aggravated, a multi-geographical synchronous monitoring system should be established to identify and monitor drought dynamics. In sandy and bare areas, ecosystems need to be restored. In agricultural farming areas, the planting structure needs to be optimized, and high-efficiency, water-saving agricultural methods need to be developed. In pastoral areas, the scale and quantity of grazing should be controlled, and the protection of grasslands should be strengthened by taking measures such as zoning and rotation grazing.

There were some aspects of our study that could be improved. In terms of the TVDI calculation, the TVDI of each area was calculated based on the geomorphological division, and then the TVDI of the whole area was obtained by mosaic. However, this method led to discontinuity of the TVDI value at the geomorphological edge, or the edge effect. Although we used the TVDI classification to weaken the edge effect, the moving window method can be tested to eliminate the edge effect in the future. In terms of index selection, the effects of natural factors, such as stratigraphic lithology, geological structure, surface water, and groundwater have not been fully considered, and the impact on human activities has not been adequately described. Future research can further explore the mechanisms of drought in the QTP and divide the regional drought risks based on drought severity and drying trends. Methods such as GWR can be used in comparison with or combined with RF methods to improve the accuracy of model interpretation.

6. Conclusions

Here, we selected the QTP as the study area. However, considering the influence of the huge spatial scale effects of the study area on TVDI inversion, we divided the QTP into eight geomorphological areas. TVDI inversion was carried out for each area and the data for the whole area were synthesized using the TVDI grading method. Partial correlation and multiple correlation analyses were used to explore the effects of precipitation and temperature on TVDI. Then, eight factors, including meteorology, geomorphology, land cover, and human activities, were selected to construct a random forest model to identify the mechanisms driving changes in TVDI. We gathered four main findings.

First, in the past 20 years drought frequency and severity in the QTP has increased, especially in the central region. The spatial distribution of TVDI was wet in the southeast and arid in the northwest, with conspicuous regional differences. The main areas with increasing TVDI were distributed in the southeast of QP, the southwest of STR, and the Kunlun Mountains. The main areas where TVDI remained unchanged and decreased were distributed in the HDM and HMLY where vegetation cover was high.

Second, the seasonal differences of drought were significant in the QTP. Spring was the driest season of the year, and there was a significant drying trend in the central and northern QTP. It was relatively dry in summer, the drought risk in the north of QTP was weakened, and the risk in the south was still very high. Autumn was relatively wet, and the region had a downward trend in drought severity. Winter was the wettest season of the year, but the risk of drought increased significantly in the northern part of the QTP.

Third, the spatial pattern of TVDI changes in the QTP was generally consistent with changes in precipitation and temperature, which had a gradual change from southeast to northwest. According to the spatial distribution and variation of aridity, we divided the study area into three categories. The first category, the HMLY and HDM in the southern region, had low drought severity and was becoming more humid. The second category, the QP, STR, and AQM in the central and northern regions, had low drought severity but had a positive trend in drought severity. The third category, the KWKM, CEKM and QHB in the north-central region, had higher drought severity and a positive trend in drought severity.

Finally, the driving effect of each factor on TVDI was significantly different in the QTP. Except in the KWKM, temperature and precipitation were the dominant factors in the variation of TVDI in the QTP, and regions with higher temperature and less precipitation had a more significant impact on TVDI. Geomorphological factors play an important role in the change of TVDI. The areas with an elevation of 3000~5200 m, slopes less than 5°, and sun-facing slopes had a more significant impact on the change of TVDI. The influence of the Euclidean distance of urban and rural built-up land and water bodies showed distance attenuation, and the sensitive areas of these two factors were <1000 m and >600 m, respectively. Among the land use types, grassland and cultivated land had a more significant impact on TVDI changes, while forest and water bodies had less impact on TVDI changes.

Author Contributions: Z.N. conceived the idea of this article and modified the manuscript, Y.L. (Yi Liu) introduced the latest technology and conducted experiments, Y.Z. explained the results and suggested revisions. G.Z., Y.L. (Yuhao Luo), S.L., D.W. and S.Z. assisted in revising the technical routes, manuscript writing issues, and provided drawing advice. All authors have read and agreed to the published version of the manuscript.

Funding: The 2021 Key Project Open Project of the Qinghai–Tibet Plateau and its Eastern Edge Human Geography Research Center, grant number RWDL2021-ZD002, the Key Research Project of National Park Research Center, grant number GJGY 2019-ZD003, the Young and Middle-aged Teacher Program of Chengdu University of Technology, grant number 10912-JXGG2021-01679, and the China Scholarship Council Fund, grant number 202008515091 supported this work.

Institutional Review Board Statement: Not applicable.

Informed Consent Statement: Not applicable.

Data Availability Statement: All data generated or analyzed during this study are included in this article.

Acknowledgments: We would like to thank Russell Doughty for the English language editing of the manuscript. Moreover, we thank the Qinghai–Tibet Plateau Science Data Center (http://data.tpdc.ac.cn, accessed on 21 April 2021) for providing open data.

Conflicts of Interest: The authors declare no conflict of interest.

Appendix A

Table A1. Full name of abbreviation in the text.

Abbreviation	Full Name
TVDI	Temperature-Vegetation Drought Index
QTP	The Qinghai–Tibet Plateau
AQM	The high valley areas of Altun–Qilian Mountains
QHB	The high mountain areas of Qaidam-Yellow River-Huangshui River Basin
KWKM	The high mountain areas of Karakorum and Western Kunlun Mountain
CEKM	The high mountain areas of Central and Eastern Kunlun Mountains
QP	The lake and basin areas of Qiangtang Plateau
STR	The mountains Sources of the Yangtze River, Yellow River, and Lancang River (Three Rivers or Sanjiangyuan) and the valley bottom of the upper reaches of Three Rivers.
HMLY	The high mountain areas of Himalayan
HDM	The high mountain and valley areas of Hengduan Mountains

Appendix B

Table A2. Basic information of each geomorphological division.

Geomorphological Division	Area (km^2)	Annual Average of TVDI	Annual Average of Temperature (°C)	Annual Precipitation (mm)	Average Elevation (m)	Average Slope (°)
AQM	182,475.56	0.55	−4.06	260.01	3768.06	5.55
QHB	231,433.86	0.66	2.18	250.69	3158.97	2.51
KWKM	207,715.72	0.55	−5.62	42.70	4765.07	8.70
CEKM	237,445.70	0.54	−4.40	262.23	4387.32	4.82
QP	511,338.20	0.59	−4.09	200.31	4988.01	2.70
STR	420,819.40	0.50	−3.59	513.88	4570.68	4.23
HMLY	513,158.23	0.59	−0.29	528.40	4492.26	8.99
HDM	316,219.62	0.52	2.07	699.36	3934.06	10.83

Table A3. Dry-edge and wet-edge equations of eight geomorphological divisions from 2000 to 2019.

Geomorphological Division	Year	Dry-Edge Equation	R^2	Wet-Edge Equation	R^2
AQM	2000	Dy = 46.4 − 9.89*X	0.27	Wy = 11.7 + 11.3*X	0.50
	2001	Dy = 48.6 − 15.2*X	0.68	Wy = 13.8 + 8.89*X	0.81
	2002	Dy = 46.0 − 15.9*X	0.78	Wy = 11.7 + 9.81*X	0.79
	2003	Dy = 43.3 − 6.96*X	0.35	Wy = 11.1 + 11.8*X	0.85
	2004	Dy = 46.5 − 12.8*X	0.69	Wy = 11.4 + 11.1*X	0.57
	2005	Dy = 46.6 − 12.6*X	0.81	Wy = 8.89 + 14.2*X	0.41
	2006	Dy = 49.3 − 19.0*X	0.93	Wy = 12.1 + 11.7*X	0.75
	2007	Dy = 50.4 − 18.7*X	0.70	Wy = 10.2 + 14.3*X	0.59
	2008	Dy = 49.3 − 21.7*X	0.78	Wy = 11.0 + 11.6*X	0.66
	2009	Dy = 42.5 − 6.92*X	0.27	Wy = 10.8 + 12.6*X	0.66
	2010	Dy = 48.4 − 16.8*X	0.85	Wy = 9.60 + 15.5*X	0.59
	2011	Dy = 43.4 − 10.1*X	0.60	Wy = 8.41 + 16.7*X	0.70
	2012	Dy = 41.8 − 6.87*X	0.39	Wy = 8.60 + 14.1*X	0.71
	2013	Dy = 47.2 − 15.2*X	0.62	Wy = 10.9 + 12.9*X	0.58
	2014	Dy = 43.8 − 8.21*X	0.45	Wy = 11.5 + 10.5*X	0.59
	2015	Dy = 47.1 − 15.9*X	0.63	Wy = 10.6 + 11.1*X	0.88
	2016	Dy = 53.1 − 20.1*X	0.92	Wy = 14.5 + 8.93*X	0.52
	2017	Dy = 50.1 − 22.7*X	0.82	Wy = 8.76 + 15.5*X	0.67
	2018	Dy = 48.5 − 18.5*X	0.75	Wy = 8.97 + 12.7*X	0.59
	2019	Dy = 45.2 − 12.1*X	0.69	Wy = 7.19 + 14.4*X	0.54
QHB	2000	Dy = 60.9 − 29.2*X	0.78	Wy = 12.8 + 7.97*X	0.52
	2001	Dy = 57.7 − 25.7*X	0.69	Wy = 10.9 + 12.4*X	0.78
	2002	Dy = 55.7 − 27.4*X	0.92	Wy = 11.3 + 10.0*X	0.57
	2003	Dy = 54.3 − 17.7*X	0.59	Wy = 12.4 + 9.58*X	0.81
	2004	Dy = 53.4 − 20.7*X	0.83	Wy = 10.9 + 11.2*X	0.90
	2005	Dy = 56.0 − 27.9*X	0.92	Wy = 10.6 + 10.5*X	0.80
	2006	Dy = 57.2 − 28.2*X	0.86	Wy = 14.6 + 7.29*X	0.74
	2007	Dy = 59.0 − 30.4*X	0.79	Wy = 12.1 + 9.59*X	0.78
	2008	Dy = 55.9 − 24.2*X	0.77	Wy = 12.5 + 9.22*X	0.71
	2009	Dy = 54.5 − 23.6*X	0.80	Wy = 12.5 + 8.88*X	0.79
	2010	Dy = 53.9 − 22.1*X	0.96	Wy = 13.0 + 5.95*X	0.14
	2011	Dy = 56.4 − 26.5*X	0.86	Wy = 11.3 + 11.7*X	0.73
	2012	Dy = 53.5 − 23.4*X	0.92	Wy = 9.76 + 12.6*X	0.91
	2013	Dy = 56.5 − 27.4*X	0.87	Wy = 14.4 + 7.99*X	0.58
	2014	Dy = 57.4 − 22.8*X	0.57	Wy = 12.4 + 9.00*X	0.61
	2015	Dy = 57.6 − 28.3*X	0.84	Wy = 12.4 + 8.87*X	0.73
	2016	Dy = 59.6 − 29.1*X	0.86	Wy = 15.2 + 6.56*X	0.63
	2017	Dy = 56.5 − 25.8*X	0.82	Wy = 11.6 + 10.0*X	0.84
	2018	Dy = 55.0 − 24.4*X	0.85	Wy = 10.3 + 9.28*X	0.75
	2019	Dy = 56.1 − 28.3*X	0.91	Wy = 12.8 + 3.60*X	0.45
KWKM	2000	Dy = 55.9 − 44.6*X	0.75	Wy = −4.8 + 36.0*X	0.34
	2001	Dy = 55.8 − 46.5*X	0.73	Wy = −6.4 + 44.8*X	0.38
	2002	Dy = 53.3 − 38.6*X	0.78	Wy = −5.7 + 41.0*X	0.44
	2003	Dy = 56.2 − 44.7*X	0.79	Wy = −5.9 + 39.0*X	0.41
	2004	Dy = 54.6 − 43.4*X	0.79	Wy = −0.2 + 19.6*X	0.11
	2005	Dy = 51.8 − 38.2*X	0.77	Wy = −7.4 + 38.5*X	0.44
	2006	Dy = 56.1 − 45.8*X	0.72	Wy = −0.8 + 31.9*X	0.39
	2007	Dy = 53.0 − 37.2*X	0.74	Wy = −0.6 + 37.0*X	0.34
	2008	Dy = 52.1 − 36.2*X	0.77	Wy = −3.1 + 36.3*X	0.42
	2009	Dy = 54.3 − 43.4*X	0.70	Wy = −4.7 + 40.6*X	0.39
	2010	Dy = 52.8 − 40.6*X	0.82	Wy = −6.5 + 32.5*X	0.41
	2011	Dy = 55.5 − 40.6*X	0.78	Wy = −2.8 + 31.4*X	0.28
	2012	Dy = 56.2 − 45.1*X	0.83	Wy = −7.1 + 33.4*X	0.46
	2013	Dy = 55.6 − 39.3*X	0.76	Wy−−3.1 + 35.8*X	0.31
	2014	Dy = 53.5 − 38.3*X	0.79	Wy = −8.3 + 45.4*X	0.51
	2015	Dy = 53.2 − 38.2*X	0.79	Wy = −7.3 + 35.4*X	0.33
	2016	Dy = 54.9 − 38.9*X	0.81	Wy = −8.9 + 45.0*X	0.50
	2017	Dy = 52.0 − 31.3*X	0.77	Wy = −2.9 + 27.2*X	0.31
	2018	Dy = 55.7 − 37.0*X	0.76	Wy = −9.2 + 43.6*X	0.53
	2019	Dy = 54.4 − 38.8*X	0.70	Wy = −6.0 + 32.0*X	0.40

Table A3. *Cont.*

Geomorphological Division	Year	Dry-Edge Equation	R^2	Wet-Edge Equation	R^2
CEKM	2000	Dy = 49.7 − 17.7*X	0.65	Wy = 10.0 + 11.4*X	0.59
	2001	Dy = 46.4 − 17.7*X	0.71	Wy = 11.6 + 8.75*X	0.67
	2002	Dy = 44.7 − 13.8*X	0.77	Wy = 9.82 + 11.6*X	0.79
	2003	Dy = 42.1 − 2.04*X	0.03	Wy = 9.56 + 11.1*X	0.77
	2004	Dy = 42.1 − 8.66*X	0.70	Wy = 10.4 + 9.08*X	0.65
	2005	Dy = 44.2 − 17.2*X	0.82	Wy = 10.9 + 6.97*X	0.39
	2006	Dy = 47.9 − 18.9*X	0.76	Wy = 11.4 + 10.5*X	0.71
	2007	Dy = 46.0 − 16.1*X	0.61	Wy = 10.5 + 9.32*X	0.71
	2008	Dy = 42.8 − 9.09*X	0.37	Wy = 10.8 + 9.99*X	0.74
	2009	Dy = 43.4 − 14.3*X	0.69	Wy = 5.75 + 14.1*X	0.59
	2010	Dy = 44.0 − 12.7*X	0.71	Wy = 7.84 + 7.88*X	0.23
	2011	Dy = 43.2 − 12.6*X	0.79	Wy = 10.2 + 10.6*X	0.74
	2012	Dy = 45.4 − 18.0*X	0.88	Wy = 8.15 + 10.6*X	0.67
	2013	Dy = 47.3 − 19.0*X	0.83	Wy = 10.4 + 10.9*X	0.67
	2014	Dy = 45.4 − 10.4*X	0.26	Wy = 8.99 + 11.8*X	0.67
	2015	Dy = 46.1 − 15.4*X	0.69	Wy = 11.1 + 9.57*X	0.72
	2016	Dy = 46.9 − 14.6*X	0.71	Wy = 11.5 + 10.0*X	0.55
	2017	Dy = 46.8 − 17.2*X	0.82	Wy = 8.92 + 10.2*X	0.62
	2018	Dy = 43.5 − 14.3*X	0.81	Wy = 7.32 + 12.1*X	0.76
	2019	Dy = 44.1 − 15.7*X	0.76	Wy = 6.38 + 10.5*X	0.52
QP	2000	Dy = 49.9 − 30.3*X	0.82	Wy = 6.68 + 14.1*X	0.61
	2001	Dy = 47.3 − 29.5*X	0.87	Wy = 6.81 + 16.8*X	0.71
	2002	Dy = 49.1 − 26.3*X	0.84	Wy = 7.04 + 16.5*X	0.69
	2003	Dy = 48.3 − 30.0*X	0.88	Wy = 5.93 + 17.1*X	0.78
	2004	Dy = 46.9 − 23.7*X	0.84	Wy = 7.46 + 15.2*X	0.68
	2005	Dy = 49.4 − 28.8*X	0.86	Wy = 6.84 + 17.3*X	0.69
	2006	Dy = 49.9 − 30.5*X	0.89	Wy = 6.90 + 17.9*X	0.66
	2007	Dy = 4.9 − 21.5*X	0.65	Wy = 6.17 + 18.2*X	0.64
	2008	Dy = 46.2 − 28.5*X	0.87	Wy = 5.29 + 17.0*X	0.74
	2009	Dy = 50.5 − 26.2*X	0.80	Wy = 7.67 + 14.1*X	0.58
	2010	Dy = 50.5 − 27.6*X	0.75	Wy = 2.17 + 22.1*X	0.68
	2011	Dy = 47.8 − 30.3*X	0.86	Wy = 6.33 + 15.7*X	0.65
	2012	Dy = 50.7 − 26.7*X	0.75	Wy = 5.47 + 17.7*X	0.65
	2013	Dy = 51.5 − 33.1*X	0.85	Wy = 8.32 + 12.4*X	0.58
	2014	Dy = 48.8 − 23.1*X	0.65	Wy = 7.07 + 12.7*X	0.57
	2015	Dy = 49.7 − 30.0*X	0.78	Wy = 7.17 + 16.5*X	0.61
	2016	Dy = 49.8 − 30.3*X	0.80	Wy = 6.71 + 14.5*X	0.71
	2017	Dy = 45.4 − 21.7*X	0.61	Wy = 6.75 + 12.9*X	0.64
	2018	Dy = 52.6 − 28.5*X	0.62	Wy = 8.08 + 7.84*X	0.67
	2019	Dy = 50.9 − 35.6*X	0.73	Wy = 8.23 + 6.56*X	0.62
STR	2000	Dy = 43.9 − 6.73*X	0.33	Wy = 4.06 + 17.0*X	0.72
	2001	Dy = 42.0 − 8.80*X	0.67	Wy = 2.24 + 20.0*X	0.78
	2002	Dy = 43.9 − 6.73*X	0.33	Wy = 4.06 + 17.0*X	0.72
	2003	Dy = 39.8 + 3.17*X	0.07	Wy = 4.70 + 14.9*X	0.66
	2004	Dy = 42.6 − 4.75*X	0.46	Wy = 5.31 + 13.6*X	0.70
	2005	Dy = 44.1 − 13.5*X	0.87	Wy = 5.95 + 12.4*X	0.86
	2006	Dy = 42.8 − 8.80*X	0.57	Wy = 6.79 + 14.2*X	0.78
	2007	Dy = 42.7 − 2.36*X	0.03	Wy = 5.65 + 16.3*X	0.79
	2008	Dy = 37.3 + 6.57*X	0.54	Wy = 2.18 + 16.1*X	0.44
	2009	Dy = 43.8 − 8.60*X	0.56	Wy = 3.96 + 18.0*X	0.87
	2010	Dy = 42.4 − 3.37*X	0.22	Wy = 0.63 + 21.1*X	0.74
	2011	Dy = 393 − 4.56*X	0.34	Wy = 5.75 + 13.8*X	0.66
	2012	Dy = 42.6 − 9.66*X	0.71	Wy − 5.89 + 8.53*X	0.27
	2013	Dy = 44.8 − 12.2*X	0.78	Wy = 7.45 + 15.2*X	0.82
	2014	Dy = 44.1 − 9.94*X	0.42	Wy = 0.85 + 19.4*X	0.81
	2015	Dy = 43.7 − 7.17*X	0.70	Wy − 6.60 + 14.8*X	0.75
	2016	Dy = 47.8 − 11.3*X	0.69	Wy = 3.75 + 19.6*X	0.79
	2017	Dy = 45.3 − 11.4*X	0.69	Wy = 5.24 + 14.2*X	0.75
	2018	Dy = 45.6 − 9.84*X	0.38	Wy = 4.46 + 15.0*X	0.76
	2019	Dy = 42.6 − 9.37*X	0.65	Wy = 4.54 + 14.0*X	0.65

Table A3. *Cont.*

Geomorphological Division	Year	Dry-Edge Equation	R^2	Wet-Edge Equation	R^2
HMLY	2000	Dy = 53.1 − 20.3*X	0.82	Wy = −12.0 + 17.6*X	0.73
	2001	Dy = 51.9 − 21.6*X	0.88	Wy = −7.3 + 16.5*X	0.82
	2002	Dy = 53.5 − 22.4*X	0.92	Wy−−8.9 + 18.2*X	0.76
	2003	Dy = 51.7 − 20.2*X	0.93	Wy = −12.0 + 16.7*X	0.73
	2004	Dy = 52.0 − 19.2*X	0.87	Wy = −9.0 + 13.9*X	0.70
	2005	Dy = 52.9 − 21.5*X	0.87	Wy = −10.0 + 14.5*X	0.75
	2006	Dy = 52.0 − 20.9*X	0.92	Wy = −3.6 + 9.67*X	0.46
	2007	Dy = 56.2 − 22.6*X	0.68	Wy = −9.1 + 18.3*X	0.73
	2008	Dy = 48.9 − 16.9*X	0.92	Wy = −11.0 + 18.3*X	0.76
	2009	Dy = 55.2 − 21.2*X	0.71	Wy = −9.2 + 19.0*X	0.81
	2010	Dy = 56.1 − 23.8*X	0.81	Wy = −9.0 + 14.2*X	0.71
	2011	Dy = 50.6 − 19.6*X	0.89	Wy = −8.3 + 11.9*X	0.69
	2012	Dy = 54.3 − 22.1*X	0.89	Wy = −13.0 + 17.6*X	0.73
	2013	Dy = 52.7 − 20.4*X	0.95	Wy = −4.7 + 11.6*X	0.48
	2014	Dy = 54.4 − 21.1*X	0.77	Wy = −12.0 + 19.6*X	0.83
	2015	Dy = 52.5 − 19.9*X	0.88	Wy = −8.8 + 14.3*X	0.66
	2016	Dy = 51.6 − 18.6*X	0.87	Wy = −3.1 + 10.0*X	0.41
	2017	Dy = 51.4 − 19.0*X	0.87	Wy = −5.7 + 14.3*X	0.74
	2018	Dy = 55.4 − 22.9*X	0.87	Wy = −8.7 + 15.7*X	0.83
	2019	Dy = 54.5 − 21.9*X	0.78	Wy = −11. + 18.3*X	0.84
HDM	2000	Dy = 47.6 − 14.8*X	0.43	Wy = −1.1 + 6.43*X	0.57
	2001	Dy = 48.3 − 16.9*X	0.60	Wy = 2.75 + 8.98*X	0.27
	2002	Dy = 48.3 − 15.9*X	0.50	Wy = 4.75 + 6.75*X	0.43
	2003	Dy = 51.4 − 17.0*X	0.42	Wy = −0.1 + 10.4*X	0.50
	2004	Dy − 45.1 − 12.1*X	0.38	Wy = 1.06 + 4.81*X	0.23
	2005	Dy = 51.0 − 18.1*X	0.43	Wy = 0.87 + 4.39*X	0.24
	2006	Dy = 50.3 − 18.2*X	0.59	Wy = 7.95 + 2.48*X	0.09
	2007	Dy = 47.8 − 14.3*X	0.48	Wy = −0.7 + 11.1*X	0.42
	2008	Dy = 48.2 − 16.4*X	0.50	Wy = −3.1 + 13.5*X	0.63
	2009	Dy = 49.3 − 14.4*X	0.38	Wy = 5.60 + 2.31*X	0.07
	2010	Dy = 47.6 − 14.3*X	0.37	Wy = 2.64 + 6.33*X	0.46
	2011	Dy = 47.9 − 16.4*X	0.65	Wy = 4.88 + 1.54*X	0.02
	2012	Dy = 49.6 − 13.1*X	0.26	Wy = 2.93 + 3.44*X	0.12
	2013	Dy = 49.3 − 14.5*X	0.33	Wy = 7.73 + 5.00*X	0.20
	2014	Dy = 49.1 − 12.0*X	0.24	Wy = 2.89 + 4.76*X	0.17
	2015	Dy = 53.2 − 17.9*X	0.40	Wy = 6.14 + 1.77*X	0.02
	2016	Dy = 46.8 − 12.8*X	0.30	Wy = 3.83 + 3.42*X	0.15
	2017	Dy = 48.6 − 14.3*X	0.37	Wy = 3.43 + 8.85*X	0.56
	2018	Dy = 47.7 − 12.6*X	0.30	Wy = −0.0 + 7.89*X	0.54
	2019	Dy = 48.9 − 12.0*X	0.23	Wy = 3.13 + 6.85*X	0.30

References

1. Du, L.T.; Tian, Q.J.; Yu, T.; Meng, Q.Y.; Jancso, T.; Udvardy, P.; Huang, Y. A comprehensive drought monitoring method integrating MODIS and TRMM data. *Int. J. Appl. Earth Obs.* **2013**, *23*, 245–253. [CrossRef]
2. Kogan, N.F. Application of Vegetation Index and Brightness Temperature for Drought Detection. *Adv. Space Res.* **1995**, *15*, 91–100. [CrossRef]
3. Kogan, F.N. Droughts of the late 1980s in the United-States as derived from NOAA polar-orbiying satellite data. *Bull. Am. Meteorol. Soc.* **1995**, *76*, 655–668. [CrossRef]
4. Liu, W.T.; Kogan, F.N. Monitoring regional drought using the Vegetation Condition Index. *Int. J. Remote Sens.* **1996**, *17*, 2761–2782. [CrossRef]
5. McFeeters, S.K. The use of the normalized difference water index (NDWI) in the delineation of open water features. *Int. J. Remote Sens.* **1996**, *17*, 1425–1432. [CrossRef]
6. Kogan, F. World droughts in the new millennium from AVHRR-based vegetation health indices. *Eos Trans. Am. Geophys. Union* **2002**, *83*, 557–563. [CrossRef]
7. Gonzalez, J.; Valdes, J.B. New drought frequency index: Definition and comparative performance analysis. *Water Resour. Res.* **2006**, *42*, 1–13. [CrossRef]
8. McKee, T.; Doesken, N.; Kleist, J. The relationship of drought frequency and duration to time scales. In Proceedings of the 8th Conference on AppliedClimatology, Anaheim, CA, USA, 17–22 January 1993; pp. 179–183.
9. Palmer, W.C. *Meteorological Drought*; US Department of Commerce, Weather Bureau: Washington, DC, USA, 1965; Volume 30.
10. Guttman, N.B. Comparing the Palmer Drought Index and the standardized precipitation index. *J. Am. Water Resour. Assoc.* **1998**, *34*, 113–121. [CrossRef]

11. Chen, S.; Zhang, L.; Liu, X.; Guo, M.; She, D. The Use of SPEI and TVDI to Assess Temporal-Spatial Variations in Drought Conditions in the Middle and Lower Reaches of the Yangtze River Basin, China. *Adv. Meteorol.* **2018**, *2018*, 1–11. [CrossRef]
12. Heim, R.R. A review of twentieth-century drought indices used in the United States. *Bull. Am. Meteorol. Soc.* **2002**, *83*, 1149–1165. [CrossRef]
13. Vicente-Serrano, S.M.; Begueria, S.; Lopez-Moreno, J.I. A Multiscalar Drought Index Sensitive to Global Warming: The Standardized Precipitation Evapotranspiration Index. *J. Clim.* **2010**, *23*, 1696–1718. [CrossRef]
14. Akinremi, O.O.; McGinn, S.M.; Barr, A.G. Evaluation of the Palmer Drought index on the Canadian prairies. *J. Clim.* **1996**, *9*, 897–905. [CrossRef]
15. Mu, Q.Z.; Zhao, M.S.; Kimball, J.S.; McDowell, N.G.; Running, S.W. A Remotely Sensed Global Terrestrial Drought Severity Index. *Bull. Am. Meteorol. Soc.* **2013**, *94*, 83–98. [CrossRef]
16. Prihodko, L.; Goward, S.N. Estimation of air temperature from remotely sensed surface observations. *Remote Sens. Environ.* **1997**, *60*, 335–346. [CrossRef]
17. Carlson, T.N.; Gillies, R.R.; Schmugge, T.J. An interpretation of methodologies for indirect measurement of soil-water content. *Agric. For. Meteorol.* **1995**, *77*, 191–205. [CrossRef]
18. Sandholt, I.; Rasmussen, K.; Andersen, J. A simple interpretation of the surface temperature/vegetation index space for assessment of surface moisture status. *Remote Sens. Environ.* **2002**, *79*, 213–224. [CrossRef]
19. Bai, J.J.; Yu, Y.; Di, L.P. Comparison between TVDI and CWSI for drought monitoring in the Guanzhong Plain, China. *J. Integr. Agric.* **2017**, *16*, 389–397. [CrossRef]
20. Zormand, S.; Jafari, R.; Koupaei, S.S. Assessment of PDI, MPDI and TVDI drought indices derived from MODIS Aqua/Terra Level 1B data in natural lands. *Nat. Hazards* **2017**, *86*, 757–777. [CrossRef]
21. Wang, C.Y.; Qi, S.H.; Niu, Z.; Wang, J.B. Evaluating soil moisture status in China using the temperature-vegetation dryness index (TVDI). *Can. J. Remote Sens.* **2004**, *30*, 671–679. [CrossRef]
22. Patel, N.; Anapashsha, R.; Kumar, S.; Saha, S.; Dadhwal, V. Assessing potential of MODIS derived temperature/vegetation condition index (TVDI) to infer soil moisture status. *Int. J Remote Sens.* **2009**, *30*, 23–39. [CrossRef]
23. Yan, H.B.; Zhou, G.; Yang, F.F.; Lu, X.J. DEM correction to the TVDI method on drought monitoring in karst areas. *Int. J. Remote Sens.* **2019**, *40*, 2166–2189. [CrossRef]
24. Liu, Y.; Yue, H. The Temperature Vegetation Dryness Index (TVDI) Based on Bi-Parabolic NDVI-T-s Space and Gradient-Based Structural Similarity (GSSIM) for Long-Term Drought Assessment Across Shaanxi Province, China (2000-2016). *Remote Sens.* **2018**, *10*, 959. [CrossRef]
25. Chen, J.; Wang, C.; Jiang, H.; Mao, L.; Yu, Z. Estimating soil moisture using Temperature-Vegetation Dryness Index (TVDI) in the Huang-huai-hai (HHH) plain. *Int. J. Remote Sens.* **2011**, *32*, 1165–1177. [CrossRef]
26. Chen, C.F.; Son, N.T.; Chen, C.R.; Chiang, S.H.; Chang, L.Y.; Valdez, M. Drought monitoring in cultivated areas of Central America using multi-temporal MODIS data. *Geomat. Nat. Hazards Risk* **2017**, *8*, 402–417. [CrossRef]
27. Shi, S.Q.; Yao, F.M.; Zhang, J.H.; Yang, S.S. Evaluation of Temperature Vegetation Dryness Index on Drought Monitoring Over Eurasia. *IEEE Access* **2020**, *8*, 30050–30059. [CrossRef]
28. Maduako, I.N.; Ndukwu, R.I.; Ifeanyichukwu, C.; Igbokwe, O. Multi-index soil moisture estimation from satellite earth observations: Comparative evaluation of the topographic wetness index (TWI), the temperature vegetation dryness index (TVDI) and the improved TVDI (iTVDI). *J. Indian Soc. Remote Sens.* **2017**, *45*, 631–642. [CrossRef]
29. Rahimzadeh-Bajgiran, P.; Omasa, K.; Shimizu, Y. Comparative evaluation of the Vegetation Dryness Index (VDI), the Temperature Vegetation Dryness Index (TVDI) and the improved TVDI (iTVDI) for water stress detection in semi-arid regions of Iran. *ISPRS J. Photogramm.* **2012**, *68*, 1–12. [CrossRef]
30. Yao, Y.; Zhang, B. Estimation of air temperature and warming effect on the Qinghai-Tibet Plateau based on MODIS data. *J. Geogr. Sci.* **2013**, *68*, 95–107. (In Chinese)
31. Wang, W.; Zhu, Y.; Xu, R.; Liu, J. Drought severity change in China during 1961-2012 indicated by SPI and SPEI. *Nat. Hazards* **2015**, *75*, 2437–2451. [CrossRef]
32. He, B.; Lü, A.; Wu, J.; Zhao, L.; Liu, M. Drought hazard assessment and spatial characteristics analysis in China. *J. Geogr. Sci.* **2011**, *21*, 235–249. [CrossRef]
33. Wang, A.; Lettenmaier, D.P.; Sheffield, J. Soil moisture drought in China, 1950–2006. *J. Clim.* **2011**, *24*, 3257–3271. [CrossRef]
34. Xu, K.; Yang, D.; Yang, H.; Li, Z.; Qin, Y.; Shen, Y. Spatio-temporal variation of drought in China during 1961-2012: A climatic perspective. *J. Hydrol.* **2015**, *526*, 253–264. [CrossRef]
35. Feng, W.; Lu, H.; Yao, T.; Yu, Q. Drought characteristics and its elevation dependence in the Qinghai–Tibet plateau during the last half-century. *Sci. Rep.* **2020**, *10*, 1–11.
36. Wang, H.J.; Chen, Y.N.; Pan, Y.P.; Chen, Z.S.; Ren, Z.G. Assessment of candidate distributions for SPI/SPEI and sensitivity of drought to climatic variables in China. *Int. J. Climatol.* **2019**, *39*, 4392–4412. [CrossRef]
37. Jain, V.K.; Pandey, R.; Jain, M.K. Spatio-temporal assessment of vulnerability to drought. *Nat. Hazards* **2015**, *76*, 443–469. [CrossRef]
38. Liu, X.; Zhu, X.; Pan, Y.; Li, S.; Liu, Y.; Ma, Y. Agricultural drought monitoring: Progress, challenges, and prospects. *J. Geogr. Sci.* **2016**, *26*, 750–767. [CrossRef]

39. Famiglietti, J.S.; Rudnicki, J.W.; Rodell, M. Variability in surface moisture content along a hillslope transect: Rattlesnake Hill, Texas. *J. Hydrol.* **1998**, *210*, 259–281. [CrossRef]
40. Reid, I. The influence of slope orientation upon the soil moisture regime, and its hydrogeomorphological significance. *J. Hydrol.* **1973**, *19*, 309–321. [CrossRef]
41. Zhang, Q.; Fan, K.; Singh, V.P.; Song, C.; Xu, C.-Y.; Sun, P. Is Himalayan-Tibetan Plateau "drying"? Historical estimations and future trends of surface soil moisture. *Sci. Total Environ.* **2019**, *658*, 374–384. [CrossRef]
42. Xia, X.; Wenbin, S.; Zhen, W. Analysis of soil moisture changes of the Buertai mining area based on TVDI. *J. Min. Sci. Technol.* **2019**, *4*, 285–291.
43. Liu, H.; Zhang, A.; Jiang, T.; Lv, H.; Liu, X.; Wang, H. The spatiotemporal variation of drought in the Beijing-Tianjin-Hebei metropolitan region (BTHMR) based on the modified TVDI. *Sustainability* **2016**, *8*, 1327. [CrossRef]
44. Breiman, L. Random forests. *Mach. Learn.* **2001**, *45*, 5–32. [CrossRef]
45. Cutler, D.R.; Edwards Jr, T.C.; Beard, K.H.; Cutler, A.; Hess, K.T.; Gibson, J.; Lawler, J.J. Random forests for classification in ecology. *Ecology* **2007**, *88*, 2783–2792. [CrossRef] [PubMed]
46. Li, T.; Tian, Y.; Wu, L. Landslide hazard zoning based on random forest method. *Geogr. Geo-Inf. Sci.* **2014**, *30*, 25–30. (In Chinese)
47. Wang, C.; Jue, A.; Zeng, Y. Population distribution pattern and influencing factors in Tibet Based on random forest model. *J. Geogr. Sci.* **2019**, *74*, 664–680. (In Chinese)
48. Zhang, Q.; Gao, W.; Su, S.; Weng, M.; Cai, Z. Biophysical and socioeconomic determinants of tea expansion: Apportioning their relative importance for sustainable land use policy. *Land Use Policy* **2017**, *68*, 438–447. [CrossRef]
49. Wang, N.; Cheng, W.; Wang, B.; Liu, Q.; Zhou, C. Geomorphological regionalization theory system and division methodology of China. *J. Geogr. Sci.* **2020**, *30*, 212–232. [CrossRef]
50. Yuan, L.; Li, L.; Zhang, T.; Chen, L.; Zhao, J.; Hu, S.; Cheng, L.; Liu, W. Soil moisture estimation for the Chinese Loess Plateau using MODIS-derived ATI and TVDI. *Remote Sens.* **2020**, *12*, 3040. [CrossRef]
51. Wu, L.; Zhang, Y.; Wang, L.; Xie, W.; Song, L.; Zhang, H.; Bi, H.; Zheng, Y.; Zhang, Y.; Zhang, X. Analysis of 22-year Drought Characteristics in Heilongjiang Province Based on Temperature Vegetation Drought Index. *Comput. Intell. Neurosci.* **2022**, *2022*, 1003243. [CrossRef]
52. Yang, F.; Duan, X.; Guo, Q.; Lu, S.; Hsu, K. The spatiotemporal variations and propagation of droughts in Plateau Mountains of China. *Sci. Total Environ.* **2022**, *805*, 150257. [CrossRef]
53. Xu, Y.; Zhang, X.; Wang, X.; Hao, Z.; Singh, V.P.; Hao, F. Propagation from meteorological drought to hydrological drought under the impact of human activities: A case study in northern China. *J. Hydrol.* **2019**, *579*, 124147. [CrossRef]
54. Holgate, C.; Van Dijk, A.; Evans, J.; Pitman, A. Local and remote drivers of southeast Australian drought. *Geophys. Res. Lett.* **2020**, *47*, e2020GL090238. [CrossRef]
55. Liang, L.; ZHAO, S.-h.; QIN, Z.-h.; HE, K.-x.; Chong, C.; LUO, Y.-x.; ZHOU, X.-d. Drought change trend using MODIS TVDI and its relationship with climate factors in China from 2001 to 2010. *J. Integr. Agric.* **2014**, *13*, 1501–1508. [CrossRef]
56. Chen, Y.; Li, X.; Shi, P. Analysis of climatic factors driving NDVI change in China from 1983 to 1992. *J. Plant Ecol.* **2001**, *25*, 716. (In Chinese)
57. Liu, J.; Wu, J.; Wu, Z.; Liu, M. Response of NDVI dynamics to precipitation in the Beijing–Tianjin sandstorm source region. *Int. J. Remote Sens.* **2013**, *34*, 5331–5350. [CrossRef]
58. Cao, X.; Feng, Y.; Wang, J. An improvement of the Ts-NDVI space drought monitoring method and its applications in the Mongolian plateau with MODIS, 2000–2012. *Arab. J. Geosci.* **2016**, *9*, 433. [CrossRef]
59. Son, N.T.; Chen, C.F.; Chen, C.R.; Molina Masferrer, M.G.; Recinos, L.E.M. Multitemporal Landsat-MODIS fusion for cropland drought monitoring in El Salvador. *Geocarto Int.* **2019**, *34*, 1363–1383. [CrossRef]
60. Han, L.; Wang, P.; Wang, J.; Liu, S. Study on characteristic space of vegetation index surface temperature composition. *Chin. Sci. Part D* **2005**, *35*, 371–377.
61. Du, L.; Song, N.; Liu, K.; Hou, J.; Hu, Y.; Zhu, Y.; Wang, X.; Wang, L.; Guo, Y. Comparison of two simulation methods of the temperature vegetation dryness index (TVDI) for drought monitoring in semi-arid regions of China. *Remote Sens.* **2017**, *9*, 177. [CrossRef]
62. Zhang, H.; Ali, S.; Ma, Q.; Sun, L.; Jiang, N.; Jia, Q.; Hou, F. Remote sensing strategies to characterization of drought, vegetation dynamics in relation to climate change from 1983 to 2016 in Tibet and Xinjiang Province, China. *Environ. Sci. Pollut. Res.* **2021**, *28*, 21085–21100. [CrossRef]
63. Li, Y.; Su, F.; Tang, Q.; Gao, H.; Yan, D.; Peng, H.; Xiao, S. Contributions of moisture sources to precipitation in the major drainage basins in the Tibetan Plateau. *Sci. China Earth Sci.* **2022**, *65*, 1088–1103. [CrossRef]
64. Yang, M.; Yao, T.; Gou, X.; Koike, T.; He, Y. The soil moisture distribution, thawing–freezing processes and their effects on the seasonal transition on the Qinghai–Xizang (Tibetan) plateau. *J. Asian Earth Sci.* **2003**, *21*, 457–465. [CrossRef]
65. Wang, S.; Liu, F.; Zhou, Q.; Chen, Q.; Niu, B.; Xia, X. Drought evolution characteristics of the Qinghai-Tibet Plateau over the last 100 years based on SPEI. *Nat. Hazards Earth Syst. Sci. Discuss.* **2021**, *73*, 1–20.
66. Zhao, S.; Cong, D.; He, K.; Yang, H.; Qin, Z. Spatial-temporal variation of drought in China from 1982 to 2010 based on a modified temperature vegetation drought index (mTVDI). *Sci. Rep.* **2017**, *7*, 1–12. [CrossRef]
67. Xie, H.; Ye, J.; Liu, X.; Chongyi, E. Warming and drying trends on the Tibetan Plateau (1971–2005). *Theor. Appl. Climatol.* **2010**, *101*, 241–253. [CrossRef]

68. Zhen, D.; Zhao, D. Alpine desert zone and cold arid core region of Qinghai Xizang Plateau. *Arid. Zone Res.* **2019**, *36*, 1–6. (In Chinese)
69. Xiangtao, W.; Xianzhou, Z.; Junhao, W.; Ben, N. Variations in the drought severity index in response to climate change on the Tibetan Plateau. *J. Resour. Ecol.* **2020**, *11*, 304–314. [CrossRef]
70. Shi, L.; Du, J.; Zhou, K.; Zhuo, G. Temporal and spatial evolution characteristics of soil moisture in Qinghai Tibet Plateau from 1980 to 2012. *J. Glaciol. Geocryol.* **2017**, *38*, 1241–1248. (In Chinese)
71. Meng, X.; Li, R.; Luan, L. Detecting hydrological consistency between soil moisture and precipitation and changes of soil moisture in summer over the Tibetan Plateau. *Clim. Dyn.* **2018**, *51*, 4157–4168. [CrossRef]
72. Shen, Q.; Liang, L.; Luo, X.; Li, Y.J.; Zhang, L.P. Analysis of the spatial-temporal variation characteristics of vegetative drought and its relationship with meteorological factors in China from 1982 to 2010. *Environ. Monit. Assess.* **2017**, *189*, 1–14. [CrossRef]
73. Sayari, N.; Bannayan, M.; Alizadeh, A.; Farid, A. Using drought indices to assess climate change impacts on drought conditions in the northeast of Iran (case study: Kashafrood basin). *Meteorol. Appl.* **2013**, *20*, 115–127. [CrossRef]
74. Dubrovsky, M.; Svoboda, M.D.; Trnka, M.; Hayes, M.J.; Wilhite, D.A.; Zalud, Z.; Hlavinka, P. Application of relative drought indices in assessing climate-change impacts on drought conditions in Czechia. *Theor. Appl. Climatol.* **2009**, *96*, 155–171. [CrossRef]
75. Naumann, G.; Alfieri, L.; Wyser, K.; Mentaschi, L.; Betts, R.; Carrao, H.; Spinoni, J.; Vogt, J.; Feyen, L. Global changes in drought conditions under different levels of warming. *Geophys. Res. Lett.* **2018**, *45*, 3285–3296. [CrossRef]
76. Cook, B.I.; Smerdon, J.E.; Seager, R.; Cook, E.R. Pan-continental droughts in North America over the last millennium. *J. Clim.* **2014**, *27*, 383–397. [CrossRef]
77. Zhen, D.; Lin, Z.; Zhang, X. Research progress of Qinghai Tibet Plateau and global environmental change. *Earth Sci. Front.* **2002**, *9*, 95–102. (In Chinese)
78. Shi, Y.; Li, J.; Li, B. The uplift of the Tibetan Plateau in the late Cenozoic and the environmental changes in East Asia. *J. Geogr. Sci.* **1999**, *54*, 10–21. (In Chinese)
79. Liu, X. Effects of the uplift of the Qinghai Tibet Plateau on the formation of Asian monsoon and global climate and environmental change. *Plateau Meterology* **1999**, *18*, 321–332. (In Chinese)
80. Zhou, X.; Zhao, P.; Chen, J. Study on the influence of thermal action of Qinghai Tibet Plateau on the climate in the Northern Hemisphere. *Chin. Sci. Part D* **2009**, 1473–1486.
81. Van Loon, A.; Laaha, G. Hydrological drought severity explained by climate and catchment characteristics. *J. Hydrol.* **2015**, *526*, 3–14. [CrossRef]
82. Qiu, Y.; Fu, B.; Wang, J.; Chen, L. Soil moisture variation in relation to topography and land use in a hillslope catchment of the Loess Plateau, China. *J. Hydrol.* **2001**, *240*, 243–263. [CrossRef]
83. Western, A.W.; Grayson, R.B.; Blöschl, G.; Willgoose, G.R.; McMahon, T.A. Observed spatial organization of soil moisture and its relation to terrain indices. *Water Resour. Res.* **1999**, *35*, 797–810. [CrossRef]
84. Jehanzaib, M.; Shah, S.A.; Yoo, J.; Kim, T.-W. Investigating the impacts of climate change and human activities on hydrological drought using non-stationary approaches. *J. Hydrol.* **2020**, *588*, 125052. [CrossRef]
85. Kim, T.-W.; Jehanzaib, M. Drought risk analysis, forecasting and assessment under climate change. *Water* **2020**, *12*, 1862. [CrossRef]
86. Li, Q.; Zhou, J.; Zou, W.; Zhao, X.; Huang, P.; Wang, L.; Shi, W.; Ma, X.; Zhao, Y.; Xue, D. A tributary-comparison method to quantify the human influence on hydrological drought. *J. Hydrol.* **2021**, *595*, 125652. [CrossRef]
87. Rangecroft, S.; Van Loon, A.F.; Maureira, H.; Verbist, K.; Hannah, D.M. An observation-based method to quantify the human influence on hydrological drought: Upstream–downstream comparison. *Hydrol. Sci. J.* **2019**, *64*, 276–287. [CrossRef]
88. Zhang, D.; Zhang, Q.; Qiu, J.; Bai, P.; Liang, K.; Li, X. Intensification of hydrological drought due to human activity in the middle reaches of the Yangtze River, China. *Sci. Total Environ.* **2018**, *637*, 1432–1442. [CrossRef]
89. Pei, y.; Jiang, g.; Cui, j. Theoretical framework of the driving mechanism of drought evolution and its key issues. *Adv. Water Sci.* **2013**, *24*, 449–456. (In Chinese)
90. Yao, T.; Bolch, T.; Chen, D.; Gao, J.; Immerzeel, W.; Piao, S.; Su, F.; Thompson, L.; Wada, Y.; Wang, L. The imbalance of the Asian water tower. *Nat. Rev. Earth Environ.* **2022**, *299*, 1–15. [CrossRef]
91. Yao, T.; Pu, J.; Lu, A.; Wang, Y.; Yu, W. Recent glacial retreat and its impact on hydrological processes on the Tibetan Plateau, China, and surrounding regions. *Arct. Antarct. Alp. Res.* **2007**, *39*, 642–650. [CrossRef]
92. Pritchard, H.D. Asia's glaciers are a regionally important buffer against drought. *Nature* **2017**, *545*, 169–174. [CrossRef]
93. Wang, M.; Zhou, C.; Wu, L.; Xu, X.; Ou, Y. Wet-drought pattern and its relationship with vegetation change in the Qinghai-Tibetan Plateau during 2001–2010. *Arid. Land Geogr.* **2013**, *36*, 49–56.
94. Hang, J.; Zhang, Y.; Wang, M.; Wang, F.; Tang, Z.; He, H. Temporal and spatial distribution characteristics and influencing factors of drought in Xinjiang in the past 17 years. *J. Ecol.* **2020**, *40*, 1077–1088.
95. Xu, M.; Zhang, T.; Zhang, Y.; Chen, N.; Zhu, J.; He, Y.; Zhao, T.; Yu, G. Drought limits alpine meadow productivity in northern Tibet. *Agric. For. Meteorol.* **2021**, *303*, 108371. [CrossRef]
96. Wang, H.; Magagi, R.; Goïta, K.; Wang, K. Soil moisture retrievals using ALOS2-ScanSAR and MODIS synergy over Tibetan Plateau. *Remote Sens. Environ.* **2020**, *251*, 112100. [CrossRef]

International Journal of
Environmental Research and Public Health

MDPI

Article

Spatial-Temporal Evolution and Driving Forces of NDVI in China's Giant Panda National Park

Mengxin Pu [1], Yinbing Zhao [1,2,3,4,*], Zhongyun Ni [1,2,4], Zhongliang Huang [1], Wanlan Peng [2], Yi Zhou [2], Jingjing Liu [2] and Yingru Gong [2]

1 College of Earth Sciences, Chengdu University of Technology, Chengdu 610059, China; pumengxin@163.com (M.P.); nizhongyun2012@mail.cdut.edu.cn (Z.N.); hzl19961002@gmail.com (Z.H.)
2 College of Tourism and Urban-Rural Planning, Chengdu University of Technology, Chengdu 610059, China; pengwanlan@stu.cdut.edu.cn (W.P.); zhouyi1@stu.cdut.edu.cn (Y.Z.); lj0011822@gmail.com (J.L.); yr852587432@163.com (Y.G.)
3 Hebei Province Key Laboratory of Wetland Ecology and Conservation, Hengshui University, Hengshui 053000, China
4 International Network for Environment and Health, School of Geography and Archaeology & Ryan Institute, National University of Ireland, Galway (NUIG), H91 CF50 Galway, Ireland
* Correspondence: zhaoyinbing06@cdut.cn

Abstract: Identifying the ecological evolution trends and vegetation driving mechanisms of giant panda national parks can help to improve the protection of giant panda habitats. Based on the research background of different geomorphological zoning, we selected the MODIS NDVI data from 2000 to 2020 to analyze the NDVI trends using a univariate linear model. A partial correlation analysis and multiple correlation analysis were used to reveal the influence of temperature and precipitation on NDVI trends. Fourteen factors related to meteorological factors, topographic factors, geological activities, and human activities were selected, and the Geographically Weighted Regression model was used to study the mechanisms driving NDVI change. The results were as follows: (1) The NDVI value of Giant Panda National Park has fluctuated and increased in the past 21 years, with an annual growth rate of 4.7%/yr. Affected by the Wenchuan earthquake in 2008, the NDVI value fluctuated greatly from 2008 to 2012, and reached its peak in 2018. (2) The NDVI in 94% of the study area improved, and the most significant improvement areas were mainly distributed in the northern and southern regions of Southwest Subalpine and Middle Mountain and the Xiaoxiangling area. Affected by the distribution of fault zones and their local activities, vegetation degradation was concentrated in the Dujiangyan–Anzhou area of Hengduan Mountain Alpine Canyon. (3) The Geographically Weighted Regression analysis showed that natural factors were dominant, with climate and elevation having a double-factor enhancement effect, the peak acceleration of ground motion and fault zone having a superimposed effect, and river density and slope having a double effect, all of which had a significant impact on the NDVI value of the surrounding area. To optimize the ecological security pattern of the Giant Panda National Park, we recommended strengthening the construction of ecological security projects through monitoring meteorological changes, preventing, and controlling geo-hazards, and optimizing the layout and intensity of human activities.

Keywords: vegetation; geographically weighted regression; climate change; Wenchuan earthquake; ecological security

Citation: Pu, M.; Zhao, Y.; Ni, Z.; Huang, Z.; Peng, W.; Zhou, Y.; Liu, J.; Gong, Y. Spatial-Temporal Evolution and Driving Forces of NDVI in China's Giant Panda National Park. *Int. J. Environ. Res. Public Health* **2022**, *20*, 6722. https://doi.org/10.3390/ijerph19116722

Academic Editor: Paul B. Tchounwou

Received: 6 May 2022
Accepted: 26 May 2022
Published: 31 May 2022

Publisher's Note: MDPI stays neutral with regard to jurisdictional claims in published maps and institutional affiliations.

1. Introduction

The giant panda is an important national protected animal in China, and changes in its habitat quality have received extensive attention from researchers. Affected by urban and rural construction activities, the fragmentation of giant panda habitats is serious, and the natural and anthropogenic threats to the habitat quality continue to intensify [1–3]. National parks are key areas for strictly protecting biodiversity in various countries [4], with the purpose of achieving effective ecological protection, rational utilization of resources and sustainable social and economic development [5,6]. In order to adapt to the multi-departmental collaborative management of giant panda reserves, the Chinese government proposed the establishment of the China Giant Panda National Park (CGPNP) in 2017 and announced its formal establishment in 2021 [7]; this balances the needs of regional ecological protection and socio-economic development and continuously improves the ability of biodiversity protection to balance the needs of regional ecological protection and socio-economic conservation [8,9]. Vegetation in CGPNP is affected by the complex interaction of ecological elements such as soil, atmosphere, and water, and is also an important medium for natural ecosystems and human production activities [10]. Among many types of remote sensing data, the Normalized Difference Vegetation Index (NDVI) is a sensitive parameter of surface vegetation coverage and vegetation growth which reflects the difference between the radiation absorption in the red spectral region caused by chlorophyll and the reflectivity of canopy structure caused by the NIR spectral region, and it can effectively characterize the vegetation environment and its changes and effects [11,12]. The NDVI generated from remote sensing data have the advantage of a long time series, wide coverage, and high spatial resolution [13], and there are many cases of NDVI being used to monitor national park vegetation, ecological environments and their changes all over the world, including passive monitoring over a long time series [14], vegetation and climate coupling characteristics research [15], vegetation phenology characteristics research [16], and so on.

Many studies have been conducted on vegetation changes in CGPNP reserves in China [17], but these study areas consisted mainly of independent reserves in CGPNP, and the data used are mostly SAR images [18–20]. Although the use of SAR image improves the accuracy of image recognition, the coverage area of SAR images is small and the period is limited, so it is not suitable for long-term dynamic monitoring of vegetation. The Moderate Resolution Imaging Spectroradiometer (MODIS) can better solve this problem and can be used for monitoring the vegetation environment and its changes in CGPNP. The research on the trends of changes in vegetation environments is the key component of dynamic monitoring, and the methods involved include the linear regression analysis [21], Mann–Kendall test [22,23], BFAST trend analysis and Theil–Sen median slope trend analysis [24]. Among these, the linear regression analysis is a more effective method in this kind of research.

The driving mechanisms of climate and other factors based on the trends of changes in NDVI form the basis for the formulation of national park protection strategies. Studying the response mechanism of climate change to NDVI changes is of great significance in predicting vegetation dynamics [25]. In the global research on the relationship between vegetation and climate, the climate-driving mechanisms in different regions show significant geographical differentiation [26–29]. The CGPNP is located in a climate transition zone, and climate drives NDVI with great complexity and uncertainty. Most previous studies on NDVI drivers in this region have only considered climatic factors. For example, Lin et al. focused on the two factors of temperature and precipitation in their research on NDVI-driving forces in North China [30], and Liu et al. also only considered the spatial characteristics of climate factors in the study of vegetation change in China [31]. The impact of geological activities, topography, and human activities was not considered sufficiently [32]. Affected by the 12 May 2008 Wenchuan earthquake and its triggered geo-hazards, the local NDVI decreased rapidly and the habitat quality deteriorated seriously [33,34]. In the more than 10 years since the earthquake, the ecological geological environment has undergone great changes, which has increased the spatial instability and complexity of the analysis of the driving forces of NDVI changes [35]. At present, there are many research methods on NDVI's driving forces, such as the enhanced regression tree model [36], the Geographically Weighted Regression (GWR) model [37], the correlation analysis method [38] and residual trend analysis [39], most of which are only used for unilateral aspects such as climate factors, ignoring the correlation and coupling between multiple factors and failing to take into account the spatial difference of the action of driving factors. To study the driving mechanisms of NDVI changes, the GWR model is a suitable choice because it enables one to change the parameter settings on the local scale, facilitates the determination of local coefficients, and can solve the problems of spatial instability and scale dependence to a certain extent in the analysis [40,41]. The GWR model can reveal the spatiotemporal variability between each driving factor and vegetation activity by studying the spatial non-stationary relationship between each driving factor and NDVI change value.

Aiming at the complex eco-geological environment of CGPNP, the MODIS NDVI data products were selected, and the Savitzky–Golay filter was used to construct NDVI serial data from 2000 to 2020. In this study, the univariate linear model was used to monitor the interannual NDVI trends in the study area based on geomorphological zoning. The GWR model was used to identify the driving mechanisms of NDVI trends by integrating the effects of natural and human factors. Finally, combined with the temporal and spatial differentiation characteristics and driving mechanisms of NDVI change, eco-geological environmental protection countermeasures were proposed according to the local conditions.

2. Study Area and Data Processing

2.1. Study Area

The CGPNP is located in the ecological barrier area of the Sichuan–Yunnan Loess Plateau in the "two screens and three belts" ecological barrier area in China's ecological security strategy [42] with the largest population, protected area type and quantity in China [43]. The establishment of CGPNP brings together various nature reserves and increases the connectivity of giant panda habitats [44–46]. From northeast to southwest, the Sichuan Area of China Giant Panda National Park (SCOCGPNP) consists of seven cities (prefectures), including 19 counties (cities and districts). It spans five areas: Qinling Mountains, Baishui River, Minshan Mountain, Qionglai Mountain–Daxiangling, and Xiaoxiangling (Figure 1a). The SCOCGPNP ranges from 102°27′ to 105°57′ E and 29°42′ to 33°34′ N, covering an area of 20,177 km^2. As the study area is located in the transition area from a subtropical zone to a warm temperate zone (Figure 1b) [47], the average annual precipitation is 830 mm, and the average temperature is 10–17 °C. In the study area, the altitude decreases from northwest to southeast, and the vertical distribution of vegetation is obvious: subtropical evergreen deciduous forest, evergreen deciduous broad-leaved mixed forest, temperate coniferous forest, cold temperate coniferous forest, shrub, and meadow.

In order to fully describe the influence of geomorphology on the basement of NDVI and its changes, the study area was divided into five geomorphological areas (Figure 1c) [48], including the Baishui River–Minshan Area of Southwest Subalpine and Middle Mountain (BSMS), the Minshan Area of Hengduan Mountain Alpine Canyon (MS), the Dujiangyan–Anzhou Area of Hengduan Mountain Alpine Canyon (DA), the Daxiangling Area of Hengduan Mountain Alpine Canyon (DXL), and the Daxiangling-Xiaoxiangling Area of Southwest Subalpine and Middle Mountain (DXLXXL).

2.2. Data Sources

2.2.1. MODIS NDVI Data

In this research, the Moderate-resolution Imaging Spectroradiometer (MODIS) spectral imager on the EOS/Terra satellite was used to obtain MOD13A1 products, among which MODIS VI products can be used to monitor the terrestrial photosynthetic vegetation activities of the earth and support the phenology and change monitoring of vegetation in national parks. The NDVI data from 2000 to 2020 in this research were downloaded from the official website of NASA (https://earthdata.nasa.gov/, accessed on 20 February 2021). The product used in this research was the collection 6 data set. Compared with the collection V data set, the 8-day surface reflection data (pre-synthesized based on the Terra and Aqua data) were used, the CV-MVC synthesis method was modified, and the necessary SDS was adjusted to reflect the new input data flow, which improves the change-detection ability of the product [49]. We selected the best available pixel value from the data set collected in 16 days with a resolution of 500 m, resulting in NDVI values with low cloud cover and low viewing angles [50].

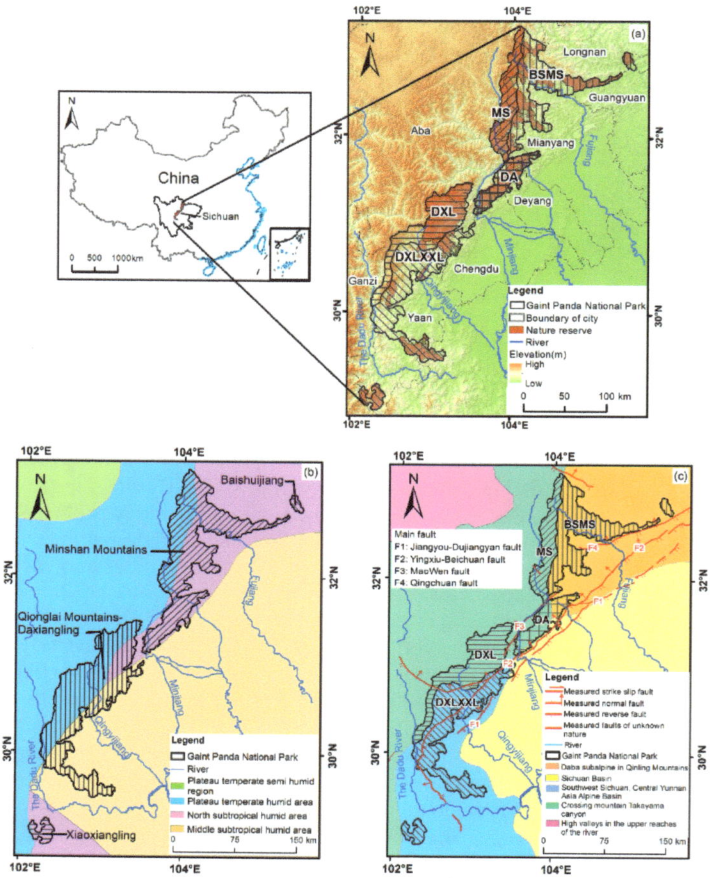

Figure 1. (**a**) Location of the study area. (**b**) Climatic zoning. (**c**) Geomorphological zoning.

2.2.2. Driving Factors

The GWR model was selected to explore the driving mechanism of various factors influencing NDVI trends, and five categories and 14 variables were selected (Table 1).

As the main controlling factor of vegetation structure, composition, and distribution, there is a strong correlation between precipitation and NDVI [51]. Since temperature can regulate the photosynthesis of vegetation [52], annual mean temperature and variability were used as analytical indicators (Figure 2a–d).

Located in a geomorphological boundary area, the study area has complex geomorphology. Among the topographic factors, elevation often determines the temperature and CO_2 content of vegetation growth (Figure 2e), slope represents soil moisture and sunlight exposure for vegetation growth (Figure 2f), and aspect also has a certain impact on vegetation lighting conditions [53] (Figure 2g).

Located near the Longmenshan fault zone, the crustal activity in the study area affects the growth of vegetation, the looseness of soil, and the stability of its growth state. The Euclidean distance from fault (Figure 2h) and the peak acceleration of ground motion (Figure 2i) were used to characterize the activity of the geological activities [54].

Table 1. Data sources.

Variable Class	Variable Name	Definition and Units	Data Sources	Spatial Resolution
Climatic	TEM_MN	Annual mean precipitation 2000–2020 (mm/yr)	National Qinghai Tibet Plateau scientific data center [a]	1 km
	PRE_MN	Annual mean temperature 2000–2020 (°C/yr)		1 km
	TEM_BT	*t*-test grading of precipitation trends (OLS) during the growing season 2000–2020		1 km
	PRE_BT	*t*-test grading of temperature trends (OLS) during the growing season 2000–2020		1 km
Geomorphological	ELEVATION	Elevation represents macroscopic geomorphology (m)	Geospatial data cloud [b]	30 m
	SLOPE	Slope represents groundcutting condition (°)		
	ASPECT	Aspect represents ground orientation		
Geological activities	ED_FAULT	Euclidean distance from fault (m)	China Geological Survey [c]	Vector
	PGA	Peak ground acceleration (g)	China earthquake administration [d]	Vector
Human activity	ED_BLAND	Euclidean distance from built-up land (m)	Data Sharing and Service Portal [e]	30 m
	DEN_ROAD	Road density (km/km^2)	National Geomatics Center of China [f]	Vector
Others	DEN_RIVER	River density (km/km^2)	National Geomatics Center of China [f]	Vector
	LUCC	Land-use change index	European Space Agency [g]	300 m

[a] http://data.tpdc.ac.cn/zh-hans/ (accessed on 25 February 2021); [b] http://www.gscloud.cn/ (accessed on 5 May 2021); [c] https://www.cgs.gov.cn/ (accessed on 20 May 2021); [d] https://www.cea.gov.cn/ (accessed on 20 May 2021); [e] http://data.casearth.cn/en/ (accessed on 16 April 2021); [f] http://www.ngcc.cn/ngcc/ (accessed on 18 April 2021); [g] https://www.esa.int/ (accessed on 25 April 2021).

Human activities around construction land, such as logging, grazing, fire, etc., which disturb the original structure of the vegetation landscape, are closely related to changes in the NDVI [55]. In addition, most of the study area is mountainous. Therefore, compared with nighttime light data and POI data, choosing the Euclidean distance for construction land can more intuitively characterize the intensity of human activities, and can exclude errors caused by terrain fluctuations and weak human activities (Figure 2j). The construction and use of roads have a devastating impact on the ecological environment, and the towns and villages connected by them imply the impact of human activities. The denser the road distribution (Figure 2k), the more serious the damage to the vegetation [31]; so the road density was selected as one influencing factor. Due to the large proportion of forest land and mountainous terrain in the study area, the dynamic change in human activities was small, so the static data of 2020 were selected to characterize the impact of human activities.

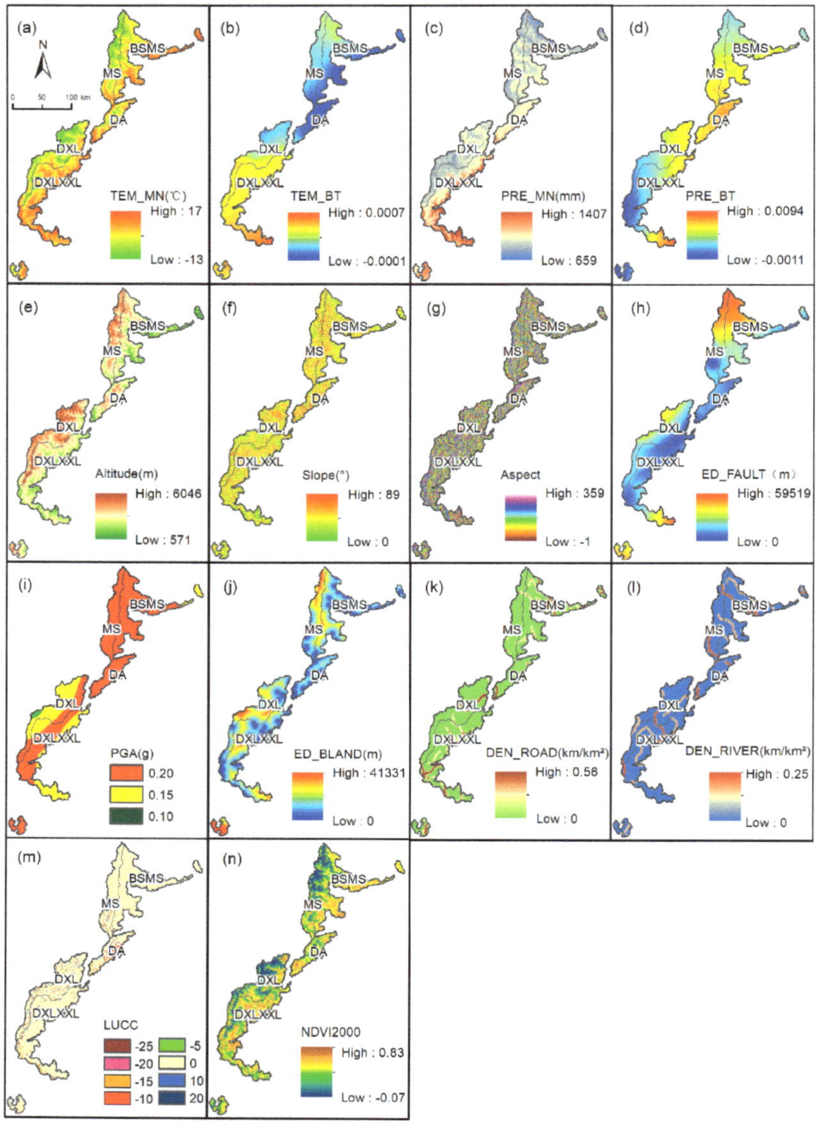

Figure 2. Driving factors used for GWR Model. (**a**) Temperature mean value; (**b**) temperature slope; (**c**) precipitation mean value; (**d**) precipitation slope; (**e**) elevation; (**f**) slope; (**g**) aspect; (**h**) Euclidean distance from fault; (**i**) seismic peak acceleration; (**j**) Euclidean distance from built-up land; (**k**) road density; (**l**) river density; (**m**) land-use change index; (**n**) NDVI in 2000.

The distribution of rivers affects the stability of regional NDVI and controls ecological changes and landscape dynamics [56]; therefore, the river density was chosen to characterize its impact (Figure 2l). Different land-use types represent different NDVI values and different NDVI change possibilities. For example, the conversion of construction land or cultivated land to grassland or forest land is beneficial to the improvement of NDVI [57]. In the same geographical environment, the order of NDVI values from high to low is forest land > cultivated land > grassland [58]; whereas a negative value of NDVI usually represents water [59]. Referring to this law, the forest land, cultivated land, grassland, construction land, water bodies, and other land types in the initial and final years were designated as 25, 15, 10, 0, −5, and 0, respectively (Figure 2m). A measure of the transformation of land-use types was obtained by subtracting the initial year from the final year. The NDVI value of the starting year in 2000 was used to characterize the vegetation trends and to assess whether the regional vegetation had reached a saturated state [60] (Figure 2n).

Different driving factors have spatial differences in different partitions (Table A1) and applying this fact to the GWR model is conducive to the discussion of the driving mechanisms of NDVI.

2.3. Method

2.3.1. Data Preprocessing

We selected the annual data synthesized from monthly data from 2000 to 2020 as the research data of long time series. There are various synthesis methods available for NDVI data, such as the average value method (AVM), maximum value composite (MVC), time series reconstruction method, and so on. To remove the NDVI outliers in multi-temporal images to a certain extent [61], the MVC method was selected to synthesize the data with a period of 16 days to obtain the monthly NDVI value from 2000 to 2020, so as to eliminate the deviation caused by atmospheric interference, solar elevation, cloud coverage, etc. To eliminate the noise caused by sensor error or cloud cover in data acquisition, Savitzky–Golay filtering was carried out on the NDVI value of continuous time series, and the NDVI value images of 21 consecutive years from 2000 to 2020 were obtained (Figure A1).

2.3.2. NDVI Change Detection

The univariate linear regression analysis model monitors the NDVI time series through the regression time function. The model can calculate the interannual variability of vegetation, which is the slope of the linear regression equation [62]. The calculation formula is as follows:

$$\theta_{slope} = \frac{n \times \sum_{i=1}^{n} i \times \mathrm{NDVI}_i - \sum_{i=1}^{n} i \sum_{i=1}^{n} \mathrm{NDVI}_i}{n \times \sum_{i=1}^{n} i^2 - \left(\sum_{i=1}^{n} i\right)^2} \quad (1)$$

where n represents the time span and i represents the NDVI value in the ith year; when $\theta_{slope} > 0$, it indicates that NDVI is in an improved state, and when $\theta_{slope} < 0$, it indicates that NDVI is degraded.

The t-test was used to obtain the significance of the temporal trend of NDVI:

$$t = \frac{\overline{x}_1 - \overline{x}_2}{s \times \sqrt{\frac{1}{n_1} + \frac{1}{n_2}}} \quad (2)$$

$$s = \sqrt{\frac{n_1 S_1^2 + n_2 S_2^2}{n_1 + n_2 - 2}} \quad (3)$$

where $\overline{x}_1$ and $\overline{x}_2$ represent the mean of the two subsamples, n_1 and n_2 represent the number of the two subsamples and S_1 and S_2 represent the standard deviation of the two subsamples.

Through the above process, the change in slope of NDVI and t-test results were obtained. The θ_{slope} was divided into improvement and degradation, with the value of 0 as the boundary. Two confidence levels of 0.01 and 0.05 were selected in the t-test to evaluate the significance of the change in NDVI. A p-value of $0.01 < p < 0.05$ was considered significant, $p < 0.01$ was extremely significant, and $p > 0.05$ was considered insignificant. The combination of the two can be used to divide the NDVI trend into six categories: extremely significant degradation, significant degradation, insignificant degradation, insignificant improvement, significant improvement, and extremely significant improvement (Table 2).

Table 2. Significance classification criteria.

θ_{slope}	p	Significance Test Classification
$\theta_{slope} < 0$	$p \leq 0.01$	Extremely significant degradation
	$0.01 < p \leq 0.05$	Significant degradation
	$p > 0.05$	Insignificant degradation
$\theta_{slope} > 0$	$p > 0.05$	No significant improvement
	$0.01 < p \leq 0.05$	Significant improvement
	$p \leq 0.01$	Extremely significant improvement

2.3.3. NDVI Climate-Driven Analysis

In order to describe the correlation between NDVI, precipitation, and temperature, the partial correlation coefficient and multiple correlation coefficient were selected as quantitative indicators to test significance, and the study area was divided into different climate-driving types according to the test results.

A partial correlation analysis can measure the correlation between two factors under the exclusion of other factors [63]. The formula for calculating the partial correlation coefficient between temperature, precipitation, and NDVI is as follows:

$$r_{123} = \frac{r_{12} - r_{13} * r_{23}}{\sqrt{(1 - r_{13}{}^2) * (1 - r_{23}{}^2)}} \tag{4}$$

$$r_{xy} = \frac{\sum_{i=1}^{n}(x_i - \overline{x})(y_i - \overline{y})}{\sqrt{\sum_{i=1}^{n}(x_i - \overline{x})^2 * \sum_{i=1}^{n}(y_i - \overline{y})^2}} \tag{5}$$

where r_{12}, r_{13}, and r_{23} are the correlation coefficients between NDVI and temperature, NDVI and precipitation, and temperature and precipitation; r_{123} is the partial correlation coefficient between the two parameters based on the third parameter.

A statistically significant confidence level of 0.05 was selected, and a *t*-test was performed on the results of partial correlation analysis to obtain the significance between climate-driving factors.

$$t = \frac{r_{123}}{\sqrt{1 - r_{123}{}^2}}\sqrt{n - m - 1} \tag{6}$$

where n is the number of samples and m is the independent variable.

A multiple correlation analysis was used to study the degree of correlation between NDVI and precipitation and temperature, thereby revealing the driving mechanism of climate on NDVI changes.

$$r_{x,yz} = \sqrt{1 - (1 - r^2{}_{xy})(1 - r^2{}_{xz,y})} \tag{7}$$

A significant F-test was performed on the results of the multiple correlation analysis:

$$F = \frac{r^2{}_{x,yz}}{1 - r^2{}_{x,yz}} \times \frac{n - k - 1}{k} \tag{8}$$

where n is the number of samples and k is the number of independent variables.

Based on the existing precipitation, temperature, and NDVI data, a partial correlation and multiple correlation analysis were carried out to test the correlation of the two coefficients. In the partial correlation analysis, the *t*-test with a confidence level of 0.05 was used to divide the precipitation-driving type and temperature-driving type pixel by pixel. In the multiple correlation analysis, the F-test with a confidence level of 0.01 was used to determine whether it was a climate-driven region. The combination of the two can be used to obtain the climate-driven zoning map in the study area (Table 3).

Table 3. Zoning criteria for climate-driving factors.

Driven Type	Zoning Criteria		
	r NDVI P, T *a	r NDVI T, P *b	r NDVI P, T *c
Precipitation-driven	$t \geq t_{0.05}$ *1		$F \geq F_{0.01}$ *2
Temperature-driven		$t \geq t_{0.05}$	$F \geq F_{0.01}$
Temperature- and precipitation-driven	$t \geq t_{0.05}$	$t \geq t_{0.05}$	$F \geq F_{0.01}$
Other driving modes			$F \leq F_{0.01}$

*a Partial correlation coefficient between NDVI and temperature. *b Partial correlation coefficient between NDVI and precipitation. *c Multiple correlation coefficient between precipitation and temperature. *1 Confidence *t*-test significance level of 0.05. *2 Confidence F-test significance level of 0.01.

2.3.4. NDVI-Driving Force Analysis Based on the GWR Model

The GWR model, which is a spatial decomposition of traditional regression models, can be extended via the estimation of local parameters. The parameters of each spatial point in the entire model are independently quantified, which is often used to test the existence of spatial non-stationarity in the relationship between dependent variables and independent variables [64]. The model can be used to characterize the effects of geology, meteorology, and human activities on vegetation coverage at different spatial locations. The GWR technique extends the traditional global regression by adding a geolocation parameter, and the formula is as follows:

$$y_i = \beta_o\,(\mu_i, v_i) + \sum_{k=1}^{p} \beta_k(\mu_i, v_i)x_{ik} + \varepsilon_i,\ i = 1, 2, \ldots n \quad (9)$$

where y_i is the dependent variable, x is the independent variable of the explanatory factor, $\beta_0(\mu_i, v_i)$ represents the intercept at position i, $\beta_k(\mu_i, v_i)$ represents the local parameter estimation of the explanatory variable x_{ik} at position i, and ε_i is the random error term at point i.

The estimated coefficients of GWR are weighted according to the observations and the spatial proximity of a particular point i. The parameters can be estimated using the rectangular equation:

$$\hat{\beta}(\mu, v) = \left(X^T W(\mu_i, v_i) X\right)^{-1} X^T W(\mu_i, v_i)\, Y \quad (10)$$

where $\hat{\beta}(\mu, v)$ represents the unbiased estimate of the regression coefficient β, $W(\mu_i, v_i)$ is the weighting matrix, and X and Y are the matrices of independent and dependent variables. $W(\mu_i, v_i)$ ensures that observations close to a specific location have greater weight, expressed using a Gaussian weighted kernel function:

$$w_{ij} = exp\left(-\frac{d_{ij}}{b^2}\right) \quad (11)$$

where w_{ij} represents the weight of observation j at position i, d_{ij} represents the Euclidean distance between regression point i and adjacent observation j, and b represents the basic width of the kernel function.

Stationarity exists when the variable x_{ik} does not vary with position i, and the GWR-based stationarity index is used to estimate spatial stationarity [65]:

$$SI = \frac{\beta_{GWR\_iqr}}{2 \times GLM\_se} \quad (12)$$

where SI is the stationarity index, β_{GWR_iqr} is the standard error interquartile range of the GWR coefficient and GLM_se is the standard error of the global regression analysis. When $SI < 1$, the explanatory variable y and the dependent variable x achieve spatial stationarity.

AIC can be used to determine the significance of the coefficients to compare relative measures of model performance, the smaller the AIC, the more reliable the model is, and *AICc* represents the limited sample size correction result of the AIC [66].

$$AICc = 2nIn(\hat{\sigma}) + nln(2\pi) + n\left(\frac{n + tr(S)}{n - 2 - tr(S)}\right) \quad (13)$$

where n is the number of samples, $\hat{\sigma}$ is the estimated value of the residual standard deviation, *tr(S)* represents the trajectory of the hat matrix, and when the *AICc* value is lower than three, the model performs better.

To study the driving mechanisms of NDVI in this study we used the local regression method in the GWR Model, taking 14 driving factors related to climate factors, terrain factors and geology and geomorphology as independent variables. In addition, the significance grading result of the NDVI value was used as the dependent variable. Among them, insignificant degradation, insignificant improvement, significant improvement, and extremely significant improvement were assigned as −1, 1, 2, and 3 respectively to represent the change in NDVI. Thus, the estimation coefficients of different factors on NDVI at more than 20,000 sampling points were obtained. To control the accuracy of the estimated coefficients, the outliers of the estimated coefficients of each driving factor were deleted, and the distribution map of the estimated coefficients was obtained through interpolation.

3. Results

3.1. Spatiotemporal Characteristics of NDVI Trends

3.1.1. Temporal Characteristics of NDVI Trends

The average value of NDVI in the whole region from 2000 to 2020 generally displayed an upward trend (Figure 3). The NDVI value changed from 0.40 to 0.51 over 21 years, roughly increasing by about 0.11, but the R^2 value was smaller, at 0.486. The variation in the mean value of NDVI had a small amount of fluctuation, and the fluctuation period was about 3 years, reaching a minimum value in 2012 and a maximum value in 2018. Affected by the Wenchuan earthquake and secondary geo-hazards, the NDVI value of the whole region was in a continuous downward trend from 2009 to 2012, and gradually increased in the following years. Although the NDVI of the whole region was generally on the rise, the variation characteristics of NDVI in each subregion were different. Among these, the NDVI value in DA generally showed a downward trend, with a serious and continuous decline from 2007 to 2012 and reached the lowest value of 0.3136 in 2012. In the past 21 years, NDVI values in other regions have been rising, but due to the influence of temperature, the NDVI values in the whole region decreased significantly in 2012.

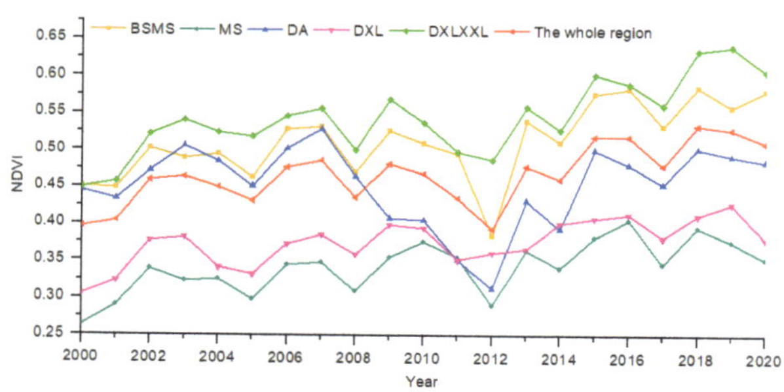

Figure 3. Average value of NDVI in different geomorphological zones from 2000 to 2020.

To obtain the correlation of the NDVI trends and average values between different zones, we conducted a Pearson analysis on the variations in the average NDVI values between different zones from 2000 to 2020 (Figure 4). Except for DA, the correlation coefficient between NDVI trends in the whole area and each area reached more than 0.85.

Affected by the surrounding crustal movement, the variation in the NDVI value in DA was unstable, and the correlation coefficient with other areas was low, ranging from 0.27 to 0.59. Due to the low altitude, suitable temperature and sufficient precipitation, the correlation coefficients of BSMS and DXLXXL with the whole region reached 0.96 and 0.97, respectively, which can better represent the NDVI trends in the whole region.

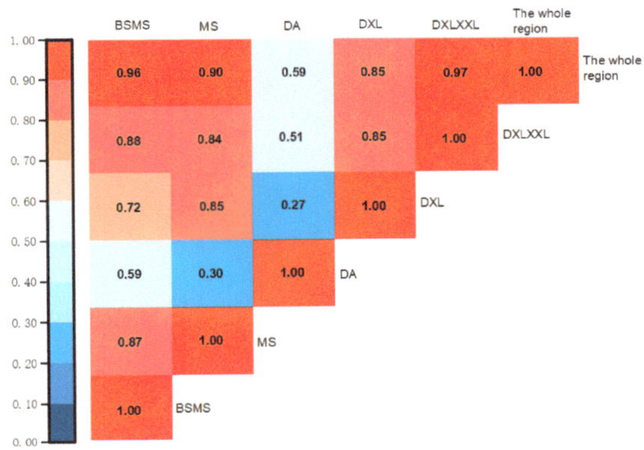

Figure 4. Correlation analysis of NDVI mean values in different geomorphological zones.

3.1.2. Spatial Characteristic of NDVI Trends

There were few human activities in the SCOCGPNP, and construction land accounted for only 0.6%, so the distribution of NDVI was mainly affected by natural factors. In the study area, the high NDVI values were mainly distributed in the BS, the eastern part of BSMS, the eastern part of DXL and other places with lower altitudes (Figure 5a). Affected by high altitude, the western region has barren vegetation, covered glaciers and low annual precipitation, with an annual precipitation level below 1000 mm all year. The average temperature in some areas was lower than 0 °C, and the basic conditions for vegetation growth were not met. Therefore, the grassland area accounted for more and the woodland accounted for less growth. The overall value of NDVI showed a decreasing trend from northwest to southeast.

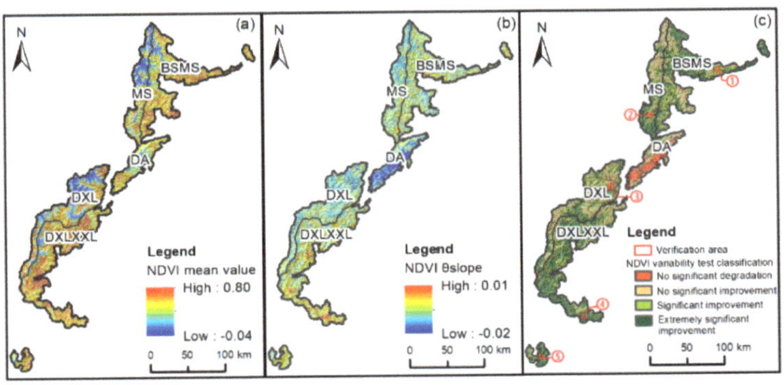

Figure 5. Annual average value of NDVI and its change from 2000 to 2020; (**a**) NDVI average value; (**b**) NDVI θ_{slope}; (**c**) NDVI trends.

To analyze the spatial improvement and degradation of NDVI values in the study area, in this study we calculated the NDVI variability (Figure 5b), and then divided the calculated NDVI variability per pixel into four levels according to the significance classification standards (Figure 5c) to obtain the spatial heterogeneity of the NDVI trends over 21 years. During the period from 2000 to 2020, the NDVI trends showed an overall trend of improvement, and a vegetation coverage rate of 94% of the study area also showed a trend of improvement with an annual growth rate of about 4.7%/yr. The proportion of non-significantly degraded areas was 6%, mainly distributed in DA. The land-use types consisted mainly of shrubs and grasslands, and these were close to the Wenchuan earthquake-generating fault zone, which is prone to secondary geo-hazards. The extremely significant improvement area accounted for 47% of the whole area, mainly distributed in the north of BSMS, the southwest of DXLXXL, and the XXL area, which had high NDVI values, abundant precipitation, and high temperature. The significant improvement area accounted for 47% of the whole area, with scattered areas of non-significant improvement and extremely significant improvement, and most of the land-use types were classed as forest land. The SCOCGPNP, from northwest to southeast, is located in a transition zone from a subtropical zone to a warm temperate zone and a transition zone from the Qinghai Tibet Plateau to the Sichuan Basin. Therefore, the NDVI trends can be roughly divided into the northwest region and southeast region, and the temperature in the southwest region is low all year, which is not conducive to the growth of vegetation, whereas the temperature in the southeast region is high all year, especially in the south.

3.1.3. Verification of NDVI Trends

To test the accuracy of the NDVI trend analysis, we selected five verification areas in different nature reserves for accuracy verification (Figures 5c and A2). Figures show that the Landsat Image of each verification area is consistent with the change trend of vegetation coverage calculated by MODIS NDVI data. In addition, due to the significant degradation of DA, the images of 2000, 2007, 2008, 2009, and 2020 were selected to verify the vegetation changes before and after the Wenchuan earthquake of 12 May 2008. The selected images are Landsat 5 images synthesized by bands 3, 2, and 1 in 2000, 2007, 2008, 2009 and Landsat 8 images synthesized by bands 4, 3, and 2 in 2020 (https://earthengine.google.com/, accessed on 10 May 2022). In DA, the NDVI value changed greatly from 2007 to 2009 (Figure 6). Due to the influence of the Wenchuan earthquake, the relatively strong vibration around the fault caused varying degrees of vegetation damage [67]. Secondary disasters such as landslides and collapses caused by the earthquake in this area gradually transformed the forest land into grassland or bare land [68,69].

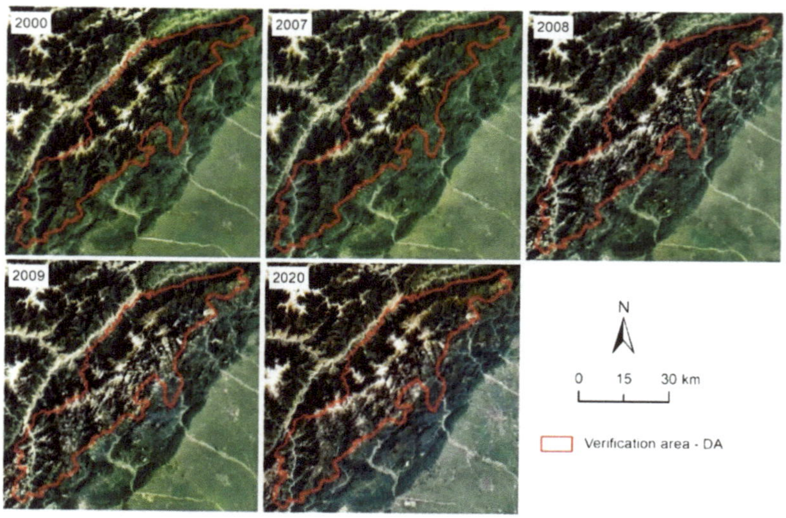

Figure 6. Remote sensing images of DA in 2000, 2007, 2008, 2009, and 2020.

3.2. NDVI Driver Analysis

3.2.1. NDVI Climate Driver Analysis

From 2000 to 2020, the average value of NDVI, annual precipitation, and annual temperature all showed an upward trend (Figure 7). The average range of NDVI was mainly between 0.4 and 0.5, showing a downward trend from 2009 to 2012, but there was a slight increase in the past 21 years. The average annual temperature range was mainly between 5.6 °C and 6.6 °C, reaching the highest value in 2006 and 2016 and the lowest value in 2000. The annual precipitation ranged from 700 mm to 900 mm, reaching the highest value in 2013 and the lowest value in 2006. The precipitation fluctuated greatly from 2000 to 2006, and the range of fluctuation in temperature was similar to that of NDVI.

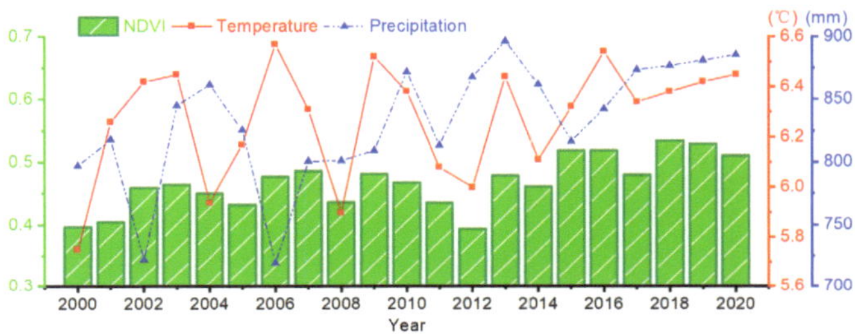

Figure 7. Statistical chart of NDVI, temperature and precipitation from 2000 to 2020.

Temperature and precipitation are two factors that directly affect the spatial distribution of vegetation. In this study, we calculated the partial correlation coefficient between NDVI, precipitation and temperature from 2000 to 2020 pixel by pixel. In general, there was a strong positive correlation between temperature and precipitation and NDVI. The partial correlation coefficient between temperature and NDVI was greater than zero in more than 90% of the regions (Figure 8a), indicating that temperature and NDVI had a basically positive correlation. The high values of the partial correlation coefficient between temperature and NDVI were mainly distributed in the BSMS and DXLXXL regions, which have a low altitude and high temperature. The partial correlation coefficient between precipitation and NDVI was greater than zero in more than 30% of the regions (Figure 8b). The places with a high correlation between precipitation and NDVI were mainly distributed near the boundaries of three geomorphological divisions in MS, the Wolong Nature Reserve, and the eastern part of XXL.

In this study, a *t*-test with a confidence level of 0.05 was carried out on the analysis results. In the partial correlation *t*-test between NDVI and temperature and precipitation the proportions of results passing the 0.05 confidence test were 51.02% and 9.14%, respectively (Figure 8c,d). After the complex correlation analysis between precipitation and temperature (Figure 8e) the analysis results were tested with a confidence level of 0.05 (Figure 8f), and the value of NDVI in 34.22% of the region was driven by climate factors, which was mainly distributed in the north and south of Southwest Subalpine and Middle Mountain. The average temperature in this region was approximately more than 12 °C, the average annual precipitation was more than 1000 mm, and the variability of precipitation and temperature was large, providing sufficient water and light for vegetation.

According to certain climate-driving factor zoning principles, the climate-driving factors of the whole region were divided into four types: temperature-driven, precipitation-driven, temperature- and precipitation-driven, and other driving modes (Figure 9). The precipitation-driven area accounted for 1.32% of the whole area, which was distributed in the middle of the Xuebaoding Nature Reserve and the Wolong Nature Reserve. The area driven by temperature accounted for 27.46% of the whole region, which was distributed in the west of MS and DXLXXL. The area driven by other driving modes accounted for 65.77% of the whole region; the elevation was higher in this region, and the land-use type was mainly bare land or grassland. Most climate-driven regions had higher NDVI values, higher temperatures, and higher average annual precipitation, which had a significant positive effect on NDVI. Other driving mode areas were generally located at higher altitudes, with lower NDVI values. In these areas, the vegetation was dominated by grasses, lichens, and mosses; the air pressure was low and the carbon dioxide content was much lower, so the driving effects of precipitation and temperature were weak. The driving mechanism in this region is very complex, so it is necessary to use the GWR model to identify the comprehensive driving mechanism of NDVI changes including climate factors based on the analysis of the correlation between climate and NDVI changes.

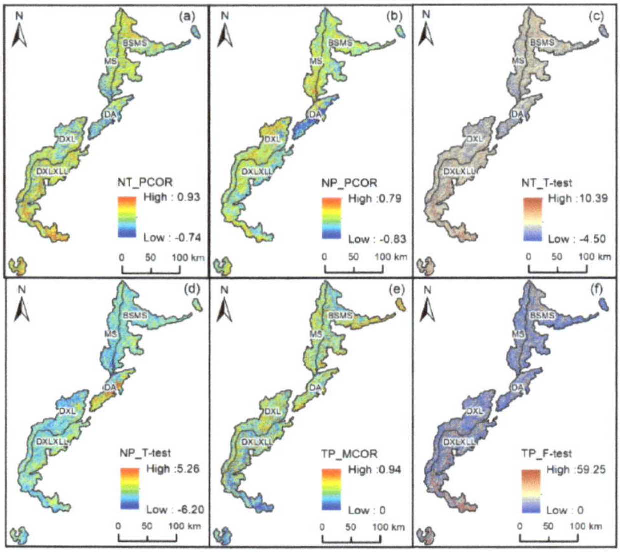

Figure 8. Partial correlation and complex correlation analysis results among NDVI, precipitation and temperature. (**a**) Partial correlation coefficient between temperature and NDVI; (**b**) partial correlation coefficient between precipitation and NDVI; (**c**) partial correlation *t*-test of temperature and NDVI; (**d**) partial correlation *t*-test of precipitation and NDVI; (**e**) multiple correlation coefficient between precipitation and temperature; (**f**) multiple correlation F-test of precipitation and temperature.

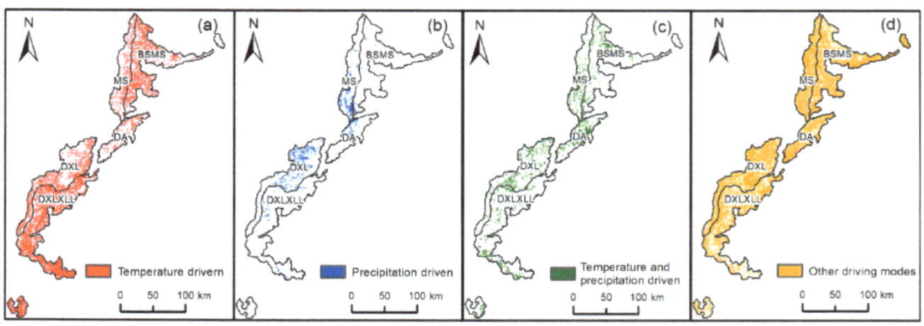

Figure 9. Climate-driven zoning map. (**a**) Temperature-driven; (**b**) precipitation-driven; (**c**) temperature- and precipitation-driven; (**d**) other driving modes.

3.2.2. NDVI Driver Analysis Based on the GWR Model

In this study we used the inverse distance weight to interpolate the spatial distribution relationship of the continuous estimated coefficients between each driving factor and NDVI (Figure 10). The average values of the estimated coefficients of each factor were sorted as follows: TEM_MN > TEM_BT > PRE_MN > PRE_BT > DEN_RIVER > PGA > LUCC >

ED_FAULT > NDVI2000 > ED_ROAD > Aspect > Elevation > ED_BLAND > Slope. This indicates that natural factors occupied a dominant position, the variability-driving effect of climate factors was strong, and the correlation between Euclidean distance from built-up land and slope was low.

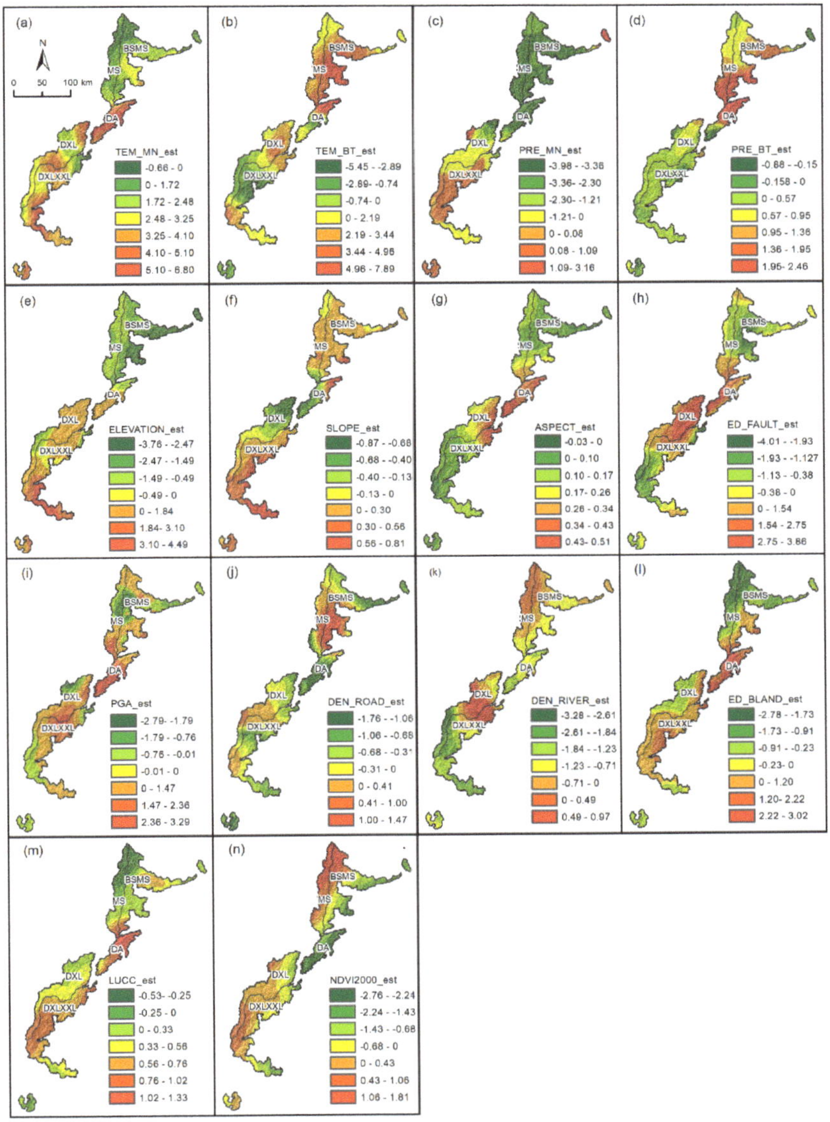

Figure 10. Driving factor estimation coefficient in GWR Model. (**a**) Temperature mean value; (**b**) temperature slope value; (**c**) precipitation mean value; (**d**) precipitation slope value; (**e**) elevation; (**f**) slope; (**g**) aspect; (**h**) Euclidean distance from fault; (**i**) peak ground acceleration; (**j**) road density; (**k**) river density; (**l**) Euclidean distance from built-up land; (**m**) land-use change index; (**n**) NDVI in 2000.

(1) Driving Effect of Climate Factors

The driving effect of climate factors on NDVI is obvious, and the climate in most areas played a driving role. High temperature and more precipitation were conducive to the growth of vegetation. The positive-driving area of the average temperature accounted for 91.95%, mainly in DA. The negative-driving effect of the area was small, and the absolute value of the estimated coefficient was only 0.66. In areas with higher altitudes, due to the limitation of the growth environment, temperature mostly drove NDVI negatively, and had no obvious promoting effect on vegetation growth. The estimation coefficient of temperature variability was mainly positive-driven and concentrated in MS, whereas the region with a large increase in temperature had less of a negative-driving effect on NDVI.

The positive-driving effect of precipitation-related factors on NDVI was smaller than that of temperature-related factors. The positive-driving area accounted for 25.91% of the estimated coefficient of the average precipitation and was mainly located in DXLXXL. The negative-driving area accounted for a large proportion, but the absolute value of the estimation coefficient was small, and was concentrated in MS.

(2) Driving Effect of Geomorphological Factors

In areas where the altitude was higher, the reduction in temperature and oxygen was not conducive to the growth of vegetation, and the driving effect of slope and NDVI was weak. The negative values of altitude estimation coefficients were mainly concentrated in high-altitude mountainous areas in BSMS and MS. The driving effect of slope on NDVI values was generally small, and the negative-driving area was mainly concentrated in Wolong Nature Reserve where (compared with the surrounding areas) the slope was relatively large. Most of the aspect estimation coefficients were positive, accounting for 87.44%, mainly concentrated in the middle of the study area.

(3) Driving Effect of Geological Activities Factors

In this study, we selected the Euclidean distance from the fault and the peak acceleration of the ground motion as the driving factors related to the geological and geomorphology, and the driving effects of the two on the NDVI were mainly positive. The positive-driving area in terms of the Euclidean distance from the fault accounted for 46.46%. It was mainly concentrated in the large area near the Longmenshan fault zone, which is sensitive to the activity of the fault zone. The negative-driving area accounted for 53.54%, which was mainly concentrated in the south and north of the study area. The positive-driving area of seismic peak acceleration accounted for 64.74%. It was mainly concentrated in DA, which was seriously affected by the secondary disaster of the earthquake. In the northeast of BSMS, the Xuebaoding nature reserve, and the southwest of DXL, the stratum was not active, so its driving effect was weak.

(4) Driving Effect of Human Activity

Human activities have a certain blocking effect on vegetation, in which the high density of roads is not conducive to vegetation growth, and the distance from construction land also has a negative impact on vegetation growth. The negative-driving area of road density exceeded 60%, mainly concentrated in the north of BSMS and DA. The positive value of Euclidean distance from built-up land accounted for 52.52%, and it was concentrated in Da and DXL. The area was close to the construction land, and the NDVI value was low, reflecting the restrictive effect of human activities on vegetation growth.

(5) Driving Effect of Other Factors

The distribution of rivers and the change in land-use types can represent the distribution of NDVI values, and the NDVI value in the initial year can indicate whether the NDVI values in some areas have reached saturation. The river density mainly had a negative impact on the NDVI trend, accounting for 76.44%, mainly in the south of DXL. The change in land-use types reflected the changes in vegetation from a dynamic point of view. The land-use change index was highly consistent with the change trends of NDVI, and the positive-driving area accounted for 85.47%, mainly concentrated in DA and DXL. The negative distribution of the estimated coefficient of NDVI in 2000 was closely related to the elevation. In the southeast of the study area and other low-altitude areas, the positive-driving area of NDVI value in 2000 accounted for 52.43%. The regions with higher altitude were limited by the environment, and the NDVI value reached a certain degree of saturation and no longer had growth potential.

(6) Gradient Variation in Each Driving Factor

To study the gradient variation in each driving factor in SCOCGPNP from northeast to southwest, the study area was divided into five areas, and the numerical distribution of each driving factor estimation coefficient in each area was counted to obtain the histogram shown in Figure A3. The estimated coefficients of the average values of temperature and precipitation tended to increase from northeast to southwest, and most of the estimated coefficients of the average values of temperature were positively correlated, especially in DXL and DXLXXL, most of which were concentrated between 2 and 3. Among the topographic factors, the elevation of BSMS in the Sichuan Basin mainly had a negative correlation with the NDVI value, and the estimation coefficient was distributed between −4 and 1; whereas the estimation coefficient in DXLXXL was mainly concentrated between zero and three, with an obvious positive-driving effect. The road density had a great impact on the NDVI trends in DA, with the slope around the road being large and close to the fault zone. In DXLXXL, the negative impact of river density on vegetation was more obvious, and the area had high annual precipitation and a large river flow. The Euclidean distance to construction land had a great impact on the vegetation growth in DA, as this area is adjacent to the urban area of Chengdu in the southeast and close to the urban area of Wenchuan in the northwest and is thus greatly affected by the development of urbanization. The low-altitude areas of DA and DXLXXL were greatly affected by human activities, and the positive-driving effect of the land-use change index on the change in the NDVI value was more obvious.

4. Discussion

4.1. Analysis of NDVI Trends

In the past two decades, mechanisms such as natural forest protection, ecological compensation, ecological protection, and regional sustainable development in the study area have all played positive roles in improving the vegetation environment. The Chinese government implemented the project of returning farmland to forest in 2002, and the area of farmland being returned to forest in Sichuan Province reached 77.6×10^4 hm^2 [70,71]. From 2000 to 2020, the NDVI trends, precipitation, and temperature showed upward trends. The years with a decreasing trend in NDVI change were similar to the years with low values of temperature. With the decreases in temperature in the study area in 2008, 2012, 2014, and 2018, the NDVI value was affected to varying degrees. Affected by the 2008 Wenchuan earthquake and its secondary disasters, the NDVI value in DA showed a continuous downward trend from 2009 to 2012. As the precipitation in the study area in 2011 was lower than that in previous years [72] due to drought events, the NDVI value of the whole area showed an obvious downward trend in 2012.

The univariate linear regression model was able to accurately and intuitively analyze the past NDVI trends and obtain the variation in vegetation characteristics of SCOCGPNP in time and space. The results showed that the variations in vegetation in the study area were mainly affected by elevation and the fault zone, with spatial heterogeneity. Affected by external forces such as crustal instability near the fault zone, the insignificant degradation areas were relatively concentrated, mainly distributed in DA. This area is located in the core area of the Longmenshan fault zone, with a basic earthquake intensity of VIII and the Wenchuan earthquake intensity of XI. In addition, this area is located in the core area of the rain screen district on the eastern edge of the Qinghai Tibet Plateau, and there are problems relating to remaining mining sites. The stability of slopes in the area has decreased sharply, and collapses, landslides, and debris flows are relatively common [73,74]. Therefore, in the past two decades, the NDVI value in this area has shown no significant degradation. On the contrary, the crust in DXL is not active, the temperature is appropriate, the precipitation is sufficient, and the vegetation growth is relatively stable, so it was found to be in a state of extremely significant improvement. In addition to the geological activities and climatic factors, NDVI is also limited by topographic factors to a certain extent [75], and the boundary between significant improvement and non-significant improvement is roughly the same as the snow line. In high-altitude areas, due to perennial snow, thin oxygen, low temperature, insufficient sunshine, and other environmental factors, vegetation generally does not improve significantly.

4.2. *Driving Force Analysis of NDVI Trends*

4.2.1. Climatic Factors

Climate affects vegetation types and spatial distribution [76]. Temperature and precipitation affect plant growth and distribution by affecting effective accumulated temperature and the amount of water available to regulate plant photosynthesis, respiration, and soil organic carbon decomposition. The climate conditions in the study area are diverse. From northwest to southeast, there are plateau temperate humid areas, northern subtropical humid areas, and middle subtropical humid areas [77]. With the increase in temperature, the annual precipitation increases to the level of more than 800 mm, and the water demands of vegetation gradually tend to be saturated, which shows that the sensitivity driven by temperature is higher than that driven by precipitation. In plateau temperate humid regions, such as DA and the Wolong Nature Reserve, due to the limitation of temperature, when NDVI reached saturation, the influence of other driving modes increases. In areas with abundant precipitation, such as the southern DXLXXL and XXL, the area of temperature-driven areas accounts for a larger proportion, because when the precipitation reaches saturation, the increase in temperature enhances the fertilization effect of CO_2, and the photosynthesis of vegetation is enhanced under high CO_2 concentration and water stress [78].

Affected by the complex topography of the study area and the transition from a subtropical zone to a warm temperate zone, there was spatial heterogeneity in the distribution of NDVI trends and climate factor regression coefficients obtained via GWR analysis. In the study area, affected by the superposition effect of climate factors, low-altitude areas are positively correlated with NDVI trends, and high-altitude areas are mostly negatively correlated [79]. In BS and the west of MS, the driving effect of average temperature on NDVI trends was mainly negative. In high-altitude, low-temperature, and relatively arid areas affected by the alpine climate the vegetation types are mainly grassland and alpine meadow [80]; the transpiration of vegetation becomes weaker and the improvement effect of temperature rise on vegetation is not significant [81,82]. The negative-driving force of annual precipitation variability on NDVI trends in the study area was mainly distributed in DXL. Because the surface of this area is deeply cut, coupled with frequent seismic activities, the increase in precipitation is accompanied by the occurrence of geo-hazards. At the same time, the presence of complex landforms and a cloudy and rainy environment can also reduce the accuracy of remote sensing results in vegetation monitoring [83].

4.2.2. Geomorphological Factors

Under natural conditions, plant growth and changes are closely related to regional-scale topographic conditions [84] and elevation, cutting depth, and aspect affect soil moisture and the solar radiation distribution. The northwest of the study area is dominated by the Hengduan Mountain Alpine Canyon, which forms a deep elevation difference with the basin in the southeast. The positive and negative-driving effect of elevation on NDVI was about 2500 m, showing vertical zoning. From low altitude to high altitude, the driving effect of elevation on NDVI gradually decreased. The Ya'an area of the subalpine basin in southwestern Sichuan Province and central Yunnan, due to its low altitude, small slope, and sufficient light and heat conditions is conducive to returning farmland to forests.

Thanks to the relatively stable geological environment, relatively balanced soil nutrients, and sufficient water conservation, the proportion of vegetation improvement in this area was relatively large [85].

Slope has an important impact on surface runoff and soil properties, and affects the intensity of human activities, so there are differences in vegetation growth conditions in areas with different slopes. In MS and DXL, some areas were affected by slope changes, and soil moisture increased [86], which is beneficial for vegetation to absorb water, and the slope had an obvious positive-driving effect on the NDVI trends. At the same time, ecological projects such as returning farmland to forests and closing mountains for afforestation since 1999 strengthened the protection of forest land [87], and the areas with significant impact were mainly concentrated in low-altitude areas such as the Zhonglongxi-Hongkou area and the Jiudingshan Nature Reserve in Aba Prefecture where there are more transitions of cultivated land to forests, and vegetation coverage increases with the increase in slope.

Aspect indicates the intensity of solar radiation received by the slope and the value and degree of changes in ground water, which affects the sunshine hours and light intensity of vegetation [88]. Compared with topographic factors such as elevation, the driving effect of aspect on NDVI was relatively small. Variations in vegetation were more positively affected by sunlight on sunny slopes. In the southern part of DXL, due to high precipitation, cloud cover and less solar radiation, and relatively wet conditions, the humidity change caused by aspect had little impact on vegetation growth, so the driving force of sunny slopes was small.

4.2.3. Geological Activities Factors

The peak acceleration of ground motion represents the differential movement of crustal fault blocks, and crustal movement is often accompanied by secondary disasters such as landslide and collapse, which affects the stability of the vegetation growth environment [89]. The terrain around DA in the Sichuan Basin is relatively large, and they are mainly considered part of the Longmenshan–Minshan strike-slip thrust seismic zones. When the crustal in situ stress caused by the earthquake exceeds the ultimate strength of the crustal rocks, the rocks fracture and cause surface damage. At the junction of mountains and basins in MS and DXL, the slope is high. Strong earthquakes not only cause a large number of co-seismic landslides, but also exacerbate slope instability for a long time after the earthquake, resulting in drastic changes in land cover in forest or shrubland areas [90], and the vegetation around the Wenchuan earthquake was affected to a certain extent.

Around the fault zone, crustal movement seriously damages the original vegetation and topsoil, forming a large area of secondary bare land [91]. The study area is pushed eastward by the western Qinghai Tibet Plateau [92], resulting in a thrust nappe structure with a sharp difference in altitude. Near the fault zone, the 2008 Wenchuan earthquake triggered rock faults and water and soil losses, which significantly reduced the species richness of woodlands and shrubs, destroyed tree roots, and reduced the density of forest crowns in the Longxi–Hongkou and the Caopo Nature Reserves near Wenchuan in the study area [61]. On the contrary, in the Wolong Nature Reserve, the NDVI showed an improving trend, indicating that after the 2008 Wenchuan earthquake, the vegetation near the fault zone was seriously damaged, resulting in a forest gap [93]. Affected by the amplification effect of slope-secondary geo-hazards after the earthquake, disasters secondary to the Wenchuan earthquake were relatively common in the middle and low mountainous areas. On the contrary, the geological environment of vegetation growth in alpine areas is relatively stable [94]. In XXL, the precipitation level is mostly more than 1000 mm. Affected by the superposition effect of secondary disasters, a large number of collapses and landslides caused by the Wenchuan earthquake provide rich loose solid materials for debris flow activities and also cause a large amount of slope instability and rock mass damage, which greatly reduces the precipitation threshold of debris flow outbreaks in earthquake-stricken areas [95].

4.2.4. Human Activity

The occupation of forest land by road construction for human activities is considered to be an important instigator of habitat fragmentation and biodiversity loss [96]. Road construction has mostly negative effects in areas with large topographic relief, especially in the north of BSMJ and DA. Road construction leads to a change in land use in the places the road passes, which can easily lead to the expansion of cultivated land and the increase in construction land, directly leading to habitat fragmentation and habitat loss, which is the most important and urgent factor threatening the habitat safety of wildlife [97]. In addition, in the south of Wolong Nature Reserve, the G317 national highway passes through areas with deep valleys and large slopes. The construction and use of roads affect the regeneration ability of vegetation to a certain extent, destroying the integrity of forest landscape, and affecting the richness and diversity of vegetation [98].

With the development of urbanization, the expansion of construction land has intensified the impact of human activities on vegetation, and human activities have brought about different degrees of ecosystem degradation. In DA, the area close to the construction land is affected by human activities, which has a high negative effect on NDVI [99], and there are residual problems of mineral development and a decline in the ecological restoration ability of giant panda habitats, which accelerates the fragmentation of protected areas [100]. For example, in the northwest of MS, the construction of the Jiuzhaigou–Mianyang expressway has affected the NDVI value since 2017. In addition, engineering construction activities related to the expansion and development of human society, such as road construction, mining, scenic spots, and water conservancy facilities, are threatening the habitats of giant pandas [101], and the contradiction between ecological environment protection and resource development and utilization is becoming more and more obvious. For example, the opening of the Jiuzhaigou–Chengdu tourism link in the north of the study area and the large-scale local hydropower development have seriously damaged the habitats of giant pandas [102]. However, China entered the scientific development stage of environmental protection in 2002 and then announced a new era of ecological civilization in 2012. During this period, policies for returning farmland to forest and grassland have been initiated as part of ecological immigration policy in extremely important areas of environmental protection [103]. Due to the large distribution of giant pandas in DXL, such as Wanglang, Wolong, and other nature reserves, the ecological migration policy of the nature reserve regulations coordinates the relationship between the community and the ecological environment [104], reducing the adverse impact of human activities. Especially around the Wolong Nature Reserve, construction land has an effect on improving the growth of the NDVI value, affected mainly by tourism, which has transformed the economic structure of the local community [105] and effectively achieved the purpose of sustainable development of the nature reserve.

4.2.5. Other Factors

The distribution of rivers has a regulating effect on the scope of human activities, and the development of agriculture and the construction of water conservancy projects are inseparable from the rivers, which have corresponding effects on the distribution and changes in vegetation. In protected areas such as Wolong in DXL, the positive-driving force of river density on NDVI changes is more obvious; benefiting from the higher elevations in these areas are the development of low-grade rivers, less human disturbance, and better vegetation preservation. However, Ya'an and other areas in DXLXXL were found to have steep slopes, long rainy seasons and concentrated rainfall, fast river velocity, frequent disasters such as floods and debris flows, and severe soil and water losses, resulting in low vegetation development. In addition, human activities around rivers in low-altitude areas are more frequent, and areas close to rivers are prone to the transformation of natural vegetation to farmland, which has a certain inhibitory effect on the improvement of NDVI [106].

Transformation between woodlands, shrubs, swamp wetlands and meadows, and built-up land imply changes in NDVI, resulting in changes in the temporal and spatial patterns of NDVI [58]. In the study area, land-use change mainly has a positive-driving effect on NDVI, which was concentrated in DA. The distribution of plant communities depends on the type of land use, and indirectly affects the vegetation coverage [107]. The transition from the main grassland in the study area to bare land or construction land shows an obvious trend of NDVI degradation. However, the study area was dominated by forest land, and the area of land-use type transformation was small, so the overall driving force of land-use change on NDVI changes was low.

The initial NDVI in the year 2000 can be used to represent the direction of variations, the degree of transition, and the spatial distribution characteristics of NDVI during the 21-year study period. The results show that the initial NDVI and the improvement of NDVI in DA showed an obvious negative correlation because the vegetation cover in this area significantly degraded from 2000 to 2020. The other negative correlations were concentrated in the northwest of BSMS, which was high in elevation and low in temperature, and was covered by ice, snow, and sparse vegetation all year, and the vegetation growth was limited and easily reached extreme values [108]. When the estimated coefficient of NDVI trends for the initial year was at a negative value, this usually indicated a constant or degrading trend in the GWR analysis.

4.3. Implications and Limitations

In this research we studied the change trends and driving mechanisms of NDVI in SCOCGPNP using univariate linear regression and the GWR model and obtained effective results, but there were some deficiencies. Due to the limitation of the spatial resolution of the NDVI data, the changes in vegetation growth at the slope scale caused by terrain factors such as surface relief, slope position, and slope may not be fully described. The Savitzky–Golay filtering was regarded as representative of annual data, and the different synthesis methods of data from multiple months in this study area are worth studying. In the GWR model, although the driving analysis considered the impact of land-use changes on NDVI changes, the main body of the study area consisted of forest land and the impact of different forest land types on NDVI changes was not subdivided.

With the continuous improvement of people's awareness of ecological protection, most of the vegetation in the study area is in a state of improvement through the implementation of measures such as ecological protection zones and artificial afforestation. In key areas of concern, ecological restoration will be gradually achieved through the combination of geological engineering and ecological engineering. At the same time, zoning management should be carried out according to the differences in different natural and human environments. In the Mianyang area of MS, the Ya'an area of DXL and XXL, precipitation is abundant, and the increase in temperature can stimulate the photosynthesis of vegetation which can further improve the vegetation coverage in these areas. To achieve the ecologically sustainable development of GPNP, the government should mobilize surrounding communities to participate in ecological protection projects and build a natural reserve system with national parks as the main body, nature reserves as the foundation, and various natural parks as supplements. During the operation of the national park, a long-term ecological public welfare forest compensation plan and monitoring mechanism should be added, so that residents can actively participate in ecological work. In ecological resource management, attention should be paid to the development of ecotourism, which is an important means of developing natural resource management by the community. At the same time, the construction of GPNP should take the impact of human activities as one of the monitoring indicators of the ecological environment, actively monitor climate change, reduce the risk of geo-hazards, optimize the layout and intensity of human activities, and actively implement measures such as ecological protection and ecological restoration.

5. Conclusions

In this study, we not only used the univariate linear model to visualize the trends of vegetation variations in the SCOCGPNP from 2000 to 2020, but also added driving factors such as topography and human activities on the basis of previous studies limited to the driving mechanisms of climate factors in GWR Model, so as to more comprehensively explain the variations in NDVI values. The main conclusions were as follows:

(1) During 2000–2020, the NDVI value showed an upward trend as a whole, with a small amount of fluctuation. Affected by the Wenchuan earthquake of 12 May 2008 and its secondary disasters, the NDVI value in DA showed a continuous downward trend from 2009 to 2012. As the precipitation in the study area in 2011 was lower than that in previous years, affected by drought events, the NDVI value of the whole area showed an obvious downward trend in 2012;

(2) The NDVI values of the study area showed an overall upward trend from 2000 to 2020, of which 94% of the areas were in an improved state, and the annual growth rate was about 4.7%/yr. The degraded area accounted for 7.94% of the total area, which was mainly concentrated in DA. This area was mainly affected by the Wenchuan earthquake, and the vegetation degradation caused by secondary geo-hazards was more serious;

(3) As the study area is located on the geomorphic boundary and climate transition zone, the NDVI trends were mainly affected by the natural environment, in which climate factors were dominant. Moreover, due to the saturation of precipitation in most areas, the driving effect of temperature was more obvious than that of precipitation, mainly concentrated in DXL and DXLXL. The superposition effect of rainfall and topographic factors means the slope had a strong influence on vegetation change, and the areas affected by this were concentrated mainly in BSMS and DA;

(4) In the protection of the ecological security patterns of the SCOCGPNP, we should closely monitor regional climate change, prevent, and control geo-hazards, optimize the vegetation growth environment, develop an ecological economy in combination with the current situation of human communities, reduce human interference in the reserve and finally realize sustainable development of the SCOCGPNP.

Author Contributions: M.P. wrote the paper with contributions from all co-authors. Y.Z. (Yinbing Zhao) and Z.N. conceived and designed the research. Z.H., W.P., Y.Z. (Yi Zhou), J.L. and Y.G. contributed to the data processing. All authors have read and agreed to the published version of the manuscript.

Funding: The National Park Research Center Project of Sichuan Key Research Base of Social Sciences (GJGY2019-ZD003), the Soft Science Project of Sichuan Provincial Department of Science and Technology (2022JDR0061), the Open Foundation of Hebei Key Laboratory of Wetland Ecology and Conservation (No. hklk201902), the Open Project of Human Geography Research Center of Qinghai Tibet Plateau and Its Eastern Edge (RWDL2021-ZD002), and the China Scholarship Council Fund (No.202108510096) supported this work.

Institutional Review Board Statement: Not applicable.

Informed Consent Statement: Not applicable.

Conflicts of Interest: The authors declare no conflict of interest.

Abbreviations

The following abbreviations are used in this manuscript:

CGPNP	China Giant Panda National Park
SCOCGPNP	Sichuan Area of China Giant Panda National Park
BS	Baishui River
DXL	Daxiangling Area
XXL	Xiaoxiangling Area
MS	Minshan Area
DA	Dujiangyan–Anzhou Area

Appendix A

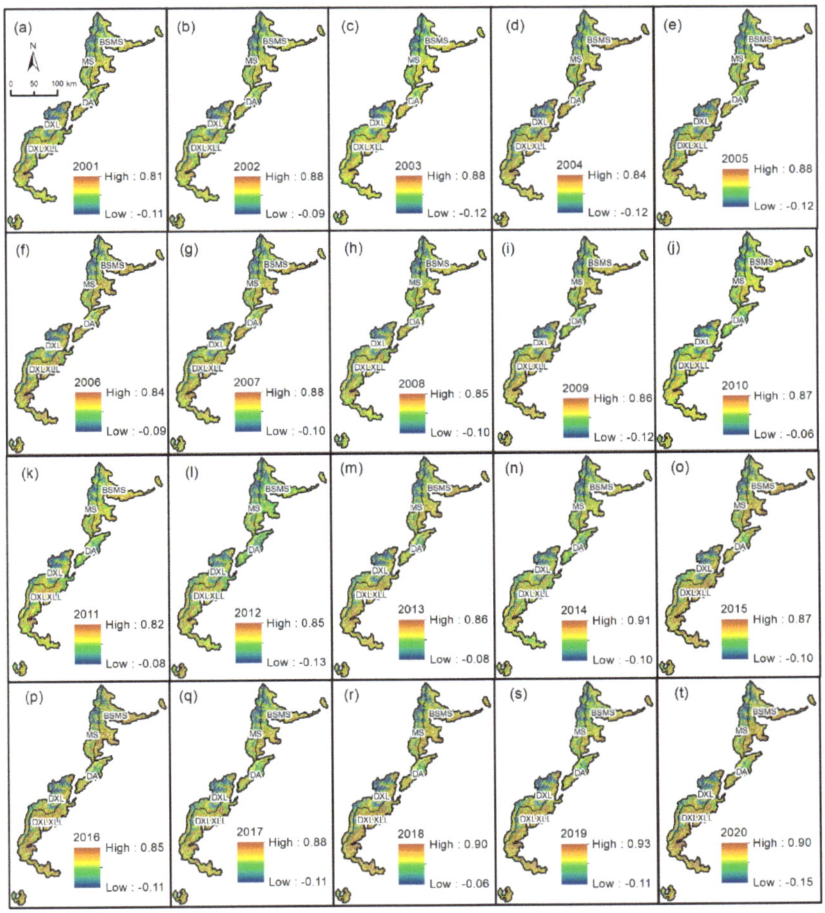

Figure A1. NDVI images from 2001 to 2020 in the study area.

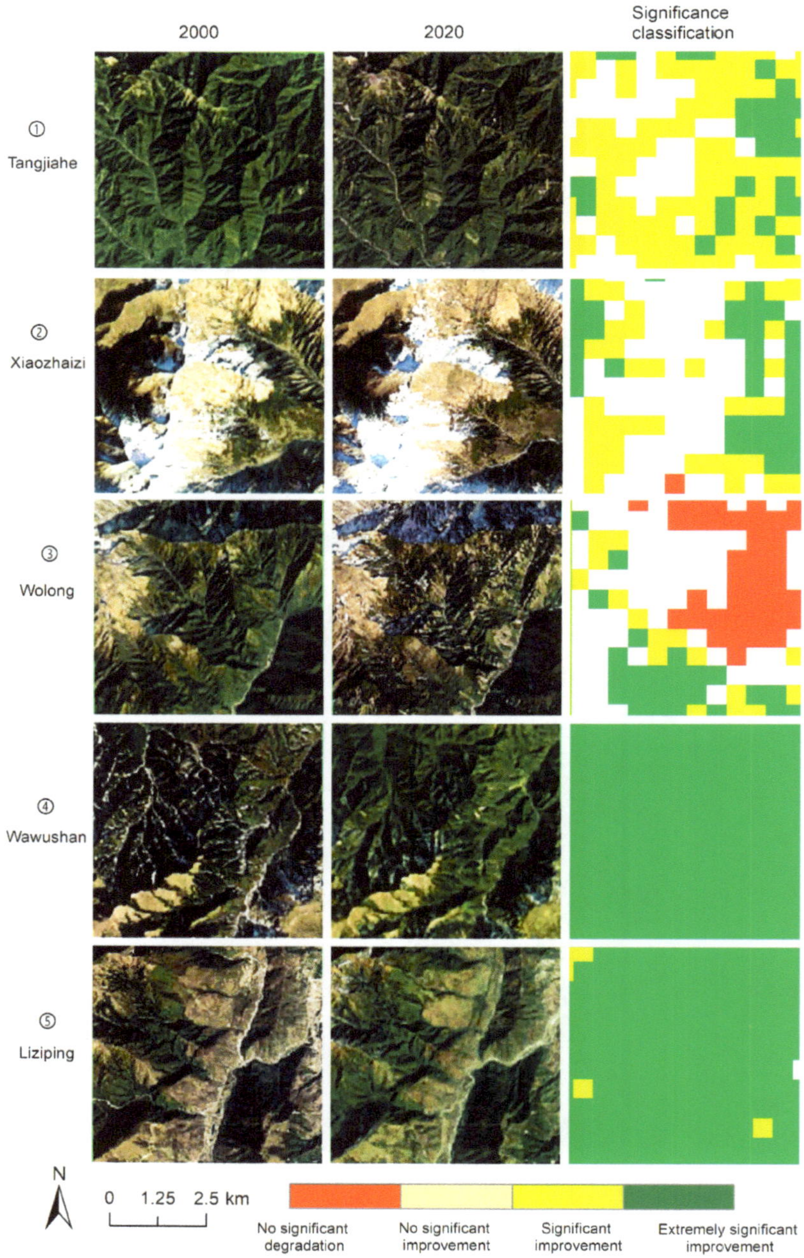

Figure A2. Remote sensing image of verification area.

Table A1. Statistics of driving factors in different geomorphological zones.

Driving Factors	Zone	MIN	MAX	MEAN	MEDIAN	STD
Temperature mean value (°C)	BSMS	−5.0829	15.7139	7.3521	70.1019	39.5631
	MS	−9.9736	13.1671	2.3445	59.9676	44.4253
	DA	−4.7338	15.3926	7.0619	61.6296	40.9538
	DXL	−12.6181	15.4079	2.2980	51.3102	46.4041
	DXLXXL	−5.7426	16.7042	8.9454	66.6759	39.5315
Temperature slope (/yr)	BSMS	−0.0001	0.0004	0.0001	0.0001	0.0001
	MS	0.0000	0.0003	0.0002	0.0001	0.0001
	DA	−0.0001	0.0002	0.0000	0.0001	0.0001
	DXL	0.0001	0.0005	0.0003	0.0002	0.0001
	DXLXXL	0.0003	0.0007	0.0005	0.0003	0.0001
Precipitation mean value (mm)	BSMS	658.9722	859.2167	782.2140	784.7940	33.5782
	MS	695.4556	873.4445	782.7805	784.2000	29.3065
	DA	773.9556	1026.3611	849.5633	793.4415	37.9322
	DXL	693.8834	1003.3723	789.6579	791.6530	31.8794
	DXLXXL	733.5555	1406.8611	897.0535	805.3220	98.8965
Precipitation slope (mm/yr)	BSMS	0.0025	0.0066	0.0045	0.0046	0.0009
	MS	0.0027	0.0066	0.0047	0.0046	0.0010
	DA	0.0056	0.0076	0.0066	0.0049	0.0004
	DXL	−0.0011	0.0062	0.0039	0.0048	0.0017
	DXLXXL	−0.0011	0.0094	0.0032	0.0045	0.0022
Elevation (m)	BSMS	540.0000	4925.0000	2384.1665	2417.0000	763.3032
	MS	1480.0000	5573.0000	3392.9168	2724.0000	692.7943
	DA	798.0000	6679.0000	2738.7173	2726.0000	784.6719
	DXL	1306.0000	6049.0000	3617.0638	3006.0000	756.6816
	DXLXXL	828.0000	5280.0000	2559.0178	2807.0000	742.5512
Slope (°)	BSMS	0.0000	87.6306	30.8077	31.2169	11.0657
	MS	0.0000	86.9273	31.7692	31.4441	11.4606
	DA	0.0000	89.2268	34.3517	31.7632	14.6286
	DXL	0.0000	85.5856	30.6938	31.5075	11.0960
	DXLXXL	0.0000	80.6044	29.6035	31.0237	11.4317
Aspect	BSMS	−1.0000	359.9654	174.0972	167.7350	99.2573
	MS	−1.0000	359.8953	172.5753	165.9640	99.2998
	DA	−1.0000	359.9691	182.2276	168.8225	106.1093
	DXL	−1.0000	359.8900	180.2346	169.4465	101.4384
	DXLXXL	−1.0000	359.8964	177.7604	168.6900	103.1578
Euclidean distance from fault (m)	BSMS	7964.65	57,340.76	33,154.21	29,525.50	13,556.48
	MS	0.00	58,590.46	30,462.28	28,367.10	17,789.54
	DA	3125.00	14,950.59	10,578.60	24,974.55	2270.77
	DXL	3334.18	44,026.04	20,923.41	22,911.85	9067.17
	DXLXXL	412.31	59,519.36	17,369.14	18,596.30	13,259.30
Peak ground acceleration (g)	BSMS	0.1500	0.2000	0.1972	0.2000	0.0116
	MS	0.2000	0.2000	0.2000	0.2000	0.0000
	DA	0.1500	0.2000	0.2000	0.2000	0.0013
	DXL	0.1000	0.2000	0.1664	0.2000	0.0277
	DXLXXL	0.1500	0.3000	0.1791	0.2000	0.0248
Road density (km/km^2)	BSMS	0.0000	0.2720	0.0179	0.0000	0.0352
	MS	0.0000	0.1675	0.0036	0.0000	0.0149
	DA	0.0000	0.5618	0.0150	0.0000	0.0474
	DXL	0.0000	0.5000	0.0173	0.0000	0.0580
	DXLXXL	0.0000	0.3815	0.0207	0.0000	0.0443

Table A1. *Cont.*

Driving Factors	Zone	MIN	MAX	MEAN	MEDIAN	STD
River density (km/km^2)	BSMS	0.0000	0.1821	0.0135	0.0000	0.0295
	MS	0.0000	0.2451	0.0166	0.0000	0.0402
	DA	0.0000	0.2138	0.0122	0.0000	0.0333
	DXL	0.0000	0.2285	0.0102	0.0000	0.0265
	DXLXXL	0.0000	0.1656	0.0099	0.0000	0.0260
Euclidean distance from built-up land (m)	BSMS	0.00	22,410.94	8355.13	7905.69	4923.44
	MS	0.00	28,517.54	12,538.66	8746.43	6845.54
	DA	0.00	12,379.42	5563.95	7632.17	2819.50
	DXL	0.00	26,504.72	11,982.42	9013.88	5840.43
	DXLXXL	0.00	41,330.98	11,434.65	9219.54	7630.49
Land-use change index	BSMS	−25.0000	20.0000	−0.1732	0.0000	2.3752
	MS	−20.0000	20.0000	−0.2752	0.0000	3.4373
	DA	−25.0000	20.0000	−1.1994	0.0000	4.7261
	DXL	−25.0000	20.0000	−0.2004	0.0000	4.2528
	DXLXXL	−25.0000	20.0000	−0.3407	0.0000	3.6004
NDVI in 2000	BSMS	−0.0564	0.7470	0.4519	0.4870	0.1651
	MS	−0.0655	0.7716	0.2638	0.4457	0.2133
	DA	−0.0373	0.7750	0.4454	0.4550	0.1594
	DXL	−0.0701	0.8263	0.3059	0.4326	0.2331
	DXLXXL	−0.0594	0.7764	0.4500	0.4488	0.1652

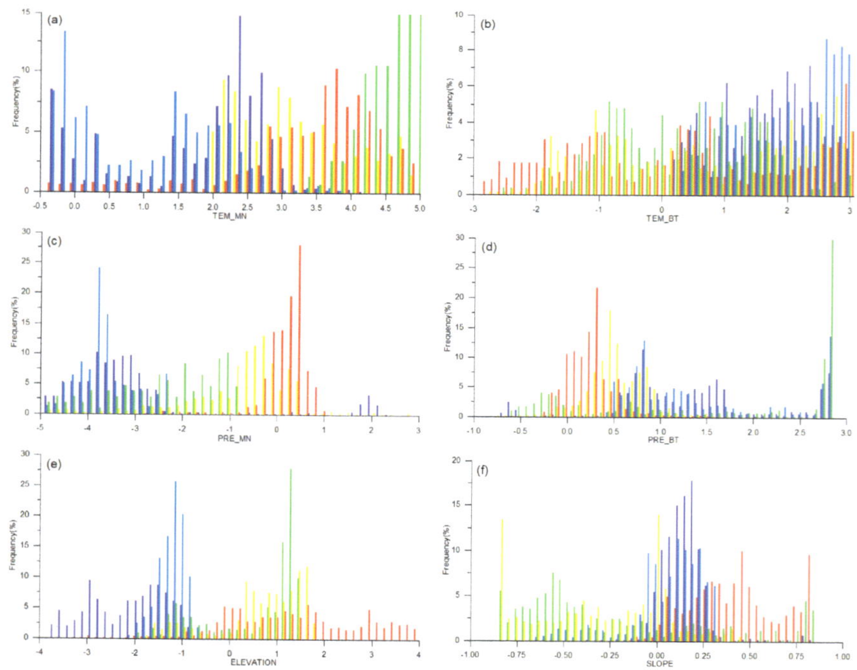

Figure A3. *Cont.*

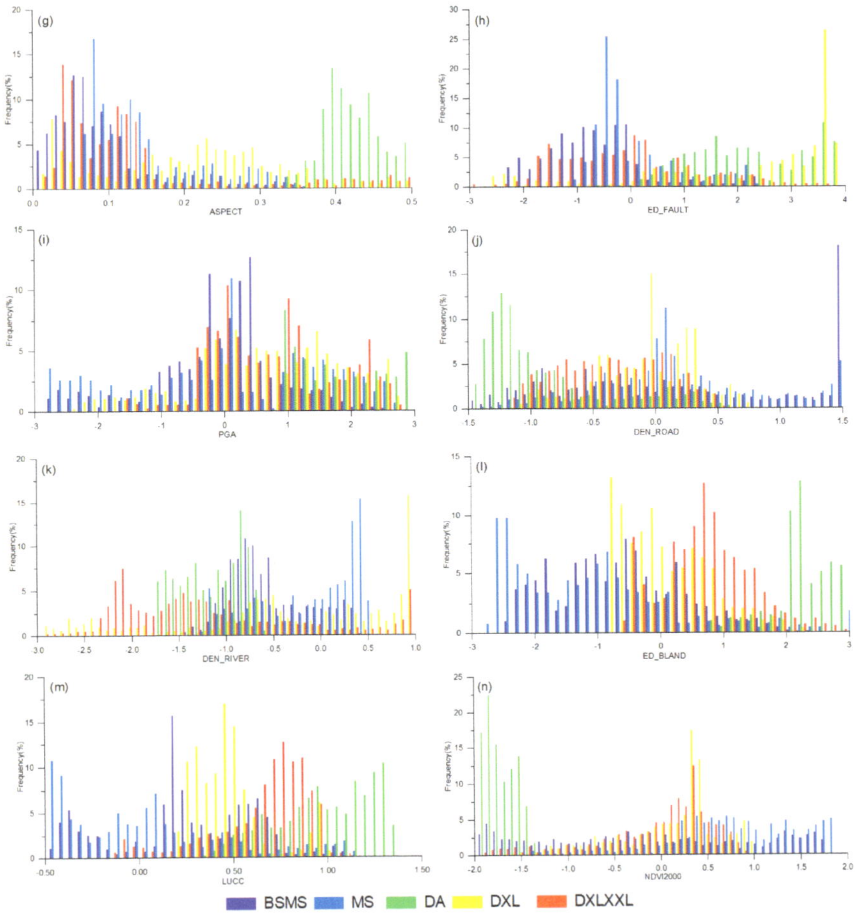

Figure A3. Histogram of estimated coefficients for each factor in different partitions. (**a**) Temperature mean value; (**b**) temperature slope value; (**c**) precipitation mean value; (**d**) precipitation slope value; (**e**) elevation; (**f**) slope; (**g**) aspect; (**h**) Euclidean distance from fault; (**i**) peak ground acceleration; (**j**) road density; (**k**) river density; (**l**) Euclidean distance from built-up land; (**m**) land-use change index; (**n**) NDVI in 2000.

References

1. Qin, Q.; Huang, Y.; Liu, J.; Chen, D.; Zhang, L.; Qiu, J.; Tan, H.; Wen, Y. The landscape patterns of the giant panda protection area in Sichuan province and their impact on giant pandas. *Sustainability.* **2019**, *11*, 5993. [CrossRef]
2. Xiao, Y.; Ouyang, Z.Y.; Zhu, C.Q.; Zhao, J.Z.; Wang, X.K. An assessment of giant panda habitat in Minshan, Sichuan, China. *Acta Ecol. Sin.* **2004**, *24*, 1373–1379. (In Chinese)
3. Li, J.; Liu, F.; Xue, Y.; Zhang, Y.; Li, D. Assessing vulnerability of giant pandas to climate change in the Qinling Mountains of China. *Ecol. Evol.* **2017**, *7*, 4003–4015. [CrossRef] [PubMed]
4. Xue, C.; Shao, C.; Gao, J. Ecological compensation strategy for SDG-based basin-type national parks: A case study of the baoxing giant panda national park. *Int. J. Environ. Res. Public Health.* **2020**, *17*, 3908. [CrossRef] [PubMed]
5. Kang, D.; Ma, Y.; Zhou, S. Comprehensive evaluation on ecological security of the habitat of the giant panda, Ailuropoda melanoleuca in the Minshan Mountains and Qionglai Mountains. *J. Biosaf.* **2020**, *29*, 6. (In Chinese)
6. Du, A.; Cui, T.; Ouyang, Z. International experiences of national park selection criteria and the inspirations to China. *Acta Ecol. Sin.* **2020**, *40*, 7231–7237. (In Chinese)
7. Cui, Q.; Zhao, M.; Tang, T.; Wang, D.; Cheng, W.; Cheng, J. The public experience perception of giant panda national park based on network text analysis. *Ecol. Econ.* **2021**, *36*, 118–124.

8. Zhao, Y.; Chen, Y.P.; Ellison, A.M.; Liu, W.G.; Chen, D. Establish an environmentally sustainable Giant Panda National Park in the Qinling Mountains. *Sci. Total Environ.* **2019**, *668*, 979–987. [CrossRef]
9. Wang, Y.; Yang, H.; Qi, D.; Songer, M.; Bai, W.; Zhou, C.; Zhang, J.; Huang, Q. Efficacy and management challenges of the zoning designations of China's national parks. *Biol. Conserv.* **2021**, *254*, 108962. (In Chinese) [CrossRef]
10. Testa, S.; Soudani, K.; Boschetti, L.; Borgogno Mondino, E. MODIS-derived EVI, NDVI and WDRVI time series to estimate phenological metrics in French deciduous forests. *Int. J. Appl. Earth Obs. Geoinf.* **2018**, *64*, 132–144. (In Chinese) [CrossRef]
11. León-Tavares, J.; Roujean, J.-L.; Smets, B.; Wolters, E.; Toté, C.; Swinnen, E. Correction of directional effects in vegetation NDVI time-series. *Remote Sens.* **2021**, *13*, 1130. [CrossRef]
12. Morawitz, D.F.; Blewett, T.M.; Cohen, A.; Alberti, M. Using NDVI to assess vegetative land cover change in central Puget Sound. *Environ. Monit. Assess.* **2006**, *114*, 85–106. [CrossRef] [PubMed]
13. Cai, Z.; Jönsson, P.; Jin, H.; Eklundh, L. Performance of smoothing methods for reconstructing NDVI time-series and estimating vegetation phenology from MODIS data. *Remote Sens.* **2017**, *9*, 1271. [CrossRef]
14. Khaznadar, M. Vegetation and land cover change in the National park of EL Kala: Application of NDVI differencing and classification analysis. *J. Biodivers. Environ. Sci.* **2015**, *7*, 231–244.
15. Plessis, W. Linear regression relationships between NDVI, vegetation and rainfall in Etosha National Park, Namibia. *J. Arid. Environ.* **1999**, *42*, 235–260. [CrossRef]
16. Wagenseil, H.; Samimi, C. Assessing spatio-temporal variations in plant phenology using Fourier analysis on NDVI time series: Results from a dry savannah environment in Namibia. *Int. J. Remote Sens.* **2007**, *27*, 3455–3471. [CrossRef]
17. Yang, B.; Qin, S.; Xu, W.; Busch, J.; Yang, X.; Gu, X.; Yang, Z.; Wang, B.; Dai, Q.; Xu, Y. Gap analysis of giant panda conservation as an example for planning China's national park system. *Curr. Biol.* **2020**, *30*, 1287–1291. [CrossRef]
18. Zhou, X.; Chen, F. Change detection of forest coverage in sichuan giant panda sanctuaries using PALSAR time series data. *Remote Sens. Technol. Appl.* **2018**, *32*, 1100–1106.
19. Chen, F.; Guo, H.; Ishwaran, N.; Zhou, W.; Yang, R.; Jing, L.; Chen, F.; Zeng, H. Synthetic aperture radar (SAR) interferometry for assessing wenchuan earthquake (2008) deforestation in the Sichuan giant panda site. *Remote Sens.* **2014**, *6*, 6283–6299. [CrossRef]
20. Tang, P.; Chen, F.; Guo, H.; Tian, B.; Wang, X.; Ishwaran, N. Large-area landslides monitoring using advanced multi-temporal InSAR technique over the giant panda habitat, Sichuan, China. *Remote Sens.* **2015**, *7*, 8925–8949. [CrossRef]
21. Fensholt, R.; Rasmussen, K.; Nielsen, T.T.; Mbow, C. Evaluation of earth observation based long term vegetation trends—Intercomparing NDVI time series trend analysis consistency of Sahel from AVHRR GIMMS, Terra MODIS and SPOT VGT data. *Remote Sens. Environ.* **2009**, *113*, 1886–1898. [CrossRef]
22. De Beurs, K.M.; Henebry, G.M. Trend analysis of the pathfinder AVHRR Land (PAL) NDVI data for the deserts of central Asia. *IEEE Geosci. Remote Sens. Lett.* **2004**, *1*, 282–286. [CrossRef]
23. Fensholt, R.; Langanke, T.; Rasmussen, K.; Reenberg, A.; Prince, S.D.; Tucker, C.; Scholes, R.J.; Le, Q.B.; Bondeau, A.; Eastman, R.; et al. Greenness in semi-arid areas across the globe 1981–2007—An Earth Observing Satellite based analysis of trends and drivers. *Remote Sens. Environ.* **2012**, *121*, 144–158. [CrossRef]
24. Huang, C.; Yang, Q.; Guo, Y.; Zhang, Y.; Guo, L. The pattern, change and driven factors of vegetation cover in the Qin Mountains region. *Sci. Rep.* **2020**, *10*, 20591. [CrossRef] [PubMed]
25. He, P.; Xu, L.; Liu, Z.; Jing, Y.; Zhu, W. Dynamics of NDVI and its influencing factors in the Chinese Loess Plateau during 2002–2018. *Reg. Sustain.* **2021**, *2*, 36–46. [CrossRef]
26. Kaufmann, R.K.; Zhou, L.; Tucker, C.J.; Slayback, D.; Shabanov, N.V.; Myneni, R.B. Reply to Comment on "Variations in northern vegetation activity inferred from satellite data of vegetation index during 1981–1999" by J. R. Ahlbeck. *J. Geophys. Res. Atmos.* **2002**, *107*, 1–3. [CrossRef]
27. Myneni, R.B.; Keeling, C.D.; Tucker, C.J.; Asrar, G.; Nemani, R.R. Increased plant growth in the northern high latitudes from 1981 to 1991. *Nature.* **1997**, *386*, 698–702. [CrossRef]
28. Mohammat, A.; Wang, X.; Xu, X.; Peng, L.; Yang, Y.; Zhang, X.; Myneni, R.B.; Piao, S. Drought and spring cooling induced recent decrease in vegetation growth in Inner Asia. *Agric. For. Meteorol.* **2013**, *178–179*, 21–30. [CrossRef]
29. Wang, J.; Rich, P.M.; Price, K.P. Temporal responses of NDVI to precipitation and temperature in the central Great Plains, USA. *Int. J. Remote Sens.* **2010**, *24*, 2345–2364. [CrossRef]
30. Lin, X.; Niu, J.; Berndtsson, R.; Yu, X.; Zhang, L.; Chen, X. NDVI dynamics and its response to climate change and reforestation in northern China. *Remote Sens.* **2020**, *12*, 4138. [CrossRef]
31. Liu, Y.; Lei, H. Responses of natural vegetation dynamics to climate drivers in China from 1982 to 2011. *Remote Sens.* **2015**, *7*, 10243–10268. [CrossRef]
32. Zou, W.C.; Xiao, G.R. Scale-location dependence of the relationship between NDVI and environmental factors in Wuyi Mountain. *Adv. Mater. Res.* **2014**, *9559*, 3828–3834. [CrossRef]
33. Cui, P.; Lin, Y.-M.; Chen, C. Destruction of vegetation due to geo-hazards and its environmental impacts in the Wenchuan earthquake areas. *Ecol. Eng.* **2012**, *44*, 61–69. [CrossRef]
34. Vina, A.; Chen, X.; McConnell, W.J.; Liu, W.; Xu, W.; Ouyang, Z.; Zhang, H.; Liu, J. Effects of natural disasters on conservation policies: The case of the 2008 Wenchuan earthquake, China. *Ambio.* **2011**, *40*, 274–284. [CrossRef]
35. Zhao, W.; Zhou, X.; Pu, H. Construction of ecological security pattern in Chengdu-Chongqing Twin-City Economic Circle. *China Environ. Sci.* **2021**, *41*, 2423–2433. (In Chinese)

36. Yang, Y.; Wang, S.; Bai, X.; Tan, Q.; Li, Q.; Wu, L.; Tian, S.; Hu, Z.; Li, C.; Deng, Y. Factors affecting long-term trends in global NDVI. *Forests.* **2019**, *10*, 372. [CrossRef]
37. Gao, J.; Jiao, K.; Wu, S. Investigating the spatially heterogeneous relationships between climate factors and NDVI in China during 1982 to 2013. *J. Geogr. Sci.* **2019**, *29*, 1597–1609. [CrossRef]
38. Li, J.; Su, Z.; Jiang, J.; Chen, W.; Yu, N.; Li, X.; Xie, J.; Wei, J. Spatial-Temporal Change in Vegetation Cover and Climate Factor Drivers of Variation in the Haihe River Basin 2003–2016. *IOP Conf. Ser. Earth Environ. Sci.* **2021**, *697*, 012005. [CrossRef]
39. He, C.; Tian, J.; Gao, B.; Zhao, Y. Differentiating climate- and human-induced drivers of grassland degradation in the Liao River Basin, China. *Environ. Monit. Assess.* **2015**, *187*, 4199. [CrossRef]
40. Propastin, P.; Kappas, M.; Erasmi, S. Application of geographically weighted regression to investigate the impact of scale on prediction uncertainty by modelling relationship between vegetation and climate. *Int. J. Spat. Data Infrastruct. Res.* **2008**, *3*, 73–94.
41. Zhao, Z.; Gao, J.; Wang, Y.; Liu, J.; Li, S. Exploring spatially variable relationships between NDVI and climatic factors in a transition zone using geographically weighted regression. *Theor. Appl. Climatol.* **2014**, *120*, 507–519. [CrossRef]
42. Wang, X.; Lesi, M.; Zhang, M. Ecosystem pattern change and its influencing factors of "two barriers and three belts". *Chin. J. Ecol.* **2019**, *38*, 11. (In Chinese)
43. Zhao, Z.; Luan, X.; Cheng, J.; Ye, J.; Li, J.; Zhang, Z.; Li, M.; Wang, H.; Yang, L. Improving management of Giant Panda National Park by quantitative assessment of ecological management. *Acta Ecol. Sin.* **2019**, *39*, 3885–3894. (In Chinese)
44. Chen, L.D.; Liu, X.H.; Fu, B.J.; LÜ, Y.H.; Qiu, J. *Identification of the potential habitat for giant panda in the Wolong Nature Reserve by using landscape ecology methodology. Landscape Ecological Applications in Man-Influenced Areas*; Springer: Dordrecht, The Netherlands, 2008; pp. 95–112.
45. Bu, H.; McShea, W.J.; Wang, D.; Wang, F.; Chen, Y.; Gu, X.; Yu, L.; Jiang, S.; Zhang, F.; Li, S. Not all forests are alike: The role of commercial forest in the conservation of landscape connectivity for the giant panda. *Landsc. Ecol.* **2021**, *36*, 2549–2564. [CrossRef]
46. Huang, Q.; Fei, Y.; Yang, H.; Gu, X.; Songer, M. Giant Panda National Park, a step towards streamlining protected areas and cohesive conservation management in China. *Glob. Ecol. Conserv.* **2020**, *22*, e00947. [CrossRef]
47. Zheng, J.; Bian, J.; Ge, Q.; Hao, Z.; Yin, Y.; Liao, Y. The climate regionalization in China for 1981–2010. *Chin. Sci. Bull.* **2013**, *58*, 3088–3099.
48. Li, B.; Pan, B.; Cheng, W.; Han, J.; Qi, D.; Zhu, C. Research on geomorphological regionalization of China. *Acta Geogr. Sin.* **2013**, *68*, 16. (In Chinese)
49. Didan, K.; Munoz, A.B.; Solano, R.; Huete, A. *MODIS Vegetation Index User's Guide (MOD13 Series)*; University of Arizona, Vegetation Index and Phenology Lab.: Tucson, AZ, USA, 2015.
50. Faisal, B.; Rahman, H.; Sharifee, N.H.; Sultana, N.; Islam, M.I.; Ahammad, T. Remotely sensed boro rice production forecasting using MODIS-NDVI: A bangladesh perspective. *AgriEngineering* **2019**, *1*, 27. [CrossRef]
51. Bolstad, P.V.; Swank, W.; Vose, J. Predicting Southern Appalachian overstory vegetation with digital terrain data. *Landsc. Ecol.* **1998**, *13*, 271–283. [CrossRef]
52. Liu, Y.; Liu, R.; Ge, Q. Evaluating the vegetation destruction and recovery of Wenchuan earthquake using MODIS data. *Nat. Hazards.* **2010**, *54*, 851–862. [CrossRef]
53. Wingate, V.R.; Phinn, S.R.; Kuhn, N. Mapping precipitation-corrected NDVI trends across Namibia. *Sci. Total Environ.* **2019**, *684*, 96–112. [CrossRef]
54. Gao, Y.; Huang, J.; Li, S.; Li, S. Spatial pattern of non-stationarity and scale-dependent relationships between NDVI and climatic factors-A case study in Qinghai-Tibet Plateau, China. *Ecol. Indic.* **2012**, *20*, 170–176. [CrossRef]
55. Hong, S.K. Changes in landscape patterns and vegetation process in the Far-Eastern cultural landscapes: Human activity on pine-dominated secondary vegetations in Korea and Japan. *Phytocoenologia* **1998**, *28*, 45–66. [CrossRef]
56. Sun, Z.; Chang, N.-B.; Opp, C.; Hennig, T. Evaluation of ecological restoration through vegetation patterns in the lower Tarim River, China with MODIS NDVI data. *Ecol. Inform.* **2011**, *6*, 156–163. [CrossRef]
57. Zhang, J.; Jia, C.; Peng, L.; Hui, T. NDVI spatial distribution and its correlation with land-use in Dan River Watershed of Shaanxi Province. *Sci. Soil Water Conserv.* **2016**, *14*, 67–73.
58. Hua, M.D.; Ming, W.Z.; Shan, S.K.; Wei, L.D.; Bai, Z.; Mei, Z.S.; Ling, L.; Hua, Z.C. The vegetation NDVI variation and its responses to climate change and LUCC from 1982 to 2006 year in northeast permafrost region. *China Environ. Sci.* **2011**, *31*, 283–292. (In Chinese)
59. Kumar, B.P.; Babu, K.R.; Ramachandra, M.; Krupavathi, C.; Swamy, B.N.; Sreenivasulu, Y.; Rajasekhar, M. Data on identification of desertified regions in Anantapur district, Southern India by NDVI approach using remote sensing and GIS. *Data Brief.* **2020**, *30*, 105560. [CrossRef]
60. Zhao, Y.; Sun, R.; Ni, Z. Identification of natural and anthropogenic drivers of vegetation change in the Beijing-Tianjin-Hebei megacity region. *Remote Sens.* **2019**, *11*, 1224. [CrossRef]
61. Pan, G.; Tang, H.J.; Chen, Z.X.; Zhang, F.R. Land cover classification based on MODIS NDVI & LST time series data in northeast China. *Resour. Sci.* **2006**, *27*, 163–169. (In Chinese)
62. Ye, Q.; Liu, H.; Lin, Y.; Han, R. Study of Maowusu sandy land vegetation coverage change based on modis Ndvi. *ISPRS Int. Arch. Photogramm. Remote Sens. Spat. Inf. Sci.* **2018**, *42*, 2127–2133. [CrossRef]
63. Xia, H.; Li, A.; Feng, G.; Li, Y.; Qin, Y.; Lei, G.; Cui, Y. The effects of asymmetric diurnal warming on vegetation growth of the Tibetan Plateau over the past three decades. *Sustainability.* **2018**, *10*, 1103. [CrossRef]

64. Tian, F.; Qiu, G.Y.; Yang, Y.H.; Xiong, Y.J.; Wang, P. Studies on the relationships between land surface temperature and environmental factors in an inland river catchment based on geographically weighted regression and MODIS data. *IEEE J. Sel. Top. Appl. Earth Obs. Remote Sens.* **2012**, *5*, 687–698. [CrossRef]
65. Brunsdon, C.; Fotheringham, A.S.; Charlton, M.E. Geographically weighted regression: A method for exploring spatial nonstationarity. *Geogr. Anal.* **2010**, *28*, 281–298. [CrossRef]
66. Leung, Y.; Mei, C.L.; Zhang, W.X. Statistical test for local patterns of spatial association. *Environ. Plan. A* **2003**, *35*, 725–744. [CrossRef]
67. Li, S.; Ni, Z.; Zhao, Y.; Hu, W.; Long, Z.; Ma, H.; Geng, C. Susceptibility analysis of geohazards in the Longmen Mountain Region after the Wenchuan Earthquake. *Int. J. Environ. Res. Public Health.* **2022**, *19*, 3229. [CrossRef]
68. Huang, R.Q.; Li, W.L. Analysis of the geo-hazards triggered by the 12 May 2008 Wenchuan Earthquake, China. *Bull. Eng. Geol. Environ.* **2009**, *68*, 363–371. [CrossRef]
69. Li, W.L.; Huang, R.Q.; Tang, C.; Xu, Q.; van Westen, C. Co-seismic landslide inventory and susceptibility mapping in the 2008 Wenchuan earthquake disaster area, China. *J. Mt. Sci.* **2013**, *10*, 339–354. [CrossRef]
70. Yang, L.; Zhang, M.; Luo, M.L.; Zhou, X. Landscape pattern change of vegetation coverage in hilly area of central Sichuan,Southwest China based on MODIS NDVI. *Chin. J. Ecol.* **2013**, *32*, 171–177. (In Chinese)
71. Qin, Q.; Liu, J.; Yu, Q.; Ma, B.; Tan, H.; Xie, L.; Wen, Y. Ecological security and spatio-temporal evolution of giant panda protected areas in Sichuan Province. *Acta Ecol. Sin.* **2020**, *40*, 7255–7266. (In Chinese)
72. Xiao, J.; Zhou, D.; Bai, X.; Qian, Q.; Yan, M. Spatial-temporal evolution of vegetation coverage and analysis of future trends in Sichuan Province. *Renmin Chang. Yangtze River.* **2018**, *49*, 16–21. (In Chinese)
73. Ge, Y.; Song, G.; Guo, Z.; Kong, Y. Characteristics and causes of 8·18 debris flow hazards at the Longmenshan Town, Pengzhou, Sichuan. *J. Hydraul. Eng.* **2012**, *43*, 147–154. (In Chinese)
74. Zeng, Q.; Xie, Q.; Xu, J. Analysis on the debris flow characteristics and effectiveness of prevention projects in Huangyang Gully in Longchi town of Dujiangyan City. *Northwestern Geol.* **2014**, *47*, 8. (In Chinese)
75. Liu, D.L.; Zhang, Y.S.; Lin, N. Association analysis of NDVI changes and topographic factors. *Appl. Mech. Mater.* **2013**, *333*, 1205–1208. [CrossRef]
76. Shi-Bo, F.; Xin-Shi, Z. Control of vegetation distribution: Climate, geological substrate, and geomorphic factors. A case study of grassland in Ordos, Inner Mongolia, China. *Can. J. Remote Sens.* **2014**, *39*, 167–174. [CrossRef]
77. Zheng, J.; Bian, J.; Ge, Q.; He, Z.; Yin, Y.; Liao, Y. China's climate zoning from 1981 to 2010. *Chin. Sci. Bull.* **2013**, *58*, 3088–3099. (In Chinese)
78. Fan, X.; Cao, X.; Zhou, H.; Hao, L.; Dong, W.; He, C.; Xu, M.; Wu, H.; Wang, L.; Chang, Z.; et al. Carbon dioxide fertilization effect on plant growth under soil water stress associates with changes in stomatal traits, leaf photosynthesis, and foliar nitrogen of bell pepper (*Capsicum annuum* L.). *Environ. Exp. Bot.* **2020**, *179*, 104203. [CrossRef]
79. Li, X.; He, B.; Quan, X.; Yin, C.; Liao, Z.; Qiu, S.; Bai, X. Recent change of vegetation growth trend and its relations with climate factors in Sichuan, China. In Proceedings of the 2015 IEEE International Geoscience and Remote Sensing Symposium (IGARSS), Milan, Italy, 26–31 July 2015; pp. 342–345.
80. Li, P.; He, Z.; He, D.; Xue, D.; Wang, Y.; Cao, S. Fractional vegetation coverage response to climatic factors based on grey relational analysis during the 2000–2017 growing season in Sichuan Province, China. *Int. J. Remote Sens.* **2019**, *41*, 1170–1190. [CrossRef]
81. Liu, Y.; Li, Y.; Li, S.; Motesharrei, S. Spatial and temporal patterns of global NDVI Trends: Correlations with climate and human factors. *Remote Sens.* **2015**, *7*, 13233–13250. [CrossRef]
82. Kong, D.; Zhang, Q.; Singh, V.P.; Shi, P. Seasonal vegetation response to climate change in the Northern Hemisphere (1982–2013). *Glob. Planet. Change.* **2016**, *148*, 1–8. [CrossRef]
83. Mao, D.; Wang, Z.; Luo, L.; Ren, C. Integrating AVHRR and MODIS data to monitor NDVI changes and their relationships with climatic parameters in Northeast China. *Int. J. Appl. Earth Obs. Geoinf.* **2012**, *18*, 528–536. [CrossRef]
84. Liu, L.; Jing, X.; Wang, J.; Zhao, C. Analysis of the changes of vegetation coverage of western Beijing mountainous areas using remote sensing and GIS. *Environ. Monit. Assess.* **2009**, *153*, 339–349. [CrossRef]
85. Zhan, Z.-Z.; Liu, H.-B.; Li, H.-M.; Wu, W.; Zhong, B. The Relationship between NDVI and Terrain Factors—A case study of Chongqing. *Procedia Environ. Sci.* **2012**, *12*, 765–771. [CrossRef]
86. Xiong, Y.; Li, Y.; Xiong, S.; Wu, G.; Deng, O. Multi-scale spatial correlation between vegetation index and terrain attributes in a small watershed of the upper Minjiang River. *Ecol. Indic.* **2021**, *126*, 107610. [CrossRef]
87. Lin, B.; Liu, Q.; You, X.; Pang, X. Engineering of Changing cultivation into tree planting and its effects to the development of rural economics in Western Sichuan, China. *J. Mt. Sci.* **2002**, *20*, 438–444. (In Chinese)
88. Bennie, J.; Hill, M.O.; Baxter, R.; Huntley, B. Influence of slope and aspect on long-term vegetation change in British chalk grasslands. *Acta Ecol. Sin.* **2006**, *94*, 355–368. (In Chinese) [CrossRef]
89. Gan, B.R.; Yang, X.G.; Zhang, W.; Zhou, J.W. Temporal and spatial evolution of vegetation coverage in the Mianyuan River basin influenced by strong earthquake disturbance. *Sci. Rep.* **2019**, *9*, 16762. [CrossRef]
90. Lin, J.; Zhou, G. *Quick Identification of Regional Earthquake-Induced Landslides Based on Sharp NDVI Change;* Springer: Berlin/Heidelberg, Germany, 2013; pp. 715–723.
91. Qing, L.; Jun, Z.; Tao, Z.; Jing, M.A.; Zhong, W.; Kai, L. The influence of vegetation restoration on soil archaeal communities in Fuyun earthquake fault zone of Xinjiang. *Acta Geogr. Sin.* **2013**, *33*, 454–463. (In Chinese) [CrossRef]

92. Zhang, Y. Seismogenic structures of the south Sichuan basin seismic zone and its neotectonic setting. *Acta Geol. Sin.* **2020**, *94*, 3161–3177. (In Chinese)
93. Meng, Q.; Miao, F.; Zhen, J.; Wang, X.; Wang, A.; Peng, Y.; Fan, Q. GIS-based landslide susceptibility mapping with logistic regression, analytical hierarchy process, and combined fuzzy and support vector machine methods: A case study from Wolong Giant Panda Natural Reserve, China. *Bull. Eng. Geol. Environ.* **2015**, *75*, 923–944. (In Chinese) [CrossRef]
94. Luo, Y.; Wang, Y. Mountain slope ground motion topography amplification effect induced by Wenchuan earthquake. *J. Mt. Sci.* **2013**, *31*, 200–210.
95. Guo, X.; Fan, J.; Cui, P.; Yan, Y. Thresholds of Rainfall Trigoerciy Debris Flows in Wenchuan Earthquake Area. *Mt. Res.* **2015**, *33*, 579–586.
96. He, G.; Chen, X.; Beaer, S.; Colunga, M.; Mertig, A.; An, L.; Zhou, S.; Linderman, M.; Ouyang, Z.; Gage, S.; et al. Spatial and temporal patterns of fuelwood collection in Wolong Nature Reserve: Implications for panda conservation. *Landsc. Urban Plan.* **2009**, *92*, 1–9. [CrossRef]
97. Gong, M.; Ouyang, Z.; Xu, W.; Song, Y.; Dai, B. The location of wildlife corridors under the impact of road disturbance: Case study of a giant panda conservation corridor. *Acta Ecol. Sin.* **2015**, *35*, 3447–3453.
98. Kang, D.; Zhao, Z.; Chen, X.; Lin, Y.; Wang, X.; Li, J. Evaluating the effects of roads on giant panda habitat at two scales in a typical nature reserve. *Sci. Total. Environ.* **2020**, *710*, 136351. [CrossRef] [PubMed]
99. Zhang, C. Using MODIS vegetation index to study urban expansion and change. *Meteorological.* **2006**, *32*, 20–26.
100. Liu, J.; Ouyang, Z.; Taylor, W.W.; Groop, R.; Tan, Y.; Zhang, H. A framework for evaluating the effects of human factors on wildlife habitat: The case of giant pandas. *Conserv. Biol.* **1999**, *13*, 1360–1370. [CrossRef]
101. Zhou, M.; Li, D.; Zhou, J. Vegetation change of giant panda habitats in Qionglai Mountains through dense Landsat Data. *Chin. J. Plant Ecol.* **2021**, *45*, 15. (In Chinese) [CrossRef]
102. Zhang, Y.; Wang, M.; Li, J. The impact of conservation projects on Giant Panda Habitat. *Acta Ecol. Sin.* **2011**, *31*, 154–163. (In Chinese)
103. Rong, Y.; Zhuang, Y.; Yang, R. Analysis on the problems and solutions of community resettlement in national parks of China. *Chin. Landsc. Archit.* **2020**, *36*, 36–40. (In Chinese)
104. Yin, W.C.; Eagles, P.F.J. Development and ranking of tourism management goals for Wolong and Wanglang Giant Panda Nature Reserves, China. *Int. J. Biodivers. Sci. Manag.* **2010**, *1*, 137–149. [CrossRef]
105. Liu, J.; Miao, H.; Zheng, H.; Ouyang, Z.; Wang, X.; Li, X.; Jiang, B. Discussion about the relationship pattern between Wolong Nature Reserve and local community. *Acta Ecol. Sin.* **2009**, *29*, 259–271. (In Chinese)
106. Zoungrana, B.J.B.; Conrad, C.; Thiel, M.; Amekudzi, L.K.; Da, E.D. MODIS NDVI trends and fractional land cover change for improved assessments of vegetation degradation in Burkina Faso, West Africa. *J. Arid. Environ.* **2018**, *153*, 66–75. [CrossRef]
107. Tappeiner, T.U. Impact of land use changes on mountain vegetation. *Appl. Veg. Sci.* **2002**, *5*, 173–184.
108. Deng, Y.; Yao, S.; Hou, M.; Zhang, T. Temporal and spatial variation of vegetation NDVI and its topographic differentiation effect in the middle and upper reaches of the Yangtze River Basin. *Resour. Environ. Yangtze Basin.* **2020**, *29*, 66–78.

International Journal of
Environmental Research and Public Health

MDPI

Article

Pollution Characteristics and Spatial Distribution of Heavy Metals in Coal-Bearing Sandstone Soil: A Case Study of Coal Mine Area in Southwest China

Dongping Deng [1,2], Yong Wu [1,2,*], Yi Sun [3], Bangzheng Ren [1,2] and Lei Song [1,2]

1 College of Environment and Civil Engineering, Chengdu University of Technology, Chengdu 610059, China; ddp900304@163.com (D.D.); rbz3796@163.com (B.R.); songlei21@mails.ucas.ac.cn (L.S.)
2 State Key Laboratory of Geohazard Prevention and Geoenvironment Protection, Chengdu 610059, China
3 School of Civil Engineering, Southwest Jiaotong University, Chengdu 610031, China; sunnys90@163.com
* Correspondence: ywu@cdut.edu.cn

Abstract: Soil pollution in coal mining areas is a serious environmental problem in China and elsewhere. In this study, surface and vertical profile soil samples were collected from a coal mine area in Dazhu, Southwestern China. Microscopic observation, concentrations, chemical speciation, statistical analysis, spatial distribution, and risk assessment were used to assess heavy metal pollution. The results show that the weathering of coal-bearing sandstone and mining activities substantially contributed to soil pollution. The concentrations of Fe, Ni, Cu, Zn, Mn, Cd, Hg, and Pb exceeded their background values. Cd caused the most intense pollution and was associated with heavily–extremely contaminated soils. The residual fraction was dominant for most metals, except Cd and Mn, for which the reducible fraction was dominant (Cd: 55.17%; Mn: 81.16%). Zn, Ni, Cd, and Cu presented similar distribution patterns, and Hg and As also shared similar distribution characteristics. Factor 1 represented anthropogenic and lithologic sources, which were affected by mining activities; Factor 2 represented anthropogenic sources, e.g., fertilizers and traffic pollution; and Factor 3 represented the contribution of metals from soil-forming parent material. More than half of the study area had high pollution risk and was not suitable for vegetable cultivation.

Keywords: coal-bearing sandstone; heavy metal pollution; soil; multivariate statistics; spatial distribution

Citation: Deng, D.; Wu, Y.; Sun, Y.; Ren, B.; Song, L. Pollution Characteristics and Spatial Distribution of Heavy Metals in Coal-Bearing Sandstone Soil: A Case Study of Coal Mine Area in Southwest China. *Int. J. Environ. Res. Public Health* **2022**, *20*, 6493. https://doi.org/10.3390/ijerph19116493

Academic Editor: José Ángel Fernández

Received: 27 April 2022
Accepted: 23 May 2022
Published: 26 May 2022

1. Introduction

Coal accounts for approximately 76% of China's primary energy consumption, and is predicted to remain the country's primary energy source for several years [1]. The coal mining industry is an important source of heavy metals in the environment, and is a major contributor to soil pollution [2,3]. The accumulation of tailings and transportation of coal, including the establishment of a large number of chemical plants, can lead to heavy metal enrichment in the soil and affect the local ecological environment [4,5]. This threatens the land productivity, ecological integrity, and ecological security of nearby areas [6,7]. Through erosion, weathering, and leaching of tailings, the metals present in the tailings can enter the surrounding groundwater, streams, sediments, and soil [8]. Moreover, in addition to causing environmental degradation, heavy metal pollution also threatens soil ecosystems and human health through food chain contamination [9,10]. High concentrations of heavy metals reduce the diversity of soil bio-communities, lead to plant toxicity, and affect agricultural productivity [11,12]. Heavy metals exist in the soil in different fractions (exchangeable, reducible, oxidizable, and residual fractions). The exchange fraction in heavy metals is easier for the plants to absorb, making it more toxic [13]). Therefore, heavy metals can threaten food security and human health through the water supply and food web [14,15].

In Southwest China, especially in the Sichuan Basin, most coal is mined in Triassic sandstone strata. The stratum is generally exposed on the surface or buried at a shallow depth and forms the unique Quaternary eluvium soil in the mining area after weathering. Weathering of parent rocks produces the soil primary and secondary minerals, such as quartz, calcite, and montmorillonite. The minerals are different in particle size, cation exchange capacity, metal species, etc., which endows soils with corresponding properties [16]. Under the influence of coal accumulation, coal and its strata in the sedimentary environment are enriched in heavy metals. With the weathering of coal-bearing sandstone, heavy metals migrate into the soil and enrich it during soil formation [17]. This soil formed by weathering of coal-bearing sandstone is widely distributed, which makes the analysis of this soil more representative. Moreover, coal mining drives the development of the local industry and agriculture, which also leads to the enrichment of heavy metals in the soil [18]. Therefore, the accumulation of heavy metals in the soil, especially in the weathered area of coal-bearing sandstone, should be investigated.

Researchers have adopted various factors and methods, including geoaccumulation index, pollution factor, and enrichment factor, to determine the degree of soil pollution [19]. The qualitative and quantitative chemical speciation of heavy metals in the soil is an important basis to clarify its migration and transformation and evaluate its potential environmental impacts [20]. Geographic information system (GIS) technology is widely used to quantify the spatial distribution of metals and identify pollution sources with low costs [21]. The availability and risks associated with soil pollution by heavy metals should be investigated for the development of reliable pollution management strategies.

To the best of our knowledge, this is the first study to investigate the characteristics of this typical soil through geological, environmental, and statistical analyses. Therefore, based on the heavy metal contents in the surface and vertical profiles of soil samples from a coal mine area in Dazhu, the objectives of this study were (1) to investigate the influence of weathering of coal-bearing sandstone on heavy metals in the soil; (2) characterize the pollution of heavy metals in the soil; and (3) analyze the correlation and spatial distribution of heavy metals in the soils of the study area. The findings can help clarify the environmental impacts of coal mining and closed mines as a basis for the development of mitigation and prevention measures by local stakeholders and authorities.

2. Materials and Methods

2.1. Site Description

The study area is located approximately 15 km northeast of Dazhu County, Dazhou City, Southwest Sichuan Province, China. It has an area of approximately 31.73 km^2 (107°20′30″–107°23′8″ E, 30°42′22″–30°45′15″ N) and is under the administration of the Xinsheng township of Dazhu County (Figure 1). The surface soil layer is covered with eluvium soil of Quaternary, with a thickness of 0–8 m. The main mineral is quartz, accompanied by typical minerals such as muscovite and calcite. The lithology includes block, gravel silt, and silty clay, which mainly derive from the weathering of coal-bearing sandstone exposed on the surface. The land-use type is farmland, and the farming method is water drought rotation. Sweet potato and rice are planted in the dry and rainy seasons, respectively. The weathered rock in the study area is upper Triassic formation, which is the coal-bearing strata of the area, with a continental sedimentary environment composed of dark gray mudstone, shale, fine sandstone, siltstone, and coal seam. Since the 1960s, more than 10 coal mines have operated in the study area. Except for Kongjiagou coal mine, which is still active, all coal mines were closed between 1990 and 2010 because of the associated pollution and economic changes. The outputs of these coal mines included mainly bright coal, followed by dark coal. Bright coal has a strong shine and a layered structure with thin charcoal. Dark coal has a milder shine, with the formation of lens-shaped specular coal and silk charcoal. The average moisture content in the raw coal is less than 0.9%, the average yield of volatile matter (V_{daf}) is 29.04–33.36%, and the CO_2 content is less than 2% [2].

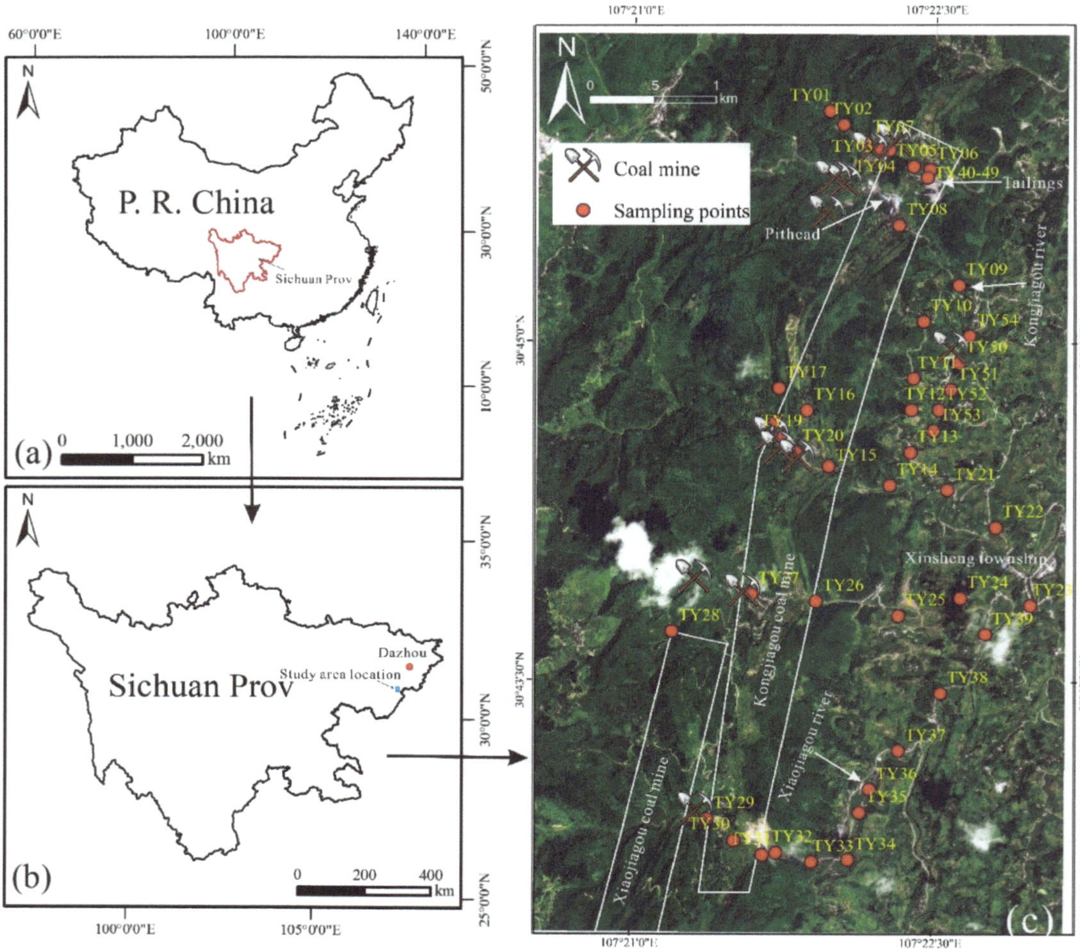

Figure 1. (**a**) The Sichuan Province in China; (**b**) location of study area in Sichuan Province; (**c**) location of soil samples in study area.

2.2. Soil Sampling and Analysis

On October 2020, 44 surface soil samples and 10 vertical profile soil samples were collected around the coal mine area in Dazhu (Figure 1). The surface soil samples were collected randomly from farmlands on both sides of the road and near the pithead of abandoned coal mines. Three soil cores collected from the top 0–15 cm were combined to produce one composite surface soil sample in the plough layer. A 1 m-deep pit was excavated at the foot of the slope of the tailings, and a vertical profile soil sample was extracted every 10 cm from top to bottom (TY40–TY49). Each sample weighed approximately 1.5–2.0 kg, and the sampling locations were recorded using GPS (Omap). All samples were stored in black polyethylene bags and immediately transported to the laboratory. After being air-dried at room temperature (15 °C) [22]), the samples were homogenized and sieved (<74 μm) for chemical analysis.

X-ray diffraction (XRD) was used for the mineralogical characterization of samples (TY02, TY40, and TY50) in a Rigaku diffractometer (Ultima IV, Akishima-shi, Tokyo, Japan). The conditions were slit fixed at 10 mm, 0.5 mm Pb monochromatic radiation, 40 mA, and

40 kV. The samples were run at a speed of 30°/min (5–80°). A microscope and scanning electron microscope (SEM, Prisma E, Thermo Scientific, Waltham, MA, USA) were used to observe the surface morphology of the soil (TY40) and tailings. The samples were gold-plated in a vacuum environment for elemental analysis by SEM using energy-dispersive X-ray (EDS) detectors.

Soil pH values were measured at a soil: water ratio of 1:2.5 (*w:v*) using potentiometry (HJ962-2018) with a pH meter (FE28-Standard, Mettler Toledo, Zurich, Switzerland). Each soil sample was divided into three parts. The first was digested using the method described in (DZ/T 0279-2016) [23]. Approximately 0.1 g of sample was digested in a Teflon crucible using a HCl: HNO_3: HF: $HClO_4$ (2 mL:2 mL:1 mL:1 mL) solution on a hot plate. Subsequently, Fe and Mn were determined using inductively coupled plasma atomic emission spectrometry (ICP–AES) (iCAP 7400, Thermo Fisher Scientific, Waltham, MA, USA), and the concentrations of Ni, Cu, Cd, and Pb were analyzed using inductively coupled plasma mass spectrometry (ICP–MS; Agilent 7700, Agilent Scientific Instruments, Palo Alto, CA, USA). The second part was digested using the method by (HJ491-2019 [24]). Approximately 0.2 g of sample was digested in a digestion tank using a HCl: HNO_3: HF (3 mL:6 mL:2 mL) solution in a microwave digestion furnace. This method was tested for Cr and Zn using a flame atomic absorption spectrophotometer (GGX-9, Beijing Haiguang Instrument Co., Beijing, China). The third part was digested using the method described in (GB/T 22105-2008) [25]. Approximately 0.2 g of sample was digested in colorimetric tubes using 10 mL aqua regia solution in a boiling water bath for 2 h. Then, 10 mL of preservation solution were added, and As and Hg were tested using an atomic fluorescence photometer (BAF-2000, Beijing Baode Instrument Co., Beijing, China). Two blank samples (one procedure blank and one reagent blank) and two standard samples (GBW07385: GSS-29) were analyzed as duplicates to ensure data reliability. The standard sample was the flood plain sediment of the main river system in China, and it was the standard substance for composition analysis, which was mainly used as the quantity value and quality control standard for sample testing of geological and geochemical investigation and mineral survey. All standard samples were free of pollution and the accuracy of the repeated analysis was less than 5% RSD. The certified and test values of standard samples and the detection limits of each element are presented in Table S1. Metal speciation was extracted sequentially using the BCR sequential extraction procedure [26,27], which includes three sequential extractions and a digestion. The obtained fractions, respectively, included: exchangeable fraction (F1, exchangeable and carbonate-associated fractions), reducible fraction (F2, fraction associated with Fe and Mn oxides), oxidizable fraction (F3, fraction bound to organic matter), and residual fraction (F4).

2.3. Evaluation of Soil Contamination

The contamination level of 44 surface soil samples could be analyzed by enrichment factor (EF) and geoaccumulation index (I_{geo}).

Enrichment factor (EF) was used to express the enrichment degree of elements in soil, and to judge and evaluate the source of elements in particulate matter [28]. EF can be calculated by [29]:

$$EF = (C_n/Fe)_{sample}/(C_n/Fe)_{background}, \quad (1)$$

In this study, where $(C_n/Fe)_{sample}$ is the ratio of the heavy metals value to the iron concentration in the sample, while $(C_n/Fe)_{background}$ is the background ratio of the heavy metals value to the iron concentration. EF numerical value can be divided into 5 grades [30]: <2 = minimal pollution; 2–5 = moderate enrichment; 5–20 = significant enrichment; 20–40 = very highly enriched; >40 = extremely enriched.

The metal contamination degree of the soil samples was obtained based on the geoaccumulation index (I_{geo}) [31], which was defined by Müller and can be calculated by [32]:

$$I_{geo} = \log_2[C_n/1.5B_n], \quad (2)$$

where C_n is the concentration of metal n in the soil, B_n is the background value of metal n, and the factor 1.5 is used to account for possible variations in background data owing to lithological variations. The background values of metals were used as references [33]. The geoaccumulation index can be divided into seven grades [34] and is shown in Table S2.

2.4. Statistical Analysis

In the field of environmental science, multivariate analysis has become a more powerful tool than the classical single variable method because it provides an easier means of data analysis [35]. Multivariate analysis methods, such as principal component analysis (PCA) and factor analysis (FA), have been successfully applied to assess soil quality and identify the chemical processes therein [19,36–39].

To characterize and compare these parameters, the soil properties were analyzed using the SPSS Statistics v22 software (International Business Machines Corporation, Armonk, NY, USA). PCA was applied to the metal variable analysis [40]. Eigenvalues were used to evaluate the number of principal components (PC), known as linear combinations of the old used factors. Factor analysis (FA) was used to determine the common latent structure among variables and reduce PC contribution through further simplification by rotating PCA-defined axes [41]. The Kaiser–Meyer–Olkin (KMO) and Bartlett's tests of sphericity have been frequently used to test the appropriateness of FA with a correlation coefficient matrix. In total, 9 soil parameters were measured, including Cr, Ni, Cu, Zn, As, Cd, Hg, Pb, and Fe, and these parameters were used in the statistical analysis.

2.5. Risk Evaluation

The pollution risk can be quantified and evaluated by the synthesis index [42]. It varies with the obtained metal concentration and the given evaluation criterion for each soil sample [43]. The synthesis index [44] can be calculated as follows:

$$P = \sqrt{\frac{(C_i/S_i)^2_{\max} + (1/n \sum_{i=1}^{n} C_i/S_i)}{2}} \quad (3)$$

where P is the synthesis index, C_i is the examined metal concentration for sample i, and S_i is the evaluation criterion of the i-th kind of metal.

The Soil Environment Quality Risk Control Standard for Soil Contamination of Agricultural Land [45] was adopted as the evaluation criterion. The risk evaluation was performed according to the method mentioned by [46]. First, the synthesis index was calculated for each soil sample based on Equation (3). Second, kriging was evaluated for all soil samples to obtain the total contamination distribution. Finally, the contamination distribution was divided into five grades based on [43,45]. All maps were created using ArcGIS version 10.2 (ESRI, Redlands, CA, USA).

3. Results and Discussion

3.1. Physical and Chemical Characterization

The samples collected in this study were mainly composed of eluvial silty clay of Quaternary. Minerals such as quartz, muscovite, albite, and kaolinite were observed in the soil samples (Figure 2). Hematite was observed in all three samples (TY02, TY40, and TY50) as secondary minerals from oxidation. This may be the reason that hematite occurs in coal and enters the soil during coal mining and transportation. As TY02 and TY40 soil samples were collected near the waste accumulation area, they contained chalcocite and cuprite. Calcite was observed in TY40 and was attributed to the leaching effect of precipitation on tailings, since calcite is found in both tailings and sandstone layers [47]. Both muscovite and quartz were found in TY40 and tailing under the microscope (Figure 3). We observed several of soil samples (TY29, TY34, and TY40) and found that they presented high similarity in the physical phase with the tailing containing coal seam [2]. This indicates that the main components of farmland soil originate from the minerals derived from sandstone weathering. In addition, carbonate minerals and siliceous rocks were observed in quartz fractures as interstitial materials (Figure 3). The SEM images show that the surfaces of soil and tailing samples were irregular (Figure 4a,c). From the figure, quartz could be observed, and pores were filled with weathered fragmental material. Figure 4b shows the EDS analysis results. The main elements in the soil were C, N, O, Si, and Ca, followed by metal elements such as Fe, Mn, Al, and Mg. This indicates that the surfaces of quartz and carbonate minerals were covered by Fe-Mn oxide. (Figure 4d) shows the EDS analysis result of tailing. The main elements were O, Si, Al, and Fe, which indicates that the sandstone layer contained metal elements such as Fe, which remain in the soil in the form of oxides after weathering. These results illustrate the environmental impact of intense mining activities on soil quality.

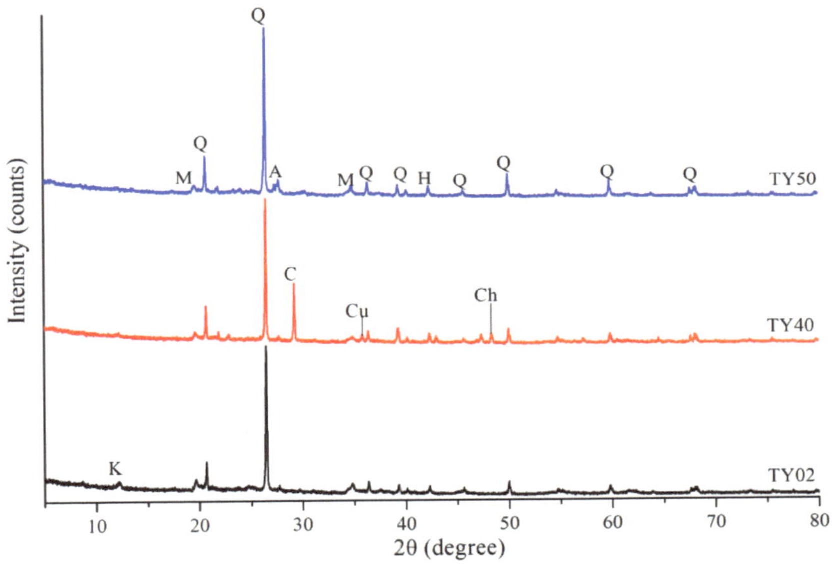

Figure 2. The X-ray diffraction patterns of soils. A—albite; C—calcite; Ch—chalcocite; Cu—cuprite; H—hematite; K—kaolinite; M—muscovite; Q—quartz.

Figure 3. Surface morphology of the soil (TY40) and tailing under microscope. M—muscovite; Q—quartz.

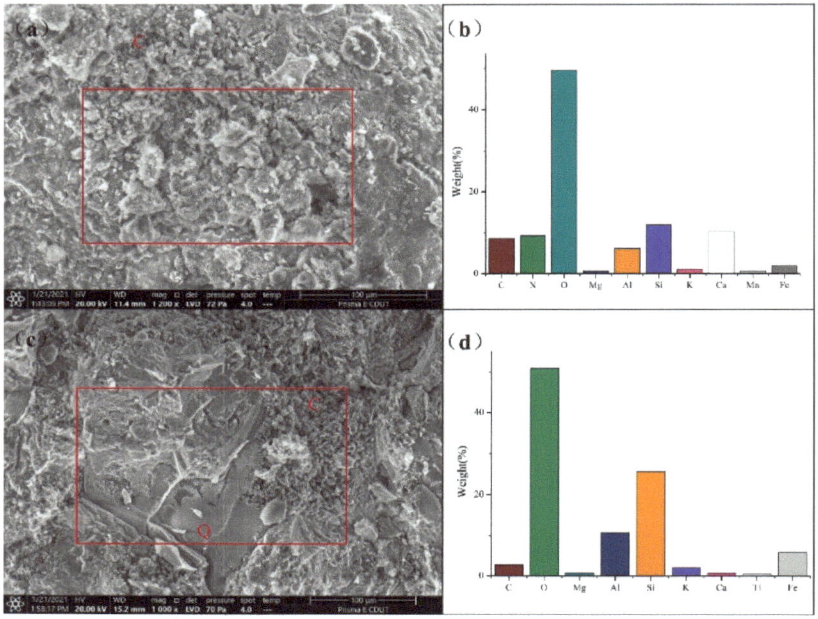

Figure 4. SEM images of the soil (**a**) and tailing (**c**), corresponding EDS spectrum (**b**,**d**). The red box indicates the scope of EDS analysis. Q—quartz; C—clay.

3.2. Concentration and Speciation of Heavy Metals

The descriptive statistical results of pH and metal concentration of soil samples are shown in Table 1 and the concentrations of these elements are listed in Table S3. The concentrations of the 10 investigated metals varied widely. Most soil samples were weakly alkaline, and their average pH was 7.22. Fe was the most abundant metal, with a mean concentration of 3.55 wt%. The mean value of Mn was 956.27 mg/kg, and that of Zn was 89.07 mg/kg. The average concentrations of Ni, Cu, and Pb were relatively similar, at 34.85, 26.38, and 28.47 mg/kg, respectively. The mean Cr and As concentrations were 58.19 and 8.51 mg/kg, respectively. The concentrations of Cd and Hg were significantly lower than those of other metals, at 0.41 and 0.14 mg/kg, respectively. Except for As and Cr, the concentrations of most metals exceeded their respective background values [33], especially those of Cd and Hg, the average concentrations of which were four and two times the background values, respectively. In the surface soil, increased concentrations of heavy metals were attributed to the dual function of secondary enrichment and parent rock inheritance. Ni and Mn presented high coefficients of variation (CV), at 194% and 218%, respectively, which exceeded 100%. Therefore, they presented a greater variation than other metals, and their high contents were strongly associated with a wide range of human activities. The CV values of Zn, As, and Cd were 52%, 54%, and 85%, respectively, which indicate a high level of spatial variation (CV > 50%) [48]. Similar to the results of most researchers [49,50], the spatial variation of coal mine pits resulted in metal data heterogeneity. The kurtosis values of Ni, Zn, Cd, Pb, and Mn were higher than 10, which indicated that there was great heterogeneity in the distribution of these elements in the soil [51]. In addition, the skewness values of all elements were greater than 0, which indicated that most metal concentrations were at relatively low values.

Table 1. pH and concentration of metals in soil samples of coal mine area.

Item	pH	Cr	Ni	Cu	Zn	As	Cd	Hg	Pb	Mn	Fe
		mg/kg	mg/kg	mg/kg	mg/kg	mg/kg	mg/kg	mg/kg	mg/kg	mg/kg	wt%
N [a]	44	44	44	44	44	44	44	44	44	44	44
Minimum [a]	4.88	34.52	15.36	7.66	39.04	5.03	0.21	0.06	16.80	258.39	2.19
Maximum [a]	9.01	85.37	471.61	59.11	346.68	29.27	2.51	0.31	91.01	14314.45	6.18
Mean [a]	7.22	58.19	34.85	26.38	89.07	8.51	0.41	0.14	28.47	956.27	3.55
Median [a]	7.28	56.20	23.39	24.26	78.77	6.96	0.33	0.13	25.15	622.01	3.49
25th [a]	6.07	47.45	20.43	20.52	72.10	6.20	0.28	0.1	22.16	460.06	3
75th [a]	8.4	64.81	28.72	32.05	94.81	8.51	0.41	0.15	28.51	780.32	3.99
Skewness [a]	−0.14	0.45	6.53	0.87	4.27	2.94	5.21	1.74	3.14	6.44	0.84
Kurtosis [a]	−1.32	−0.43	43.10	1.66	22.64	9.89	30.43	4.84	10.75	42.2	1.55
SD [a]	1.27	12.51	67.70	9.72	46.44	4.60	0.35	0.04	13.99	2080.77	0.8
CV [a]	0.18	0.22	1.94	0.37	0.52	0.54	0.85	0.33	0.49	2.18	0.22
TY40-49 [b]	8.8	66.3	36.81	35.02	79.43	18.48	0.41	0.28	30.15	676.18	3.56
Background values [c]	-	61	26.9	22.60	74.2	11.2	0.1	0.065	26	583	2.94
Limit value [d]	-	200	100	100	250	30	0.3	2.4	120	-	-

[a] Item of surface soil. [b] Item of profile soil. [c] Background values for soils in China [33]. [d] Soil Environment Quality Risk Control Standard for Soil Contamination of Agricultural Land [45].

The EF index values are presented in Figure 5 and in Table S4. According to the EF index values, heavy metals showed a wide range of enrichment at each sampling location. Cd had the highest degree of enrichment, belonging to moderate enrichment and significant enrichment. The Hg and Pb enrichment level of most soil samples was minimal pollution, but a small part showed significant enrichment. The lower enrichment degree was found for Cr and Ni, except in TY01. Briefly, the order of the average values of the EF index was Cd > Hg > Pb > As > Zn > Cu > Ni > Cr. The high contents of Fe/Al/Mn oxides, carbonaceous species, and clay (Al_2/SiO_2) were closely related to the enrichment of heavy metals in the soil. Human activities such as coal combustion, waste incineration, and transportation can release a large amount of dust containing Cd and Hg into the atmosphere, which is then enriched through natural sedimentation and rain [52]. In addition, the heavy application of chemical fertilizers and pesticides in agricultural production can lead to Cd pollution [53]. These results illustrate the effects of weathering of primary minerals for soil formation and the impacts of mining activities on the concentrations of heavy metals in the soil of the mining area.

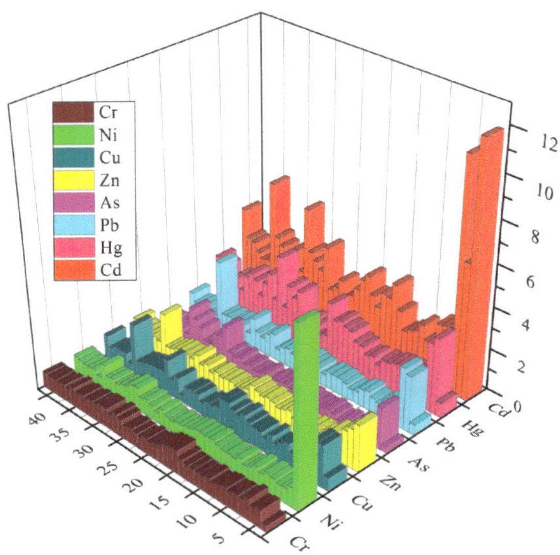

Figure 5. Three-dimensional plots of enrichment factor (EF).

The mean Cd concentration exceeded the standard value given in [45] by 36% (Table 1). The results of the geoaccumulation index (I_{geo}) are shown in Figure 6 and listed in Table S5. Moreover, Figure 6 indicated that soils in the coal mine area were significantly polluted by Cd, which presented Class 5 (heavily to extremely contaminated, TY01) and Class 3 (moderately–heavily contaminated, TY02 and TY03). In addition, the I_{geo} of Cd in most soil samples were classified as Class 2 (moderately contaminated). Hg, Pb, and Zn presented lower I_{geo} values, dominated by Classes 2 (moderately contaminated) and 1 (uncontaminated–moderately contaminated). Because the TY01 and TY02 mixed soil samples were collected near the pithead of an abandoned coal mine, they were more strongly affected by the deposition of primary minerals and immersional wetting of tailings, which resulted in extremely high I_{geo} values for Ni, Pb, and Zn. In addition, As and Cu were associated with I_{geo} Classes 1 and 0 (practically uncontaminated). Cr presented the lowest pollution, and it was the only contaminant with I_{geo} Class 0 for all samples. According to the I_{geo} results, heavy metal pollution in the study area was at a safe level, except for Hg and Cd. The main pollution sources are likely the tailings, dust piled up in mining areas, and settlement of industrial coal. The high concentration of Cd caused heavy metal pollution in the agricultural soils of the study area. Accordingly, long-term consumption of rice, vegetables, fruits, and water seriously polluted by Cd are likely to lead to chronic poisoning [54,55]. In addition, Cd has become the most serious heavy metal soil pollutant in China and substantially affects the quality and yield of crops [4,56].

Figure 7 shows the chemical speciation percentages of the metals in the soil sample (TY40), which was at the foot of the slope of the tailings. In addition, a large number of crops were planted near the site from which this soil was sampled. In general, the residual fraction was dominant in most metals (F4), except for Cd and Mn, in which the reducible fraction was dominant (F2) (Cd: 55.17%; Mn: 81.16%). Owing to their detection limit, only residual fractions were observed for As and Hg (100%). Cd yielded the highest exchangeable fraction (F1), up to 17.24%, whereas the exchangeable fraction of As, Pb, Hg, and Fe were not detected. Pb presented the highest oxidizable fraction (F3) (14.98%).

The percentages of different Cd fractions for F1, F2, F3, and F4 were 17.24%, 55.17%, 6.9%, and 20.69%, respectively. The high proportion exchangeable fraction of Cd indicated that its bioavailability was high, and its migration ability in the soil was strong [57]. In addition, the release of exchangeable elements can cause a large amount of cation replacement in soil through precipitation (acid rain), which leads to substantial nutrient deficiency and toxicity to plants [58]. The reducible fraction generally exists in the outer capsule of minerals and fine powder particles, with strong exclusive adsorption, and it is easily released when the redox potential of the water body decreases or the water lacks oxygen [59]. Both exchangeable and reducible fractions of Cd showed great bioavailability, which indicated that Cd presented a greater pollution to the environment than other elements in the study area.

The proportions of reducible and residual fractions of Pb were similar, at 37.65% and 47.37%, respectively. Generally, the release of Pb in reducible species is difficult, but under anoxic conditions, generated by a soil water saturation (for example) with no water transfer, Pb will be released, thereby causing secondary pollution [60]. In addition, the proportions of oxidizable fraction of Pb was 14.98%, which indicated that Pb easily formed complexes or chelates with humic acid and other organic matter in sediments, and it then coprecipitated with sulfide. The oxidable fraction of heavy metals reflects the aquatic activities and effects of the discharge of organic-rich sewage. Heavy metals in this form are relatively stable in the soil, but under strong oxidation conditions, their mobility can increase, and they can enter the water [61].

The high ratio of residual fraction in As, Hg, Cr, and Fe indicated a strong combination with the crystal structure of minerals, which was stable under natural conditions with a low transferability [62]. Because these elements mainly existed in the crystal lattice of the minerals, a non-anthropogenic source was suggested for the metals in the residual fraction [63]. Similar to the conclusion of many researchers [64], Cu and Zn appeared mostly in the residual and reducible fractions. In this study, soil particles were mainly composed of silty clay, and the adsorption of heavy metals increased with the decrease in particle size [65]. Mn is mainly found in the reducible fraction (from Fe-Mn oxides) and less in the oxidizable fraction (bound to OM) [66]. Many studies have shown that Ni can preferentially combine with aluminosilicate minerals (e.g., kaolinite and muscovite (Figure 2) [67], resulting in its residual fraction being high (78.06%).

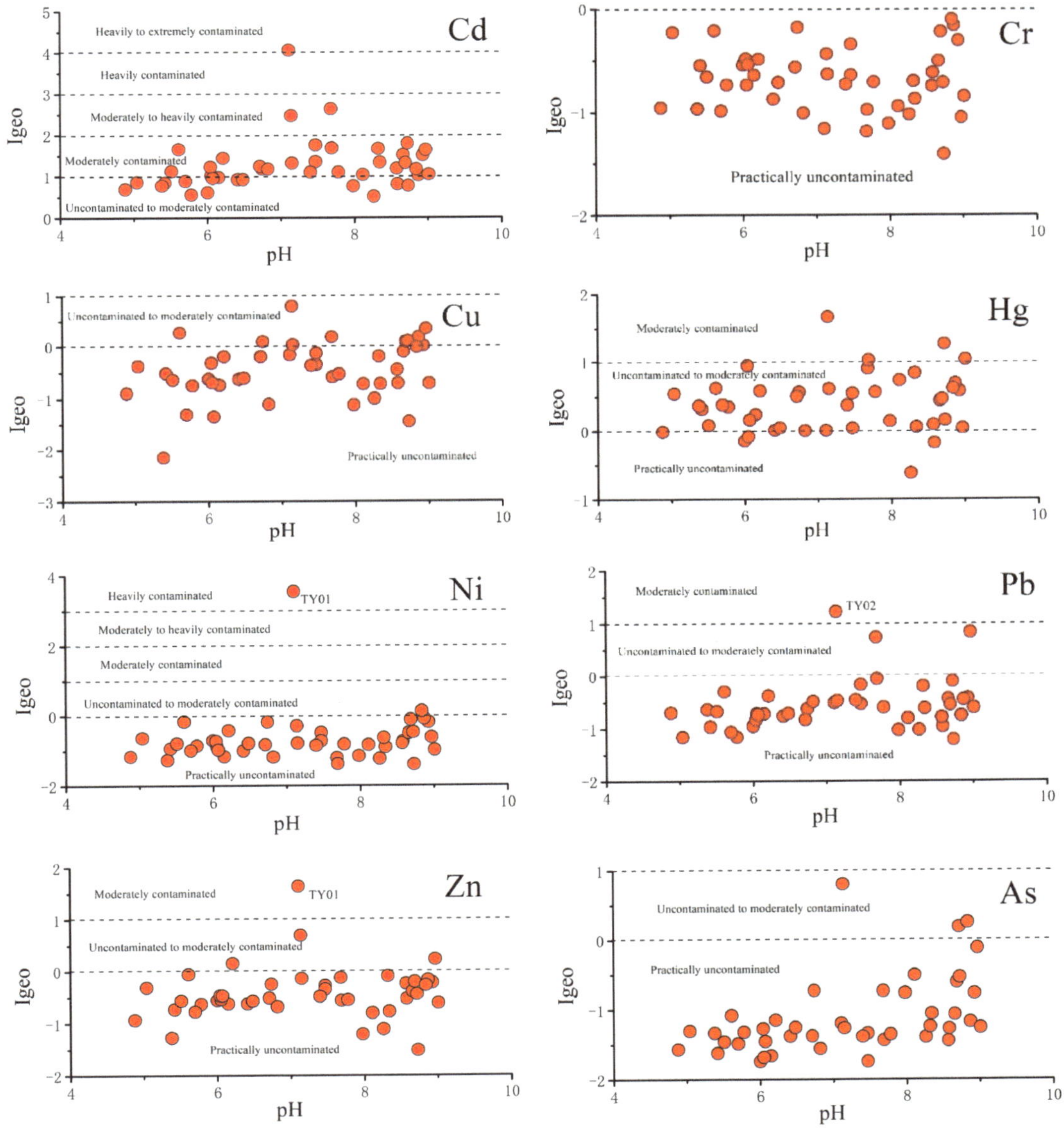

Figure 6. The index of geoaccumulation (I_{geo}) of soils in the study area.

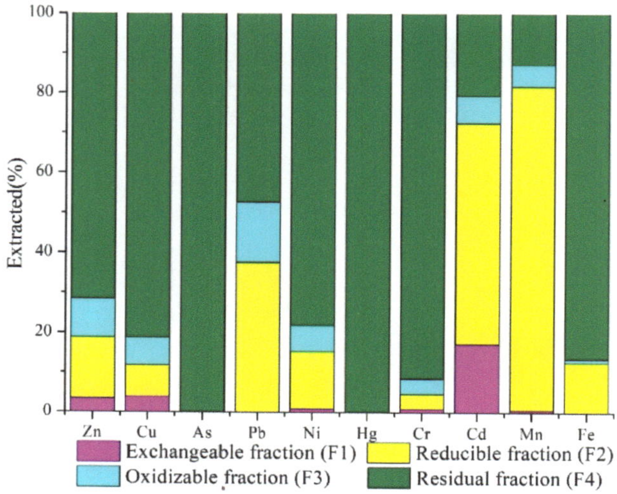

Figure 7. The percentages of metals chemical speciation in soils.

3.3. Statistical Analysis

To further investigate the relationship between heavy metals in surface soils, the Pearson correlation coefficients were calculated, and the results are shown in Table 2. Ni was positively ($p < 0.01$) correlated with Zn, Cd, and Fe, whereas Cu was positively correlated with Cr, Zn, As, Hg, and Pb. In addition, Zn exhibited positive significant correlations with Cd, Pb, and Fe.

Table 2. The correlation matrix of metals and pH in soil.

	Cr	Ni	Cu	Zn	As	Cd	Hg	Pb	Fe
Cr	1								
Ni	−0.128	1							
Cu	0.506 **	0.132	1						
Zn	0.055	0.882 **	0.511 **	1					
As	0.166	0.007	0.633 **	0.259	1				
Cd	−0.183	0.921 **	0.340 *	0.927 **	0.18	1			
Hg	0.251	−0.094	0.604 **	0.186	0.636 **	0.143	1		
Pb	−0.031	0.005	0.727 **	0.389 **	0.659 **	0.326 *	0.596 **	1	
Fe	0.543 **	0.568 **	0.331 *	0.606 **	−0.022	0.473 **	−0.025	−0.05	1

* Correlation is significant at the 0.05 level; ** Correlation is significant at the 0.01 level.

In this study, the sphericity (0.6) was larger than 0.5, and the KMO result was less than 0.001, which indicated that the data was suitable for FA [68]. Based on eigenvalues (eigenvalue > 1), three main factors explained 87.918% of the total variance. The variance contribution rate of Factor 1 (F1) was 43.106%, and was positively correlated with Ni, Cd, Zn, and Fe (0.975, 0.964, 0.934, and 0.589), respectively (Table 3). The interrelationships between Ni, Zn, and Cd suggest the influence of local human activities (domestic waste and fertilization) and lithology (weathering of parent rocks) on soil samples [41]. F2, which explained 27.906% of the total variance, was highly positively correlated with Pb, As, Hg, and Cu (0.885, 0.856, 0.831, and 0.8), respectively. Many researchers [69,70] believe that the main reason for the enrichment of Cu and As in soils is the application of chemical fertilizers and pesticides. The high correlation between Hg and Pb was attributed to the automobile emissions from coal transportation and the fly ash produced by coal combustion [71]). F3 presented a variance contribution rate of 16.906%, and Cr and Fe presented high loads (0.954 and 0.734, respectively). The high ratio of residual fraction in Cr and Fe indicated a high correlation with soil-forming parent material. Therefore, the analysis indicates that F1 represents anthropogenic and lithologic sources, which are affected by mining activities; F2 represents anthropogenic sources, such as fertilizers and traffic pollution; and F3 represents the contribution of metals from soil-forming parent material.

Table 3. Factor loadings in soils.

	F1	F2	F3	Communalities
Ni	**0.975**	−0.102	0.019	0.962
Cd	**0.964**	0.192	−0.084	0.962
Zn	**0.934**	0.275	0.146	0.879
Pb	0.167	**0.885**	−0.147	0.969
As	0.043	**0.856**	0.024	0.735
Hg	−0.047	**0.831**	0.112	0.972
Cu	0.233	**0.8**	0.43	0.705
Cr	−0.152	0.169	**0.954**	0.832
Fe	**0.589**	−0.099	**0.734**	0.896
Eigenvalues	0.975	−0.102	0.019	
% of variance explained	43.106	27.906	16.906	
Cumulative % of variance	43.106	71.012	87.918	

Loading values for the PC axis higher than +0.5 and lower than −0.5 are given in bold.

3.4. Spatial Distribution and Risk Assessment

Figure 8 shows the distribution of metal concentrations in the vertical profile of the soils under the tailings. The mean concentrations of all metals exceeded the background values [33], especially those of Cd and Hg, which were four times the background values. Only the concentration of Cd exceeded the limit value [45]. The metal concentrations in the vertical profile of the soils (TY40–49) are listed in Table S3.

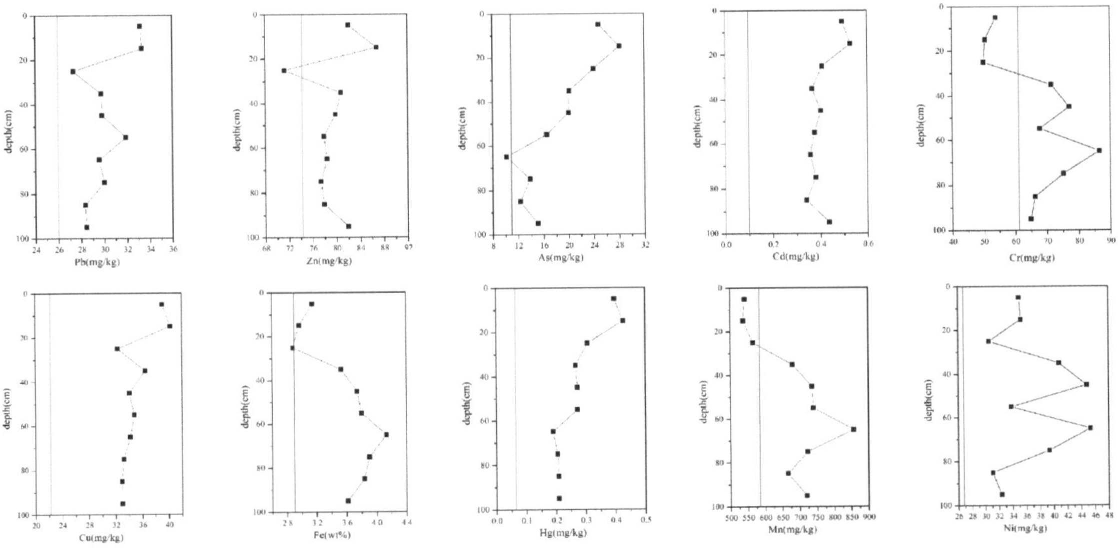

Figure 8. Distribution of metal concentrations in vertical profile soils under the tailings. The red line indicates background value.

In general, the concentrations of most metals decreased with depth, which indicated that most metals in the farmland were still in the plow layer. The average deviation of Mn was the highest (79.1°), followed by Cr (9.2°). The average deviations of other metals were found to be within ±5%. In Section 3.1, calcite and cuprite were found in TY40, which was sampled from the surface of the deep pit at the foot of the slope of the tailings. This may be because the atmospheric precipitation would flow through the tailings before entering the soil, bringing these minerals in the tailings into the surface soil. Therefore, heavy metals were easier to enrich in surface layer. Fe and Mn concentrations were lower at 0–20 cm, and higher at 60–80 cm, which was attributed to their predominant reducible (F2) and residual fractions (F4). The weak acid water produced by tailing leaching reacted with Fe and Mn oxides in the surface layer, which led to their enrichment in deeper soil layers. The Cr concentration was low in the surface soil but rich at 70 cm. With the pH increase (mean pH of 8.8), the Cr adsorption on soil was clearly weakened. This occurred because higher pH values lead to more negative charges on the soil surface. This increases the probability of the formation of complexes with organic acids, which decrease the adsorption capacity and enhance the mobility of heavy metals in the soil [72]. In this study, the surface–bottom soil pH showed a decreasing trend (from 9.05 to 8.54), which led to more Cr being adsorbed on the clay in a deeper position than on the surface. The vertical distribution of Ni concentration in the soil first decreased and then increased. The highest concentration of Ni was at 70 cm. When the permeability of shallow soil was good, the density of deep soil was high, the water retention was good, and the ability of Ni for downward migration would increase [73]. In addition, the accumulation of Pb was likely related to the transportation of leaded gasoline [67].

The spatial analysis of metals in the soils of the study area are shown in Figure 9, with a clear spatial distribution pattern of heavy metals in the soil. Pb was mainly distributed in the north and middle parts of the study area. The highest Pb values were observed for TY02 (91.01 mg/kg), which was close to the entrance and exit of the transportation center. This suggested that industrial activities substantially affected the enrichment of Pb in the soil. The concentrations of Ni and Zn presented similar spatial distribution patterns. High concentrations of Ni and Zn were mainly observed in areas where human activities were concentrated, and they were observed throughout the entire study area, from north to south. The high Ni and Zn contents in these other areas were mainly caused by geochemical-related industrial activities, such as application of pesticides and chemical fertilizers. The spatial distribution of Cu concentration was similar to those of Ni and Zn, but it was more intense, which indicated that human activities had a more severe impact on Cu.

The spatial distribution patterns of Cr were higher in the north and lower in the south. High Cd concentration was mainly distributed in the north of the study area, where pulverized fuel ash pipelines and tailings were located. A large number of pesticides and chemical additives are used in agricultural activities both north and south of the study area, resulting in the enrichment of As in the soils. High Hg concentrations were observed southwest of the Xiaojiagou coal mine, which characterized a decrease in spatial distribution from northeast to southwest. The Xiaojiagou coal mine, located south of the study area, had been closed, whereas the Kongjiagou coal mine in the north was still operative, so the northern road was the main coal transportation road. Moreover, there was a large number of coal companies and tailings in the north of the research area, which likely justified the enrichment of Zn, Ni, Cr, Cu, and Cd in the northern region. This also showed that the mining operations exerted a more severe impact on the environment, whereas the closure of mines was conducive to the recovery of the local environment.

We conducted a risk assessment for the study area to provide a basis for effective suggestions for policy makers and farmers. The synthesis index was adopted for the assessment. The results are shown in Table 4 and Figure 10, and they indicate that there were few safe areas, accounting for only 0.06 km^2 (0.21%). This was mainly attributed to the concentration of Cd, which exceeded the limit value in most areas. Most of the soil (56.90%, 18.06 km^2) belonged to guard grade ($0.7 < p \leq 1$), including central and southern parts of the study area, where human activities were concentrated. Owing to the use of chemical fertilizers and pesticides, these concentrated areas belonged to low pollution grade. Considering the pithead of Kongjiagou coal mine as the boundary, several factors in the northern area caused them to belong to Levels 3–5, including coal transportation, tailings accumulation, and chemical plants. These areas were classified under moderate (1.96 km^2, 6.17%) and severe (0.36 km^2, 1.13%) with high concentrations of Cd (2.51 mg/kg) and Pb (91.01 mg/kg). Some of the physiological effects of chronic exposure to waterborne cadmium at sub-lethal concentrations are manifested in the form of reduction in growth and changes in hematology and enzyme activity [74]. Lead at sub-lethal levels destroys normal metabolic processes by disrupting calcium and sodium homeostasis [75]. In view of such risks, protective measures are required to avoid heavy metal contamination of vegetables and ensure food safety for human consumption. The risk assessment results can be used as a basis to estimate the environmental cleaning costs of coal mine areas. The classification suggests that vegetables should not be planted in high-risk areas, whereas in low-risk areas, the cultivation of crops should consider the economic and environmental impacts.

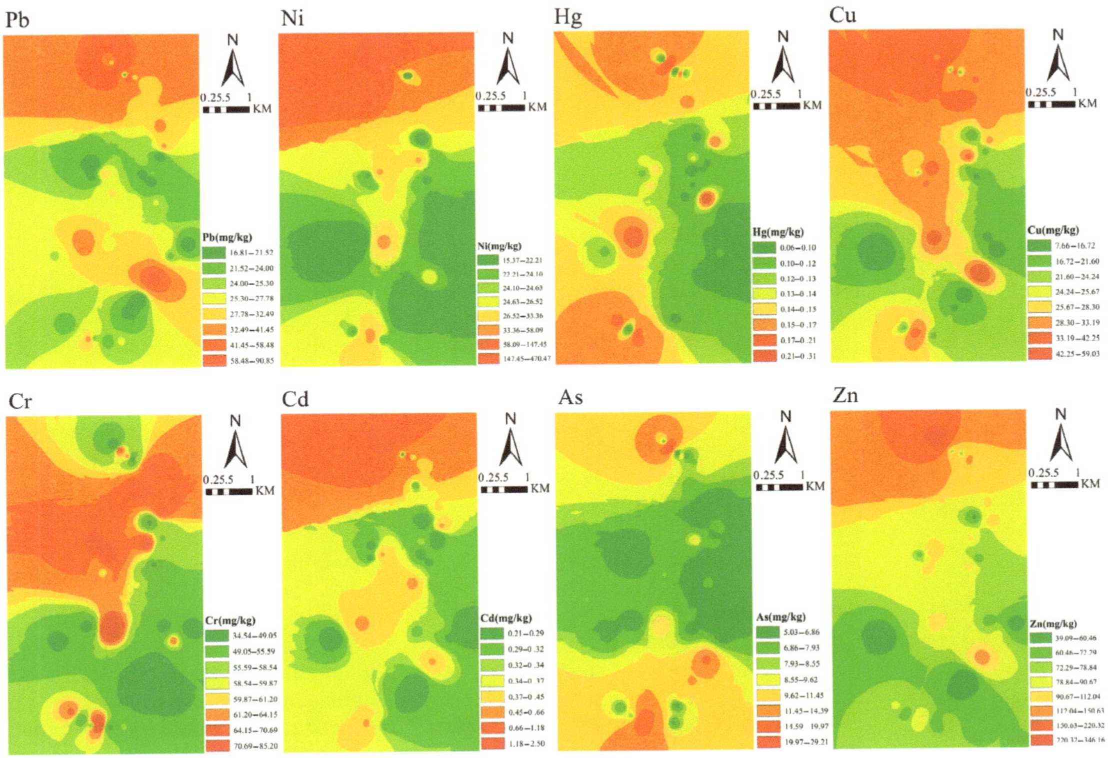

Figure 9. Spatial distributions of metals in farmland soils.

Table 4. The evaluation standard and results in study area.

Level	p [a]	Grade	Area (km^2)	Percent (%)
1	$p \leq 0.7$	Safety	0.06	0.21
2	$0.7 < \partial p \leq 1$	Guard	18.06	56.90
3	$1 < p \leq 2$	Low pollution	11.29	35.59
4	$2 < p \leq 3$	Moderate pollution	1.96	6.17
5	$p < 3$	Severe pollution	0.36	1.13

[a] The standard based on [43,45].

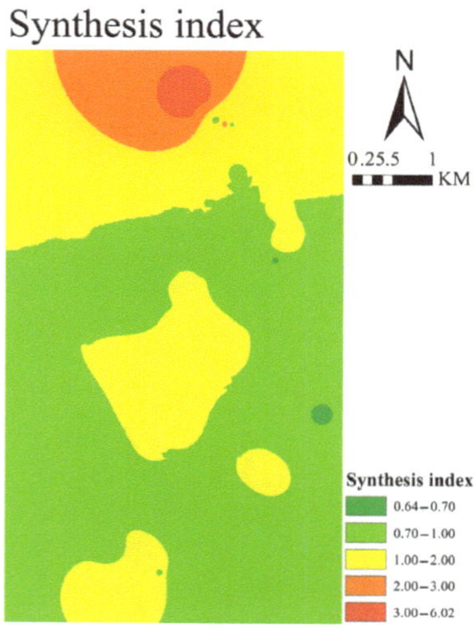

Figure 10. The synthesis index map of metals in study area.

4. Conclusions

In this study, we investigated the mineralogical characterization, concentrations, speciation, statistical analysis, spatial distribution, and risk assessment of metals in a coal mine area in Dazhu, China. Except for As and Cr, the concentrations of all metals (Fe, Ni, Cu, Zn, Mn, Cd, Hg, and Pb) exceeded the background values [33], which indicated that the weathering of primary minerals in soil formation and coal mining activities substantially affected soil quality. Cd pollution was the most intense, and it exceeded the limit by 36% [45]. Cd was classified under I_{geo} Class 5 (heavily–extremely contaminated) or Class 6 (extremely contaminated). The XRD and SEM analyses indicated the presence of many secondary minerals, which likely influenced the concentrations of heavy metals. The residual fraction was dominant for most metals, except Cd and Mn, for which the reducible fraction was dominant (Cd: 55.17%; Mn: 81.16%). The concentrations of most metals decreased with depth, which indicated that most metals in the farmlands originated from human activities and remained in the plow layer. The statistical analysis showed that Factor 1 can represent anthropogenic and lithologic sources, which are affected by mining activities, whereas Factor 2 can represent anthropogenic sources such as fertilizers and traffic pollution. Factor 3 represented the contribution of metals from soil-forming parent material. The heavy metal concentration in the mining area was high and presented high risk, so the area is not suitable for agriculture. The mining activities had a severe impact on the environment, whereas areas with closed mines were associated with the recovery of the local environment.

Supplementary Materials: The following supporting information can be downloaded at: https://www.mdpi.com/article/10.3390/ijerph19116493/s1, Table S1: The certified and test values of standard samples and the detection limits of each element, Table S2: Six classes of the geoaccumulation index, Table S3: pH and concentrations of soil samples, Table S4: The EF index values, Table S5: The geoaccumulation index (I_{geo}).

Author Contributions: Conception and design of sampling, D.D.; investigation, D.D., Y.S., B.R. and L.S.; chemical analysis, D.D. and Y.S.; statistical analysis and modeling, D.D.; writing—original draft preparation, D.D. and Y.W.; writing—review and editing, D.D. and Y.W. All authors have read and agreed to the published version of the manuscript.

Funding: This work was funded by key research and development project of Sichuan Province (No. 2018JY0425 and 2018SZ0290).

Institutional Review Board Statement: Not applicable.

Informed Consent Statement: Not applicable.

Data Availability Statement: Not applicable.

Acknowledgments: The authors wish to thank Deng Xiaoquan and Gao Yuping for their test and financial support.

Conflicts of Interest: The authors declare no conflict of interest.

References

1. Du, P.P. *Research on Geologic Environment Effects Induced by Coal Mining in Ecological Weakness Area and Assessment Techniques: A Case Study Taking Yushenfu Mining District in Northern Shanxi Province*; China University of Mining and Technology: Xuzhou, China, 2011.
2. Li, M.B.; Wu, H. Coal seam physical property and coal quality characteristics of Jingang coal mine in Dazhou city, Sichuan province. *Mod. Min.* **2018**, *34*, 41–45.
3. Niu, S.P.; Gao, L.M.; Zhao, J.J. Distribution and risk assessment of heavy metals in the Xinzhuangzi reclamation soil from the Huainan coal mining area, China. *Hum. Ecol. Risk Assess. Int. J.* **2015**, *21*, 900–912. [CrossRef]
4. Li, Z.; Ma, Z.; van der Kuijp, T.J.; Yuan, Z.; Huang, L. A review of soil heavy metal pollution from mines in China: Pollution and health risk assessment. *Sci. Total Environ.* **2014**, *468–469*, 843–853. [CrossRef] [PubMed]
5. Beane, S.J.; Comber, S.D.W.; Rieuwerts, J.; Long, P. Abandoned metal mines and their impact on receiving waters: A case study from Southwest England. *Chemosphere* **2016**, *153*, 294–306. [CrossRef] [PubMed]
6. Zuzolo, D.; Cicchella, D.; Catani, V.; Giaccio, L.; Guagliardi, I.; Esposito, L.; De Vivo, B. Assessment of potentially harmful elements pollution in the Calore River basin (Southern Italy). *Environ. Geochem. Health* **2016**, *39*, 531–548. [CrossRef] [PubMed]
7. Wang, J.; Liu, W.; Yang, R.; Zhang, L.; Ma, J. Assessment of the potential ecological risk of heavy metals in reclaimed soils at an opencast coal mine. *Disaster. Adv.* **2013**, *6*, 366–377.
8. Kumar, V.; Sharma, A.; Minakshi; Bhardwaj, R.; Thukral, A.K. Temporal distribution, source apportionment, and pollution assessment of metals in the sediments of Beas river, India. *Hum. Ecol. Risk Assess.* **2018**, *24*, 2162–2181. [CrossRef]
9. Dong, J.; Yang, Q.; Sun, L.; Zeng, Q.; Liu, S.; Pan, J.; Liu, X. Assessing the concentration and potential dietary risk of heavy metals in vegetables at a Pb/Zn mine site, China. *Environ. Earth Sci.* **2011**, *64*, 1317–1321. [CrossRef]
10. Park, B.; Lee, J.; Ro, H.; Kim, Y.H. Effects of heavy metal contamination from an abandoned mine on nematode community structure as an indicator of soil ecosystem health. *Appl. Soil Ecol.* **2011**, *51*, 17–24. [CrossRef]
11. Zhao, F.J.; Ma, Y.; Zhu, Y.G.; Tang, Z.; McGrath, S.P. Soil contamination in China: Current status and mitigation strategies. *Environ. Sci. Technol.* **2015**, *49*, 750–759. [CrossRef]
12. Luo, J.; Cheng, H.; Ren, J.; Davison, W.; Zhang, H. Mechanistic insights from DGT and soil solution measurements on the uptake of Ni and Cd by radish. *Environ. Sci. Technol.* **2014**, *48*, 7305–7313. [CrossRef] [PubMed]
13. Cai, L.M.; Wang, Q.S.; Luo, J.; Chen, L.G.; Zhu, R.L.; Wang, S.; Tang, C.H. Heavy metal contamination and health risk assessment for children near a large Cu-smelter in central China. *Sci. Total Environ.* **2019**, *650*, 725–733. [CrossRef] [PubMed]
14. Khan, S.; Reid, B.J.; Li, G.; Zhu, Y.-G. Application of biochar to soil reduces cancer risk via rice consumption: A case study in Miaoqian village, Longyan, China. *Environ. Int.* **2014**, *68*, 154–161. [CrossRef] [PubMed]
15. Rai, P.K.; Lee, S.S.; Zhang, M.; Tsang, Y.F.; Kim, K.H. Heavy metals in food crops: Health risks, fate, mechanisms, and management. *Environ. Int.* **2019**, *125*, 365–385. [CrossRef]
16. He, S.; Lu, Q.; Li, W.; Ren, Z.; Zhou, Z.; Feng, X.; Zhang, Y.; Li, Y. Factors controlling cadmium and lead activities in different parent material-derived soils from the Pearl River Basin. *Chemosphere* **2017**, *182*, 509–516. [CrossRef]
17. Liu, X.; Jing, M.; Bai, Z. Heavy Metal Concentrations of Soil, Rock, and Coal Gangue in the Geological Profile of a Large Open-Pit Coal Mine in China. *Sustainability* **2022**, *14*, 1020. [CrossRef]
18. Sun, Z.; Xie, X.; Wang, P.; Hu, Y.; Cheng, H. Heavy metal pollution caused by small-scale metal ore mining activities: A case study from a polymetallic mine in South China. *Sci. Total Environ.* **2018**, *639*, 217–227. [CrossRef]
19. Barbieri, M. The importance of enrichment factor (EF) and geoaccumulation index (igeo) to evaluate the soil contamination. *J. Geol. Geophys.* **2016**, *5*, 2. [CrossRef]
20. Qiu, Y.; Guan, D.S.; Song, W.W.; Huang, K.Y. Capture of heavy metals and sulfur by foliar dustin urban Huizhou, Guangdong Province, China. *Chemosphere* **2009**, *75*, 447–452. [CrossRef]
21. Zhang, X.Y.; Lin, F.F.; Wong, M.T.; Feng, X.L.; Wang, K. Identification of soil heavy metal sources from anthropogenic activities and pollution assessment of Fuyang County, China. *Environ. Monit. Assess.* **2009**, *154*, 439–449. [CrossRef]
22. Gee, G.W.; Bauder, J.W. Particle-size analysis. In *Methods of Soil Analysis. Part 1: Physical and Mineralogical Methods*, 2nd ed.; Klute, A., Ed.; American Society of Agronomy and Soil Science Society of America: Madison, WI, USA, 1986; pp. 383–410.
23. *DZ/T 0279-2016*; Geological and Mineral Industry Standards. Ministry of Land and Resources, PRC: Beijing, China, 2016; p. 4.
24. *HJ491-2019*; National Environmental Protection Standards. Ministry of Ecological Environment, PRC: Beijing, China, 2019; p. 4.
25. *GB/T 22105-2008*; National Standard-Method for Determination of Total Mercury, Total Arsenic and Total Lead by Atomic Fluorescence. Standardization Administration of China: Beijing, China, 2008; p. 2.

26. Hu, C.; Yang, X.; Dong, J.; Zhang, X. Heavy metal concentrations and chemical fractions in sediment from Swan Lagoon, China: Their relation to the physiochemical properties of sediment. *Chemosphere* **2018**, *209*, 848–856. [CrossRef] [PubMed]
27. Liu, G.N.; Yu, Y.J.; Hou, J.; Xue, W.; Liu, X.; Liu, Y.; Wang, W.; Alsaedi, A.; Hayat, T.; Liu, Z. An ecological risk assessment of heavy metal pollution of the agricultural ecosystem near a lead-acid battery factory. *Ecol. Indic.* **2014**, *47*, 210–218. [CrossRef]
28. Bourennane, H.; Douay, F.; Sterckeman, T.; Villanneau, E.; Ciesielski, H.; King, D.; Baize, D. Mapping of anthropogenic trace elements inputs in agricultural topsoil from Northern France using enrichment factors. *Geoderma* **2010**, *157*, 165–174. [CrossRef]
29. Chen, Z.L.; Huang, Y.; Cheng, X.; Ni, S.; Wang, J.J.; Ren, B.Z.; Yu, Q.L. Assessment of toxic elements in road dust from Hutou Village, China: Implications for the highest incidence of lung cancer. *Environ. Sci. Pollut. Res.* **2021**, *28*, 1850–1865. [CrossRef]
30. Sutherland, R.A. Bed sediment-associated trace metals in an urban stream, Oahu, Hawaii. *Environ. Geol.* **2000**, *39*, 611–627. [CrossRef]
31. Tian, K.; Huang, B.; Xing, Z.; Hu, W.Y. Geochemical baseline establishment and ecological risk evaluation of heavy metals in greenhouse soils from Dongtai, China. *Ecol. Indic.* **2017**, *72*, 510–520. [CrossRef]
32. Karim, Z.; Qureshi, B.A.; Mumtaz, M. Geochemical baseline determination and pollution assessment of heavy metals in urban soils of Karachi. *Pakistan. Ecol. Indic.* **2015**, *48*, 358–364. [CrossRef]
33. China National Environmental Monitoring Center (CNEMC). *The Background Values of Soil Elements in China*; China Environmental Science Press: Beijing, China, 1990. (In Chinese)
34. Ji, Y.Q.; Feng, Y.C.; Wu, J.H.; Wu, J.H.; Zhu, T.; Bai, Z.P.; Duan, C.Q. Using geoaccumulation index to study source profiles of soil dust in China. *J. Environ. Sci.* **2008**, *20*, 571–578. [CrossRef]
35. Hou, D.Y.; O'Connor, D.; Nathanail, P.; Li, T.; Ma, Y. Integrated GIS and multivariate statistical analysis for regional scale assessment of heavy metal soil contamination: A critical review. *Environ. Pollut.* **2017**, *231 Pt 1*, 1188–1200. [CrossRef]
36. Micó, C.; Recatalá, L.; Peris, M.; Sánchez, J. Assessing heavy metal sources in agricultural soils of an European Mediterranean area by multivariate analysis. *Chemosphere* **2006**, *65*, 863–872. [CrossRef]
37. Rodríguez Martín, J.A.; López Arias, M.; Grau Corbí, J.M. Heavy metals contents in agricultural topsoils in the Ebro basin (Spain). Application of the multivariate geoestatistical methods to study spatial variations. *Environ. Pollut.* **2006**, *144*, 1001–1012. [PubMed]
38. Franco-Uría, A.; López-Mateo, C.; Roca, E.; Fernández-Marcos, M.L. Source identification of heavy metals in pastureland by multivariate analysis in NW Spain. *J. Hazard. Mater.* **2009**, *165*, 1008–1015. [CrossRef]
39. Cai, L.M.; Xu, Z.C.; Ren, M.Z.; Guo, Q.W.; Hu, X.B.; Hu, G.C.; Wang, H.F.; Peng, P.A. Source identification of eight hazardous heavy metals in agricultural soils of Huizhou, Guangdong Province, China. *Ecotoxicol. Environ. Saf.* **2012**, *78*, 2–8. [CrossRef] [PubMed]
40. Cai, L.M.; Xu, Z.; Bao, P.; He, M.; Dou, L.; Chen, L.; Zhou, Y.; Zhu, Y.G. Multivariate and geostatistical analyses of the spatial distribution and source of arsenic and heavy metals in the agricultural soils in Shunde, Southeast China. *J. Geochem. Explor.* **2015**, *148*, 189–195. [CrossRef]
41. Khosravi, Y.; Zamani, A.A.; Parizanganeh, A.H.; Yaftian, M.R. Assessment of spatial distribution pattern of heavy metals surrounding a lead and zinc production plant in Zanjan Province, Iran. *Geoderma Reg.* **2018**, *12*, 10–17. [CrossRef]
42. Cai, S.; Zhou, S.; Wu, P.; Zhao, J. Speciation and bioavailability of metals in sediments from a stream impacted by abandoned mines in maoshi town, Southwest of China. *Bull. Environ. Contam. Toxicol.* **2019**, *103*, 302–307. [CrossRef]
43. Yan, W.; Mahmood, Q.; Peng, D.; Fu, W.; Chen, T.; Wang, Y.; Li, S.; Chen, J.; Liu, D. The spatial distribution pattern of heavy metals and risk assessment of moso bamboo forest soil around lead–zinc mine in Southeastern China. *Soil Tillage Res.* **2015**, *153*, 120–130. [CrossRef]
44. Wang, L.; Guo, Z.; Xiao, X.; Chen, T.; Liao, X.; Song, J.; Wu, B. Heavy metal pollution of soils and vegetables in the midstream and downstream of the Xiangjiang River, Hunan Province. *J. Geogr. Sci.* **2008**, *18*, 353–362. [CrossRef]
45. *GB15618-2018*; Soil Environment Quality Risk Control Standard for Soil Contamination of Agricultural Land. PRC: Beijing, China, 2018; p. 2.
46. Li, Y.; Wang, Y.; Gou, X.; Su, Y.; Wang, G. Risk assessment of heavy metals in soils and vegetables around non-ferrous metals mining and smelting sites, Baiyin, China. *J. Environ. Sci.* **2006**, *18*, 1124–1134. [CrossRef]
47. Li, Y.J. *Sequence-Palaeogeography and Coal Accumulation of the Late Triassic Xujiahe Formation in the Sichuan Basin*; China University of Mining and Technology: Beijing, China, 2014.
48. Liu, X.; Shi, H.; Bai, Z.; Zhou, W.; Liu, K.; Wang, M.; He, Y. Heavy metal concentrations of soils near the large opencast coal mine pits in China. *Chemosphere* **2019**, *244*, 125360. [CrossRef]
49. Amari, K.E.; Valera, P.; Hibti, M.; Pretti, S.; Marcello, A.; Essarraj, S. Impact of mine tailings on surrounding soils and ground water: Case of Kettara old mine, Morocco. *J. Afr. Earth Sci.* **2014**, *100*, 437–449. [CrossRef]
50. Neiva, A.M.R.; Carvalho, P.C.S.; Antunes, I.M.H.R.; Silva, M.M.V.G.; Santos, A.C.T.; Cabral Pinto, M.M.S.; Cunha, P.P. Contaminated water, stream sediments and soils close to the abandoned Pinhal do Souto uranium mine, Central Portugal. *J. Geochem. Explor.* **2014**, *136*, 102–117. [CrossRef]
51. Zhao, K.; Fu, W.; Liu, X.; Huang, D.; Zhang, C.; Ye, Z.; Xu, J. Spatial variations of concentrations of copper and its speciation in the soil–rice system in Wenling of southeastern China. *Environ. Sci. Pollut. Res.* **2014**, *21*, 7165–7176. [CrossRef] [PubMed]
52. Wu, X.L.; Huang, X.F.; Li, C.C.; Hu, J.W.; Tang, F.H.; Zhang, Z.D. Soil heavy metal pollution degrees and metal chemical forms around the coal mining area in western Guizhou. *Res. Soil Water Conserv.* **2018**, *25*, 335–341.

53. Wang, H.; Wu, Q.; Hu, W.; Huang, B.; Dong, L.; Liu, G. Using multi-medium factors analysis to assess heavy metal health risks along the Yangtze River in Nanjing, Southeast China. *Environ. Pollut.* **2018**, *243*, 1047–1056. [CrossRef]
54. Ge, F.F.; Wang, X.F.; Fu, W.J.; Wang, L.L.; Jin, C.X.; Lv, J.H.; Cao, Z.G.; Yan, G.X.; Zhang, X. A Review of Cadmium Polluted Agricultural Soils and Phytoremediation in China. *Environ. Prot. Sci.* **2017**, *43*, 105–110.
55. Ren, B.Z.; Wu, Y.; Deng, D.P.; Tang, X.F.; Li, H.T. Effect of multiple factors on the adsorption of Cd in an alluvial soil from Xiba, China. *J. Contam. Hydrol.* **2020**, *232*, 103605. [CrossRef]
56. Bin, H.; Zhaojun, Y.; Jianbo, S.; Guibin, J. Research progress of heavy metal pollution in China: Sources, analytical methods, status, and toxicity. *Chin. Sci. Bull.* **2013**, *58*, 134–140.
57. Guo, J.K.; Zhao, J.; Wei, T.; Li, Y.P.; Lü, X.; Ren, Q.; Wang, X.H. Speciation and pollution assessment of cadmium and lead in vegetable greenhouse soil from a Xi'an Suburb with different cultivating years. *J. Agro-Environ. Sci.* **2018**, *37*, 2570–2577.
58. Kassir, L.N.; Darwish, T.; Shaban, A.; Olivier, G.; Ouaini, N. Mobility and bioavailability of selected trace elements in Mediterranean red soil amended with phosphate fertilizers: Experimental study. *Geoderma* **2012**, *189–190*, 357–368. [CrossRef]
59. Huang, Y.; Li, Y.X.; Gao, F.W.; Xu, M.M.; Sun, B.; Wang, N.; Yang, J. Speciation and Risk Assessment of Heavy Metals in Surface Sediments from the Heavily Polluted Area of Xiaoqing River. *Environ. Sci.* **2015**, *36*, 2046–2053.
60. Zhang, K.; Chai, F.; Zheng, Z.; Yang, Q.; Zhong, X.; Fomba, K.W.; Zhou, G. Size distribution and source of heavy metals in particulate matter on the lead and zinc smelting affected area. *J. Environ. Sci.* **2018**, *71*, 188–196. [CrossRef] [PubMed]
61. Davidson, C.M.; Urquhart, G.J.; Ajmone-Marsan, F.; Biasioli, M.; da Costa Duarte, A.; Díaz-Barrientos, E.; Grčman, H.; Hossack, I.; Hursthouse, A.S.; Madrid, L.; et al. Fractionation of potentially toxic elements in urban soils from five European cities by means of a harmonised sequential extraction procedure. *Anal. Chim. Acta* **2006**, *565*, 63–72. [CrossRef]
62. Ma, X.L.; Zuo, H.; Tian, M.J.; Zhang, L.Y.; Meng, J.; Zhou, X.; Min, N.; Chang, X.; Liu, Y. Assessment of heavy metals contamination in sediments from three adjacent regions of the Yellow River using metal chemical fractions and multivariate analysis techniques. *Chemosphere* **2016**, *144*, 264–272. [CrossRef]
63. Acosta, J.A.; Gabarrón, M.; Faz, A.; Martínez-Martínez, S.; Zornoza, R.; Arocena, J.M. Influence of population density on the concentration and speciation of metals in the soil and street dust from urban areas. *Chemosphere.* **2015**, *134*, 328–337. [CrossRef]
64. Adriano, D. Trace Elements in Terrestrial Environments. In *Biogeochemistry, Bioavailability and Risks of Metals*; Springer: New York, NY, USA, 2001; 866p.
65. Zong, Y.T.; Xiao, Q.; Lu, S.G. Distribution, bioavailability, and leachability of heavy metals in soil particle size fractions of urban soils (northeastern China). *Environ Sci Pollut Res.* **2016**, *23*, 14600–14607.
66. Huang, Q.; Tang, S.; Huang, X.; Zhang, F.; Yi, Q.; Li, P.; Fu, H. Influence of rice cultivation on the abundance and fractionation of Fe, Mn, Zn, Cu, and Al in acid sulfate paddy soils in the Pearl River Delta. *Chem. Geol.* **2017**, *448*, 93–99. [CrossRef]
67. Li, H.X.; Ji, H.B. Chemical speciation, vertical profile and human health risk assessment of heavy metals in soils from coal-mine brownfield, Beijing, China. *J. Geochem. Explor.* **2017**, *183*, 22–32. [CrossRef]
68. Davis, J.C.; Sampson, R.J. *Statistics and Data Analysis in Geology*; John Wiley & Sons, Inc.: New York, NY, USA, 1986; pp. 166–171.
69. Jiao, W.; Chen, W.; Chang, A.C.; Page, A.L. Environmental risks of trace elements associated with long-term phosphate fertilizers applications: A review. *Environ. Pollut.* **2012**, *168*, 44–53. [CrossRef]
70. Wang, Y.; Duan, X.; Wang, L. Spatial distribution and source analysis of heavy metals in soils influenced by industrial enterprise distribution: Case study in Jiangsu Province. *Sci. Total Environ.* **2020**, *710*, 134953. [CrossRef]
71. Lv, J.; Liu, Y.; Zhang, Z.; Dai, J. Factorial kriging and stepwise regression approach to identify environmental factors influencing spatial multi-scale variability of heavy metals in soils. *J. Hazard Mater.* **2013**, *261*, 387–397. [CrossRef] [PubMed]
72. Dou, W.Q.; An, Y.; Qin, L.; Lin, D.S.; Dong, M.M. Characteristics of vertical distribution and migration of heavy metals in farmland soils and ecological risk assessment. *Environ. Eng.* **2021**, *39*, 166–172.
73. Shi, R.; Yue, R.; Zhang, H. Research on Vertical Distribution of Heavy Metal in Soil around Non-ferrous Metal Industry Area. *Chin. J. Soil Sci.* **2016**, *47*, 186–191.
74. Heydarnejad, M.S.; Khosravian-Hemamai, M.; Nematollahi, A. Effects of cadmium at sub-lethal concentration on growth and biochemical parameters in rainbow trout (Oncorhynchus mykiss). *Ir. Vet. J.* **2013**, *66*, 11. [CrossRef]
75. Van Ginneken, M.; Blust, R.; Bervoets, L. How lethal concentration changes over time: Toxicity of cadmium, copper, and lead to the freshwater isopod Asellus aquaticus. *Environ. Toxicol. Chem.* **2017**, *36*, 10. [CrossRef]

remote sensing

MDPI

Article

The Identification and Driving Factor Analysis of Ecological-Economi Spatial Conflict in Nanjing Metropolitan Area Based on Remote Sensing Data

Ji Cao [1], Weidong Cao [1,*], Xianwei Fang [1], Jinji Ma [1], Diana Mok [2] and Yisong Xie [3]

[1] School of Geography and Tourism, Anhui Normal University, Wuhu 241002, China
[2] Department of Management and Organizational Studies, The University of Western Ontario, London, ON N6A 5C2, Canada
[3] Environment Protection Key Laboratory of Satellite Remote Sensing, Aerospace Information Research Institute, Chinese Academy of Sciences, Beijing 100094, China
* Correspondence: weidong1@ahnu.edu.cn

Abstract: The rapid socio-economic development of the metropolitan area has led to the continuous deterioration of the ecological environment. This leads to intense competition and conflict between different spatial use types. Spatial conflict research is essential to achieve ecological-economic coordination and high-quality development. However, existing studies lack comprehensive and direct ecological-economic spatial conflicts, especially those on the spatial-temporal evolution and potential drivers of spatial conflict. In this study, we identified the ecological-economic spatial conflicts in the Nanjing metropolitan area in 2010, 2015, and 2020. This study used the random forest to analyze the factors that influenced the change of spatial conflict. Results show that: (1) From 2010 to 2020, the ecological-economic spatial conflict in the Nanjing metropolitan area changed significantly. (2) Land use change has an important effect on spatial conflicts, which are easily triggered by uncontrolled urban expansion, but ecological land can mitigate spatial conflicts. (3) Relevant driving factors of spatial conflicts show multi-level features, so the development of conflict reconciliation countermeasures needs to be tailored to local conditions. This study provides a significant foundation for the high-quality development of the Nanjing metropolitan area and provides a reference for the planning and management of the territorial space.

Keywords: spatial conflict; conflict identification; analysis of driving factor; the Nanjing metropolitan area

Citation: Cao, J.; Cao, W.; Fang, X.; Ma, J.; Mok, D.; Xie, Y. The Identification and Driving Factor Analysis of Ecological-Economi Spatial Conflict in Nanjing Metropolitan Area Based on Remote Sensing Data. *Remote Sens.* **2022**, *14*, 5864. https://doi.org/10.3390/rs14225864

Academic Editor: Javier Marcello

Received: 9 October 2022
Accepted: 16 November 2022
Published: 19 November 2022

1. Introduction

In sustainable development, the development of human society should not only meet the needs of contemporary people, but also should not damage the ability of future generations to meet their own needs. "Space" is the collective term for the Earth's surface area that is suitable for human economic and social activities [1]. As an objective geographical phenomenon caused by spatial resource scarcity and spatial function overflow, spatial conflict is caused by the competition for spatial resources in human activities [2]. Although there are relatively few studies related to spatial conflicts, scholars have noticed the widespread phenomenon of spatial conflicts in society and have put forward the concepts of "regional deprivation" and "spatial competition", which are similar to spatial conflicts. Those concepts' connotations are all contradictory phenomena in the process of spatial conflict. For example, the collision between urban space and rural space, the encroachment of arable land by urban expansion, the competition between green ecological space and urban construction land, and the decline of ecological environment quality due to the disorderly expansion of urban construction land are all important manifestations of the phenomenon of spatial conflict. Along with the continuous development of the social economy, human activities now have far more influence on the Earth's surface than any period in history. The friction generated by human activity space and natural ecological space also exceeds the limit that the local environment can bear. The degradation of the ecological environment caused by rapid urban expansion makes the ecological-economic space conflict increasingly fierce [3]. The ecological-economic space conflict has also become an important reason hindering sustainable development, which has attracted the continuous attention of scholars [4].

Previous studies have provided various identification and evaluation methods of ecological-economic space conflicts, but most existing studies on ecological-economic conflict are based on panel data and lack spatial analysis. In the past few decades, rapid advances in remote sensing (RS) and geographical information systems (GIS) technology have provided the basis for spatial data acquisition and analysis, which provides an accurate source of data for monitoring and detecting land use, ecological changes, and human activity intensity [5,6]. The spatial data provided by remote sensing can reflect the spatial variation in regions better than traditional statistical surveys. As the carrier of ecological environment and economic activities, land use conflict is the most direct manifestation of ecological-economic spatial conflict and the earliest research involving ecological-economic space conflict [7]. Scholars determine the types of space they belong to according to different land use patterns, and they have found that potential spatial conflicts may arise from different land use patterns overlapping in space [8–10]. Based on the actual situation in China, Chinese scholars have put forward the theory of "ecology-production-life" to study spatial conflict, which is based on the three pillar theories of sustainable development. In the ecological-production-life framework, space is divided into three independent and interrelated spaces: ecological, production, and living space, which makes it easier to distinguish conflicts between different spaces [11]. Meanwhile, this theory also coordinates the relationship between ecological, production, and living space and offers countermeasures and suggestions for the sustainable development of society [12].

In addition, domestic and foreign scholars also analyze the ecological-economic space conflict from the perspective of landscape [13]. The spatial conflict caused by the unreasonable spatial structure of land use can be accurately identified by the landscape pattern theory [14,15]. Scholars use landscape pattern theory to build a spatial conflict model to describe the impact of human activity or natural environment change on the landscape composition, structure, and function and also use relevant landscape pattern models to measure spatial conflict [16,17]. As attention to ecological protection has increased, a conflict analysis method that can directly link ecology and the economy has also made great progress [18–20]. The theory of ecosystem service is considered the bridge between natural environmental systems and social-economic systems [21]. Based on the "land use mode-ecological process-ecological service system" [22], the contradiction between regional ecological and economic space can be directly analyzed through the contradiction between the supply and demand of ecosystem service theory [23–26].

In a word, the existing studies can basically reflect the ecological-economic space conflict, but they are mostly specific ecological aspects, such as for wetlands, water, protected natural areas, and other social and economic space conflicts, and there is still a lack of overall analysis of the ecological-economic space conflict [27–29]. Moreover, the identification of spatial conflicts relies too much on land use change, neglecting other socio-economic and natural environmental factors. The existing related methods also have their disadvantages. The "ecology-production-life" theory divides space into three separate spaces, thus separating the integrity of space, and cannot well analyze the space conflict in complex situations. Due to its theoretical characteristics, the landscape pattern theory considers economic factors less [30]. Influenced by dynamic changes in time and space, the supply and demand studies of ecosystem services are greatly influenced by the size of the study area [31]. Thus, the existing related methods have their shortcomings. Some scholars have identified the relevant shortcomings and tried to combine multiple methods to study the regional spatial conflict situation, such as by analyzing the regional ecosystem service value change under the framework of the "ecology-production-life" theory [32], using the landscape pattern index to quickly and accurately identify the regional "ecology-production-life" spatial evolution [33], or studying the response of ecosystem service function to the landscape pattern change caused by land use transformation [34]. However, there is no study combining these three theories, leading to one-sided results in the identification of ecological-economic space conflicts.

Most of the existing studies take cities and counties as the research units or take the whole country as the subject investigated. Few studies have researched the spatial conflict situation in metropolitan areas, which is a novel and advanced form of territorial spatial organization [4,33,34]; the role of the metropolitan is becoming apparent in regional economic development. Along with the rapid development of metropolitan areas, there are frequent transfers between various land types [35,36]. The original spatial pattern breaks, leading to many spatial conflict problems. Metropolitan areas have become an important driving force behind China's economic growth and contribute to the coordinated development of regional space, which is increasingly receiving national attention [37].

Located in the Yangtze River Delta, the Nanjing metropolitan area is China's first inter-provincial metropolitan area. In February 2021, the Nanjing metropolitan area became the first metropolitan area plan in China to be officially approved by the National Development and Reform Commission, marking a further increase in the strategic status of the Nanjing metropolitan area. Owing to its superior natural geographical conditions, the Nanjing metropolitan area economy developed rapidly. At the same time, conflicts caused by environmental protection and economic development are increasingly intensified. Therefore, there is an urgent need to study the actual situation and the relevant influencing factors of ecological-economic space conflicts [38]. To explore the conflict caused by the disharmony and inconsistency between ecological and economic space, this study used the comprehensive evaluation model of ecological-economic space conflict to identify the severity using the random forest method. This study provides an important realistic basis for the high-quality development of the Nanjing metropolitan area and new ideas for studying the sustainable development path of metropolitan areas in China.

2. Materials and Methods

2.1. Study Area

Situated at the lower reaches of the Yangtze River (Figure 1), the Nanjing metropolitan area has a warm and humid climate and is blessed with abundant natural resources. The southern part of the metropolitan area has a high forest cover and rich forest resources. The northern and central parts are mainly plains, which are suitable for agricultural development. The metropolitan area has well-developed river systems with two important drainage systems, the Yangtze River and the Huai River, and large lakes such as Hung-tse Lake and Gaoyou Lake. Now, the Nanjing metropolitan area has 8 cities and 2 districts, including Nanjing, Zhenjiang, Yangzhou, Huaian, Maanshan, Chuzhou, Wuhu, Xuancheng, Liyang, and Jintan in Changzhou. The total area of the Nanjing metropolitan is over 65,000 square km and the population is more than 35 million. Its GDP is above 4.6 trillion yuan, making the Nanjing metropolitan area one of the most important economic centers in the Yangtze River Delta. The rapid development of the social economy in the Nanjing metropolitan area has brought great pressure and challenges to the local environment, making the ecological-economic space conflict increasingly fierce.

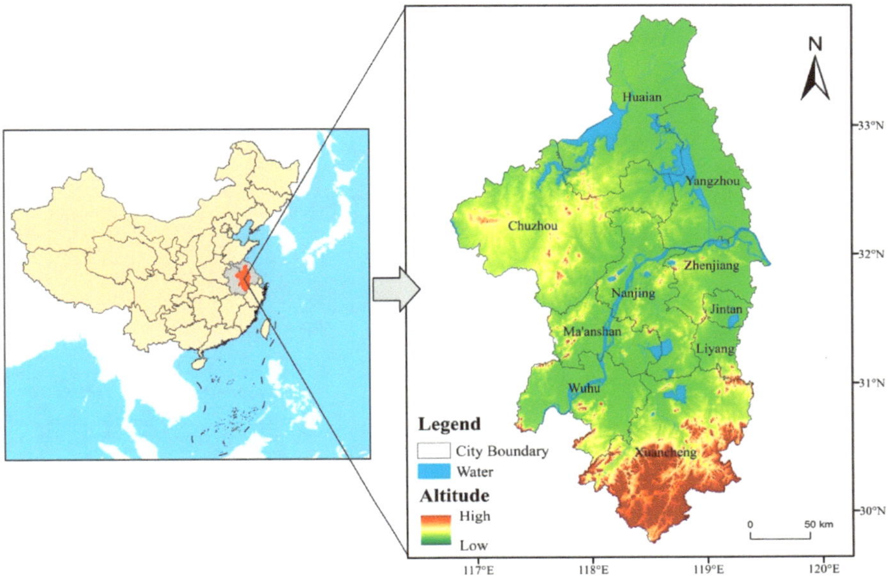

Figure 1. Administrative division and terrain of the study area.

2.2. Research Framework

The essence of spatial conflict is the game of spatial resource possession between the conflicting parties. Along with the development and use of spatial resources, the original spatial pattern will also change, leading to changes in spatial functions and thus changes in the spatial carrying capacity of the region. On this basis, we divide the spatial use mode into three different use types: spatial resource use, spatial function use, and spatial capacity use and construct a spatial conflict identification system by combining relevant previous studies [39]. Then, we classified spatial conflicts into five levels to identify and analyze the ecological-economic conflicts in the Nanjing metropolitan area. Finally, the study used the random forest method to measure the contribution of the relevant driving factors to analyze their importance. Our study research framework is shown in Figure 2.

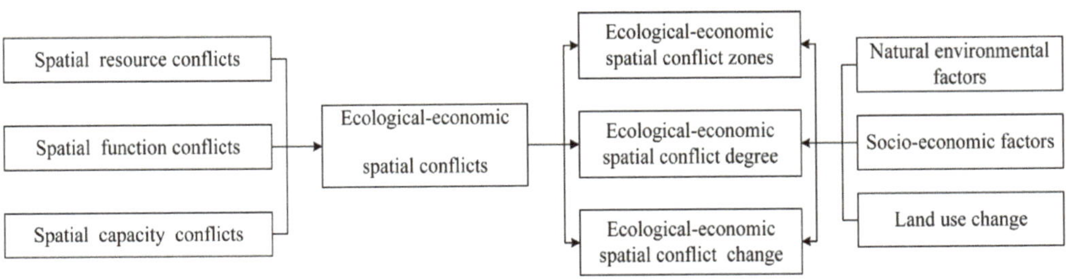

Figure 2. Research framework diagram.

2.3. Data Collection

Table 1 shows the source of the data; our data can be divided into three main categories. First, the study divided the land use data into six types according to LUCC standards [40]; the specific situation of land use is shown in Figure 3. Second, the natural environment data are mainly from relevant satellite remote-sensing image data. Third, the spatial resolution of the socioeconomic data is 1 km and was mostly provided by the Resource and Environment Data Sharing Center of the Chinese Academy of Sciences.

Table 1. Data sources.

Data	Resolution Data	Available Time Interval	Source
Land Use Data	1 km × 1 km	2010, 2015, 2020	Resource and Environmental Science and Data Center (https://www.resdc.cn (accessed on 25 January 2020))
Net primary productivity (NPP)	0.5 km × 0.5 km	2010–2020	Product of MOD17A3H estimated by moderate resolution imaging spectroradiometer (MODIS) images (http://www.noaa.gov/ (accessed on 26 January 2020))
Normalized difference vegetation index (Ndvi)	1 km × 1 km	2010–2020	MYDND1M China 500M (http://www.noaa.gov/ (accessed on 11 January 2021))
Fine particulate matter (PM2.5)	1 km × 1 km	2010–2020	https://doi.org/10.5281/zenodo.6372847 (accessed on 18 March 2022)
Nighttime light (NtL)	1 km × 1 km	2010–2020	NOAA (https://ngdc.noaa.gov/eog/dmsp/downloadV4 composites.html (accessed on 7 April 2022))
Gross domestic product (GDP)	1 km × 1 km	2010, 2015, 2019	Resource and Environmental Science and Data Center (https://www.resdc.cn (accessed on 17 April 2022))
Population data (Pop)	1 km × 1 km	2010–2020	Worldpop (https://www.worldpop.org/ (accessed on 27 March 2021))

Since the Chinese Academy of Sciences has not yet given the spatial data of China's GDP in 2020, we choose to use the spatial data of China's GDP in 2019 instead.

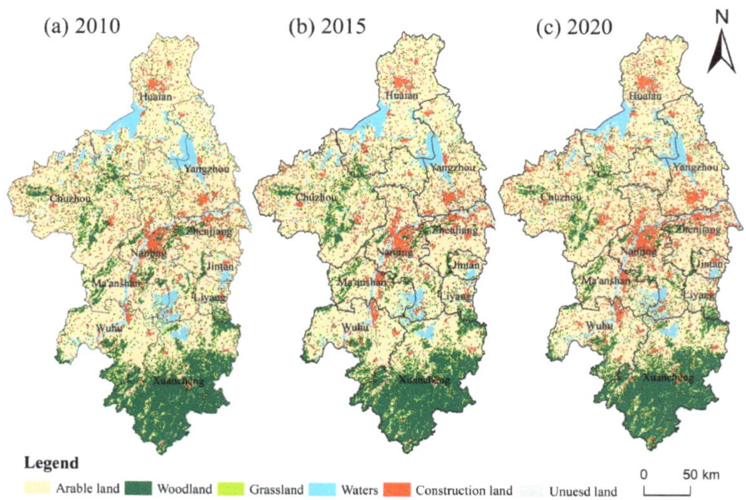

Figure 3. Land use map of Nanjing metropolitan area. (**a**) 2010 land use map; (**b**) 2015 land use map; (**c**) 2020 land use map.

2.4. Indicator Construction

In this study, spatial use was divided into three types, corresponding to the spatial resource development process, the spatial function change situation, and the spatial carrying capacity change situation. We then selected the corresponding evaluation indexes from the natural factors and socio-economic factors to construct an ecological-economic spatial conflict evaluation index system. We selected relevant data from 2010, 2015, and 2020 and conducted standardized processing before the calculation to prevent the possible uncertain impact of different data values on the overall operation. The composition of the indicator system is shown in Figure 4. The entropy weighting method was used to calculate the weights of resource conflict (RC), function conflict (FC), and capacity conflict (CC) to derive the spatial conflict value for that year [39]. All of these are shown in Table 2. The entropy weight method is an objective empowerment way to calculate the weight through the information entropy, which is through the dispersion degree of the original data of each index. It can effectively avoid the deviation caused by subjective factors and improve the credibility and accuracy of the index weight value [41,42]. We used the weighted average of the 3 option weights as the final weight of the study, with resource conflict (RC), function conflict (FC), and capacity conflict (CC) having final weight values of 0.32, 0.42, and 0.25. Finally, the study used ArcGIS 10.7 and applied equal interval classification to classify the Nanjing metropolitan area 2010–2020 ecosystem service demand index into five classes: highest-conflict, high-conflict, medium-conflict, low-conflict, and lowest-conflict.

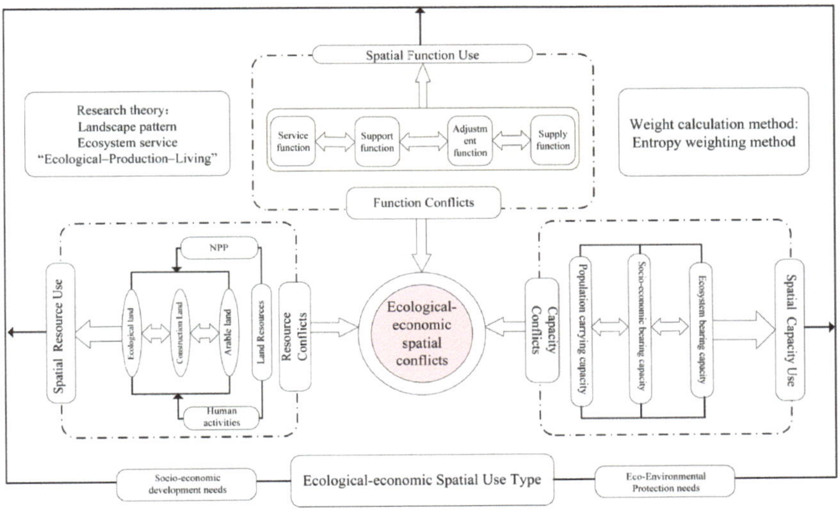

Figure 4. Indicator system diagram.

Table 2. Ecological-economic spatial conflict evaluation Indicator system.

Indicator Type	Standard Layer	Indicator Layer	Indicator Attribute	Indicator Description
Resource Conflict (RC)	Land use conflict (RC1)	Landscape aggregation index (AI)	Negative	Reflects the conflict between cultivated land resources and construction land
		Landscape sprawl index (Contag)	Negative	Reflects the conflict between ecological land resources and construction land
	Human activities clash with natural resources (RC2)	NPP data	Negative	Reflects the vegetation regeneration capacity
		Construction–land density reaction	positive	Reaction to the consumption of land resources
Function Conflict (FC)	Supply and demand of ecosystem services conflict (FC1)	Supply of ecosystem services	Negative	Reflects the supply and demand of ecosystem services
		Demand of ecosystem services	positive	
	Carbon-fixing capacity conflicts with carbon emissions (FC2)	Carbon emissions	positive	Reflect carbon emissions and carbon storage
		Carbon sequestration	Negative	
Capacity Conflict (CC)	Biodiversity conflict (CC1)	Habitat quality	Negative	Reflects species richness through biodiversity
	Economic and environmental conflict (CC2)	GDP	positive	Reflects the economic development situation
		PM2.5	positive	Reflects the air environmental quality situation
		Pop	positive	Reflects the size of the population
		Ndvi data	Negative	Reflects the vegetation coverage situation

2.5. Interpretation and Calculation of Indicators

Resource conflict (RC) is mainly expressed as the conflicts between human-led economic activities, food production, and other processes on other biological and resource supplies. Among them, RC1 uses AI to measure the conflicts between arable land and construction land, the degree of cultivated land fragmentation to reflect the conflict between construction land and arable land, Contag to measure the degree of ecological land and construction land, and the degree of fragmentation to respond to the conflicts between construction land and ecological land. The degree of ecological land fragmentation responds to the spatial connectivity of ecological land by the encroachment of construction land [43,44]. RC2 uses NPP data and the density of built-up land to reflect the influence of urban development on vegetation regeneration capacity. The space occupied by human socio-economic activities undoubtedly affects the growth of the original ecological vegetation. The NPP data were used to reflect the spatial situation of vegetation growth activities on the surface, whereas the density of built-up land reflects the distribution intensity of human economic activities on the surface [45,46].

$$AI = \left[\sum_{i=1}^{m}\left(\frac{g_{ii}}{\max g_{ii}}\right)p_i\right] \times 100 \tag{1}$$

where g is the number of nodes between image elements of patch type i based on the single-fold method and $\max g_{ii}$ is the maximum number of nodes between image elements of patch type i based on the single-fold method. P_i is the area proportion of patch type i in landscape.

$$CONTAG = \left\{1 + \frac{\sum_{i=1}^{m}\sum_{k=1}^{m}\left[p_i\left(g_{ik}/\sum_{k=1}^{m} g_{ik}\right)\right] \times \left[\ln p_i\left(g_{ik}/\sum_{k=1}^{m} g_{ik}\right)\right]}{2\ln(m)}\right\} \times 100 \tag{2}$$

where P_i is the proportion of area of patch type i in the landscape, g_{ik} is the number of nodes between patch type i and patch type k on the basis of the doubling method, and m is the number of patch types in the landscape, including those in the landscape boundary.

Functional conflict (FC) is mainly reflected in the overlapping area of the economic activity area and ecological function area. The overlap results in conflict caused by social and economic production to ecological function disturbance. According to a previous study, we used ecosystem service supply and demand to judge the functional conflict of ecological-economic space [47]. The ecosystem service supply was calculated by the value equivalent method, while the demand for ecosystem service was calculated according to previous studies [25]. Based on previous studies, this study uses nighttime light (NtL) to estimate carbon emissions [48]. Then we used the invest model carbon storage module to calculate carbon reserves. The invest model carbon storage module specific formula is

$$C_{x,t} = \sum_{j=1}^{J} A_{xj}\left(C_{aj} + C_{bj} + C_{sj} + C_{dj}\right) \tag{3}$$

C_x is the carbon stock of region x in t, A_{xj} is the area of land cover type j in region x, and C_{aj}, C_{bj}, C_{sj}, and C_{dj} represent the above-ground carbon density, below-ground carbon density, soil carbon density, and dead organic matter carbon density of land cover type j, respectively [49].

Capacity conflict (CC) mainly occurs in the process of space evolution between human activities and the ecological environment. CC1 uses the invest model of the habitat quality module to calculate the impact of human activities on the bearing capacity of the natural environment [50]. Based on previous studies, CC2 uses GDP, PM2.5, Pop, and Ndvi to reflect the conflict between environmental protection and economic development [39].

GDP and Pop represent the capacity of economic development while the Ndvi and PM2.5 represent the capacity of ecological protection.

The habitat quality module formula is

$$Q_{xj} = H_j\left(\frac{K^z}{\left(D_{xj}^z + K^z\right)}\right) \tag{4}$$

where Q_{xj} is the habitat quality index of raster cell x in land use type j and H_j is the habitat suitability of land use type j. The value range is [0, 1]. The closer the value is to 1, the higher the habitat quality. D_{xj} is the degradation degree of raster cell x in land use/cover type j. K is the half-saturation constant, which is usually half of the maximum degradation degree; the default value is 0.5. z is the normalization constant, which is the default parameter of the model, and takes the model definition value of 2.5 [51].

2.6. Data Sources and Methods of Driving Factor

Eight potential driving factors were selected for analysis as potential causes of ecological-economic spatial conflicts that may affect the Nanjing metropolitan area [52]. These driving factors include two main aspects: (1) For natural environmental factors, the study selected DEM, distance to water, average annual temperature, and soil as driving factors in the natural environment. DEM, distance to water, and soil type data were obtained from the Chinese Academy of Sciences Resource Environment Data Sharing Center (https://www.resdc.cn/(accessed on 11 March 2022)). The annual average temperature data were obtained from the site data interpolation of China Meteorological Network (http://www.cma.gov.cn/ (accessed on 11 January 2022)). (2) For socio-economic factors such as distance to the major highways, distance to the Nanjing, industrial density (nuclear density of industrial parks and development zones), and distance to the railroad, the data were crawled from Amap.

The random forest algorithm has excellent performance for establishing the nonlinear relationship between input variables and output variables [53]. Based on the principle of random forest, Liang proposed a patch-generated land-use simulation model (PLUS model), which has been successfully applied to dynamic simulation and prediction of land-use change and can analyze the contribution of related drivers to land use change [54]. The role of the LEAS model is to transform the mining of transition rules of each land use type in the PLUS model into a binary classification problem. This is specifically done to calculate the relationship between the growth of each land use type and the associated drivers based on the random forest algorithm and finally output the growth probability $P_{i,k}^d$ of land use type k at cell i. The random forest algorithm formula is

$$P_{i,k}^d(x) = \frac{\sum_{n=1}^{M} I(h_n(x) = d)}{M} \tag{5}$$

The value of d is either 0 or 1; when the value of d was 1, there were other land use types changed to land use type k, while 0 represents other transitions; x is a vector that consists of multiple driving factors; $I(\cdot)$ is the indicative function of the decision tree set; $h_n(x)$ is the prediction type of the n-th decision tree for vector x; and M is the total count of decision trees.

Jiang used similar principles to analyze the impact of related factors on potential pollution sites in the Yangtze River Delta [55]. On this basis, we performed spatial superpositions on the ecological-economic spatial conflict zone data in different periods, subtracted the previous spatial conflict zone data from the spatial conflict zone data in the latter period, and extracted the changes to represent the change areas of each spatial conflict level. Then, we used the LEAS module of the PLUS model to mine the influence of each driving factor on the change of each level of conflict area based on the random forest algorithm to evaluate the relative importance of each driver on different types of spatial conflict area and provide stable and accurate classification results. For a detailed description of the PLUS model, please refer to https://github.com/HPSCIL/Patch-generating_Land_Use_Simulation_Model (accessed on 27 January 2022).

3. Results

3.1. Spatial Distribution Characteristics of Ecological-Economic Space Conflict

This study used the constructed index system to obtain the ecological-economic spatial conflict distribution characteristics of the Nanjing metropolitan area for the three periods of 2010, 2015, and 2020. Figure 5 and Table 3 show that the ecological-economic spatial conflict in the Nanjing metropolitan area changes substantially from 2010 to 2020. The spatial conflict state in the Nanjing metropolitan area was dominated by low conflict. The proportion of low-conflict areas increased from 58.94% in 2010 to 66.15% in 2020. Meanwhile, the proportion of highest-conflict and high-conflict areas were low. These situations together show that the degree of ecological-economic space conflict in the Nanjing metropolitan area is not serious. However, we found some features of its space-time evolution: First, numerous medium-conflict areas were spread along the edges of the city in 2010. Medium-conflict areas accounted for 24.49% of the period, which gradually stabilized at around 18% over the next decade. Second, the high-conflict and highest-conflict areas were distributed in the city and its surrounding areas, and two larger core conflict areas formed in the central region along the river and northern region; the area proportion of these two conflicts is increasing. Finally, the lowest-conflict areas were mainly in the mountainous zone of Xuancheng and the hilly areas of Chuzhou, but the proportion of low-conflict areas is shrinking.

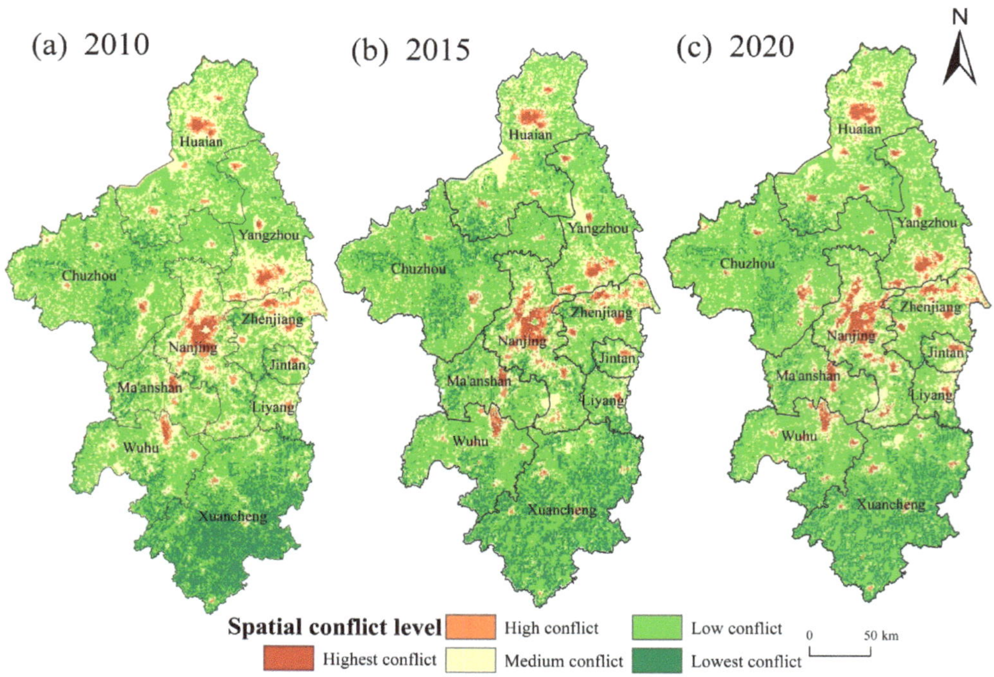

Figure 5. Distribution of ecological-economic spatial conflict in Nanjing metropolitan area. (**a**) 2010 spatial conflict map; (**b**) 2015 spatial conflict map; (**c**) 2020 spatial conflict map.

Table 3. Area composition of different conflict types.

Conflict Types	Highest Conflict	High Conflict	Medium Conflict	Low Conflict	Lowest Conflict
2010	1.07%	2.95%	24.49%	58.94%	12.56%
2015	1.24%	3.33%	18.22%	64%	13.21%
2020	1.94%	3.69%	18.98%	66.15%	9.24%

3.2. Spatial-Temporal Evolution of Ecological-Economic Spatial Conflict

To investigate the spatial and temporal changes of ecological-economic spatial conflicts in the Nanjing metropolitan area, we created a map of ecological-economic spatial conflict change zone transfer in the Nanjing metropolitan area (Figure 6) between 2010 and 2020. Figure 6 shows that transfers between spaces of different conflict levels were more frequent during 2010–2015, whereas the frequency of transfers decreased remarkably during 2015–2020. Specifically, highest-conflict and high-conflict areas had the least probability of spatial shifts. Shifts between medium-conflict areas, low-conflict areas, and lowest-conflict areas were frequent, especially between medium-conflict areas and low-conflict areas.

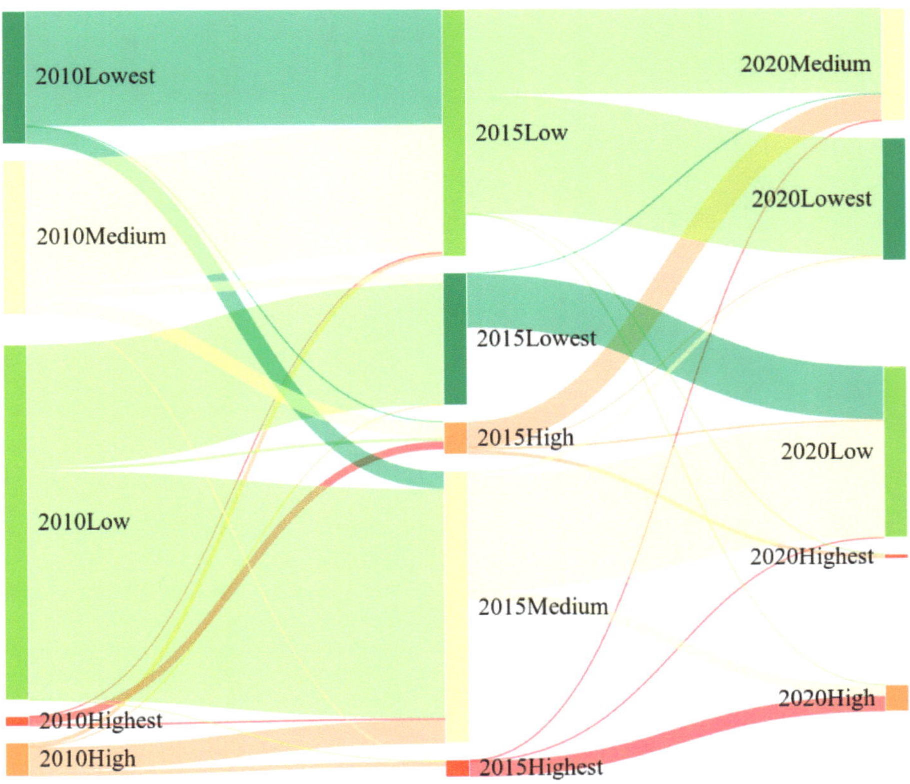

Figure 6. Transfer of ecological-economic space conflict change zone in Nanjing metropolitan area from 2010 to 2020 (the color of the highest, high, medium, low, and lowest five different spatial conflict levels is consistent with those in Figure 5).

Hotspot analysis (Figure 7) can fully show the clustering of ecological-economic spatial conflicts in the Nanjing metropolitan area: the deeper the red color is, the higher the degree of clustering in areas with high-conflict values, and vice versa [56]. Figure 7 shows that the spatial agglomeration characteristic of cold and hot spots in the Nanjing metropolitan area is remarkable. The cold spots in the Nanjing metropolitan area are mostly concentrated in the hilly areas of Chuzhou and the mountainous region of Xuancheng. The overall change is large, with an obvious decrease in cold spots over the last 10 years. The proportion of cold points in 2020 decreased by 22% compared with 2010. The hot spot areas were mainly concentrated around the central part of the metropolitan area along the Yangtze river and the urban area of Huai'an in the north, showing a trend of concentration to the city and its surrounding areas, and the hot spot value has significantly increased.

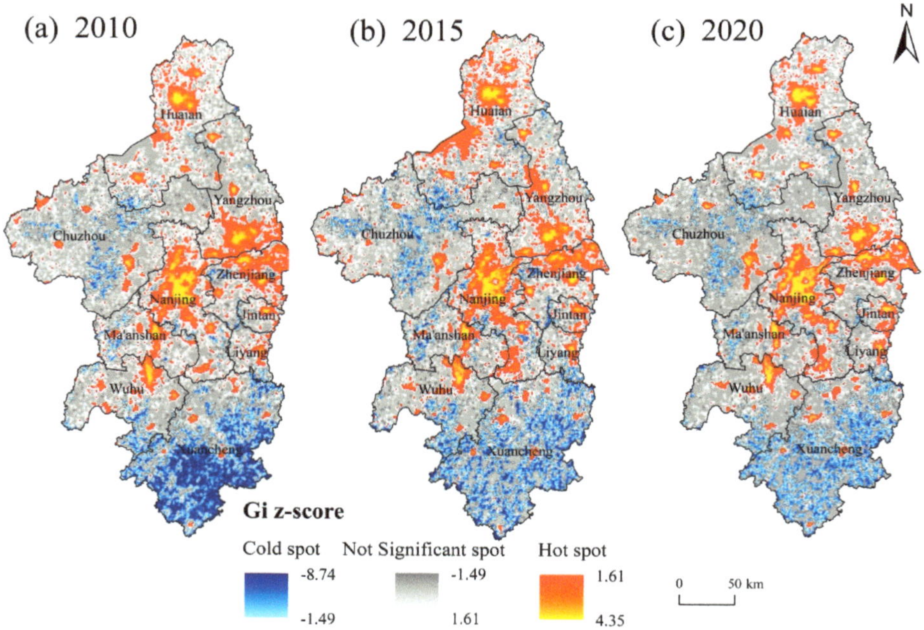

Figure 7. Cold−hot spot map of ecological-economic space conflict for 2010−2020. (**a**) 2010 cold−hot spot map; (**b**) 2015 cold−hot spot map; (**c**) 2020 cold−hot spot map.

3.3. Factors Influencing the Change of Ecological-Economic Spatial Conflict in Nanjing Metropolitan Area

3.3.1. Effect of Land Use on Ecological-Economic Space Conflict

Based on the analysis of the proportion of land use type in different conflict levels (Figure 8), construction land is the main land use type in the highest-conflict zone and high-conflict zone in Nanjing metropolitan area from 2010 to 2020. Construction land in 2010 only accounted for 60% of the high-conflict area, while this proportion rose to 79% by 2020. In the highest-conflict areas, the proportion of construction land has been above 90%, and reached 98% in 2020. The medium-conflict areas are mostly located around various cities and mainly show spatial conflict between arable land and construction land; that is, the proportion of arable land and construction land is the highest. As an important carrier of many socio-economic activities, the coastal parts of lake areas are disturbed by human activities and their ecological function is weakened, which makes them prone to medium spatial conflicts. As arable land is the biggest land type in the Nanjing metropolitan area, arable land is also the main type of land in the low-conflict zones, where its share is consistently around 70%. According to the current situation, the areas with the lowest conflict are mainly ecological lands such as forest and grassland, with the highest percentage of woodlands consistently being around 70% (Table 4).

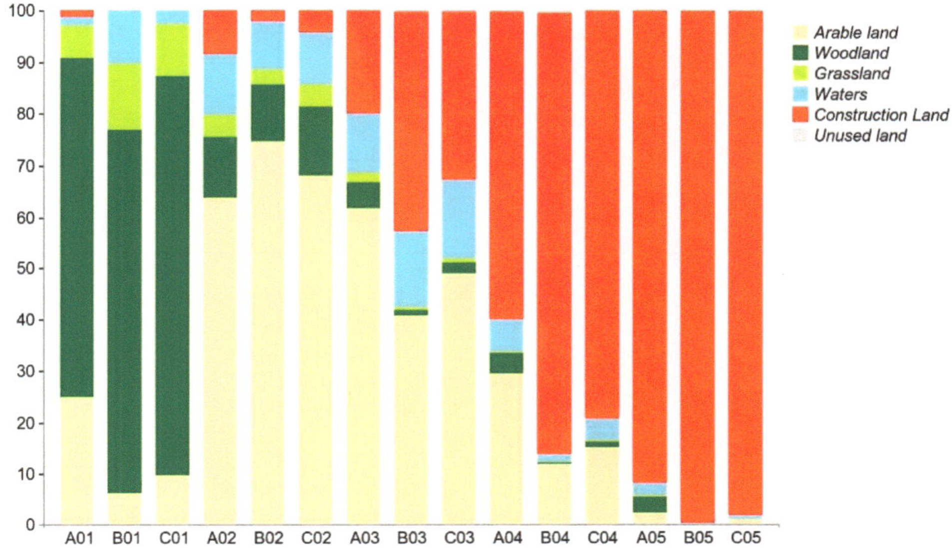

Figure 8. Composition of land use in different years of each conflict types (A, B, and C refer to 2010, 2015, and 2020; 01, 02, 03, 04, 05 refer to lowest, low, medium, high, highest spatial conflict levels).

Table 4. The proportion of different land use types in each conflict level.

Time	Land Use Type	Lowest Conflict	Low Conflict	Medium Conflict	High Conflict	Highest Conflict
2010	Arable land	24.97	63.97	61.76	29.31	2.34
	Woodland	65.73	11.68	5.11	3.97	3.21
	Grassland	6.46	4.26	1.83	0.26	0.15
	Waters	1.57	11.57	11.52	6.52	2.63
	Construction Land	1.20	8.44	19.68	59.57	91.68
	Unused land	0.07	0.08	0.10	0.37	0
2015	Arable land	6.3	74.71	40.90	11.80	0
	Woodland	70.51	10.86	1.12	0.33	0
	Grassland	12.89	3.08	0.53	0.09	0
	Waters	10.20	9.23	14.79	1.42	0.25
	Construction Land	0.06	2.09	42.36	85.88	99.75
	Unused land	0.04	0.03	0.29	0.47	0
2020	Arable land	9.67	68.17	48.93	15.12	0.98
	Woodland	77.54	13.20	2.28	1.07	0
	Grassland	10.22	4.18	0.77	0.17	0
	Waters	2.45	10.07	15.46	4.33	0.98
	Construction Land	0.10	4.31	32.35	79.22	98.04
	Unused land	0.02	0.06	0.21	0.09	0

3.3.2. Driving Factor of Changes in the Ecological-Economic Space Conflict

This study was based on previous studies that used similar principles to the LEAS model, which is based on the random forest method to rank the ecological-economic spatial conflict driving factors in the Nanjing metropolitan area. Then, we analyzed the contribution of different driving factors to different conflict rank regions. The number of decision trees was set as 50, the sampling rate set as 0.1, and the total sample size was 8939; also, this study used RMSE to reflect the accuracy of random forests. In general, the precision values of the random forest were inversely proportional to the RMSE values (Table 5).

Table 5. Random forest accuracy by RMSE.

Time	Highest Conflict	High Conflict	Medium Conflict	Low Conflict	Lowest Conflict
2010–2015	0.03	0.07	0.14	0.17	0.12
2015–2020	0.07	0.15	0.17	0.13	0.09
2010–2020	0.07	0.14	0.16	0.13	0.09

Just as shown in Figure 9 and Table 6, the different influencing factors affect each conflict state to different degrees, and the influence of DEM on each conflict-level area is substantial, especially for the lowest-conflict areas. In the 2010–2015 period, the contribution weight of DEM to the lowest-conflict region and low-conflict region both exceeded 0.2. In the 2015–2020 period, the contribution weight of DEM to the lowest-conflict region reached 0.39. As a whole, the contribution weights of DEM to the lowest conflict region and low conflict region during 2010–2020 were 0.18 and 0.24, both of which are at the top of the contribution scale. The distance to water is also an important driving factor for lowest-conflict and low-conflict areas. Overall, industrial density had the greatest impact on the expansion of high and highest conflict areas in the 2010–2020 period, with contribution weights of 0.2 and 0.23, which were above the other driving factors. However, between 2010 and 2015, the largest contributor to the expansion of highest-conflict areas was distance to railroads with a weight of 0.22, slightly higher than the industrial density (0.18), while the distance to highways contributed much less than the distance to railroads. The weight of the contribution of industrial density to the highest-conflict areas was 0.23 for the 2015–2020 period, which was at the top of each driving factor. The degree of contribution of soil type, the distance to the Nanjing, and the annual average temperature were also not remarkable, indicating that the development of the Nanjing metropolitan area relies more on the remaining characteristics.

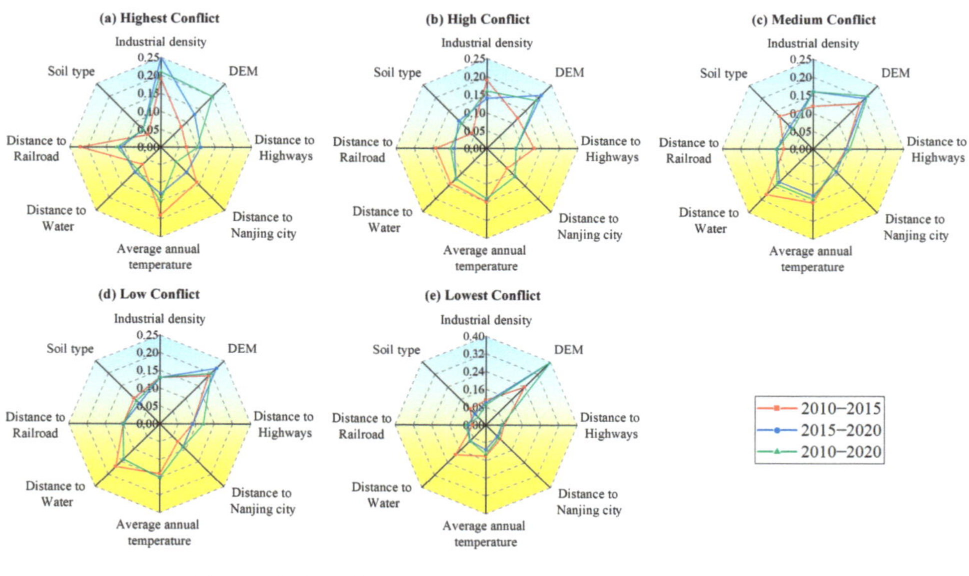

Figure 9. Driving factors of the ecological-economic spatial change.

Table 6. Contribution weight values of related driving factors.

Time	Factors	Highest Conflict	High Conflict	Medium Conflict	Low Conflict	Lowest Conflict
2010–2015	Industrial density	0.18	0.18	0.12	0.13	0.12
	DEM	0.09	0.13	0.19	0.2	0.24
	Distance to highways	0.06	0.13	0.09	0.1	0.09
	Distance to Nanjing city	0.13	0.08	0.07	0.07	0.09
	Average annual temperature	0.19	0.17	0.14	0.13	0.13
	Distance to water	0.07	0.13	0.18	0.16	0.19
	Distance to railroad	0.22	0.12	0.08	0.11	0.05
	Soil type	0.05	0.07	0.12	0.1	0.09
2015–2020	Industrial density	0.23	0.15	0.16	0.11	0.11
	DEM	0.15	0.19	0.19	0.21	0.39
	Distance to highways	0.11	0.1	0.09	0.12	0.06
	Distance to Nanjing city	0.07	0.1	0.1	0.1	0.09
	Average annual temperature	0.13	0.14	0.13	0.15	0.11
	Distance to water	0.1	0.13	0.12	0.14	0.1
	Distance to railroad	0.14	0.1	0.1	0.1	0.08
	Soil type	0.07	0.09	0.1	0.08	0.05
2010–2020	Industrial density	0.2	0.23	0.13	0.15	0.13
	DEM	0.12	0.1	0.22	0.18	0.24
	Distance to highways	0.09	0.09	0.08	0.1	0.08
	Distance to Nanjing city	0.09	0.08	0.08	0.08	0.08
	Average annual temperature	0.17	0.17	0.16	0.14	0.17
	Distance to water	0.12	0.11	0.12	0.15	0.11
	Distance to railroad	0.15	0.13	0.1	0.11	0.09
	Soil type	0.07	0.09	0.11	0.09	0.1

4. Discussion

In the context of ecological civilization construction, the identification of ecological-economic spatial conflict and research on influencing factors have become an important prerequisite for high-quality development, which may help policymakers and stakeholders to conduct investigation and adjustment work. In this study, all the data were derived from remote sensing satellite data and their deductive data to ensure accuracy and authenticity. Researchers have carried out many related studies on space conflict identification, which provided us with the relevant theoretical basis and methods for the identification of ecological-economic space conflicts. However, due to the defects of relevant theories, there are still some deficiencies in the identification of ecological-economic space conflicts. Based on previous studies, we tried to create a comprehensive, integrated index system to reflect the ecological-economic space conflict in the Nanjing metropolitan area as truthfully and thoroughly as possible. Furthermore, we analyze the extent of different drivers of the ecological-economic spatial conflict in the Nanjing metropolitan area through the latest random forest methods [55].

4.1. The overall Pattern of Ecological-Economic Spatial Conflicts Is Stabilizing, but the Lowest Conflict Areas Will Gradually Shrink

Our study shows that the situation of ecological-economic spatial conflict in the Nanjing metropolitan area has changed significantly over the past decade. This is reflected in the continuous increase of the highest-conflict and high-conflict zones, while the area of the lowest-conflict zone continues to decline. The percentage of area in the lowest conflict region in 2020 was only 9.24%. The area proportion of low-conflict zones increased from 58.94% in 2010 to 66.15%, and the area share of medium-conflict zones gradually decreased from 24.49% in 2010 before stabilizing at about 18% in 2020. In terms of the transfer of spatial conflict regions, the frequency of transfer is decreasing, indicating that the overall spatial conflict patterns gradually stabilize. Among them, the probability of the highest-conflict and high-conflict areas undergoing spatial transfer is minimal, and the highest-conflict and high-conflict areas will be difficult to change once formed. Medium-conflict and low-conflict areas are the most likely to be affected and change. In particular, the transfer of medium-conflict zones from 2010 to 2015 was very frequent. The spatial structure of medium-conflict zones and low-conflict zones was still not stable enough, the spatial structure of these areas was not stable enough, and changes in relevant policies or new development projects can change the local spatial conflict situation. From the change on the cold–hot spot maps, we can further see that the hot spots eventually contracted to be near the city, especially around the central part of the metropolitan area along the Yangtze river and the urban area of Huai'an in the north. This further proves that the ecological-economic space conflict situation in the Nanjing metropolitan area is effectively controlled and that the high-conflict level areas are restricted to the city zone. At the same time, we also note that with the development of the economy, the proportion of the cold spot areas will continue to decrease; in particular, the Xuancheng area decreased at a significant rate, which means that the pressure on ecological and environmental protection will continue to increase.

4.2. Different Land Use Types Have Different Effects on Changes in Ecological-Economic Spatial Conflicts

The spatial overlap of different functional requirements leads to spatial conflict. As the most common and direct influencing factor of spatial conflicts, land use patterns have always had a profound effect on the evolution of spatial conflicts [57]. Based on the proportion of different land use types at different conflict levels, construction land is undoubtedly the most important trigger of ecological-economic conflicts, because it must carry out most socio-economic activities, which further weakens the ecological functions of this land use type. The medium-conflict and low-conflict areas are mainly made up of arable land and water areas. Arable and water areas are mostly on the edge area of human activities. By rational planning of the use of these two types of land, industrial development can be achieved without damaging the ecosystem. However, they are inherently less stable and vulnerable to the influence of surrounding areas, so strict spatial boundary planning control of these land use types is needed to reduce the impact of adjacent units on them [12]. Pure ecological land types such as woodland and grassland are the main land types in the lowest-conflict areas, and strict protection measures should be taken so that ecological land can rebuild the regional ecological security pattern. Developing ecotourism industries and other industries that have less effect on the ecological environment while improving land use efficiency as much as possible is also important [58].

4.3. The Development of Eco-Economic Spatial Conflict Mitigation Measures Needs to Be Tailored to Local Conditions

According to the analysis of the relevant factors, we found that the relevant driving factors of spatial conflicts showed multi-level features, meaning that the development of conflict reconciliation countermeasures should also be adjusted to local conditions [59]. The distribution of industries largely influences the proliferation of both high-conflict and highest-conflict areas, so the study can conclude that the construction of industrial parks will quickly change the local conflict state and will lead to the rapid enhancement of socio-economic functions as well as the decline of ecological functions [60]. Hence, industrial parks must be built in and around cities as much as possible, away from ecological reserves or ecologically fragile areas. This was also confirmed in a study by Wang et al. [61]. DEM is the factor that contributes most to the lowest-conflict and low-conflict areas. Given that the Nanjing metropolitan high-altitude areas are mountainous forests, they are not amenable to socio-economic activities. Moreover, most of these areas belong to ecological protection zones, so the distribution of altitude highly overlaps with the distribution of lowest-conflict and low-conflict areas. At the same time, the low-elevation Nanjing metropolitan areas are mostly plains, which are suitable for socio-economic activities and agricultural production. Therefore, the spatial utilization of the Nanjing metropolitan area should be developed rationally according to local conditions. In addition, relevant policies or major constructions can have a remarkable effect on the conflict situation in that year. For example, railroads contributed the most to the expansion of high-conflict areas during 2010–2015. High-speed railway stations were mostly established near suburbs in the past and the establishment of high-speed railway stations could drive the rapid development of the surrounding area, which made the ecological-economic spatial conflicts near high-speed railway stations intensify rapidly.

4.4. Findings and Policy Suggestions

In sum, we found that due to better ecological background conditions, the degree of ecological-economic spatial conflict in the Nanjing metropolitan area is not serious and mainly dominated by low conflict. However, the Nanjing metropolitan area's ecological environment pressure will gradually increase with the continuous development of the social economy, so relevant ecological environment protection measures, such as set natural protection and water protection, should be implemented to ensure that the low-conflict areas no longer drop. Along with the continuous development of the city, the high-conflict and highest-conflict areas will inevitably increase. Based on the analysis of the changes in spatial conflict over the past 10 years and related influencing factors, we find that the most advisable way to alleviate the ecological-economic space conflict in the Nanjing metropolitan area is to, first, limit the disorderly growth of the urban area and strictly implement the requirements of ecological civilization construction and land space planning, second, protect the relevant ecological areas, and third, lock the eco-economic space conflict area in the city and its surrounding areas. This can maximize the avoidance of disorderly expansion similar to 2010 to ensure the normal operation of other ecological spaces and achieve the overall ecological-economic harmony state. In so doing, the Nanjing metropolitan circle can finally achieve the goal of sustainable development. Our conclusion is consistent with the findings of Zhang and Xu [62,63], as well as with the current mainstream solution view of spatial conflict. Scholars now believe that since spatial conflict is inevitable, the solution to spatial conflict should shift from traditional confrontation and elimination to guidance and weakening and should give priority to the legitimate needs of human social development [64–66].

4.5. Implications and Limitations

To diagnose Nanjing metropolitan area ecological-economic spatial conflicts, identify them, and determine their intensity, several problems remain to be solved. First, the theoretical understanding of spatial conflict must be strengthened. Most existing studies consider the overlap of different functions of space to be a conflict [67,68]. Most existing studies also believe that an important cause of spatial conflicts is the overlap of different spatial functions. However, a space with only a single function is rare in reality. Therefore, subsequent studies need to systematically discuss the forms and connotations of spatial conflicts. Second, although this study attempted to use spatial data to establish a comprehensive system of indicators to measure ecological-economic spatial conflicts to compensate for the shortcomings of existing studies, we have only analyzed spatial conflicts from an ecological-economic perspective. Space also has cultural, aesthetic, and other social functions, and analyzing these social spatial functions requires the reuse of panel data or through methods such as questionnaires. Our future research may explore spatial conflict from the perspective of civil rights protection and social spatial equity [69]. However, the method for considering these functions in the study of spatial conflicts remains to be explored.

5. Conclusions

This study identified and diagnosed the intensity of ecological-economic spatial conflicts in the Nanjing metropolitan area in 2010, 2015, and 2020. The degree of contribution of selected natural environmental and socio-economic factors to the spatial conflict changes was analyzed via the random forest method. The main conclusions of this paper are as follows:

(1) From 2010 to 2020, the ecological-economic space conflict in the Nanjing metropolitan area changed considerably. The spatial conflict status of the Nanjing metropolitan area was dominated by low conflict areas, and the lowest-conflict areas were mainly concentrated in the hilly areas of Chuzhou and the mountainous areas of Xuancheng. High-conflict and highest-conflict areas had the lowest proportion and were mainly concentrated in urban areas, while two large conflict areas formed in the central and northern regions of the metropolitan area. The proportion of medium-conflict areas is larger and mainly concentrated in the urban periphery.

(2) The change in land use has a substantial effect on spatial conflicts. In general, the main land use types in the lowest-conflict zone are forest land and arable land, the main land use types in the low-conflict and medium-conflict zones are arable land, and the high-conflict and highest-conflict zones consist mainly of construction land. Therefore, spatial conflicts are easily triggered or intensified by disorderly urban expansion, whereas the presence of ecological land can mitigate spatial conflicts.

(3) The relevant driving factors of spatial conflicts showed multi-level features. The factor that contributed most to the lowest-conflict and low-conflict areas was DEM, and the factor that contributed most to the highest-conflict and high-conflict areas was industrial density. However, the situation varied from year to year and from region to region, so the development of conflict reconciliation countermeasures needs to be tailored to local conditions.

Finally, this paper offers suggestions to help the Nanjing metropolitan area achieve sustainable development. As a typical metropolitan area in China, the sustainability and spatial conflict situation of the Nanjing metropolitan area needs more attention.

Author Contributions: Conceptualization, W.C.; Formal analysis, J.C.; Methodology, J.C. and J.M.; Supervision, Y.X.; Validation, W.C.; Visualization, J.C. and X.F.; Writing—original draft, J.C.; Writing—review and editing, J.C. and D.M. All authors have read and agreed to the published version of the manuscript.

Funding: This work was funded by the National Natural Science Foundation of China (41571124).

Data Availability Statement: Not applicable.

Conflicts of Interest: The authors declare no conflict of interest.

References

1. Zhou, G.H.; Peng, J.J. The evolution characteristics and influence effect of spatial conflict: A case study of Changsha-Zhuzhou-Xiangtan urban agglomeration. *Prog. Geogr.* **2012**, *31*, 717–723.
2. Peng, J.; Zhou, G.; Tang, C.; He, Y. The analysis of spatial conflict measurement in fast urbanization region based on ecological security: A case study of Changsha-Zhuzhou-Xiangtan urban agglomeration. *J. Nat. Resour.* **2012**, *27*, 1507–1519.
3. Fan, J.; Wang, Y.; Wang, C.; Chen, T.; Jin, F.; Zhang, W.; Li, L.; Xu, Y.; Dai, E.; Tao, A.; et al. Reshaping the Sustainable Geographical Pattern: A Major Function Zoning Model and Its Applications in China. *Earth's Future* **2019**, *7*, 25–42. [CrossRef]
4. Zhou, D.; Xu, J.; Lin, Z. Conflict or coordination? Assessing land use multi-functionalization using production-living-ecology analysis. *Sci. Total Environ.* **2017**, *577*, 136–147. [CrossRef] [PubMed]
5. Svoboda, J.; Štych, P.; Laštovička, J.; Paluba, D.; Kobliuk, N. Random Forest Classification of Land Use, Land-Use Change and Forestry (LULUCF) Using Sentinel-2 Data—A Case Study of Czechia. *Remote Sens.* **2022**, *14*, 1189. [CrossRef]
6. Khavarian Nehzak, H.; Aghaei, M.; Mostafazadeh, R.; Rabiei-Dastjerdi, H. Chapter 5—Assessment of machine learning algorithms in land use classification. In *Computers in Earth and Environmental Sciences*; Pourghasemi, H.R., Ed.; Elsevier: Amsterdam, The Netherlands, 2022; pp. 97–104.
7. Carr, M.; Zwick, P. Using GIS suitability analysis to identify potential future land use conflicts in North Central Florida. *J. Conserv. Plan.* **2005**, *1*, 58–73.
8. Dong, G.; Ge, Y.; Jia, H.; Sun, C.; Pan, S. Land Use Multi-Suitability, Land Resource Scarcity and Diversity of Human Needs: A New Framework for Land Use Conflict Identification. *Land* **2021**, *10*, 1003. [CrossRef]
9. Geoghegan, J. The value of open spaces in residential land use. *Land Use Policy* **2002**, *19*, 91–98. [CrossRef]
10. Shan, L.; Yu, A.T.; Wu, Y. Strategies for risk management in urban–rural conflict: Two case studies of land acquisition in urbanising China. *Habitat Int.* **2017**, *59*, 90–100. [CrossRef]

11. Zhang, Y.; Long, H.; Tu, S.; Ge, D.; Ma, L.; Wang, L. Spatial identification of land use functions and their tradeoffs/synergies in China: Implications for sustainable land management. *Ecol. Indic.* **2019**, *107*, 105550. [CrossRef]
12. Zou, L.; Liu, Y.; Wang, J.; Yang, Y. An analysis of land use conflict potentials based on ecological-production-living function in the southeast coastal area of China. *Ecol. Indic.* **2020**, *122*, 107297. [CrossRef]
13. Sun, Y.; Ge, X.; Liu, J.; Chang, Y.; Liu, G.-J.; Chen, F. Mitigating Spatial Conflict of Land Use for Sustainable Wetlands Landscape in Li-Xia-River Region of Central Jiangsu, China. *Sustainability* **2021**, *13*, 11189. [CrossRef]
14. Ma, G.; Li, Q.; Yang, S.; Zhang, R.; Zhang, L.; Xiao, J.; Sun, G. Analysis of Landscape Pattern Evolution and Driving Forces Based on Land-Use Changes: A Case Study of Yilong Lake Watershed on Yunnan-Guizhou Plateau. *Land* **2022**, *11*, 1276. [CrossRef]
15. Jiao, M.; Hu, M.; Xia, B. Spatiotemporal dynamic simulation of land-use and landscape-pattern in the Pearl River Delta, China. *Sustain. Cities Soc.* **2019**, *49*, 101581. [CrossRef]
16. Gao, Y.; Wang, J.; Zhang, M.; Li, S. Measurement and prediction of land use conflict in an opencast mining area. *Resour. Policy* **2021**, *71*, 101999. [CrossRef]
17. Qiu, G.; Wang, Y.; Guo, S.; Niu, Q.; Qin, L.; Zhu, D.; Gong, Y. Assessment and Spatial-Temporal Evolution Analysis of Land Use Conflict within Urban Spatial Zoning: Case of the Su-Xi-Chang Region. *Sustainability* **2022**, *14*, 2286. [CrossRef]
18. Costanza, R.; Arge, A.; Groot, R.; Farberk, S.; Grasso, M.; Bruce, H.; Limburg, K.; Naeem, S.; O'Neill, R.V.; Paruelo, J.; et al. The value of the world's ecosystem services and natural capital. *Ecol. Econ.* **1997**, *25*, 3–15. [CrossRef]
19. Wolff, S.; Schulp, C.; Verburg, P. Mapping ecosystem services demand: A review of current research and future perspectives. *Ecol. Indic.* **2015**, *55*, 159–171. [CrossRef]
20. Wei, H.; Fan, W.; Wang, X.; Lu, N.; Dong, X.; Zhao, Y.; Ya, X.; Zhao, Y. Integrating supply and social demand in ecosystem services assessment: A review. *Ecosyst. Serv.* **2017**, *25*, 15–27. [CrossRef]
21. Assessment, M.E. *Ecosystems and Human Well-Being: Wetlands and Water*; World Resources Institute: Washington, DC, USA, 2005.
22. Tian, J.-L.; Peng, Y.; Huang, Y.-H.; Bai, T.; Liu, L.-L.; He, X.-A.; Luo, S.-Q. Identifying the priority areas for enhancing the ecosystem services in hilly and mountainous areas of southern China. *J. Mt. Sci.* **2022**, *19*, 338–349. [CrossRef]
23. Fu, J.; Zhang, Q.; Wang, P.; Zhang, L.; Tian, Y.; Li, X. Spatio-Temporal Changes in Ecosystem Service Value and Its Coordinated Development with Economy: A Case Study in Hainan Province, China. *Remote Sens.* **2022**, *14*, 970. [CrossRef]
24. Elliot, T.; Goldstein, B.; Gómez-Baggethun, E.; Proença, V.; Rugani, B. Ecosystem service deficits of European cities. *Sci. Total Environ.* **2022**, *837*, 155875. [CrossRef] [PubMed]
25. Sun, R.; Jin, X.; Han, B.; Liang, X.; Zhang, X.; Zhou, Y. Does scale matter? Analysis and measurement of ecosystem service supply and demand status based on ecological unit. *Environ. Impact Assess. Rev.* **2022**, *95*, 106785. [CrossRef]
26. Wu, X.; Liu, S.; Zhao, S.; Hou, X.; Xu, J.; Dong, S.; Liu, G. Quantification and driving force analysis of ecosystem services supply, demand and balance in China. *Sci. Total Environ.* **2018**, *652*, 1375–1386. [CrossRef] [PubMed]
27. Wang, A.; Liao, X.; Tong, Z.; Du, W.; Zhang, J.; Liu, X.; Liu, M. Spatial-temporal dynamic evaluation of the ecosystem service value from the perspective of "production-living-ecological" spaces: A case study in Dongliao River Basin, China. *J. Clean. Prod.* **2021**, *333*, 130218. [CrossRef]
28. Wang, H.; Liu, X.; Zhao, C.; Chang, Y.; Liu, Y.; Zang, F. Spatial-temporal pattern analysis of landscape ecological risk assessment based on land use/land cover change in Baishuijiang National nature reserve in Gansu Province, China. *Ecol. Indic.* **2021**, *124*, 107454. [CrossRef]
29. Zhang, S.; Zhong, Q.; Cheng, D.; Xu, C.; Chang, Y.; Lin, Y.; Li, B. Coupling Coordination Analysis and Prediction of Landscape Ecological Risks and Ecosystem Services in the Min River Basin. *Land* **2022**, *11*, 222. [CrossRef]
30. Liu, X.; Zhang, Z.; Li, M.; Fu, Y.; Hui, Y. Spatial conflict simulation of land-use based on human-land-landscape elements intercoordination: A case study in Tianjin, China. *Environ. Monit. Assess.* **2022**, *194*, 1–17. [CrossRef]
31. Tao, Y.; Wang, Q. Quantitative Recognition and Characteristic Analysis of Production-Living-Ecological Space Evolution for Five Resource-Based Cities: Zululand, Xuzhou, Lota, Surf Coast and Ruhr. *Remote Sens.* **2021**, *13*, 1563. [CrossRef]
32. Liu, Y.; Wang, S.; Chen, Z.; Tu, S. Research on the Response of Ecosystem Service Function to Landscape Pattern Changes Caused by Land Use Transition: A Case Study of the Guangxi Zhuang Autonomous Region, China. *Land* **2022**, *11*, 752. [CrossRef]
33. Wang, Q.; Wang, H. Dynamic simulation and conflict identification analysis of production–living–ecological space in Wuhan, Central China. *Integr. Environ. Assess.* **2022**. [CrossRef] [PubMed]
34. Hong, W.; Guo, R.; Wang, W. A diagrammatic method for the identification and resolution of urban spatial conflicts. *J. Environ. Manag.* **2022**, *316*, 115297. [CrossRef] [PubMed]
35. Din, S.U.; Mak, H.W.L. Retrieval of Land-Use/Land Cover Change (LUCC) Maps and Urban Expansion Dynamics of Hyderabad, Pakistan via Landsat Datasets and Support Vector Machine Framework. *Remote Sens.* **2021**, *13*, 3337. [CrossRef]
36. Shih, H.-C.; Stow, D.A.; Chang, K.-C.; Roberts, D.A.; Goulias, K.G. From land cover to land use: Applying random forest classifier to Landsat imagery for urban land-use change mapping. *Geocarto Int.* **2022**, *37*, 5523–5546. [CrossRef]
37. Zeng, C.; He, J.; He, Q.; Mao, Y.; Yu, B. Assessment of Land Use Pattern and Landscape Ecological Risk in the Chengdu-Chongqing Economic Circle, Southwestern China. *Land* **2022**, *11*, 659. [CrossRef]
38. Liu, Y.; Yang, R.; Sun, M.; Zhang, L.; Li, X.; Meng, L.; Wang, Y.; Liu, Q. Regional sustainable development strategy based on the coordination between ecology and economy: A case study of Sichuan Province, China. *Ecol. Indic.* **2021**, *134*, 108445. [CrossRef]
39. Han, B.; Jin, X.; Sun, R.; Li, H.; Liang, X.; Zhou, Y. Land use sustainability evaluation based on conflict-adaptation perspective. *Acta Geogr. Sin.* **2021**, *76*, 1763–1777.

40. Zhang, T.; Zhang, S.; Cao, Q.; Wang, H.; Li, Y. The spatiotemporal dynamics of ecosystem services bundles and the social-economic-ecological drivers in the Yellow River Delta region. *Ecol. Indic.* **2022**, *135*, 108573. [CrossRef]
41. Pan, G.; Xu, Y.; Yu, Z.; Song, S.; Zhang, Y. Analysis of river health variation under the background of urbanization based on entropy weight and matter-element model: A case study in Huzhou City in the Yangtze River Delta, China. *Environ. Res.* **2015**, *139*, 31–35. [CrossRef]
42. Wen, D.; Liu, M.; Yu, Z. Quantifying Ecological Landscape Quality of Urban Street by Open Street View Images: A Case Study of Xiamen Island, China. *Remote Sens.* **2022**, *14*, 3360. [CrossRef]
43. Zhang, N.; Ye, H.; Wang, M.; Li, Z.; Li, S.; Li, Y. Response Relationship between the Regional Thermal Environment and Urban Forms during Rapid Urbanization (2000–2010–2020): A Case Study of Three Urban Agglomerations in China. *Remote Sens.* **2022**, *14*, 3749. [CrossRef]
44. Liu, J.; Chen, J.J.; Zhao, X.N.; Zhang, H.Z. Spatial differentiation characteristics of alpine grassland patches in the source region of the yellow river and its ecological significance. *ISPRS-Int. Arch. Photogramm. Remote Sens. Spat. Inf. Sci.* **2020**, *42*, 755–762. [CrossRef]
45. Peng, J.; Shen, H.; Wu, W.; Liu, Y.; Wang, Y. Net primary productivity (NPP) dynamics and associated urbanization driving forces in metropolitan areas: A case study in Beijing City, China. *Landsc. Ecol.* **2015**, *31*, 1077–1092. [CrossRef]
46. Chen, Y.; Wang, J.; Xiong, N.; Sun, L.; Xu, J. Impacts of Land Use Changes on Net Primary Productivity in Urban Agglomerations under Multi-Scenarios Simulation. *Remote Sens.* **2022**, *14*, 1755. [CrossRef]
47. Zhai, T.; Wang, J.; Jin, Z.; Qi, Y.; Fang, Y.; Liu, J. Did improvements of ecosystem services supply-demand imbalance change environmental spatial injustices? *Ecol. Indic.* **2020**, *111*, 106068. [CrossRef]
48. Lin, X.; Ma, J.; Chen, H.; Shen, F.; Ahmad, S.; Li, Z. Carbon Emissions Estimation and Spatiotemporal Analysis of China at City Level Based on Multi-Dimensional Data and Machine Learning. *Remote Sens.* **2022**, *14*, 3014. [CrossRef]
49. Yang, H.; Huang, J.; Liu, D. Linking climate change and socioeconomic development to urban land use simulation: Analysis of their concurrent effects on carbon storage. *Appl. Geogr.* **2020**, *115*, 102135. [CrossRef]
50. Xiao, P.; Zhou, Y.; Li, M.; Xu, J. Spatiotemporal patterns of habitat quality and its topographic gradient effects of Hubei Province based on the InVEST model. *Environ. Dev. Sustain.* **2022**. [CrossRef]
51. Tang, F.; Fu, M.; Wang, L.; Song, W.; Yu, J.; Wu, Y. Dynamic evolution and scenario simulation of habitat quality under the impact of land-use change in the Huaihe River Economic Belt, China. *PLoS ONE* **2021**, *16*, e0249566. [CrossRef]
52. Zhou, D.; Lin, Z.; Lim, S.H. Spatial characteristics and risk factor identification for land use spatial conflicts in a rapid urbanization region in China. *Environ. Monit. Assess.* **2019**, *191*, 677. [CrossRef]
53. Wei, J.; Li, Z.; Lyapustin, A.; Sun, L.; Peng, Y.; Xue, W.; Su, T.; Cribb, M. Reconstructing 1-km-resolution high-quality PM2.5 data records from 2000 to 2018 in China: Spatiotemporal variations and policy implications. *Remote Sens. Environ.* **2020**, *252*, 112136. [CrossRef]
54. Liang, X.; Guan, Q.; Clarke, K.C.; Liu, S.; Wang, B.; Yao, Y. Understanding the drivers of sustainable land expansion using a patch-generating land use simulation (PLUS) model: A case study in Wuhan, China. *Comput. Environ. Urban Syst.* **2020**, *85*, 101569. [CrossRef]
55. Jiang, Y.; Huang, M.; Chen, X.; Wang, Z.; Xiao, L.; Xu, K.; Zhang, S.; Wang, M.; Xu, Z.; Shi, Z. Identification and risk prediction of potentially contaminated sites in the Yangtze River Delta. *Sci. Total Environ.* **2021**, *815*, 151982. [CrossRef]
56. Tang, Y.; Gao, C.; Wu, X. Urban Ecological Corridor Network Construction: An Integration of the Least Cost Path Model and the InVEST Model. *ISPRS Int. J. Geo-Information* **2020**, *9*, 33. [CrossRef]
57. Liu, J.-Y.; Deng, X.-Z.; Liu, M.-L.; Zhang, S.-W. Study on the spatial patterns of land-use change and analyses of driving forces in Northeastern China during 1990–2000. *Chin. Geogr. Sci.* **2002**, *12*, 299–308. [CrossRef]
58. Cui, S.; Han, Z.; Yan, X.; Li, X.; Zhao, W.; Liu, C.; Li, X.; Zhong, J. Link Ecological and Social Composite Systems to Construct Sustainable Landscape Patterns: A New Framework Based on Ecosystem Service Flows. *Remote Sens.* **2022**, *14*, 4663. [CrossRef]
59. Wei, S.; Pan, J.; Liu, X. Landscape ecological safety assessment and landscape pattern optimization in arid inland river basin: Take Ganzhou District as an example. *Hum. Ecol. Risk Assessment: Int. J.* **2018**, *26*, 782–806. [CrossRef]
60. Cao, Y.; Liu, M.; Cao, Y.; Chen, C.; Zhang, D. Change Pattern and Driving Mechanism of Construction Land in China's Undertaking Industrial Transfer Demonstration Area: Taking the Wanjiang City Belt along the Yangtze River as an Example. *Earth Sci. Res. J.* **2020**, *24*, 215–223. [CrossRef]
61. Wang, Y.; Duan, X.; Wang, L.; Zou, H. Spatial temporal patterns and driving factors of industrial pollution and structures in the Yangtze River Economic Belt. *Chemosphere* **2022**, *303*, 134996. [CrossRef]
62. Zhang, Y.; Liu, Y.; Kong, X.; Jing, Y.; Cai, E.; Zhang, L.; Liu, Y.; Wang, Z.; Liu, Y. Spatial Patterns and Driving Forces of Conflicts among the Three Land Management Red Lines in China: A Case Study of the Wuhan Urban Development Area. *Sustainability* **2019**, *11*, 2025. [CrossRef]
63. Xu, W.; He, M.; Meng, W.; Zhang, Y.; Yun, H.; Lu, Y.; Huang, Z.; Mo, X.; Hu, B.; Liu, B.; et al. Temporal-spatial change of China's coastal ecosystems health and driving factors analysis. *Sci. Total Environ.* **2022**, *845*, 157319. [CrossRef] [PubMed]
64. Saarikoski, H.; Raitio, K.; Barry, J. Understanding 'successful' conflict resolution: Policy regime changes and new interactive arenas in the Great Bear Rainforest. *Land Use Policy* **2013**, *32*, 271–280. [CrossRef]
65. Bircol, G.A.C.; de Souza, M.P.; Fontes, A.T.; Chiarello, A.G.; Ranieri, V.E.L. Planning by the rules: A fair chance for the environment in a land-use conflict area. *Land Use Policy* **2018**, *76*, 103–112. [CrossRef]

66. Moomen, A. Strategies for managing large-scale mining sector land use conflicts in the global south. *Resour. Policy* **2017**, *51*, 85–93. [CrossRef]
67. Lovell, S.T.; DeSantis, S.; Nathan, C.A.; Olson, M.B.; Méndez, V.E.; Kominami, H.C.; Erickson, D.L.; Morris, K.S.; Morris, W.B. Integrating agroecology and landscape multifunctionality in Vermont: An evolving framework to evaluate the design of agroecosystems. *Agric. Syst.* **2010**, *103*, 327–341. [CrossRef]
68. Stockdale, A.; Barker, A. Sustainability and the multifunctional landscape: An assessment of approaches to planning and management in the Cairngorms National Park. *Land Use Policy* **2009**, *26*, 479–492. [CrossRef]
69. Schirpke, U.; Timmermann, F.; Tappeiner, U.; Tasser, E. Cultural ecosystem services of mountain regions: Modelling the aesthetic value. *Ecol. Indic.* **2016**, *69*, 78–90. [CrossRef]

remote sensing

Article

Water-Vapour Monitoring from Ground-Based GNSS Observations in Northwestern Argentina

Nikolaos Antonoglou [1,2,*], Kyriakos Balidakis [2], Jens Wickert [2,3], Galina Dick [2], Alejandro de la Torre [4] and Bodo Bookhagen [1]

1 Institute of Geosciences, University of Potsdam, Karl-Liebknecht-Str. 24-25, 14476 Potsdam, Germany
2 German Research Centre for Geosciences GFZ, Telegrafenberg, 14473 Potsdam, Germany
3 Institute of Geodesy and Geoinformation Science, Technische Universität Berlin, Straße des 17. Juni 135, 10623 Berlin, Germany
4 Facultad de Ingeniería, Universidad Austral, Mariano Acosta 1611, Pilar B1630, Argentina
* Correspondence: nikolaos.antonoglou@gfz-potsdam.de

Abstract: The Central Andes in northwestern Argentina are characterized by steep topographic and climatic gradients. The humid foreland areas at 1 km asl elevation rapidly rise to over 5 km in the eastern Cordillera, and they form an orographic rainfall barrier on the eastern windward side. This topographic setting combined with seasonal moisture transport through the South American monsoon system leads to intense rainstorms with cascading effects such as landsliding and flooding. In order to better quantify the dynamics of water vapour transport, we use high-temporal-resolution global navigation satellite system (GNSS) remote sensing techniques. We are particularly interested in better understanding the dynamics of high-magnitude storms with high water vapour amounts that have destructive effects on human infrastructure. We used an existing GNSS station network with 12 years of time series data, and we installed two new ground stations along the climatic gradient and collected GNSS time series data for three years. For several stations we calculated the GNSS signal delay gradient to determine water vapour transport direction. Our statistical analysis combines in situ rainfall measurements and ERA5 reanalysis data to reveal the water vapour transport mechanism for the study area. The results show a strong relationship between altitude and the water vapour content, as well as between the transportation pathways and the topography.

Keywords: GNSS meteorology; GNSS remote sensing; intense rain events; water vapour; Central Andes; orographic barrier; South American monsoon system

Citation: Antonoglou, N.; Balidakis, K.; Wickert, J.; Dick, G.; de la Torre, A.; Bookhagen, B. Water-Vapour Monitoring from Ground-Based GNSS Observations in Northwestern Argentina. *Remote Sens.* **2022**, *14*, 5427. https://doi.org/10.3390/rs14215427

Academic Editors: Zhengqiang Li, Zhongwei Huang, Chi Li, Kai Qin, Han Wang, Tianhe Wang and Jie Luo

Received: 14 September 2022
Accepted: 22 October 2022
Published: 28 October 2022

1. Introduction

Strong rainfall events repeatedly lead to natural disasters in steep mountain regions, e.g., [1–5]. Especially along the eastern Andes, intense hydro-meteorological events cause landsliding and flooding that impact population and infrastructures [6–8]. Previous studies indicate that hydro-meteorological extreme events are often a consequence of several additive climatic and topographic factors [9]. For example, the availability of high water vapour transported through the South American low-level jet (SALLJ) from the north along the eastern Andes, the advancement of cold fronts from the south, and the steep topography lead to unstable atmospheric conditions and heavy cloudbursts in the eastern Central Andes [4,6].

A core component for understanding these hydrometeorologic processes are highly dynamic observations. While satellite-based observations have advanced our understanding of large-scale orographic effects at windward sides of mountain ranges, e.g., [10,11], these data often do not provide the temporal resolution to decipher dynamic processes, including the vertical and horizontal components of water vapour transport within storms. Similarly, reanalysis and numerical weather prediction data are very useful for understanding the dynamics of large-scale processes, but they often do not allow for the reliable measurement of spatially small (<2 km) and temporally short (<1 h) processes, e.g., [9,12]. With the recent advancement of global navigation satellite system (GNSS)-based observation and the increase in availability of these data, an alternative meteorological observation method is readily available, e.g., [13–15].

The study area in northwestern Argentina is on the eastern side of the second largest orogenic plateau—the Altiplano–Puna plateau. The plateau is rich in mineral and georesources, especially lithium. Maintaining infrastructural networks in this area is important, but the seasonal South American monsoon system (SAMS) heavily impacts the road network connecting the high-elevation and resource-rich areas with the low-elevation foreland areas. In addition, global warming leads to changing rainfall and discharge patterns, shifting the distributions of low frequency and high magnitude hydrometeorological events [4,9,16,17].

In this study, we explore the potential of GNSS-based observations to track water vapour transport along and across the south-central Andes. For this purpose, we used available station data since 2010, but also installed additional monitoring sites in medium and high elevation areas for the duration of three years (cf. Figure 1). We use GNSS time series data to group stations into clusters and use their high temporal resolution to analyse water vapour distributions. Furthermore, we implement an analysis of water vapour gradients using GNSS delays to determine direction of water vapour transport. Our investigation is complemented by in situ weather stations and ERA5 reanalysis data to validate results and provide a hydrometeorological context.

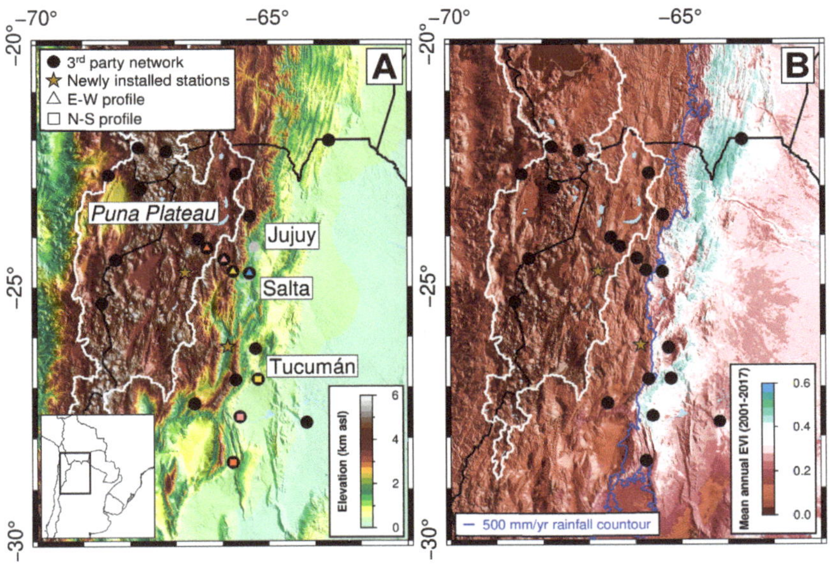

Figure 1. Topographic setting of the south-central Andes with GNSS station network location. We have selected a subset of these stations for an analysis of water vapour transport: across the Andes in east–west (E-W) direction (UNSA, GOLG, SRSA, and SALC) and along the Andes in north–south (N-S) direction (UNSA, TUCU, JBAL, and CATA). UNSA, GOLG/TUCU, SRSA/JBAL, and SALC/CATA are represented with blue, yellow, pink, and orange colours, respectively. The white line outlines the internally drained Altiplano–Puna plateau, also called the Central Andean plateau. Black lines are international borders (**A**). (**B**) shows enhanced vegetation index (EVI) information for the area of interest between 2001 and 2017. The blue line points out 500 mm annual rainfall (Topographic data obtained from ETOPO1 [18], EVI data obtained from MODIS/Terra [19], rainfall information retrieved from the Tropical Rainfall Measuring Mission (TRMM) [20], as cited in Bookhagen and Strecker [10]).

2. Climatic Setting of the Study Region in Northwestern Argentina

The climatic setting of the study area in the Central Andes in northwestern Argentina is controlled by several interfering factors. In short, these are: (1) the moisture transport from tropical regions; (2) unstable atmospheric conditions at the boundary between tropical and subtropical air masses; (3) the steep topographic gradient leading to orographic processes; and (4) cold temperature excursion from the southern polar region. This leads to strong climatic gradients in the east–west (E-W) and north–south (N-S) directions.

The conveyor belt transporting water vapour from north to south east of the Andes is the South American low-level jet (SALLJ). The authors in Montini et al. [21] showed that moisture transport associated with the jet is greater during summer, when the SALLJ is influenced by warm, moisture-rich air masses from tropical South America. During the austral summer (DJF), it mainly transports moist air from the latitudes of the Amazon southward [22,23]. Reanalysis data show that the inter-annual variability of the jet's strength and frequency is significantly modulated by the El Niño southern oscillation, especially during spring [24].

Moreover, the SALLJ is controlled by the dynamics of the SAMS. From a synoptic and seasonal perspective, the SAMS is highly dependent on the surface temperature difference between the ocean and surrounding land masses [23,25,26]. During the SAMS activity in the DJF, more than 50% of annual rainfall occurs along the Andes. The same region receives less than 10% of the annual rainfall in JJA season, which leads to distinct warm-wet and cold-dry seasons, e.g., [17,23].

The Central Andes in northwestern Argentina are located near the end of the SALLJ and also receive moisture from mesoscale convective systems [27–29]. To the east of the study area exists one of the global hotspots for mesoscale convective systems. The complex topography and interaction of unstable air masses leads to some of the heaviest rainfall and largest cloudbursts on the South American continent, e.g., [4,6,16,30]. An additional important component influencing heavy cloudbursts in this area is the interplay of high moisture availability through the SALLJ and cold surges and frontal systems propagating from the south [4,29]. This is an important process, especially in the generation of areal-extensive heavy rainfall leading to major flooding in the eastern Central Andes and downstream areas. Previous studies have shown that 80% of the 40 largest discharge events of the past 40 years are associated with propagating cold fronts [4]. The interplay of these factors lead to complex rainfall patterns on the South American continent and in the study area [31].

In addition to the atmospheric conditions, the topographic setting of the study area with deeply incised valleys funneling moisture to the higher-elevation area lead to pronounced orographic barriers on the eastern, windward sides of the mountain ranges [10,15,32]. The authors of Castino et al. [17] suggest three climate zones that follow the topographic and rainfall gradient from east to west: a low-elevation and low-slope sector, a medium-elevation sector dissected by rivers with steep hill slopes, and a high-elevation sector with moderate to steep slopes. The climatic contrast between these areas is large: the foreland zone receives more than 1.5 m/yr rainfall, whereas the high-elevation areas receive less than 0.2 m/yr [10]. The intermediate elevations show high topographic relief and force orographic lifting and convection resulting in pronounced rainfall on the windward slopes [6,9,33]. The convective cells have lifetimes of several hours and observations with high temporal resolution are required [23,34,35]. In addition to the spatial and inter-annual rainfall distribution patterns, there is also an inter-diurnal variation. According to meteorological data from Salta [36], we observe a nocturnal rainfall peak during DJF between 21:00 and 4:00 local time [37].

3. Data

We processed GNSS and reanalysis data from ERA5 [38], and our analysis focuses on the years between 2010 and 2021 [39]. This time frame is mainly constrained by the availability of the GNSS data.

3.1. GNSS Network Description

We analysed data from 23 GNSS stations listed in Table 1. A total of 16 stations are located in northwestern Argentina, four stations in Chile, and three stations in Bolivia. The elevation range of the station spans more than 5 km, and the network extends 450 km in the E-W and 700 km in the N-S direction. This area is impacted by the moisture transport associated with the SALLJ from north to south, but also the westward moisture transport across the orographic barrier.

Table 1. Geographic coordinates of the stations that were used for water vapour analysis. The institutions that were responsible for the installation of the facilities are listed in the column **Source** and are the National Geographic Institute of Argentina (Instituto Geográfico Nacional—IGN), UNAVCO, the University of Potsdam (UP), and the German Research Centre for Geosciences (Deutsches GeoForschungsZentrum—GFZ). The column **Analysis Centre** shows where the data processing was carried out: either at the GFZ or the Nevada Geodetic Laboratory (NGL).

Station Name	Latitude	Longitude	Height (m)	Source	Analysis Centre
ABRA	22°43′19.32″S	65°41′50.31″W	3530.10	IGN	NGL
ALUM	27°19′24.33″S	66°35′47.86″W	2736.94	IGN	NGL
CAFJ	26°10′51.22″S	65°52′49.17″W	1702.36	UP/GFZ	GFZ
CATA	28°28′15.54″S	65°46′26.83″W	547.15	IGN	NGL
CBAA	22°44′46.92″S	68°26′53.33″W	3514.84	UNAVCO	NGL
CJNT	23°01′38.96″S	67°45′38.06″W	5074.05	UNAVCO	NGL
COLO	22°10′02.57″S	67°48′14.32″W	4376.93	UNAVCO	NGL
GOLG	24°41′26.11″S	65°45′38.80″W	2381.15	UNAVCO	NGL
JBAL	27°35′03.86″S	65°37′21.89″W	409.16	IGN	NGL
LCEN	25°19′33.81″S	68°36′09.36″W	4270.94	UNAVCO	NGL
PUNJ	24°42′46.96″S	66°47′37.27″W	3802.58	UP/GFZ	GFZ
SALC	24°12′47.11″S	66°19′20.83″W	3841.62	UNAVCO	NGL
SOCM	24°27′16.60″S	68°17′42.59″W	3969.45	UNAVCO	NGL
SRSA	24°26′59.24″S	65°57′11.85″W	3153.80	UNAVCO	NGL
TAVA	26°51′10.72″S	65°42′36.02″W	2036.74	IGN	NGL
TERO	27°41′57.30″S	64°10′42.17″W	222.63	IGN	NGL
TIL2	23°34′37.70″S	65°23′42.26″W	2517.78	IGN	NGL
TRNC	26°13′48.77″S	65°16′55.82″W	816.08	IGN	NGL
TUCU	26°50′35.71″S	65°13′49.26″W	485.02	IGN	NGL
TUZG	24°01′53.82″S	66°30′59.56″W	4338.67	UNAVCO	NGL
UNSA	24°43′38.84″S	65°24′27.51″W	1257.79	IGN	NGL
UTUR	22°14′31.21″S	67°12′19.94″W	5184.09	UNAVCO	NGL
YCBA	22°01′01.56″S	63°40′47.94″W	659.66	IGN	NGL

The station data are maintained by different data providers, and they have been compiled and prepared for this study. The stations CAFJ and PUNJ were installed by the University of Potsdam (UP) and the German Research Centre for Geosciences (Deutsches GeoForschungsZentrum—GFZ), while their observations were processed by the GFZ. All other stations were installed by UNAVCO [40–52] and by the National Geographic Institute of Argentina (Instituto Geográfico Nacional—IGN) [53]. The Nevada Geodetic Laboratory (NGL) provides access to these data [54].

Due to the heterogeneous development of the GNSS network, observations do not always overlap. Generally, the network spans the years from 2010 to 2021, but only six time series extend through the entire range. Furthermore, all stations have undergone data loss for limited periods of time due to technical reasons.

For the analysis of the topographic impact, we selected four stations across the central Andes in an E-W direction (UNSA, GOLG, SRSA, and SALC, cf. Figure 1). Moreover, four stations were chosen (UNSA, TUCU, JBAL, and CATA) perpendicular to this and along a N-S direction. Those two subsets have one station in common, they cover multiple years with simultaneous observations, and they reflect the diverse climate conditions of the region. The largest elevation difference exists between the stations UNSA at 1224 m and SALC at

3799 m asl. In this research, we use these two stations to assess data measurements from different climatic conditions. In addition, we rely on the stations UNSA, CAFJ, and PUNJ for in situ comparisons because these are accompanied by rain gauge sensors in short distance. The water vapour readings in those locations is directly related to high-precision rainfall information.

3.2. ERA5 Data

We rely on ERA5 hourly data on pressure levels from 2010 to 2021 [38], and we use the native temporal and spatial resolution (0.25°). ERA5 data span several decades, and they are continually updated with a minimal latency of a few days. Furthermore, ERA5 analyses 37 pressure levels, reaching a height of 80 km. This way we can calculate refraction and wind speed information, which is vital for the methodology that we follow. Finally, we linearly resample ERA5 data (with a temporal resolution of one hour) to the 5 min temporal resolution of the GNSS data.

4. GNSS Meteorology

GNSS meteorology is a methodology for acquiring neutral atmosphere information by employing GNSS measurements. The primary output of this technique is the slant total delay between the receiver and the satellite. In this section, we discuss the translation of this product into the zenith total delay and its gradients along the E-W and N-S directions, and subsequently, the calculation of the zenith hydrostatic delay, the zenith wet delay, and the water vapour amount. We also describe the approach to derive the gradients of the zenith hydrostatic and wet delay.

4.1. Slant Delay Decomposition

The atmospheric delays of the GNSS signals, both in the slant and the zenith directions, are composed of the hydrostatic (or dry) and the wet counterparts [55]. In order to decompose the slant total delay, we project these to the vertical using mapping functions. The simplest versions of the mapping functions assumes a uniform atmosphere and mainly depends on the elevation (ε), whereas the more advanced versions take into account azimuthal asymmetry [56,57]. The latter approximations yield significantly better results because they better reflect the reality by introducing gradients along the E-W and N-S directions. According to Kačmařík et al. [58], a complete expression of the observation equation of the slant total delay can be written as follows:

$$S_{total} = m_{dry}Z_{dry} + m_{wet}Z_{wet} + m_{grad}(G_{NS}\cos(a) + G_{EW}\sin(a)) \tag{1}$$

where:

a	azimuth
Z_{dry}, Z_{wet}	zenith hydrostatic and wet delay
m_{dry}, m_{wet}	mapping functions for the dry and wet component
m_{grad}	mapping function of the gradient parts
G_{NS}, G_{EW}	gradients in the N-S and E-W directions

In a later step, the water vapour is directly calculated by the slant total delay. The formula for this conversion is implemented by Bevis et al. [59], and it can be written as follows:

$$WV = Z_{wet}\Pi \tag{2}$$

$$\Pi = \frac{10^6}{\varrho R_u[(k_3/T_m) + k_2']} \tag{3}$$

$$k_2' = k_2 - mK_1 \tag{4}$$

where:

ϱ	density of liquid water
R_u	specific gas constant of water
m	ratio of molar masses of water vapour and air
k_1, k_2, k_3	physical constants

4.2. Data Processing

Because the GNSS data were not obtained from a homogenized network (cf. Section 3.1), various software packages are used for the calculation of the zenith total delay and its gradients. More specifically, Receiver Independent Exchange Format files are processed with Earth Parameter and Orbit System (EPOS) [60] and GipsyX [61] software packages. Although both programs are very robust and they rely on the same fundamentals, each application follows its own strategy, and it is important to know the detailed differences for a proper assessment of the analysis results.

4.2.1. EPOS

EPOS is a GNSS analysis software that was developed in the 1990s by the scientific team of the GFZ [60]. Even though it was initially designed for space applications (e.g., precise orbit determination), it can be also used for terrestrial applications. More specifically, EPOS estimates the slant total delay in near real-time mode using precise point positioning algorithms [62,63]. For each set of epochs, the zenith hydrostatic and wet delay, and the gradients of the zenith total delay are estimated in a least squares adjustment where the functional model is Equation (1). Additionally, the utilized mapping function for the zenith components is the global mapping function [64], while for the azimuthal component it is a mapping function described by Bar-Sever et al. [57].

4.2.2. GipsyX

Gipsy is a multi-purpose navigation software that was developed in the 1980s by the Jet Propulsion Laboratory [61]. Its last version is GipsyX and it does not only allow for GNSS data processing, but for other space geodetic techniques, such as satellite laser ranging and Doppler orbitography and radiopositioning integrated by satellite. The processing strategy of the GNSS data is similar to EPOS. The only difference is that GipsyX employs the Vienna Mapping Function 1 [65,66] instead of a global mapping function [54,61].

4.3. Ray Tracing

To obtain a signal delay while propagating through the atmosphere, we adopted the geometrical optics approximation. First, we calculated the index of refraction based on hourly ERA5 data on equiangular 0.25° grids up to the maximum height of 80 km; we rely on pressure, geopotential height, temperature, and specific humidity of all ERA5 levels.

To perform ray tracing, we adopt a variational approach (Euler–Lagrange equations) employing an implicit finite difference scheme [67–69]. The atmospheric delays are calculated by integrating refractivity along the ray path. The ray-traced delays describe how atmospheric delay varies with elevation and azimuth. Because it is not practical to employ geodetic observations, delay is often described as a continuous function of elevation and azimuth as shown in Equation (1). The weather model-derived zenith delays and gradients are comparable to those from space geodetic data analysis.

While zenith delays are not affected by the choice of the parametric model describing directional delay variations, the gradient components that describe azimuthal delay variations do. In particular, the choice of the scheme describing the elevational decay of the asymmetric delay with increasing elevation is crucial. Although the latest International Earth Rotation Service [70] recommends a first-order continued fraction form for $\sin(\varepsilon)\cot(\varepsilon)$ $((\sin(\varepsilon)\tan(\varepsilon) + 0.0032)^{-1})$ [56], it is popular to use a gradient mapping function that involves the wet mapping function ($m_{wet}\cot(\varepsilon)$) [57]. As demonstrated in Baldidakis [71] and Kačmařík et al. [58], these two sets of gradients are incompatible, and they cannot be accurately transformed to be used with other gradient mapping functions. Because the latter gradient mapping function has been adopted during the estimation of first-degree gradient components employing GipsyX [54] and EPOS [60], we adopt consistent elevation-decay modelling to estimate gradients from our ray-traced delays. That is, in the station- and epoch-wise least squares adjustment of ray-traced delays, we constrained the estimates of the dry and wet symmetric mapping function to the values given by the gridded version of Vienna Mapping Function 1 so that we may directly compare ray-traced gradients with GNSS-derived gradients.

5. Analysis Methods

We performed spatio-temporal analyses by applying several methods: First, we identified temporal correlations between GNSS time series through k-means clustering. Second, we used spectral analysis to identify recurrence of specific event magnitudes. Third, we related GNSS-based water vapour observations with rainfall measurements for several in situ meteorologic stations, and we analysed the frequency–magnitude distributions of water vapour and rainfall. Furthermore, fourth, we determined latitude moisture transport through zenith delay gradient measurements.

5.1. k-Means Clustering of GNSS Time Series

We used k-Means clustering of monthly averaged water vapour measurements to group similar station data. More specifically, we employed the TimeSeriesKMeans algorithm within the Tslearn library [72], which is especially implemented for time series. Using this algorithm, we relied on Euclidean distances, because the seasonal meteorological of various locations data differ in amplitude and not in frequency, and they are not subject to time shifts. We set the number of clustering classes to three, because the topographic and climatic setting creates three zones in low-, medium-, and high-elevation areas [17] (cf. Figure 1 and Section 2).

Due to the lack of simultaneous and overlapping observations (cf. Figure 2), the clustering for all years cannot be realized. Instead, we selected only the measurements from 20 stations during 2014, when there were no major data interruptions. Even though this is a fraction of the available data, it is adequate for the classification because it covers an entire annual cycle, and the selected stations are distributed over various altitudes along the E-W and N-S cross-sections.

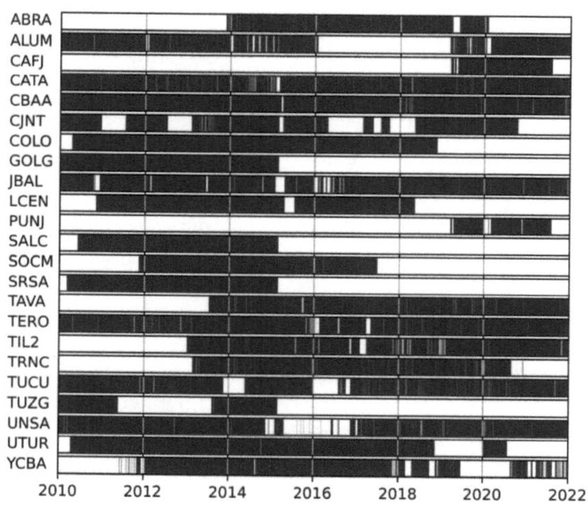

Figure 2. Temporal coverage of the stations used in this study. Blue areas indicate times with data availability.

5.2. Spectral Analysis

The GNSS water vapour signal was generated by overlapping periodic oscillations and we decomposed the signals by analysing their spectral behaviour. The annual time series depict the water vapour observations in a low- (UNSA) and a high-altitude (SALC) station during 2012 with a daily temporal resolution (Figure 3). In this characteristic overview, one can clearly identify the annual oscillation in both stations. Moreover, there are shorter oscillations with frequencies of about one week, which correspond to the synoptic-scale water vapour cycles in the atmosphere. We also show the 5 min water vapour time series illustrating the sub-daily cycles for two periods in summer and winter.

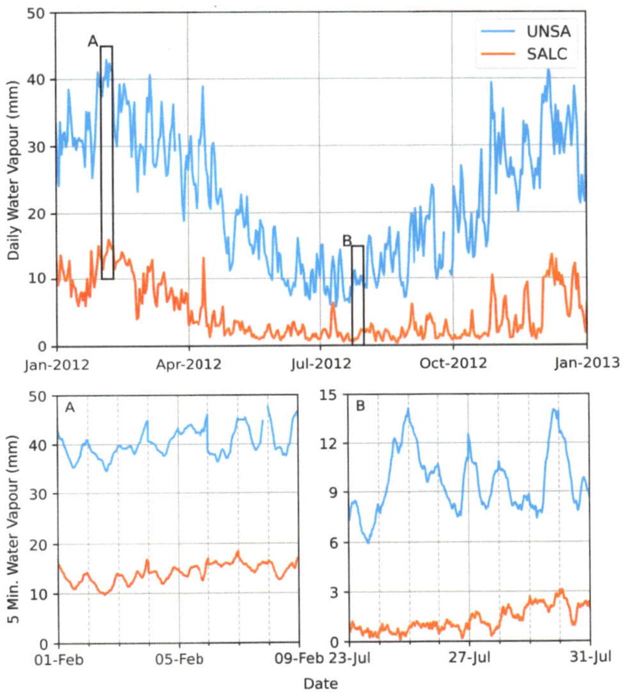

Figure 3. Top: Water vapour observations with 1 day sampling rate for the UNSA and SALC stations during 2012. **Bottom:** Detailed view with 1 h sampling rate during two incidents in February (**A**) and July (**B**) 2012. The panels have a different scales in their y-axes because of the contrasting water vapour levels during the wet and dry season and their different temporal resolutions.

The frequency domain analysis is accomplished by generating 3D graphs of the signal responses at various frequencies over time (or spectrograms) with a series of Fourier transforms [73,74]. Considering the non-simultaneous observations at all stations, we selected data between 2010 and 2014, when the stations along the E-W and N-S cross-sections were functional. We set the sampling window to seven days because it is aligned with the period of the synoptic events, and we forecast data discontinuities smaller than two weeks using the Prophet algorithm. This approximation employs an additive model that is sensitive to periodic fluctuations. It decomposes the signal into a trend, seasonality, and irregular occurrences, and it does not require regularlyspaced data as input [75].

For the next step, we quantified the spectral signals. Transient water vapour variations y were approximated as the sum of four different signal groups:

$$y = y_p + y_h + y_a + y_n \quad (5)$$

where y_p, y_h, y_a, and y_n indicate the polynomial term, the harmonic variations, the synoptic term, and the noise, respectively.

The polynomial counterpart ($y_p = \sum_i x_i t^i$) is time (t) dependent, and it yields largely non-significant estimates for coefficient terms $i > 1$. It will not be discussed herein given the relatively short data duration. The harmonic variations that are of main interest are described as follows:

$$y_h(t) = \sum_j A_j \cos(\chi_j(t) - \phi_j) \tag{6}$$

where j is a certain wave with a particular frequency, A_j denotes the amplitude, ϕ_j denotes the phase, and χ_j denotes the astronomical argument. To build the latter, we adopt Doodson multipliers from Hartmann and Wenzel [76]. We estimate in-phase and quadrature components for all waves whose speed spans from the Nyquist frequency to one cycle per half-time series length, making sure that the Rayleigh criterion is fulfilled for all possible pairs, while water vapour features marked modulation in certain spectral lines (e.g., S_1 and S_2) it is not explicitly considered, because the purpose of the this step is the estimation of the power spectral density (PSD) of the post-fit residual time series, and eventually to perform a scaling analysis using power-law fitting [77]. We refer the interested reader to Balidakis et al.[78] and the accompanying supplementary material for further details on the estimation of harmonic amplitudes from meteorological time series. We prefer this approach to adopting a bandstop filter (as in [79]) because the gap filling strategy may introduce artificial spectral signatures. A power-law distribution is suitable for describing natural phenomena because we assume that the probability of an event is inversely proportional to the power of its magnitude as it has been documented in several other natural science datasets, e.g., [80].

To estimate the PSD given the water vapour post-fit residuals, we utilize the multitaper method, e.g., [81]. Two sets of data are employed: raw post-tidal-fit residuals with their uncertainty estimates and a normalized version of the former by its standard deviation to facilitate the direct comparison of the PSD estimates from several data types stemming either from the weather model (ERA5) or the GNSS observations themselves. The motivation behind employing post-fit residuals instead of the raw water vapour time series is that the sharp spectral lines at frequencies associated with radiative forcing (S_a and S_1, as well as overtones thereof) bias the estimation of the spectral indices in the power-law approximation of the PSD.

5.3. Water Vapour and Rainfall Relation

We analysed the relation between liquid precipitation and water vapour to better understand their relation. The magnitude of water vapour varies throughout the season and during rainstorm events, but water vapour is always present in the atmosphere. On the other hand, rainfall occurs during events or in the form of episodes, when water vapour reaches the peak relative humidity level (100%), and it forms water particles in various forms [82]. In an initial analysis, we selected all daily averaged water vapour values that exceed the 90th percentile and the corresponding daily summed rainfall, and we analysed their relationship in a power-law framework. This data subset also includes zero-rainfall days and we exclude those epochs from the power-law relation analysis. We repeated the analysis by selecting daily rainfall exceeding the 75th percentile and the associated daily averaged water vapour, and we examined the relationship of their linearly binned readings in a quantile–quantile (Q-Q) plot. For this comparison analysis, we selected the UNSA, CAFJ, and PUNJ stations for the period between August 2019 and July 2021. UNSA is a long-term station, and the other two sites were specifically installed for this study. All three stations are accompanied by in situ rain-gauges.

5.4. Latitudinal Moisture Gradient Transport

In order to better understand the dynamics of the moisture transported by the SALLJ and the impact of the strong orographic effect in our study area, we examined the zenith wet delay gradients (or wet gradients). In order to detect correlation between wind vectors and wet gradients, we firstly plotted their azimuthal distributions. We removed potential wind shear effects with the surface by only considering pressure levels exceeding the altitude by 1 km above the surface. Subsequently, we focused on the variation of the wet gradients per epoch. We analysed the seasonal distributions and, apart from the number of events per azimuthal segment, we also showed the 90th to 50th percentile ratio for the corresponding segment. This assessment assists us in the detection of the directions towards which the strongest—compared to the median average—events take place. Those occurrences indicate large increases in the zenith wet delay gradient that points in the direction of incoming moisture. Finally, we selected zenith wet delay gradient readings for the corresponding epochs, when the greater 75th percentile rainfall amount occurs. According to those epochs, we plotted the azimuthal distributions of the wet gradients to detect key directions. Similarly to the previous step, we used the stations UNSA, CAFJ, and PUNJ because they are complimented by precise rainfall data.

6. Results

In this section, we present the results for each analysis method following the same order as in Section 5.

6.1. k-Means Clustering of GNSS Time Series

Our k-means clustering analysis shows that the separation of the water vapour time series into three classes divides them according to their geographic position and seasonality (cf. Figure 4). The clustering is based on mean monthly time series and their seasonal gradients determine the cluster generation. The time series of the highest seasonal amplitude (cluster 3) shows an oscillating signal with a central value 27.5 mm and an amplitude of 12.5 mm. The baseline of cluster 2 is shifted by 10 to 15 mm. Cluster 1 represents the stations with the lowest seasonal amplitude.

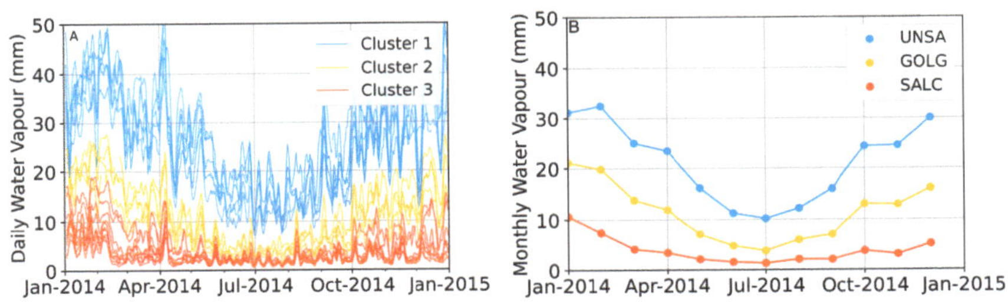

Figure 4. Results of clustering the water vapour readings showing all time series within a cluster with daily temporal resolution (**A**), but station clustering was performed on mean monthly values shown in (**B**). (**B**) shows individual mean monthly values along the topographic gradient exemplified for UNSA at 1224 m, GOLG at 2343 m, and SALC at 3799 m asl station elevation.

We observe the highest values in the austral summer during the SAMS seasons (January and February), whereas the lowest values take place in June. Moreover, October yields high magnitudes—in both the daily and mean monthly water vapour time series—followed by a slightly lower magnitude in November for 2014.

The map view of the station clusters show their expected geographic separation (cf. Figure 5). We emphasize that the clustering did not take into account the spatial location, but only the time series data. We separate stations by elevation and climatic conditions and use the 500 mm annual rainfall contour.

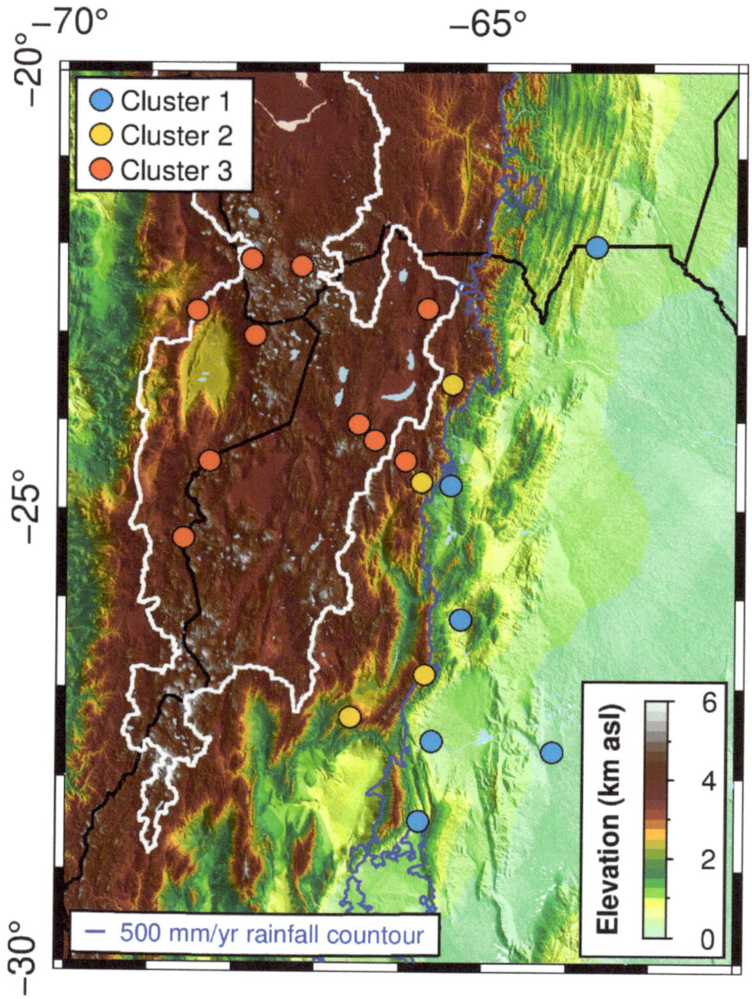

Figure 5. Map view of the clustered stations using the monthly mean values during 2014. Circle colours are similar to Figure 4. The stations with elevations up to 1225 m asl are located in the foothill zone (blue points), the sites with elevation between 2000 m and 2700 m asl are located in the transition zone (yellow points), and almost all stations above 3115 m asl are situated on the Altiplano–Puna plateau (red points). (Topographic data obtained from ETOPO1 [18], rainfall information retrieved from Tropical Rainfall Measuring Mission (TRMM) [20], as cited in Bookhagen and Strecker [10]).

6.2. Spectral Analysis

We analysed the spectrograms for all stations along the E-W and N-S directions (Figure 6). In our analysis, we only evaluated periods up to seven days, because this is the maximum range of the analysis window. The comparison between the UNSA, GOLG, SRSA, and SALC stations shows that the response strength is inversely proportional to the altitude. On the other hand, the stations along the N-S direction show a relatively homogeneous signal. Moreover, there is a notable difference in magnitude between the weaker winter and stronger summer signals. We also observe a diurnal signal independently of the station.

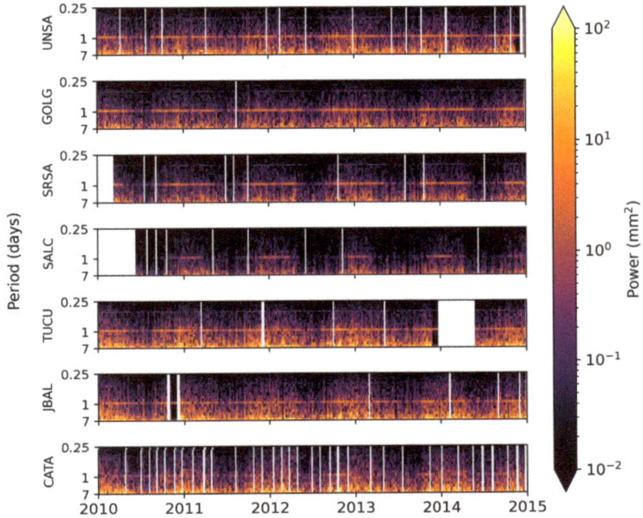

Figure 6. Spectral analysis of the stations along the E-W and N-S directions (Altitude asl: UNSA—1224 m, GOLG—2343 m, SRSA—3113 m, SALC—3799 m, TUCU—456 m, JBAL—381 m, and CATA—518 m; cf. Figure 1 and Table 1 for station location). A seven-day sampling window was used.

In Figure 7, we observe the PSDs of the UNSA and SALC stations. In addition, we highlight the power responses of the annual (An), semi-annual (S-An), monthly (M), weekly (W), and diurnal (D) periods. The strongest power signals at both stations are the annual, semi-annual, and diurnal periods. In a next step, we compare power signal strength between the stations: the higher elevations show significantly lower power for lower frequencies, in some cases, by one order of magnitude. The only exception is the half-year period, which is relatively higher in the SALC station.

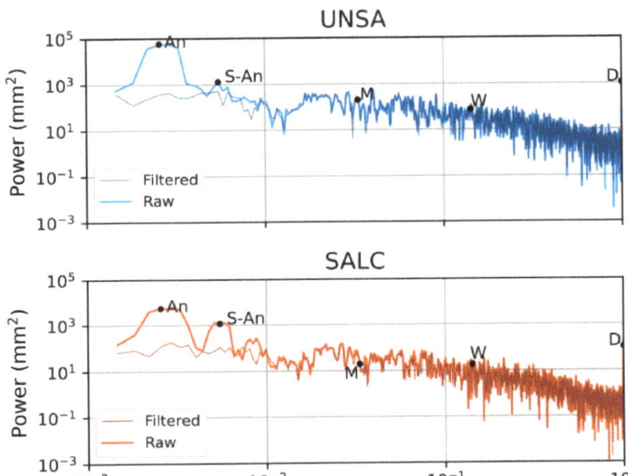

Figure 7. Power spectral densities (PSDs) of the low-elevation UNSA (upper panel) and the high-elevation SALC stations (lower panel). The thick and light-coloured lines represent the raw PSDs, while the thinner lines show the filtered values.

We note that the stations along the N-S cross-section are relatively homogeneous, while the stations along the topographic gradient alter significantly (Table 2). The decrease in the amplitude is at the level of 70–80%, which is in accordance with the observations made in Figure 6. In terms of the lower power signals, it appears that there is no physical relationship between those periods and the data series. Moreover, the normalized amplitudes show homogeneous behaviour for the daily and semi-annual signals, but we observe a decrease in the annual cycle at higher altitudes.

Table 2. Amplitude estimation of various harmonics for the stations along the E-W and N-S profiles.

Period	Absolute and Std.-Normalized Amplitude (mm/—)						
	UNSA	GOLG	SRSA	SALC	JBAL	TUCU	CATA
1 day	1.85/0.19	1.82/0.25	1.44/0.26	0.54/0.14	1.63/0.14	1.90/0.16	0.83/0.18
1 week	0.30/0.03	0.08/0.01	0.05/0.01	0.11/0.03	0.37/0.03	0.54/0.05	0.53/0.05
1 month	0.51/0.05	0.43/0,06	0.32/0.06	0.24/0.06	0.94/0.08	1.27/0.11	0.68/0.06
6 months	10.13/1.01	7.22/1.00	5.39/0.97	3.12/0.80	11.77/0.99	11.73/0.98	10.16/0.96
1 year	21.86/2.19	12.60/1.74	7.99/1.44	4.89/1.25	27.55/2.31	28.22/2.37	23.37/2.22

The power-law fitting of the binned filtered PSDs signals show a general agreement between the behaviours of the two stations (cf. Figure 8). The power-law exponents are comparable, but their roll-over magnitudes differ. With the exception of the low-magnitude spectrum, the log-binned datasets match well with this distribution. The estimated alpha values for UNSA and SALC stations are 1.50 and 1.46, respectively. Taking into account the power-law exponents, their standard deviations, and the degrees of freedom (18), the t-score equals 2.898. This value is lower than the threshold point of a t-distribution with a confidence interval of 99.9%, and we can assume that the two lines have comparable power-law slopes. Because of the different seasonal magnitudes of these two stations, we expect different roll-over magnitudes.

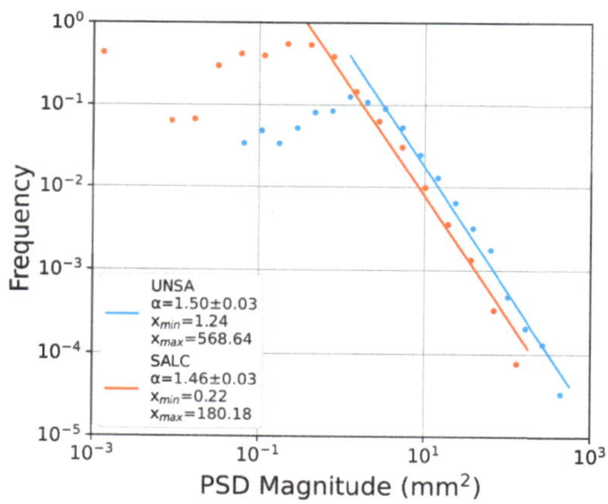

Figure 8. Power-law fitting of the filtered PSDs of the UNSA and SALC stations. The dots indicate the input data that are logarithmically binned using 20 classes. The power-law exponent is described by α, and both exponents are comparable within their standard deviations. The fitting is constrained by the minimum value (x_{min}) that corresponds to the point/bin, where the frequency magnitude starts to decrease.

6.3. Water Vapour and Rainfall Relation

We analysed the differences between power-law fits and water vapour and rainfall amount (Figure 9). For our analysis, we filtered the greater than 90th percentile daily mean water vapour values, and we selected the cumulative rainfall readings for the corresponding days. The water vapour series suggest that a power-law distribution is appropriate for modelling the log-binned observations. The exponent (α), which describes the slope of the line, is similar for the low- and medium-elevation stations, but it changes significantly for the high-elevation stations on the Altiplano–Puna plateau. On the contrary, the maximum value (x_{max}), which is related to the shift of the slope on the x-axis, is negatively correlated with the altitude. With respect to the rainfall values, we observe that the differences between the slopes are less pronounced, and the PUNJ station shows a slightly steeper relation. This indicates a lower variation between the rainfall events in the UNSA station. Even though there is a trend of heavier events in lower altitudes, the CAFJ time series yields lower peak values than the higher-elevation PUNJ station. We observe that the standard deviations of power-law exponents for the water vapour and rainfall relation varies: exponent uncertainties are large for water vapour relation and uncertainties are reduced to 8–22% for the rainfall relation (Table 3).

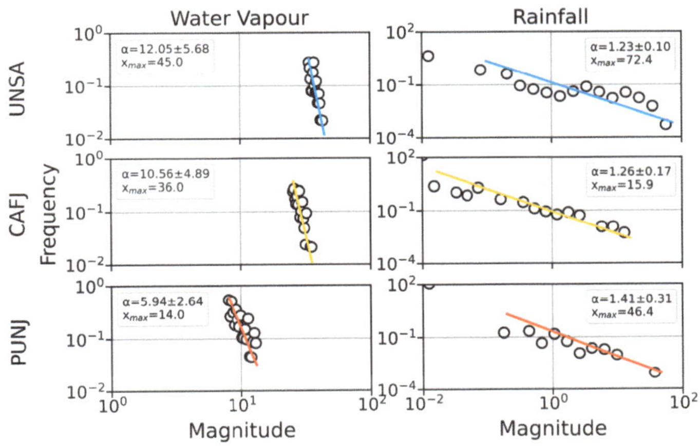

Figure 9. Power-law fitting of greater than the 90th percentile daily mean water vapour values (**left**) and the cumulative rainfall for the corresponding epochs (**right**) of the UNSA, CAFJ, and PUNJ stations. In both cases, the inputs (white dots) are grouped into 20 bins using a logarithmic scale. The minimum value (x_{min}) for the water vapour modelling was set to the lowest-magnitude bin, whereas the daily sums lower than 0.01 mm were omitted for the rainfall fitting.

Table 3. Statistical attributes for α and its standard deviation and x_{max} of the power-law fitted lines in Figure 9.

Station Name	Water Vapour		Rainfall	
	α	x_{max}	α	x_{max}
UNSA	12.05 ± 5.68	45.0	1.23 ± 0.10	72.40
CAFJ	10.56 ± 4.89	36.0	1.26 ± 0.17	15.90
PUNJ	5.94 ± 2.64	14.0	1.41 ± 0.31	46.40

In the next step, we reverse the reference dataset, and we examine the relation between rainfall and their corresponding water vapour values. We exemplify this by selecting the 75th percentile cumulative rainfall events on daily basis and their corresponding mean water vapour values during those epochs. Taking into account the non-continuous presence of this scalar, we set a lower percentile threshold in order to retrieve sufficient data from all stations.

While we do not observe a relation between all rainfall events and water vapour amounts, we detect certain location-related tendencies when comparing stronger rainfall events and water vapour (cf. Figure 10). The scattering in the y-axis is broader in the lower-elevation stations. On the other hand, the spreading in the x-axis follows the same trend but at a lower rate. However, the relations are very dynamic, and they cannot be described by a linear function.

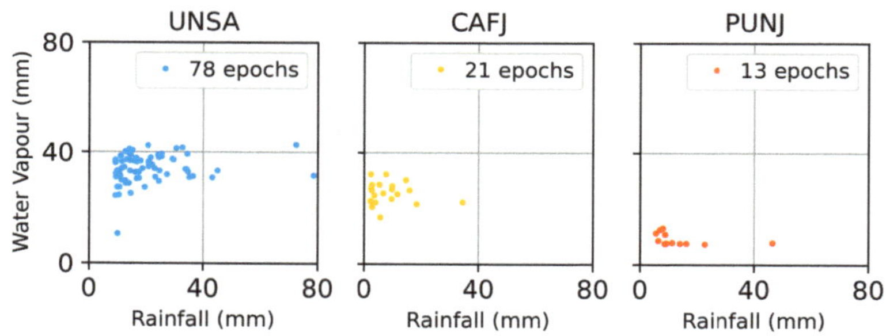

Figure 10. Quantile–quantile (Q-Q) plots of the greater than the 75th percentile daily-summed rainfall against the corresponding daily mean water vapour values of UNSA, CAFJ, and PUNJ stations. The lower number of measurements (epochs) in the higher-altitude stations is explained by the lower frequency of rainfall occurrences.

6.4. Zonal Moisture Gradient Transport

We compared the calculated zonal moisture gradient with ERA5-derived wind vectors (Figure 11). We observe annual wet gradients pointing in the general moisture directions to the north-west for UNSA and to the east for the high-elevation PUNJ station. This corresponds to the main wind direction for UNSA, which is toward the south-west. The observed eastward wind speeds in PUNJ are much higher, and they reflect the generally higher wind velocities following a high-to-low elevation gradient.

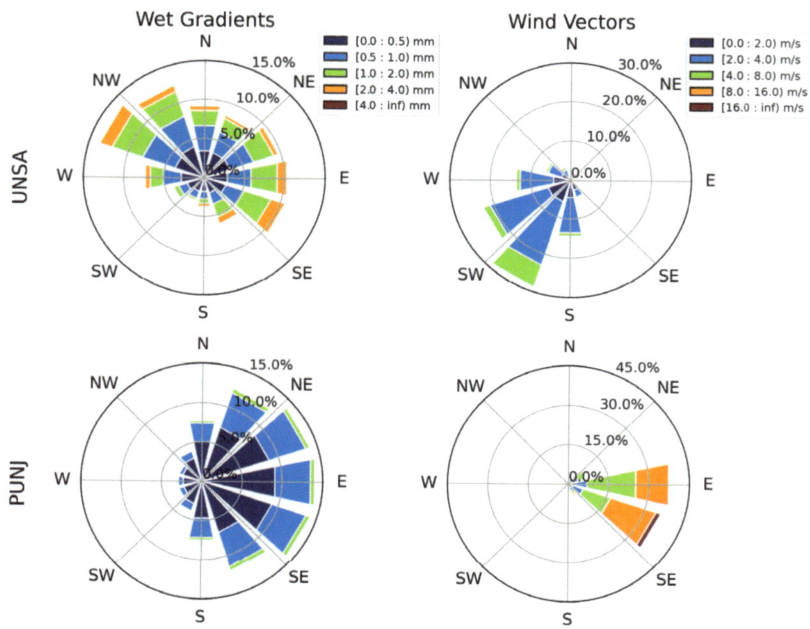

Figure 11. Azimuthal and magnitude distributions of the wet gradients and wind vectors of the UNSA and PUNJ stations. Wet gradients show the direction of moisture from the station location and

wind vectors show direction of wind transport. The azimuths are separated into 12 angular bins of 30°, and the magnitudes are arranged into five non-equal-width classes, in order to better illustrate distinct features for each station. The radius of the major influence is approximately 20 km. The wind directions are calculated from ERA5 hourly data on pressure levels [38] by extracting the median wind components between the station and the pressure level of 200 hPa. Because of the wind surface friction, the first km above the station is ignored. In both datasets we utilize a temporal resolution of 5 min, and the temporal coverage is adjusted to the shorter extent of the GNSS measurements.

We performed a seasonal analysis of wet gradient directions to highlight their strong seasonal dependence (Figure 12). For the low-elevation UNSA station located in the SALLJ we observe two peaks, and the majority of the azimuths point either to the east (90°) or to the west (270°) during austral spring/summer and fall/winter, respectively. During the fall season, there are higher values pointing to the west, which is reversed in the spring. The high-elevation PUNJ station shows different patterns: there is a peak azimuth direction from the north-east (45°) to the south-east (135°) throughout the year.

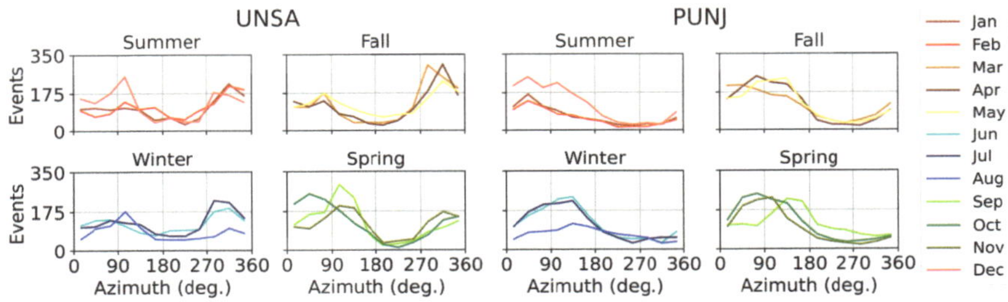

Figure 12. Azimuthal distribution of monthly-based wet gradients for the UNSA and PUNJ stations. Seasons are given as austral seasons (i.e., Summer is DJF), and the angular bins are defined as in Figure 11. The events indicate the number of hourly-sampled wet gradients that occur for each angular bin during each month.

Subsequently, we examined the ratios between the 90th to 50th percentile for all directions and seasons (Figure 13). In other words, we normalized the 90th percentile values by their medians and show the relation to the higher percentile: if the ratio is high, the 90th is much higher than the median. This analysis focuses on the wet gradient hotspots that are significantly larger than the mean, both in the spatial and temporal domain. Those occurrences are particularly interesting, because they indicate changing boundary conditions fluctuations. The ratios are high for the UNSA station and suggest a wide directional range. The PUNJ station shows a more homogeneous signal. There are no significant intra-seasonal variations of the distribution of the wet gradients in both cases. We also observe that moisture–gradient ratios vary by their directions. At the low-elevation UNSA station, the largest ratios occur during the summer and are directed towards the east-north-east (15° to 75° azimuth).

In the last step, we analysed the wet gradient direction during rainfall events exceeding the 75th percentile rainfall amount (Figure 14) in order to focus on those episodes. We observe that moisture gradients during rainfall events are spread out with a dominant direction from the south-east. In contrast, the high-elevation PUNJ station only shows five main direction that reflect local topographic shielding patterns and main moisture directions. We note that the majority of the events point towards the east in an azimuth range between 105–165° and 15–135° for UNSA and PUNJ, respectively.

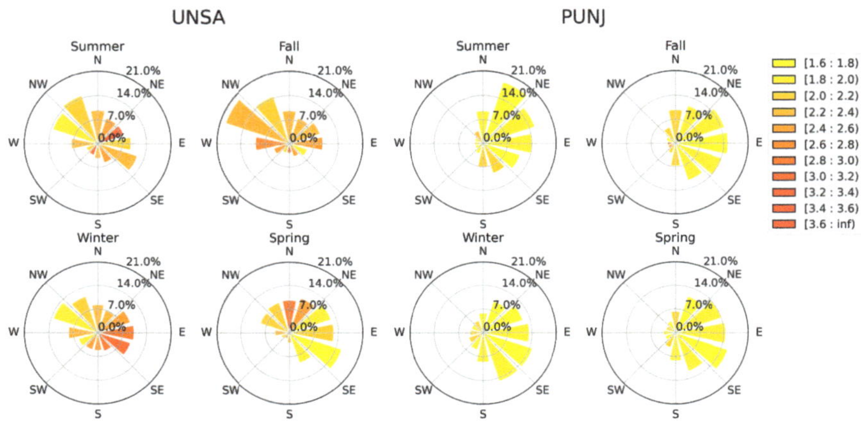

Figure 13. 90th to 50th percentile ratio of the wet gradients for the UNSA and PUNJ stations. The temporal and azimuthal separation is done similarly to Figures 11 and 12. The percentiles are extracted from the hourly averaged observations that occur in each directional and seasonal segment.

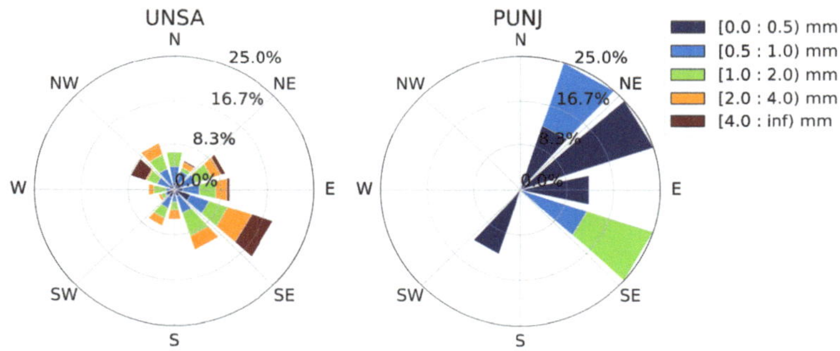

Figure 14. Summer season wet gradients for the larger than the 75th percentile rainfall events for the UNSA and PUNJ stations. The seasonal and azimuthal segmentation coincides with Figures 11–13, and the rainfall events are identified by the cumulative rainfall on 1 h basis that is measured with in situ rain gauges.

7. Discussion

The discussion follows the same organization of the Results sections.

7.1. k-Means Clustering of GNSS Time Series

We analysed the time series of the monthly water vapour during 2014 (Figure 4). The results of the clustering and the spatial water vapour distribution show the impact of topography and climate: the higher-elevation stations with low seasonal amplitudes are located on the arid Altiplano–Puna plateau, and the low-elevation densely vegetated areas are characterized by high seasonality. The transitional zone between these two end members shows an intermediate behaviour. This finding confirms our initial hypothesis and previous observations that orography plays a significant role on the local climate [10,32]. We further calculated the water vapour readings of UNSA at various altitudes, and we directly compare them with other stations (Figure 15). According to [83,84], the water vapour of a site can be projected to a higher altitude with the following equation:

$$wv = wv_0 \exp \frac{C_2 \Delta h}{1000} \tag{7}$$

where C_2 is a constant equal to 0.439, and Δh is the height difference in meters.

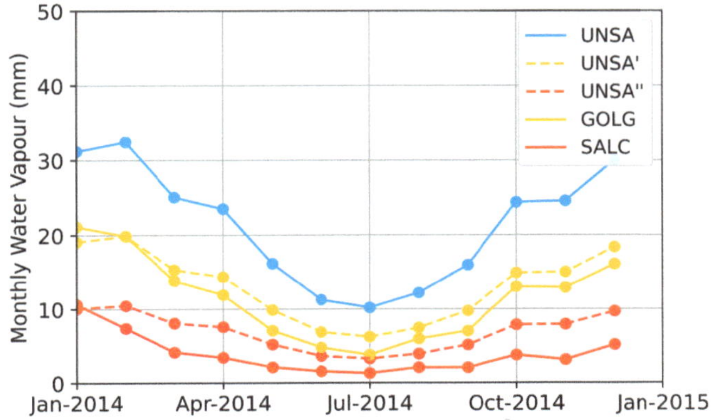

Figure 15. Projection of the mean monthly water vapour values of UNSA station (1224 m asl) at higher altitudes for the direct comparison of the readings with actual measurements at those points. UNSA′ is the water vapour content at the corresponding height of the GOLG station at 2343 m asl, and UNSA″ corresponds to the SALC station at 3799 m asl. In most cases, the water vapour contents at UNSA at the corresponding heights are higher than at the measured station.

Additionally, all stations show a seasonal signal, independent of altitude, and we observe a near-continuous signal from austral winter to spring (September to November). In order to examine this signal, we plot the monthly means over a longer period (Figure 16). The transitional seasons are characterized by larger 1 sigma standard deviations and show a larger variability in atmospheric water vapour.

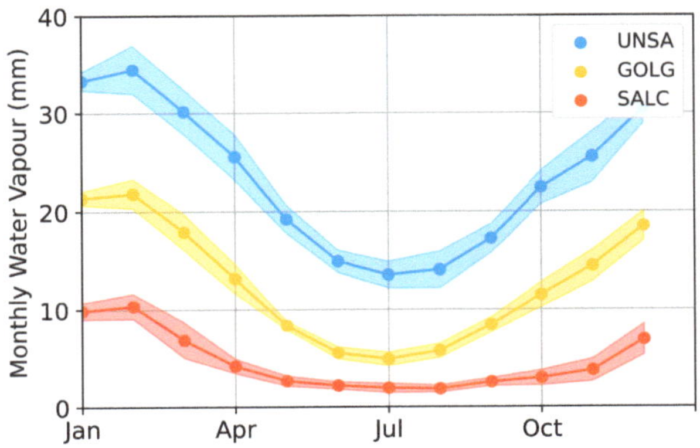

Figure 16. Mean monthly water vapour values for the stations along the topographic gradient from low to high elevations (cf. Figure 4). We analysed water vapour readings from 2010 to 2021 derived from ray tracing that only utilizes ERA5 meteorological data on pressure levels [38]. The semi-transparent colouring indicates the per-month standard deviation.

7.2. Spectral Analysis

We compare the signal responses at various frequencies of the GNSS-derived water vapour estimates against the ERA5 reanalysis data (Figure 17). We observe similar behaviour between the two spectrograms, which indicates high coincidence in the seasonal signals. In addition, the majority of the relative differences are less than 1%, showing that the GNSS observations are equally reliable. One advantage of GNSS data is that they can achieve a significantly higher temporal resolution of five minutes, and detect features in this region on short time scales. In contrast, the reanalysis or mesoscale models have a temporal resolution of one hour and capture synoptic-scale processes.

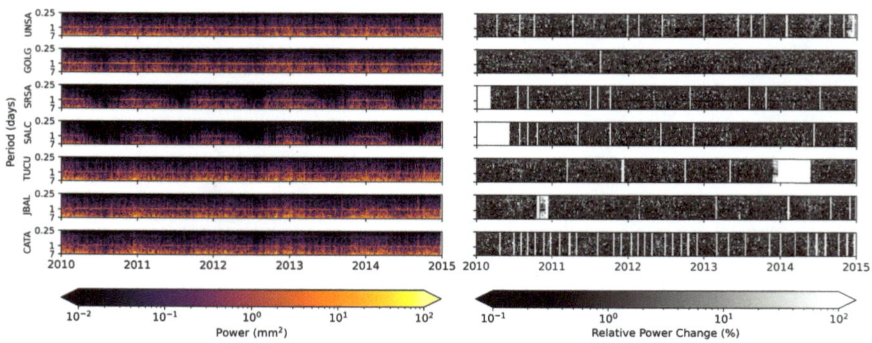

Figure 17. Left: Spectral analysis of the station along the E-W and N-S directions using a seven-day sampling window. This figure can be directly compared to Figure 6. In this case, we employed the ray-tracing derived water vapour using only meteorological information from ERA5 hourly data on pressure levels [38], instead of the GNSS-derived water vapour. **Right:** Difference in relative response power between the spectrograms of the GNSS- and the ERA5 ray-tracing derived water vapour series.

The spectral behaviour of the water vapour shows that all signals are primarily tuned at the annual, semi-annual, and diurnal periods, which correspond to the seasonal cycles of the moist air masses. The shape of the signal affects the ratio between period responses. In the case of the higher-elevation stations, the semi-annual periods are pronounced. This is due to the flattening of the water vapour series, which leads to a time series that can be better characterized by a harmonic equation with two oscillations. The water vapour values of the lower-elevation stations are reflected both in the spectrograms and the PSDs, showing the influence of the topography on the regional climate. Lastly, the frequency–magnitude relation of the filtered PSDs reveals relatively analogous responses to the seasonal fluctuations, regardless of location (cf. Figure 8).

7.3. Water Vapour and Rainfall Relation

The power-law fitting of the water vapour observations reveals higher values in lower altitudes on the one hand and a lower decay ratio for the PUNJ station. The prior is expected, and it has been noted in the previous sections (e.g., Figures 3 and 4). The latter is interesting, and we note that the high-percentile readings are more equally distributed in this station. Additionally, the fitting of the rainfall readings demonstrates that strong events will also take place at high elevations, but less frequently. However, this interpretation is not well applicable to the CAFJ station, because this station is part of an inter-mountain valley; thus the lower cloud coverage results in different temperature conditions. The high standard deviations of the exponents indicate skewed datasets that diverge significantly from a normal distribution. Despite the high uncertainty of the water vapour fitting, we observe clear trends. The Q-Q plots also show that extreme rainfall events are observed at high elevations, but they occur in a narrower water vapour peak range. This demonstrates the direct relationship between the required amount of water vapour to produce atmospheric saturation and the elevation. Additionally, the saturation is also dependent on the tropospheric temperature above the examined locations. In this case, there is a notable difference because of the within-the-layer temperature decrease along the altitude and the complex terrain that impacts cloud coverage.

7.4. Zonal Moisture Gradient Transport

We observe that the wind directions in the low-elevation stations are associated with the SALLJ and show moisture transport from the north-east and east towards the southerly directions (cf. Figure 11). Wind directions on the Altiplano–Puna plateau show only a minimal correlation with the SALLJ. There is a large difference in the wind speed between the low- and high-elevation stations, because the measurements do not take place at the same height. When considering a fixed pressure level (e.g., 500 hPa), the wind speed is homogeneous over the area. The distribution of the per-season-separated wet gradients shows dominant patterns for every station that slightly change through the year. This suggests a major influence of the topography and the altitude.

The higher 90th to 50th percentile ratios of the wet gradients in the low-elevation station are associated with the SALLJ that transports moist masses over the foothill zone of the south-central Andes. Moreover, the direction of the strongest occurrences (in terms of ratio) in this station (east-north-east) reveal important information. This is the direction of the topographic barrier which intersects with the SALLJ. The analogy between the wet gradients and the most intense rainfall occurrences suggests that most rainfall events reaching Salta are transported across the orographic barriers to the east and south-east of the city. In conjunction with our prior findings, this indicates good correlation between the wet gradients and the location of the wet air masses. Moreover, some rainfall events occur in the opposite direction, pointing to the orographic barrier west of Salta (cf. Figures 12 and 18). For the arid PUNJ station, we observe weaker wet gradients, but a strong correlation with topography. The correlation is pronounced because gradient generation is only triggered by the moisture transport on topographic barriers.

Figure 18. Topography setting of the low-elevation UNSA (1224 m asl) and high-elevation PUNJ (3760 m asl) stations. The dark- and light-blue circles show the area of influence of the 50th and 90th percentile of the water vapour above each site, respectively. The black vectors indicate the main moisture directions associated with high rainfall, as calculated in Figure 14. Topographic data obtained from ETOPO1 [18].

8. Conclusions

In this study, we have used GNSS time series data to better understand the climatic dynamics of the central Andes in northwestern Argentina. We have compiled 23 GNSS observations from 2001 to 2021 and have installed two GNSS stations that collected data between 2019 and 2021. The GNSS signal is used to measure water vapour content in the atmosphere at high temporal resolutions of 5 min. We use these data to make the following key observations:

(1) The GNSS time series data show distinctive climatic behaviour for the Central Andes that was analysed using a clustering analysis: GNSS stations from the low-elevation area in the pathway of the SALLJ show similar behaviour and cluster into the same group. Stations from intermediate elevations at the transition from low- to high-

elevation areas show a distinctive signal and cluster into the same group. Stations from the high elevations located on the Altiplano–Puna plateau behave similarly and have the lowest absolute (not relative) seasonal component.

(2) A frequency analysis depicts the seasonal signals, and it illustrates the impact of the orographic uplift. The annual, semi-annual, and diurnal periods can be clearly identified, but there are also spectral differences across the time series. The most prominent variations between the stations are found in the magnitude of the water vapour levels, where the readings are inversely proportional to the station elevations.

(3) The association between water vapour and rainfall reveals a general correlation of stronger water vapour amounts. We observe that high water vapour episodes are less frequent at higher altitudes, but strong events still occur. We note that the rainfall–water vapour relation varies along the topographic gradient. At lower altitudes, rainfall occurs across a wide water vapour peak range. In contrast, at high elevations only a narrow band of water vapour amounts can be associated with rainfall events.

(4) We have used wet gradients to identify moisture transport for two sites: the low-elevation UNSA station at 1224 m and the high-elevation PUNJ station at 3760 m. The wet gradients allow us to document that local topographic effects strongly impact the characteristics of the GNSS and hydrologic stations. Even though the moisture fluxes' magnitude is subject to the circulation of the SALLJ and the mesoscale convective systems, nearby topography controls the circulation of atmospheric water vapour and controls the moisture pathways.

In comparison to reanalysis data, there are several advantages of GNSS meteorology techniques: a good accuracy of water vapour measurements, the ability to measure water vapour in three dimensions, and the high sampling rate of seconds to minutes. A network of homogenized and reliable GNSS stations will allow for an improved weather prediction. The National Geographic Institute of Argentina continuously operates a GNSS network that is very dense in urban zones, but lacks facilities in remote areas. Considering the ability of GNSS to measure the atmospheric moisture gradient, the aggregation of more stations and the integration of those data into meteorological applications would allow for short-term predictions of heavy rainfall, and would improve the weather prediction. This knowledge may help in reducing the damage from natural hazards, and it would benefit the agriculture sector, which is crucial for the local economy.

Author Contributions: Conceptualization, N.A., B.B. and K.B.; methodology, N.A. and B.B.; software, N.A. and K.B.; validation, N.A. and B.B.; formal analysis, N.A., K.B. and G.D.; investigation, N.A., B.B. and K.B.; resources, J.W. and B.B.; data curation, N.A.; writing—original draft preparation, N.A.; writing—review and editing, B.B., J.W., K.B., A.d.l.T. and G.D.; visualization, N.A. and B.B.; supervision, B.B., J.W. and A.d.l.T.; project administration, B.B. and J.W.; funding acquisition, B.B. and J.W. All authors have read and agreed to the published version of the manuscript.

Funding: This research was funded by the German Research Foundation (Deutsche Forschungsgemeinschaft—DFG) and the federal state of Brandenburg through the International Research Training Group-StRATEGy (DFG IRTG2018). Kyriakos Balidakis is funded by the DFG through the TerraQ Project (ID 434617780—SFB 1464). The publication costs were covered by the DFG within the funding programme "Open-Access-Publikationskosten" and the German Research Centre for Geosciences (Deutsches GeoForschungsZentrum—GFZ).

Data Availability Statement: The ETOPO1 dataset was provided by the National Oceanic and Atmospheric Administration (NOAA). The MODIS/Terra dataset was provided by the National Aeronautics and Space Administration (NASA). The Tropical Rainfall Measuring Mission (TRMM) dataset was provided by NASA and the Japanese Aerospace Exploration Agency (JAXA). The ERA5 dataset was provided by the European Centre for Medium-Range Weather Forecasts (ECMWF). The water vapour estimates and the zenith total delays, as well as their gradients, were provided by the German Research Centre for Geosciences (Deutsches GeoForschungsZentrum—GFZ) and the Nevada Geodetic Laboratory (NGL) (http://geodesy.unr.edu/, accessed on 3 December 2020). The entire time series of the GNSS and ray-tracing products are provided by the GFZ Data Services (https://dataservices.gfz-potsdam.de/panmetaworks/showshort.php?id=8fd77904-2f86-11

ed-9732-32bb1430b4f7, accessed on 9 September 2022). Precise rainfall information was provided by an in situ rain-gauge station, the Nationa Institute of Agriculture Technology (Instituto Nacional de Tecnología Agropecuaria—INTA) of Argentina (http://siga.inta.gob.ar/, accessed on 16 December 2021), and Meteostat (https://meteostat.net, accessed on 16 December 2021).

Acknowledgments: We would like to thank Fernando Hongn and Luis Alvarado from the Institute of Bio- and Geosciences of Northeast Argentina (Instituto de Bio y Geociencias del NOA—IBIGEO) for their assistance in installing the ground stations.

Conflicts of Interest: The authors declare no conflict of interest.

Abbreviations

The following abbreviations are used in this manuscript:

EPOS	Earth Parameter and Orbit System
EVI	enhanced vegetation index
E-W	east–west
GFZ	Deutsches GeoForschungsZentrum (German Research Centre for Geosciences—ger)
GNSS	global navigation satellite system
GPS	global positioning system
IGN	Instituto Geográfico Nacional (National Geographic Institute of Argentina—sp)
NGL	Nevada Geodetic Laboratory
N-S	north–south
PDF	probability density function
PSD	power spectral density
Q-Q	quantile–quantile
SALLJ	South American low-level jet
SAMS	South American monsoon system
UP	University of Potsdam

References

1. Bookhagen, B.; Thiede, R.C.; Strecker, M.R. Abnormal monsoon years and their control on erosion and sediment flux in the high, arid northwest Himalaya. *Earth Planet. Sci. Lett.* **2005**, *231*, 131–146. [CrossRef]
2. Bookhagen, B. Appearance of extreme monsoonal rainfall events and their impact on erosion in the Himalaya. *Geomat. Nat. Hazards Risk* **2010**, *1*, 37–50. [CrossRef]
3. Wulf, H.; Bookhagen, B.; Scherler, D. Seasonal precipitation gradients and their impact on fluvial sediment flux in the Northwest Himalaya. *Geomorphology* **2010**, *118*, 13–21. [CrossRef]
4. Castino, F.; Bookhagen, B.; de la Torre, A. Atmospheric dynamics of extreme discharge events from 1979 to 2016 in the southern Central Andes. *Clim. Dyn.* **2020**, *55*, 3485–3505. [CrossRef]
5. Jones, J.N.; Boulton, S.J.; Stokes, M.; Bennett, G.L.; Whitworth, M.R.Z. 30-year record of Himalaya mass-wasting reveals landscape perturbations by extreme events. *Nat. Commun.* **2021**, *12*, 6701. [CrossRef]
6. Boers, N.; Bookhagen, B.; Barbosa, H.M.; Marwan, N.; Kurths, J.; Marengo, J.A. Prediction of extreme floods in the eastern Central Andes based on a complex networks approach. *Nat. Commun.* **2014**, *5*, 5199. [CrossRef]
7. Castino, F.; Bookhagen, B.; Strecker, M.R. River-discharge dynamics in the Southern Central Andes and the 1976–77 global climate shift. *Geophys. Res. Lett.* **2016**, *43*, 11–679. [CrossRef]
8. Poveda, G.; Espinoza, J.C.; Zuluaga, M.D.; Solman, S.A.; Garreaud, R.; Van Oevelen, P.J. High impact weather events in the Andes. *Front. Earth Sci.* **2020**, *8*, 162. [CrossRef]
9. Schumacher, V.; Fernández, A.; Justino, F.; Comin, A. WRF high resolution dynamical downscaling of precipitation for the Central Andes of Chile and Argentina. *Front. Earth Sci.* **2020**, *2020*, 328. [CrossRef]
10. Bookhagen, B.; Strecker, M.R. Orographic barriers, high-resolution TRMM rainfall, and relief variations along the eastern Andes. *Geophys. Res. Lett.* **2008**, *35*, 6. [CrossRef]
11. Bookhagen, B.; Burbank, D.W. Toward a complete Himalayan hydrological budget: Spatiotemporal distribution of snowmelt and rainfall and their impact on river discharge. *J. Geophys. Res. Earth Surf.* **2010**, *115*, F3. [CrossRef]
12. Cannon, F.; Carvalho, L.M.V.; Jones, C.; Norris, J.; Bookhagen, B.; Kiladis, G.N. Effects of topographic smoothing on the simulation of winter precipitation in High Mountain Asia. *J. Geophys. Res. Atmos.* **2017**, *122*, 1456–1474. [CrossRef]
13. Yao, Y.; Shan, L.; Zhao, Q. Establishing a method of short-term rainfall forecasting based on GNSS-derived PWV and its application. *Sci. Rep.* **2017**, *7*, 12465. [CrossRef]
14. Rohm, W.; Guzikowski, J.; Wilgan, K.; Kryza, M. 4DVAR assimilation of GNSS zenith path delays and precipitable water into a numerical weather prediction model WRF. *Atmos. Meas. Tech.* **2019**, *12*, 345–361. [CrossRef]

15. Ramezani Ziarani, M.; Bookhagen, B.; Schmidt, T.; Wickert, J.; de la Torre, A.; Deng, Z.; Calori, A. A Model for the Relationship between Rainfall, GNSS-Derived Integrated Water Vapour, and CAPE in the Eastern Central Andes. *Remote Sens.* **2021**, *13*, 3788. [CrossRef]
16. Boers, N.; Bookhagen, B.; Marengo, J.; Marwan, N.; von Storch, J.S.; Kurths, J. Extreme rainfall of the South American monsoon system: A dataset comparison using complex networks. *J. Clim.* **2015**, *28*, 1031–1056. [CrossRef]
17. Castino, F.; Bookhagen, B.; Strecker, M.R. Rainfall variability and trends of the past six decades (1950-2014) in the subtropical NW Argentine Andes. *Clim. Dyn.* **2017**, *48*, 1049–1067. [CrossRef]
18. Amante, C. ETOPO1 1 Arc-Minute Global Relief Model: Procedures, Data Sources and Analysis. 2009. Available online: https://www.ncei.noaa.gov/access/metadata/landing-page/bin/iso?id=gov.noaa.ngdc.mgg.dem:316 (accessed on 15 November 2021).
19. Didan, K. MOD13A3 MODIS/Terra Vegetation Indices Monthly L3 Global 1 km SIN Grid V006. 2015. Available online: https://lpdaac.usgs.gov/products/mod13a3v006/ (accessed on 20 November 2021).
20. Tropical Rainfall Measuring Mission (TRMM). TRMM Combined Precipitation Radar and Microwave Imager Rainfall Profile L2 1.5 hours V7. 2011. Available online: https://disc.gsfc.nasa.gov/datacollection/TRMM_2B31_7.htmls (accessed on 30 May 2022).
21. Montini, T.L.; Jones, C.; Carvalho, L.M.V. The South American low-level jet: A new climatology, variability, and changes. *J. Geophys. Res. Atmos.* **2019**, *124*, 1200–1218. [CrossRef]
22. Marengo, J.A.; Soares, W.R.; Saulo, C.; Nicolini, M. Climatology of the low-level jet east of the Andes as derived from the NCEP–NCAR reanalyses: Characteristics and temporal variability. *J. Clim.* **2004**, *17*, 2261–2280.<2261:COTLJE>2.0.CO;2. [CrossRef]
23. Vera, C.; Higgins, W.; Amador, J.; Ambrizzi, T.; Garreaud, R.; Gochis, D.; Gutzler, D.; Lettenmaier, D.; Marengo, J.; Mechoso, C.; et al. Toward a unified view of the American monsoon systems. *J. Clim.* **2006**, *19*, 4977–5000. 10.1175/JCLI3896.1. [CrossRef]
24. Jones, C. Recent changes in the South America low-level jet. *NPJ Clim. Atmos. Sci.* **2019**, *2*, 20. [CrossRef]
25. Marengo, J.A.; Liebmann, B.; Grimm, A.; Misra, V.; Silva Dias, P.d.; Cavalcanti, I.; Carvalho, L.M.V.; Berbery, E.; Ambrizzi, T.; Vera, C.S.; et al. Recent developments on the South American monsoon system. *Int. J. Climatol.* **2012**, *32*, 1–21. [CrossRef]
26. Silva, V.B.; Kousky, V.E. The South American monsoon system: Climatology and variability. *Mod. Climatol.* **2012**, *123*, 152. [CrossRef]
27. Salio, P.; Nicolini, M.; Zipser, E.J. Mesoscale Convective Systems over Southeastern South America and Their Relationship with the South American Low-Level Jet. *Mon. Weather Rev.* **2007**, *135*, 1290–1309. [CrossRef]
28. Durkee, J.D.; Mote, T.L.; Shepherd, J.M. The contribution of mesoscale convective complexes to rainfall across subtropical South America. *J. Clim.* **2009**, *22*, 4590–4605. [CrossRef]
29. Boers, N.; Barbosa, H.M.; Bookhagen, B.; Marengo, J.A.; Marwan, N.; Kurths, J. Propagation of strong rainfall events from southeastern South America to the central Andes. *J. Clim.* **2015**, *28*, 7641–7658. [CrossRef]
30. Carvalho, L.M.V.; Jones, C.; Posadas, A.N.; Quiroz, R.; Bookhagen, B.; Liebmann, B. Precipitation characteristics of the South American monsoon system derived from multiple datasets. *J. Clim.* **2012**, *25*, 4600–4620. [CrossRef]
31. Boers, N.; Bookhagen, B.; Marwan, N.; Kurths, J.; Marengo, J. Complex networks identify spatial patterns of extreme rainfall events of the South American Monsoon System. *Geophys. Res. Lett.* **2013**, *40*, 4386–4392. [CrossRef]
32. Bookhagen, B.; Strecker, M.R. Spatiotemporal trends in erosion rates across a pronounced rainfall gradient: Examples from the southern Central Andes. *Earth Planet. Sci. Lett.* **2012**, *327*, 97–110. [CrossRef]
33. Romatschke, U.; Houze, R.A. Characteristics of precipitating convective systems accounting for the summer rainfall of tropical and subtropical South America. *J. Hydrometeorol.* **2013**, *14*, 25–46. [CrossRef]
34. Houze, R.A., Jr. Orographic effects on precipitating clouds. *Rev. Geophys.* **2012**, *50*. [CrossRef]
35. Houze, R.A., Jr.; Rasmussen, K.L.; Zuluaga, M.D.; Brodzik, S.R. The variable nature of convection in the tropics and subtropics: A legacy of 16 years of the Tropical Rainfall Measuring Mission satellite. *Rev. Geophys.* **2015**, *53*, 994–1021. [CrossRef]
36. Lamprecht, C. Meteostat. 2022. Available online: https://meteostat.net (accessed on 16 December 2021).
37. Hierro, R.; Fonseca, Y.B.; Ziarani, M.R.; Llamedo, P.; Schmidt, T.; de la Torre, A.; Alexander, P. On the behavior of rainfall maxima at the eastern Andes. *Atmos. Res.* **2020**, *234*, 104792. [CrossRef]
38. Hersbach, H.; Bell, B.; Berrisford, P.; Biavati, G.; Horányi, A.; Muñoz Sabater, J.; Nicolas, J.; Peubey, C.; Radu, R.; Rozum, I.; et al. ERA5 hourly data on pressure levels from 1959 to present. Copernicus Climate Change Service (C3S) Climate Data Store (CDS). 2018. Available online: https://cds.climate.copernicus.eu/cdsapp#!/dataset/10.24381/cds.bd0915c6?tab=overview (accessed on 29 November 2021).
39. Antonoglou, N.; Balidakis, K.; Dick, G.; Wickert, J.; Bookhagen, B. Hydrostatic and Wet Signal Delays Calculated from a Network of 23 GNSS Stations in Northwestern Argentina between 2010–2021. 2022. Available online: https://dataservices.gfz-potsdam.de/panmetaworks/showshort.php?id=8fd77904-2f86-11ed-9732-32bb1430b4f7 (accessed on 9 September 2022).
40. Simons, M.; Galetzka, J.E.; Genrich, J.F.; Ortega, F.; Comte, D.; Glass, B.; Gonzalez, G. Central Andean Tectonic Observatory Geodetic Array—CBAA-Cordon Barros Arana P.S., The GAGE Facility Operated by UNAVCO, Inc., GPS/GNSS Observations Dataset. 2010. Available online: https://www.unavco.org/data/doi/10.7283/T52J68XN (accessed on 9 September 2022).
41. Simons, M.; Galetzka, J.E.; Genrich, J.F.; Ortega, F.; Comte, D.; Glass, B.; Gonzalez, G. Central Andean Tectonic Observatory Geodetic Array—CJNT-Chajnantor P.S., The GAGE Facility Operated by UNAVCO, Inc., GPS/GNSS Observations Dataset. 2010. Available online: https://www.unavco.org/data/doi/10.7283/T5B27S9T (accessed on 9 September 2022).

42. Pritchard, M. Plutons GPS Network—COLO-Laguna Colorada P.S., The GAGE Facility Operated by UNAVCO, Inc., GPS/GNSS Observations Dataset. 2010. Available online: https://www.unavco.org/data/doi/10.7283/T5Q23XBT (accessed on 9 September 2022).
43. Pritchard, M. Plutons GPS Network—UTUR-Uturuncu Volcano P.S., The GAGE Facility Operated by UNAVCO, Inc., GPS/GNSS Observations Dataset. 2010. Available online: https://www.unavco.org/data/doi/10.7283/T5K935MG (accessed on 9 September 2022).
44. Pritchard, M. Plutons GPS Network—LCEN-Lazufre Center of Deformation, The GAGE Facility Operated by UNAVCO, Inc., GPS/GNSS Observations Dataset. 2011. Available online: https://www.unavco.org/data/doi/10.7283/T53776T2 (accessed on 9 September 2022).
45. Pritchard, M. Plutons GPS Network—SOCM-Socompa P.S., The GAGE Facility Operated by UNAVCO, Inc., GPS/GNSS Observations Dataset. 2012. Available online: https://www.unavco.org/data/doi/10.7283/T5TT4P27 (accessed on 9 September 2022).
46. Smalley R., Jr.; Bennett, R. CAP Andes GPS Network—LVRA-Liviara P.S., The GAGE Facility Operated by UNAVCO, Inc., GPS/GNSS Observations Dataset. 2015. Available online: https://www.unavco.org/data/doi/10.7283/T5TB1596 (accessed on 9 September 2022).
47. Smalley R., Jr.; Bennett, R. CAP Andes GPS Network—SCTC-Santa Catalina P.S., The GAGE Facility Operated by UNAVCO, Inc., GPS/GNSS Observations Dataset. 2015. Available online: https://www.unavco.org/data/doi/10.7283/T52V2DHH (accessed on 9 September 2022).
48. Smalley R., Jr.; Bennett, R. CAP Andes GPS Network—YAVC-Yavi P.S., The GAGE Facility Operated by UNAVCO, Inc., GPS/GNSS Observations Dataset. 2015. Available online: https://www.unavco.org/data/doi/10.7283/T5Z31X1J (accessed on 9 September 2022).
49. Smalley R., Jr.; Bennett, R. Puna-Andes GPS Network—GOLG-GOLG P.S., The GAGE Facility Operated by UNAVCO, Inc., GPS/GNSS Observations Dataset. 2015. Available online: https://www.unavco.org/data/doi/10.7283/T59C6VTS (accessed on 9 September 2022).
50. Smalley R., Jr.; Bennett, R. Puna-Andes GPS Network—SALC-SALC P.S., The GAGE Facility Operated by UNAVCO, Inc., GPS/GNSS Observations Dataset. 2015. Available online: https://www.unavco.org/data/doi/10.7283/T5PN9413 (accessed on 9 September 2022).
51. Smalley R., Jr.; Bennett, R. Puna-Andes GPS Network—SRSA-SRSA P.S., The GAGE Facility Operated by UNAVCO, Inc., GPS/GNSS Observations Dataset. 2015. Available online: https://www.unavco.org/data/doi/10.7283/T5F47MJH (accessed on 9 September 2022).
52. Smalley R., Jr.; Bennett, R. Puna-Andes GPS Network—TUZG-TUZG P.S., The GAGE Facility Operated by UNAVCO, Inc., GPS/GNSS Observations Dataset. 2015. Available online: https://www.unavco.org/data/doi/10.7283/T5JW8C9C (accessed on 9 September 2022).
53. Piñón, D.A.; Gómez, D.D.; Smalley, R., Jr.; Cimbaro, S.R.; Lauría, E.A.; Bevis, M.G. The history, state, and future of the Argentine continuous satellite monitoring network and its contributions to geodesy in Latin America. *Seismol. Res. Lett.* **2018**, *89*, 475–482. [CrossRef]
54. Blewitt, G.; Hammond, W.C.; Kreemer, C. Harnessing the GPS data explosion for interdisciplinary science. *Eos* **2018**, *99*, 485. [CrossRef]
55. Teunissen, P.; Montenbruck, O. *Springer Handbook of Global Navigation Satellite Systems*; Springer: Cham, Switzerland, 2017; ISBN 978-3-319-42928-1.
56. Chen, G.; Herring, T. Effects of atmospheric azimuthal asymmetry on the analysis of space geodetic data. *J. Geophys. Res. Solid Earth* **1997**, *102*, 20489–20502. [CrossRef]
57. Bar-Sever, Y.E.; Kroger, P.M.; Borjesson, J.A. Estimating horizontal gradients of tropospheric path delay with a single GPS receiver. *J. Geophys. Res. Solid Earth* **1998**, *103*, 5019–5035. [CrossRef]
58. Kačmařík, M.; Douša, J.; Zus, F.; Václavovic, P.; Balidakis, K.; Dick, G.; Wickert, J. Sensitivity of GNSS tropospheric gradients to processing options. *Ann. Geophys.* **2019**, *37*, 429–446. [CrossRef]
59. Bevis, M.; Businger, S.; Chiswell, S.; Herring, T.A.; Anthes, R.A.; Rocken, C.; Ware, R.H. GPS meteorology: Mapping zenith wet delays onto precipitable water. *J. Appl. Meteorol.* **1994**, *33*, 379–386.<0379:GMMZWD>2.0.CO;2. [CrossRef]
60. GFZ German Research Centre for Geosciences. Earth Parameter and Orbit System Software. Software EPOS, 2021. Available online: https://www.gfz-potsdam.de/en/section/global-geomonitoring-and-gravity-field/topics/earth-system-parameters-and-orbit-dynamics/earth-parameter-and-orbit-system-software-epos (accessed on 17 February 2022).
61. Bertiger, W.; Bar-Sever, Y.; Dorsey, A.; Haines, B.; Harvey, N.; Hemberger, D.; Heflin, M.; Lu, W.; Miller, M.; Moore, A.W.; et al. GipsyX/RTGx, a new tool set for space geodetic operations and research. *Adv. Space Res.* **2020**, *66*, 469–489. j.asr.2020.04.015. [CrossRef]
62. Gendt, G.; Dick, G.; Reigber, C.; Tomassini, M.; Liu, Y.; Ramatschi, M. Near real time GPS water vapor monitoring for numerical weather prediction in Germany. *J. Meteorol. Soc. Japan. Ser. II* **2004**, *82*, 361–370. [CrossRef]
63. Douša, J.; Václavovic, P.; Zhao, L.; Kačmařík, M. New adaptable all-in-one strategy for estimating advanced tropospheric parameters and using real-time orbits and clocks. *Remote Sens.* **2018**, *10*, 232. [CrossRef]

64. Böhm, J.; Niell, A.; Tregoning, P.; Schuh, H. Global Mapping Function (GMF): A new empirical mapping function based on numerical weather model data. *Geophys. Res. Lett.* **2006**, *33*, 7. [CrossRef]
65. Böhm, J.; Werl, B.; Schuh, H. Troposphere mapping functions for GPS and very long baseline interferometry from European Centre for Medium-Range Weather Forecasts operational analysis data. *J. Geophys. Res. Solid Earth* **2006**, *111*, B2. [CrossRef]
66. Kouba, J. Implementation and testing of the gridded Vienna Mapping Function 1 (VMF1). *J. Geod.* **2008**, *82*, 193–205. [CrossRef]
67. Zus, F.; Bender, M.; Deng, Z.; Dick, G.; Heise, S.; Shang-Guan, M.; Wickert, J. A methodology to compute GPS slant total delays in a numerical weather model. *Radio Sci.* **2012**, *47*, 1–15. [CrossRef]
68. Zus, F.; Dick, G.; Douša, J.; Heise, S.; Wickert, J. The rapid and precise computation of GPS slant total delays and mapping factors utilizing a numerical weather model. *Radio Sci.* **2014**, *49*, 207–216. [CrossRef]
69. Zus, F.; Balidakis, K.; Dick, G.; Wilgan, K.; Wickert, J. Impact of tropospheric mismodelling in GNSS precise point positioning: A simulation study utilizing ray-traced tropospheric delays from a high-resolution NWM. *Remote Sens.* **2021**, *13*, 3944. [CrossRef]
70. Luzum, B.; Petit, G. The IERS Conventions (2010): Reference systems and new models. *Proc. Int. Astron. Union* **2012**, *10*, 227–228. [CrossRef]
71. Balidakis, K. On the Development and Impact of Propagation Delay and Geophysical Loading on Space Geodetic Technique Data Analysis. Doctoral Thesis, Technische Universität Berlin, Berlin, Germany, 2019. [CrossRef]
72. Tavenard, R.; Faouzi, J.; Vandewiele, G.; Divo, F.; Androz, G.; Holtz, C.; Payne, M.; Yurchak, R.; Rußwurm, M.; Kolar, K.; et al. Tslearn, A Machine Learning Toolkit for Time Series Data. *J. Mach. Learn. Res.* **2020**, *21*, 1–6.
73. Oppenheim, A.V.; Schafer, R.W.; Buck, J.R. *Discrete-Time Signal Processing*; Prentice Hall: Upper Saddle River, NJ, USA, 1999.
74. Virtanen, P.; Gommers, R.; Oliphant, T.E.; Haberland, M.; Reddy, T.; Cournapeau, D.; Burovski, E.; Peterson, P.; Weckesser, W.; Bright, J.; et al. SciPy 1.0: Fundamental Algorithms for Scientific Computing in Python. *Nat. Methods* **2020**, *17*, 261–272. [CrossRef]
75. Taylor, S.J.; Letham, B. Forecasting at scale. *Am. Stat.* **2018**, *72*, 37–45. [CrossRef]
76. Hartmann, T.; Wenzel, H.G. The HW95 tidal potential catalogue. *Geophys. Res. Lett.* **1995**, *22*, 3553–3556. 95GL03324. [CrossRef]
77. Clauset, A.; Shalizi, C.R.; Newman, M.E. Power-law distributions in empirical data. *SIAM Rev.* **2009**, *51*, 661–703. [CrossRef]
78. Balidakis, K.; Sulzbach, R.; Shihora, L.; Dahle, C., Dill, R.; Dobslaw, H. Atmospheric Contributions to Global Ocean Tides for Satellite Gravimetry. *Journal of Advances in Modeling Earth Systems* **2022**. e2022MS003193. [CrossRef]
79. Butterworth, S. On the theory of filter amplifiers. *Wirel. Eng.* **1930**, *7*, 536–541.
80. White, E.P.; Enquist, B.J.; Green, J.L. On estimating the exponent of power-law frequency distributions. *Ecology* **2008**, *89*, 905–912. [CrossRef]
81. Di Matteo, S.; Viall, N.M.; Kepko, L. Power Spectral Density Background Estimate and Signal Detection via the Multitaper Method. *J. Geophys. Res. Space Phys.* **2021**, *126*, e2020JA028748. [CrossRef]
82. Glickman, T.S.; Zenk, W. *Glossary of Meteorology*; AMS (American Meteorological Society): Boston, MA, USA, 2000; ISBN 978-1-87-822034-9.
83. Leckner, B. The spectral distribution of solar radiation at the Earth's surface—elements of a model. *Sol. Energy* **1978**, *20*, 143–150. [CrossRef]
84. Wang, S.; Xu, T.; Nie, W.; Jiang, C.; Yang, Y.; Fang, Z.; Li, M.; Zhang, Z. Evaluation of precipitable water vapor from five reanalysis products with ground-based GNSS observations. *Remote Sens.* **2020**, *12*, 1817. [CrossRef]

remote sensing

MDPI

Article

Automated Small River Mapping (ASRM) for the Qinghai-Tibet Plateau Based on Sentinel-2 Satellite Imagery and MERIT DEM

Xiangan Liang [1], Wei Mao [2], Kang Yang [2,3,4] and Luyan Ji [5,6,*]

1 Ministry of Education Key Laboratory for Earth System Modeling, Department of Earth System Science, Tsinghua University, Beijing 100084, China
2 School of Geography and Ocean Science, Nanjing University, Nanjing 210023, China
3 Jiangsu Provincial Key Laboratory of Geographic Information Science and Technology, Nanjing 210023, China
4 Southern Marine Science and Engineering Guangdong Laboratory (Zhuhai), Zhuhai 519080, China
5 Aerospace Information Research Institute, Chinese Academy of Sciences, Beijing 100094, China
6 Key Laboratory of Technology in Geo-Spatial Information Processing and Application System, Chinese Academy of Sciences, Beijing 100094, China
* Correspondence: jily@mail.ustc.edu.cn

Abstract: The dynamic variation in the water surfaces of the river networks within the Qinghai-Tibet Plateau affects the water resource availability for downstream ecosystems and human activities. Small rivers (with a river width less than 30 m) are an important component of this network, but are difficult to map in the Qinghai-Tibet Plateau. Firstly, the width of most rivers is very narrow, at around 20 m, which appears as only one or two pixels in Sentinel-2 images and thus is susceptible to salt-and-pepper noise. Secondly, local mountain shadows, cloud shadows, and snow pixels have spectral characteristics similar to those of rivers, leading to misclassification. Therefore, we propose an automated small river mapping (ASRM) method based on Sentinel-2 imagery to address these two difficulties. A preprocessing procedure was designed to remove the salt-and-pepper noise and enhance the linear characteristic of rivers with specific widths. A flexible digital elevation model (DEM)-based post-processing was then imposed to remove the misclassifications caused by mountain shadows, cloud shadows, and snow pixels. The ASRM results achieved an overall accuracy of 87.5%, outperforming five preexisting remote sensing-derived river network products. The proposed ASRM method has shown great potential for small river mapping in the entire Qinghai-Tibet Plateau.

Keywords: small rivers; Sentinel-2; Google Earth Engine; Gabor filtering; HAND index

Citation: Liang, X.; Mao, W.; Yang, K.; Ji, L. Automated Small River Mapping (ASRM) for the Qinghai-Tibet Plateau Based on Sentinel-2 Satellite Imagery and MERIT DEM. *Remote Sens.* **2022**, *14*, 4693. https://doi.org/10.3390/rs14194693

Academic Editor: Chang-Keun Song

Received: 31 July 2022
Accepted: 13 September 2022
Published: 20 September 2022

1. Introduction

Terrestrial river networks store and transport large amounts of water and surface materials, and are an important part of the Earth's biochemical cycle [1–3]. They influence global climate change processes and ecosystem nutrient balances [4,5]. Rivers with channel patterns can be clustered into heterogeneous types based on their morphological characteristics and flow behaviors [6], and can be analyzed through automated and semi-automated approaches [7] using various data sources such as historical maps [8]. With the rapid development of earth observation technology, satellite remote sensing images have become an important data source for monitoring the dynamic changes of terrestrial river networks [9–11].

In recent years, many studies used satellite remote sensing imagery to produce remote sensing data for terrestrial river networks. Yamazaki et al. (2015) produced a Global 3 arc-second Water Body Map (G3WBM, 90 m) dataset using long time series, multi-seasonal Landsat imagery. The Global 3 arc-second Water Body Map (G3WBM, 90 m) dataset used long time series of multi-seasonal Landsat imagery to distinguish permanent from seasonal water bodies based on their frequency of occurrence in the images [12]. Gong et al. (2017) produced a 10-m spatial resolution global land cover product based on Sentinel-2 (10 m) imagery from

2017, and a multiseasonal, multiscale (from 30 m × 30 m to 500 m × 500 m) training sample of global land cover types sampled by visual interpretation using a supervised classification approach, named the Finer Resolution Observation and Monitoring-Global Land Cover (FROM-GLC10) [13], which labelled the global surface water distribution extent with high accuracy. Pekel et al. (2016) used nearly three million Landsat images to produce the Global Surface Water (GSW, 30 m) dataset [14], which contains month-by-month global surface water extraction results from the past 37 years and classifies permanent and seasonal water bodies according to their frequency of occurrence, supporting studies on water conservation management and decision-making, biodiversity conservation, and climate change. GSW records the spatial distribution and dynamics of global water bodies from 1984 to 2020 (37 years). Allen and Pavelsky (2018) produced the Global River Widths from Landsat (GRWL, 30 m) dataset containing river centerline and river width attributes using 7376 Landsat images and field measurements from 3963 hydrographic stations [15]. The total global river area estimated based on the GRWL dataset is more than 44% larger than the results of previous studies. The existing river network remote sensing data products are mainly based on Landsat imageries with a spatial resolution of 30 m. As a result, small rivers with a width of 30 m or less are ignored [16,17]. However, compared with large rivers, small rivers have active ecosystems and frequent land-air interactions [18]. Moreover, ignoring small rivers can cause underestimation of the role of terrestrial river networks in geochemical cycles [19]. Therefore, it is important to use high-resolution satellite imagery to monitor the dynamics of the water surfaces of the small rivers [20].

The Qinghai-Tibet Plateau in China, which accounts for 1/4 of its land area, stores a large number of water resources in the form of glaciers and snow, and is the birthplace of China's major rivers, such as the Yangtze, Yellow, Lantsang, and Tarim rivers [21]. It has a dense network of rivers with different river morphologies, and the spatial and temporal distribution of its rivers dynamically affects the downstream ecosystems and human activities [22–24]. Therefore, water monitoring on the Qinghai-Tibet Plateau is crucial for China's water security. However, the Qinghai-Tibet Plateau has steep terrain, narrow river channels, and widespread small rivers. Additionally, shadows from mountains and clouds and snow cover inevitably influence the remote sensing monitoring of river networks [25,26]. Therefore, developing an advanced method for mapping small rivers on the Qinghai-Tibet Plateau is necessary to improve our understanding of the spatial and temporal dynamic characteristics in water surfaces of the river networks on the Qinghai-Tibet Plateau and quantify their impacts on downstream ecosystems.

2. Materials and Methods

2.1. Study Area

Zagunao Basin (about 2391 km^2) is a sub-basin of the Yangtze River Basin within the Tibetan Plateau (Figure 1) and was selected as the study area. The Yangtze River is one of the five major rivers in the Tibetan Plateau outflow area. The terrain of the study area is steep. The river networks are complex, and the rivers are mostly narrow and small, with an average width of 20 m (Figure 1d) as observed from the existing remote sensing-derived product Finer Resolution Observation and Monitoring-Global Land Cover (FROM-GLC10). Additionally, the study area has an extensive snow cover, and the surface water recharge is snowmelt runoff, which is regulated by temperature. The flood season is usually in summer (June–August), while the dry season is usually in winter (December–February) [27].

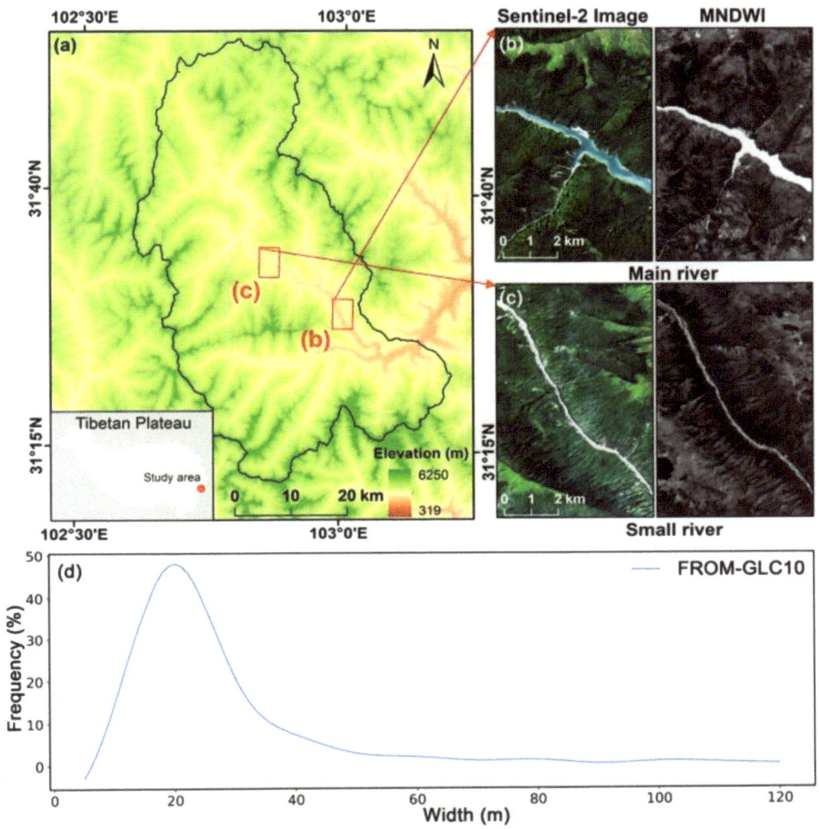

Figure 1. Location of the study area. (**a**) Location of the study area; (**b**) Main river channel (the confluence of more than two tributaries with wider river width) within the study area; (**c**) Small river (<30 m) channel within the study area; (**d**) Distribution of river widths in the study area. The river widths were extracted from FROM-GLC10.

2.2. Data

2.2.1. Satellite Data

Optical remote sensing images of the Sentinel-2 satellite with 10-m spatial resolution were used for small river monitoring. Sentinel-2 acquires images with a Multispectral Imager (MSI) and is mainly used for terrestrial observation [28]. There are two Sentinel-2 satellites: Sentinel-2A (launched in 2015) and Sentinel-2B (launched in 2017). The Sentinel-2 MSI includes visible and shortwave-infrared bands: Band 2 (blue, 10 m), Band 3 (green, 10 m), Band 4 (red, 10 m), and Band 11 (SWIR, 20 m) [29]. Sentinel-2 has a relatively high spatial resolution among the currently freely available satellite images and has a broad application prospect in the global remote sensing monitoring of environmental resources [30]. Sentinel-2 imagery from May 2017 to October 2020 was acquired and mosaiced to a monthly time step using the GEE platform to produce monthly cloud-free synthetic images for the study area, which were then processed for the extraction and analysis of the river networks.

2.2.2. Digital Elevation Model (DEM)

The MERIT DEM (Multi-Error-Removed Improved-Terrain DEM) is a hydrologically adjusted DEM produced from the MERIT Hydro dataset with a spatial resolution of 30 m that is particularly suitable for hydrological analysis [31]. We selected MERIT DEM to build the river ROI because it is specially developed for hydrological analysis by reducing error components from existing DEM products, in which river networks and hill-valley structures are clearly represented. The vertical error of MERIT DEM is ±12 m (90th percentile of the error range) [32]. Therefore, we suggest it possesses sufficient accuracy to create river AOIs.

The Height Above the Nearest Drainage Network (HAND) is a topographic index based on topographic data used to simulate the possible occurrence of water bodies [33]. It is widely used in the fields of soil moisture simulation, flood prediction, and assessment [33]. The HAND index has also been used for Arctic river mapping [34]. The higher the difference in elevation, the higher the gravitational potential energy, and the more work is required to transport water from the river to a location, so a lower soil-water content indicates a lower possibility of the presence of a water body [33]. The HAND data produced from the MERIT DEM were obtained from the MERIT Hydro dataset on the GEE platform and used as auxiliary data to remove mountain and cloud shadows as well as snow cover.

The *HAND* index is calculated as follows:

$$HAND = E_i - E_{nearest} \quad (1)$$

where E_i denotes the elevation of each pixel in DEM. The $E_{nearest}$ is the elevation of the nearest pixel of the DEM-modeled river network.

2.2.3. Existing Remote Sensing River Network Products

Five existing river network remote sensing products were obtained for comparison with ASRM (Table 1).

Table 1. Existing river network remote sensing products.

Name	Data Source	Resolution	Introduction	Reference
FROM-GLC10	Sentinel-2 MSI	10 m	Global land cover data products	Gong et al., 2019 [13]
GSW	Landsat MSS/TM/ETM+/OLI	30 m	Long time series monthly surface water cover	Pekel et al., 2016 [14]
GRWL	Landsat MSS/TM/ETM+/OLI	30m	Vector data product with river width	Allen et al., 2018 [15]
OSM	Aerial, satellite imagery, GPS devices and in situ observation data	None	Vector data product based on opensource community contributions	https://www.openstreetmap.org/, accessed on 1 January 2021
HydroSHEDS	SRTM DEM	450, 900 m	Continuous river networks generated by DEM	Lehner et al., 2008 [35]

(1) FROM-GLC10: 10-m spatial resolution global land cover product

Finer Resolution Observation and Monitoring-Global Land Cover (FROM-GLC10) is a 10-m spatial resolution global land cover product that includes water bodies identified from Sentinel-2 imagery [13]. The overall accuracy of all land cover types tested was 72.76%.

(2) GSW: Long-time-series remote sensing dataset

Global Surface Water (GSW) is global surface water remote sensing dataset produced based on nearly three million Landsat images [14]. The dataset contains the global monthly surface water extents for the past 37 years. This dataset contains sub-datasets of monthly surface water extents, the annual frequencies of water bodies, and the maximum water body distribution range, all with a spatial resolution of 30 m. In this paper, the "maximum water body distribution range" data in the GSW dataset were selected for comparison with ASRM.

(3) GRWL: A global river network dataset with river width

Global River Widths from Landsat (GRWL) is a global river network dataset with river width attributes [15]. The dataset is based on over 7000 cloud-free Landsat images acquired during the normal water flow period and field measurements from nearly 4000 hydrological stations, combined with statistical models to model the centerline locations and width of the global rivers. GRWL is the first global river network dataset with river width attributes.

(4) OSM: Real-time update of a global geographic dataset

OpenStreetMap (OSM) is free open-source dataset. Every member of the OSM community can upload and maintain the data, including rivers, lakes, roads, railroads, etc. Contributors combine aerial and satellite imagery, GPS devices, and field work to verify and correct the data to ensure accuracy and timeliness. The river network dataset in vector format from the OSM community was acquired for this paper.

(5) HydroSHEDS: A global river network dataset produced from DEM

HydroSHEDS is a global river network dataset produced using Shuttle Radar Topography Mission (SRTM) DEM data [35]. Since the HydroSHEDS dataset is generated from DEM data modelling, it contains only river centerline information, and lacks information on the actual extents of the rivers.

2.2.4. Validation Data

We extracted samples from various months (Figure 2b) and manually classified the pixels as rivers or background (i.e., non-river pixels) referring to the original Sentinel-2 image. A total of 137,357 pixels were selected as validation samples, 73,254 of which corresponded to river pixels.

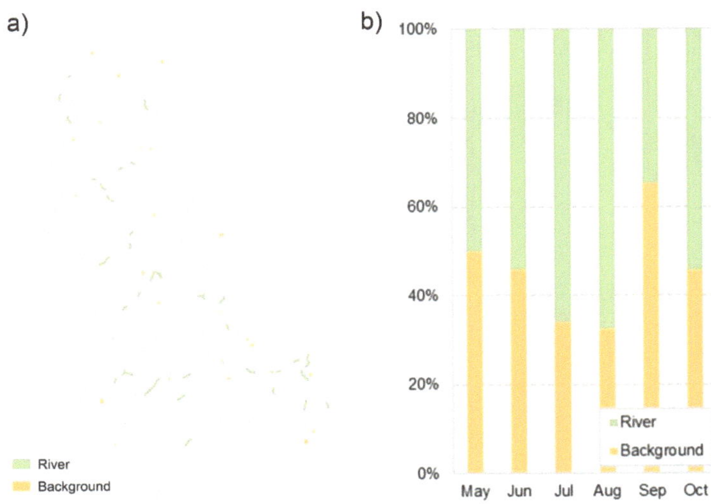

Figure 2. Validation samples. (**a**) Distribution of validation samples collected in July 2018; (**b**) Ratio of river pixels to background pixels in samples from various months.

2.3. *Automated Small River Mapping*

Due to the unique climatic and topographical conditions of the Tibetan Plateau, there are two main difficulties in mapping small rivers. Firstly, the widths of the rivers are very narrow (around 20 m) (Figure 1d). They appear in the images as one or two pixels and are susceptible to salt-and-pepper noise from unexpected atmospheric conditions and quality issues with the electronic sensor during the image production process. Secondly, local mountain shadows, cloud shadows and snow cover have similar spectral characteristics to those of rivers. To address these two problems, a preprocessing procedure was imposed to remove the salt-and-pepper noise and enhance the linear characteristics of the rivers. Additionally, a DEM-based post-processing was imposed to crop the misclassifications caused by mountain shadows, cloud shadows, and snow cover. Then, the river network map generated by ASRM was compared with the existing five products.

2.3.1. Water Index

The first step is to calculate the MNDWI (Modified Normalized Difference Water Index) (Figure 3b), which is an improved normalized water index based on the NDWI (Normalized Difference Water Index) [36], both of which take advantage of the spectral characteristics of water bodies with very low reflectance in the short-wave infrared band and high reflectance in the visible bands. The MNDWI is constructed by adjusting the combination of the bands that comprise the NDWI [37]. In particular, MNDWI can better distinguish between shadows and water bodies, and is calculated as follows:

$$\mathrm{MNDWI} = \frac{\mathrm{Green} - \mathrm{SWIR}}{\mathrm{Green} + \mathrm{SWIR}} \tag{2}$$

where Green and SWIR denote the reflectance in the green and shortwave infrared bands, corresponding to Band 3 and Band 11 of Sentinel-2 images, respectively.

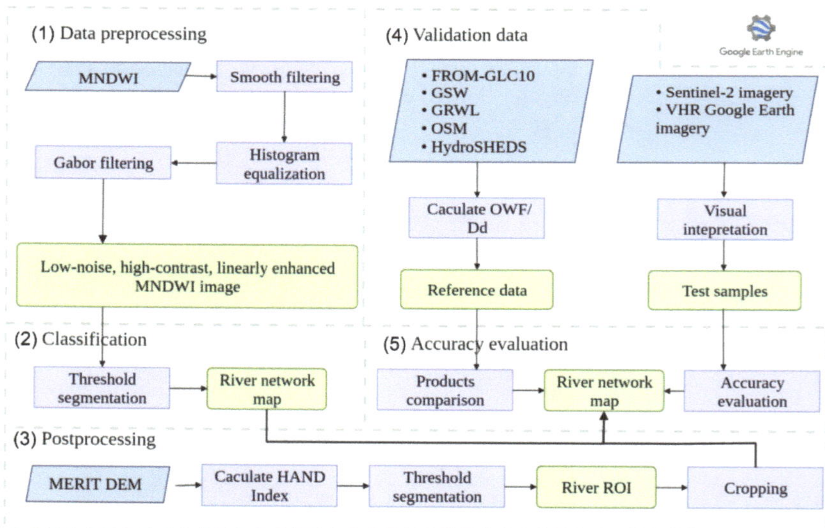

Figure 3. Workflow of remote sensing river extraction.

2.3.2. River Enhancement

The salt-and-pepper effect increases the difficulty of distinguishing small rivers. Therefore, mean smooth filtering with a window size of 3 × 3 is imposed to suppress the salt-and-pepper noise (Figure 4). However, the smoothing process also blurs river boundaries. Therefore, we use a histogram equalization algorithm to enhance the local contrast between the water boundary and the background. Then, a Gabor filter is applied to enhance the linear characteristics.

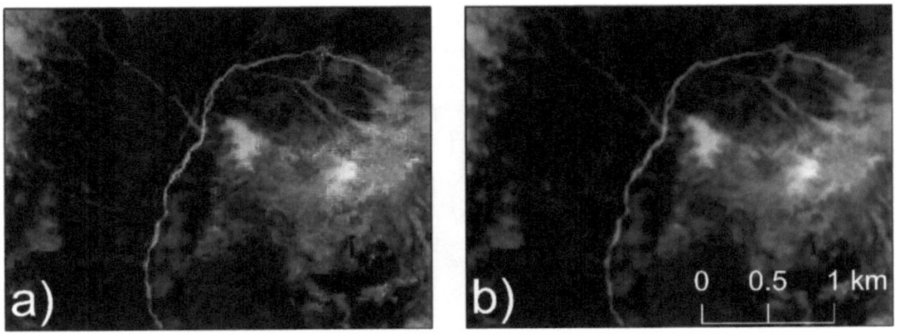

Figure 4. Salt-and-pepper noise removal using smooth filtering. (**a**) MNDWI before smoothing. (**b**) MNDWI after smoothing.

A Gabor filter is a band-pass filter that enhances the linear characteristics of a specific width (20 m for this study) while suppressing irrelevant linear elements (e.g., feature edges). It is often used in the delineation of linear blood vessels in medical imaging [38]. Since rivers and blood vessels have similar linear features, it is also used for river identification [39]. Here, Gabor filtering is imposed to enhance the linear characteristics of small rivers in MNDWI images so that the contrast between small linear rivers and background is enhanced.

The Gabor filter kernel with an angle of $\theta = -\pi/2$ can be expressed as:

$$g(x,y) = \frac{1}{2\pi\sigma_x\sigma_y}\exp\left[-\frac{1}{2}\left(\frac{x^2}{\sigma_x^2}+\frac{y^2}{\sigma_y^2}\right)\right]\cos(2\pi f_0 x) \quad (3)$$

where σ_x, σ_y is the standard deviation in the x, y-direction, and f_0 is the frequency of the modulated sinusoidal curve. The setting of these three parameters can be adjusted to the specific width of the linear features to be enhanced. Assuming that W is the width of the river to be enhanced, the bandwidth w of the Gabor filter and W satisfy Equation (3). The values of σ_x, σ_y, and f_0 satisfy Equations (4) and (5).

$$W = 2w + 1. \quad (4)$$

$$\sigma_x = \sigma_y = \frac{w}{2\sqrt{2ln2}} \quad (5)$$

$$f_0 = \frac{1}{w} \quad (6)$$

Since Gabor filtering is performed in the null domain, it is necessary to place $g(x,y)$ according to a certain angle θ to enhance the river lines in different directions. The rotation, ($\theta \in [-\pi/2, \pi/2]$) equation is as follows:

$$\begin{cases} g^{\theta}(x',y') = g(x,y) \\ x' = xcos\theta + ysin\theta \\ y' = ycos\theta - xsin\theta \end{cases} \quad (7)$$

Depending on the rotation angle, the filter can observe the river in the direction that is not used. For the input image $f(x,y)$, the filter will respond with the following equation:

$$r^{\theta}(x,y) = g^{\theta}(x,y) * f(x,y) \quad (8)$$

The final maximum response in all directions is expressed as:

$$r(x,y) = \max\left(r^{\theta}(x,y)\right), \theta \in \left[-\frac{\pi}{2}, \frac{\pi}{2}\right] \quad (9)$$

After experiments, the filter performs most successfully for rivers with a width of 20 m when w = 2, θ = 15°. The result after applying the Gabor filter is shown in Figure 5.

The river is more clearly differentiated from the background, and the connectivity of the river is further improved. Moreover, the range of pixel values is set to 0–255.

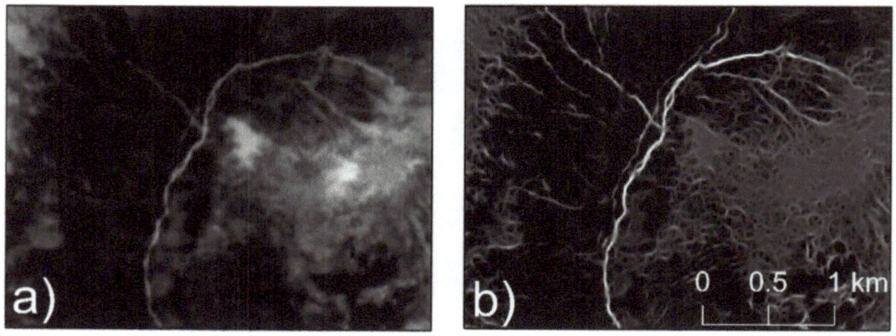

Figure 5. Performance of Gabor filtering. (**a**) MNDWI before filtering. (**b**) MNDWI after filtering.

After manual testing at 0.5 intervals, a threshold of 10 was applied on the low-noise high-contrast and on the linear enhanced MNDWI image for river segmentation. Pixels with values higher than 10 were classified as river pixels, while others were classified as non-river pixels. However, the interferences from mountain shadows, cloud shadows, and snow cover remain unremoved and will be handled in the post-processing.

2.3.3. Post-Processing

The preliminary remote sensing river networks have been obtained after threshold segmentation of the Gabor-filtered MNDWI (Figure 6b). However, there is still extensive noise caused by mountain shadows, cloud shadows, and snow cover. The spectrum of shadows will vary depending on the underlying surface. The shadow overlying on the vegetation and snow cover would have a quite low reflectance in all bands with a higher reflectance in the visible bands than in the SWIR band, which is an effect similar to that of water. While the shadow overlying on the bare land has a lower reflectance in the visible bands than in the SWIR band, snow has a much higher reflectance in the visible bands than in the SWIR band. Although the reflectance value of snow is greater than that of water (especially the visible part), the MNDWI of snow and water are similar. Since we use the MNDWI index in this paper, both snow and shadows can be misclassified as rivers. Therefore, DEM data were used as auxiliary data to post-process the river extraction results and to remove the noises caused by shadows and snow cover.

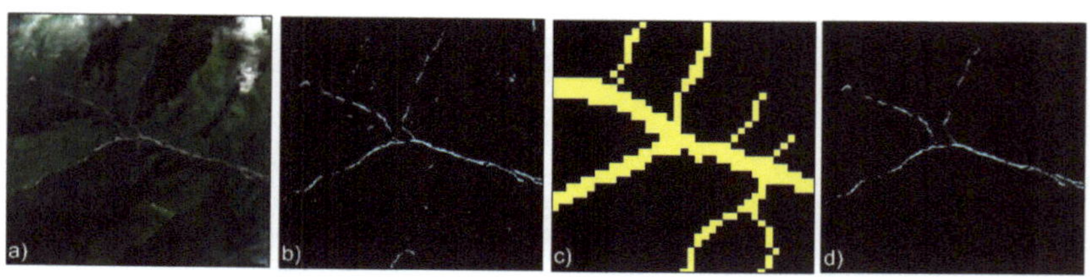

Figure 6. Classification removal based on DEM-modelled area of interest (AOI). (**a**) Original Sentinel-2 image. RGB: band 4 (Red), band 3 (Green), band 2 (Blue). (**b**) Results before cropping. (**c**) AOI generated from HAND index. (**d**) Cropped results.

Lu et al. [40] used a DEM to model the central line of a river channel and generated a 50-m buffer on each side of the river. The water pixels outside the AOI were considered to be misclassified. However, the method did not consider different river widths. Creating buffers with uniform distances along different parts of the rivers may cause the partial cropping out of wider tributaries while retaining the noise around the small tributaries.

To overcome this limitation, the HAND index was introduced, which considers the possibility of a river's presence (Figure 6) to create river AOIs of different widths [38]. We used a relatively high threshold of 50 m for the segmentation of HAND imaging to ensure all river pixels are within the AOI, while the pixels with a value of lower than 50 m were considered as a part of river AOI. The AOIs were used to crop the preliminary results of the extraction (Figure 6b) to obtain the final images. After cropping, the noise caused by shadows and snow cover were removed, while the river networks were kept intact (Figure 6d).

2.4. Accuracy Evaluation

We constructed a validation sample set by randomly and uniformly sampling a total of 137,357 pixels from different months (Figure 2b), identifying the river pixels by manual visual interpretation and by referring to the original Sentinel-2 image.

Based on the manually interpreted validation data, we calculated the user accuracy, producer accuracy, and overall accuracy to evaluate the reliability of ASRM results. The formulae are as follows:

$$\text{User accuracy} = \frac{Number\ of\ Correct\ river\ pixels}{Number\ of\ Correct{+}misclassified\ pixels} \tag{10}$$

$$\text{Producer accuracy} = \frac{Number\ of\ Correct\ river\ pixels}{Number\ of\ Total\ river\ pixels} \tag{11}$$

$$\text{Overall accuracy} = \frac{Number\ of\ Correctly\ predicted\ pixels}{Number\ of\ Total\ pixels} \tag{12}$$

The Kappa coefficient is calculated as:

$$\text{Kappa coefficient} = \frac{N\sum_{i=1}^{n} m_{i,i} - \sum_{i=1}^{n}(G_i C_i)}{N^2 - \sum_{i=1}^{N}(G_i C_i)} \tag{13}$$

where i is the number of classes, N is the total number of pixels, $m_{i,i}$ is the number of pixels correctly predicted as class I, C_i is the total number of pixels predicted as class I, and G_i is the total number of truth values belonging to class i.

3. Results

3.1. Results of ASRM

The overall accuracy of the ASRM reached 87.5%, and the Kappa coefficient was 0.75. The producer accuracy of river classification was 79.25%, and the user accuracy was 96.82% (Table 2), which indicates that there is more leakage of river pixels than misclassification. Most of the unidentified river sections were small rivers with a width less than or equal to the spatial resolution (10 m) of the Sentinel-2 images. These small rivers were represented as mixed pixels in the images, and their spectral characteristics differed less from the background features, rendering them difficult to identify. In addition, we validated the performance per month using the samples collected in the corresponding months (Figure 7).

Table 2. Validation of the accuracy of the proposed method (Unit: pixels).

	Background	River	Overall	User Accuracy
Background	62,194	15,197	77,391	80.36%
River	1909	58,057	59,966	96.82%
Overall	64,103	73,254	137,357	
Producer Accuracy	97.02%	79.25%		
Total accuracy: 87.5% Kappa coefficient: 0.75				

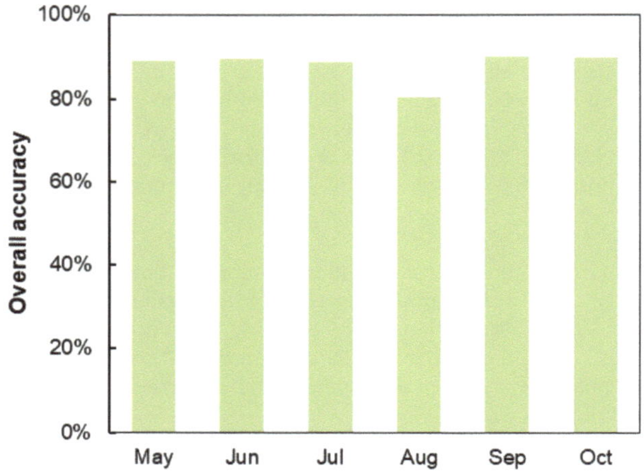

Figure 7. Validation of the accuracy of the proposed method per month using the samples collected in the corresponding months.

Further, we evaluated the performance of the ASRM method in different situations. When there were residential areas around the river (Figure 8), the residential areas would potentially affect river mapping, as the edge of the buildings would be enhanced through the Gabor filtering. ASRM identified the intact river channels and removed the residential areas with non-linear characteristics.

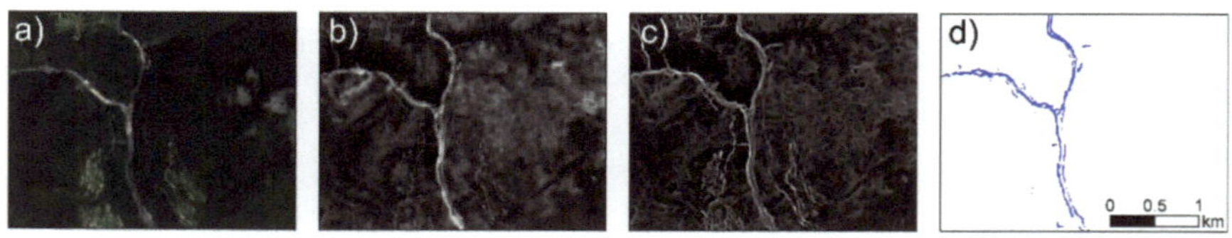

Figure 8. ASRM results in the presence of residential areas. (**a**) Original Sentinel-2 image. RGB: band 4 (Red), band 3 (Green), band 2 (Blue). (**b**) MNDWI image. (**c**) Gabor-filtered MNDWI image. (**d**) ASRM results.

To eliminate the influence of mountain shadows (Figure 9), Gabor filtering can enhance the contrast between linear rivers and the nearby pixel clusters of mountain shadows. The post-processing further cropped the mountain shadows far from the central line of the river channel.

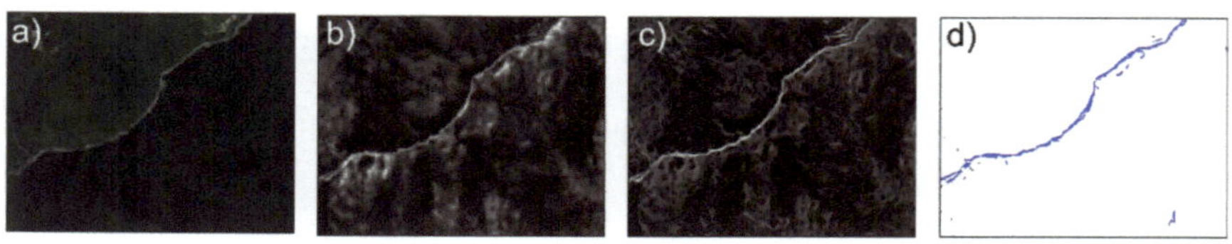

Figure 9. ASRM results in the presence of mountain shadow. (**a**) Original Sentinel-2 image. RGB: band 4 (Red), band 3 (Green), band 2 (Blue). (**b**) MNDWI image. (**c**) Gabor filtered MNDWI image. (**d**) ASRM results.

In the presence of cloud cover (Figure 10), although the spectral characteristics of clouds and cloud shadows are similar to those of rivers, the result contained few misclassifications. The misclassifications caused by cloud cover were cropped by the DEM-modelled AOI because they were far from the central line of the river.

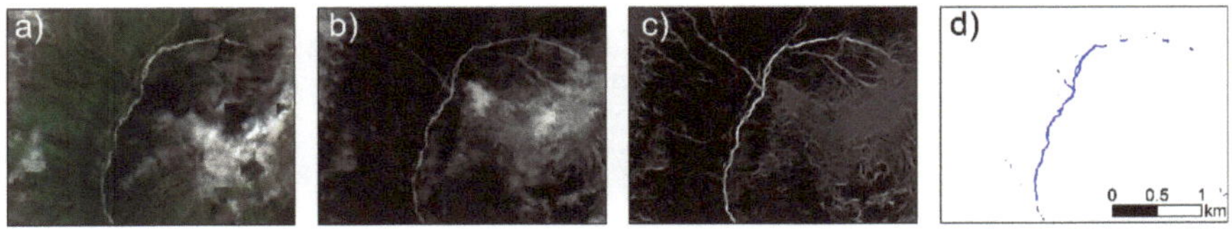

Figure 10. ASRM results in the presence of cloud cover. (**a**) Original Sentinel-2 image. RGB: band 4 (Red), band 3 (Green), band 2 (Blue). (**b**) MNDWI image. (**c**) Gabor filtered MNDWI image. (**d**) ASRM results.

3.2. Assessment against Existing River Network Products

Five existing remote sensing data products of river network (FROM-GLC10, GSW, GRWL, HydroSHEDS, OSM) were quantitatively compared with ASRM (Figure 11). The results demonstrate that the accuracy of ASRM is superior to those of the other data products (Table 3). In addition, to compare the completeness and continuity of river networks in different data products, drainage density (Dd, the ratio of river length to watershed area) and Open Water Fraction (OWF, the ratio of water body area to watershed area) (Figure 12) were calculated for each product. For the GSW product, we utilized the maximum water extent to form the comparison. For other products with only one period of results, we consider them to represent the maximum extent of water bodies detected in these datasets. We compared them with the ASRM results of 2018.08, which has the highest OWF among results of all months.

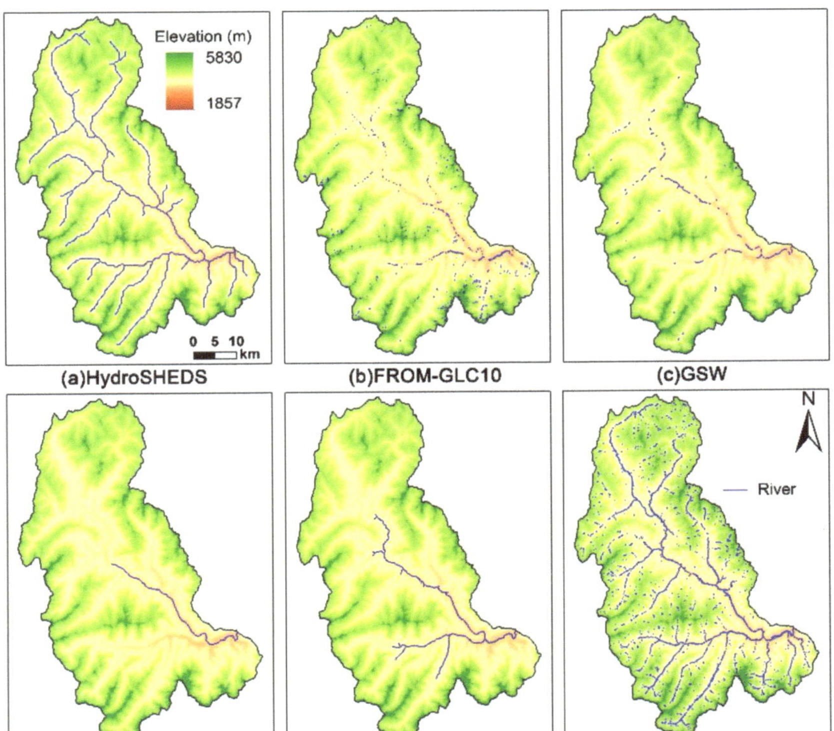

Figure 11. Comparison of ASRM with other remote sensing data products.

Table 3. Comparison of accuracy of various products.

Product	Producer Accuracy	User Accuracy	Total Accuracy
ASRM [1]	79.25%	96.82%	87.5%
FROM-GLC10	4.8%	92.9%	59.0%
GSW	37.1%	93.2%	69.6%
GRWL	11.6%	100.0%	57.0%

[1] Result of the proposed method.

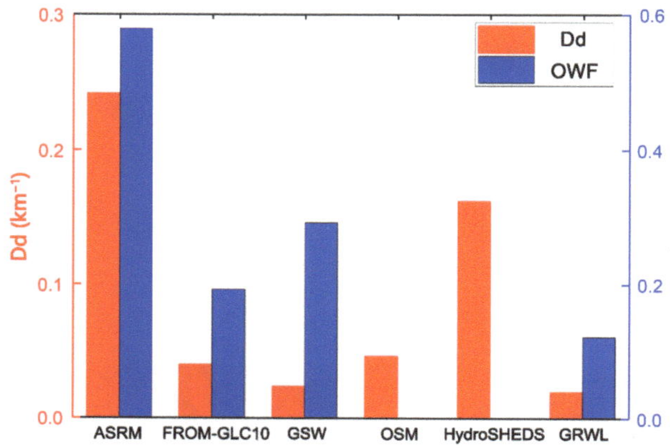

Figure 12. Comparison of ASRM with other data products for Dd and OWF.

Compared with FROM-GLC10, GSW, and GRWL, the river networks in ASRM are more complete and more consistent. The Dd of ASRM is 6 times higher than that of FROM-GLC10, 10.17 times higher than that of GSW, and 12.14 times higher than that of GRWL, while only 0.48 times higher than that of HydroSHEDS.

The OWF of ASRM is 6.84 times higher than that of FROM-GLC10, 1.98 times higher than that of GSW, and 12.81 times higher than that of GRWL.

The river width of both ASRM and the FROM-GLC10 dataset reached the peak frequency around 20 m and were mostly distributed in the range of 10–30 m (Figure 13). The peak frequency of river width in the GSW dataset was around 30 m because Landsat images identify rivers less than 30 m. As for the GRWL dataset, the peak frequency of river width is around 70 m.

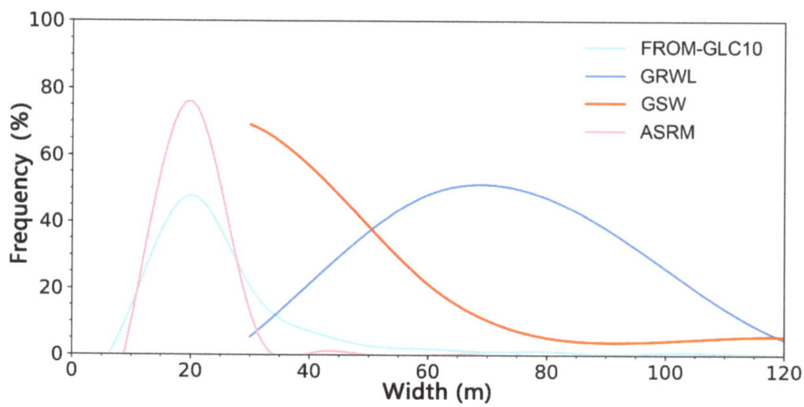

Figure 13. Comparison of ASRM with other data products on river width.

4. Discussion

4.1. Advantages of the Proposed Method over Existing Datasets

Considering the special geographical condition of the Qinghai-Tibet Plateau, there are two important advances in our approach. Firstly, to address the problem of narrow river channels caused by the steep terrain, we used Gabor filtering to enhance the linear features of small rivers. Secondly, we introduced the HAND index to build AOIs and mitigate the effects of mountain and cloud shadows and snow cover.

Gabor filtering enhanced the linear characteristics of small and medium-sized rivers in the image. Additionally, Sentinel-2 images (10 m) have higher spatial resolution compared with Landsat images (30 m), based on which the GSW and GRWL were produced. Therefore, the Dd of ASRM is much higher than that of GSW, GRWL and FROM-GLC10 (Figure 12). HydroSHEDS is produced from a DEM, which continuity ensures that its rivers features show fewer interruptions than remote sensing results, and for this reason, it should have a higher Dd. However, because it lacks the details of small tributaries, its Dd is still smaller than that of ASRM.

The OWF difference between ASRM and the GSW dataset is relatively small (1.98 times) compared to the Dd difference (Figure 12). The low spatial resolution of the Landsat imagery results in the overestimation of OWF due to the identification of rivers less than 30 m wide as 30 m, and the mixed pixels of river boundaries as rivers.

The study compared the river width attributes of the Sentinel-2 extraction results and other five global river network remote sensing data products (Figure 11) and found that Sentinel-2 could extract more rivers with river widths of less than 30 m, which compensated for the shortcoming of the Landsat imagery, which could only extract rivers wider than 30 m. For both the ASRM and the FROM-GLC10 extraction results, the peak frequency was around 20 m, while the peak frequency of GSW was around 30 m as a result of the Landsat images pixel 30-m resolution, assigning all rivers less than 30 m to a 30-m width. The peak frequency of the river width of GRWL was around 70 m, which is likely because the GRWL data product only identified the main streams with the largest flow connected to the pour point in the study area, while the small tributaries with smaller river widths upstream were not identified.

However, there are still several limitations of this method. First, the producer accuracy of river classification was 79.25% (Table 2), which indicates that a large portion of river pixels were not identified. Compared to the Sentinel-2 image, it is found that most of the unidentified river sections were small rivers with a width less than or equal to the spatial resolution (10 m) of the Sentinel-2 images. These small rivers were represented as mixed pixels in the images, and their spectral characteristics differed less from the background features, rendering them difficult to identify. Second, this method is based on a pixel-wise algorithm, leading to the disconnection in the river networks. To create a real river network, those river pixels must be interconnected to form a continuity. Third, our method can only monitor dynamic variation in the water surface of the rivers, instead of the transformation of the river courses. Therefore, our analysis is limited to the variation of the water surface rather than the geomorphological characteristics of the river network.

4.2. Dynamic Monitoring in Water Surface of the River Networks in the Study Area

The main source of surface water recharge in the basin is snowmelt runoff, which is seasonal. The proposed method automatically extracted the study area's river network from May to October during the 2017–2020 period (20 extractions in total, as there were no high-quality images in May, June, and July 2017, and October 2018) (Figure 14). Furthermore, the spatial distribution characteristics, as well as the seasonal and annual dynamic change characteristics of the variation in water surface of the river networks in the study area, were analyzed based on the extraction results.

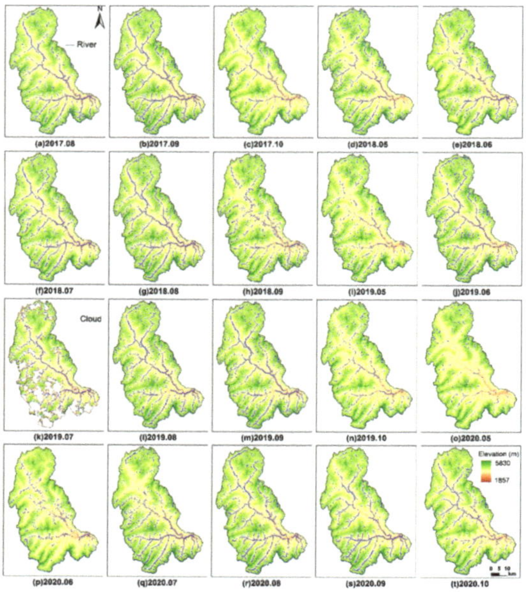

Figure 14. Remote sensing river network of the study area, 2017–2020. The cloud area of 2019.07 was cropped, and the remaining area was used to calculate the Dd and OWF.

4.2.1. Spatial Distribution Characteristics of the River Networks

The highest elevation along the river network is about 4800 m. The farthest distance from the pour point is 66.4 km (Figure 15). The study areas have steep terrain, and a large proportion (>15%) of the river networks are in areas with slopes greater than 30°. The tributary streams in the upper reaches are narrow and merge into the wider mainstream in the lower reaches. The mainstream with the highest flow and best continuity is distributed near the downstream pour point in the southeast part of the study area. All the water in the basin runs off through the pour point.

4.2.2. Dynamic Variation in Water Surface of the River Networks

Since our method can only monitor dynamic variation in water surface of the rivers instead of the transformation of the river courses, our analysis is limited to the variation of the water surface rather than the geomorphological characteristics of the river network.

(1) Annual Dynamic Variation

From 2017 to 2020, the Dd and OWF of the river networks were stable, exhibiting small annual differences (Table 4, Figure 16). The annual peak of OWF was highest in 2018 (0.58%) and lowest in 2020 (0.52%). In contrast, the peak of Dd in each year did not differ significantly.

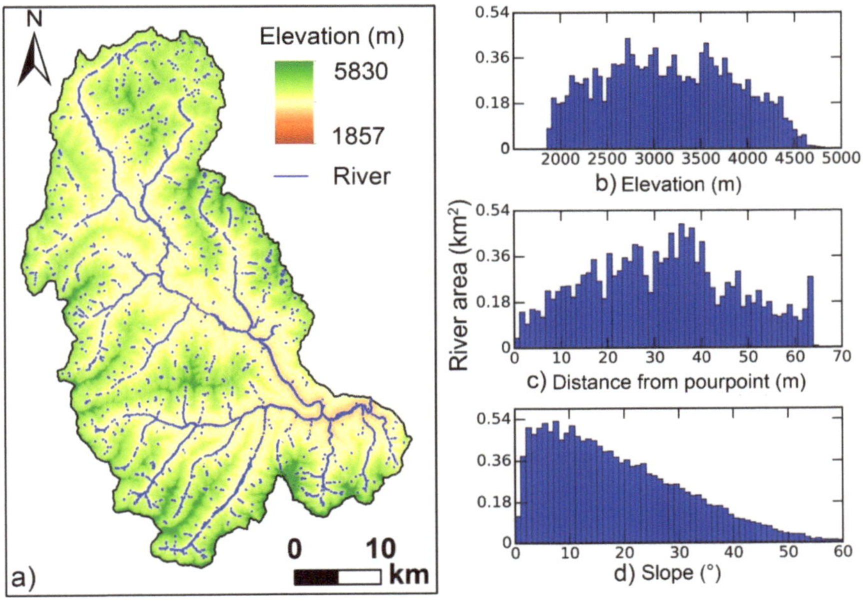

Figure 15. Spatial distribution of river networks. (**a**) River mask; (**b**) Elevation of river pixels; (**c**) Distance from water pixels to pour point; (**d**) Slope of river pixels.

Table 4. Statistics of OWF, Dd of river networks.

Date	OWF (%)	Dd (km^{-1})
2017/08	0.51	0.21
2017/09	0.56	0.25
2017/10	0.18	0.08
2018/06	0.37	0.16
2018/07	0.52	0.22
2018/08	0.58	0.24
2018/09	0.43	0.17
2019/05	0.24	0.11
2019/06	0.44	0.20
2019/07	0.40	0.17
2019/08	0.54	0.22
2019/09	0.26	0.22
2019/10	0.17	0.07
2020/05	0.12	0.05
2020/06	0.22	0.08
2020/07	0.49	0.20
2020/08	0.53	0.23
2020/09	0.49	0.21
2020/10	0.30	0.13

The trends of Dd and OWF are consistent, both experiencing an increase followed by a decrease during the summer (May to October), with the flood period in July-August and the dry period in May, September, and October.

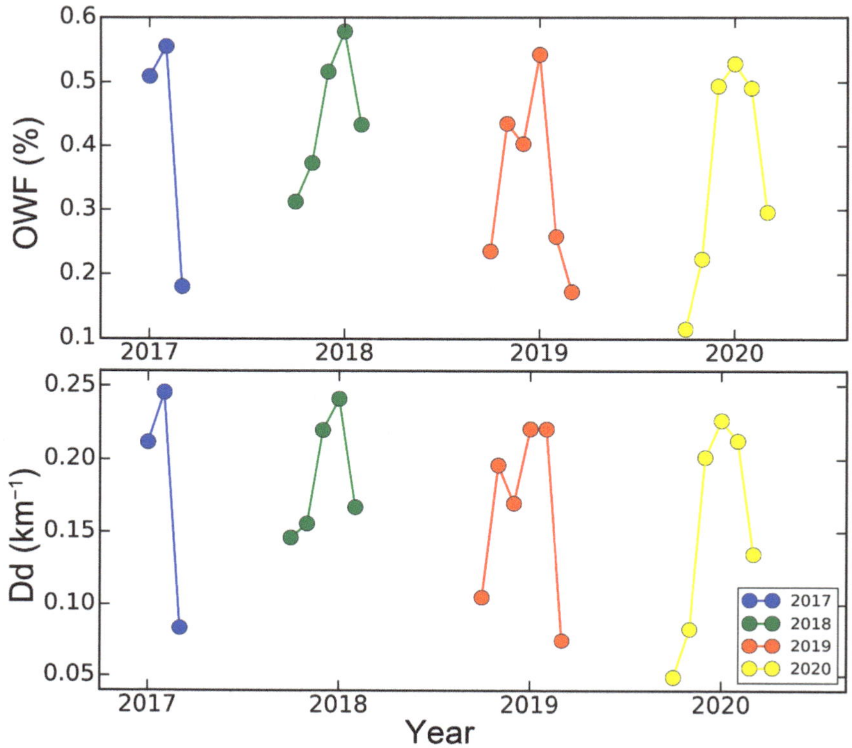

Figure 16. Seasonal variation in the water surfaces of the river networks in the study area.

(2) Seasonal Variation

Based on the proposed method, the dynamic variations in the water surfaces of the river networks in the study area were monitored and well-characterized during each summer between 2017–2020 (Table 5, Figure 17).

Table 5. Mean values of OWF and Dd from 2017 to 2020.

Month	OWF (%)	Dd (km^{-1})
May	0.22	0.10
June	0.34	0.15
July	0.47	0.20
August	0.54	0.23
September	0.43	0.21
October	0.22	0.10

May is the transition period from low to abundant water in the study area, and the lowest Dd and OWF are recorded in May. June is the period of rapid development of water surface, when the OWF is 56% higher than that of May, while Dd is 45% higher. In July and August, the high-water period, both the Dd and OWF reach their peak. September is the transition period from high-water to normal flow; finally, rivers gradually enter the low-flow period in October, and the OWF and Dd decrease by 0.22% and 0.098 km, respectively.

Sentinel-2 imagery displays successful performance in mapping the small rivers within the Qinghai-Tibet Plateau.

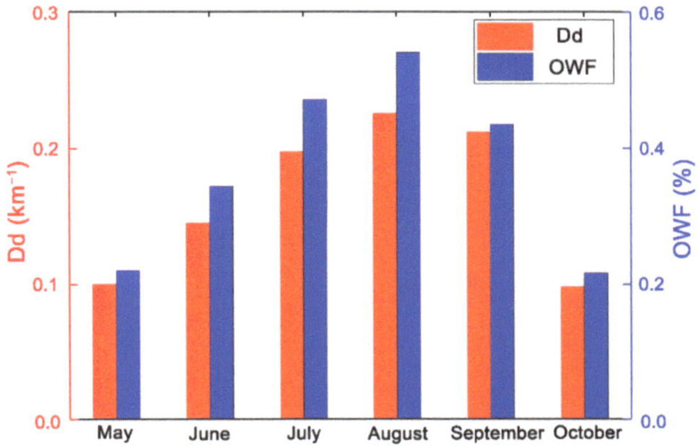

Figure 17. Monthly averages of OWF and Dd from 2017 to 2020.

5. Conclusions

In this study, we proposed ASRM, an advanced method used for small river mapping using 10-m Sentinel-2 imagery that can overcome the mapping challenges created by the special geographic conditions of the Qinghai-Tibet Plateau. The approach was evaluated with validation samples collected from different months and compared with five existing remote sensing products. Finally, the method was applied to a basin (~2391 km^2) within the Qinghai-Tibet Plateau, and the spatial distribution and dynamic variation in the water surfaces of the river networks in the study area from May to October in the period from 2017 to 2020 were analyzed. The results reveal that:

(1) ASRM achieved an overall accuracy of 87.5%, with more leakage of river pixels than misclassifications. The ASRM performed well in the presence of residential areas, mountain shadows, and cloud cover.

(2) Compared to five existing remote sensing products, ASRM identified more small rivers, providing more detailed and consistent maps. The Drainage density (Dd) of ASRM was more than six times that of other datasets, and the Open Water Fraction (OWF) was more than 1.9 times that of other datasets.

(3) From the perspective of interannual variations, the annual variations of maximum Dd and OWF of the river networks in the study area were less than 15% from 2017 to 2020. In terms of the seasonal variations, both Dd and OWF increased from May to August, and decreased monthly after August.

Author Contributions: Conceptualization, X.L., L.J. and K.Y.; methodology, X.L. and W.M.; software, W.M.; validation, L.J. and X.L.; formal analysis, X.L. and L.J.; investigation, X.L.; resources, K.Y.; data curation, W.M.; writing—original draft preparation, X.L.; writing—review and editing, L.J.; visualization, X.L.; supervision, L.J. and K.Y.; project administration, L.J.; funding acquisition, L.J. All authors have read and agreed to the published version of the manuscript.

Funding: This research was supported by the Second Tibetan Plateau Scientific Expedition and Research Program (STEP), Grant No. 2019QZKK0206.

Data Availability Statement: The data presented in this study are available on request from the corresponding author.

Conflicts of Interest: The authors declare no conflict of interest.

References

1. Benstead, J.P.; Leigh, D.S. An expanded role for river networks. *Nat. Geosci.* **2012**, *5*, 678–679. [CrossRef]
2. Raymond, P.A.; Hartmann, J.; Lauerwald, R.; Sobek, S.; McDonald, C.; Hoover, M.; Butman, D.; Striegl, R.; Mayorga, E.; Humborg, C.J.N. Global carbon dioxide emissions from inland waters. *Nature* **2013**, *503*, 355–359. [CrossRef] [PubMed]

3. Sjögersten, S.; Black, C.R.; Evers, S.; Hoyos-Santillan, J.; Wright, E.L.; Turner, B.L. Tropical wetlands: A missing link in the global carbon cycle? *Glob. Biogeochem. Cycles* **2014**, *28*, 1371–1386. [CrossRef] [PubMed]
4. Jung, M.; Reichstein, M.; Ciais, P.; Seneviratne, S.I.; Sheffield, J.; Goulden, M.L.; Bonan, G.; Cescatti, A.; Chen, J.; De Jeu, R.J.N. Recent decline in the global land evapotranspiration trend due to limited moisture supply. *Nature* **2010**, *467*, 951–954. [CrossRef] [PubMed]
5. Gudmundsson, L.; Boulange, J.; Do, H.X.; Gosling, S.N.; Grillakis, M.G.; Koutroulis, A.G.; Leonard, M.; Liu, J.; Müller Schmied, H.; Papadimitriou, L.J.S. Globally observed trends in mean and extreme river flow attributed to climate change. *Science* **2021**, *371*, 1159–1162. [CrossRef]
6. Buffington, J.M.; Montgomery, D.R. Geomorphic Classification of Rivers. In *Treatise on Geomorphology*; Shroder, J.F., Ed.; Academic Press: San Diego, CA, USA, 2013; pp. 730–767. [CrossRef]
7. Nardini, A.; Brierley, G. Automatic river planform identification by a logical-heuristic algorithm. *Geomorphology* **2021**, *375*, 107558. [CrossRef]
8. Witkowski, K. Reconstruction of Nineteenth-Century Channel Patterns of Polish Carpathians Rivers from the Galicia and Bucovina Map (1861–1864). *Remote Sens.* **2021**, *13*, 5147. [CrossRef]
9. Allen, G.H.; Pavelsky, T.M. Patterns of river width and surface area revealed by the satellite-derived North American River Width data set. *Geophys. Res. Lett.* **2015**, *42*, 395–402. [CrossRef]
10. Alsdorf, D.E.; Rodríguez, E.; Lettenmaier, D.P. Measuring surface water from space. *Rev. Geophys.* **2007**, *45*, RG2002. [CrossRef]
11. Gleason, C.J.; Smith, L.C. Toward global mapping of river discharge using satellite images and at-many-stations hydraulic geometry. *Proc. Natl. Acad. Sci. USA* **2014**, *111*, 4788. [CrossRef]
12. Yamazaki, D.; Trigg, M.A.; Ikeshima, D. Development of a global ~90 m water body map using multi-temporal Landsat images. *Remote Sens. Environ.* **2015**, *171*, 337–351. [CrossRef]
13. Gong, P.; Liu, H.; Zhang, M.; Li, C.; Wang, J.; Huang, H.; Clinton, N.; Ji, L.; Li, W.; Bai, Y.; et al. Stable classification with limited sample: Transferring a 30-m resolution sample set collected in 2015 to mapping 10-m resolution global land cover in 2017. *Sci. Bull.* **2019**, *64*, 370–373. [CrossRef]
14. Pekel, J.-F.; Cottam, A.; Gorelick, N.; Belward, A.S. High-resolution mapping of global surface water and its long-term changes. *Nature* **2016**, *540*, 418–422. [CrossRef] [PubMed]
15. Allen, G.H.; Pavelsky, T.M.J.S. Global extent of rivers and streams. *Science* **2018**, *361*, 585–588. [CrossRef] [PubMed]
16. Hossain, A.A.; Mathias, C.; Blanton, R.J.R.S. Remote sensing of turbidity in the Tennessee River using Landsat 8 satellite. *Remote Sens.* **2021**, *13*, 3785. [CrossRef]
17. Huang, J.; Zhang, Y.; Bing, H.; Peng, J.; Dong, F.; Gao, J.; Arhonditsis, G.B.J.W.R. Characterizing the river water quality in China: Recent progress and on-going challenges. *Water Res.* **2021**, *201*, 117309. [CrossRef]
18. Meyer, J.L.; Strayer, D.L.; Wallace, J.B.; Eggert, S.L.; Helfman, G.S.; Leonard, N.E. The contribution of headwater streams to biodiversity in river networks. *Jawra J. Am. Water Resour. Assoc.* **2010**, *43*, 86–103. [CrossRef]
19. Allen, G.H.; Pavelsky, T.M.; Barefoot, E.A.; Lamb, M.P.; Butman, D.; Tashie, A.; Gleason, C.J. Similarity of stream width distributions across headwater systems. *Nat. Commun.* **2018**, *9*, 610. [CrossRef]
20. Mao, W.; Yang, K.; Zhang, W.; Wang, Y.; Li, M. High-resolution global water body datasets underestimate the extent of small rivers. *Int. J. Remote Sens.* **2022**, *43*, 4315–4330. [CrossRef]
21. Wang, S. Progresses in Variability of Snow Cover over the Qinghai-Tibetan Plateau and Its Impact on Water Resources in China. *Plateau Meteorol.* **2017**, *36*, 1153–1164.
22. Jiang, W.; Niu, Z.; Wang, L.; Yao, R.; Gui, X.; Xiang, F.; Ji, Y.J.R.S. Impacts of Drought and Climatic Factors on Vegetation Dynamics in the Yellow River Basin and Yangtze River Basin, China. *Remote Sens.* **2022**, *14*, 930. [CrossRef]
23. Krause, S.; Abbott, B.W.; Baranov, V.; Bernal, S.; Blaen, P.; Datry, T.; Drummond, J.; Fleckenstein, J.H.; Velez, J.G.; Hannah, D.M.J.W.R.R. Organizational principles of hyporheic exchange flow and biogeochemical cycling in river networks across scales. *Water Resour. Res.* **2022**, *58*, e2021WR029771. [CrossRef]
24. Wei, H.; Xue, D.; Huang, J.; Liu, M.; Li, L.J.R.S. Identification of Coupling Relationship between Ecosystem Services and Urbanization for Supporting Ecological Management: A Case Study on Areas along the Yellow River of Henan Province. *Remote Sens.* **2022**, *14*, 2277. [CrossRef]
25. Luo, J.; Niu, F.; Lin, Z.; Liu, M.; Yin, G. Thermokarst lake changes between 1969 and 2010 in the Beilu River Basin, Qinghai–Tibet Plateau, China. *Sci. Bull.* **2015**, *60*, 556–564. [CrossRef]
26. Li, D.; Wang, G.; Qin, C.; Wu, B.J.R.S. River extraction under bankfull discharge conditions based on sentinel-2 imagery and DEM data. *Remote Sens.* **2021**, *13*, 2650. [CrossRef]
27. Chinese Academy of Sciences. *China Natrual Geoscience: Groundwater*; Chinese Academy of Sciences: Beijing, China, 1981.
28. Drusch, M.; Del Bello, U.; Carlier, S.; Colin, O.; Fernandez, V.; Gascon, F.; Hoersch, B.; Isola, C.; Laberinti, P.; Martimort, P.; et al. Sentinel-2: ESA's Optical High-Resolution Mission for GMES Operational Services. *Remote Sens. Environ.* **2012**, *120*, 25–36. [CrossRef]
29. Bertini, F.; Brand, O.; Carlier, S.; Bello, U.D.; Pieiro, J. Sentinel-2 ESA's Optical High-Resolution Mission for GMES Operational Services. *Remote Sens. Environ.* **2012**, *120*, 25–36.
30. Meer, F.; Werff, H.; Ruitenbeek, F. Potential of ESA's Sentinel-2 for geological applications. *Remote Sens. Environ.* **2014**, *148*, 124–133. [CrossRef]

31. Yamazaki, D.; Ikeshima, D.; Sosa, J.; Bates, P.D.; Allen, G.H.; Pavelsky, T.M. MERIT Hydro: A High-Resolution Global Hydrography Map Based on Latest Topography Dataset. *Water Resour. Res.* **2019**, *55*, 5053–5073. [CrossRef]
32. Yamazaki, D.; Ikeshima, D.; Tawatari, R.; Yamaguchi, T.; O'Loughlin, F.; Neal, J.C.; Sampson, C.C.; Kanae, S.; Bates, P.D. A high-accuracy map of global terrain elevations. *Geophys. Res. Lett.* **2017**, *44*, 5844–5853. [CrossRef]
33. Nobre, A.D.; Cuartas, L.A.; Hodnett, M.; Rennó, C.D.; Rodrigues, G.; Silveira, A.; Saleska, S. Height Above the Nearest Drainage–a hydrologically relevant new terrain model. *J. Hydrol.* **2011**, *404*, 13–29. [CrossRef]
34. Lu, X.; Yang, K.; Bennett, M.M.; Liu, C.; Mao, W.; Li, Y.; Zhang, W.; Li, M. High-resolution satellite-derived river network map reveals small Arctic river hydrography. *Environ. Res. Lett.* **2021**, *16*, 054015. [CrossRef]
35. Lehner, B.; Verdin, K.; Jarvis, A. New global hydrography derived from spaceborne elevation data. *Eos Trans. Am. Geophys. Union AGU* **2008**, *89*, 93–94. [CrossRef]
36. McFeeters, S.K. The use of the Normalized Difference Water Index (NDWI) in the delineation of open water features. *Int. J. Remote Sens.* **1996**, *17*, 1425–1432. [CrossRef]
37. Xu, H. Modification of normalised difference water index (NDWI) to enhance open water features in remotely sensed imagery. *Int. J. Remote Sens.* **2006**, *27*, 3025–3033. [CrossRef]
38. Liu, J.-L.; Feng, D.-Z. Two-dimensional multi-pixel anisotropic Gaussian filter for edge-line segment (ELS) detection. *Image Vis. Comput.* **2014**, *32*, 37–53. [CrossRef]
39. Yang, K.; Li, M.; Liu, Y.; Cheng, L.; Huang, Q.; Chen, Y. River detection in remotely sensed imagery using Gabor filtering and path opening. *Remote Sens.* **2015**, *7*, 8779–8802. [CrossRef]
40. Lu, X.; Yang, K.; Lu, Y.; Gleason, C.J.; Li, M. Small Arctic rivers mapped from Sentinel-2 satellite imagery and ArcticDEM. *J. Hydrol.* **2020**, *584*, 124689. [CrossRef]

atmosphere

MDPI

Article

Case Study of the Aerosol Optical Properties in the Atmosphere over Wuhan, China

Qianjun Mao *, Gangzheng Hu and Xin Nie

School of Urban Construction, Wuhan University of Science and Technology, Wuhan 430065, China
* Correspondence: author: maoqianjun@wust.edu.cn

Abstract: The research on regional aerosol optical properties is of great significance for exploring climate regulation mechanisms and controlling atmospheric pollution. Based on the solar radiation observation platform, a three-month optical observation of atmospheric aerosols was conducted in Wuhan, China. The daily and monthly variation characteristics of aerosol optical depth (AOD_{550}), Angstrom parameter ($\alpha_{440-870}$), and turbidity coefficient (β) were revealed, and the interrelations between the three optical parameters were fitted. Then, the potential relationships between atmospheric particulate matter ($PM_{2.5}$, PM_{10}) with AOD_{550} and β were discussed. The results show that the average values of AOD_{550}, $\alpha_{440-870}$, and β in this case study are 0.42, 1.32, and 0.20, respectively. The frequency distribution patterns of the three optical parameters are all unimodal. AOD_{550} has a good linear correlation system with β, and the Pearson correlation coefficient reaches 0.94, while its correlation with $\alpha_{440-870}$ is not significant. The daily variation in AOD_{550} and β both show an increasing trend, and their monthly increases are more than 50%. However, the daily variation in $\alpha_{440-870}$ is relatively stable, and the fitted line is a nearly horizontal line with no significant monthly variation. The fluctuation of particulate matter concentration affects the aerosol optical properties to some extent, among which β has a prominent effect on the response to the change in $PM_{2.5}$ concentration with a linear correlation coefficient of 0.861. As the concentration of particulate matter increases, the proportion of fine particulate matter in the atmosphere increases monthly, and the ratio of PM_{10} to $PM_{2.5}$ concentrations decreases from 1.8:1 to 1.2:1. Atmospheric pollution conditions are frequent during this observation period, mainly at mildly turbid levels. Atmospheric turbidity shows an increasing trend month by month, and the concentration of particulate matter increases rapidly. The response of atmospheric aerosol optical properties to the changes in fine particulate matter concentration is significant, and controlling the particulate matter content in the atmosphere is an effective means to mitigate aerosol pollution.

Keywords: TBS-4 solar spectrometer; aerosol optical depth; Angstrom parameter; turbidity coefficient; particulate matter

Citation: Mao, Q.; Hu, G.; Nie, X. Case Study of the Aerosol Optical Properties in the Atmosphere over Wuhan, China. *Atmosphere* **2023**, *14*, 1108. https://doi.org/10.3390/atmos14071108

Academic Editors: Jean-Christophe Raut and Matteo Rinaldi

Received: 23 May 2023
Revised: 23 June 2023
Accepted: 29 June 2023
Published: 4 July 2023

1. Introduction

Aerosols are multiphase systems of solid particles and liquid droplets suspended in the atmosphere, with diameters usually ranging from 0.001 to 100 μm. Aerosols change the regional climate environment by participating in the formation of clouds and rain in the atmosphere, and their absorption and scattering effects on solar radiation also affect the radiation balance of the Earth-atmosphere system [1,2]. Aerosols can be classified into various categories according to different classification bases, and multiple types of aerosols can exist simultaneously in the atmosphere [3,4]. Various types of aerosol particles participate together in the evolution of the atmosphere, and their interactions lead to the production of more aerosol particles, diversifying the composition of the atmosphere [5–7]. Aerosols are present in the human living environment all the time, and the occurrence of atmospheric pollution events such as acid rain, haze, and dust are closely related to aerosols [8,9]. Harmful substances such as viruses and bacteria carried in aerosols can cause respiratory damage and pose a serious threat to human health [10–12]. Some studies have shown that transmissible diseases such as COVID-19 can be transmitted through aerosols [13]. As an important component of the atmosphere, aerosols have an inescapable role in the regional climate environment, the energy balance of the Earth-atmosphere system, and human life [14,15]. Therefore, it is of great application and scientific significance to actively extend the research of aerosol optical properties of multi-regional and deeply analyze their response mechanisms to the atmospheric environment.

The more pronounced spatial and temporal variability of aerosols and their smaller amount in the atmosphere makes it difficult to obtain the optical properties of aerosols than other gases [16,17]. The current methods of aerosol observation are ground-based observations and satellite remote sensing [18,19]. The advantages of satellite remote sensing include wide spatial coverage, lengthy time series, high resolution, and regional projection imaging [20,21]. In recent years, a large number of observations and analyses of aerosol optical properties have been carried out by domestic and foreign scientists using satellite remote sensing [22–25]. Instruments such as Moderate-resolution Imaging Spectroradiometer (MODIS) and Multi-angle Imaging Spectrometer (MISR) have become common tools for satellite remote sensing observations [26,27]. Satellite remote sensing technology allows real-time aerosol observations in most regions of the world by combining different inversion algorithms. Therefore, it is a good solution to the challenge of aerosol research across regions [28,29]. However, the drawbacks of satellite remote sensing cannot be ignored, such as fewer aerosol product bands, longer intervals between remote sensing observations, and insufficient data accuracy [30,31].

Continuous long-term series of aerosol characterization is indispensable in resolving the response of the atmospheric environment to aerosol changes. In order to make up for the shortage of satellite remote sensing observations, ground-based observations are favored by many scientists because of their high accuracy and time continuity [32–34]. Gong W. et al. analyzed columnar aerosol volume size distributions from March 2012 to February 2013 in Wuhan, China [35]. Ground-based observation methods have also been used to verify the accuracy of satellite remote sensing data [36–38]. Wang W. et al. evaluated the AOD of VIIRS and MODIS in the Wuhan area using photometric measurements [39]. In fact, most ground-based observational studies rely on the global Aerosol Robotic Network (AERONET) established by NASA. However, there are few ground-based sites in China, and many of them are not publicly authorized for use. Therefore, in regions where AERONET stations are not located, self-established aerosol ground-based observation platforms are of significant importance for aerosol characterization research [40,41].

Domestic and foreign scientists have conducted extensive studies worldwide through ground-based measurements and remote-sensing observations. However, regional research on aerosol optical properties, especially for regions without AERONET sites, still needs to be supplemented. Therefore, in this study, aerosol optical depth (AOD_{550}), Angstrom parameter ($\alpha_{440-870}$), and turbidity coefficient (β) were measured for the period from November 2021 to January 2022 using a ground-based observation platform equipped with the TBS-4 solar spectrometer built in Wuhan. Its applicability as an aerosol observation instrument was verified by comparison with the aerosol products obtained from satellite remote sensing. The variation characteristics of aerosol optical parameters and their interrelationships were then analyzed from the perspective of daily and monthly variation, respectively. Finally, the effects of particulate matter ($PM_{2.5}$, PM_{10}) on aerosol optical properties were revealed based on air quality data.

2. Experimental Principles and Methods

2.1. Experimental Instruments and Data

The TBS-4 solar spectrometer used in this paper is manufactured by Jinzhou Sunshine Meteorological Technology Co., Ltd. (Jinzhou, China). It uses synchronous measurement technology and GPS satellite location tracking technology to simultaneously measure solar spectral radiation in 9 bands (340, 380, 440, 500, 675, 870, 936, 1020, and 1640 nm) with a measurement accuracy of less than 1%. The experimental site is located at Wuhan University of Science and Technology (114°26′ E, 30°44′ N), with an altitude of 40 m above sea level. There is no shade over the location of the solar spectrometer, and the instrument is powered by solar photovoltaic panels, with a power storage device that can meet the low light operation for 3 days. Figure 1 shows the location and physical presence of the TBS-4 sun photometer. The observation period is from November 2021 to January 2022, with spectral radiation collected every 5 min. Valid data selection criteria: Cumulative sunshine hours must be greater than 5 h, or the data for that day are rejected. Mean values were taken for each half-hour observation, excluding offsets greater than 20% and 550 band radiative values less than 10 $W \cdot m^{-2}$ from the observations. To take three adjacent data in chronological order, the intermediate moment data need not be smaller than the previous and next data values; otherwise, the data are not available. The remaining daily data amounts larger than 50 were used as the final radiation data. The data were filtered to leave 38 available observation days.

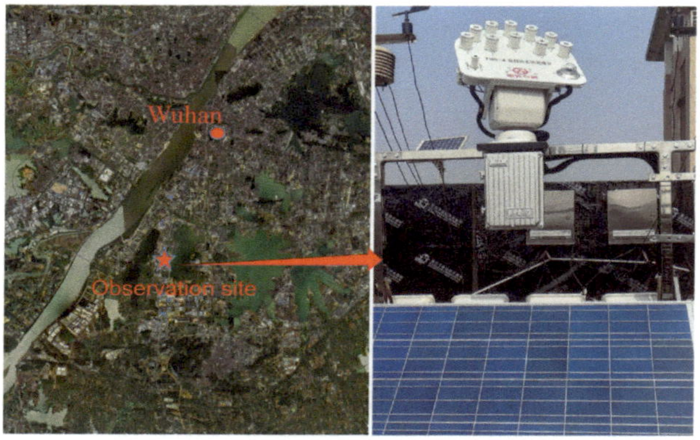

Figure 1. The TBS-4 solar photometer and its location.

$PM_{2.5}$ and PM_{10} are pollution concentration indicators that are divided based on the equivalent diameter of particles, and their values directly affect atmospheric quality. The concentration data of $PM_{2.5}$ and PM_{10} are sourced from the Air Quality Online Monitoring and Analysis Platform (https://www.aqistudy.cn/ (accessed on 21 May 2023)). The daily and monthly average PM concentrations were calculated separately for use. The satellite remote sensing data used in this study were obtained from 3 km resolution AOD_{550} products provided by the Terra satellite of the National Aeronautics and Space Administration (NASA). Remote sensing image processing and analysis of the study area were carried out by ENVI + IDL software.

2.2. Principles and Methods of Calculation

According to Bouguer–Lamber law, in the non-water vapor absorption channel, the direct solar radiation $E(\lambda)$ of the surface incident on the earth with a wavelength of λ is:

$$E(\lambda) = E_0(\lambda)R^2 T_g exp[-m(\theta)\tau(\lambda)] \quad (1)$$

$E_0(\lambda)$ is the solar spectral irradiance at the solar-terrestrial mean distance at the upper boundary of the atmosphere at a wavelength of λ. R^2 is the solar-terrestrial distance correction factor. Neglecting the absorption of solar radiation by the absorbing gas in the above bands, $T_g = 1$. $m(\theta)$ is the atmospheric optical mass at a zenith angle of θ. $\tau(\lambda)$ is the total atmospheric optical depth of the spectrum.

$$R^2 = 1 + 0.033 \cos \frac{2\pi \cdot n}{365} \quad (2)$$

$$m(\theta) = \frac{1}{\sin(h) + 0.15(h + 3.885)^{-1.253}} \quad (3)$$

where n is the number of day sequences, with January 1st as the first day, and h is the solar altitude angle.

Take the logarithm at both ends of Equation (1) to obtain the following formula:

$$\ln\left(\frac{E_\lambda}{R^2}\right) = \ln E_{0\lambda} - m\tau_\lambda \quad (4)$$

A simple linear regression was performed with m as the independent variable and $\ln(E_\lambda/R^2)$ as the dependent variable. The total atmospheric optical depth $\tau(\lambda)$ is represented by the slope of the fit result in absolute terms, and the instrument calibration value is represented by the intercept $\ln E_{0\lambda}$.

In order to ensure calibration accuracy, clear weather is chosen for the calibration of the spectrometer. When the atmosphere is clear and cloudless, the aerosol state is stable, and the AOD changes slightly. In the linear regression fitting, the variation range of m should be larger than 3.5. The sample points with the largest fitting residuals should be excluded when the correlation coefficient is less than 0.99. The number of excluded samples should be less than 30%, and the remaining samples should be more than 15.

The total atmospheric optical depth $\tau(\lambda)$ is mainly composed of Rayleigh scattering optical depth $\tau_r(\lambda)$, aerosol optical depth $\tau_a(\lambda)$ and absorbing gas optical depth $\tau_o(\lambda)$.

$$\tau_a(\lambda) = \tau(\lambda) - \tau_r(\lambda) - \tau_o(\lambda) \quad (5)$$

$$\tau_r(\lambda) = \frac{P}{P_0} 0.0088\lambda^{-4.05} \quad (6)$$

$$\tau_0(\lambda) = k_0 \frac{U}{1000} \quad (7)$$

where the main components of the absorbing gas are ozone, water vapor, carbon dioxide, and oxygen, among others. For the 440 nm and 870 nm bands chosen for the calculations in this paper, only the absorption effect of ozone needs to be taken into account; P and P_0 are the atmospheric pressure and standard atmospheric pressure (1013.25 mbar) at the observation site, respectively; λ is the wavelength of the spectrum, calculated in μm; k_0 is the absorption coefficient of ozone; and U is the ozone content.

The formula for calculating the ozone content is as follows:

$$U = 235 + \{150 + 40\sin[0.9865(n - 30)] + 20\sin[3(\xi + 20)]\}\sin^2(1.28\varphi) \quad (8)$$

where ξ and φ are the longitude and latitude of the observation location.

The relationship between aerosol optical depth $\tau_a(\lambda)$ and the Angstrom parameter α and turbidity coefficient β is expressed as:

$$\tau_a(\lambda) = \beta \times \lambda^{-\alpha} \quad (9)$$

AOD, α, and β are three basic dimensionless parameters to characterize the atmospheric aerosol optical properties, which are key factors used to assess aerosol content, determine the types of aerosols and measure the degree of atmospheric pollution [42–44]. AOD is defined as the integral of the atmospheric extinction coefficient in the vertical direction, which reflects the extinction characteristics of aerosols. α reflects the particle size of aerosol particles, and its value generally ranges from 0 to 2. The closer the value of α is to 2, the smaller the particle size of the main particles of aerosols, and vice versa. β is numerically equal to the aerosol optical depth at a wavelength of 1 μm and can be used to indicate the degree of atmospheric turbidity. According to the common classification basis, the atmosphere is generally classified into four classes: clean atmosphere ($\beta \leq 0.1$), mildly turbid atmosphere ($0.1 < \beta \leq 0.2$), turbid atmosphere ($0.2 < \beta \leq 0.4$), and heavily turbid atmosphere ($\beta > 0.4$) [45,46].

To match the 550 nm band aerosol products provided by MODIS, the AOD_{550} are obtained by substituting $\alpha_{440\text{–}870}$ and β values into Equation (10) using the interpolation method.

$$\tau(550) = \beta \times 0.55^{-\alpha_{440-870}} \quad (10)$$

2.3. *Accuracy Verification of the Measured AOD_{550}*

In order to verify the feasibility of observing aerosol optical properties by this model of the spectrometer, the accuracy of ground-based measurement of AOD_{550} was verified using AOD_{550} of MOD04_3K. The average value of ground-based observations before and after the satellite transit for half an hour is calculated as the measured value. The image element of the remote sensing image where the ground-based site is located, its surroundings are selected as the study area, and its mean value is used as the satellite remote sensing value. The ground-based measurement values and the satellite remote sensing values were spatially and temporally matched according to the above method and then presented as data points in Figure 2, with linear regression processing. The linear fit of data points results in $Y = 0.08 + 0.89 \times X$, which shows a small angle with the 1:1 reference line. In the low-value region of AOD_{550}, the ground-based measurements are somewhat underestimated compared with the satellite remote sensing values. The experimentally measured AOD_{550} was calculated to have a good linear correlation with the remotely sensed observed AOD_{550}. Their Pearson correlation coefficients reached 0.96, and the root mean square error was 0.076. It is evident that the inversion of AOD_{550} using the TBS-4 model solar spectrometer is workable.

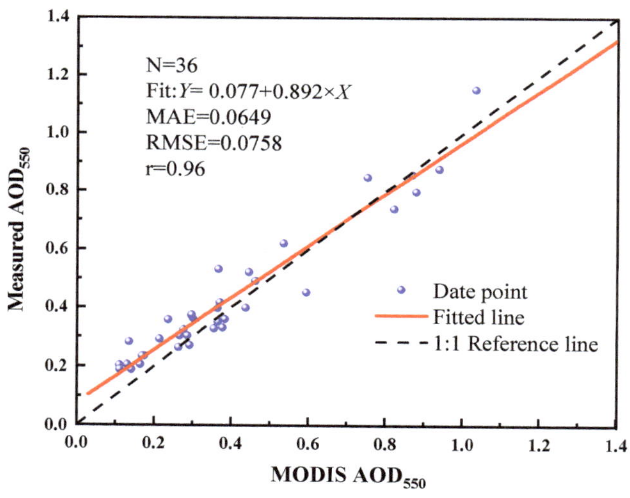

Figure 2. Accuracy verification of ground-based measured AOD_{550}.

3. Results and Discussion

3.1. *Aerosol Optical Properties*

3.1.1. General Characteristics of Aerosol Optical Parameters

The daily average values of the three optical parameters (AOD_{550}, $\alpha_{440\text{–}870}$, and β) were calculated daily, and their daily variations were presented as dotted line plots in Figure 3. As seen in the figure, the AOD_{550} on November 30, 2021, is 0.19, and the β coefficient is only 0.09, which is the lowest value during the observation period. According to the criteria for classifying atmospheric turbidity levels by beta coefficients, the atmosphere was in the best clean state on that day. The AOD_{550} values on 9 December 2021 and 19 January 2022 showed anomalously high values of 1.16 and 1.26, respectively, which far exceeded the values of other observed days. Moreover, the turbidity coefficients exceed 0.4, and the atmosphere is in a heavily turbid state on these two days due to severe aerosol pollution events. The daily mean values of the optical parameters are fitted linearly, as shown by the dashed lines in Figure 3. AOD_{550} and β show an increasing trend, and the slope of the fitted line is 0.0067 and 0.0029, respectively, which indicates that the atmospheric pollution condition is worsening. As a whole, when the AOD_{550} value increases or decreases, the β value also produces a certain magnitude of isotropic fluctuations. The trends of AOD_{550} and β are consistent, but the growth rate of AOD_{550} is higher than that of β, and the linear fit of $\alpha_{440-870}$ results in an approximate horizontal straight line. The data points show irregular and limited fluctuations above and below the fitted line, indicating that the particle size of the main atmospheric aerosol particles changes steadily during the observation period.

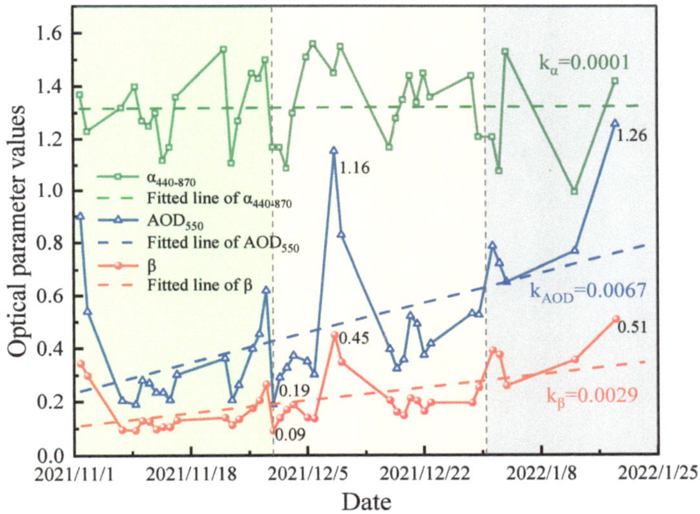

Figure 3. Daily variation in the atmospheric aerosol optical properties.

The experimental AOD_{550}, $\alpha_{440-870}$, and β were numerically counted, and their means and standard deviations were calculated for the whole experimental period. The detailed values are shown in Table 1. The intervals of AOD_{550}, $\alpha_{440-870}$, and β were divided by 0.1 steps, and the frequency ratios of each interval were counted, and the results are shown in Figure 4. The frequency ratios for all three optical parameters show unimodal distribution. The percentage of frequencies of AOD_{550} in the low-value range ($AOD_{550} < 0.4$) reached 55.7%. Its peak occurs in the interval of 0.3–0.4, and the frequency in this interval reaches 23.0%. The mean value of AOD_{550} is 0.42, and the standard deviation is 0.24, which shows that the attenuation effect of aerosols on solar radiation is not significant. The mean value of $\alpha_{440-870}$ is 1.32 with a standard deviation of 0.20. Its frequency proportion in the peak interval of 1.3–1.4 reaches 23.4%, indicating that the atmosphere is mainly composed of fine size aerosol particles. The mean value of β is 0.20 with a standard deviation of 0.09. According to the distribution of the frequency proportion of β in each interval, it can be seen that the frequency proportion of clean atmosphere is only 12.9%, while the frequency in the interval of 0.1–0.2 is as high as 52.6%. Based on the criteria for classifying atmospheric turbidity levels based on β values, the atmospheric pollution conditions in the Wuhan area during this observation period were frequent, mainly at mild turbidity degrees, and some states reached heavy turbidity levels.

Table 1. Mean value and standard deviation of aerosol optical parameters.

Optical Parameter	Mean Value	Standard Deviation
AOD_{550}	0.42	0.24
$\alpha_{440-870}$	1.32	0.20
β	0.20	0.09

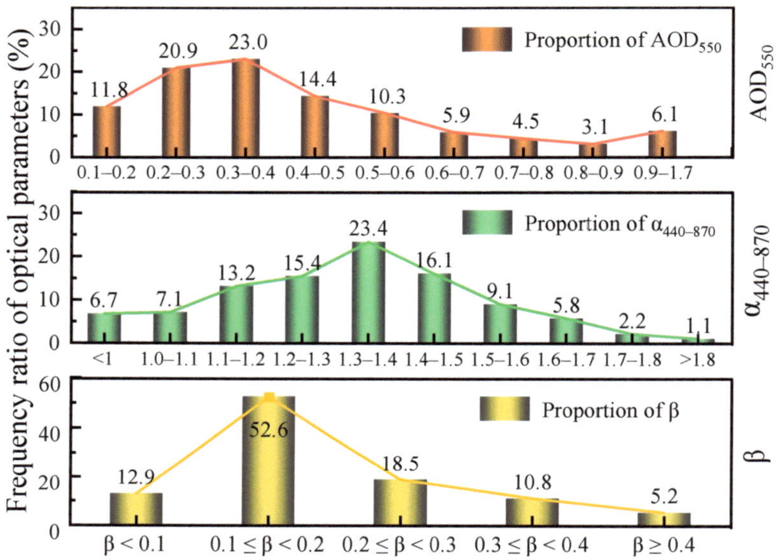

Figure 4. Frequency distribution of atmospheric aerosol optical properties.

3.1.2. Correlation of Aerosol Optical Parameters

Figure 5 shows the scatter plot of AOD_{550} with $\alpha_{440-870}$ and β, respectively, using density mapping to show the data distribution. The solid line in the figures is the result of the linear fit. The red color represents the high degree of sample aggregation, while the blue color represents the small sample size and low aggregation. AOD_{550} and β show a regular linear distribution, and their fitting equation is $Y = 2.34 \times X - 0.02$. With the increase in atmospheric turbidity, the attenuation effect of aerosol particles on radiation is enhanced, which makes AOD_{550} increase. Their linear correlation was extremely high, with a positive correlation coefficient of 0.94. In the low-value region where AOD_{550} was less than 0.4 and β was less than 0.2, the proportion of samples reached 57.6%, while in the high-value region, it was relatively scattered. Once again, it shows that the main state of atmospheric aerosols during this observation period was mildly turbid degree with low aerosol content. The sample distribution of AOD_{550} and $\alpha_{440-870}$ was relatively scattered, and the linear fit correlation coefficient was only 0.36. The scattered distribution results of AOD_{550} and $\alpha_{440-870}$ showed that they were mainly gathered in the region of AOD_{550} less than 0.6 but did not show an obvious linear correlation. With the increase in AOD_{550}, $\alpha_{440-870}$ and β mainly showed increasing trends, among which the increasing trend of $\alpha_{440-870}$ was relatively slight. This phenomenon indicated that the particle size of the main aerosol particles showed a small decrease when the atmospheric turbidity increased.

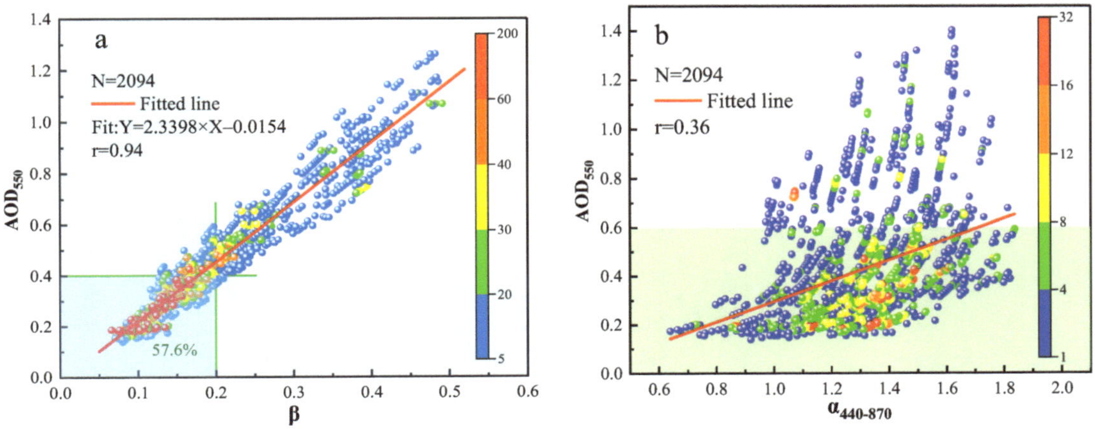

Figure 5. Scatter distribution of the combination of AOD_{550} with β (**a**) and $\alpha_{440–870}$ (**b**), respectively.

3.1.3. Monthly Variation in Aerosol Optical Parameters

The mean values of AOD_{550}, $\alpha_{440–870}$, and β by month are counted, and the results are shown in Table 2. Moreover, Figure 6 shows the numerical variations in the three optical parameters in the form of a boxplot. Compared with November 2021, the mean value of $\alpha_{440–870}$ in December 2021 increased by 0.1, indicating a small decrease in the aerosol particle size. The standard deviation of $\alpha_{440–870}$ in each month was close, among which the standard deviation of $\alpha_{440–870}$ in December 2021 was the smallest at 0.19, and the best concentration of particle size can be found in that month combined with the boxplot. The mean value of $\alpha_{440–870}$ fell back to 1.26 in January 2022. However, the particle size distribution of aerosols was more dispersed in that month, with a maximum standard deviation of 0.25. The aerosols in each month of the observation period mainly consisted of fine-sized particles with a more concentrated particle size distribution. The mean value of β showed an increasing trend monthly, from 0.14 in November 2021 to 0.22 in December 2021, reflecting the increase in the atmospheric from mild turbidity level to turbidity level. The mean value of β increased again to 0.36 in January 2022, which was 63.64% higher compared to the previous month, and the degree of atmospheric pollution further increased. The standard deviation of β for all months was less than 0.1, indicating the high degree of aggregation of β distribution. AOD_{550} increased rapidly at a rate of more than 0.20 per month. Its average value was 0.30 in November 2021, and by January 2022, the average AOD_{550} value reached 0.76, an increase of 153.33%. It revealed that the atmospheric turbidity in the Wuhan region gradually increased during the study period, and the attenuation of solar radiation by atmospheric aerosol particles also gradually increased.

Table 2. Monthly means and standard deviations of aerosol optical parameters.

Optical Parameters	November 2021	December 2021	January 2022
$\alpha_{440–870}$	1.28 ± 0.21	1.38 ± 0.19	1.26 ± 0.25
β	0.14 ± 0.06	0.22 ± 0.09	0.36 ± 0.08
AOD_{550}	0.30 ± 0.15	0.50 ± 0.22	0.76 ± 0.18

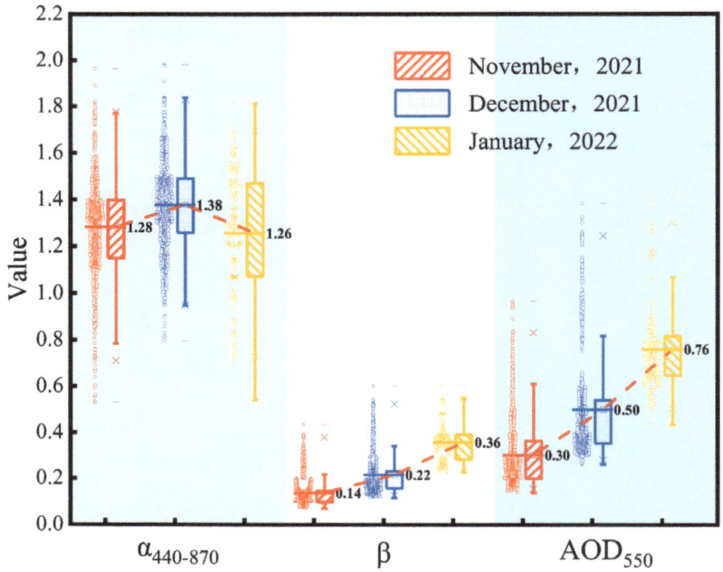

Figure 6. Boxplot of monthly variation in AOD_{550}, $\alpha_{440-870}$, and β.

3.2. Atmospheric Particulate Matter

The mean values of particulate matter concentrations ($PM_{2.5}$, PM_{10}) during the observation period were calculated by month and shown in Figure 7, along with the monthly variations in aerosol optical depth and turbidity coefficient. By comparing with the aerosol optical parameters, it can be found that the concentration of particulate matter rises rapidly with the aggravation of atmospheric turbidity, showing an increasing trend from month to month. The monthly relative increases in $PM_{2.5}$, AOD_{550}, and β all exceeded 50%, while the monthly relative increase in PM_{10} was near 30%. The monthly increase in fine particulate matter content far exceeded the increase in coarse particulate matter over the same period, as detailed in Table 3. Regarding the closest monthly relative increase in $PM_{2.5}$ and β, the increase in $PM_{2.5}$ content from November 2021 to December 2021 was 52%, and the increase in β during the same period was 57%. By January 2022, the monthly increase in $PM_{2.5}$ increased to 76%, and the increase in β value in that period was 64%. This indicates that there was a relatively strong correlation between the fine particulate matter content and the degree of atmospheric turbidity. In November 2021, the PM_{10} and $PM_{2.5}$ concentrations were 74.47 $\mu g \cdot m^{-3}$ and 41.82 $\mu g \cdot m^{-3}$, respectively, with a ratio of 1.8:1, and in January 2022, the ratio decreased to 1.2:1. As atmospheric turbidity increases during the observation period, the particulate matter content increases significantly, with fine particulate matter playing a major role in this process.

Table 3. Statistics of monthly average growth of particulate matter and optical parameters.

Project	November 2021	Monthly Increase	December 2021	Monthly Increase	January 2022
$PM_{2.5}$ ($\mu g \cdot m^{-3}$)	41.82	52%	63.38	76%	111.60
PM_{10} ($\mu g \cdot m^{-3}$)	74.47	30%	96.88	33%	129.20
AOD_{550}	0.30	67%	0.50	52%	0.76
β	0.14	57%	0.22	64%	0.36

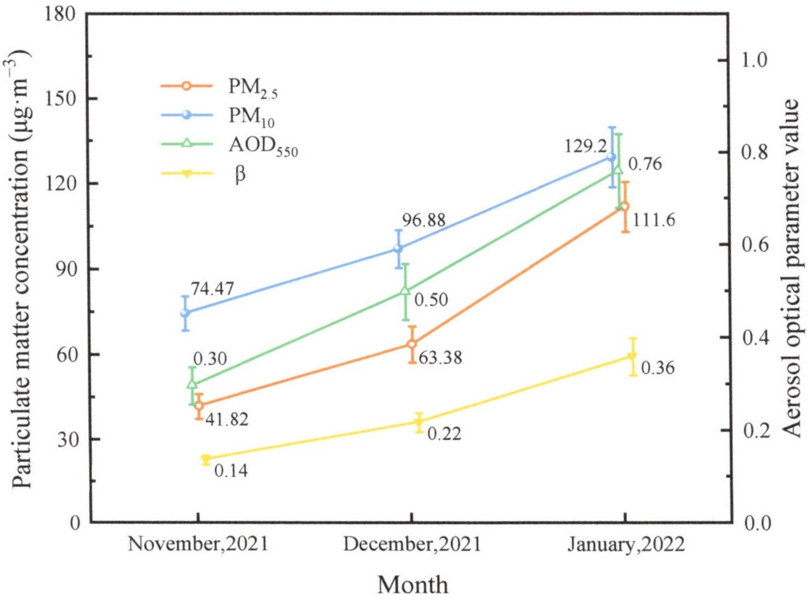

Figure 7. Variation in monthly mean and standard error of particulate matter, AOD_{550}, and β.

The interrelation between atmospheric particulate matter and aerosol optical properties was further investigated. Firstly, the daily average concentration values of particulate matter were calculated. Then scatter plots were drawn with AOD_{550} and β as horizontal coordinates and particulate matter concentrations as vertical coordinates, respectively. The results of the linear fit based on scatter plots are shown in Figure 8. The fitting results of PM concentration and AOD_{550} were similar to those of PM concentration and β, and their scatter distribution had some similarity, which was caused by the high linear correlation between AOD_{550} and β. With the increase in AOD_{550} or β, the PM concentration showed an obvious positive correlation trend. Among them, $PM_{2.5}$ concentration showed the strongest linear correlation with β, and the Pearson correlation coefficient reached 0.861. When the concentration of particulate matter suspended in the atmosphere increases, the attenuation effect of aerosols on solar radiation is enhanced, leading to an increase in aerosol optical depth, along with an increase in atmospheric turbidity. During the fluctuation of aerosol optical properties caused by the change in particulate matter concentration, the response effect of AOD_{550} and β to the change in fine particulate matter concentration is more prominent. From the above analysis, it can be seen that reducing the particulate matter content in the atmosphere is an effective way to reduce atmospheric turbidity, of which controlling the fine particulate matter content is the most important.

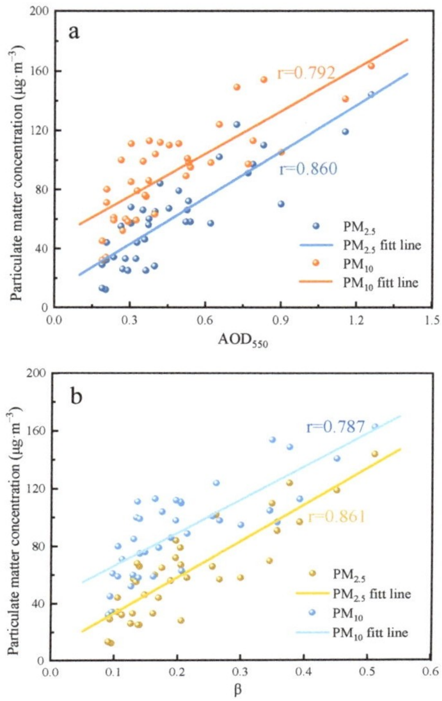

Figure 8. The correlation between particulate matter concentration with AOD_{550} (**a**) and β (**b**).

4. Conclusions

Experimental observations of aerosols were carried out in Wuhan from January 2021 to January 2022 using the TBS-4 solar spectrometer, and the variation characteristics of aerosol optical properties are analyzed in detail. By monitoring the air quality during the observation period, the influence of atmospheric particulate matter content on the aerosol optical properties is revealed to some extent. The main conclusions are as follows:

(1) During this case study, the mean value of AOD_{550} was 0.42, and the atmospheric aerosol content was at a low level; the mean value of β was 0.20, and the atmosphere was lightly turbid. Their daily variations show an increasing trend. The mean value of $\alpha_{440\text{–}870}$ was 1.32, and the fitting result of its daily variation indicates that the fluctuation of aerosol particle size is more stable, which is mainly fine particle size aerosol. The frequency ratios of AOD_{550}, $\alpha_{440\text{–}870}$, and β are unimodal distributions. The results of the frequency distribution of the aerosol optical parameters show the frequency of atmospheric pollution conditions. The atmosphere was mainly in mild turbidity degree. The linear correlation coefficient between AOD_{550} and β reached 0.94. With the increase in atmospheric turbidity, the attenuation effect of aerosol particles on solar radiation was enhanced, resulting in an increase in AOD_{550}. The data points consisting of AOD_{550} and $\alpha_{440\text{–}870}$ presented a weak positive correlation growth trend, and their linear fit correlation was poor.

(2) Atmospheric turbidity increased on a monthly basis, from a mild turbidity level in November 2021 to a high turbidity level in December 2021. β increased again in January 2022, and atmospheric turbidity increased further. Along with the increase in β, AOD_{550} rapidly increased, and the monthly increase in both exceeded 50%. The monthly variation in $\alpha_{440\text{–}870}$ was relatively small, and its value had a small proportional increase in December 2021 and decreased in January 2022. The mean value of $\alpha_{440\text{–}870}$ in December 2021 was higher than the remaining two months, the standard deviation of $\alpha_{440\text{–}870}$ in that month was the smallest, and the aerosol particle size distribution was the most concentrated.

(3) Particulate matter concentrations show a good linear correlation with both AOD_{550} and β. A sharp increase in the concentration of particulate matter was observed when the aerosol content increased, and the turbidity level increased. Among them, the response of β to the change in fine particulate matter concentration is most effective, with their linear correlation coefficient reaching 0.861 and their monthly relative increases being close. The monthly relative increase in $PM_{2.5}$ exceeds 50%, while the monthly relative increase in PM_{10} is near 30%. As atmospheric turbidity increases month by month, the ratio of PM_{10} to $PM_{2.5}$ concentration shrinks from 1.8:1 in November 2021 to 1.2:1 in January 2022, with the proportion of fine particulate matter in the atmosphere increasing.

Author Contributions: Conceptualization, Q.M. and G.H.; methodology and validation, G.H.; writing—original draft preparation, G.H.; editing and review, Q.M. and X.N.; funding acquisition, Q.M. All authors have read and agreed to the published version of the manuscript.

Funding: This work was supported by the National Natural Science Foundation of China (Grant No. 51876147).

Institutional Review Board Statement: Not applicable.

Informed Consent Statement: Not applicable.

Data Availability Statement: Data not available due to privacy restrictions.

Acknowledgments: A very special acknowledgment is made to the editors and referees who make important comments to improve this paper.

Conflicts of Interest: The authors declare no conflict of interest.

References

1. Allen, R.J.; Landuyt, W.; Rumbold, S.T. An increase in aerosol burden and radiative effects in a warmer world. *Nat. Clim. Chang.* **2015**, *6*, 269–274. [CrossRef]
2. He, Z.; Dong, C.; Liang, D.; Mao, J. A weighted-sum-of-gray soot-fractal-aggregates model for nongray heat radiation in the high temperature gas-soot mixture. *J. Quant. Spectrosc. Radiat. Transf.* **2021**, *260*, 107431. [CrossRef]
3. Xin, J.; Gong, C.; Wang, S.; Wang, Y. Aerosol direct radiative forcing in desert and semi-desert regions of northwestern China. *Atmos. Res.* **2016**, *171*, 56–65. [CrossRef]
4. Yu, X.; Kumar, K.R.; Lu, R.; Ma, J. Changes in column aerosol optical properties during extreme haze-fog episodes in January 2013 over urban Beijing. *Environ. Pollut.* **2016**, *210*, 217–226. [CrossRef]
5. Mao, Q.; Cheng, F.; Chen, M. Experimental Study on the Chemical Characterization of Atmospheric Aerosols in Wuhan, China. *Atmosphere* **2021**, *12*, 1393. [CrossRef]
6. Ji, D.; Li, L.; Wang, Y.; Zhang, J.; Cheng, M.; Sun, Y.; Liu, Z.; Wang, L.; Tang, G.; Hu, B.; et al. The heaviest particulate air-pollution episodes occurred in northern China in January, 2013: Insights gained from observation. *Atmos. Environ.* **2014**, *92*, 546–556. [CrossRef]
7. Zhao, H.; Che, H.; Zhang, X.; Ma, Y.; Wang, Y.; Wang, H.; Wang, Y. Characteristics of visibility and particulate matter (PM) in an urban area of Northeast China. *Atmos. Pollut. Res.* **2013**, *4*, 427–434. [CrossRef]
8. Zheng, Y.; Che, H.; Xia, X.; Wang, Y.; Yang, L.; Chen, J.; Wang, H.; Zhao, H.; Li, L.; Zhang, L.; et al. Aerosol optical properties and its type classification based on multiyear joint observation campaign in north China plain megalopolis. *Chemosphere* **2020**, *273*, 128560. [CrossRef]
9. Liu, L.; Tan, H.; Fan, S.; Cai, M.; Xu, H.; Li, F.; Chan, P. Influence of aerosol hygroscopicity and mixing state on aerosol optical properties in the Pearl River Delta region, China. *Sci. Total Environ.* **2018**, *627*, 1560–1571. [CrossRef]
10. Kampa, M.; Castanas, E. Human health effects of air pollution. *Environ. Pollut.* **2008**, *151*, 362–367. [CrossRef]
11. Lelieveld, J.; Klingmuller, K.; Pozzer, A.; Burnett, R.T.; Haines, A.; Ramanathan, V. Effects of fossil fuel and total anthropogenic emission removal on public health and climate. *Proc. Natl. Acad. Sci. USA* **2019**, *116*, 7192–7197. [CrossRef] [PubMed]
12. Alam, K.; Trautmann, T.; Blaschke, T.; Majid, H. Aerosol optical and radiative properties during summer and winter seasons over Lahore and Karachi. *Atmos. Environ.* **2012**, *50*, 234–245. [CrossRef]
13. Wang, H.; Miao, Q.; Shen, L.; Yang, Q.; Wu, Y.; Wei, H. Air pollutant variations in Suzhou during the 2019 novel coronavirus (COVID-19) lockdown of 2020: High time-resolution measurements of aerosol chemical compositions and source apportionment. *Environ. Pollut.* **2021**, *271*, 116298. [CrossRef] [PubMed]
14. Reddington, C.L.; Butt, E.W.; Ridley, D.A.; Artaxo, P.; Morgan, W.T.; Coe, H.; Spracklen, D.V. Air quality and human health improvements from reductions in deforestation-related fire in Brazil. *Nat. Geosci.* **2015**, *8*, 768–771. [CrossRef]
15. Logan, T.; Xi, B.; Dong, X.; Li, Z.; Cribb, M. Classification and investigation of Asian aerosol absorptive properties. *Atmos. Chem. Phys.* **2013**, *13*, 2253–2265. [CrossRef]

16. Patel, P.N.; Dumka, U.C.; Babu, K.N.; Mathur, A.K. Aerosol characterization and radiative properties over Kavaratti, a remote island in southern Arabian Sea from the period of observations. *Sci. Total Environ.* **2017**, *599–600*, 165–180. [CrossRef]
17. Shaik, D.S.; Kant, Y.; Mitra, D.; Babu, S.S. Assessment of Aerosol Characteristics and Radiative Forcing Over Northwest Himalayan Region. *IEEE J. Sel. Top. Appl. Earth Obs. Remote Sens.* **2017**, *10*, 5314–5321. [CrossRef]
18. Levy, R.C.; Remer, L.A.; Dubovik, O. Global aerosol optical properties and application to Moderate Resolution Imaging Spectroradiometer aerosol retrieval over land. *J. Geophys. Res. Atmos.* **2007**, *112*, 7815. [CrossRef]
19. Mao, Q.; Huang, C.; Zhang, H.; Chen, Q.; Yuan, Y. Performance of MODIS aerosol products at various timescales and in different pollution conditions over eastern Asia. *Sci. China Technol. Sci.* **2020**, *64*, 774–784. [CrossRef]
20. Wei, J.; Peng, Y.; Guo, J.; Sun, L. Performance of MODIS Collection 6.1 Level 3 aerosol products in spatial-temporal variations over land. *Atmos. Environ.* **2019**, *206*, 30–44. [CrossRef]
21. Gui, K.; Che, H.; Wang, Y.; Xia, X.; Holben, B.N.; Goloub, P.; Cuevas-Agulló, E.; Yao, W.; Zheng, Y.; Zhao, H.; et al. A global-scale analysis of the MISR Level-3 aerosol optical depth (AOD) product: Comparison with multi-platform AOD data sources. *Atmos. Pollut. Res.* **2021**, *12*, 101238. [CrossRef]
22. Jose, S.; Mishra, A.K.; Singh, S. A correlational study on size differentiated aerosols on monsoonal and pre-monsoonal cloud properties over the Indo Gangetic Basin. *Atmos. Res.* **2021**, *262*, 105796. [CrossRef]
23. Sabetghadam, S.; Alizadeh, O.; Khoshsima, M.; Pierleoni, A. Aerosol properties, trends and classification of key types over the Middle East from satellite-derived atmospheric optical data. *Atmos. Environ.* **2021**, *246*, 118100. [CrossRef]
24. Mao, Q.; Huang, C.; Chen, Q.; Zhang, H.; Yuan, Y. Satellite-based identification of aerosol particle species using a 2D-space aerosol classification model. *Atmos. Environ.* **2019**, *219*, 117057. [CrossRef]
25. Tan, Y.; Wang, Q.; Zhang, Z. Assessing spatiotemporal variations of AOD in Japan based on Himawari-8 L3 V31 aerosol products: Validations and applications. *Atmos. Pollut. Res.* **2022**, *13*, 101439. [CrossRef]
26. Ma, Y.; Liu, B.; Gong, W.; Shi, Y.; Jin, S. Impact of environmental pollution on the retrieval of AOD products from Visible Infrared Imaging Radiometer Suite (VIIRS) over wuhan. *Atmos. Pollut. Res.* **2019**, *10*, 2063–2071. [CrossRef]
27. Xie, Q.; Sun, Q. Monitoring the Spatial Variation of Aerosol Optical Depth and Its Correlation with Land Use/Land Cover in Wuhan, China: A Perspective of Urban Planning. *Int. J. Environ. Res. Public Health* **2021**, *18*, 1132. [CrossRef]
28. Sayer, A.M.; Munchak, L.A.; Hsu, N.C.; Levy, R.C.; Bettenhausen, C.; Jeong, M.J. MODIS Collection 6 aerosol products: Comparison between Aqua's e-Deep Blue, Dark Target, and "merged" data sets, and usage recommendations. *J. Geophys. Res. Atmos.* **2014**, *119*, 13965–913989. [CrossRef]
29. Levy, R.C.; Mattoo, S.; Munchak, L.A.; Remer, L.A.; Sayer, A.M.; Patadia, F.; Hsu, N.C. The Collection 6 MODIS aerosol products over land and ocean. *Atmos. Meas. Tech.* **2013**, *6*, 2989–3034. [CrossRef]
30. Tao, M.; Wang, Z.; Tao, J.; Chen, L.; Wang, J.; Hou, C.; Wang, L.; Xu, X.; Zhu, H. How Do Aerosol Properties Affect the Temporal Variation of MODIS AOD Bias in Eastern China? *Remote Sens.* **2017**, *9*, 800. [CrossRef]
31. Zhu, J.; Xia, X.; Wang, J.; Che, H.; Chen, H.; Zhang, J.; Xu, X.; Levy, R.; Oo, M.; Holz, R.; et al. Evaluation of aerosol optical depth and aerosol models from VIIRS retrieval algorithms over North China Plain. *Remote Sens* **2017**, *9*, 432. [CrossRef]
32. Zhao, H.; Che, H.; Zhang, X.; Ma, Y.; Wang, Y.; Wang, X.; Liu, C.; Hou, B.; Che, H. Aerosol optical properties over urban and industrial region of Northeast China by using ground-based sun-photometer measurement. *Atmos. Environ.* **2013**, *75*, 270–278. [CrossRef]
33. Abdullah, N.A.; Latif, M.T.; Juneng, L.; Uning, R.; Hassan, H.; Azhari, A.; Tuch, T.; Wiedensohler, A. Aerosol particle properties at a remote tropical rainforest in Borneo. *Atmos. Pollut. Res.* **2022**, *13*, 101383. [CrossRef]
34. Vicente-Luis, A.; Tremblay, S.; Dionne, J.; Chang, R.Y.W.; Fogal, P.F.; Leaitch, W.R.; Sharma, S.; Kolonjari, F.; Hayes, P.L. In situ optical and microphysical properties of tropospheric aerosols in the Canadian High Arctic from 2016 to 2019. *Atmos. Environ.* **2021**, *250*, 118254. [CrossRef]
35. Gong, W.; Zhang, S.; Ma, Y. Aerosol Optical Properties and Determination of Aerosol Size Distribution in Wuhan, China. *Atmosphere* **2014**, *5*, 81–91. [CrossRef]
36. Bibi, S.; Alam, K.; Chishtie, F.; Bibi, H. Characterization of absorbing aerosol types using ground and satellites based observations over an urban environment. *Atmos. Environ.* **2017**, *150*, 126–135. [CrossRef]
37. Boiyo, R.; Kumar, K.R.; Zhao, T. Statistical intercomparison and validation of multisensory aerosol optical depth retrievals over three AERONET sites in Kenya, East Africa. *Atmos. Res.* **2017**, *197*, 277–288. [CrossRef]
38. Liu, B.; Ma, Y.; Gong, W.; Zhang, M.; Wang, W.; Shi, Y. Comparison of AOD from CALIPSO, MODIS, and Sun Photometer under Different Conditions over Central China. *Sci. Rep.* **2018**, *8*, 10066. [CrossRef] [PubMed]
39. Wang, W.; Mao, F.; Pan, Z.; Du, L.; Gong, W. Validation of VIIRS AOD through a Comparison with a Sun Photometer and MODIS AODs over Wuhan. *Remote Sens.* **2017**, *9*, 403. [CrossRef]
40. Chen, Q.-X.; Shen, W.-X.; Yuan, Y.; Tan, H.-P. Verification of aerosol classification methods through satellite and ground-based measurements over Harbin, Northeast China. *Atmos. Res.* **2019**, *216*, 167–175. [CrossRef]
41. Jin, S.; Ma, Y.; Zhang, M.; Gong, W.; Lei, L.; Ma, X. Comparation of aerosol optical properties and associated radiative effects of air pollution events between summer and winter: A case study in January and July 2014 over Wuhan, Central China. *Atmos. Environ.* **2019**, *218*, 117004. [CrossRef]
42. Ångström, A. On the Atmospheric Transmission of Sun Radiation and on Dust in the Air. *Geogr. Ann.* **2017**, *11*, 156–166. [CrossRef]

43. Xu, X.; Xie, L.; Yang, X.; Wu, H.; Cai, L.; Qi, P. Aerosol optical properties at seven AERONET sites over Middle East and Eastern Mediterranean Sea. *Atmos. Environ.* **2020**, *243*, 117884. [CrossRef]
44. Breon, F.M.; Tanre, D.; Generoso, S. Aerosol effect on cloud droplet size monitored from satellite. *Science* **2002**, *295*, 834–838. [CrossRef]
45. Ångström, A. The parameters of atmospheric turbidity. *Tellus* **2016**, *16*, 64–75. [CrossRef]
46. Zakey, A.S.; Abdelwahab, M.M.; Makar, P.A. Atmospheric turbidity over Egypt. *Atmos. Environ.* **2004**, *38*, 1579–1591. [CrossRef]

remote sensing

MDPI

Technical Note

Machine Learning to Identify Three Types of Oceanic Fronts Associated with the Changjiang Diluted Water in the East China Sea between 1997 and 2021

Dae-Won Kim [1], So-Hyun Kim [1,2] and Young-Heon Jo [1,2,*]

1 Department of Oceanography, Pusan National University, Busan 46241, Korea; daewon@pusan.ac.kr (D.-W.K.); hyun3460@pusan.ac.kr (S.-H.K.)
2 BK21 School of Earth Environmental System, Pusan National University, Busan 46241, Korea
* Correspondence: joyoung@pusan.ac.kr; Tel.: +82-51-510-3372

Abstract: Long-term sea surface salinity (SSS) in the East China Sea (ECS) was estimated based on Ocean Color Climate Change Initiative (OC-CCI) data using machine learning during the summer season (June to September) from 1997 to 2021. Changjiang diluted water (CDW) in the ECS propagates northeastward and forms longitudinally-oriented ocean fronts. To determine the CDW's distribution, three fronts were investigated: (1) a CDW front based on chlorophyll-a concentration (Chl), SSS, and sea surface temperature (SST); (2) a CDW front based on sea surface density (SSD); and (3) a CDW front for nutrient distribution. The Chl fronts matched well with the SSS fronts, suggesting that Chl variation in the ECS is highly correlated with the CDW. Furthermore, the SSD fronts spatially matched well with nitrogen concentration. Sea level anomaly (SLA) variation with SSD was also detected, indicating that CDW had sufficiently large effects on SLA so that they may be detectable by altimeter measurements. This result suggests that the influence of steric height changes and the inflow from rivers are significant in the ECS. Additionally, the continuous long-term SSD developed in this study enables researchers to detect the CDW front and its influence on the ECS marine environment.

Keywords: Changjiang diluted water; sea surface salinity estimation; ocean front; long-term ocean color; OC-CCI

Citation: Kim, D.-W.; Kim, S.-H.; Jo, Y.-H. Machine Learning to Identify Three Types of Oceanic Fronts Associated with the Changjiang Diluted Water in the East China Sea between 1997 and 2021. *Remote Sens.* **2022**, *14*, 3574. https://doi.org/10.3390/rs14153574

Academic Editor: SeungHyun Son

Received: 26 May 2022
Accepted: 22 July 2022
Published: 25 July 2022

1. Introduction

Because Changjiang diluted water (CDW) includes freshwater, nutrients, and sediments, understanding the variations in CDW systems is crucial for analyzing material transport and biogeochemical processes in the East China Sea (ECS) [1–3]. CDW detection has been attemptI confirmed using satellite measurements of sea surface salinity (SSS), chlorophyll-a concentration (Chl), and turbidity; however, studies on long-term SSS distribution patterns are scarce, possibly because of limitations in continuous measurements [2,4].

Several recent studies have demonstrated the potential of ocean color satellite remote sensing for estimating the SSS in the ECS [2,5,6]. This is because CDW contains colored dissolved solids or pollutant matter that reflects light at specific wavelengths, such as colored dissolved organic matter (CDOM) and terrestrial particles; the optical feature of CDW is indirectly related to SSS. Based on this principle, Kim et al. [6] estimated the SSS using Geostationary Ocean Color Imager (GOCI, 500 m resolution) data from 2015 to 2020. The application of the same approach also detected SSS variation due to river discharge near the estuaries of the mid-Atlantic [7] and Gulf of Mexico [8], suggesting that SSS variation may be monitored globally using ocean color satellite sensors.

Kim et al. [4] defined CDW based on statistical analysis with the K-means clustering technique using Sea-viewing Wide Field-of-view Sensor (SeaWiFS) Chl in the summer in the ECS from 1998 to 2007. The study results revealed the interannual variation in the CDW indicated by the high satellite Chl and correlated it with the interannual variation in the Changjiang summer freshwater discharge. Bai et al. [2] analyzed summertime CDW variations from 1998 to 2010 using the remote sensing reflectance (Rrs) of SeaWiFS. This study identified three types of plume shapes: (1) the commonly known northeastward transportation, (2) a case in which most of the CDW is to the north of Jeju Island, and (3) a rare case in which the CDW front is transported southeastward. The plume shapes were related to discharge, wind speed in the 45° and 60° directions, and typhoons. Although these studies analyzed the CDW distribution over 10 years, they mainly focused on the variation in CDW distribution and did not investigate the impact of CDW fronts on the biogeochemical environment in the ECS.

Because the Changjiang River is one of the primary ECS nutrient sources, plume water is rich in nutrients compared with that of ambient seawater [9–11]. In particular, dissolved inorganic nitrogen (DIN) and phosphate (DIP) fluxes have increased owing to the widespread use of chemical fertilizers in the Changjiang River basin [12]. Kwon et al. [13] analyzed DIN and DIP concentrations from the Changjiang River Estuary to the Korean Strait. The concentrations of DIN and DIP decreased sharply from the river mouth and were depleted within 200 km, and subsequently maintained approximately 0.28 μM L^{-1} and 0.07 μM L^{-1}, respectively, until 800 km. However, the analysis of DIN and DIP concentrations depending on CDW extension was conducted in previous studies because of the limitation of the spatiotemporal coverage of shipboard surveys. Therefore, satellite-based CDW monitoring and analysis with in situ chemical measurements are required.

Therefore, the objective of this study was to estimate the long-term SSS distribution using 25 years of satellite observations and to determine the biogeochemical influence of CDW. Because oceanic fronts offer habitats for marine biology, detecting the principal front is an important task. This study presents the differences in the location of the SSS, Chl, sea surface temperature (SST), and sea surface density (SSD) fronts. In addition, the nitrogen and phosphate distributions were compared to the CDW front locations.

2. Materials and Methods

2.1. Materials

2.1.1. Satellite Data

To estimate the long-term SSS in the ECS, Rrs measurements derived from the Ocean Color Climate Change Initiative (OC-CCI) version 5.0 were used [14]. The OC-CCI products were developed by merging observations from the SeaWiFS, moderate resolution imaging spectroradiometer onboard the Aqua Earth Observing System (MODIS-Aqua), and Medium Resolution Imaging Spectrometer (MERIS). The products were constructed based on SeaWiFS datasets, and the atmospherically corrected Rrs values of MODIS and MERIS data were shifted to the nearest SeaWiFS bands (412, 443, 490, 510, 560, and 665 nm) based on a bio-optical model [15]. This process generates a set of Rrs values in six bands, which increases the inter-sensor consistency and reduces the bias values. In addition, the OC-CCI Chl dataset was used. OC-CCI uses the OC4V6 algorithm to retrieve Chl. The spatial and temporal resolutions of the OC-CCI used in this study are 4 km and daily, respectively.

MODIS-Aqua global level 3 mapped SST version 2019.0 data from 2002 to 2021 were obtained from the Physical Oceanography Distributed Active Archive Center (PO.DAAC) of National Aeronautics and Space Administration (NASA). Because SST data were available from 2002 to 2021 (20 years), SST front detection and SSD calculations were conducted for 20 years. In addition, altimeter satellite-gridded SLA data from 1997 to 2020 were computed with respect to a twenty-year (1993–2012) mean. These data were estimated by optimal interpolation by merging the measurements from the different available altimeter missions: Jason-3, Sentinel-3A, Haiyang-2A (HY-2A), Saral/AltiKa, Cryosat-2, Jason-2, Jason-1, Topex/Poseidon, Environmental Satellite (ENVISAT), Geostat follow-on radar altimeter (GFO), European Remote Sensing Satellite (ERS) 1 and 2. The spatial and temporal resolutions were 0.25 × 0.25° and monthly mean, respectively.

2.1.2. Observation Data

Repeated shipboard measurements conducted by the National Institute of Fisheries Science (NIFS) from 1997 to 2021 were used to develop and evaluate the SSS model. Observations were conducted near the Korean Peninsula, including the Yellow (YS) and East/Japan Sea (EJS), South Sea of Korea (SSK), and ECS. This shipboard observation investigates SST, SSS, dissolved oxygen, phosphate, nitrite, nitrate, and silica bi-monthly in the YS, EJS, and SSK, and at three-month intervals in the ECS. Because the model was designed to detect CDW, the southern part of the YS and EJS and the total SSK and ECS station datasets were collected (Figure 1). The data for these observations in various regions are presented in Table 1. All datasets, including satellite measurements, were obtained during the summer season (June to September) when the CDW is a major factor in SSS variation in the ECS.

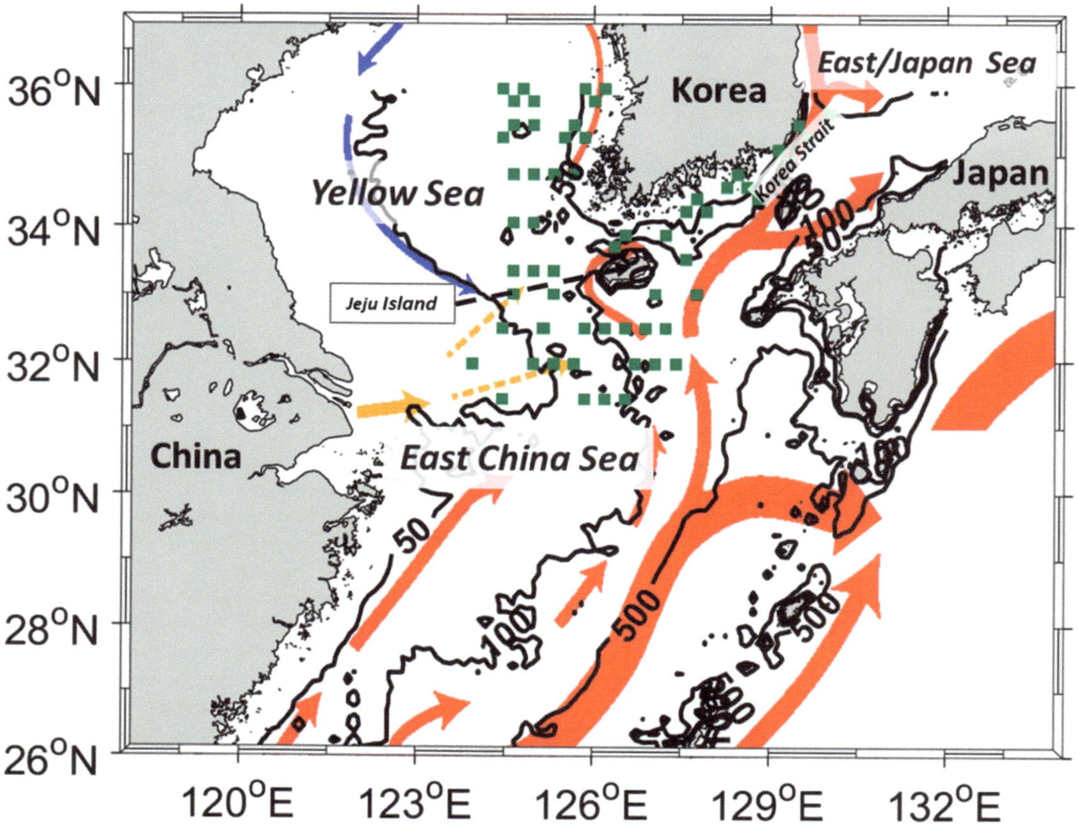

Figure 1. Research region and matching stations for the in situ and OC-CCI data from 1997 to 2021. The serial shipboard observation stations were marked in green squares. The red, blue, and yellow arrows indicate warm, cold, and Changjiang outflow ocean currents, respectively, during summer season from 2002 to 2005 in this region (provided by Korea Hydrographic and Oceanographic Agency). Three bathymetric isobaths (50, 100, and 500 m) are also overlaid.

Table 1. The number of serial shipboard observations by Korean National Institute of Fisheries Science at the four regions from summertime (June to September) of 1998 to 2021. Among total 5488 data, about half (2421) data were matched with OC-CCI.

Year	ECS [a]	SSK [b]	YC [c]	ES [d]
1998	26	122	77	34
1999	22	108	68	19
2000	32	105	68	20
2001	33	102	68	20
2002	32	98	67	18
2003	61	108	68	20
2004	32	108	68	20
2005	32	108	69	20
2006	33	108	68	20
2007	32	110	68	20
2008	32	110	68	20
2009	32	108	68	20
2010	32	108	48	20
2011	32	108	68	20
2012	32	108	68	20
2013	32	108	68	20
2014	32	108	68	20
2015	34	108	68	20
2016	32	108	68	20
2017	32	108	68	20
2018	32	108	68	20
2019	32	108	68	20
2020	32	108	68	20
2021	32	108	68	20

[a] East China Sea. [b] Southern Sea of Korea. [c] Yellow Sea. [d] East Sea.

2.2. MPNN Model for SSS Estimation

Because the multi-layer perceptron neural network (MPNN) has been used to estimate SSS based on ocean color measurements in various regions [5–8], this method was adopted to develop an SSS estimation model for the OC-CCI. Because the shipboard survey stations are far from the Changjiang River mouth, the SSS range of in situ data lacking in the low-salinity conditions (under 25 psu). Thus, the model overestimates the Changjiang River mouth region, where the SSS is extremely low (<20 psu). Figure 2 shows an MPNN schematic diagram including six OC-CCI Rrs bands in the input layer and in situ SSS data in the output layer. To assemble two different datasets, the temporal resolution of in situ measurements was regarded as daily, and OC-CCI data were selected at the nearest location of the in situ survey stations. Thereafter, the datasets were randomly divided into training (80%) and validation (20%) datasets during the training period. The training was repeated 1000 times for each step while changing the number of hidden layers and neurons.

Figure 3a shows the performance of the MPNN model based on the validation dataset independent of the training dataset. The scatters were concentrated from 30 to 33 psu and R^2 and root mean square error (RMSE) were 0.56 and 1.06 psu, respectively. Although the RMSE was acceptable, the SSS was corrected by the slope and bias owing to overestimation in the low SSS range (under 30 psu). After linear fitting, the overestimation and underestimation in the low and high SSS ranges were reduced despite the RMSE increasing to 0.34 psu. As the KNIFS serial shipboard survey stations are far from the Changjiang River mouth, an SSS under 25 psu was not obtained. Therefore, the accuracy decreased for low SSS values (under 30 psu). Given that the isohaline of the CDW in previous studies was 30 to 32 over the middle ECS [2,4,16,17], detection of the CDW fronts using our model near Jeju Island and the Korean Peninsula was possible. Moreover, the OC-CCI SSD was validated with KNIFS in situ measurements from 1997 to 2021 (Figure 3b). The R^2 and RMSE between in situ and OC-CCI-based SSD were 0.67 and 1.27 kg m^{-3}, respectively.

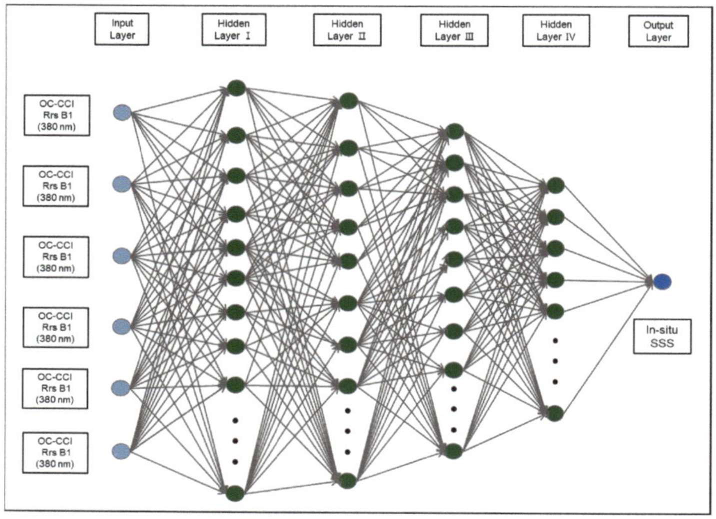

Figure 2. The MPNN model architecture based on the OC-CCI Rrs six bands data and KNIFS serial shipboard measurements. Among the 2421 datasets, 80% were used to train, and 20% were used to validate model. The input datasets were fixed, and the number of hidden layers and neurons in hidden layers were varied to derive the best performance.

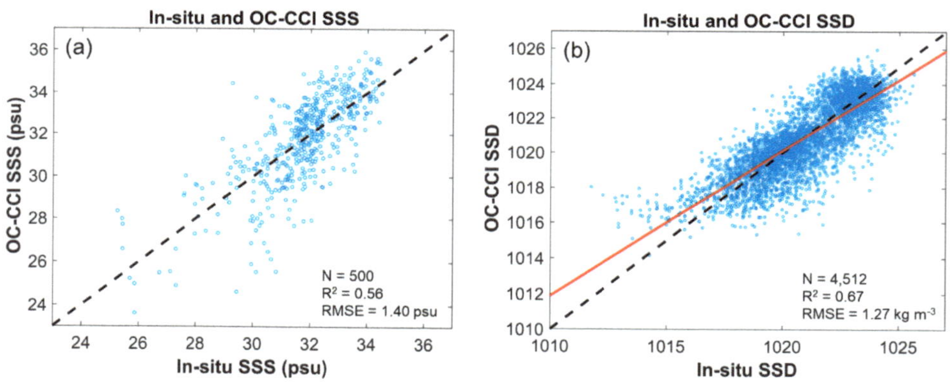

Figure 3. (**a**) Linear fitting by slope and bias scatter diagrams between NIFS serial shipboard measurements and OC-CCI SSS. (**b**) Scatter plot between in situ and OC-CCI SSD. Red line indicates a linear regression line.

3. Results

3.1. CDW Front Based on SSS, Chl, and SST

Because CDW rapidly evolves, factors such as gaps from frequent cloud cover, sun glint, and thick aerosols make it difficult to produce a monthly SSS distribution. However, it is possible to combine daily isohaline locations to produce cumulative isohaline footprints. Figure 4 shows the monthly cumulative isohaline footprints and monthly mean Chl concentrations from the June to September 2016 case as an example. The same figures from 1997 to 2021 are provided in the Supplementary Figures. The background color displays Chl; black dotted and solid lines and gray solid lines indicate 30, 31, and 32 isohaline footprints, respectively. The plume was generally directed northeastward, reaching Jeju Island, the Korean Peninsula, and sometimes the EJS through the Tsushima–Korea Strait.

The isohalines are nearly parallel to the isobaths, especially the 50 m isobaths in June and the 100 m isobaths in the other three months (Figure 1). The plume shapes were in good agreement with the results of Bai et al. [2]. They presented the SSS distribution in July and August from 1998 to 2010 and identified three major types of extending shapes of the CDW based on the 31 isohalines. In August (Figure 4b and Supplementary Figures), the most apparent shape appeared as a type 2 plume in Bai et al. [2], which extends entirely northeastward, with the majority of the plume water traveling northward to the southern YS and then advected eastward through the Jeju Strait, with less low-salinity water remaining on the middle shelf of the southern ECS. However, the classification of the three types was not conducted owing to the ambiguous criteria. For example, mixed features appeared in August 2016 and the plume extended northeastward and southward.

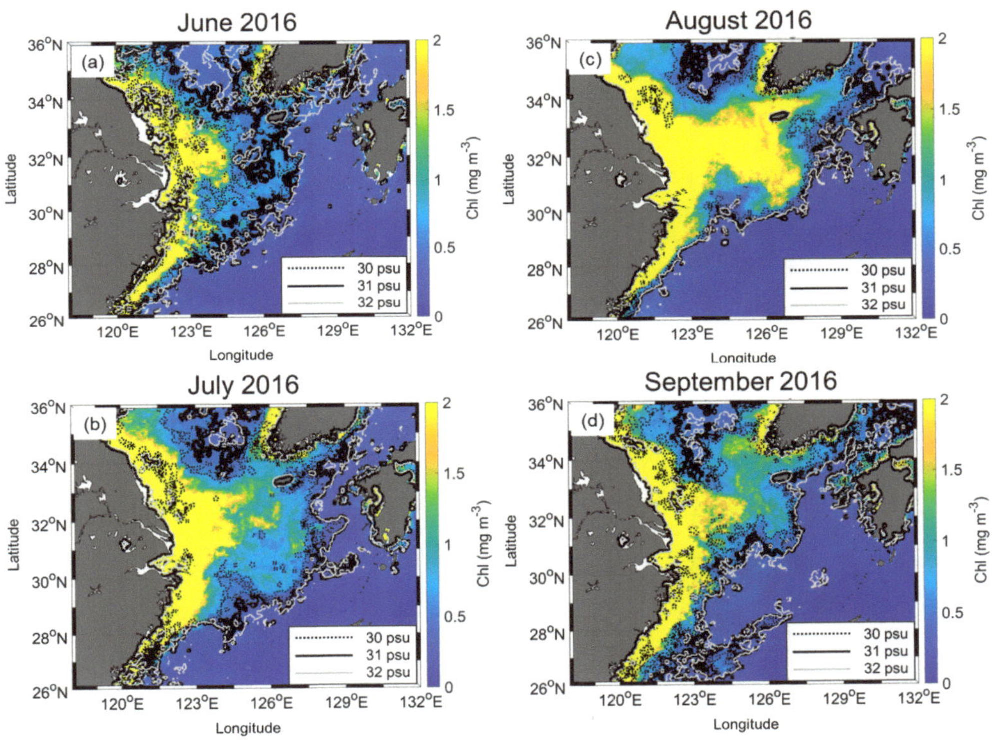

Figure 4. Monthly Chl (shown with color map) and cumulative SSS footprint (shown with contours) from (**a**–**d**) June to September 2016. The dotted and solid black and gray lines indicate 30, 31, and 32 isohalines, respectively.

The Chl distribution corresponded well with the isohaline footprints, despite not accumulating Chl. A probable reason for this is that the interaction between Chl and CDW had a time lag. Chl directly changes near the river mouth from the river origin Chl, but the rich nutrients in the CDW influence Chl within offshore regions [1,4,18]. This indicates that CDW could affect Chl for approximately one month. In addition, Chl rapidly decreased with increasing distance from the river mouth, unlike the isohalines pattern. This is because of complex situations, such as nutrient consumption, subsurface mixing, and the phytoplankton life cycle, which are difficult to demonstrate quantitatively in this study.

Figure 5 shows the climatology of SSS, Chl, and SST from June to September. The SSS and Chl were averaged from 1997 to 2021 (25 years), and SST was averaged from 2002 to 2021 (20 years) because the available data periods were different. Similar to CDW, the SSS and Chl fronts were distributed longitudinally. Chl and SSS showed river plume fronts, whereas SST did not. Near the Changjiang River mouth, SST formed ocean fronts longitudinally but was maintained up to 123° E. Except for this area, SST generated fronts latitudinally because of an insolation gradient. In addition, the CDW water mass, distinguished by 30 isohalines, was detached from the river mouth and transported to the Jeju Island. This detachment can be generated by the interaction between tide-induced vertical mixing and horizontal wind-driven movement of the CDW [19]. They determined the mechanism for the offshore detachment of the CDW using a three-dimensional numerical model. The detached patches in this study showed a pattern similar to the results of Moon et al. [19].

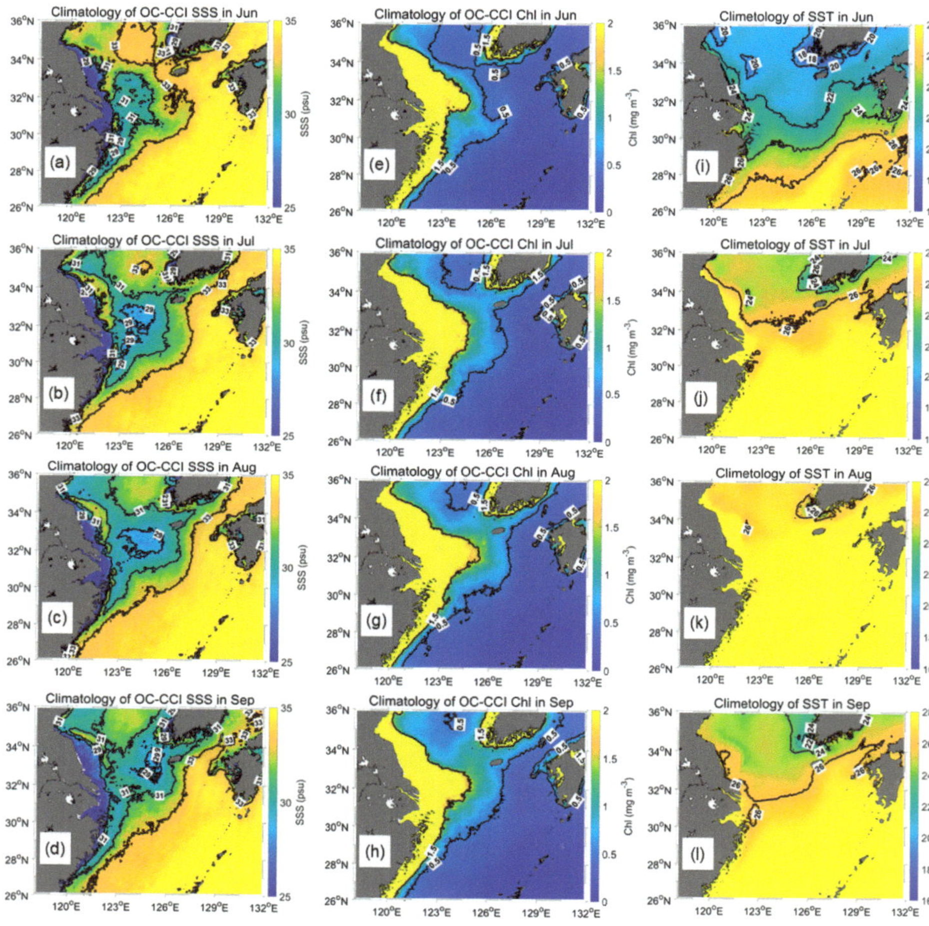

Figure 5. Climatology of OC-CCI (**a–d**) SSS and (**e–h**) Chl from June to September between 1997 and 2021. (**i–l**) Climatology of MODIS SST from June to September between 2002 and 2021.

3.2. CDW Front Based on Surface Density

The different patterns of the various fronts suggest a complicated environment in the study area. Therefore, SSD was calculated using SST and SSS to distinguish CDW. Figure 6 shows the climatology of SSD from 2002 to 2021. The overall pattern seems similar to SSS (Figure 5a–d), but over the southeast of the CDW, it is different. This region is known as the high SST and SSS region because the Kuroshio Current is strong in summer. Because

SSS is almost sustained in this area, SST increases from June to September, decreasing SSD by approximately 2 kg m^{-3}. However, in the northern area, isopycnic lines follow isohalines, although the isothermal lines are obviously different. In September, the area of the 1019 isopycnic line (Figure 6a) was smaller than the area of 30 isohalines (Figure 5d) west of Jeju Island. However, detached patches were still clearly identified in the SSD distribution. During the summer, SSD decreased with increasing SST in the entire study region, but a noticeable reduction was observed in the CDW-dominated area.

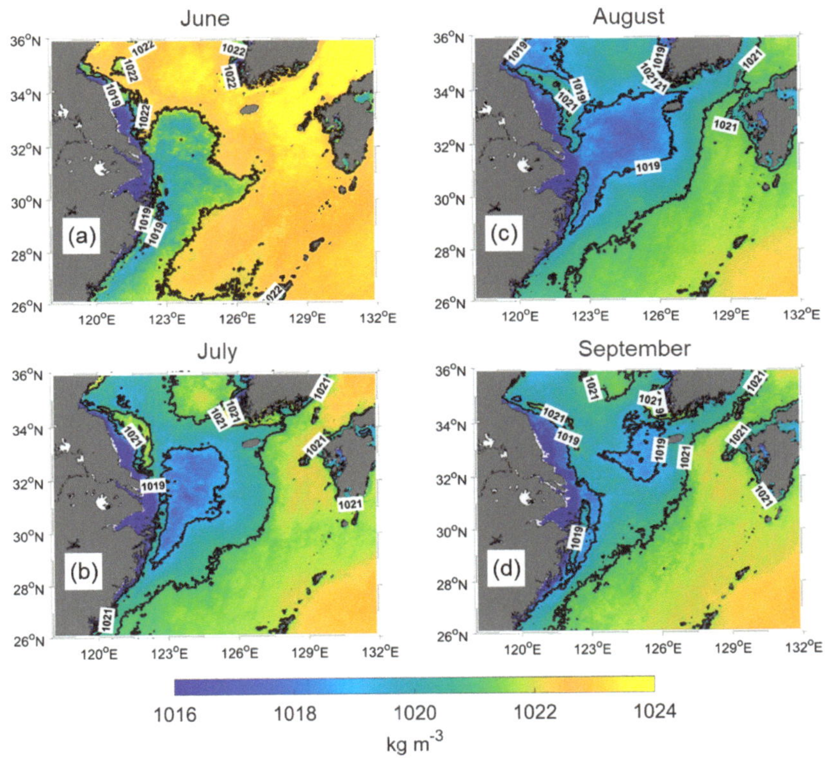

Figure 6. SSD climatology from June to September (**a**–**d**) between 2002 and 2021. The CDW identified by 1019 isopycnic lines in the central ECS.

Figure 7 shows the climatology of SSD in September for 20 years (2002 to 2021) and the SSD, SSS, SST, Chl, and SLA values along with a line originating from the Changjiang River mouth to the northeastern offshore region of the Jeju Island (AB line, presented in Figure 7a). The SSD shows a similar pattern to the SSS rather than the SST, indicating that SSS is a major factor driving SSD during September at this region. The detached SSD patch is also observed in Figure 7. The detached patch (high SSD and high SSS) was formed by tidal mixing with the subsurface, and ambient seawater covering due to the wind [19]. The Chl is at maximum at the river mouth and decreases toward B. However, the second peak of Chl appears around 124.3°E, where the SSD phase changes from high to low. This signifies that the Chl increased near the CDW front. Moreover, the SLA showed patterns similar to SSD. From the steric height variation perspective, the SLA evolved not only as water was added but also as SST or SSS evolved. In this case, the SSS variation could change the SLA since the SST is almost constant at 26 °C. When the SSD is high (low), SLA is low (high) because the water volume has a negative relationship with water density. However, the exact locations of the SSD and SLA peaks were slightly different. There are three possible causes: (1) the observed values were diminished by the 20-year averaging, (2) the spatial resolution of the SSD and SLA observations varied, and (3) the time lag between the SSD evolutions and the resulting SLA reaction. However, these assumptions are not clear. Consequently, additional studies are required to understand the relationship between SSD and SLA.

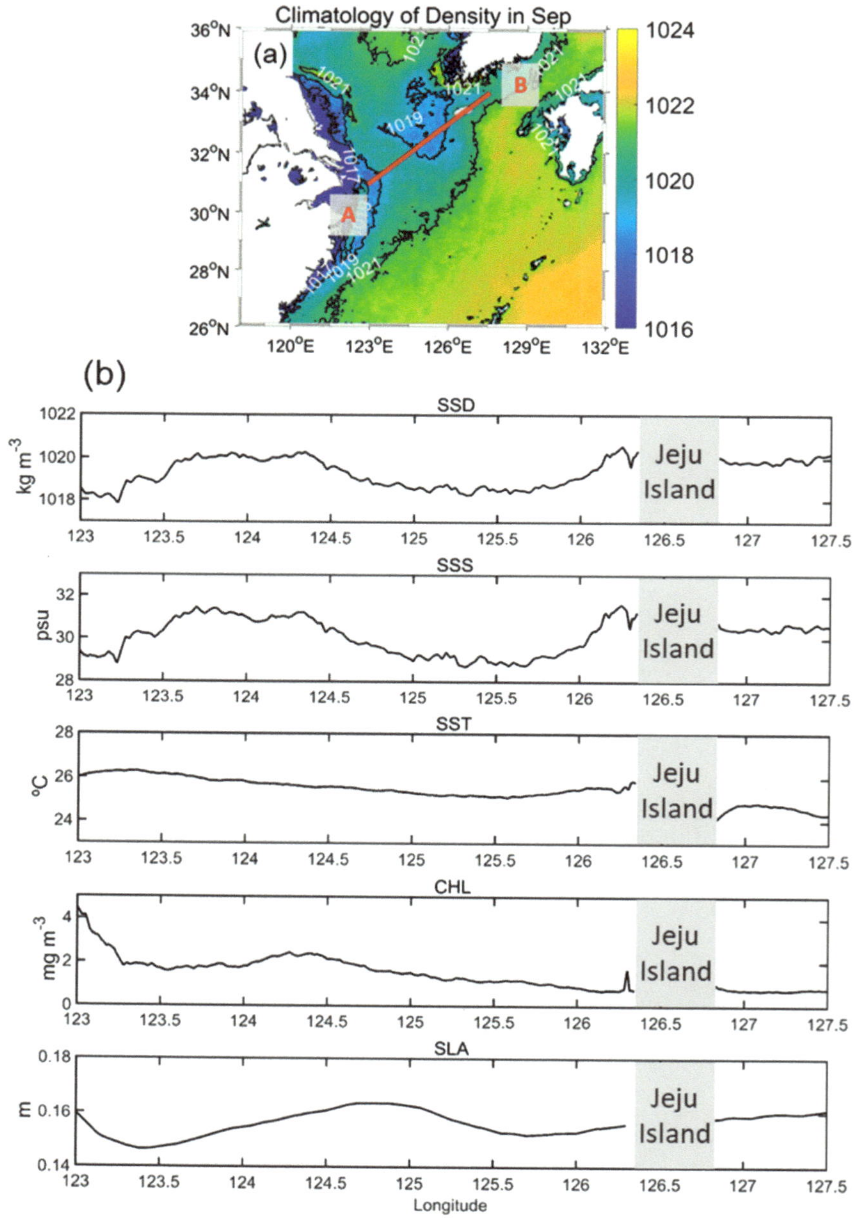

Figure 7. (**a**) SSD climatology during September for 20 years (from 2002 to 2021). (**b**) Climatology for SSD, SSS, SST, Chl, and SLA along the A–B line (red line in (**a**)) during September.

In addition, the correlation coefficient map and R^2 map showed the regions where Chl was affected by SSD (Figure 8). A relatively high R^2 (negative relationship) between Chl and SSD was observed near Jeju Island, the Yellow Sea, and the Tsushima–Korea Strait ($p < 0.05$). However, the other regions (particularly further south than 30°N) showed a low R^2 at insignificant levels. Therefore, the SSD variation caused by SST changes did not influence Chl; however, the SSS changes contributed to regional Chl variation.

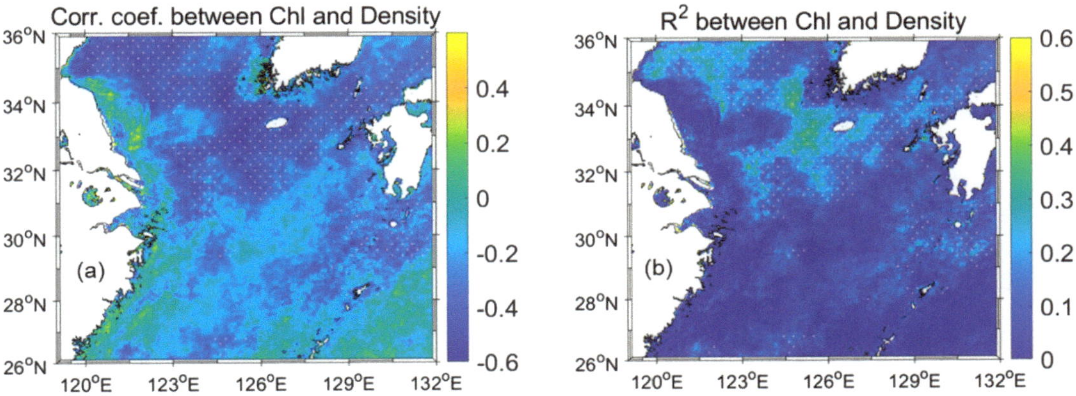

Figure 8. (**a**) Correlation coefficient and (**b**) R^2 between Chl and SSD from 2002 to 2021. The gray cross symbols mark a p-value < 0.05.

3.3. CDW Front for Nutrients' Distribution

Because the CDW is rich in nutrients, the CDW extension transports nutrients over the ECS. Therefore, the nutrient conditions differed across the CDW front. In this study, the shipboard observed nitrogen (N) and phosphorus (P) concentrations, and the SSD fronts from satellite data were compared. Figures 9 and 10 show the N and P levels with isopycnic lines in August for nine years (2002, 2005, 2006, 2007, 2009, 2011, 2014, 2019, and 2020) because the shipboard observations were conducted only in August at the ECS stations. In general, N was relatively high (4 to >10 μM L^{-1}) at a low SSD, close to the Changjiang River mouth. N showed a different pattern across the 1019 isopycnic lines, indicating that the CDW had more N than ambient seawater. Because N was obtained only once at each station, in contrast with SSD, which was a monthly average, the comparison of the spatial distribution had some limitations. Nevertheless, clear differences were revealed in August 2005, 2007, 2011, 2014, and 2020, except for a few stations that showed relatively high (low) N inside (outside) of each 1019 isopycnic line. This indicates that the CDW is relatively rich in N from the river plume, and that the satellite-driven SSD can distinguish the front of N.

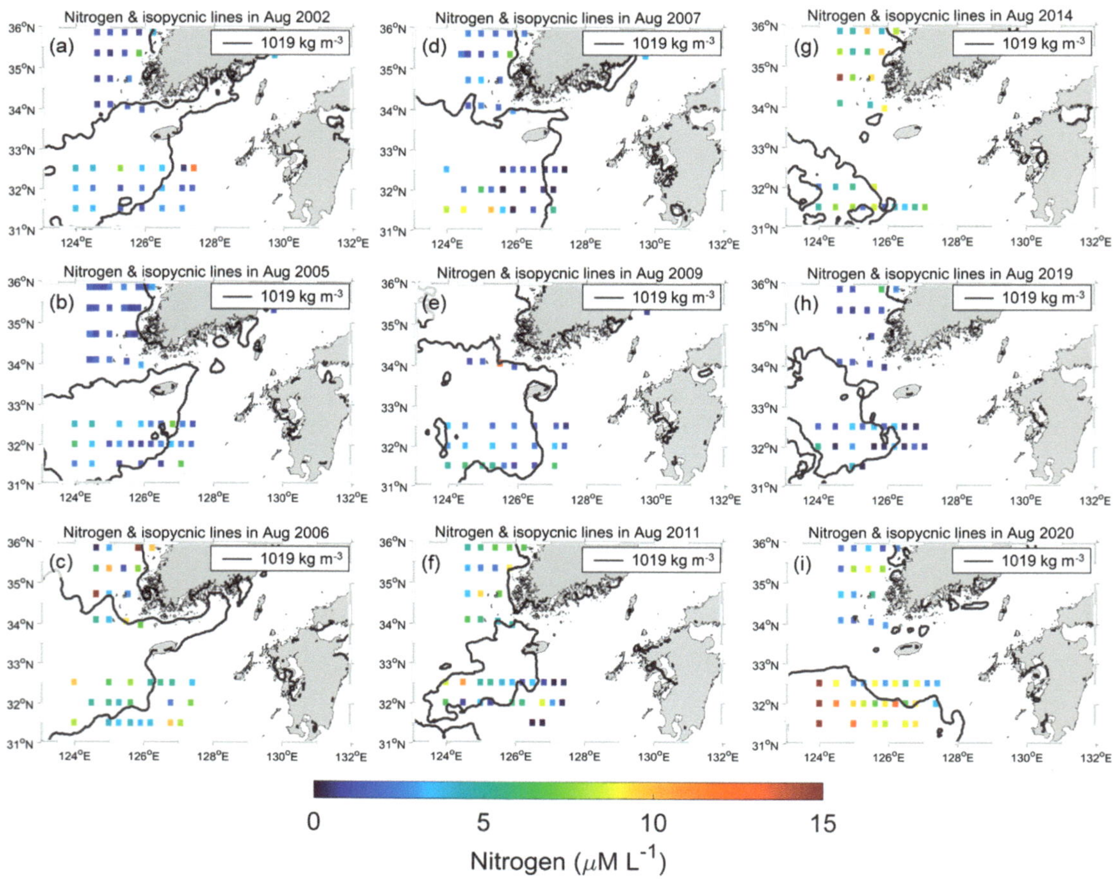

Figure 9. (**a–i**) In situ nitrogen concentration (colored square) and 1019 isopycnic line during August 2002, 2005, 2006, 2007, 2009, 2011, 2014, 2019, and 2020. Nitrogen resulted from NIFS serial shipboard observations.

However, P did not follow the N and SSD spatial distributions over the nine year period. A high P inside the CDW cases was observed in 2002, 2006, and 2007 (Figure 10). In particular, P was relatively lower in the CDW interior than in the exterior during 2020, which was the opposite phenomenon in the case of N. Li et al. [20] determined the increasing N trend during the past 50 years in the ECS due to the Changjiang River plume; however, P has been nearly stable since 2000. Therefore, the N/P ratio increased, which contributed to changes in the phytoplankton community's composition. Jiang et al. [21] obtained the same results in August 2009. They determined the factors that control the summer phytoplankton community in the ECS. This study examined phytoplankton blooms that caused enhanced eastward and southward CDW extension in August 2009. The diatoms and dinoflagellates bloomed under high N and low P conditions off the Changjiang Estuary. In addition, Zhou et al. [22] suggested that CDW plays an important role in diatom blooms by enhancing the pycnocline and P stress. Moreover, the coastal area adjacent to the Changjiang River Estuary is a high-risk region for harmful algal blooms [20,23,24]. Although the analysis of the phytoplankton community's composition depending on the CDW front was not conducted in this study, satellite-driven SSD can be useful for the prediction of N and P distributions or major phytoplankton species before shipboard surveys.

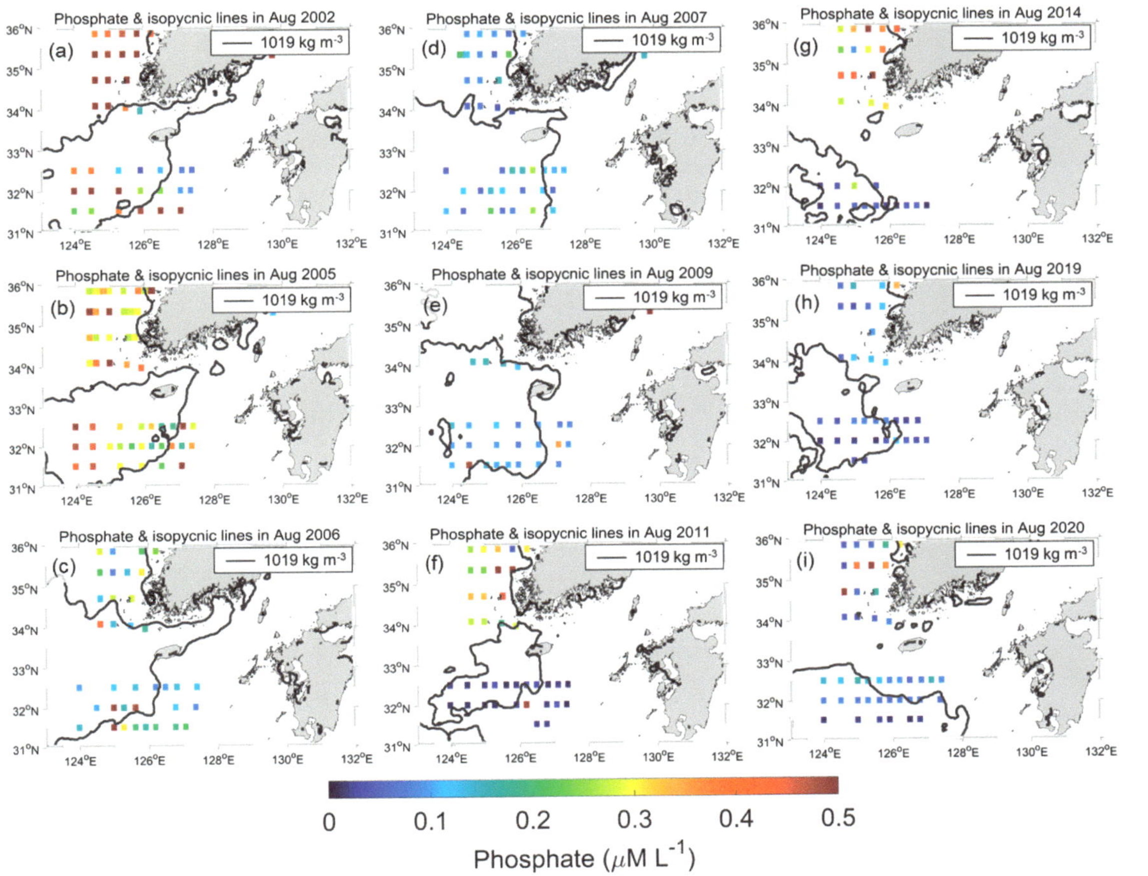

Figure 10. (**a–i**) Same as Figure 9 but phosphate during August 2002, 2005, 2006, 2007, 2009, 2011, 2014, 2019, and 2020.

4. Discussion

The SSS estimation model based on the OC-CCI ocean color measurements was developed using NIFS serial shipboard observation data. Although the in situ observation stations cover large regions, including the SSK and central ECS, the lowest SSS in these data was higher than 25 psu (see Figure 3a) because the stations were far from the Changjiang River mouth. Therefore, the model seldom estimated low SSS conditions (<20 psu) (Figure 5) despite being revealed near the Changjiang River mouth. Hence, the SSD in this area is underestimated. Bai et al. [2] also revealed this limitation. They used a simple analytical algorithm to estimate the SSS using the absorption of CDOM. The range of the in situ SSS was 26–36, and a low SSS (<28 psu) was not observed. However, such handling was not excessive owing to their interest in the variation in the CDW extension between 28, 29, 30, and 31 isohalines. Similarly, in this study, the middle range of SSS (29–32 psu) was used to analyze CDW identification and its impact; thus, the overestimation of low SSS was not significant. The training dataset contained abundant data in the middle SSS range, for which the model performance was relatively accurate (Figure 5). As shown in Figure 4, the monthly cumulative 31 isohalines were in good agreement with Chl fronts. As the cumulative SSS footprint was not influenced by low SSS (<20 psu), the results were reliable. However, the detached patches in July and August may have impacted the SSS (SSD) overestimation (underestimation) (Figures 5 and 6). Thirty isohalines and 1019 isopycnic lines were separated from the Changjiang River mouth at 122.5°E, where the uncertainty was high. The overestimation (underestimation) of the SSS (SSD) produced a westward extension of the detached patches. Therefore, offshore CDW detachment occurred on the east side, and the area of the patch might have been much smaller than indicated.

To overcome these model limitations, more training datasets near the Changjiang River mouth region are required. Rarely obtained shipboard measurements are valuable for validating satellite-driven SSS products; however, they are insufficient for developing an SSS estimation model. Microwave satellite sensor-based SSS data can solve this problem. Kim et al. [5] and Kim et al. [6] avoided such limitations by using soil moisture active passive (SMAP) SSS data. Their model could estimate an extremely low SSS (<20 psu) near the Changjiang River mouth. However, it was difficult to develop the SSS estimation model for 25 years because SMAP data are available from 2015. Although the Soil Moisture and Ocean Salinity (SMOS) satellite has observed SSS since 2009, the low spatiotemporal resolutions and imprecision in coastal regions (a result of land interference) result in difficulties for its utilization in the ECS. Improving microwave SSS data quality in the ECS and assimilation with in situ measurements are recommended for further research.

In general, oceanic fronts are detected using specific algorithms, such as the histogram-based Cayula–Cornillon algorithm (CCA) and gradient-based Belkin–O'Reilly algorithm (BOA) [25]. However, these algorithms are unsuitable for investigating continuous front detection because they are highly sensitive to different parameter ranges. Therefore, the gradient-based fronts were inconsistent at different times. Therefore, the CDW fronts based on SSS and SSD were identified using thresholds (31 psu for SSS and 1019 kg m^{-1} for SSD). Although the thresholds were simply determined, previous research [2] supports the value of the threshold (31 psu), and effectively produced CDW fronts for 25 years.

This study was mainly conducted based on qualitative analysis because we focused on the spatial variation in marine environments caused by CDW. The spatial distribution of the CDW fronts explains the influence of CDW on Chl and DIN. However, to understand the mechanism and major factors of CDW behavior, more quantitative analyses are required in further studies. To conduct a quantitative analysis, more accurate SSS estimation and small-scale analyses are required. GOCI and GOCI-II measurements are valuable for small-scale analysis. Using high spatial (<500 m) and temporal (hourly) resolutions, more detailed analyses are possible, such as tides, small-scale eddies, wind, and typhoons.

5. Conclusions

In this study, an MPNN model for summertime SSS estimation in the ECS was developed based on satellite and in situ data from 1997 to 2021. Using time-series satellite-derived data, the monthly SSS, Chl, SST, and SSD fronts were presented and examined. The concurrence of SSS and Chl fronts indicated that CDW contributed significantly to Chl distribution in the ECS. In contrast, the SST fronts formed different patterns with Chl, indicating that the nutrients in the CDW were a major factor for Chl compared to the SST conditions during the summer months. CDW was identified in 31 isohaline and 1019 isopycnic lines. Detached patches were clearly revealed in the SSS and SSD distributions. In addition, N differences across the SSD fronts were observed. Although such an N distribution is possibly a consequence of CDW extension, more studies are required to explain the interaction between CDW extension and N.

Supplementary Materials: The following supporting information can be downloaded at: https://www.mdpi.com/article/10.3390/rs14153574/s1.

Author Contributions: Conceptualization, D.-W.K., S.-H.K. and Y.-H.J.; methodology, D.-W.K. and S.-H.K.; data curation, D.-W.K. and S.-H.K.; writing—original draft preparation, D.-W.K.; writing—review and editing, D.-W.K. and Y.-H.J.; visualization, D.-W.K.; supervision, Y.-H.J.; funding acquisition, Y.-H.J. All authors have read and agreed to the published version of the manuscript.

Funding: This research was supported by the Korea Institute of Marine Science & Technology Promotion (KIMST) funded by the Ministry of Oceans and Fisheries (20220546) and "Technology development for Practical Applications of Multi-Satellite data to maritime issues" funded by the Ministry of Ocean and Fisheries, Korea.

Data Availability Statement: Data sharing is not applicable to this article.

Acknowledgments: The authors acknowledge the Ocean Color Climate Change Initiative dataset, Version 5.0, European Space Agency, available online at http://www.esa-oceancolour-cci.org/ (accessed on 25 May 2022), for providing access to remote sensing reflectance products. We also thank the National Institute of Fisheries Science (NIFS) for providing in situ measurements, the Copernicus Marine Environment Monitoring Service for providing SLA data, and the Physical Oceanography Distributed Active Archive Center (PO.DAAC) of the National Aeronautics and Space Administration for providing SST data.

Conflicts of Interest: The authors declare no conflict of interest.

References

1. He, X.; Bai, Y.; Pan, D.; Chen, C.T.; Cheng, Q.; Wang, D.; Gong, F. Satellite views of the seasonal and interannual variability of phytoplankton blooms in the eastern China seas over the past 14 yr (1998–2011). *Biogeosciences* **2013**, *10*, 4721–4739. [CrossRef]
2. Bai, Y.; He, X.; Pan, D.; Chen, C.T.A.; Kang, Y.; Chen, X.; Cai, W.J. Summertime Changjiang River plume variation during 1998–2010. *J. Geophys. Res. Ocean.* **2014**, *119*, 6238–6257. [CrossRef]
3. Chen, Y.; Hu, C.; Yang, G.P.; Gao, X.C.; Zhou, L.M. Variation and reactivity of organic matter in the surface sediments of the Changjiang Estuary and its adjacent East China Sea. *J. Geophys. Res. Biogeosci.* **2021**, *126*, e2020JG005765. [CrossRef]
4. Kim, H.C.; Yamaguchi, H.; Yoo, S.; Zhu, J.; Okamura, K.; Kiyomoto, Y.; Tanaka, K.; Kim, S.W.; Park, T.; Ishizaka, J. Distribution of Changjiang diluted water detected by satellite chlorophyll-a and its interannual variation during 1998–2007. *J. Oceanogr.* **2009**, *65*, 129–135. [CrossRef]
5. Kim, D.W.; Park, Y.J.; Jeong, J.Y.; Jo, Y.H. Estimation of Hourly Sea Surface Salinity in the East China Sea Using Geostationary Ocean Color Imager Measurements. *Remote Sens.* **2020**, *12*, 755. [CrossRef]
6. Kim, D.W.; Kim, S.H.; Jo, Y.H. A Development for Sea Surface Salinity Algorithm Using GOCI in the East China Sea. *Korean J. Remote Sens.* **2021**, *37*, 1307–1315. (In Korean)
7. Geiger, E.F.; Grossi, M.D.; Trembanis, A.C.; Kohut, J.T.; Oliver, M.J. Satellite-derived coastal ocean and estuarine salinity in the Mid-Atlantic. *Cont. Shelf Res.* **2013**, *63*, S235–S242. [CrossRef]
8. Chen, S.; Hu, C. Estimating Sea surface salinity in the northern Gulf of Mexico from satellite ocean color measurements. *Remote Sens. Environ.* **2017**, *201*, 115–132. [CrossRef]
9. Yan, W.; Zhang, S. The composition and bioavailability of phosphorus transport through the Changjiang (Yangtze) River during the 1998 food. *Biogeochemistry* **2003**, *65*, 179–194. [CrossRef]
10. Zhang, J.; Liu, S.M.; Ren, J.L.; Wu, Y.; Zhang, G.L. Nutrient gradients from the eutrophic Changjiang (Yangtze River) Estuary to the oligotrophic Kuroshio waters and re-evaluation of budgets for the East China Sea Shelf. *Prog. Oceanogr.* **2007**, *74*, 449–478. [CrossRef]
11. Gao, L.; Li, D.; Ishizaka, J.; Zhang, Y.; Zong, H.; Guo, L. Nutrient dynamics across the river-sea interface in the Changjiang (Yangtze River) estuary—East China Sea region. *Limnol. Oceanogr.* **2015**, *60*, 2207–2221. [CrossRef]
12. Dai, Z.; Du, J.; Zhang, X.; Su, N.; Li, J. Variation of riverine material loads and environmental consequences on the Changjiang (Yangtze) Estuary in recent decades (1955−2008). *Environ. Sci. Technol.* **2011**, *45*, 223–227. [CrossRef] [PubMed]
13. Kwon, H.K.; Kim, G.; Hwang, J.; Lim, W.; Park, J.W.; Kim, T.H. Significant and conservative long-range transport of dissolved organic nutrients in the Changjiang diluted water. *Sci Rep* **2018**, *8*, 12768. [CrossRef]
14. Sathyendranath, S.; Grant, M.; Brewin, R.J.W.; Brockmann, C.; Brotas, V.; Chuprin, A. *ESA Ocean Colour Climate Change Initiative (Ocean_Colour_cci): Version 3.1 Data*; Centre for Environmental Data Analysis: Oxfordshire, UK, 2018.
15. Mélin, F.; Sclep, G. Band shifting for ocean color multi-spectral reflectance data. *Opt. Express* **2015**, *23*, 2262–2279. [CrossRef]
16. Su, Y.-s.; Weng, X.-c. Water masses in China seas. In *Oceanology of China Seas*; Springer: Dordrecht, The Netherlands, 1994; pp. 3–16.
17. Chang, P.H.; Isobe, A. A numerical study on the Changjiang diluted water in the Yellow and East China Seas. *J. Geophys. Res. Ocean.* **2003**, *108*, 3299. [CrossRef]
18. Chen, C.T.A.; Wang, S.L. Carbon, alkalinity, and nutrient budgets on the East China Sea. *J. Geophys. Res.* **1999**, *104*, 20675–20686. [CrossRef]
19. Moon, J.H.; Hirose, N.; Yoon, J.H.; Pang, I.C. Offshore detachment process of the low-salinity water around Changjiang Bank in the East China Sea. *J. Phys. Oceanogr.* **2010**, *40*, 1035–1053. [CrossRef]
20. Li, H.M.; Tang, H.J.; Shi, X.Y.; Zhang, C.S.; Wang, X.L. Increased nutrient loads from the Changjiang (Yangtze) River have led to increased harmful algal blooms. *Harmful Algae* **2014**, *39*, 92–101. [CrossRef]

21. Jiang, Z.; Chen, J.; Zhou, F.; Shou, L.; Chen, Q.; Tao, B.; Wang, K. Controlling factors of summer phytoplankton community in the Changjiang (Yangtze River) Estuary and adjacent East China Sea shelf. *Cont. Shelf Res.* **2015**, *101*, 71–84. [CrossRef]
22. Zhou, Z.X.; Yu, R.C.; Sun, C.; Feng, M.; Zhou, M.J. Impacts of Changjiang River discharge and Kuroshio intrusion on the diatom and dinoflagellate blooms in the East China Sea. *J. Geophys. Res. Ocean.* **2019**, *124*, 5244–5257. [CrossRef]
23. Zhou, Z.X.; Yu, R.C.; Zhou, M.J. Resolving the complex relationship between harmful algal blooms and environmental factors in the coastal waters adjacent to the Changjiang River estuary. *Harmful Algae* **2017**, *62*, 60–72. [CrossRef] [PubMed]
24. Qi, L.; Hu, C.; Liu, J.; Ma, R.; Zhang, Y.; Zhang, S. Noctiluca blooms in the East China Sea bounded by ocean fronts. *Harmful Algae* **2022**, *112*, 102172. [CrossRef] [PubMed]
25. Belkin, I.M. Remote sensing of ocean fronts in marine ecology and fisheries. *Remote Sens.* **2021**, *13*, 883. [CrossRef]

atmosphere

Review

The Widespread Use of Remote Sensing in Asbestos, Vegetation, Oil and Gas, and Geology Applications

Leydy K. Torres Gil, David Valdelamar Martínez and Manuel Saba *

Civil Engineering Program, Universidad de Cartagena, Cartagena 130001, Colombia
* Correspondence: msaba@unicartagena.edu.co

Abstract: Remote sensing is the technique of acquiring data from the earth's surface from sensors installed on satellites or on manned or unmanned aircrafts. Its use is common in dozens of sectors of science and technology, agriculture, atmosphere, soil, water, land surface, oceans and coasts, snow and ice, and natural disasters, among others. This article focuses on an in-depth literature review of some of the most common and promising disciplines, which are asbestos–cement roof identification, vegetation identification, the oil and gas industry, and geology, with the aim of having clarity on the trends in research on these issues at the international level. The most relevant problems in each sector have been highlighted, evidencing the need for future research in the area in light of technological advances in multi- and hyperspectral sensors and the availability of satellite images with more precise spatial resolution. A bibliometric analysis is proposed for each discipline and the network of related keywords is discussed. Finally, the results suggest that policymakers, urban planners, mine, and oil and gas companies should consider remote sensing as primary tool when planning comprehensive development strategies and in field parameter multitemporal analysis.

Keywords: remote sensing review; vegetation index; oil spill detection; soil identification

Citation: Torres Gil, L.K.; Valdelamar Martínez, D.; Saba, M. The Widespread Use of Remote Sensing in Asbestos, Vegetation, Oil and Gas, and Geology Applications. *Atmosphere* **2023**, *14*, 172. https://doi.org/10.3390/atmos14010172

Academic Editor: Filomena Romano

Received: 13 December 2022
Revised: 3 January 2023
Accepted: 9 January 2023
Published: 13 January 2023

1. Introduction

Remote sensing is the process of detecting and monitoring the physical characteristics of an area by measuring its reflected and emitted radiation at a distance [1–5]. Electromagnetic (EM) energy, produced by the vibration of charged particles, travels as waves through the atmosphere and the vacuum of space [6–8]. These waves have different wavelengths (the distance from wave crest to wave crest) and frequencies; a shorter wavelength means a higher frequency. Some waves, such as radio, micro-, and infrared waves, have longer wavelengths, while others, such as ultraviolet rays, X-rays, and gamma rays, have much shorter wavelengths. Visible light is in the middle of that range of long- to short-wave radiation. This small portion of energy is all that the human eye is capable of detecting. Instrumentation is needed to detect all other forms of electromagnetic energy (see Figure 1 taken from [2]).

Some waves are absorbed or reflected by atmospheric components, such as water vapor and carbon dioxide, while some wavelengths allow unimpeded movement through the atmosphere; visible light has wavelengths that can be transmitted through the atmosphere. Microwave energy has wavelengths that can pass through clouds, an attribute used by many weather and 5rftcommunication satellites. All things on earth reflect, absorb, or transmit energy, the amount of which varies by wavelength, creating a spectral fingerprint, called a spectral signature, unique to each object [9]. Remote sensing sensors sense the EM waves reflected/emitted by the objects on the Earth's surface. The sensors record these waves as images depending on their spectral capabilities.

Spectral sensors collect remote sensing images taken via drone, plane, or satellite. They can be used in the environmental field to monitor changes in temperature in the land surface, oceans and in the topography of the ocean floor [10,11], map great forest fires [12–14], track clouds to help predict the weather [15,16], or observe volcanoes erupt and help monitor dust storms [17,18]. Additionally, among its most common applications are the monitoring of the growth of a city through multitemporal analysis [19,20], the identification of objects on the ground, military, intelligence, commercial, planning, and humanitarian applications, among others [21].

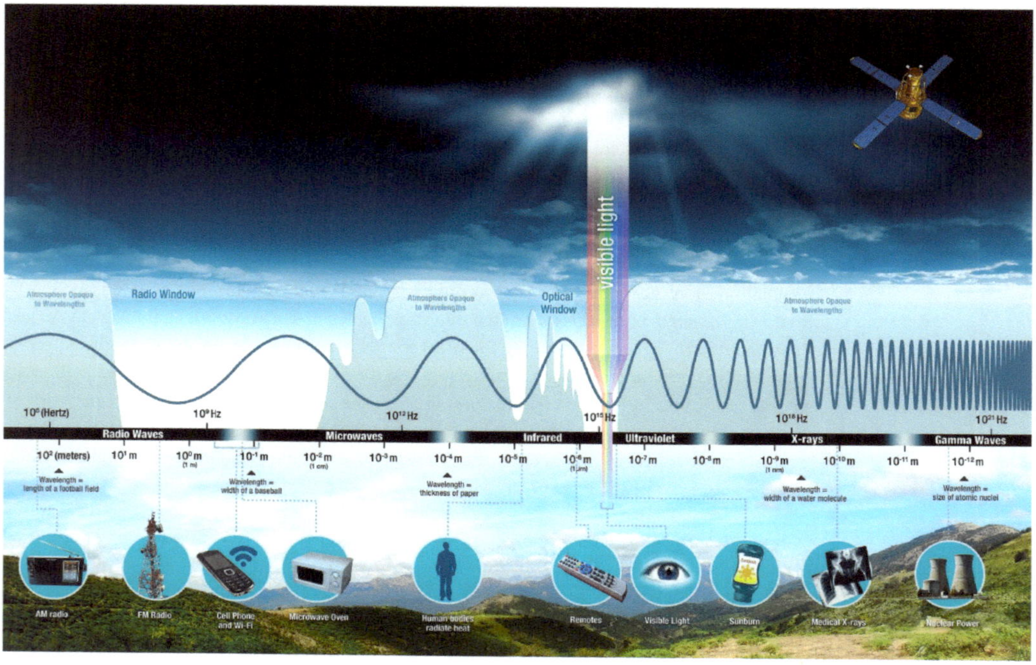

Figure 1. Diagram of the electromagnetic spectrum. Source: NASA Science [2].

As remote technologies continue to develop, their use is becoming more widespread and multidisciplinary [22–29]. Often, review works focus mainly on one highly specific topic [22–29], while a comparison between remote sensing applications is neglected.

The present study focuses particularly on the selected applications of remote sensing for civil and environmental engineering, such as the identification of asbestos–cement roofs, vegetation and its health status, oil and gas, and geological applications. The authors acknowledge the fact that many other applications of remote sensing exist and are of primary interest in the academic world; nevertheless, in this work, the topics were narrowed for extension. This review includes the most relevant papers on the matter in the last decades found mainly in Science Direct, Scopus, and SpringerLink databases. A historical cross section of research on the subject aims to detect past and present trends, thinking about possible developments in the coming years.

2. Methodology

The thirty most relevant articles of each topic treated were identified considering the number of citations and the relevance of the journal. They were read in order to be filtered, tabulated, and finally included in the discussion of this document. This review reports a chapter for each main application. Where relevant, the keywords network was analyzed. Finally, an original discussion of the spectral signature is presented, which is useful for the reader while comparing different applications for similar materials to identify a pattern that is useful for future policy makers' decisions and future research.

3. Remote Sensing and Asbestos–Cement Roofs

In the 1940s to 1990s, asbestos was widely used as an additive in construction materials [30–32]. This mineral is formed by a group of fibrous microcrystalline hydrated silicates, among which Chrysotile stands out in the form of serpentines. On the other hand, thermolite, crocidolite, actinolite, anthophyllite, and amosite are classified as part of the amphibole's family. All of them are considered carcinogenic to humans by the International Agency for Cancer Research (IACR) and by the World Health Organization (WHO) [33–35]. The risk of this mineral to humans has been known approximately since the first decades of the 20th century. The danger is mainly due to the inhalation of fibers that, penetrating into the respiratory system, cause permanent damage in humans, producing diseases such as asbestosis, pleural thickening, neoplastic diseases such as lung cancer, mesothelioma of the pleura, and peritoneum, among others [36]. The use of asbestos on a large scale began around World War II, with an annual world production of 1 million metric tons in 1950 up to 5 million in 1975. In 2000, a cumulative world production of approximately 173 millions of tons was estimated [37]. More recently, in 2021, independent agencies estimated that the annual world production of asbestos was still at 1.29 million metric tons [38], mainly in Russia, Kazakhstan, China, and Brazil.

Due to the great impact on public health that occurs due to the inhalation of asbestos fibers, it is vitally important to invest in methodologies that seek to estimate the amount of this mineral in the urban environment, where people are more exposed. This is especially the case for countries that have recently prohibited the use and commercialization of the material, as it would allow the competent authorities to formulate strategies to mitigate the environmental and public health problem.

One of the first steps to mitigate the public health problem [37,39–42] is the identification of the distribution of asbestos–cement roofs through remote sensing from spectral images. As evidenced in the literature [30,43,44], for the capture of this type of image, multispectral sensors are particularly used, with a band range between 400 nm and 2400 nm. The sensor can be mounted on an unmanned aircraft (drone), on a manned aircraft, and on satellites.

The relationship between the spatial resolution (SRE) and the economic factor is one of the most relevant factors when choosing between the three modalities mentioned above. The unmanned aircraft guarantees the SRE of the order of centimeters, with a flight height between 100 m and 300 m [45,46]. However, when it comes to identifying large areas, such as cities or entire regions, this tool has strong logistical limitations; companies providing this service are scarce and the costs are very high. On the other hand, the overflight with a manned aircraft could reach an SRE between 0.4 m and 2.0 m with a flight height between 500 m and 1500 m. The relationship between the SRE and cost is acceptable, although it costs between 600 and 1000 USD per km^2, which makes it a common tool in developed countries and scarcely used in developing countries such as those in South America [47]. The third option, through satellite images, has been highly debated in the international scientific community in recent years. This is because until 2020, the SRE of WorldView-3 images had 7.5 m pixels (56.25 m^2) for eight short-wave infrared (SWIR) bands [48]. Considering that most roofs have a dimension equal to or less than 56.25 m^2, this method is not effective according to the literature [44,49]. However, to date, thanks to technological advances, the satellite images that are on the market offer an SRE of 3.70 m (13.69 m^2) in the SWIR band range between 1000 nm and 2400 nm [50], and an SRE of 0.30 m (0.09 m^2) in the band range between 400 nm and 1000 nm (eight bands in the visible and near-infrared (VNIR) region). This makes them attractive since they have limited costs and do not require flight permits or particular logistics to purchase them directly from official suppliers.

In the last two decades, the use of remote sensing has proven to be a good instrument to identify and evaluate the condition and material of roofs [30,51,52] (Table A1). There are different methodologies for classifying asbestos–cement roofs through the use of multispectral and hyperspectral images, which focus on algorithms such as object classification or object-based image analysis (OBIA) [53], Spectral Feature Fitting (SFF) [54], Spectral Angle

Mapper (SAM) [43], Support Vector Machine (SVM) [55], decision trees and Random Forest (RF), discriminant function analysis (DFA) [56], and the maximum likelihood method (MLC), among others. The use of Convolutional Neural Networks (CNNs) has also been implemented [44] for the identification of asbestos–cement tiles with the use of aerial RGB and color–infrared (CIR) imagery.

OBIA is characterized by simulating the way in which human beings perceive and recognize objects in the real world. The analysis begins by segmenting an image into homogeneous regions or objects that roughly represent real-world objects. This approach is based on the idea that the information in an image should be interpreted as significant objects rather than individual pixels, classified from spectral, spatial, textual, and contextual data [53]. On the other hand, the SFF classification method is responsible for comparing the image pixel spectrum with a reference spectrum from a spectral library or a field/laboratory spectrum. [54]. Something similar is conducted by the SAM and SVM algorithms. In the case of SAM, this allows one to quickly map the similarities between the image spectra and the reference spectra by calculating the angle formed between the spectra. This algorithm is used in the ENVI® software [57]. Similarly, the SVM is a supervised machine learning statistical algorithm that analyzes the data and recognizes patterns based on a decision plane that defines the decision boundary that separates objects with different class memberships [53]. On the other side, RF is a machine learning algorithm that works based on the bagging method and the classification and regression tree, in which each tree contributes a single vote to determine the most frequent class in a set of variables data [58,59].

In attempts to identify asbestos–cement roofs, the trend is to increase the overall accuracy of the classification by reducing the error of the algorithm through the delimitation of the construction areas. Abriha et al. (2018) [56] carried out the identification of asbestos–cement tiles in the city of Debrecen in Hungary through a mask with the normalized digital surface model derived from a Digital Terrain Model and a digital surface model using a LiDAR study, in addition to another mask with the values obtained from the Normalized Difference Vegetation Index (NDVI). At the time of classification, the discrimination of tiles was made by those that were shaded and sunny. However, the classification accuracy was 6–7 percent worse compared to the simple approach that did not discriminate, so it is not particularly efficient to perform this procedure.

In another study carried out in the same city mentioned above by Szabó et al. (2014) [51], the same procedure was carried out for the delimitation of the construction areas without including the LiDAR model, giving a global precision of almost 80% against 85% where the model was applied. On the other hand, it was shown that the DFA and RF classification methods that were applied by [56] showed better results compared to SVM, SAM, and MLC [51].

Tommasini et al. (2019) [30] filtered areas in image that were not covered by buildings using a vector mask with a layer of cadastral forms. This involved a topographic map where all the buildings related to the selected area were described using a vector graphics editor. The clipping operation was performed using functions offered by the free software QGIS. Something similar was performed in the study carried out by Cilia et al. (2015) [52] in five municipalities located in northern Italy, where the cadastre map was superimposed on a multispectral infrared visible imaging spectrometer (MIVIS) to improve the classification. In addition, roofs with surfaces less than 36 m^2 were excluded due to the 3×3 m resolution of the images. However, Frassy et al. (2014) [43] established that MIVIS data from the Valle d'Aosta region, Italy, with a spatial resolution of 4×4 m, assume that at least a 3×3 pixel window containing asbestos–cement roofing is needed for correct detection to occur, classifying roofs larger than 144 m^2 with reasonable confidence. These assumptions can influence the accuracy of the classification. The authors found a 43% correct classification independent of the roofs size and 75% when only roofs bigger than 3×3 pixels were considered. Krówczyńska et al. (2016) [44] found similar results with an accuracy of 60% for surfaces up to 300 m^2 and 20% for roof surfaces smaller than

3 × 3 pixels. On the contrary, Cilia et al. (2015) [52] obtained better results by applying the same algorithm with precisions of 86% and 89%, respectively, showing the impact of spatial resolution when making the classification. Hence, the finer the spatial resolution, the better the performance of the classification algorithms.

Methodologies to improve spatial resolution are available, such as pansharpening, which is an improvement in the geometric resolution of multispectral bands with the finer resolution of the panchromatic band. This method tends to affect the spectral profiles of the objects; nevertheless, it improves the spatial characteristics. This method is applied when a pixel is larger than a possible house, so its values are mixed with those of the environment, making the spectral profiles not accurate. Pansharpening has been used in different works found in the literature, such as Abriha et al. (2018) [56], where this technique improved the classification by 2–3%. It was also used in the cities of Kajang and Bangi, Malaysia, for the determination of asbestos–cement tiles in buildings, achieving an accuracy of up to 93.10%, [53].

These methods from hyperspectral and multispectral images present in some cases low resolutions or little information within the spectral range. To ensure that they cover a spectral range and a high spatial resolution, the costs increase significantly. For this reason, over time, other methodologies have been studied that have high spatial resolutions, do not need spectral information, and are cheaper. Among these is the use of CNNs, which were introduced theoretically for the first time by LeCun et al. (1989) [60]. However, the first practical use was probably by Krizhevsky in 2012 [61]. The network was named AlexNet and was used to classify images from the ImageNet library. Neural networks are characterized by having greater precision than other methods used. Nevertheless, to classify asbestos–cement tiles, it was used for the first time by Krówczyńska et al. (2020) [44] in Checiny, Polonia. RGB and color–infrared (CIR) images were used and global accuracies of 89% were obtained. According to the authors, this could be a more economical and practical method for the identification of both roof types and other areas of interest.

Remote sensing has been widely studied worldwide, mainly for the identification of vegetation, minerals, and oil spills. However, in the case of asbestos–cement tiles, studies are still limited in the literature. The first part of Table A1 shows an overview of the most recent and cited works found in the literature in the matter of asbestos identification. A comparison between the classification results in WV3 satellite images (multispectral) and hyperspectral images is not found in the literature. This would allow for an assessment of the technical–economic feasibility of the two options, favoring greater awareness in decision making by authorities and professionals.

The words *Remote sensing* and *Asbestos* were searched for in *Scopus*, resulting in 708 works; then, through the free platform VOSviewer, the network of related keywords and their distribution per year were obtained (Figure 2). It is clear that in general, the issue of the remote sensing of asbestos–cement has had its greatest relevance in the last 20 years, due to strict regulation, especially in European countries [62], which has led to a greater sensitivity in the authorities and the academic world on the subject. The use of the words airborne remote sensing, MIVIS, multispectral infrared marked the 2000s [54], especially in studies in Italy, which also appeared as a keyword; however, in the last 5 years, the most trending keywords have been hyperspectral, machine learning, satellite data, and satellite imagery, showing an evolution in technology towards more precise sensors with higher bands both in aircrafts and at a satellite level [56,63]. The need to have greater spatial resolution and the costs of these images are the two contrasting factors that dominate the academic and professional scene. The dominant countries in these investigations are the United States, Italy, and China; however, adding the studies of European countries, it is noted that this continent represents the true center of research on asbestos and remote sensing worldwide. Asia also plays an important role as China, India, and Malaysia are dominant countries in these investigations, while Latin America has a deep lack of studies.

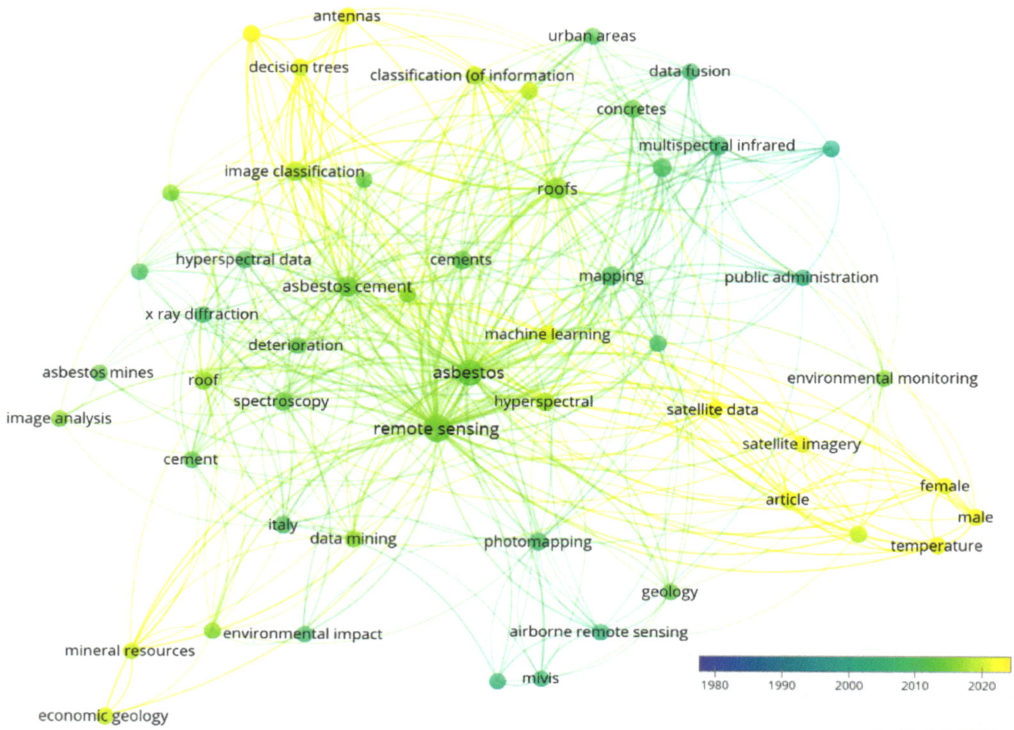

Figure 2. Relationship between keywords found in Scopus per year. Graph made with VOSviewer.

4. Remote Sensing and Vegetation

Vegetation studies are carried out for the planning and management of land use through studies of vegetation cover, the determination of changes caused by fires, the evaluation of water stress in plants, or the evaluation of stress in the vegetation caused by water or atmosphere contaminants, among others [64–66]. Remote sensing allows one to see places that are difficult to access and allows for clarity when estimating changes or anomalies in the vegetation. For multispectral imagery, images from Landsat, Sentinel, or QuickBird satellites are commonly used [67,68]. Hyperspectral images are hardly ever used for this purpose according to the literature found.

Table A2 shows the development of the most common vegetation indices (VIs) over the last 50+ years; more IVs can be found at [69]. Among these, one of the first to be conceived was the Normalized Difference Vegetation Index (NDVI), which is attractive for its ability to rapidly delineate vegetation and vegetative stress. It is widely used in commercial agriculture and in land-use studies. For reasons related to its long history, its simplicity, and its reliance on readily available multispectral bands, the NDVI has become the most popular index used for vegetation assessment [70]. This index measures the relationship between the energy absorbed and emitted by plant covers through intensity values of the greenness of the area, the amount of vegetation present on a surface, and its state of health or vegetative vigor [71]. The NDVI relates the information acquired in the red and near-infrared (NIR) bands with the state and characteristics of the vegetation covers through the normalized difference of the two bands whose range of variation is between -1 and 1. Negative values (-1) to zero (0) are bare surfaces, while values from zero (0) to one (1) show the presence of plants [72]. Dense vegetation is given from values between 0.5 and 0.7 [73]. Another index used for vegetation is the Simple Relation Index

(SR), which reduces or eliminates the influence of the soil on solar reflectance values. It is considered as a structural index that allows for the estimation of leaf area index values [71]. On the other hand, the Soil-Adjusted Vegetation Index (SAVI) minimizes the effect of the soil on the characterization of the vegetation, especially on partially covered surfaces [71].

Different combinations of bands were made in order to analyze the most appropriate combinations for the study of the detection of pests in the vegetation between the Misantla and Coatepec coffee regions in the state of Veracruz, Mexico [73]. For this, Landsat satellite images with 11 bands were taken, with each one used to record the characteristics of surface objects such as soil, vegetation, and water. The methodology consisted of three main steps: obtaining digital images by flying a drone equipped with a multispectral sensor, characterizing the levels of the disease in a field-monitoring plot, and the digital processing of the images to obtain eleven vegetation indices compared in different levels of severity. These were evaluated from the Shapiro–Wilk, Levene, ANOVA, Kruskal–Wallis, and Wilcoxon statistical tests [74].

On the other hand, for studies carried out in areas affected by fires [75,76] as in Las Peñuelas in Moguer, Spain, vegetation indices were calculated for the evaluation of the recovery of plant vigor. In this case, cartographic maps were made that allowed one to observe the levels of recovery or retreat of the affected vegetation. Images from the Sentinel 2 and Pléaides satellites and images obtained in flyovers were used. Imagery was corrected for dark pixels produced by atmospheric scattering and then they compared with information obtained in the field to assess the reliability of the classification.

In general, Table A2 shows that IVs have their origins in the '60s–'70s and had their great development during the '90s and '00s. Although, in the last decade, new methodologies and advances have also been proposed. However, some fundamental aspects are evident. The first is that with technological advances, the concentration of researchers has focused on improving the quality of images with a greater number of bands in satellite sensors and smaller pixels on the ground. The second is that almost all the new indices have been developed in the United States, denoting a predominance of interest from the large universities and companies of this country on the subject, probably due to dominance over technology, the large extensions of vegetation in the country, and attention to environmental issues. Additionally, it was observed that the great concern was initially the reduction in atmospheric disturbance; however, with the arrival of specialized software on the subject, IVs have focused more on the preliminary identification of the state of the vegetation to correctly identify the use, timing, and dosage of treatments to intensify crop production, especially corn [77].

It is felt that the development of new indices is not as essential as in the past and that agreement has been reached on this. The academy and companies in the last decade have focused mainly on multitemporal studies of vegetation, evaluating water stress, and deforestation [78].

Other methodologies for the delimitation and calculation of vegetation areas are found with the use of spectral libraries that allow for the incorporation of new representative spectral signatures. For example, in Manaus, Brazil, Landsat satellite images were used, and the spectral library of urban materials based on categories of urban land cover components was created. The spectra that had a high probability of confusion with other classes of materials were eliminated and the most representative member of each class was chosen. To identify the most representative spectra of each class of material, the root mean square error (RMSE) was applied [79,80]. The end product of this analysis was a set of fractional abundance maps for each material class (i.e., vegetation, impervious surfaces, soil, and water). A similar methodology was applied in Yellowstone National Park, USA, which aimed at mapping the vegetation cover. In this study, a spectral library of reflectance signatures was created by pixel averaging over the known occurrences of 27 vegetation cover types in the study area [81].

Scopus showed that there were 43,651 works that included the words Remote sensing and Vegetation in the keywords, abstract, or title. Figure 3 shows the most common results of the associated keywords. In terms of work frequency, the subject has had its greatest relevance in the last 20 years, probably due to greater attention to climate–environmental issues by institutions and companies. China, the United States, and Europe are the areas where the subject is most studied. Asia also plays an important role as China, India, and Japan are the dominant countries in these investigations, while Latin America again showed a trend that was similar to the case of asbestos.

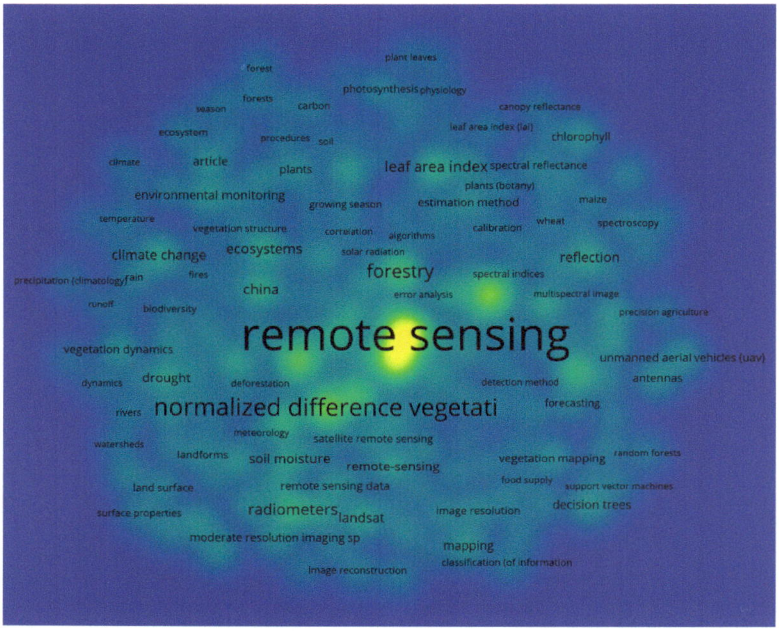

Figure 3. Keywords network found in Scopus searching with remote sensing and vegetation. Graph elaborated with VOSviewer.

5. Remote Sensing and Oil and Gas

Multispectral remote sensing is an emerging technology for the oil and gas industry. It has experienced an enormous advancement in relatively new fields such as geosciences for the exploration of hydrocarbons [82]. Remote sensing is a support tool for hydrocarbon exploration as it allows access to areas of physical inaccessibility, guaranteeing control of the threats caused by oil spills. Oil is an essential asset for societies; however, its transfer turns out to be difficult, since it is necessary to manage several transfers to take the hydrocarbons from the deposits to the refineries, the industries, and later the final clients [82]. During these transfers, spills may occur on land or in the water. In the latter case, waves, wind, and ocean currents can spread large slicks across open water in a matter of hours [83]. The early detection of anthropogenic oil slicks can enable the timely protection of critical habitats and limit economical damage [84].

In San Juan Capistrano, California, a remote sensing study was carried out in which the authors presented thermal infrared spectra of oil slicks made from five oil samples of different compositions. Different thicknesses of oil in water was tested in a laboratory to determine the reflectance spectra [85]. A similar methodology was applied by [86] in which three annual experiments of artificial mesocosms with oil in ice were carried out at the Sea Ice Environmental Research Facility (SERF) of the University of Manitoba, Canada, during 2016–2018 in order to see the behavior of oil in the ice.

On the other hand, in the Niger Delta (Nigeria) [87,88], Landsat images were acquired for the period of 2000–2018, and the NDVI was applied to determine the changes in vegetation caused by oil spills. Another similar procedure was the case study in Deep Horizon (2012) and Campo Basin (2011), in which they used resolution images to monitor oil slicks using the fluorescence/emissivity index and image analysis, using a multispectral sensor 36-band with a resolution of 250 m to 1000 m on the EOS AM (Terra) and EOS PM (Aqua) satellites [89]. The study did not show acceptable results since oil does not have specific spectral characteristics that can be used for direct detection. However, in some circumstances, oil can have a silvery appearance with a reflectance higher than that of the background [82].

Another use of satellite images is that applied to the coast of Lake Albert (Africa), in which time series of multisensory satellites with images from 1999 to 2008 were used. These images were corrected for the atmosphere and radiometric calibration; subsequently, a comparison was made between the anomaly map generated by multisensor satellite images and the map with superimposed oil and gas fields. In the comparison, it was shown that all the anomalies are located in areas with high gravimetric gradients, so it was concluded that microleakage maps can provide new high-quality data, complementary to those of traditional geophysics, at an affordable cost and with no need for exploration licenses to help the oil and gas industry reduce exploration risk [82]. This methodology was also applied in Lake Turkana (Africa), in which satellite multispectral data were used to detect microfiltration signals. These were grouped into spectral anomalies according to land use, geographic, geological, and subsoil variables. The study gave good results since an underground accumulation of hydrocarbons was evidenced [90].

Scopus showed that works with the keywords remote sensing and oil and gas were not very frequent as 1176 documents were found, which was probably because most of the works were carried out by private companies that do not publish their results due to issues related to copyright, patents, and industrial secrets, as this is a strategic economic sector. Appendix A reports some of the most relevant papers on the matter, confirming that most of them are related to hydrocarbon seepage. Once again, the United States and China led the investigations, followed by the Russian Federation.

6. Geology Applications

Remote sensing has become a very important instrument, being used in various investigations related to the identification of minerals [91–93], lithology mapping [94–96], and environmental geology due to contamination by mining areas [97]. Mineralogical studies and lithological mapping have been performed in different climatic and tectonic conditions where hyperspectral imaging for VNIR and SWIR spectral ranges is common [98]. Nevertheless, according to [99], most minerals are identifiable in the SWIR and Long Wavelength Infrared (LWIR) ranges.

The basis on which a spectral processing technique requires a priori reference data or not is used to establish a categorization scheme. In the case of no reference data, the method is usually able to directly use the available spectral patterns in a pixel (or measured spectra). Additionally, there are techniques that try to describe the spectral content of a pixel according to some predefined representative facts, known as reference data or endmembers. This initial difference gives rise to two different categories: the knowledge-based approach and the data-driven approach, which contain different classifiers [100].

Other forms for the identification of mineral assemblages is from relation indices, similar to the case of vegetation indices [101]. The Index Data Base (IDB) is a tool for working with remote sensing indices available on the web [69,102]. It provides a quick overview of which indices are usable for a specific sensor and a specific topic. It is a valuable tool for indices developed before 2012 based on the Advanced Spaceborne Thermal Emission and Reflection Radiometer sensor (ASTER), 15 bands from 520–11,650 μm, resolution of 15–90 m. Carbonate, clay, ferric iron, and SiO_2, among many other indexes to identify minerals, are reported in the literature. Clearly, for this type of large-scale application, high resolution is not needed. The use of hyperspectral images with a spatial resolution of less than 30 cm is possible thanks to modern technology; however, the information requires extraordinary storage and a consequent computational capacity that is not very accessible at a commercial level.

An applied step in methodologies for geology that can be diligent within other disciplines would be the use of Linear Spectral Unmixing (LSU) classifiers, which is a spectral unmixing tool that decomposes a reflectance source spectrum into a set of end-member spectra. This classifier showed better results compared to others that were used. According to [91], the LSU indicates a better technique for estimating the distribution of pure and impure pixels compared to methodologies such as SAM and SFF that classify all pixels as pure when they are actually impure. Another technique that performs a partial unmixing of spectra is the CEM algorithm, applied by [103], which is considered a powerful sub-pixel demixing analysis tool for analyzing ASTER reflectance data.

Machine learning algorithms are known because they work from training data sets. In a study carried out by [104] to obtain a lithological map in the greenstone belt of the Hutti area, India, the impact of the quantity on the performance of these algorithms was evaluated, for which a reduction of 15%, 30%, and 45% was made of the total samples, showing that there was a slight reduction in the global accuracies. In the case of LDA, the precision was reduced by 5%, in RF by 2%, and in SVM by 1%, thus showing that the least sensitive method to the size of data sets was SVM compared to the other two. This statement was validated in other studies [105], where SVM showed better results than other methodologies such as SAM using a relatively low number of sampling data. However, the SVM method is not widely used for lithology mapping; it is an effective algorithm for remote predictive mapping for remote areas as well as for updating existing lithology maps. It generates a high precision of up to 85% as in the lithological cartography of the Souk Arbaa Sahel region belonging to the Sidi Ifni located in southern Morocco [106].

Over time, new methodologies that are different from the conventional ones have been applied, such as the Convolutional Neural Networks (CNNs), which have presented better results. The authors of [96] presented a comparison of this methodology based on other conventional methodologies such as SAM, SID, FCLSU, SVM, and RF, showing that two-dimensional CNNs and three-dimensional CNNs were approximately 2.5–12% higher than that of SVM and RF and approximately 12−25% higher than that of SAM, SID, and FCLSU. Therefore, the CNN 2D and CNN 3D algorithms improve the classification of hyperspectral TIR remote sensing images, as it offers a better classification performance, higher noise immunity, and more accurate boundary classification. It should be noted that with this method, the global accuracies were up to 98.56%. Despite other classification methodologies being better for the identification of minerals and the generation of lithological maps such as SVM, RF, and CNN, the most used classifier is SAM which, as mentioned above, does not work very well when there are not enough training data [28,96,107]. In geological studies, it is quite difficult to obtain a lot of training data due to difficult access areas as in the case of [105] where a lithological map of an area formed by the Teide-Pico Viejo stratovolcano was made.

In unsupervised classification techniques, where classes are created purely based on spectral information and not manual visual interpretation, methods such as K-means and the Iterative self-organizing method (ISODATA) are most common. However, this topic is outside the scope of this work.

Geological applications of remote sensing are among those that have generated interest in researchers and companies since the 1980s, aimed at characterizing and monitoring large extensions of soil, the search for precious minerals, and lithological studies. Appendix A shows a summary of the most relevant and mentioned works in the literature on the subject. It can be noted that this field has the greatest variety of applications, mostly focused on the identification of minerals; however, there are also more advanced applications, such as in structural geology, crustal deformation, and geological lineaments. Scopus shows that there are 16,326 works to date with the words "Remote sensing" and "Geology". Figure 4 shows the most common results of the associated keywords. In terms of the frequency of keywords, the same frequencies per country are evident as in the case of asbestos, vegetation, and oil and gas. On the other hand, relationships between geological and geotechnical studies with climate change, glacial geology, risk, and landslides are noted.

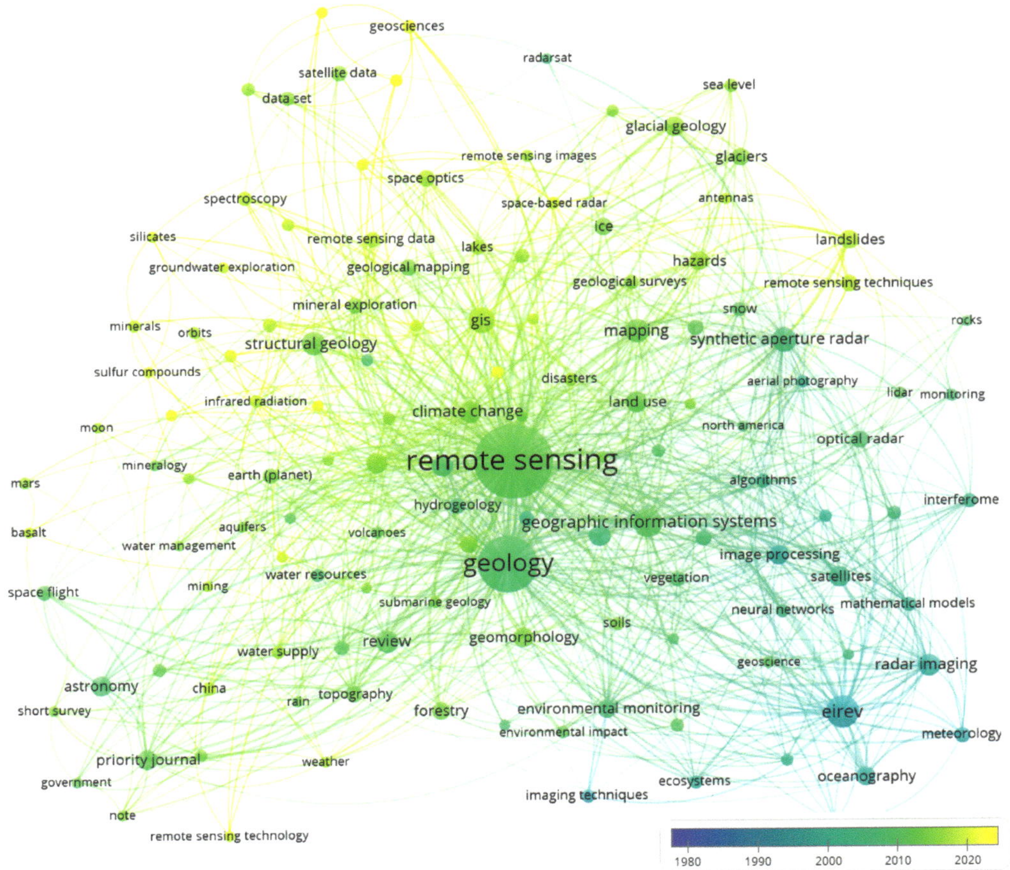

Figure 4. Keywords found in Scopus while looking for works with the keywords "Remote sensing" and "Geology". Graphic made with VOSviewer.

7. Spectral Signature

One of the key points in remote sensing studies in any of its fields of application is the response of materials to the incidence of a light ray and their identification through the spectral signature [49]. Each material has a different spectral signature, which varies depending on its physical–chemical and morphological characteristics and its capacity to absorb, transmit, or reflect the energy received. The latter is measurable through reflectance as a function of wavelength [108]. The most reflective part of the electromagnetic spectrum is in the wavelength range of 350 nm to 2500 nm, so it is within these values that most remote sensing research and applications are carried out [109].

Figure 5 shows some of the spectral signatures that can be found in the fields of application mentioned above: asbestos, vegetation, oil and gas, and soil. Most of these spectral signatures can be found in the U.S. Geological Survey (USGS) spectral library, made with laboratory, field, or aircraft-mounted spectroradiometers and covering the spectrum from 200 nm to 200,000 nm depending on the measurement performed [110].

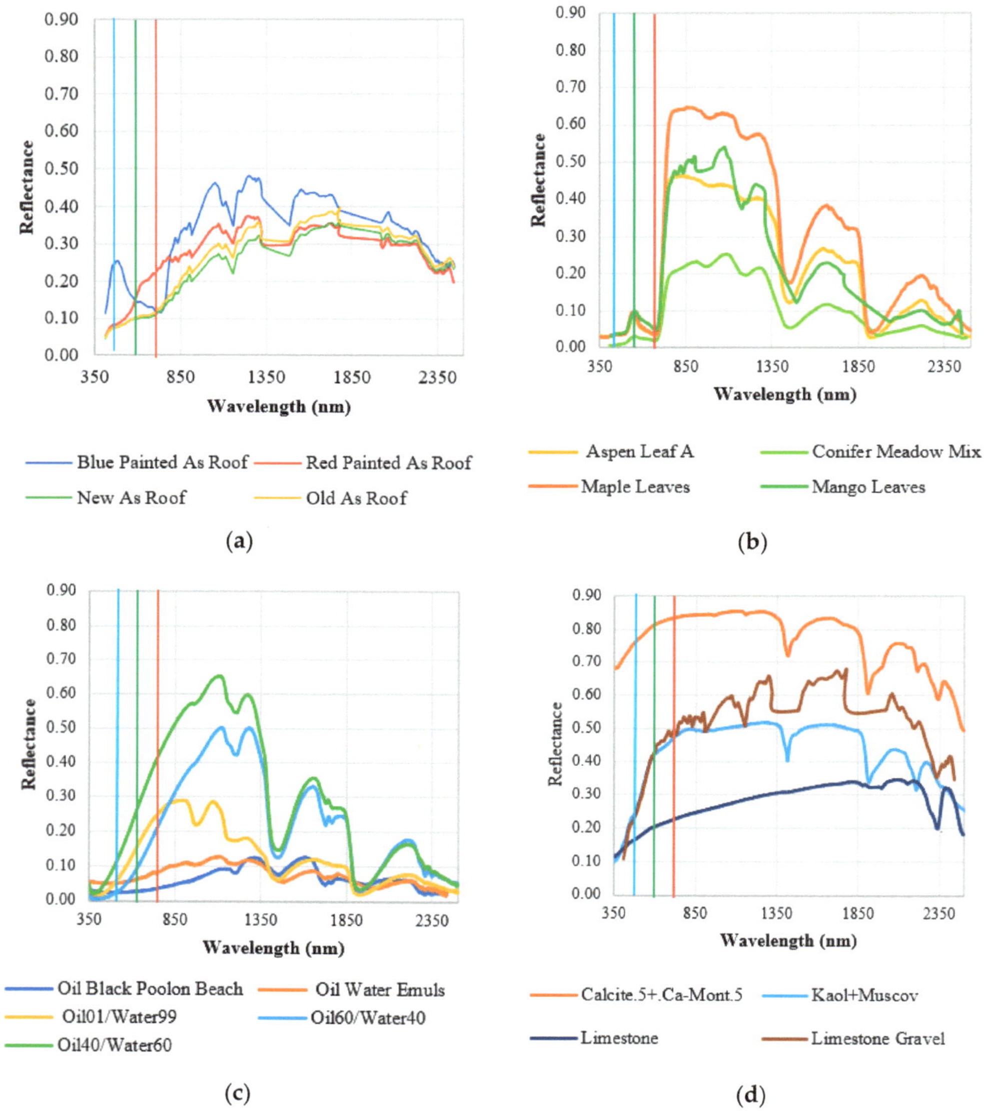

Figure 5. Spectral signatures of some materials within the categories: (**a**) asbestos, (**b**) oil and gas, (**c**) vegetation, and (**d**) soil. Vertical lines represent blue, green and red area of the spectrum. Reflectance can vary from 0 to 1.

The spectral signatures of asbestos are part of an ongoing investigation by the authors, where an aircraft mounted with a HySpex V620 hyperspectral sensor was used to collect the information. Asbestos, having different types of fibers, can present different spectral signatures according to the material under study; however, the greatest variations in reflectance occurred near the SWIR wavelengths between 1195 to 1415 nm and between 1742 and 1974 nm, which makes this range quite useful in its identification by means of classification tools such as SAM or SVM [111]. The state of deterioration and type of asbestos fibers significantly modify the spectral signature [63,112]. Asbestos–cement roofs with a nondeteriorated cement matrix may present greater energy reflection; on the contrary, those painted or protected with waterproofing usually have lower energy reflection [63,112]. Figure 5a shows some spectral signatures of asbestos–cement roofs in the city of Cartagena, Colombia. Higher reflectance values in the NIR and SWIR zones were observed, with some reflectance peaks near 1080, 1250, and 1550 nm. However, low spectral values in the visible range could differ significantly, while the spectral values for longer wavelengths are similar. Some asbestos–cement ceilings were painted different colors, and this produced an effect on the spectral signature of this material, reflecting a high reflectance near the ranges corresponding to the color; in the case of a blue ceiling, there was a modification that was significant near 490 nm, and for the red ceiling, an increase in reflectance was seen from 620 nm to 900 nm, passing from the green band to the first part of the near infrared. In both cases, the spectral signature maintained its shape, but with higher reflectance values.

In the case of the oil and gas industry, the application of remote sensing for the identification of contamination from oil spills has been of the utmost importance. During these events, it is necessary to know where the contaminant moves in the water and the magnitude of the places affected. In an investigation performed by [113] on an oil spill from the Deepwater Horizon platform in Barataria Bay, Louisiana, the authors found that the oil spectral signature was affected according to the materials that were in the background, such as water, sediment, or vegetation. Figure 5b shows the series "Oil Black Poolon Beach" and "Oil Water Emuls", where the difference between them was that the first one was on beach sand and the second one was in the water. Another important factor is the mixing ratio and the thickness of the oil layer, both of which can cause the oil to emulsify and therefore change the spectral signature. In the same event as the Deepwater Horizon spill, the authors of [114] found that the proportion of oil:water that yielded the highest reflectance values was 40:60 and that the thickness of the layers, even if they varied by a few millimeters, caused the reflectance to vary considerably.

The vegetation, for its part, presented low values of reflectance in the visible range, except for a peak near 500 nm where the green band is located; this was due to the photosynthesis processes of the plants that absorbed most of the energy in the other wavelengths. However, advancing in the spectrum, near 700 nm where it is known as the "red-edge region", the spectral signature increased the reflectance values, maintaining a similar trend until crossing the midinfrared, where there was a large absorption of energy due to the water content in the plants. Again the reflectance values rose in the SWIR area [115]. It is for this reason that the vegetation indices are mostly related to red and near-infrared wavelengths and some more recent ones use SWIR for applications of nonphotosynthetic processes [116]. In Figure 5c, the typical behavior of vegetation in four different species, i.e., Aspen Leaf A (Denver, CO [110]), Conifer Meadow Mix (Yellowstone Park [81]), Maple Leaves (Golden, CO [110]), and Mango Leaves (Cartagena, Colombia) is observed, in which the reflectance values vary but the trend and shape of the spectral signatures are similar to each other. These changes in reflectance allow one to know the type of vegetation under study and contribute to analyzing changes within the vegetation, such as the stress levels, leaves' health, and overall quality of the vegetation [117,118].

The soils presented higher reflectance values than the other categories in the entire spectrum. They tended to reflect electromagnetic energy in a greater proportion. Most of the time it was related to the chemical composition of the soil under study and its relationship with the ecosystem in which it was found [118–120]. The characteristics of each type of soil made it possible to find some absorption peaks in the spectral signature. In Figure 5d, it is observed that the greatest energy absorption was found in the SWIR part; however, the highest reflectance values were maintained in this area. In the case of the Calcite.5+.Ca-Mont.5 (mixture of Calcite and Montmorillonite) and Limestone signatures, the authors of [110] describe that the absorptions in the SWIR are related to the interaction of C-O within the chemical composition and some organic impurities of the materials, giving the signature a negative slope in this short wave zone. A similar effect also occurred for Kaol + Muscov (mixture of Kaolinite and Muscovite), where it was observed that a band was highly influenced by Muscovite in the 2200 nm zone. These patterns can explain the behavior of the Limestone Gravel series, from the city of Cartagena, which presented combinations of the first two materials mentioned.

8. Conclusions

In this work, world trends on remote sensing related to some of the most relevant issues at the global level were studied. In particular, the present manuscript focused on public health problems in the case of asbestos; environmental issues in the case of vegetation and oil and gas, in addition to hydrocarbon exploration; and geological applications in the case of the identification and characterization of soils and minerals. Several aspects that are worth highlighting were evidenced. Remote sensing studies related to vegetation were the studies most frequently found in the literature, with a marked tendency for researchers in the 1990s and 2000s to create new vegetation indices (Figure 6). More than 30 indices were found among the most relevant and mentioned studies in the literature, which are useful for different applications. However, no recent studies were found that would allow all the indices to be compared with each other in the same multispectral or hyperspectral image. In the latter case, it is probably due to the scarcity of satellite hyperspectral sensors and their cost/logistics for use in drones or manned aircraft.

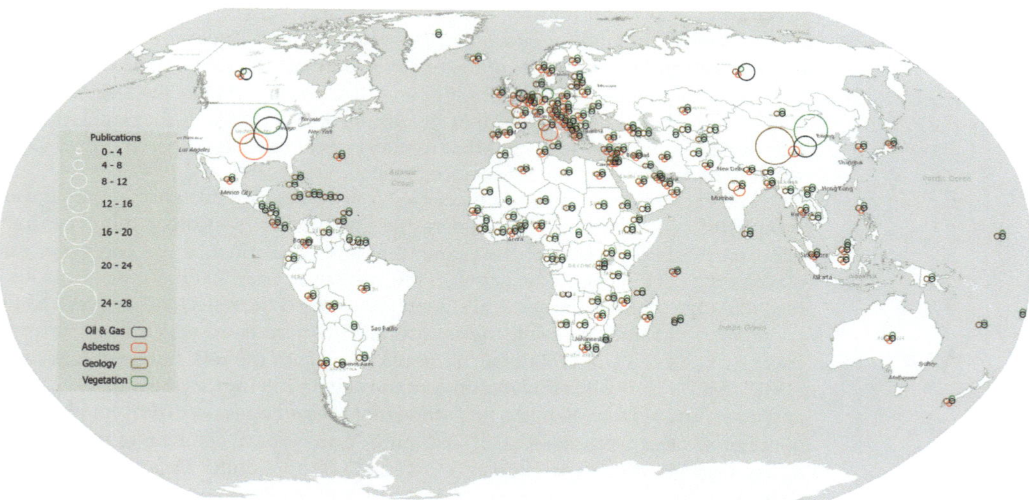

Figure 6. Summary of publications by country on oil and gas (black), asbestos (red), geology (brown), and vegetations (green).

In terms of public health, the applications for the detection of asbestos–cement roofs are interesting. Developing countries that are barely banning the use of this material have multispectral satellite images available, such as those from Word View 3 (WV3), to be able to identify asbestos–cement roofing. Since 2020, these have had a reasonable SRE for SWIR (~3.7 m). However, with this tool, countries that have also banned the material decades ago can track the progress of removal. It is evident that this technology of the WV3, with its 8 VNIR bands and 8 SWIR bands, is also useful for other applications. The main limitation in this case is cloudiness, which in some areas is constant throughout the year; however, images can be taken upon request with a limit to the percentage of cloudiness. More studies are needed to investigate the efficiency of cover identification through multispectral satellite imagery and hyperspectral flyby imagery. Since these studies are on urban areas, the overflights have the limitation of urban airports, or military areas, with restrictions in the landing cone and surrounding areas. Finally, remote sensing applied to oil and gas is a subject that has not been studied much; however, it has great potential as a tool to support preliminary studies for the exploration of wells to find hydrocarbons. As mentioned, this lack of published studies probably reflects the need for companies in the sector to keep information and methodologies confidential.

9. Recommendations

The results highlighted in the present review are relevant for environmental policymakers. Remote sensing should be applied in problem identification, policy formulation, policy implementation, and policy control and evaluation to strengthen governance and improve policy efficiency and effectiveness. For instance, in the case of asbestos–cement roofs, better strategies for removing and wasting can be tailored if roof distribution has been previously categorized. On the other hand, great economic development has characterized entire economic sectors after asbestos prohibition, and private companies should encourage remote sensing in urban areas to customize roof replacement options and optimize position of the disposal sites.

Air quality and quality of life in urban environments are in part related to the presence of green areas. Policymakers and environmental engineering companies should promote multitemporal remote sensing analysis to control and improve vegetation in urban environments, especially in underdeveloped countries where the population density and building speculations are extreme. Similar to the case of oil slick identification, permanent control with remote sensing will help prevent the spread of environmental disasters, especially in remote areas of developing counties.

Finally, remote sensing should change paradigms in policymakers and companies since inversions in this technology and these images may generate high returns in terms of quality of life and environmental quality, maximize the resources to explore the raw materials, and minimize the costs of mitigating environmental problems.

Author Contributions: Conceptualization, M.S. and L.K.T.G.; methodology, M.S.; formal analysis, L.K.T.G., D.V.M. and M.S.; investigation, L.K.T.G., D.V.M. and M.S.; resources, L.K.T.G., D.V.M. and M.S.; data curation, M.S.; writing—original draft preparation, L.K.T.G. and M.S; writing—review and editing, L.K.T.G., D.V.M. and M.S.; visualization, L.K.T.G., D.V.M. and M.S; supervision, M.S.; project administration, M.S.; funding acquisition, M.S. All authors have read and agreed to the published version of the manuscript.

Funding: This research was funded by General System of Royalties of Colombia (SGR) grant number BPIN 2020000100366. And the APC was funded by the University of Cartagena, project approved by the resolution N. 01385 od 2021.

Institutional Review Board Statement: Not applicable.

Informed Consent Statement: Not applicable.

Data Availability Statement: Not applicable.

Acknowledgments: This article is considered a product in the framework of the project "Formulation of an integral strategy to reduce the impact on public and environmental health due to the presence of asbestos in the territory of the Department of Bolivar", financed by the General System of Royalties of Colombia (SGR) and identified with the code BPIN 2020000100366. This project was executed by the University of Cartagena, Colombia, and the Asbestos-Free Colombia Foundation. Finally, the authors thank Daniela Velez Clavijo and María Angélica Márquez de León for their invaluable support.

Conflicts of Interest: The authors declare no conflict of interest.

Appendix A

Table A1. The most relevant studies for asbestos roofing identification, oil and gas, and geology related to remote sensing found in literature in the last decade, identified by relevance and organized in increasing order by year.

Material	Location	Image Taking Tool	Type of Sensor/Satellite	Number of Bands	Ground Resolution	Methodology	Year	Ref.
Asbestos	Follonica and Rimini, Italy	Overflight	MIVIS	102	3.0–4.0 m	Spectral Angle Mapper (SAM)	2008	[118]
Asbestos	Rome, Italia	Overflight	MIVIS	102	4.0 m	Spectral Angle Mapper (SAM)	2012	[57]
Asbestos	Hyderabad, India	Satellite	QuickBird	3	Panchromatic = 0.61–072 m; VNIR = 2.44–2.88 m	PCA-based; line-detection-based	2012	[119]
Asbestos	Barcelona, Spain	Overflight	Hyperspectral	32	2.0–2.4 m	Integration of rooftop greenhouses	2017	[120,121]
Asbestos	Lombardía, Italy	Overflight	MIVIS	102	3.0 m	Spectral Angle Mapper (SAM)	2018	[54]
Asbestos	Debrecen, Hungary	Satellite	WorldView-2	8	Panchromatic = 2 m VNIR= 0.5 m	LDFA = Linear Discriminant Function Analysis; QDFA = Quadratic Discriminant Function Analysis; RF = Random Forest;	2018	[56]
Asbestos	Prato, Italy	Satellite	WorldView-3	16	Panchromatic = 0.31 m; VNIR = 1.24 m; SWIR= 3.70 m	QGIS Plugin named RoofClassify	2019	[30]
Asbestos	Chęciny, Poland	Overflight	Orthophotomap	3	0.25 m	Convolutional Neural Networks (CNNs)	2020	[44]
Asbestos	São José do Rio Preto, Brazil	Satellite	WorldView-3	16	Panchromatic = 0.31 m; VNIR = 1.24 m	Maximum likelihood, mahalanobis distance, and minimum distance.	2020	[122]
Asbestos	Chęciny and Baranów, Poland	Overflight	Orthophotomap	3	0.25 m	Convolutional Neural Networks (CNNs)	2022	[123]
Asbestos	Paldal-dong, Daegu, South Korea	Overflight	Orthophotomap	NA	NA	Visual counting method	2022	[124]

Table A1. *Cont.*

Material	Location	Image Taking Tool	Type of Sensor/Satellite	Number of Bands	Ground Resolution	Methodology	Year	Ref.
Exploration of oil	Southern Tunisia	Satellite	Landsat Enhanced Thematic Mapper (ETM+); ASTER Red–Green–Blue (RGB) radar (RADARSAT)	See the reference	10 m to 100 m	Interpretation of the Shuttle Radar Topography Mission (SRTM) Digital Elevation Models (DEMs)	2006	[125]
Hydrocarbon seepages	Campos Basin, Brazil and Bay of Campeche, SE Gulf of Mexico	Satellite	ASTER	9	VNIR= 15 m; SWIR= 30 m; Thermal Infrared (TIR) = 90 m.	Spectral processing of the data; selection and preprocessing (e.g., atmospheric compensation) of ASTER imagery containing seepage records; mapping the extension of oil over water through some classification scheme (e.g., Fuzzy Clustering); selection of representative spectra from seepage pixels extracted from ASTER imagery; integration of multivariate statistics processing.	2012	[126]
Exploration of oil	Louisiana (USA) (Deep Horizon) and Campo Basin, Brazil	Satellite	EOS AM (Terra) and EOS PM (Aqua) Moderate-Resolution Imaging Spectro-radiometer (MODIS)	36	250–1000 m	Object-based image analysis (OBIA)	2014	[87]

Table A1. *Cont.*

Material	Location	Image Taking Tool	Type of Sensor/Satellite	Number of Bands	Ground Resolution	Methodology	Year	Ref.
Gas leak	Kelowna, Canada	Unmanned aerial vehicle (UAV)	Laser Methane mini-G SA3C50A	NA	Not specified. Flight altitudes: 25–30 m	Off-the-shelf laser-based methane detector into a multirotor UAV	2017	[127]
Hydrocarbon seepages	Gulf of Mexico	Satellite	RADARSAT-2; ASTER and WorldView-2	See the reference	1–15 m	Oil/emulsion thickness classification using Satellite Synthetic Aperture Radar (SAR)	2020	[128]
Gas leak	Katowice, Poland	Unmanned aerial vehicle (UAV)	LaserMethane mini SA3C321-BE	NA	Not specified. Flight altitudes: 3.5 m, 6 m, 9 m, 12 m, 15 m, 18 m, 21 m, and 25 m	Data cleaning; back-ground/leakage gas concentration determination; location of the leakage estimation.	2021	[129]
Hydrocarbon seepages	Sudd Wetlands in South Sudan	Satellite	Sentinel-1; Sentinel-2	13	10–60 m	Random Forest (RF)	2021	[130]
Hydrocarbon seepages	Louisiana (USA) (Deep Horizon)	Satellite	Sentinel-1 and RADARSAT-2	Not specified.	Sentinel-1 = 20m; RADARSAT-2 = 50 m	Faster Region-based Convolutional Neural Network (Faster R-CNN) model	2022	[131]
Hydrocarbon seepages	Matruh Basin, Egypt	Satellite y Hyperspectral	EO-1 (ALI) and Landsat-7; EO-1 (HYPERION)	6; 4; 49; 49	30 m for all sensors	Spectral Angle Mapper (SAM)	2022	[132]
Gold exploration	Southeastern Desert of Egypt	Satellite	ASTER and ETM+	6	Not specified	Band ratioing, principal component analysis (PCA), false-color composition (FCC), and frequency filtering (FFT-RWT)	2012	[133]

Table A1. *Cont.*

Material	Location	Image Taking Tool	Type of Sensor/Satellite	Number of Bands	Ground Resolution	Methodology	Year	Ref.
General mineral identification	Girón, Colombia	Satellite	Hyperion-Satellite EO-1	220	30 m	Spectral Angle Mapper (SAM)	2015	[134]
Geologic mapping	Edembo area, Algerian Sahara	Satellite	Multispectral ASTER	9	VNIR = 15 m; SWIR = 30 m; thermal infrared (TIR) = 90 m.	Maximum likelihood classifier method (MLC)	2016	[135]
Map alteration minerals	Southeast Spain	Satellite	WorldView-3 imagery and ASTER TIR	See the reference	Panchromatic = 0.31 m; VNIR = 1.24 m; SWIR = 3.70 m TIR =90 m.	Spectral Angle Mapper (SAM)	2019	[48]
Groundwater exploration	Gongola Basin, Nigeria	Satellite	Landsat 8	3	Not specified	Detection of lineaments through geophysical gravity	2020	[136]
Mineral exploration	Semna region, Eastern Desert (ED) of Egypt	Satellite	Multispectral ASTER	9	VNIR = 15 m; SWIR = 30 m; thermal infrared (TIR) = 90 m.	The ASTER data were enhanced in terms of mapping lithological units and the hydrothermal zones	2010	[137]
Kimberlite exploration	Kimberlite Province, Lesotho	Satellite	ASTER, Shuttle Radar Topography Mission (SRTM) Digital Elevation Model (DEM) and Google Earth	9	VNIR = 15 m; SWIR = 30 m	Spectral Angle Mapper (SAM)	2021	[138]
Iron mineral	Çankırı Province, Turkie	Satellite	Sentinel-2	13	5, 30 and 60 m	Spectral Angle Mapper (SAM)	2021	[139]
Structural framework and mineral occurrences	Nimas-Khadra, Southern Arabian	Satellite	ASTER	14	VNIR = 15 m SWIR = 30 m TIR = 90 m	Geophysical and image analyses to identify the tectonic framework and establish the relationship of the lithology and tectonic features with the known and prospective mineral occurrences	2022	[140]

Table A1. *Cont.*

Material	Location	Image Taking Tool	Type of Sensor/Satellite	Number of Bands	Ground Resolution	Methodology	Year	Ref.
Structural geology measurements of lava flows	Lake Assal, Djibouti	Satellite	Pleiades	5	0.5 m	Mouse Mode (MM) and Virtual Reality (VR) approaches	2022	[141]
Crustal deformation	Niger Delta Basin	Satellite	Landsat 8, Advanced Land Observation Satellite (ALOS), World 3D DEM	See the reference	30 m	Integration of satellite images	2022	[142]
Geological lineaments	Central Turkey	Satellite	Landsat 8; Advanced Land Observing Satellite (ALOS)	8	30 m; 10–100 m	Preprocessing of both optical and radar images, the image enhancement, and the determination of optimal parameter values employed in the extraction of lineaments from the data sets and the verification and the interpretation of the resultant lineament maps	2022	[143]

MIVIS = multispectral infrared visible imaging spectrometer; ASTER = Advanced Spaceborne Thermal Emission and Reflection Radiometer.

Table A2. The most relevant vegetation indices found in literature.

Vegetation Indices (VIs)	Formulas	Study Area	Observations	Year	Ref.
Simple Ratio	SR = NIR/Red	Marysville, USA	A two-wavelength reflectance ratio R745/R675 was developed for an objective index of turf color	1968	[144]
Normalized Difference Vegetation Index	NDVI = (NIR − Red)/(NIR + Red)	Texas, USA	Multispectral satellite images are used. A method has been developed for quantitative measurement of vegetation conditions over broad regions using ERTS-1 MSS data	1974	[145]
Green Vegetation Index	GVI = (−0.2848 TM1) + (−0.2435 TM2) + (−0.5436 TM3) + (0.7243 TM4) + (0.0840 TM5) + (−0.18 TM7)	Worldwide	This index minimizes the effects of background soil while emphasizing green vegetation. It uses global coefficients that weigh the pixel values to generate new transformed bands. It is also known as the Landsat TM Tasseled Cap green vegetation index	1976	[146]
Difference Vegetation Index	DVI = NIR-Red	Maryland, USA	In situ collected spectrometer data were used	1979	[147]
Soil-Adjusted Vegetation Index	SAVI = (1.5 (NIR − Red))/(NIR + Red + 0.5)	Arizona, USA	Similar to NDVI; nevertheless, it is a proposed index that minimizes soil brightness influences involving red and near-infrared (NIR) spectra	1988	[148]
Infrared Percentage Vegetation Index	IPVI = NIR/(NIR + Red)	Worldwide	The near-infrared (NIR) versus red "infrared percentage vegetation index," NIR/(NIR + Red), is functionally and linearly equivalent to the Normalized Difference Vegetation Index, (NIR-Red/(NIR + Red). Advantageously, it is both computationally faster and never negative	1990	[149]
Global Environmental Monitoring Index	GEMI = η (1 − 0.25 η) + (Red − 0.125)/(1 − Red) $\eta = 2(NIR^2 - Red^2) + 1.5NIR + 0.5Red)/(NIR + Red + 0.5)$	Worldwide	Designed specifically to reduce the relative effects of these undesirable atmospheric perturbations	1992	[150]

Table A2. *Cont.*

Vegetation Indices (VIs)	Formulas	Study Area	Observations	Year	Ref.
Atmospherically Resistant Vegetation Index Difference Vegetation	ARVI = (NIR − (Red − γ(Blue − Red)))/(NIR + (Red − γ(Blue − Red)))	Worldwide	MODIS sensor. ARVI has a similar dynamic range to the NDVI but is on average four times less sensitive to atmospheric effects than the NDVI	1992	[151]
Modified Soil-Adjusted Vegetation Index 2	MSAVI2 = (2 NIR + 1-√((2 IR + 1)- 8 (NIR-Red)))/2	Tucson, USA	This index is a simpler version of the MSAVI proposed by Qi et al. (1994), which improves upon the Soil-Adjusted Vegetation Index (SAVI). It reduces soil noise and increases the dynamic range of the vegetation signal. MSAVI2 is based on an inductive method that does not use a constant L value (as with SAVI) to highlight healthy vegetation [152]	1994	[153]
Nonlinear Index	NLI = (NIR^2 − Red)/(NIR^2 + Red)	Detroit, USA	Multispectral satellite images are used. A comparison between 3D crop model and several VIs is proposed focusing on soil brightness, optical properties of canopy elements, leaf angle distribution, and spacing, among others. The authors found that VIs using off-nadir reflectances are more informative and useful than those based on nadir reflectances; the optimal VI and sun/view geometries are usually different for inferring different parameters, depending on canopy architecture; and LAI can be practically estimated by VI only for homogeneous canopies	1994	[154]

Table A2. *Cont.*

Vegetation Indices (VIs)	Formulas	Study Area	Observations	Year	Ref.
Renormalized Difference Vegetation Index	RDVI = (NIR − Red)/√(NIR + Red)	Toulouse, France	Similar to NDVI; nevertheless, a VI to minimize soil effects is proposed	1995	[155]
Structurally Independent Pigment	SIPI = (NIR–Blue)/(NIR–Red)	Barcelona, Spain	The index minimizes the confounding effect of leaf surface and mesophyll structure	1995	[156]
Optimized Soil-Adjusted Vegetation Index	OSAVI = (NIR − Red)/(NIR + Red + 0.16)	Nottingham, UK	Similar to NDVI; nevertheless, the value of the parameter X is critical in the minimization of soil effects. A value of 0.16 is proposed	1996	[157]
Green Atmospherically Resistant Index	GARI = (NIR – [Green-γ(Blue – Red)])/(NIR + [Green-γ(Blue – Red)])	Worldwide	MODIS sensor. GARI is tailored to the concept of ARVI. Resistant to atmospheric effects as ARVI but more sensitive to a wide range of Chl-a concentrations. While NDVI and ARVI are sensitive to vegetation fraction and to rate of absorption of photosynthetic solar radiation, a green vegetation index such as GARI should be added to sense the concentration of chlorophyll, to measure the rate of photosynthesis, and to monitor plant stress.	1996	[158]
Modified Simple Ratio	MSR = ((NIR/Red) − 1)/(√(NIR/Red) + 1)	Ottawa, Canada	Multispectral satellite images are used to classify boreal forests. They evaluate several vegetation indices against experimental data sets for their performance in terms of the ability to minimize the error induced by noise in remote sensing data. The authors propose a nonlinear index that has the advantage of both low noise effects and good linearity with biophysical parameters	1996	[159]

Table A2. *Cont.*

Vegetation Indices (VIs)	Formulas	Study Area	Observations	Year	Ref.
Green Normalized Difference Vegetation Index	GNDVI = (NIR − Green)/NIR + Green)	Worldwide	Satellite images for remote sensing of chlorophyll concentration	1998	[160]
Green Leaf Index (GLI)	GLI = ((Green – Red) + (Green – Blue))/ (2*Green + Red + Blue)	Oregon, USA	This index was originally designed for use with a digital RGB camera to measure wheat cover, where the red, green, and blue digital numbers (DNs) range from 0 to 255 GLI values range from −1 to +1. Negative values represent soil and nonliving features, while positive values represent green leaves and stems [152].	2001	[161]
Enhanced Vegetation Index	EVI = 2.5 (NIR-Red)/(NIR + (6 Red) − (7.5 Blue) + 1)	Worldwide	The study was performed using the Moderate Resolution Imaging Spectroradiometer (MODIS), which is a 36-band imaging radiometer, on the NASA Earth Observing System (EOS) satellites Terra [162]	2002	[163]
Leaf Area Index	LAI = 3.618 × EVI – 0.118	Denmark	Multispectral data were acquired with the Compact Airborne Spectral Imager (CASI). The results allowed for the evaluation of the spatial variations in the photosynthetic light, nitrogen, and water use efficiencies. While photosynthesis was linearly related to transpiration, the light use efficiency (LUE) was found to be dependent on nitrogen concentrations	2002	[164]

Table A2. *Cont.*

Vegetation Indices (VIs)	Formulas	Study Area	Observations	Year	Ref.
Visible Atmospherically Resistant Index	VARI = (Green − Red)/(Green + Red − Blue)	Nebraska, USA	The goal of this study was to investigate the information content of reflectance spectra of crops in the visible and near-infrared range of the spectrum and develop a technique for remote estimation of vegetation fraction	2002	[165]
Transformed Difference Vegetation Index	TDVI = 1.5((NIR-Red)/$\sqrt{}$(NIR^2 + Red + 0.5))	Ottawa, Canada	This index shows the same sensitivity as the Soil-Adjusted Vegetation Index (SAVI) to the optical properties of bare soil subjacent to the cover. It does not saturate like NDVI and SAVI and it shows an excellent linearity as a function of the rate of vegetation cover	2002	[166]
Green Chlorophyll Index	GCI = (NIR + Green) − 1	Lincoln, USA	This index is used to estimate leaf chlorophyll content across a wide range of plant species. Having broad NIR and green wavelengths provides a better prediction of chlorophyll content while allowing for more sensitivity and a higher signal-to-noise ratio [152]	2003	[167]
Sum Green Index	SGI = Green	California, USA	SGI is the mean of reflectance across the 500 nm to 600 nm portion of the spectrum. The sum is then normalized by the number of bands to convert it back to units of reflectance. The value of this index ranges from 0 to more than 50 (in units of % reflectance). The common range for green vegetation is 10 to 25 percent reflectance [152].	2003	[168]

Table A2. *Cont.*

Vegetation Indices (VIs)	Formulas	Study Area	Observations	Year	Ref.
Wide Dynamic Range Vegetation Index	WDRVI = (a NIR-Red)/(a NIR + Red)	Lincoln, NE, USA	This index is similar to NDVI, but it uses a weighting coefficient (a) to reduce the disparity between the contributions of the near-infrared and red signals to the NDVI. The WDRVI is particularly effective in scenes that have moderate-to-high vegetation density when NDVI exceeds 0.6. NDVI tends to level off when vegetation fraction and leaf area index (LAI) increase, whereas the WDRVI is more sensitive to a wider range of vegetation fractions and to changes in LAI. The weighting coefficient (a) can range from 0.1 to 0.2. ENVI uses a value of 0.2, as recommended by Henebry, Viña, and Gitelson (2004) [152]	2004	[169]
Green Optimized Soil-Adjusted Vegetation Index	GOSAVI = (NIR-Green)/(NIR + Green + 0.16)	North Carolina, USA	This index was originally designed with color–infrared photography to predict nitrogen requirements for corn. It is similar to OSAVI, but it substitutes the green band for red [152]	2005	[170]
Green Difference	GDVI = NIR - Green	North Carolina Coastal Plain, USA	Aerial photography used for nitrogen requirements in corn	2006	[171]
Green Ratio Vegetation Index	GRVI = NIR/Green	North Carolina Coastal Plain, USA	Aerial photography used for nitrogen requirements in corn	2006	[171]

Table A2. *Cont.*

Vegetation Indices (VIs)	Formulas	Study Area	Observations	Year	Ref.
Modified Nonlinear Index	MNLI = ((NIR^2 − Red) × (1 + L))/(NIR^2 + Red + L)	Colorado, USA	Multispectral satellite images. The impact of using band ratio and vegetation indices of the AWIFS sensor images to the crop classification accuracy is empirically investigated via supervised classification. The research indicates that appropriately used vegetation indices and image ratios can potentially improve crop classification accuracy	2008	[172]
MERIS terrestrial chlorophyll index	MTCI = (R740 − R705)/(R705 − R665)	Southampton, UK	This paper reports on the design and indirect evaluation of a surrogate REP index for use with spectral data recorded at the standard band settings of the Medium Resolution Imaging Spectrometer (MERIS). This index, termed the MERIS terrestrial chlorophyll index (MTCI), was evaluated using model spectra, field spectra, and MERIS data	2010	[173]
Normalized Area Over Reflectance Curve	$NAOC = 1 - \frac{\int_a^b \rho d\lambda}{\rho_{max}(b-a)}$ Where ρ is the reflectance; λ is the wavelength; ρmax is the maximum far-red reflectance, corresponding to reflectance at the wavelength "b"; and "a" and "b" are the integration limits surrounding the chlorophyll well centered at ∼670 nm.	Valencia, Spain	The Normalized Area Over Reflectance Curve (NAOC) is proposed as a new index for remote sensing estimation of the leaf chlorophyll content of heterogeneous areas with different crops, different canopies, and different types of bare soil. This index is based on the calculation of the area over the reflectance curve obtained by high spectral resolution reflectance measurements determined from the integral of the red–near-infrared interval and divided by the maximum reflectance in that spectral region	2010	[75]

Table A2. *Cont.*

Vegetation Indices (VIs)	Formulas	Study Area	Observations	Year	Ref.
Triangular Greenness Index	TGI = ((λRed − λBlue)(ρRed − ρGreen) − (λRed − λGreen)(ρRed − ρBlue))/2	Maryland, USA	This index approximates the area of a triangle bounding a leaf reflectance spectrum, where the vertices are in the red, green, and blue wavelengths. The Lambda (λ) terms represent the center wavelengths of the respective bands. The Rho (ρ) terms represent the pixel values of those bands. The original TGI equation (Hunt et al., 2011) used 670 nm, 550 nm, and 480 nm for the red, green, and blue wavelength centers, with a 10 nm band width [152]	2011	[174]
WorldView Improved Vegetative Index	WV-VI = (NIR^2-Red)/(NIR^2 + Red)	Maryland, USA	This index uses WorldView-2 bands to compute NDVI. The value of this index ranges from −1 to 1. The common range for green vegetation is 0.2 to 0.8 [152]	2012	[175]
Enhanced Normalized Difference Vegetation Index	ENDVI = ((NIR + Green) − (2 × Blue))/((NIR + Green) + (2 × Blue))	Carlstadt, USA	The blue channel for NDVI can be used equally as well for the visible absorption channel as the Kodak film using red as the visible absorption channel Maxar found that better results are achieved using red and green as the reflective channels while using blue as the absorption channel	2015	[176–179]

References

1. Furukawa, K. Resource Base: Global Distribution and Characteristics of Estuaries and Associated Coastal Shores. *Treatise Estuar. Coast. Sci.* **2012**, *1*, 207–228. [CrossRef]
2. Earthdata What Is Remote Sensing? 2022. Available online: https://www.earthdata.nasa.gov/learn/backgrounders/remote-sensing (accessed on 23 September 2022).
3. Horning, N. Remote Sensing. *Encycl. Ecol.* **2019**, *7*, 404–413. [CrossRef]
4. Xiong, X.; Butler, J. Volume 1 Overview. *Compr. Remote Sens.* **2018**, *1*, 1–6. [CrossRef]
5. Calvet, J.-C.; Albergel, C.; Baghdadi, N.; Leroux, D.; Munier, S. Chapter 6—Remote sensing. *Water Resour. Mediterr. Reg.* **2020**, *1*, 137–156. [CrossRef]
6. Haus, H.A.; Melcher, J.R. *Electromagnetic Fields and Energy*; Prentice Hall: Cambridge, MA, USA, 1989.
7. Lorrain, P. *Fundamentals of Electromagnetic Phenomena*; W.H. Freeman: New York, NY, USA, 2000.
8. Adler, R.B.; Chu, L.J.; Fano, R.M. *Electromagnetic Energy Transmission and Radiation*; The MIT Press: Cambridge, MA, USA, 1968.
9. Ting, D. Thermal radiation. *Thermofluids* **2022**, 357–372. [CrossRef]
10. Rani, M.; Masroor, M.; Kumar, P. Remote sensing of Ocean and Coastal Environment—Overview. *Remote Sens. Ocean Coast. Environ.* **2021**, *1*, 1–15. [CrossRef]
11. Krug, L.A.; Platt, T.; Sathyendranath, S.; Barbosa, A.B. Ocean surface partitioning strategies using ocean colour remote Sensing: A review. *Prog. Oceanogr.* **2017**, *155*, 41–53. [CrossRef]
12. Gale, M.G.; Cary, G.J.; Van Dijk, A.I.J.M.; Yebra, M. Forest fire fuel through the lens of remote sensing: Review of approaches, challenges and future directions in the remote sensing of biotic determinants of fire behaviour. *Remote Sens. Environ.* **2021**, *255*, 112282. [CrossRef]
13. Yu, Y.; Fu, L.; Cheng, Y.; Ye, Q. Multi-view distance metric learning via independent and shared feature subspace with applications to face and forest fire recognition, and remote sensing classification. *Knowledge-Based Syst.* **2022**, *243*, 108350. [CrossRef]
14. de Jesus, C.S.L.; Delgado, R.C.; Wanderley, H.S.; Teodoro, P.E.; Pereira, M.G.; Lima, M.; de Ávila Rodrigues, R.; da Silva Junior, C.A. Fire risk associated with landscape changes, climatic events and remote sensing in the Atlantic Forest using ARIMA model. *Remote Sens. Appl. Soc. Environ.* **2022**, *26*, 100761. [CrossRef]
15. Wetherley, E.B.; Roberts, D.A.; Tague, C.L.; Jones, C.; Quattrochi, D.A.; McFadden, J.P. Remote sensing and energy balance modeling of urban climate variability across a semi-arid megacity. *Urban Clim.* **2021**, *35*, 100757. [CrossRef]
16. Avand, M.; Moradi, H.; lasboyee, M.R. Using machine learning models, remote sensing, and GIS to investigate the effects of changing climates and land uses on flood probability. *J. Hydrol.* **2021**, *595*, 125663. [CrossRef]
17. Ganci, G.; Cappello, A.; Bilotta, G.; Del Negro, C. How the variety of satellite remote sensing data over volcanoes can assist hazard monitoring efforts: The 2011 eruption of Nabro volcano. *Remote Sens. Environ.* **2020**, *236*, 111426. [CrossRef]
18. McAlpin, D.; Meyer, F.J. Multi-sensor data fusion for remote sensing of post-eruptive deformation and depositional features at Redoubt Volcano. *J. Volcanol. Geotherm. Res.* **2013**, *259*, 414–423. [CrossRef]
19. Fu, X.; Yao, L.; Xu, W.; Wang, Y.; Sun, S. Exploring the multitemporal surface urban heat island effect and its driving relation in the Beijing-Tianjin-Hebei urban agglomeration. *Appl. Geogr.* **2022**, *144*, 102714. [CrossRef]
20. Bovolo, F.; Bruzzone, L.; Solano-Correa, Y.T. Multitemporal Analysis of Remotely Sensed Image Data. *Compr. Remote Sens.* **2018**, *2*, 156–185. [CrossRef]
21. Wu, X.; Xiao, Q.; Wen, J.; You, D.; Hueni, A. Advances in quantitative remote sensing product validation: Overview and current status. *Earth-Sci. Rev.* **2019**, *196*, 102875. [CrossRef]
22. Rivera-Marin, D.; Dash, J.; Ogutu, B. The use of remote sensing for desertification studies: A review. *J. Arid Environ.* **2022**, *206*, 104829. [CrossRef]
23. Veettil, B.K.; Hong Quan, N.; Hauser, L.T.; Doan Van, D.; Quang, N.X. Coastal and marine plastic litter monitoring using remote sensing: A review. *Estuar. Coast. Shelf Sci.* **2022**, *279*, 108160. [CrossRef]
24. Wang, P.; Bayram, B.; Sertel, E. A comprehensive review on deep learning based remote sensing image super-resolution methods. *Earth-Sci. Rev.* **2022**, *232*, 104110. [CrossRef]
25. Ali, A.M.; Abouelghar, M.; Belal, A.A.; Saleh, N.; Yones, M.; Selim, A.I.; Amin, M.E.S.; Elwesemy, A.; Kucher, D.E.; Maginan, S.; et al. Crop Yield Prediction Using Multi Sensors Remote Sensing (Review Article). *Egypt. J. Remote Sens. Space Sci.* **2022**, *25*, 711–716. [CrossRef]
26. Liu, C.; Xing, C.; Hu, Q.; Wang, S.; Zhao, S.; Gao, M. Stereoscopic hyperspectral remote sensing of the atmospheric environment: Innovation and prospects. *Earth-Sci. Rev.* **2022**, *226*, 103958. [CrossRef]
27. Chen, Z.; Deng, L.; Luo, Y.; Li, D.; Marcato Junior, J.; Nunes Gonçalves, W.; Awal Md Nurunnabi, A.; Li, J.; Wang, C.; Li, D. Road extraction in remote sensing data: A survey. *Int. J. Appl. Earth Obs. Geoinf.* **2022**, *112*, 102833. [CrossRef]
28. Shirmard, H.; Farahbakhsh, E.; Müller, R.D.; Chandra, R. A review of machine learning in processing remote sensing data for mineral exploration. *Remote Sens. Environ.* **2022**, *268*, 112750. [CrossRef]
29. Asadzadeh, S.; de Oliveira, W.J.; de Souza Filho, C.R. UAV-based remote sensing for the petroleum industry and environmental monitoring: State-of-the-art and perspectives. *J. Pet. Sci. Eng.* **2022**, *208*, 109633. [CrossRef]
30. Tommasini, M.; Bacciottini, A.; Gherardelli, M. A QGIS Tool for Automatically Identifying Asbestos Roofing. *ISPRS Int. J. Geo-Inf.* **2019**, *8*, 131. [CrossRef]

31. Stevulova, N.; Estokova, A.; Holub, M.; Singovszka, E. Demolition waste contaminated with asbestos. *Adv. Toxic. Constr. Build. Mater.* **2022**, *1*, 261–283. [CrossRef]
32. Obmiński, A. Asbestos cement products and their impact on soil contamination in relation to various sources of anthropogenic and natural asbestos pollution. *Sci. Total Environ.* **2022**, *848*, 157275. [CrossRef]
33. OMS. Eliminación de las Enfermedades Relacionadas con el Asbestos. 2020. Available online: https://www.who.int/es/news-room/fact-sheets/detail/asbestos-elimination-of-asbestos-related-diseases (accessed on 2 December 2022).
34. Frank, A.L.; Joshi, T.K. The global spread of asbestos. *Ann. Glob. Heal.* **2014**, *80*, 257–262. [CrossRef]
35. Villamizar, G.; Camero, G. *Asbesto en Colombia. Fundamentos Para el Debate*; Universidad Nacional de Colombia: Bogotá, Colombia, 2019.
36. Książek, J. Methods for Detection of Asbestos-Cement Roof ng Sheets. *Geomatics Environ. Eng.* **2014**, *8*, 59–76. [CrossRef]
37. Hemminki, K.; Försti, A.; Chen, T.; Hemminki, A. Incidence, mortality and survival in malignant pleural mesothelioma before and after asbestos in Denmark, Finland, Norway and Sweden. *BMC Cancer* **2021**, *21*, 1189. [CrossRef] [PubMed]
38. Statista. Major Countries in Worldwide Asbestos Mine Production in 2021. 2021. Available online: https://www.statista.com/statistics/264923/world-mine-production-of-asbestos/ (accessed on 7 October 2022).
39. Carmonaa, R.E.; Rivera Rosalesa, R.M. Asbestosis y mesotelioma pleural maligno. *Rev. Fac. Med.* **2013**, *52*, 5–17.
40. Selikoff, I.J.; Lee, D.H.K. Asbestos and disease. *Asbestos Dis.* **1979**, *36*, 157–158. [CrossRef]
41. Navarro-Vargas, J.R.; Villamizar, G.A. Artículo de Reflexión El largo y sinuoso camino de la enfermedad laboral en Colombia. El caso de la asbestosis. *Rev. De La Acad. Nac. De Med.* **2019**, 231–240.
42. Neitzel, R.L.; Sayler, S.K.; Demond, A.H.; d'Arcy, H.; Garabrant, D.H.; Franzblau, A. Measurement of asbestos emissions associated with demolition of abandoned residential dwellings. *Sci. Total Environ.* **2020**, *722*, 137891. [CrossRef] [PubMed]
43. Frassy, F.; Candiani, G.; Rusmini, M.; Maianti, P.; Marchesi, A.; Nodari, F.R.; Via, G.D.; Albonico, C.; Gianinetto, M. Mapping asbestos-cement roofing with hyperspectral remote sensing over a large mountain region of the Italian western alps. *Sensors* **2014**, *14*, 15900–15913. [CrossRef] [PubMed]
44. Krówczyńska, M.; Raczko, E.; Staniszewska, N.; Wilk, E. Asbestos-cement roofing identification using remote sensing and convolutional neural networks (CNNs). *Remote Sens.* **2020**, *12*, 408. [CrossRef]
45. Suomalainen, J.; Oliveira, R.A.; Hakala, T.; Koivumäki, N.; Markelin, L.; Näsi, R.; Honkavaara, E. Direct reflectance transformation methodology for drone-based hyperspectral imaging. *Remote Sens. Environ.* **2021**, *266*, 112691. [CrossRef]
46. Pfitzner, K.; Bartolo, R.; Whiteside, T.; Loewensteiner, D.; Esparon, A. Multi-temporal spectral reflectance of tropical savanna understorey species and implications for hyperspectral remote sensing. *Int. J. Appl. Earth Obs. Geoinf.* **2022**, *112*, 102870. [CrossRef]
47. Scopus. Scopus Research. 2020. Available online: https://unicartagena.elogim.com:2085/results/results.uri?sort=plf-f&src=s&st1=Asbestos+&nlo=&nlr=&nls=&sid=ab9cb7ceaa0a84c920417ae66c0e2036&sot=b&sdt=cl&cluster=scoaffilctry%2C%22Colombia%22%2Ct&sl=24&s=TITLE-ABS-KEY%28Asbestos+%29&cl=t&offset=21&origin=r (accessed on 7 October 2020).
48. Bedini, E. Application of WorldView-3 imagery and ASTER TIR data to map alteration minerals associated with the Rodalquilar gold deposits, southeast Spain. *Adv. Space Res.* **2019**, *63*, 3346–3357. [CrossRef]
49. Carlos Chicaíza-Rojas, L.; Francisco Lopez-Parra, J. Uso de imágenes satelitales para detección y cuantificación de asbesto. *DYNA* **2014**, *81*, 1–2.
50. The European Space Agency. WorldView-3 Instruments. 2022. Available online: https://earth.esa.int/eogateway/missions/worldview-3 (accessed on 13 December 2022).
51. Szabó, S.; Burai, P.; Kovács, Z.; Szabó, G.; Kerényi, A.; Fazekas, I.; Paládi, M.; Buday, T.; Szabó, G. Testing algorithms for the identification of asbestos roofing based on hyperspectral data. *Environ. Eng. Manag. J.* **2014**, *143*, 512900. [CrossRef]
52. Cilia, C.; Panigada, C.; Rossini, M.; Candiani, G.; Pepe, M.; Colombo, R. Mapping of Asbestos Cement Roofs and Their Weathering Status Using Hyperspectral Aerial Images. *ISPRS Int. J. Geo-Inf.* **2015**, *4*, 928–941. [CrossRef]
53. Gibril, M.B.A.; Shafri, H.Z.M.; Hamedianfar, A. New semi-automated mapping of asbestos cement roofs using rule-based object-based image analysis and Taguchi optimization technique from WorldView-2 images. *Int. J. Remote Sens.* **2017**, *38*, 467–491. [CrossRef]
54. Bassani, C.; Cavalli, R.M.; Cavalcante, F.; Cuomo, V.; Palombo, A.; Pascucci, S.; Pignatti, S. Deterioration status of asbestos-cement roofing sheets assessed by analyzing hyperspectral data. *Remote Sens. Environ.* **2007**, *109*, 361–378. [CrossRef]
55. Norman, M.; Mohd Shafri, H.Z.; Idrees, M.O.; Mansor, S.; Yusuf, B. Spatio-statistical optimization of image segmentation process for building footprint extraction using very high-resolution WorldView 3 satellite data. *Geocarto Int.* **2019**, *35*, 1124–1147. [CrossRef]
56. Abriha, D.; Kovács, Z.; Ninsawat, S.; Bertalan, L.; Balázs, B.; Szabó, S. Identification of roofing materials with Discriminant Function Analysis and Random Forest classifiers on pan-sharpened WorldView-2 imagery—A comparison. *Hungarian Geogr. Bull.* **2018**, *67*, 375–392. [CrossRef]
57. Fiumi, L.; Campopiano, A.; Casciardi, S.; Ramires, D. Method validation for the identification of asbestos–cement roofing. *Appl. Geomatics* **2012**, *4*, 55–64. [CrossRef]
58. Wu, P.-Y.; Sandels, C.; Mjörnell, K.; Mangold, M.; Johansson, T. Predicting the presence of hazardous materials in buildings using machine learning. *Build. Environ.* **2022**, *213*, 108894. [CrossRef]

59. Ashtari, A.; Alizadeh, B. A comparative study of machine learning classifiers for secure RF-PUF-based authentication in internet of things. *Microprocess. Microsyst.* **2022**, *93*, 104600. [CrossRef]
60. LeCun, Y.; Boser, B.; Denker, J.S.; Henderson, D.; Howard, R.E.; Hubbard, W.; Jackel, L.D. Backpropagation Applied to Handwritten Zip Code Recognition. *Neural Comput.* **1989**, *1*, 541–551. [CrossRef]
61. Krizhevsky, B.A.; Sutskever, I.; Hinton, G.E. ImageNet Classification with Deep Convolutional Neural Networks. *Commun. ACM* **2012**, *60*, 84–90. [CrossRef]
62. Kazan-Allen, L. *Asbestos Profile: European Union*; International Asbestos Ban Secretariat: Bruxelles, Belgium, 2019; Available online: http://www.ibasecretariat.org/prof_eu.php (accessed on 23 September 2022).
63. Norman, M.; Shafri, H.Z.M.; Mansor, S.; Yusuf, B.; Radzali, N.A.W.M. Fusion of multispectral imagery and LiDAR data for roofing materials and roofing surface conditions assessment. *Int. J. Remote Sens.* **2020**, *41*, 1–22. [CrossRef]
64. García-Pardo, K.A.; Moreno-Rangel, D.; Domínguez-Amarillo, S.; García-Chávez, J.R. Remote sensing for the assessment of ecosystem services provided by urban vegetation: A review of the methods applied. *Urban For. Urban Green.* **2022**, *74*, 127636. [CrossRef]
65. Neinavaz, E.; Schlerf, M.; Darvishzadeh, R.; Gerhards, M.; Skidmore, A.K. Thermal infrared remote sensing of vegetation: Current status and perspectives. *Int. J. Appl. Earth Obs. Geoinf.* **2021**, *102*, 102415. [CrossRef]
66. Pérez-Cabello, F.; Montorio, R.; Alves, D.B. Remote sensing techniques to assess post-fire vegetation recovery. *Curr. Opin. Environ. Sci. Heal.* **2021**, *21*, 100251. [CrossRef]
67. Andreatta, D.; Gianelle, D.; Scotton, M.; Dalponte, M. Estimating grassland vegetation cover with remote sensing: A comparison between Landsat-8, Sentinel-2 and PlanetScope imagery. *Ecol. Indic.* **2022**, *141*, 109102. [CrossRef]
68. Meusburger, K.; Bänninger, D.; Alewell, C. Estimating vegetation parameter for soil erosion assessment in an alpine catchment by means of QuickBird imagery. *Int. J. Appl. Earth Obs. Geoinf.* **2010**, *12*, 201–207. [CrossRef]
69. Henrich, V.; Krauss, G.; Götze, C.; Sandow, C. Index DataBase. A Database for Remote Sensing Indices. 2012. Available online: https://www.indexdatabase.de/db/i.php (accessed on 29 September 2022).
70. Huang, S.; Tang, L.; Hupy, J.P.; Wang, Y.; Shao, G. A commentary review on the use of normalized difference vegetation index (NDVI) in the era of popular remote sensing. *J. For. Res.* **2021**, *32*, 1–6. [CrossRef]
71. Beltrán Hernández, D.H. *Aplicación de Indices de Vegetación Para Evaluar Procesos de Restauración Ecológica en el Parque Forestal Embalse del Neusa*; Universidad Militar Nueva Granada: Neusa, Colombia, 2017.
72. Song, Y.; Chen, B.; Ho, H.C.; Kwan, M.-P.; Liu, D.; Wang, F.; Wang, J.; Cai, J.; Li, X.; Xu, Y.; et al. Observed inequality in urban greenspace exposure in China. *Environ. Int.* **2021**, *156*, 106778. [CrossRef]
73. Sánchez-Méndez, A.G.; Arguijo-Hernández, S.P. Análisis de imágenes multiespectrales para la detección de cultivos y detección de plagas y enfermedades en la producción de café. *Res. Comput. Sci.* **2018**, *147*, 309–317. [CrossRef]
74. Pedrali, L.D.; Borges Júnior, N.; Pereira, R.S.; Tramontina, J.; Alba, E.; Marchesan, J. Multispectral remote sensing for determining dry severity levels of pointers in *Eucalyptus* spp. *Sci. For.* **2019**, *122*, 224–234.
75. Vales, J.J.; Pino, I.; Granado, L.; Prieto, R.; Méndez, E.; Rodríguez, M.; Giménez de Azcárate, F.; Ortega, E.; Moreira, J.M. Cartografía de la afección y recuperación vegetal del incendio de Las Peñuelas en Moguer (Huelva) con imágenes satelitales. *Rev. Teledetección* **2020**, *57*, 79. [CrossRef]
76. Michalijos, M.P.; Uboldi, J. Propuesta metodológica para la evaluación de áreas afectadas por incendios mediante el uso de imágenes satelitales (Sierra de la Ventana, Argentina). *Rev. Geogr. Norte Gd.* **2013**, *56*, 223–234. [CrossRef]
77. Delegido, J.; Alonso, L.; González, G.; Moreno, J. Estimating chlorophyll content of crops from hyperspectral data using a normalized area over reflectance curve (NAOC). *Int. J. Appl. Earth Obs. Geoinf.* **2010**, *12*, 165–174. [CrossRef]
78. Saito, H.; Uchiyama, S.; Teshirogi, K. Rapid vegetation recovery at landslide scars detected by multitemporal high-resolution satellite imagery at Aso volcano, Japan. *Geomorphology* **2022**, *398*, 107989. [CrossRef]
79. Dennison, P.E.; Roberts, D.A. The effects of vegetation phenology on endmember selection and species mapping in southern California chaparral. *Remote Sens. Environ.* **2003**, *87*, 295–309. [CrossRef]
80. Dennison, P.E.; Roberts, D.A. Endmember selection for multiple endmember spectral mixture analysis using endmember average RMSE. *Remote Sens. Environ.* **2003**, *87*, 123–135. [CrossRef]
81. Kokaly, R.F.; Despain, D.G.; Clark, R.N.; Livo, K.E. Mapping vegetation in Yellowstone National Park using spectral feature analysis of AVIRIS data. *Remote Sens. Environ.* **2003**, *84*, 437–456. [CrossRef]
82. Frassy, F.; Maianti, P.; Marchesi, A.; Nodari, F.R.; Dalla Via, G.; De Paulis, R.; Biffi, P.G.; Gianinetto, M. Satellite remote sensing for hydrocarbon exploration in new venture areas. In Proceedings of the 2015 IEEE International Geoscience and Remote Sensing Symposium (IGARSS), Milan, Italy, 26–31 July 2015; pp. 2884–2887. [CrossRef]
83. Fingas, M.; Wang, Z.; Fieldhouse, B.; Smith, P. Environment Canada Arctic and Marine Oil Spill Program Technical Seminar (AMOP) Proceedings. 2003. Available online: https://www.researchgate.net/publication/256925779_The_correlation_of_chemical_characteristics_of_an_oil_to_dispersant_effectiveness (accessed on 2 December 2022).
84. Salisbury, J.W.; D'Aria, D.M.; Sabins, F.F. Thermal infrared remote sensing of crude oil slicks. *Remote Sens. Environ.* **1993**, *45*, 225–231. [CrossRef]
85. Sabins, F.F. Remote sensing for petroleum exploration, Part 1: Overview of imaging systems. *Lead. Edge* **2012**, *17*, 467. [CrossRef]
86. Desmond, D.S.; Crabeck, O.; Lemes, M.; Harasyn, M.L.; Mansoori, A.; Saltymakova, D.; Fuller, M.C.; Rysgaard, S.; Barber, D.G.; Isleifson, D.; et al. Investigation into the geometry and distribution of oil inclusions in sea ice using non-destructive X-ray

microtomography and its implications for remote sensing and mitigation potential. *Mar. Pollut. Bull.* **2021**, *173*, 112996. [CrossRef] [PubMed]

87. Aweda, E.D.; Adeyewa, Z.D. Inter annual variation of vegetation anomaly over Nigeria using satellite-derived index. *Adv. Appl. Sci. Res.* **2011**, *2*, 468–475.
88. Obida, C.B.; Blackburn, G.A.; Whyatt, J.D.; Semple, K.T. Counting the cost of the Niger Delta's largest oil spills: Satellite remote sensing reveals extensive environmental damage with >1million people in the impact zone. *Sci. Total Environ.* **2021**, *775*, 145854. [CrossRef]
89. Maianti, P.; Rusmini, M.; Tortini, R.; Dalla Via, G.; Frassy, F.; Marchesi, A.; Rota Nodari, F.; Gianinetto, M. Monitoring large oil slick dynamics with moderate resolution multispectral satellite data. *Nat. Hazards* **2014**, *73*, 473–492. [CrossRef]
90. Gianinetto, M.; Frassy, F.; Marchesi, A.; Maianti, P.; De Paulis, R.; Biffi, P.G.; Nodari, F.R. Mapping large-scale microseepage signals for supporting oil and gas exploration in new ventures. *Int. Geosci. Remote Sens. Symp.* **2016**, *2016*, 5430–5433. [CrossRef]
91. Bhadra, B.K.; Pathak, S.; Nanda, D.; Gupta, A.; Rao, S.S. Spectral characteristics of talc and mineral abundance mapping in the Jahazpur Belt of Rajasthan, India using AVIRIS-NG data. *Int. J. Remote Sens.* **2020**, *41*, 8757–8777. [CrossRef]
92. Camacho-Velasco, A.; Vargas-García, C.A.; Rojas-Morales, F.A.; Castillo-Castelblanco, S.F.; Arguello-Fuentes, H. Aplicaciones y retos del sensado remoto hiperespectral en la geología colombiana. *Rev. Fac. Ing.* **2015**, *24*, 17–29. [CrossRef]
93. Carrino, T.A.; Crósta, A.P.; Toledo, C.L.B.; Silva, A.M. Hyperspectral remote sensing applied to mineral exploration in southern Peru: A multiple data integration approach in the Chapi Chiara gold prospect. *Int. J. Appl. Earth Obs. Geoinf.* **2018**, *64*, 287–300. [CrossRef]
94. Thannoun, R.G. Mapping lithological and mineralogical units using hyperspectral imagery. *Malaysian J. Sci.* **2021**, *40*, 93–106. [CrossRef]
95. Fan, Y.; Wan, Y.; Wang, H.; Yang, X.; Liang, M.; Pan, C.; Zhang, S.; Wang, W.; Tan, F. Application of an airborne hyper-spectral survey system CASI/SASI in the gold-silver-lead-zinc ore district of Huaniushan, Gansu, China. *Geol. Croat.* **2021**, *74*, 73–83. [CrossRef]
96. Lyu, P.; He, L.; He, Z.; Liu, Y.; Deng, H.; Qu, R.; Wang, J.; Zhao, Y.; Wei, Y. Research on remote sensing prospecting technology based on multi-source data fusion in deep-cutting areas. *Ore Geol. Rev.* **2021**, *138*, 104359. [CrossRef]
97. Dkhala, B.; Mezned, N.; Gomez, C.; Abdeljaouad, S. Hyperspectral field spectroscopy and SENTINEL-2 Multispectral data for minerals with high pollution potential content estimation and mapping. *Sci. Total Environ.* **2020**, *740*, 140160. [CrossRef] [PubMed]
98. Peyghambari, S.; Zhang, Y. Hyperspectral remote sensing in lithological mapping, mineral exploration, and environmental geology: An updated review. *J. Appl. Remote. Sens.* **2021**, *15*, 031501. [CrossRef]
99. Coulter, D.W.; Harris, P.D.; Wickert, L.M.; Zhou, X. Advances in Spectral Geology and Remote Sensing: 2008–2017. In Proceedings of the Exploration 17: Sixth Decennial International Conference on Mineral Exploration, Toronto, ON, Canada, 21–25 October 2017; pp. 23–50.
100. Asadzadeh, S.; de Souza Filho, C.R. A review on spectral processing methods for geological remote sensing. *Int. J. Appl. Earth Obs. Geoinf.* **2016**, *47*, 69–90. [CrossRef]
101. Henrich., V.; Götze, C.; Jung, A.; Sandow, C.; Thürkow, D.; Glaesser, C. Development of an Online Indices Database: Motivation, Concept and Implementation | Request PDF. 2009. Available online: https://www.researchgate.net/publication/259802556_Development_of_an_online_indices_database_Motivation_concept_and_implementation (accessed on 29 September 2022).
102. Ukrainski, P. IDB: A Remote Sensing Indices Database. 2022. Available online: http://www.50northspatial.org/idb-remote-sensing-indices-database/ (accessed on 29 September 2022).
103. Zhang, X.; Pazner, M.; Duke, N. Lithologic and mineral information extraction for gold exploration using ASTER data in the south Chocolate Mountains (California). *ISPRS J. Photogramm. Remote Sens.* **2007**, *62*, 271–282. [CrossRef]
104. Kumar, C.; Chatterjee, S.; Oommen, T.; Guha, A. Automated lithological mapping by integrating spectral enhancement techniques and machine learning algorithms using AVIRIS-NG hyperspectral data in Gold-bearing granite-greenstone rocks in Hutti, India. *Int. J. Appl. Earth Obs. Geoinf.* **2020**, *86*, 102006. [CrossRef]
105. Amici, S.; Piscini, A.; Buongiorno, M.F.; Pieri, D. Geological classification of Volcano Teide by hyperspectral and multispectral satellite data. *Int. J. Remote Sens.* **2013**, *34*, 3356–3375. [CrossRef]
106. Bachri, I.; Hakdaoui, M.; Raji, M.; Teodoro, A.C.; Benbouziane, A. Machine Learning Algorithms for Automatic Lithological Mapping Using Remote Sensing Data: A Case Study from Souk Arbaa Sahel, Sidi Ifni Inlier, Western Anti-Atlas, Morocco. *ISPRS Int. J. Geo-Inf.* **2019**, *8*, 248. [CrossRef]
107. Caruso, A.S.; Clarke, K.D.; Tiddy, C.J.; Lewis, M.M. Airborne hyperspectral characterisation of hydrothermal alteration in a regolith-dominated terrain, southern Gawler Ranges, South Australia. *Aust. J. Earth Sci.* **2020**, *68*, 590–608. [CrossRef]
108. Sandoval, P.J.M.; González, J.A.C. *Principios y Aplicaciones de la Percepción Remota en el Cultivo de la Caña de Azúcar en Colombia*; Centro de investigación de la >caña de azúcar de Colombia: Cali, Colombia, 2012.
109. Negi, H.S.; Shekhar, C.; Singh, S.K. Snow and glacier investigations using hyperspectral data in the Himalaya. *Curr. Sci.* **2015**, *108*, 892–902.
110. Kokaly, R.F.; Clark, R.N.; Swayze, G.A.; Livo, K.E.; Hoefen, T.M.; Pearson, N.C.; Wise, R.A.; Benzel, W.M.; Lowers, H.A.; Driscoll, R.L.; et al. *USGS Spectral Library Version 7: U.S. Geological Survey Data Series 1035*; U.S. Geological Survey: Reston, VA, USA, 2017.

111. Bonifazi, G.; Capobianco, G.; Serranti, S. Asbestos containing materials detection and classification by the use of hyperspectral imaging. *J. Hazard. Mater.* **2018**, *344*, 981–993. [CrossRef] [PubMed]
112. Samsudin, S.H.; Shafri, H.Z.M.; Hamedianfar, A. Development of spectral indices for roofing material condition status detection using field spectroscopy and WorldView-3 data. *J. Appl. Remote Sens.* **2016**, *10*, 025021. [CrossRef]
113. Kokaly, R.F.; Couvillion, B.R.; Holloway, J.A.M.; Roberts, D.A.; Ustin, S.L.; Peterson, S.H.; Khanna, S.; Piazza, S.C. Spectroscopic remote sensing of the distribution and persistence of oil from the Deepwater Horizon spill in Barataria Bay marshes. *Remote Sens. Environ.* **2013**, *129*, 210–230. [CrossRef]
114. Clark, R.N.; Swayze, G.A.; Leifer, I.; Livo, K.E.; Lundeen, S.; Eastwood, M.; Green, R.O.; Kokaly, R.F.; Hoefen, T.; Sarture, C.; et al. A Method for Qualitative Mapping of Thick Oil Spills Using Imaging Spectroscopy. *U.S. Geol. Surv.* **2010**, *2010*, 51.
115. De Jong, S.M. Chapter 1 Basics of Remote Sensing. In *Remote Sensing Image Analysis: Including the Spatial Domain*; Springer: Dordrecht, The Netherlands, 2007; pp. 1–15. [CrossRef]
116. Liu, J.; Fan, J.; Yang, C.; Xu, F.; Zhang, X. Novel vegetation indices for estimating photosynthetic and non-photosynthetic fractional vegetation cover from Sentinel data. *Int. J. Appl. Earth Obs. Geoinf.* **2022**, *109*, 102793. [CrossRef]
117. Ma, X.Q.; Leng, P.; Liao, Q.Y.; Geng, Y.J.; Zhang, X.; Shang, G.F.; Song, X.; Song, Q.; Li, Z.L. Prediction of vegetation phenology with atmospheric reanalysis over semiarid grasslands in Inner Mongolia. *Sci. Total Environ.* **2022**, *812*, 152462. [CrossRef]
118. Bai, Y.; Li, S.; Liu, M.; Guo, Q. Assessment of vegetation change on the Mongolian Plateau over three decades using different remote sensing products. *J. Environ. Manag.* **2022**, *317*, 115509. [CrossRef]
119. Sun, W.; Liu, S.; Zhang, X.; Zhu, H. Performance of hyperspectral data in predicting and mapping zinc concentration in soil. *Sci. Total Environ.* **2022**, *824*, 153766. [CrossRef]
120. Ge, X.; Ding, J.; Teng, D.; Xie, B.; Zhang, X.; Wang, J.; Han, L.; Bao, Q.; Wang, J. Exploring the capability of Gaofen-5 hyperspectral data for assessing soil salinity risks. *Int. J. Appl. Earth Obs. Geoinf.* **2022**, *112*, 102969. [CrossRef]
121. Zambrano Prado P., L. Assessing Building's Rooftops Potencial to Integrate Food, Water and Energy Sysitems: The Use of Remote Sensing Technology and Percemptual Aspects in a Mediterranean Region. Universitat Autonoma de Barcelona. 2021. Available online: https://www.tdx.cat/bitstream/10803/674530/1/plzp1de1.pdf (accessed on 2 September 2022).
122. Kit, O.; Lüdeke, M.; Reckien, D. Texture-based identification of urban slums in Hyderabad, India using remote sensing data. *Appl. Geogr.* **2012**, *32*, 660–667. [CrossRef]
123. Nadal, A.; Alamús, R.; Pipia, L.; Ruiz, A.; Corbera, J.; Cuerva, E.; Rieradevall, J.; Josa, A. Urban planning and agriculture. Methodology for assessing rooftop greenhouse potential of non-residential areas using airborne sensors. *Sci. Total Environ.* **2017**, *601*, 493–507. [CrossRef] [PubMed]
124. Zini, E. Esperienze ed Esigenze di ARPA Lombardia di Utilizzazione del Telerilevamento Iperspettrale nei Monitoraggi e nei Controlli Ambientali; San Piero a Grado (Pisa). 2018. Available online: http://conferenzecisam.it/convegni/c-i-s-a-m-2018-1/documenti/Zini_ARPALombardia.pdf (accessed on 28 November 2022).
125. Lorenz, C.; Chiaravalloti-Neto, F.; de Oliveira Lage, M.; Quintanilha, J.A.; Parra, M.C.; Dibo, M.R.; Fávaro, E.A.; Guirado, M.M.; Nogueira, M.L. Remote sensing for risk mapping of Aedes aegypti infestations: Is this a practical task? *Acta Trop.* **2020**, *205*, 105398. [CrossRef] [PubMed]
126. Raczko, E.; Krówczyńska, M.; Wilk, E. Asbestos roofing recognition by use of convolutional neural networks and high-resolution aerial imagery. Testing different scenarios. *Build. Environ.* **2022**, *217*, 109092. [CrossRef]
127. Zhang, Y.L.; Hong, W.H.; Kim, Y.C. Methods for assessing asbestos-containing roofing slate distribution in an area with poor dwelling conditions. *J. Clean. Prod.* **2022**, *358*, 132032. [CrossRef]
128. Sobral, B.S.; de Oliveira-Júnior, J.F.; Alecrim, F.; Gois, G.; Muniz-Júnior, J.G.; de Bodas Terassi, P.M.; Pereira-Júnior, E.R.; Lyra, G.B.; Zeri, M. PERSIANN-CDR based characterization and trend analysis of annual rainfall in Rio De Janeiro State, Brazil. *Atmos. Res.* **2020**, *238*, 104873. [CrossRef]
129. Lammoglia, T.; de Souza Filho, C.R. Mapping and characterization of the API gravity of offshore hydrocarbon seepages using multispectral ASTER data. *Remote Sens. Environ.* **2012**, *123*, 381–389. [CrossRef]
130. Emran, B.J.; Tannant, D.D.; Najjaran, H. Low-altitude aerial methane concentration mapping. *Remote Sens.* **2017**, *9*, 823. [CrossRef]
131. Garcia-Pineda, O.; Staples, G.; Jones, C.E.; Hu, C.; Holt, B.; Kourafalou, V.; Graettinger, G.; DiPinto, L.; Ramirez, E.; Streett, D.; et al. Classification of oil spill by thicknesses using multiple remote sensors. *Remote Sens. Environ.* **2020**, *236*, 111421. [CrossRef]
132. Iwaszenko, S.; Kalisz, P.; Słota, M.; Rudzki, A. Detection of natural gas leakages using a laser-based methane sensor and uav. *Remote Sens.* **2021**, *13*, 510. [CrossRef]
133. Löw, F.; Stieglitz, K.; Diemar, O. Terrestrial oil spill mapping using satellite earth observation and machine learning: A case study in South Sudan. *J. Environ. Manag.* **2021**, *298*, 113424. [CrossRef] [PubMed]
134. Huang, X.; Zhang, B.; Perrie, W.; Lu, Y.; Wang, C. A novel deep learning method for marine oil spill detection from satellite synthetic aperture radar imagery. *Mar. Pollut. Bull.* **2022**, *179*, 113666. [CrossRef]
135. El-Hadidy, S.M.; Alshehri, F.; Sahour, H.; Abdelmalik, K.W. Detecting hydrocarbon micro-seepage and related contamination, probable prospect areas, deduced from a comparative analysis of multispectral and hyperspectral satellite images. *J. King Saud Univ.-Sci.* **2022**, *34*, 102192. [CrossRef]
136. Zoheir, B.; Emam, A. Integrating geologic and satellite imagery data for high-resolution mapping and gold exploration targets in the South Eastern Desert, Egypt. *J. Afr. Earth Sci.* **2012**, *66*, 22–34. [CrossRef]

137. Ghulam, A.; Amer, R.; Kusky, T.M. Mineral exploration and alteration zone mapping in eastern desert of Egypt using aster data. In Proceedings of the ASPRS 2010 Annual Conference, San Diego, CA, USA, 26–30 April 2010.
138. Lamri, T.; Djemaï, S.; Hamoudi, M.; Zoheir, B.; Bendaoud, A.; Ouzegane, K.; Amara, M. Satellite imagery and airborne geophysics for geologic mapping of the Edembo area, Eastern Hoggar (Algerian Sahara). *J. Afr. Earth Sci.* **2016**, *115*, 143–158. [CrossRef]
139. Epuh, E.E.; Okolie, C.J.; Daramola, O.E.; Ogunlade, F.S.; Oyatayo, F.J.; Akinnusi, S.A.; Emmanuel, E.O.I. An integrated lineament extraction from satellite imagery and gravity anomaly maps for groundwater exploration in the Gongola Basin. *Remote Sens. Appl. Soc. Environ.* **2020**, *20*, 100346. [CrossRef]
140. Eldosouky, A.M.; El-Qassas, R.A.Y.; Pour, A.B.; Mohamed, H.; Sekandari, M. Integration of ASTER satellite imagery and 3D inversion of aeromagnetic data for deep mineral exploration. *Adv. Space Res.* **2021**, *68*, 3641–3662. [CrossRef]
141. Muavhi, N.; Tessema, A. Identification of potential targets for kimberlite exploration using satellite imagery and map combination approach in the Lesotho Kimberlite Province. *Ore Geol. Rev.* **2021**, *132*, 104001. [CrossRef]
142. Soydan, H.; Koz, A.; Düzgün, H.Ş. Secondary Iron Mineral Detection via Hyperspectral Unmixing Analysis with Sentinel-2 Imagery. *Int. J. Appl. Earth Obs. Geoinf.* **2021**, *101*, 102343. [CrossRef]
143. Zouaghi, T.; Harbi, H. Airborne geophysics and remote sensing of an Nimas-Khadra area, southern Arabian shield: New insights into structural framework and mineral occurrences. *Adv. Space Res.* **2022**, *70*, 3649–3673. [CrossRef]
144. Jaud, M.; Geoffroy, L.; Chauvet, F.; Durand, E.; Civet, F. Potential of a virtual reality environment based on very-high-resolution satellite imagery for structural geology measurements of lava flows. *J. Struct. Geol.* **2022**, *158*, 104569. [CrossRef]
145. Epuh, E.E.; Moshood, A.I.; Okolie, C.J.; Daramola, O.E.; Akinnusi, S.A.; Arungwa, I.D.; Orji, M.J.; Olanrewaju, H.O.; Fatoyinbo, A.A. Integration of satellite gravimetry, multispectral imagery and digital elevation model for investigating crustal deformation in the Niger Delta Basin. *Geosyst. Geoenviron.* **2022**, *1*, 100067. [CrossRef]
146. Tözün, K.A.; Özyavaş, A. Automatic detection of geological lineaments in central Turkey based on test image analysis using satellite data. *Adv. Space Res.* **2022**, *69*, 3283–3300. [CrossRef]
147. Birth, G.S.; McVey, G.R. Measuring the Color of Growing Turf with a Reflectance Spectrophotometer. *Agron. J.* **1968**, *60*, 640–643. [CrossRef]
148. Rouse, J.W.J.; Haas, R.H.; Schell, J.A.; Deering, D.W.; Haas, R.H.; Schell, J.A.; Deering, D.W. Monitoring vegetation systems in the great plains with erts. Conference Paper, NASA. In Proceedings of the Goddard Space Flight Center 3d ERTS-1 Symp, Houston, TX, USA, 1974.
149. Kauth, R.J.; Thomas, G.S.P. *The Tasselled Cap—A Graphic Description of the Spectral-Temporal Development of Agricultural Crops as Seen by Landsat*; Pordue University: West Lafayette, IN, USA, 1976.
150. Tucker, C.J. Red and photographic infrared linear combinations for monitoring vegetation. *Remote Sens. Environ.* **1979**, *8*, 127–150. [CrossRef]
151. Huete, A.R. A soil-adjusted vegetation index (SAVI). *Remote Sens. Environ.* **1988**, *25*, 295–309. [CrossRef]
152. Crippen, R.E. Calculating the vegetation index faster. *Remote Sens. Environ.* **1990**, *34*, 71–73. [CrossRef]
153. Pinty, B.; Verstraete, M.M. GEMI: A non-linear index to monitor global vegetation from satellites. *Vegetatio* **1992**, *101*, 15–20. [CrossRef]
154. Kaufman, Y.J.; Tanre, D. Atmospherically resistant vegetation index (ARVI) for EOS-MODIS. *IEEE Trans. Geosci. Remote Sens.* **1992**, *30*, 261–270. [CrossRef]
155. L3Harris. Broadband Greenness. 2022. Available online: https://www.l3harrisgeospatial.com/docs/broadbandgreenness.html#:~{}:text=State%20University%2C%202005.-,Green%20Vegetation%20Index%20(GVI),range%20from%20-1%20to%201 (accessed on 27 September 2022).
156. Qi, J.; Chehbouni, A.; Huete, A.R.; Kerr, Y.H.; Sorooshian, S. A modified soil adjusted vegetation index. *Remote Sens. Environ.* **1994**, *48*, 119–126. [CrossRef]
157. Goel, N.S.; Qin, W. Influences of canopy architecture on relationships between various vegetation indices and LAI and Fpar: A computer simulation. *Remote Sens. Rev.* **1994**, *10*, 309–347. [CrossRef]
158. Roujean, J.L.; Breon, F.M. Estimating PAR absorbed by vegetation from bidirectional reflectance measurements. *Remote Sens. Environ.* **1995**, *51*, 375–384. [CrossRef]
159. Penuelas, J.; Baret, F.; Filella, I. Semi-empirical indices to assess carotenoids/chlorophyll a ratio from leaf spectral reflectance. *Photosynthetica* **1995**, *31*, 221–230.
160. Rondeaux, G.; Steven, M.; Baret, F. Optimization of soil-adjusted vegetation indices. *Remote Sens. Environ.* **1996**, *55*, 95–107. [CrossRef]
161. Gitelson, A.A.; Kaufman, Y.J.; Merzlyak, M.N. Use of a green channel in remote sensing of global vegetation from EOS-MODIS. *Remote Sens. Environ.* **1996**, *58*, 289–298. [CrossRef]
162. Chen, J.M. Evaluation of Vegetation Indices and a Modified Simple Ratio for Boreal Applications. *Can. J. Remote Sens.* **1996**, *22*, 229–242. [CrossRef]
163. Gitelson, A.A.; Merzlyak, M.N. Remote sensing of chlorophyll concentration in higher plant leaves. *Adv. Space Res.* **1998**, *22*, 689–692. [CrossRef]
164. Louhaichi, M.; Borman, M.M.; Johnson, D.E. Spatially Located Platform and Aerial Photography for Documentation of Grazing Impacts on Wheat. *Geocarto Int.* **2001**, *16*, 65–70. [CrossRef]
165. Minnett, P.J. Satellite Remote Sensing of Sea Surface Temperatures. *Encycl. Ocean Sci. Second Ed.* **2001**, *1*, 91–102. [CrossRef]

166. Huete, A.; Didan, K.; Miura, T.; Rodriguez, E.P.; Gao, X.; Ferreira, L.G. Overview of the radiometric and biophysical performance of the MODIS vegetation indices. *Remote Sens. Environ.* **2002**, *83*, 195–213. [CrossRef]
167. Boegh, E.; Soegaard, H.; Broge, N.; Schelde, K.; Thomsen, A.; Hasager, C.B.; Jensen, N.O. Airborne multispectral data for quantifying leaf area index, nitrogen concentration, and photosynthetic efficiency in agriculture. *Remote Sens. Environ.* **2002**, *81*, 179–193. [CrossRef]
168. Gitelson, A.A.; Stark, R.; Grits, U.; Rundquist, D.; Kaufman, Y.; Derry, D. Vegetation and soil lines in visible spectral space: A concept and technique for remote estimation of vegetation fraction. *Int. J. Remote Sens.* **2002**, *23*, 2537–2562. [CrossRef]
169. Bannari, A.; Asalhi, H.; Teillet, P.M. Transformed difference vegetation index (TDVI) for vegetation cover mapping. *IEEE Int. Geosci. Remote Sens. Symp.* **2002**, *5*, 3053–3055.
170. Gitelson, A.A.; Gritz, Y.; Merzlyak, M.N. Relationships between leaf chlorophyll content and spectral reflectance and algorithms for non-destructive chlorophyll assessment in higher plant leaves. *J. Plant Physiol.* **2003**, *160*, 271–282. [CrossRef]
171. Lobell, D.B.; Asner, G.P. Hyperion studies of crop stress in Mexico. In Proceedings of the 12th JPL Airborne Earth Science Workshop, Pasadena, CA, USA, 12–16 January 2004; pp. 1–6.
172. Gitelson, A.A. Wide Dynamic Range Vegetation Index for Remote Quantification of Biophysical Characteristics of Vegetation. *J. Plant Physiol.* **2004**, *161*, 165–173. [CrossRef]
173. Sripada, R.P. *Determining In-Season Nitrogen Requirements for Corn Using Aerial Color-Infrared Photography*; North Carolina State University: Raleigh, NC, USA, 2005.
174. Sripada, R.P.; Heiniger, R.W.; White, J.G.; Meijer, A.D. Aerial Color Infrared Photography for Determining Early In-Season Nitrogen Requirements in Corn. *Agron. J.* **2006**, *98*, 968–977. [CrossRef]
175. Yang, Z.; Willis, P.; Mueller, R. Impact of Band-Ratio Enhanced AWiFS Image to Crop Classification Accuracy. *Proc. Pecora.* **2008**, *17*, 1–11.
176. Dash, J.; Curran, P.J. The MERIS terrestrial chlorophyll index. *Int. J. Remote Sens.* **2004**, *25*, 5403–5413. [CrossRef]
177. Raymond Hunt, E.; Daughtry, C.S.T.; Eitel, J.U.H.; Long, D.S. Remote Sensing Leaf Chlorophyll Content Using a Visible Band Index. *Agron. J.* **2011**, *103*, 1090–1099. [CrossRef]
178. Wolf, A.F. Using WorldView-2 Vis-NIR multispectral imagery to support land mapping and feature extraction using normalized difference index ratios. In Proceedings of the Algorithms and Technologies for Multispectral, Hyperspectral, and Ultraspectral Imagery XVIII, Baltimore, MD, USA, 14 May 2012; Society of Photo-Optical Instrumentation Engineers (SPIE): Bellingham, DC, USA; Volume 8390, pp. 188–195. [CrossRef]
179. MaxMax. Enhanced Normalized Difference Vegetation Index (ENDVI). 2015. Available online: https://www.maxmax.com/endvi.htm (accessed on 26 September 2022).

geosciences

Article

Monitoring the Ambient Seismic Field to Track Groundwater at a Mountain–Front Recharge Zone

Quinn Buzzard [1], Jeff B. Langman [2,*], David Behrens [2] and James G. Moberly [3]

1 Water Resources Program, University of Idaho, Moscow, ID 83844, USA
2 Department of Earth and Spatial Sciences, University of Idaho, Moscow, ID 83844, USA
3 Department of Chemical and Biological Engineering, University of Idaho, Moscow, ID 83844, USA
* Correspondence: jlangman@uidaho.edu

Abstract: The heterogeneity of the fractured-basalt and interbedded-sediment aquifer along the eastern margin of the Columbia Plateau Regional Aquifer System has presented challenges to resource managers in quantifying recharge and estimating sustainable withdrawals. Previous studies indicated recharge pathways in alluvial sediments atop a mountain–front interface upgradient of the basalt flows. In this sedimentary zone, six seismic stations were deployed for one year to detect velocity changes in low-frequency seismic waves that could be correlated to changes in groundwater recorded by a well transducer near the center of the seismic station network. Waveforms in the 1−5 Hz range were recorded at each station to determine changes in wave velocities between station pairs and correlate these velocity changes to changes in groundwater levels. The velocity–groundwater relation allowed for estimation of daily groundwater levels beneath the seismic station network. Existing hydrogeologic information was used to estimate hydraulic gradients and hydraulic conductivities, which allowed for the calculation of the daily volume of recharge passing beneath the seismic stations and into the confined aquifer system. The daily recharge volumes across the seismic station network were summed for comparison of the total annual recharge calculated from the change in seismic wave velocities (154,660 m^3) to a flow model calculation of recharge based on areal precipitation and infiltration (26,250 m^3). The 6× greater recharge estimated from the seismic wave velocity changes for this portion of the recharge zone is attributed to preferential pathways of high hydraulic conductivity and greater depth associated with paleochannels beneath the seismic station network.

Keywords: groundwater recharge; ambient seismic field; passive monitoring

Citation: Buzzard, Q.; Langman, J.B.; Behrens, D.; Moberly, J.G. Monitoring the Ambient Seismic Field to Track Groundwater at a Mountain–Front Recharge Zone. *Geosciences* **2023**, *13*, 9. https://doi.org/10.3390/geosciences13010009

Academic Editors: Nicolò Colombani and Jesus Martinez-Frias

Received: 27 August 2022
Revised: 20 December 2022
Accepted: 23 December 2022
Published: 28 December 2022

1. Introduction

Groundwater is an important resource for municipal, agricultural, and industrial uses across Idaho, the United States, and the globe [1–4]. Since 1935, water levels have declined in the multi-aquifer system in the South Fork Palouse River Basin (Figure 1) located in the Palouse geographic region and eastern margin of the Columbia Plateau Regional Aquifer System [5–7]. The South Fork Palouse River Basin aquifer system is contained in the fractured basalts of the Columbia River Basalt Group (CRBG) and interbedded sediments of the Latah Formation (Figure 2) that compose the eastern portion of the basin, designated as the Moscow–Pullman Basin (MPB) [8–10]. Groundwater in the local basin provides a primary source for drinking water and irrigation [11] and is the sole source of municipal water in the MPB [12]. Extrapolation of current trends in declining groundwater levels indicates the possibility of insufficient groundwater resources to meet future community needs [13]. Quantification of recharge to the MPB aquifer system is necessary to evaluate sustainable withdrawals or potential water storage/recovery systems. This study was conducted to evaluate groundwater changes and quantify the annual recharge along a portion of a theorized recharge zone by passively monitoring the ambient seismic field and correlating changes in seismic wave velocities to changes in groundwater levels.

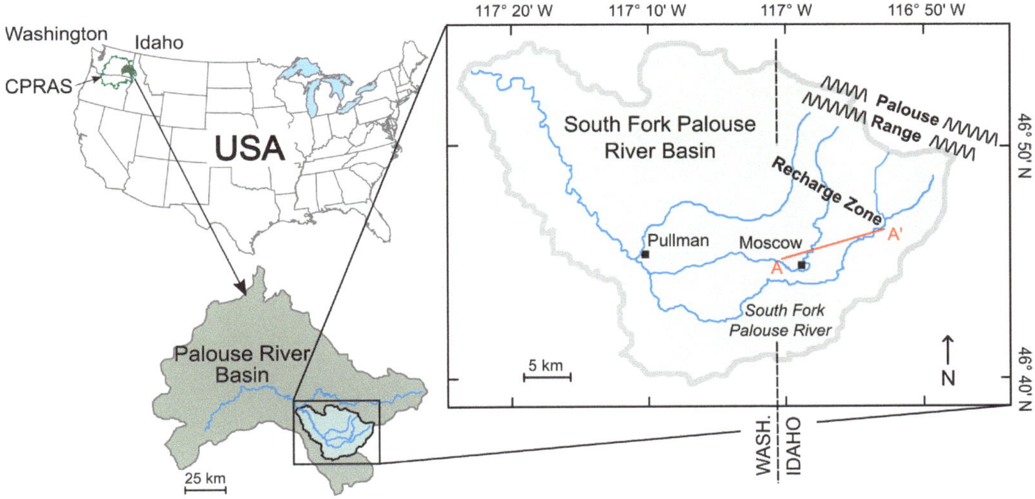

Figure 1. Location of the South Fork Palouse River Basin in the Palouse River Basin within the Columbia Plateau Regional Aquifer System (CPRAS) (modified from Behrens et al. [14]).

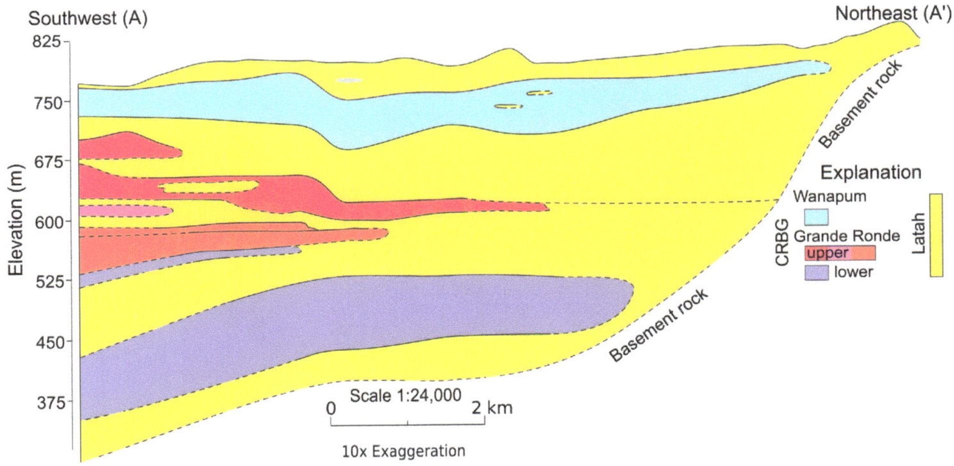

Figure 2. Southwest-to-northeast cross section (A–A′, Figure 1) of the eastern South Fork Palouse River Basin near Moscow, Idaho, USA (modified from Bush et al. [9]).

Past modeling efforts to predict future declines in groundwater levels of the MPB have produced mixed results due to a limited understanding of recharge processes [15–19]. The variable permeability and discontinuity of basalt flows and interbedded sediments create heterogeneous and anisotropic aquifer matrices in the basin [10,20,21]. Resource management entities across the northwestern United States continue to struggle to model and predict recharge in such terrains [22]. An interstate, multi-agency committee of water providers in the MPB, Palouse Basin Aquifer Committee (PBAC), implemented a study to develop a new groundwater flow model to assist in understanding the continued decline in groundwater levels. As part of the modeling effort, recharge to the aquifer system was estimated by assigning a higher areal precipitation and infiltration rate to the foothill/mountainous region across the eastern portion of the basin (aligns with the recharge zone in Figure 1) and a lower rate for the lowlands of the basin [20]. This current study was conducted to compare the annual recharge of the PBAC groundwater model in a portion of the recharge zone to the annual recharge calculated from groundwater levels derived from changes in the velocity of low frequency seismic waves recorded in the same portion of the recharge zone.

1.1. Recharge Zone

Previous studies have indicated that groundwater recharge (e.g., snowmelt) is entering the aquifer system through sediments of the Latah Formation [14,23–25] at the mountain–front interface along the eastern margin of the MPB [15,25–28]. These sediments overly the granitic basement rock at the mountain front of the Palouse Range (Figure 3). The sediments of the Latah Formation can range from permeable alluvial/colluvial deposits to clayey wetland deposits emplaced during damming of streams with the intrusion of CRBG flows [10]. Additionally, coarse paleochannel sediments are interspersed throughout the Latah Formation because of the continued rerouting of the paleostream network with the intrusion of at least 25 basalt flows [9,10]. The uppermost sediments of the Latah Formation can be clay rich but also contain coarser material that corresponds to the current stream network [9,10].

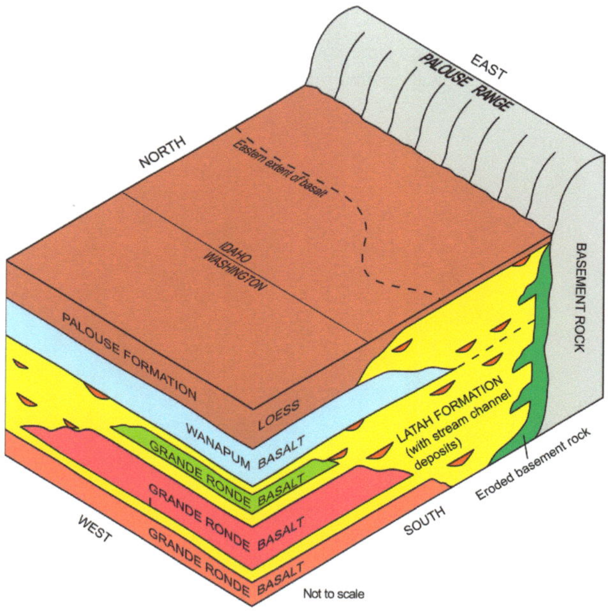

Figure 3. Theorized mountain–front interface of the Palouse Range and sedimentary units of the Latah Formation that contain paleochannel deposits from prior iterations of the stream network draining the Palouse Range (updated from Bush et al. [29]).

Downgradient of the theorized recharge zone, Duckett et al. [25] were able to discriminate two primary groundwater sources that originated from snowmelt moving either quickly into the subsurface ("fast pathway") or snowmelt and/or rainfall that stayed in the surface-water network and entered the subsurface further downgradient ("slow pathway"). Behrens et al. [14] were able to refine the fast and slow pathway concept through an isotopic analysis of snowpack, snowmelt, runoff, creek, and groundwater samples collected from the mountain top to the recharge zone. The fastest recharge pathways appear to be located within the central portion of the recharge zone and slower pathways are located along the western and eastern peripheries [14]. These pathway types have some overlap with the existing stream network but are not fully aligned, and the higher conductivity flowpaths in the recharge zone likely are associated with paleochannels [14].

1.2. Passive Seismic Monitoring for Estimating Groundwater Levels

Passive seismic monitoring can be used to interpret near-surface conditions [30–34], such as changes in groundwater levels in unconfined and non-compartmentalized alluvial aquifers [35–37]. The scattering of seismic waves in the Earth's crust allows for an averaged and volumetric view of changes in groundwater where the velocity of scattered seismic waves is sensitive to changes in pore pressure (e.g., grain-to-grain contact) [37–39]. Seismometers can be deployed to passively record low frequency waves of the ambient seismic field generated by natural or anthropogenic earth movements [40,41]. These low-frequency waves are influenced by the elastic properties of near surface materials and properties, such as changes in saturated thickness/pore pressure [37,38,40,42–46]. The velocity of scattered seismic waves in an aquifer will respond to changes in pore pressure caused by increased water levels and decreased grain contacts [36,38]. The recharge zone at the mountain front in the MPB is composed of unconsolidated sediments of the Latah Formation outside of the furthest extent of the basalts, which allowed for deployment of a temporary network of seismic stations to enhance the limited groundwater monitoring in this area (one well transducer).

2. Materials and Methods

To quantify the annual recharge along a portion of the mountain–front recharge zone, six seismic stations were installed as a transect perpendicular to groundwater flow. This temporary seismic network was used to collect seismic spectra from October 2020 through September 2021 to correlate changes in seismic wave velocities to changes in groundwater levels. Available geologic data (e.g., well logs and local geologic reports) were used to interpret hydraulic gradients and hydraulic conductivities. The combination of groundwater levels/saturated thicknesses, hydraulic gradients, and hydraulic conductivities, allowed for estimating the volume of water passing beneath the seismic network and entering the MPB confined aquifer system during the study period.

2.1. Seismometer and Station Construction

The Raspberry Shake® 1D was used for construction of the seismic stations. The Raspberry Shake® 1D contains a 4.5-Hz vertical geophone and internal memory for datalogging of up to 80 days. The geophone has the potential to resolve the low frequency range (0.1–5 Hz) that constitute the portion of the ambient seismic field that has previously been used to detect changes in seismic velocity because of changes in pore pressure/groundwater levels [31,37,45,47]. The seismometers were fitted with GPS units for an accurate record of time because of the need for cross-correlation analysis between stations for identifying changes in wave velocities [37,41,48,49]. The seismometer vaults (Figure 4) consisted of a weather-proof sealable container (action packer) for containing the seismometer in a weatherproof case and a deep cycle marine battery for power. The weatherproof case containing the seismometer was bolted to a granitic rock plinth and placed on a sand bed inside the action packer to ensure connection of the seismometer to the surrounding earth. The battery was connected to a solar panel (Figure 4) to reduce the need for battery replacement during the deployment period. Each seismic station was placed 1 m below land surface to connect with the surrounding earth and allow access to the seismometer. A data retrieval cable was paired with the power cable connecting the solar panel to the seismic station.

Figure 4. Seismic station composed of the buried, sealable container with solar panel and an inner view of the container with the marine battery and weatherproof case containing the seismometer.

2.2. Seismic Station Locations

Seismic station locations (Table 1) were based on proximity to the mountain front and outside the extent of the Wanapum basalt (Figure 3). Local drilling logs indicated that the selected sites likely had relatively shallow groundwater (<100 m) and relatively shallow basement rock (<500 m). One site was pre-selected because of an existing well containing an hourly recording Aqua4Plus 1.9.10 transducer (Figure 5) to which the seismic spectra were correlated for estimating groundwater across the seismic station network. The transducer well is 77 m deep and set in a mixed alluvium consisting of alternating clay- or gravel-rich layers that are part of the sediments of Bovill [9,24]. For quality control purposes, each seismic station was visited monthly for data downloading to ensure data preservation and identification of possible recording/power issues. If abnormal data output or power levels were detected, the vault was opened, and the instrumentation checked on-site.

Table 1. Seismic station location description.

Station ID	Latitude [1]	Longitude [1]	Elevation (m) [2]
1	46.78935	−117.010	848
2	46.78417	−116.987	853
3	46.77367	−116.975	824
4	46.77975	−116.972	848
5	46.77078	−116.951	846
6	46.76875	−116.936	863

[1] North American Datum of 1983 (NAD 83); [2] North American Vertical Datum of 1988 (NAVD 88).

2.3. Seismic Station Network and Quantifying Recharge

The seismic stations constituted a network of points overlying the non-compartmentalized sedimentary units composing the recharge zone, which connects the primary source water (e.g., infiltrated snowmelt at the mountain front) to the confined portion of the aquifer system. To correlate changes in seismic wave velocities and groundwater levels, the seismic station network was divided into station pairs and associated segments (Table 2 and Figure 5). Stations were paired by closest neighbor (west to east) for cross-correlation analysis of the waveforms recorded at each station. If a station could be paired to multiple stations (correlatable waveform distributions), each available pair was included in the analysis and recharge volumes from overlapping station pairs were averaged across the intersected area. From the station pairs and given sufficient ambient waveforms, a change in velocity relative to velocity (dv/v) can be used to determine groundwater levels [37]. Such a correlation is possible because the velocity of ambient waves is sensitive to changes in pore pressure with increasing or decreasing groundwater levels [38]. The velocity comparison (dv/v) is a relativistic determination of waveform velocity differences recorded across the paired stations and represents a perturbation in the waveform velocity due to a change in groundwater levels that influence grain-to-grain contact from changes in pore pressure [37].

Table 2. Station pairs and associated network segments.

Station pairs	1–2	2–3	2–4	3–5	4–5	5–6
Recharge segments	A	B [1]	C [1]	D [1]	E [1]	F

[1] Overlapping station pairs were averaged for recharge calculations.

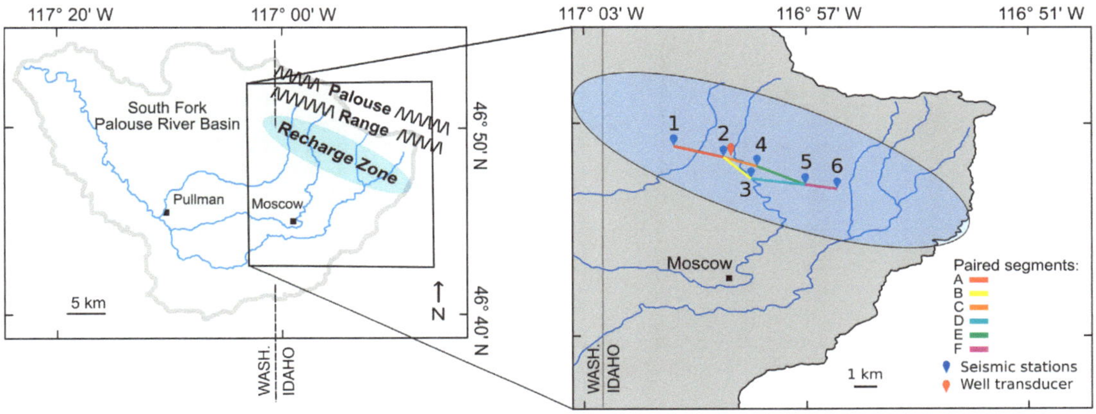

Figure 5. Seismic station locations and station pairs (paired segments) in the study area that is part of the theorized recharge zone along the Palouse Range.

2.4. Identifying Applicable Waveforms in the Ambient Seismic Field

The waveforms from each seismic station were evaluated in ObsPy [50] with probabilistic power spectral density (PPSD) plots [51], which provided a view of smoothed and binned power spectral densities. These plots assisted in determining if low frequency waves were consistently detected by each seismometer. The 1–5 Hz range proved to be the most consistent waveform range at each station, which is within the applicable range for detecting changes in saturated thickness/pore pressure [37]. Small periods (hours to a few days) of data loss occurred at most seismic stations because of data corruption, but these short periods were linearly interpolated using the preceding and following changes in velocity. The percent of missing data ranged from 0% (segment F) to 11.8% (segment A) with an average data loss of 4.4%.

The cross-correlation function of MsNoise [52] was used to identify similar waveforms recorded between stations to create a proxy of Green's function. A whitening filter from 1–5 Hz was applied to correct for frequency attenuation of the recorded waves in this target range [53,54]. The cross-correlation functions between each station pair were computed at 1 h intervals with a 30 min overlap [37]. A 14-day stack of cross-correlation functions was used to maximize temporal resolution while minimizing spurious oscillations. A moving window cross spectral (MWCS) technique [37,55] was used to evaluate the delay in arrival times (change in time relative to time or dt/t) for waveforms in the 1–5 Hz target range. It is assumed that there is a linear relation between relative time lags and seismic wave velocity changes (change in velocity relative to velocity or dv/v), or $-dt/t = dv/v$ [37,41,54,56].

2.5. Velocity Changes to Groundwater Levels

The groundwater level (GWL) between each station pair was derived through correlation of station pair dv/v and groundwater levels recorded by the well transducer near the center of the network. A single transducer recording groundwater levels can be reflective of changes in saturated thickness across a seismic station network overlying a non-compartmentalized alluvial aquifer (no barriers to recharge and GWL changes across the sedimentary units composing the aquifer) [37]. Estimates of GWL were calculated for each day at each station pair to produce daily groundwater values across the seismic station network for estimating the annual recharge volume. To correlate dv/v to groundwater, the relative changes had to be correlated within distinct seasonal periods. The dv/v-groundwater relations were assumed linear during the seasonal periods (rising or falling on a seasonal basis) and reflective of the elastic properties of the aquifer [37,40,57]. The study time frame was divided into four periods that correlate with periods of seasonal precipitation and infiltration or the lack of precipitation and infiltration: the end of the dry season and return of rainfall (October or period 1), winter snowfall/snowmelt (November through May or period 2), spring/summer snowmelt (June or period 3), and the dry summer season (July through September or period 4). These divisions align with basin precipitation patterns and subsequent streamflow and mountain–front groundwater response as monitored by the Natural Resources Conservation Service [58] and identified by past research in the basin [1,14,20,59]. The linear relation of groundwater changes (ΔGWL) and dv/v changes (Δdv/v) were calculated from the period difference (maximum value − minimum value) of each seasonal period to determine the applicable correlation constant (C_{period}):

$$\frac{GWL_{max} - GWL_{min}}{dv/v_{max} - dv/v_{min}} = \frac{\Delta GWL}{\Delta dv/v} = C_{period} \tag{1}$$

The daily dv/v change ($\Delta dv/v_{day}$) was calculated by the difference between the initial dv/v of the period and a specific day dv/v:

$$dv/v_{initial} - dv/v_{day} = \Delta dv/v_{day} \tag{2}$$

The daily change in groundwater level (ΔGWL_{day}) was derived from the C_{period} and the $\Delta dv/v_{day}$):

$$\Delta dv/v_{day} \times C_{period} = \Delta GWL_{day} \tag{3}$$

The ΔGWL_{day} was added to the initial period groundwater level ($GWL_{initial}$) measured by the transducer to obtain the daily groundwater level (GWL_{day}) for each station pair:

$$\Delta GWL_{day} + GWL_{initial} = GWL_{day} \tag{4}$$

2.6. Interpretations of Hydraulic Conductivity, Gradient, and Recharge

By discriminating sedimentary layer composition beneath seismic stations from local well logs and geologic reports [9,10,60,61], a composite hydraulic conductivity (K, m/d) was assigned for each station pair according to accepted K values for such alluvium types [62,63]. Given the unconfined alluvial aquifer of the recharge zone, hydraulic gradients (Δh/L) were assumed to correspond to basement rock gradients beneath each station pair. The Δh/L of groundwater passing beneath each station pair were estimated from well logs above and below each station pair (depth to bedrock and linear interpretation of depth perpendicular to the station pair) and checked against the bedrock gradient derived by Bush et al. [10]. The Δh/L values ranged from 0.03 to 0.08 and correspond to the land surface gradient with the transition from the steeper mountain slope of the Palouse Range to the basin floor [10]. With calculation of daily groundwater levels from dv/v and depth to basement rock from the well logs, the daily saturated thickness could be calculated for each network segment (depth × segment length = area (A) in m^2). Given K, Δh/L, and A, the daily volume of recharge (Q, m^3/d) passing beneath each network segment was calculated using Darcy's law (Q = A × K × Δh/L).

3. Results

3.1. Velocity Changes and Relation to Groundwater

Changes in seismic velocity varied between station pairs (Figure 6) and ranged from a dv/v high of +0.45% (period 4) to a dv/v low of −0.3% (period 2). The velocity changes inversely reflected the seasonal changes (rising or falling during the seasonal period) in groundwater elevation that ranged between 791 m and 795 m (Figure 6). The recorded changes in groundwater levels were representative of historical annual changes recorded at the transducer well location near the center of the transducer network. The dv/v values were lowest during periods of higher groundwater elevation (period 2 or the winter/spring snowmelt season) and highest during the dry periods (periods 1 and 4) that produced lower groundwater elevations (Figure 6). This inverse relation of dv/v and groundwater elevation corresponds to the expected changes in low-frequency wave velocities with changes in saturated thickness [37,40,45]. The seasonal flux of groundwater at the transducer represents the expected seasonal flux of recharge to the aquifer that is primarily driven by fall rainfall and winter/spring snowmelt [1,14,59].

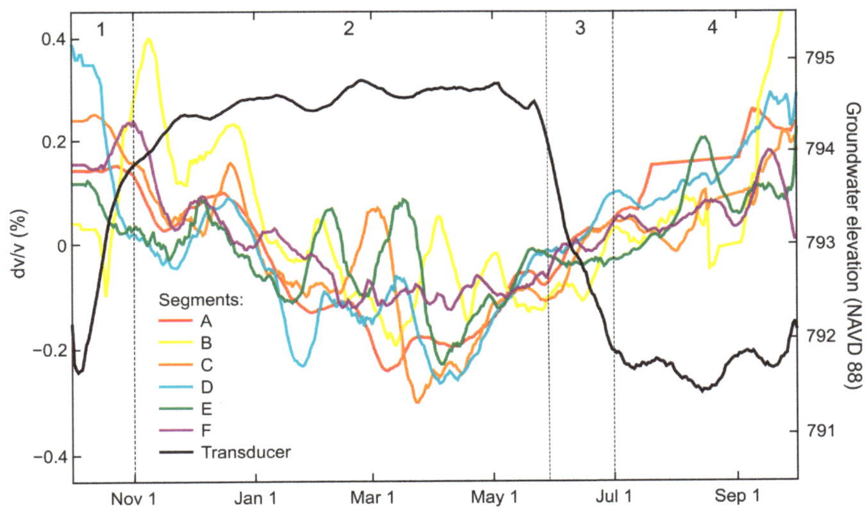

Figure 6. Changes in seismic velocity (dv/v) at each network segment (Figure 5) and groundwater elevation recorded by the well transducer. Temporal periods (1–4) are seasonal divisions used to develop correlations between dv/v and groundwater changes for each seasonal period.

3.2. Converting Seismic Velocity to Groundwater

The distinct seasonal periods in groundwater levels recorded by the transducer provided the necessary temporal periods for correlating dv/v and groundwater as separate seasonal relations (Table 3). The assumption of seasonal dv/v-groundwater relations parallels the seasonal flux of recharge that corresponds to surface hydrological processes of the basin [1,14,28,59]. The change in C_{period} (range of 4.6 to 37.3) reflects the high variability of groundwater levels/recharge during the 1-year study period (Table 3). Although the seasonal discrimination of the dv/v-groundwater relation provided a more refined correlation compared to an annual relation, groundwater elevations derived from dv/v tended to underestimate groundwater elevation during periods of increasing groundwater and overestimate groundwater elevation during periods of decreasing groundwater (Figure 7). These underestimation/overestimation periods represent a lag in the dv/v-groundwater relation following substantial changes in aquifer recharge (Figure 7). It is assumed that the lag period is an adjustment of overall grain-to-grain contact to the pore pressure changes reflective of the change in groundwater level.

Table 3. Seasonal periods and associated changes in groundwater (ΔGWL) and seismic wave velocity (Δdv/v) for correlating (C_{period}) the data sets and estimating groundwater levels.

Period	Date Range (2020–2021)	ΔGWL (m)	Δdv/v (%)	C_{period}
1	October	+2.19	−0.07	31.2
2	November–May	+0.93	−0.20	4.6
3	June	−1.89	+0.05	37.3
4	July–September	−0.62	+0.12	5.3

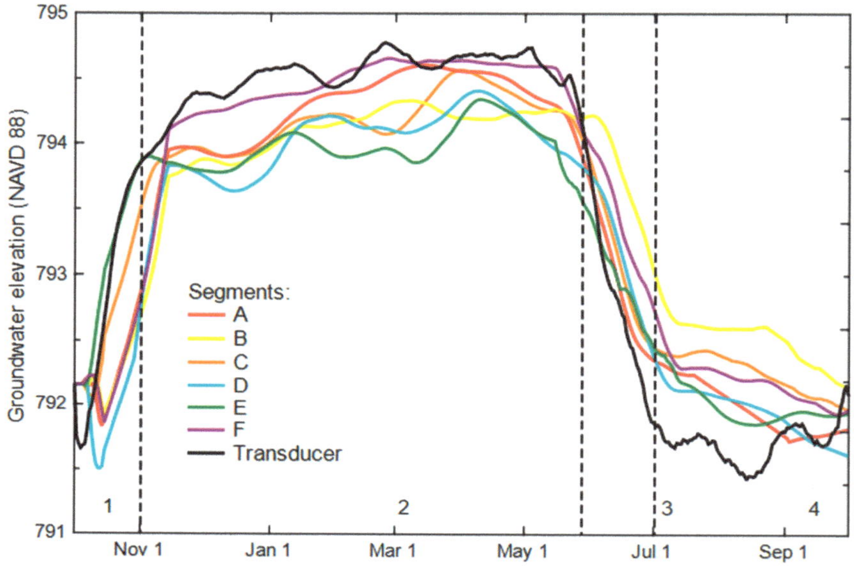

Figure 7. Groundwater elevations derived from dv/v for each network segment (Figure 5) compared to the groundwater elevation measured by the well transducer. Temporal periods (1–4) were seasonal divisions used to develop correlations between dv/v and groundwater changes.

3.3. *Recharge Volumes by Network Segment*

Interpretation of the well logs and geologic reports for evaluation of hydraulic conductivity (K) by network segment produced a range of segment composite K values from a low of 0.024 m/d (more clayey sediments of Bovill that are part of the Latah Formation [8]) to a high of 0.052 m/d (more paleochannel sand) (Table 4). These interpreted K values were calculated by the proportion of different sediment types estimated beneath each station or the mixture of lower conductivity alluvium (clayey) with paleochannel deposits (sand) and the presence of eroded basement rock (granular) [24]. The segment K values were smaller towards the west and largest on the east end of the seismic network. Hydraulic (bedrock) gradients also varied from low to high moving west to east with a corresponding increase in saturated thickness (Table 4). With the available groundwater levels (saturated thicknesses) across the seismic network and associated hydraulic conductivities and hydraulic gradients at each station pair, daily recharge volumes (example in Table 4) were calculated for each network segment and the overall seismic network (Figure 8). The average recharge volume was 422 m^3/d with the largest recharge during period 2 (435 m^3/d) and smallest during period 1 (404 m^3/d) (Figure 8). Recharge was largest after a 10-day snowmelt period in early spring when approximately 15% of the mountain-snowpack water equivalent was lost [59]. Recharge volumes were smallest in period 1 following the summer dry season when <4 cm precipitation occurred in the preceding 3 months [59]. The total annual recharge for the recharge zone beneath the seismic network was estimated at 154,660 m^3.

Table 4. Example of recharge calculations at each network segment and total recharge across the network for 1 October 2020. Overlapping segments were averaged for an adjusted recharge value.

Network Segment	Hydraulic Conductivity (m/d)	Saturated Thickness (m)	Station Distance (m)	Hydraulic Gradient	Potential Recharge (m^3/d)	Adjusted Recharge [1] (m^3/d)
A	0.024	8.0	1812	0.030	10.3	10.3
B [1]	0.033	17.8	1253	0.031	22.4	43.0
C [1]	0.033	23.5	1500	0.055	63.5	
D [1]	0.042	32.9	1883	0.080	210.6	189.8
E [1]	0.042	47.6	1927	0.044	169.0	
F	0.052	44.9	1130	0.063	164.6	164.6
					Network sum (m^3/d):	407.7

[1] Average recharge for overlapping network segments (Figure 5).

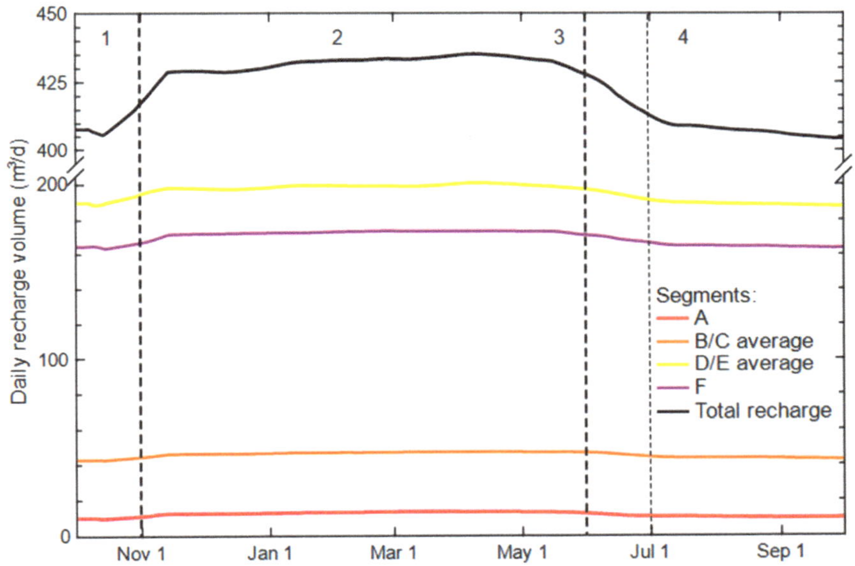

Figure 8. Daily recharge passing beneath the segments of the seismic network (Figure 5) and their summation from dv/v-derived groundwater elevations. Temporal periods (1–4) were seasonal divisions used to develop correlations between dv/v and groundwater changes.

4. Discussion

Recharge volumes were spatially variable across the seismic network with the largest volumes occurring in the central to eastern portion of the network because of greater saturated thicknesses (deeper bedrock), higher hydraulic conductivities (coarser grains from the presence of paleochannels), and steeper hydraulic gradients. This portion of the network (segments D/E and F) constituted 86% of the annual recharge volume while comprising about 50% of the network. The coarser grains, larger hydraulic conductivities, and steeper gradients of the eastern portion of the seismic network suggest faster recharge pathways, which aligns with the theorized fast recharge pathway identified for this area by Behrens et al. [14]. This faster pathway was assumed to be dominated by a greater concentration of paleochannels, which aligns with the review of sedimentary layer composition beneath this portion of the seismic network. The deeper bedrock of this area suggests greater erosion of the mountain front and correlates with the greater concentration of paleochannels and steeper gradients.

To compare the dv/v-derived recharge volume and the recharge volume derived by the PBAC groundwater model, the model aerial infiltration rate used for the foothills/mountainous region (105 mm/yr) was applied to the area from the seismic network to the upgradient watershed boundary for an annual estimate of 26,250 m^3/yr. The larger estimate of recharge derived from the dv/v data (154,660 m^3) is a reflection of greater saturated thicknesses and higher hydraulic conductivities paired with steeper hydraulic gradients, which align with the fast pathway concept of Duckett et al. [25] and Behrens et al. [14].

5. Conclusions

Discrimination of recharge pathways and quantification of recharge to the Moscow–Pullman Basin aquifer system in the Columbia Plateau Regional Aquifer System has posed challenges to resource managers due to the unique geology of the basin and limited well drilling in the theorized recharge zone. Such limitations have made it difficult to determine sustainable withdrawals from the aquifer system, which has undergone groundwater mining for a century. A recent groundwater modeling effort to assist with interpreting the effects of water conservation and withdrawal practices used an aerial infiltration method to estimate recharge along the eastern margin of the basin in a primary recharge zone. Six seismic stations were temporarily installed to enhance groundwater monitoring in a portion of the recharge zone and calculate an annual recharge to the confined aquifer system for comparison to recharge estimates from the groundwater model. Sufficient low-frequency seismic waves were recorded at the six seismic stations composing the seismic network for correlation to groundwater levels recorded by a well transducer located in the center of the network. Estimates of groundwater changes from changes in seismic wave velocities and estimates of hydraulic conductivities and hydraulic gradients from local well logs and geologic reports allowed for estimation of daily recharge volumes passing beneath the seismic network. Summation of the daily recharge estimates produced an annual recharge volume of 154,660 m^3, which is six times greater than the model estimate of 26,250 m^3 for the same area. The larger estimate of recharge derived from the dv/v data is a reflection of a perceived faster pathway of recharge underlying a substantial portion of the seismic network. This faster pathway area highlights the variability of recharge pathways across the mountain front and the difficulty in modeling recharge in the basin.

Author Contributions: Conceptualization, J.B.L., D.B. and Q.B.; methodology, J.B.L. and Q.B.; software, J.G.M., Q.B., J.B.L. and D.B.; validation, Q.B., J.B.L. and J.G.M.; formal analysis, Q.B., J.B.L. and J.G.M.; investigation, Q.B. and J.B.L.; resources, Q.B., J.B.L. and J.G.M.; writing—original draft preparation, Q.B. and J.B.L.; writing—review and editing, Q.B., J.B.L. and J.G.M.; visualization, Q.B. and J.B.L.; supervision, J.B.L.; project administration, J.B.L.; funding acquisition, J.B.L. All authors have read and agreed to the published version of the manuscript.

Funding: This project was funded in part by the Palouse Basin Aquifer Committee.

Data Availability Statement: The data presented in this study are available upon request.

Acknowledgments: We acknowledge the support of the Palouse Basin Aquifer Committee to make this research possible. Special thanks to all the private landowners around Moscow who kindly allowed the installation of seismometers on their property.

Conflicts of Interest: The funding agency had no role in the design of the study; in the collection, analyses, or interpretation of data; in the writing of the manuscript, or in the decision to publish the results.

References

1. Candel, J.; Brooks, E.; Sánchez-Murillo, R.; Grader, G.; Dijksma, R. Identifying Groundwater Recharge Connections in the Moscow (USA) Sub-Basin Using Isotopic Tracers and a Soil Moisture Routing Model. *Hydrogeol. J.* **2016**, *24*, 1739–1751. [CrossRef]
2. Custodio, E. *Trends in Groundwater Pollution: Loss of Groundwater Quality & Related Services*; Global Environment Facility and the Food and Agriculture Organization of the United Nations: Barcelona, Spain, 2014.
3. Llamas, M.; Custodio, E. *Intensive Use of Groundwater: Challenges and Opportunities*; A.A. Balkema Publishers: Madrid, Spain, 2002.

4. Gleeson, T.; Wada, Y.; Bierkens, M.F.P.; van Beek, L.P.H. Water Balance of Global Aquifers Revealed by Groundwater Footprint. *Nature* **2012**, *488*, 197–201. [CrossRef] [PubMed]
5. Foxworthy, B.L.; Washburn, R.L. *Ground Water in the Pullman Area, Whitman County, Washington*; U.S. Geological Survey: Washington, DC, USA, 1963.
6. Robischon, S. *Annual Water Use Report 2016*; Palouse Basin Aquifer Committee: Moscow, ID, USA, 2016.
7. Burns, E.R.; Snyder, D.T.; Haynes, J.V.; Waibel, M.S. *Groundwater Status and Trends for the Columbia Plateau Regional Aquifer System, Washington, Oregon, and Idaho*; U.S. Geological Survey: Washington, DC, USA, 2012.
8. Bush, J.H.; Pierce, J.L.; Potter, G.N. *Bedrock Geologic Map of the Moscow East Quadrangle, Latah County, Idaho*; Idaho Geological Survey: Moscow, ID, USA, 2010.
9. Bush, J.H.; Dunlap, P.; Reidel, S.P.; Kobayashi, D. *Geologic Cross Sections Across the Moscow-Pullman Basin, Idaho and Washington*; Idaho Geological Survey: Moscow, ID, USA, 2018.
10. Bush, J.H.; Dunlap, P.; Reidel, S.P. *Miocene Evolution of the Moscow-Pullman Basin, Idaho and Washington*; Idaho Geological Survey: Moscow, ID, USA, 2018.
11. Reidel, S.P.; Tolan, T.L.; Hooper, P.R.; Beeson, M.H.; Fecht, K.R.; Bentley, R.D.; Anderson, J.L. The Grande Ronde Basalt, Columbia River Basalt Group; Stratigraphic Descriptions and Correlations in Washington, Oregon, and Idaho. In *Volcanism and Tectonism in the Columbia River Flood-Basalt Province*; Geological Society of America: McLean, VA, USA, 1989; Volume 239, pp. 21–54.
12. Beall, A.; Fiedler, F.; Boll, J.; Cosens, B. Sustainable Water Resource Management and Participatory System Dynamics. Case Study: Developing the Palouse Basin Participatory Model. *Sustainability* **2011**, *3*, 720–742. [CrossRef]
13. Dhungel, R.; Fiedler, F. Water Balance to Recharge Calculation: Implications for Watershed Management Using Systems Dynamics Approach. *Hydrology* **2016**, *3*, 13. [CrossRef]
14. Behrens, D.; Langman, J.B.; Brooks, E.S.; Boll, J.; Waynant, K.; Moberly, J.G.; Dodd, J.K.; Dodd, J.W. Tracing $\delta^{18}O$ and $\delta^{2}H$ in Source Waters and Recharge Pathways of a Fractured-Basalt and Interbedded-Sediment Aquifer, Columbia River Flood Basalt Province. *Geosciences* **2021**, *11*, 400. [CrossRef]
15. Hernandez, H.P. Observations of Recharge to the Wanapum Aquifer System in the Moscow Area, Latah County, Idaho. Master's Thesis, University of Idaho, Moscow, ID, USA, 2007.
16. Kopp, W.P. Hydrogeology of the Upper Aquifer of the Pullman-Moscow Basin at the University of Idaho Aquaculture Site. Ph.D. Thesis, University of Idaho, Moscow, ID, USA, 1994.
17. Leek, F. Hydrogeological Characterization of the Palouse Basin Basalt Aquifer System, Washington and Idaho. Master's Thesis, Washington State University, Pullman, WA, USA, 2006.
18. Tong, L. Hydrogeologic Characterization of a Multiple Aquifer Fractured Basalt System. Ph.D. Thesis, University of Idaho, Moscow, ID, USA, 1991.
19. Lum II, W.E.; Smoot, J.L.; Ralston, D.R. *Geohydrology and Numerical Model Analysis of Ground-Water Flow in the Pullman-Moscow Area, Washington and Idaho*; US Geological Survey: Washington, DC, USA, 1990; p. 79.
20. Medici, G.; Engdahl, N.B.; Langman, J.B. A Basin-Scale Groundwater Flow Model of the Columbia Plateau Regional Aquifer System in the Palouse (USA): Insights for Aquifer Vulnerability Assessment. *Int. J. Environ. Res.* **2021**, *15*, 299–312. [CrossRef]
21. Medici, G.; Langman, J.B. Pathways and Estimate of Aquifer Recharge in a Flood Basalt Terrain; A Review from the South Fork Palouse River Basin (Columbia River Plateau, USA). *Sustainability* **2022**, *14*, 11349. [CrossRef]
22. Ryu, J.H.; Contor, B.; Johnson, G.; Allen, R.; Tracy, J. System Dynamics to Sustainable Water Resources Management in the Eastern Snake Plain Aquifer Under Water Supply Uncertainty1: System Dynamics to Sustainable Water Resources Management in the Eastern Snake Plain Aquifer Under Water Supply Uncertainty. *J. Am. Water Resour. Assoc.* **2012**, *48*, 1204–1220. [CrossRef]
23. Provant, A.P. Geology and Hydrogeology of the Viola and Moscow West Quadrangles, Latah County, Idaho and Whitman County, Washington. Master's Thesis, University of Idaho, Moscow, ID, USA, 1995.
24. Bush, J.H.; Dunlap, P. *Geologic Interpretations of Wells and Important Rock Outcrops in the Moscow-Pullman Basin and Vicinity, Idaho and Washington*; Idaho Geological Survey: Moscow, ID, USA, 2018.
25. Duckett, K.A.; Langman, J.B.; Bush, J.H.; Brooks, E.S.; Dunlap, P.; Welker, J.M. Isotopic Discrimination of Aquifer Recharge Sources, Subsystem Connectivity and Flow Patterns in the South Fork Palouse River Basin, Idaho and Washington, USA. *Hydrology* **2019**, *6*, 15. [CrossRef]
26. Ackerman, D.J. *Transmissivity of the Snake River Plain Aquifer at the Idaho National Engineering Laboratory, Idaho*; U.S. Geological Survey: Washington, DC, USA, 1991.
27. Duckett, K.A.; Langman, J.B.; Bush, J.H.; Brooks, E.S.; Dunlap, P.; Stanley, J.R. Noble Gases, Dead Carbon, and Reinterpretation of Groundwater Ages and Travel Time in Local Aquifers of the Columbia River Basalt Group. *J. Hydrol. Amst.* **2020**, *581*, 124400. [CrossRef]
28. Langman, J.B.; Martin, J.; Gaddy, E.; Boll, J.; Behrens, D. Snowpack Aging, Water Isotope Evolution, and Runoff Isotope Signals, Palouse Range, Idaho, USA. *Hydrology* **2022**, *9*, 94. [CrossRef]
29. Bush, J.H.; Dunlap, P.; Kodayashi, D. *A Collection of Geologic Maps, Cross Sections, and Schematic Diagrams That Illustrate the Subsurface Geology of the Moscow-Pullman Basin and Vicinity*; Palouse Basin Aquifer Committee: Moscow, ID, USA, 2019.
30. Bohnhoff, M.; Dresen, G.; Ellsworth, W.L.; Ito, H. Passive Seismic Monitoring of Natural and Induced Earthquakes: Case Studies, Future Directions and Socio-Economic Relevance. In *New Frontiers in Integrated Solid Earth Sciences*; Cloetingh, S., Negendank, J., Eds.; International Year of Planet Earth; Springer: Dordrecht, The Netherlands, 2010; pp. 261–285. ISBN 978-90-481-2737-5.

31. Mordret, A.; Courbis, R.; Brenguier, F.; Chmiel, M.; Garambois, S.; Mao, S.; Boué, P.; Campman, X.; Lecocq, T.; Van der Veen, W.; et al. Noise-Based Ballistic Wave Passive Seismic Monitoring—Part 2: Surface Waves. *Geophys. J. Int.* **2020**, *221*, 692–705. [CrossRef]
32. Meier, U.; Shapiro, N.M.; Brenguier, F. Detecting Seasonal Variations in Seismic Velocities within Los Angeles Basin from Correlations of Ambient Seismic Noise. *Geophys. J. Int.* **2010**, *181*, 985–996. [CrossRef]
33. Campillo, M.; Roux, P. Crust and Lithospheric Structure—Seismic Imaging and Monitoring with Ambient Noise Correlations. In *Treatise on Geophysics*; 2015; pp. 391–417, ISBN 978-0-444-53803-1.
34. McNamara, D.E.; Boaz, R.I. Visualization of the Seismic Ambient Noise Spectrum. In *Seisimic Ambient Noise*; Nakata, N., Gualtieri, L., Fichtner, A., Eds.; Cambridge University Press: Cambridge, UK, 2019; ISBN 978-1-108-41708-2.
35. Grêt, A.; Snieder, R.; Scales, J. Time-Lapse Monitoring of Rock Properties with Coda Wave Interferometry. *J. Geophys. Res. Solid Earth* **2006**, *111*. [CrossRef]
36. Christensen, N.I.; Wang, H.F. The Influence of Pore Pressure and Confining Pressure on Dynamic Elastic Properties of Berea Sandstone. *Geophysics* **1985**, *50*, 207–213. [CrossRef]
37. Clements, T.; Denolle, M.A. Tracking Groundwater Levels Using the Ambient Seismic Field. *Geophys. Res. Lett.* **2018**, *45*, 6459–6465. [CrossRef]
38. Grêt, A.; Snieder, R.; Aster, R.C.; Kyle, P.R. Monitoring Rapid Temporal Change in a Volcano with Coda Wave Interferometry. *Geophys. Res. Lett.* **2005**, *32*. [CrossRef]
39. Christensen, N.I. The Influence of Pore Pressure on Oceanic Crustal Seismic Velocities. *J. Geodyn.* **1986**, *5*, 45–48. [CrossRef]
40. Lecocq, T.; Longuevergne, L.; Pedersen, H.A.; Brenguier, F.; Stammler, K. Monitoring Ground Water Storage at Mesoscale Using Seismic Noise: 30 Years of Continuous Observation and Thermo-Elastic and Hydrological Modeling. *Sci. Rep.* **2017**, *7*, 14241. [CrossRef]
41. Shapiro, N.M.; Campillo, M. Emergence of Broadband Rayleigh Waves from Correlations of the Ambient Seismic Noise. *Geophys. Res. Lett.* **2004**, *31*. [CrossRef]
42. Wang, Q.-Y.; Brenguier, F.; Campillo, M.; Lecointre, A.; Takeda, T.; Aoki, Y. Seasonal Crustal Seismic Velocity Changes Throughout Japan. *J. Geophys. Res. Solid Earth* **2017**, *122*, 7987–8002. [CrossRef]
43. Gassenmeier, M.; Sens-Schönfelder, C.; Delatre, M.; Korn, M. Monitoring of Environmental Influences on Seismic Velocity at the Geological Storage Site for CO_2 in Ketzin (Germany) with Ambient Seismic Noise. *Geophys. J. Int.* **2015**, *200*, 524–533. [CrossRef]
44. Tsai, V.C. Understanding the Amplitudes of Noise Correlation Measurements. *J. Geophys. Res. Solid Earth* **2011**, *116*. [CrossRef]
45. Voisin, C.; Guzmán, M.A.R.; Réfloch, A.; Taruselli, M.; Garambois, S. Groundwater Monitoring with Passive Seismic Interferometry. *J. Water Resour. Prot.* **2017**, *9*, 1414–1427. [CrossRef]
46. Garambois, S.; Voisin, C.; Romero Guzman, M.A.; Brito, D.; Guillier, B.; Réfloch, A. Analysis of Ballistic Waves in Seismic Noise Monitoring of Water Table Variations in a Water Field Site: Added Value from Numerical Modelling to Data Understanding. *Geophys. J. Int.* **2019**, *219*, 1636–1647. [CrossRef]
47. Prieto, G.A.; Denolle, M.; Lawrence, J.F.; Beroza, G.C. On Amplitude Information Carried by the Ambient Seismic Field. *Comptes Rendus Geosci.* **2011**, *343*, 600–614. [CrossRef]
48. Berryman, J.G. Seismic Wave Attenuation in Fluid-Saturated Porous Media. In *Scattering and Attenuations of Seismic Waves, Part I*; Aki, K., Wu, R.-S., Eds.; Pageoph Topical Volumes; Birkhäuser: Basel, Switzerland, 1988; pp. 423–432, ISBN 978-3-0348-7722-0.
49. Nakata, N.; Gualtieri, L.; Fichtner, A. *Seisimic Ambient Noise*; Nakata, N., Gualtieri, L., Fichtner, A., Eds.; Cambridge University Press: Cambridge, UK, 2019; ISBN 978-1-108-41708-2.
50. Krischer, L.; Megies, T.; Barsch, R.; Beyreuther, M.; Lecocq, T.; Caudron, C.; Wassermann, J. ObsPy: A Bridge for Seismology into the Scientific Python Ecosystem. *Comput. Sci. Discov.* **2015**, *8*, 014003. [CrossRef]
51. McNamara, D.E.; Buland, R.P. Ambient Noise Levels in the Continental United States. *Bull. Seismol. Soc. Am.* **2004**, *94*, 1517–1527. [CrossRef]
52. Lecocq, T.; Caudron, C.; Brenguier, F. MSNoise, A Python Package for Monitoring Seismic Velocity Changes Using Ambient Seismic Noise. *Seismol. Res. Lett.* **2014**, *85*, 715–726. [CrossRef]
53. Sajid, M.; Ghosh, D. A Fast and Simple Method of Spectral Enhancement. *Geophysics* **2014**, *79*, V75–V80. [CrossRef]
54. Schimmel, M.; Stutzmann, E.; Ventosa, S. Low-Frequency Ambient Noise Autocorrelations: Waveforms and Normal Modes. *Seismol. Res. Lett.* **2018**, *89*, 1488–1496. [CrossRef]
55. Clarke, D.; Zaccarelli, L.; Shapiro, N.M.; Brenguier, F. Assessment of Resolution and Accuracy of the Moving Window Cross Spectral Technique for Monitoring Crustal Temporal Variations Using Ambient Seismic Noise. *Geophys. J. Int.* **2011**, *186*, 867–882. [CrossRef]
56. Sánchez-Sesma, F.J.; Campillo, M. Retrieval of the Green's Function from Cross Correlation: The Canonical Elastic Problem. *Bull. Seismol. Soc. Am.* **2006**, *96*, 1182–1191. [CrossRef]
57. Nanda, N.C. Seismic Wave and Rock-Fluid Properties. In *Seismic Data Interpretation and Evaluation for Hydrocarbon Exploration and Production: A Practitioner's Guide*; Nanda, N.C., Ed.; Advances in Oil and Gas Exploration & Production; Springer International Publishing: Cham, Switzerland, 2021; pp. 3–23, ISBN 978-3-030-75301-6.

58. National Water and Climate Center, NWCC Report Generator—Idaho SNOTEL Moscow Mountain Site, Average Precipitation Accumulation (1981–2010). Available online: https://wcc.sc.egov.usda.gov/reportGenerator/view/customGroupByMonthReport/daily/989:id:SNTL\T1\textbar{}id=%22%22\T1\textbar{}name/1980-10-01,1981-09-30/PREC::average_1981 (accessed on 9 January 2019).
59. Sánchez-Murillo, R.; Brooks, E.S.; Elliot, W.J.; Boll, J. Isotope Hydrology and Baseflow Geochemistry in Natural and Human-Altered Watersheds in the Inland Pacific Northwest, USA. *Isotopes Environ. Health Stud.* **2015**, *51*, 231–254. [CrossRef]
60. Bush, J.H. *Bedrock Geologic Map of the Viola Quadrangle, Latah County, Idaho, and Whitman County, Washington*; Idaho Geological Survey: Moscow, ID, USA, 1998.
61. Pierce, J.L. Geology and Hydrology of the Moscow East and Robinson Lake Quadrangles, Latah County, Idaho. Master's Thesis, University of Idaho, Moscow, ID, USA, 1998.
62. Domenico, P.A.; Schwartz, F.W. *Physical and Chemical Hydrogeology*, 2nd ed.; Wiley & Sons: New York, NY, USA, 1997; ISBN 978-0-471-59762-9.
63. Freeze, R.A.; Cherry, J.A. *Groundwater*; Prentice-Hall: Englewood Cliffs, NJ, USA, 1979; ISBN 978-0-13-365312-0.

atmosphere

Article

Comparison of Planetary Boundary Layer Height Derived from Lidar in AD-Net and ECMWFs Reanalysis Data over East Asia

Zhijuan Zhang *, Ling Mu and Chen Li

College of Environmental Science and Engineering, Taiyuan University of Technology, Taiyuan 030024, China
* Correspondence: zhangzhijuan@tyut.edu.cn

Abstract: The planetary boundary layer height is a very important parameter in the atmosphere because it determines the range where the most effective dispersion processes take place, and it serves as a medium for the vertical transport of heat, moisture, and pollutants. The accurate estimation of boundary layer height (BLH) is vital for air pollution prediction. In this paper, the BLH estimated by AD-Net was compared with that from the ECMWFs over East Asia from September 2015 to August 2018. A continuous 24 h BLH estimation from AD-Net generally matched with the aerosol vertical structures. Diurnal and seasonal variation and spatial variation of BLH can also be shown, suggesting the good performance of AD-Net BLH. The comparison of seasonal mean BLH between AD-Net and ECMWFs was conducted at 20 lidar sites. On average, there was an underestimation of the ECMWFs, mostly in summer and winter. A significant disagreement between AD-Net and the ECMWFs was noted, especially over coastal areas and mountain areas. In order to investigate the difference between them, two BLHs were compared under different land cover types and climate conditions. In general, the BLH of the ECMWFs was less than that of AD-Net over most of the land cover types in summer and winter. The smallest differences (0.26 km) existed over water surfaces in winter compared with AD-Net, and the largest underestimation (1.42 km) occurred over grassland surfaces in summer. Similarly, all the BLHs of the ECMWFs were lesser than those of AD-Net under different climatological conditions in summer and winter. The mean difference between AD-Net BLH and ECMWFs BLH was 1.05, 0.71, and 0.48 km for arid regions, semi-arid and semi-wet regions, and wet regions, respectively. The largest underestimation occurred over arid regions in winter, with a value of 1.42 km. The smallest underestimation occurred over wet regions, with a value of 0.27 km. The present research provides better insight into the BLH performance in the ECMWFs reanalysis data. The new continuous PBL dataset can be used to improve the model parameterization of PBL and our understanding of the atmospheric transport of pollutants which affect air quality and human health.

Keywords: planetary boundary layer; boundary layer height; lidar; ECMWFs; aerosol

Citation: Zhang, Z.; Mu, L.; Li, C. Comparison of Planetary Boundary Layer Height Derived from Lidar in AD-Net and ECMWFs Reanalysis Data over East Asia. *Atmosphere* **2022**, *13*, 1976. https://doi.org/10.3390/atmos13121976

Academic Editors: Zhengqiang Li, Zhongwei Huang, Chi Li, Kai Qin, Han Wang, Tianhe Wang and Jie Luo

Received: 20 October 2022
Accepted: 23 November 2022
Published: 26 November 2022

1. Introduction

The planetary boundary layer (PBL) is the lowest part of the troposphere and is directly influenced by the Earth's surface through the exchange of heat, momentum, and moisture [1,2]. It determines the vertical extent of turbulent mixing, vertical diffusion, and convective transport within the PBL, and it is also a primary determinant of cloud type and coverage that affects the Earth's radiation budget [3]. The important roles of the PBL in weather and climate have long been recognized. In order to better characterize the structure of the PBL, the concept of boundary layer height (BLH) is commonly used [4,5]. Therefore, it is critical to obtain an accurate BLH, since it is important for air quality, weather forecasting, and the assessment of regional climate change.

Traditionally, the BLH is diagnosed using the vertical profiles of temperature, humidity, and wind from radiosonde observation [4–6]. The bulk Richardson number (Ri) method [7]

can also be used to estimate the BLH. New data sources such as ground-based lidar [8,9] instruments, sodar [10], ceilometer [11], aircraft sounding [12], and space-borne lidar [13–15] can be relied on to retrieve the BLH. In addition, the structure of turbulence can also be determined from optical measurements performed on ground-based telescopes [16–19].

It is noted that products on the global climatology of BLHs are available freely from the European Centre for Medium-Range Weather Forecasts (ECMWFs). A comparison of seasonally averaged BLHs derived from fine-resolution sounding observations and ECMWFs reanalysis over China showed good agreement on average, despite the pronounced inconsistencies in some regions [6]. However, the ECMWFs reanalysis products perform significantly worse than those from the Cloud-Aerosol Lidar and Infrared Pathfinder Satellite Observations (CALIPSOs) lidar-based BLH over the oceans and coastal areas in China. The BLHs from the ECMWFs are greater than those from CALIPSOs over North China in spring and summer. The majority of the ECMWFs BLHs are within 25% of the estimates derived from CALIPSOs in autumn. The comparison of CALIPSOs and ECMWFs BLHs under different land-cover conditions reveals that the BLH estimated by the CALIPSOs backscatter climatology is higher over ocean and forest surfaces and lower over grassland and bare land surfaces in spring and summer [14]. Although many researchers have used the ECMWFs BLH, its precision is still poor because the verification of the BLH by direct observations is rare. Additionally, climate modelers are constantly searching for new opportunities to verify model outputs.

BLH derived from ground-based lidar can be used for comparison and validation with others. Ground-based lidar has the advantage of continuous tracing and thus avoids confusion with elevated layers. BLH is determined using a combination of the averaging variance method and the high-resolution gradient method from lidar backscatter measurements [8]. Additionally, it also can be detected from micropulse lidar (MPL)-measured backscatter [9]. In addition, AD-Net is used for the validation of EarthCARE satellite observation and data assimilation to evaluate emissions of air pollution and dust aerosols in East Asia [20,21]. Therefore, AD-Net BLHs can be used for comparison with the ECMWFs reanalysis data in this study.

Here, the seasonal mean BLH derived from the National Institute for Environmental Studies' (NIESs') ground-based lidar in the Asian Dust and Aerosol Lidar Observation Network (AD-Net) is directly used to perform comparisons and further validations with the BLH from the ECMWF ERA-Interim fields from September 2015 to August 2018 in this study. This paper is arranged as follows. The details of the datasets are given in Section 2. In Section 3, a comparison of BLHs obtained from AD-Net and the ECMWF in summer and winter and under different land cover types and climate conditions is presented. The conclusion and discussion are presented in Sections 4 and 5, respectively.

2. Materials and Methods

2.1. BLH Derived from AD-Net

AD-Net is a lidar network for the continuous observation of vertical distributions of Asian dust and other aerosols (including industrial, forest fire, volcanic) in East Asia, and for studying the impacts of aerosols on the air quality and atmosphere environment. Gaining a quantitative understanding of Asian dust emissions, transport, and deposition, the long-term variations of Asian dust and anthropogenic aerosols, and the effects of aerosols on human health, vegetation, and radiation is objective of this network. Officially, there are 20 sites with automatic lidars over East Asia, mainly situated in Japan. The standard lidar in AD-Net is a two-wavelength polarization-sensitive (532 nm) Mie-scattering lidar [22]. The measured data are transferred to the AD-Net in real-time and processed automatically. Data from AD-Net are strictly managed and quality-assured (http://www-lidar.nies.go.jp/AD-Net/ncdf/, accessed on 23 November 2022).

Figure 1 shows the topography of 20 AD-Net lidar stations in AD-Net [23,24], i.e., Chiba (CHB), Fukue (FKE), Fukuoka (FKO), Hedo (HED), Gosan Jeju (JEJ), Matsue (MTS), Nagasaki (NGS), Niigata (NIG), Osaka (OSK), Phimai (PHM), Seoul (SEO), Sendai (SND),

Sainshand (SNS), Sapporo (SPR), Tsukuba (TKB), Tokyo (TKO), Toyama (TYM), Ulaanbaatar (ULN), Ulsan (ULS), and Zamynuud (ZMY). The basic parameters derived are the attenuated backscattering coefficients (ABCs) at 532 nm and 1064 nm, and the volume depolarization ratio at 532 nm. Extinction coefficient estimates for non-spherical and spherical aerosols are also derived with this method using the backscattering and depolarization ratio [25,26].

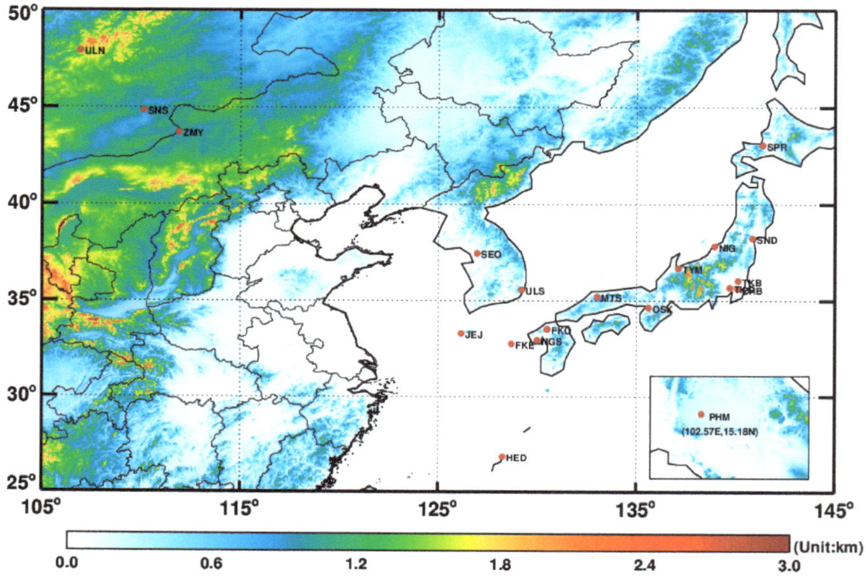

Figure 1. The location of AD-Net lidar stations (the red solid circle). The color represents surface elevation.

BLH is identified using the vertical gradient of the ABC at 532 nm with the method used in Sugimoto et al. [22]. Due to the fact that the aerosols have the largest loading in the PBL, the gradient of the backscattering signal decreases rapidly at the top of the PBL. Therefore, the BLH is defined as the height at which the gradient of the ABC at 532 nm reaches a minimum value. The minimum value was set empirically. When this criterion is satisfied, the layer is identified as an aerosol layer, and the top of the layer is identified as the BLH [20]. AD-Net provides the near-real-time mixing layer height (MLH) since 2001 at 20 sites. The vertical and temporal resolution of AD-Net lidar is 30 m and 15 min, respectively. Here, the MLH was regarded as the BLH and was compared with the ECMWFs BLH.

2.2. BLHs from ECMWFs

The ECMWFs' core mission is to produce numerical weather forecasts and monitor the Earth system, as well as to carry out scientific and technical research to improve forecasting skills and to maintain an archive of meteorological data. It provides air quality analysis, atmospheric composition monitoring, and climate monitoring. Global climatology BLHs are a small part of their products and are important for understanding the physical process in the boundary layer. The ECMWFs models produce BLHs by means of the bulk Richardson number method. The bulk Richardson number is defined as the ratio of stability to vertical wind shear, which is expressed as:

$$Ri(z) = \frac{\left(\frac{g}{\theta_{vs}}\right)(\theta_{vz} - \theta_{vs})(z - z_s)}{(u_z - u_s)^2 + (v_z - v_s)^2 + (bu_*^2)} \quad (1)$$

where g is the acceleration due to gravity, θ is the potential temperature, θvs and θvz are the virtual potential temperature at surface and at height z above ground level, respectively, u and v are the components of wind speed, and u∗ is the surface friction velocity, while b is a coefficient to be determined [7]. Previous theoretical and laboratory studies (e.g., [2]) suggested that when Ri is smaller than the critical value (~0.25), the laminar flow becomes unstable. Thus, the lowest-level z at which the interpolated Ri crosses the critical value of 0.25 is referred to as the BLH in this study, similar to the criteria used by Seidel et al. [12,27–31]. This method works well for both stable and convective boundary layers [27]. However, it may reach this critical value at a height somewhat below the BLH defined by other means [32].

The ECMWFs has provided monthly mean BLHs with 0.125×0.125 latitude–longitude resolution at the global scale from 1979 to the present. Here, the BLH from September 2015 to August 2018 over East Asia was used for comparison with the AD-Net BLH.

3. Results

It was proved that the BLH from CALIPSO is strongly correlated with that from ground-based lidar [14]. Therefore, the BLH derived using the maximum standard variance method is reliable. Figure 2 shows a series of cases of BLH estimation using the ABCs profile and extinction profile in August 2018. The BLHs from the AD-Net product and the BLHs derived using the maximum standard variance method [14] are shown as a dotted line and dashed line, respectively. The two methods are both based on the aerosol backscatter. Close agreement was observed between them, which suggested the good performance of the AD-Net BLH.

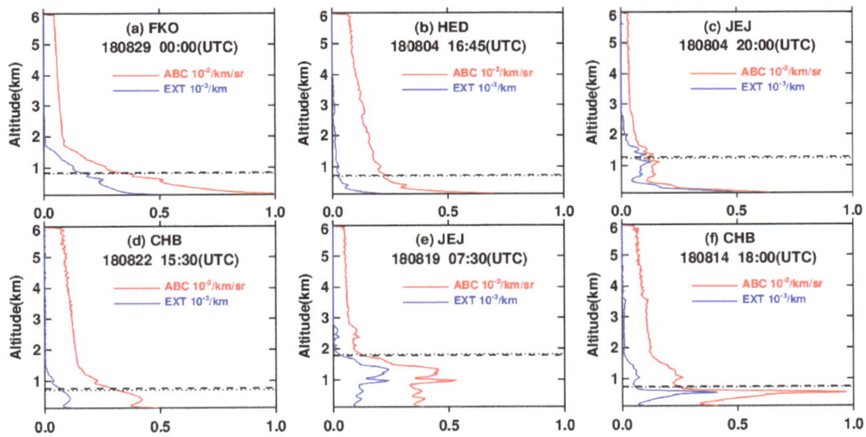

Figure 2. Profile of the attenuated backscatter coefficient (ABC) at 532 nm and extinction coefficient profiles (EXT) using the Fernald inversion on (**a**) 29 August 2018 at the site of FKO, (**b**) 4 August 2018 at the site of HED, (**c**) 4 August 2018 at the site of JEJ, (**d**) 22 August 2018 at the site of CHB, (**e**) 19 August 2018 at the site of JEJ and (**f**) 14 August 2018 at the site of CHB. The BLH from CALIPSO and Lidar is shown in dotted line and dashed line, respectively. The vertical and time resolution is 30 m and 15 min, respectively.

Figure 3 shows a case study of a continuous 24 h AD-Net ground-based lidar profile, which contains the ABC at 532 nm (a), the ABC at 1064 nm (b), the volume depolarization ratio (d), and the color ratio (d) distributions observed from AD-Net lidar at the MTS station, with the BLH (gray solid circle) on 7 August 2018. The depolarization ratio, which is a parameter sensitive to the nonsphericity of the scatters, is useful for identifying mineral dust aerosols and in discriminating between water and ice clouds [22]. The color ratio between 1064 nm and 532 nm backscatter is efficient in detecting falling droplets which evaporate before reaching the surface [33]. Clear aerosol signature layers extending from 120 m to 1 km and the boundary layer cloud at the top of the boundary layer are shown. In order to minimize the influence of the clouds on the BLH determination, all of the lidar measurements with clouds were excluded for further analyses. The BLH derived from AD-Net lidar generally matched with the aerosol vertical structures. In addition, a diurnal variation of BLH can be clearly seen. During a diurnal cycle, the BLH is typically shallow (about 900 m) at night due to the strong near-surface stability, and well-developed and reaching a maximum of 1.18 km in the afternoon.

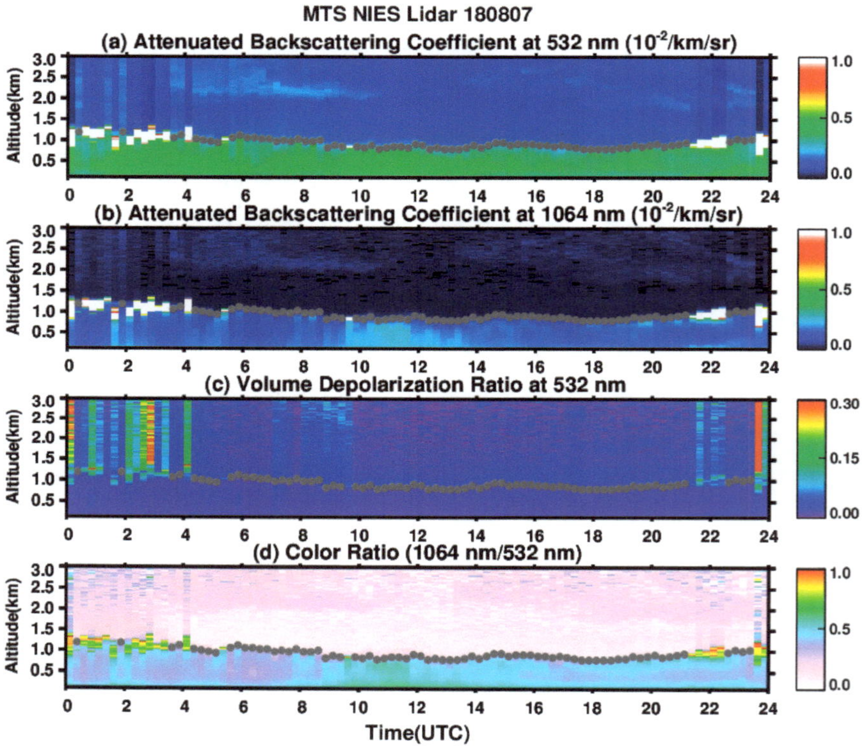

Figure 3. (**a**) Attenuated backscatter coefficient at 532 nm, (**b**) attenuated backscatter coefficient at 1064 nm, (**c**) volume depolarization ratio at 532 nm, and (**d**) color ratio (1064 nm/532 nm) observed from AD–Net ground-based lidar on 7 August 2018. The gray solid circle indicates AD–Net lidar BLH.

Figure 4 shows the mean seasonal geographic distribution of the BLH concerning the location of each lidar site. It can be clearly seen that the BLHs exhibit large spatial and seasonal variations. On average, more intense solar radiation reaching the surface in summer favors the PBL development on the land, but the development of the PBL is typically suppressed due to the smaller amount of solar radiation received at the surface [2]. For ocean areas, the situation is entirely different due to its relatively large heat capacity. The mean BLH over inland areas (including the ULN, SNS, ZMY, and PHM sites) was larger than that over ocean areas (including the JEJ, FKE, and HED sites) in summer (see Table 1). However, the opposite distribution was observed in winter. Additionally, this indicated that over coastal areas, the BLH in winter was greater than that in summer, which was consistent with Liu et al. [14]. A possible explanation is that the sea–land wind contributes to the development of the PBL height over the coastal areas. The highest BLH (2.23 km) was seen at the ZMY site (at the border of Mongolia, surface elevation of 962 m) in summer

(June, July, and August), and the lowest BLH value (0.77 km) occurred at the TYM site (on the west coast of Japan) in winter (December, January, and February).

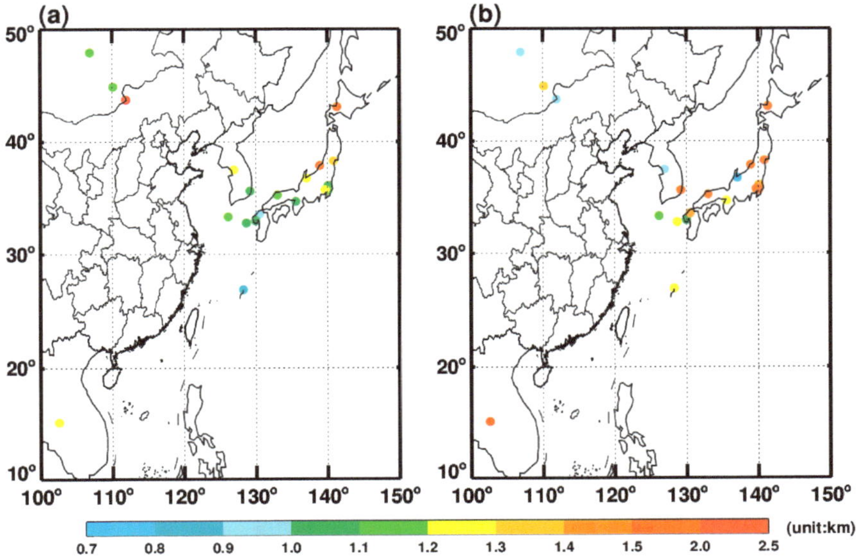

Figure 4. Spatial patterns of 3-year seasonal mean BLH for 20 sites in AD-Net over East Asia from the AD-Net for summer (**a**) and winter (**b**).

Table 1. BLH derived from AD-Net over inland and ocean area.

	Summer	Winter
Inland	1.44	1.19
Ocean	1.01	1.22

A comparison between BLHs from AD-Net and the ECMWFs was conducted at 20 lidar sites, as shown in Figure 5. In terms of similar spatial distribution of BLH, discrepancies between them still existed. Compared with AD-Net BLH, there was an underestimation of the ECMWFs BLH both in summer and winter, except over the ocean in winter. The ECMWFs BLH was in close agreement with the AD-Net BLH at the JEJ, MTS, and NGS sites in winter with a relative difference smaller than 10% (Figure 6b). The spatial distribution of the BLH revealed a tendency for higher BLHs over high-elevation regions, consistent with the dependence on elevation reported in the United States [27]. However, the BLH from the ECMWFs was severely underestimated at high-surface elevation sites (including ULN, ZMY, and SNS in Mongolia), especially in winter, with a relative difference around 80% (Figure 6b).

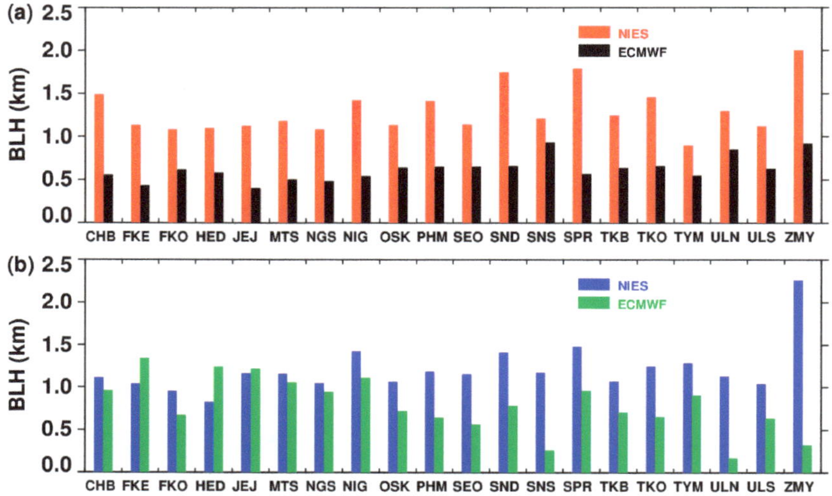

Figure 5. Comparison of the AD-Net PBL and ECMWFs PBL for 20 sites from AD-Net for summer (**a**) and winter (**b**).

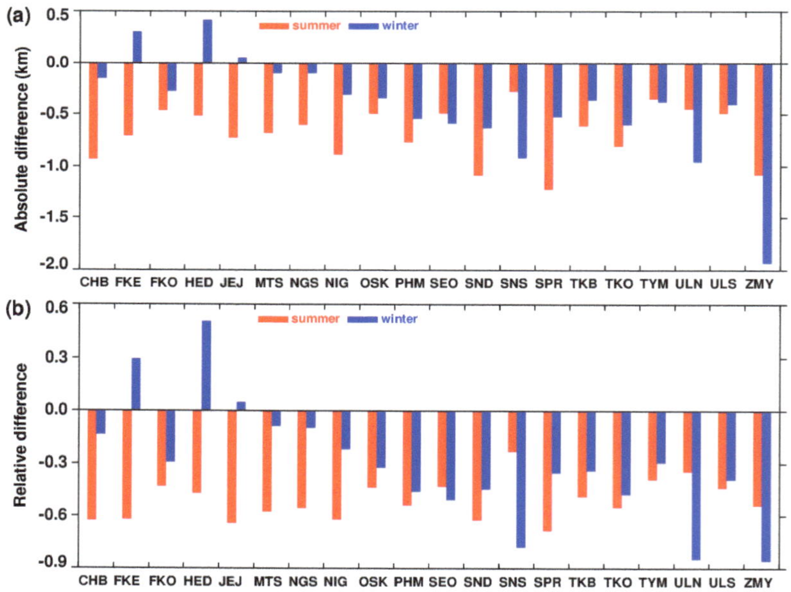

Figure 6. (**a**) Absolute difference, (**b**) relative difference in the BLH from AD-Net and ECMWFs for 20 sites in AD-Net.

As noted above, a seasonal disagreement was found in BLH climatology. Here, we continue to explore these seasonal biases quantitatively. The seasonally averaged absolute differences and relative differences between the AD-Net BLH and the ECMWFs BLH are presented in Figure 6a,b. The mean absolute difference was 0.68 km and 0.42 km in summer and in winter, respectively. Additionally, the relative difference was 0.51 and 0.30 in summer and in winter, respectively. The larger absolute difference in BLH was in summer because the BLH from the ECMWFs was uniformly underestimated compared to the AD-Net BLH. The poor estimation of BLH can mainly be attributed to the lower

BLH over land, especially for mountainous areas (see Table 2). However, large biases of the seasonal BLHs were most likely due to the different methods utilized based on radiosonde, ground-based lidar, and CALIPOs observations over one site in South Africa [34].

Table 2. Mean difference in BLH derived from AD-Net and ECMWFs (ECMWFs BLH minus AD-Net BLH) in summer and winter.

	Absolute Difference (km)		Relative Difference	
	Summer	**Winter**	**Summer**	**Winter**
Inland	−0.64	−1.09	−0.41	−0.73
Ocean	−0.65	0.26	−0.58	0.28
Coastal areas	−0.70	−0.37	−0.53	−0.30
Mountain	−0.60	−1.27	−0.37	−0.83

Next, we will analyze which land cover type and climate condition has the largest influence on the differences in the seasonal variation between the ECMWFs BLH and the AD-Net BLH. The vegetation types in China, as determined by MODIS at 0.05 degrees of spatial resolution, are plotted in Figure 7a, including 17 different surface types. Precipitation data from the University of East Anglia Climate Research Unit (CRU) global climate dataset at 0.5° latitude and longitude resolution were used as a proxy for climate state [35]. In this study, the monthly mean climatology was calculated relative to the average for the period of 1961–1990. The spatial distribution of precipitation is shown in Figure 7b. The annual mean precipitation ranges from 0 to 200, from 200 to 800, and over 800 mm yr^{-1} in arid regions, semi-arid and semi-wet regions, and wet regions, respectively.

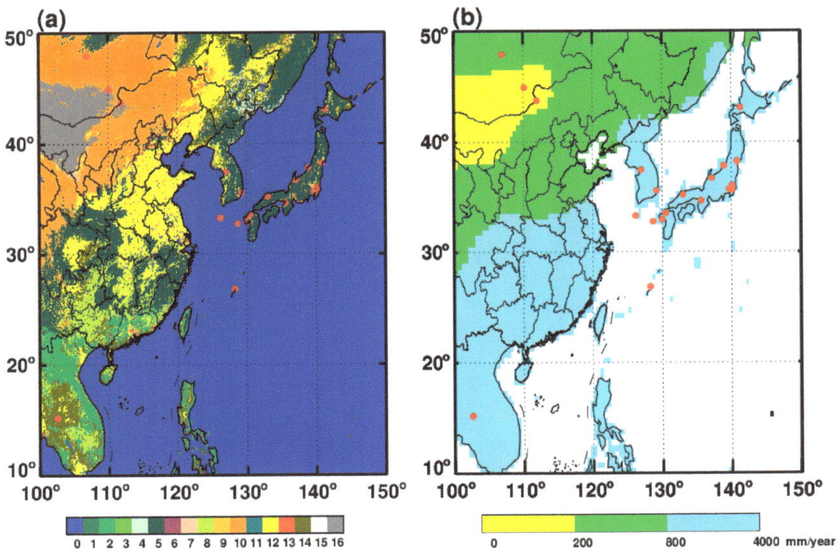

Figure 7. (**a**) Land cover distribution in 2011 from MODIS (0.05 degree) over East Asia. (**b**) Precipitation distribution averaged from 1961 to 1990. (0 = water, 1 = evergreen needleleaf, 2 = evergreen broadleaf, 3 = deciduous needleleaf, 4=deciduous broadleaf, 5 = mixed forests, 6 = closed shrubland, 7 = open shrublands, 8 = woody savannas, 9 = savannas, 10 = grasslands, 11=permanent wetlands, 12 = croplands, 13 = urban and built-up, 14 = cropland mosaic, 15 = snow and ice, 16 = bare or sparsely vegetated).

The seasonal mean values of the AD-Net BLH and the ECMWFs BLH under the condition of different land cover types are presented in Figure 8, and detailed differences in summer and winter are listed in Table 3. We found that, in general, the BLHs of the ECMWFs were lower than those of AD-Net over most of the land cover types in summer and winter, except for over ocean areas in winter. The smallest differences existed over water surfaces in winter and bare surfaces in summer. Additionally, the largest differences occurred over grassland surfaces in winter, with the largest underestimation of 1.93 km compared to the AD-Net BLH, followed by 1.08 km over grassland surfaces in summer. Overall, the mean minimum and maximum differences occurred over water surfaces and grassland surfaces, respectively. These results are different from those of Liu et al. [14], who suggested that the largest differences occurred over bare land surfaces in summer with the largest overestimate (0.97 km), and over water surfaces in spring with the greatest underestimate (−0.91 km) compared with CALIPSOs. There are more land cover types in China due to its vast area, and different meteorological conditions along with its complex pollution conditions, which may lead to the above results.

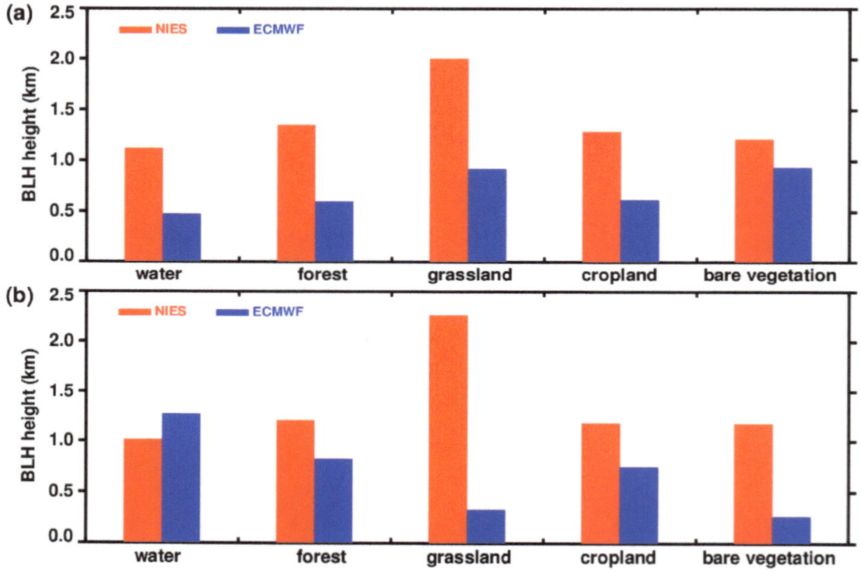

Figure 8. Comparison of AD-Net (red) and ECMWFs (blue) PBL in the case of different land cover types (water, forest, grassland, cropland, and bare vegetation) for summer (**a**) and winter (**b**).

Table 3. Seasonal differences in BLH between AD-Net and ECMWFs (ECMWFs PBL minus AD-Net PBL) given different land cover conditions.

	Water	Forest	Grassland	Cropland	Bare Vegetation
Summer	−0.65	−0.79	−1.08	−0.67	−0.28
Winter	0.26	−0.52	−1.93	−0.43	−0.91

The seasonal mean values of the AD-Net BLH and the ECMWFs BLH under different climatological conditions are presented in Figure 9, and detailed differences in summer and winter are listed in Table 4. We found that all the BLHs of the ECMWFs were lower than those of AD-Net under different climatological conditions in summer and winter. The largest underestimation occurred over arid regions in winter, with a value of 1.42 km. The smallest underestimation occurred over wet regions, with a value of 0.27 km. The mean difference between the AD-Net BLH and the ECMWFs BLH was 1.05, 0.71, and 0.48 km for arid regions, semi-arid and semi-wet regions, and wet regions, respectively. In wet regions, the number of lidar stations was the largest, which may contribute to the minor difference between the AD-Net BLH and the ECMWFs BLH. Special attention should be paid to the fact that it is in arid regions that the relative difference in summer and in winter (0.74 km) was the largest, which meant that this region was the most sensitive in the related physical parameterization in the ECMWFs model.

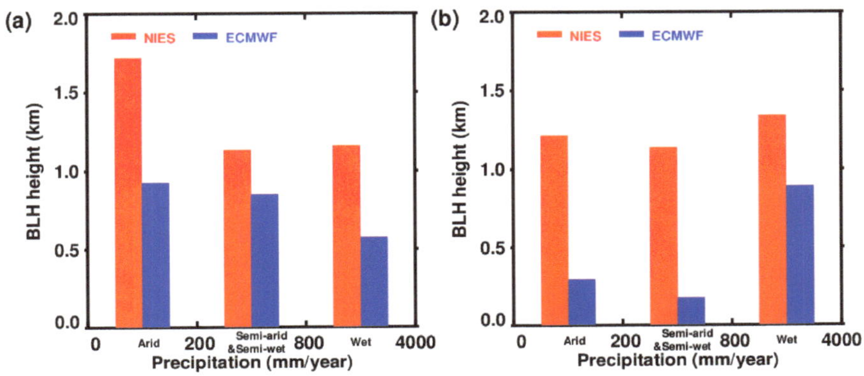

Figure 9. Comparison of AD-Net (red) and ECMWFs (blue) PBL in the case of different climate conditions (arid, semi-arid and semi-wet, and wet) for summer (**a**) and winter (**b**).

Table 4. Mean differences in BLH between AD-Net and ECMWFs (ECMWFs PBL minus AD-Net PBL) in summer and winter when given different climate conditions.

	Arid	Semi-Arid and Semi-Wet	Wet
Summer	−0.68	−0.45	−0.69
Winter	−1.42	−0.96	−0.27

4. Conclusions

In this paper, we compared the BLHs estimated by AD-Net and the ECMWFs over East Asia from September 2015 to August 2018. The present research provides better insight into the BLH performance in the ECMWFs reanalysis data.

The BLH determined by AD-Net was in close agreement with that calculated by Liu et al. [14], using the methods based on the aerosol backscatter. Additionally, a continuous 24 h BLH estimation from AD-Net generally matched with the aerosol vertical structures and the diurnal variation of the BLH is clearly shown, which suggests the good performance of the AD-Net BLH.

A comparison between seasonal mean BLH from AD-Net and the ECMWFs was conducted at 20 lidar sites. On average, there was an underestimation from the ECMWFs compared with AD-Net, both in summer and winter, except for over ocean areas in winter. A significant disagreement between the AD-Net BLH and the ECMWFs was noted over coastal areas and mountain areas. The absolute difference in BLH was larger in summer because the BLH from the ECMWFs was uniformly underestimated.

In general, the BLH of the ECMWFs was lower than that of AD-Net over most of the land cover types in summer and winter, except for over ocean areas in winter. The smallest differences existed over water surfaces in winter and bare surfaces in summer, and the largest differences occurred over grassland surfaces in summer, with the largest underestimation of 1.93 km compared with AD-Net. Overall, the mean minimum and maximum differences occurred over water surfaces and grassland surfaces, respectively.

All the BLHs of the ECMWFs were lower than those of AD-Net under different climatological conditions in summer and winter. The largest underestimation occurred over arid regions in winter, with a value of 1.42 km. The smallest underestimation occurred over wet regions, with a value of 0.27 km. The mean difference between the AD-Net BLH and the ECMWFs BLH was of 1.05, 0.71, and 0.48 km for arid regions, semi-arid and semi-wet regions, and wet regions, respectively. The difference in absolute uncertainly of the two BLHs in summer and in winter (0.74 km) was the largest in arid regions, which meant that this region was the most sensitive in the related physical parameterization in the ECMWFs model.

5. Discussion

To the best of our knowledge, there have been a number of studies that have used remote sensing to generate the BLH. These studies have used different techniques to identify the top of the PBL. The first set of studies used the temperature difference at the top marine stratus or stratocumulus clouds and sea surface to estimate PBL depth [36]. Height-resolved temperature observations from limb sounders, however, have the potential for PBL depth detection. Due to the fact that the PBL top is often marked with a sharp decrease in aerosol scattering signals that can be detected from space [37], the second approach is to use these signals to estimate the BLH [14]. Additionally, refractivity profiles with high vertical resolution provide, by means of a space-borne global positioning system, radio occultation, making this approach attractive for PBL detection [38,39]. Their strengths and weaknesses are all different, owing to the different detection methods. It is noted that the BLH is not produced directly. Therefore, the quality of the related physical variables is vital for the accurate estimation of the BLH.

Many researchers have shown that dust events seem to be a possible reason for the great difference between the different estimations of the BLH [40,41]. Thus, it is necessary to screen the dust cases when a comparison is conducted. In order to identify the dust cases, it is the best solution for us to use the lidar observations. The volume linear depolarization ratio (VLDR), which can indicate the shape of atmospheric particles, is a key parameter to distinguish the aerosol types. The different aerosol types exhibit a clear difference in the VLDR. Due to this, the VLDR can be considered to identify the dust aerosols, together with the lidar return signal [28,42–46]. The identification of the dust case can be achieved by using the VLDR; that is, when the value of the layer is larger than the threshold of dust, the profile is considered to be affected by the dust aerosols, and then the profile is screened for the comparison of BLHs. On the other hand, during a diurnal cycle, the atmosphere boundary layer structure can be classified into three major regimes [2]: convective boundary layer (CBL), stable boundary layer (SBL), and residual layer (RL). Lidar has been used for tracking the evolution of the BLH by using aerosol backscatter as a tracer, assuming aerosol is generally well-mixed in the planetary boundary layer. However, the validity of this assumption in fact varies with atmospheric stability [47]. For example, the relationship of the heights of intense turbulent exchange and the corresponding temperature gradients in the boundary layer of the atmosphere are analyzed with the ground values of wind speed and vertical turbulent heat flow [48]. Therefore, it is necessary to compare the BLH under different atmosphere stratifications.

In coastal regions, the formation of a thermal internal boundary layer (TIBL) is a common boundary layer phenomenon. In the sunny daytime, the sea breeze will blow the stable or neutrally stratified air over the sea toward the land. The surface-heating effect and dynamic-disturbance effect intensify the turbulence in the lowest atmospheric layer to form an unstable layer, which develops into the TIBL. Notably, the TIBL progressively grows from the coastline to the inland areas. The structure and dynamics of the planetary boundary layer in the ocean–continent zone during the summer period was reconstructed from lidar sounding data [49]. The TIBL is always associated with a sea breeze, so the meteorological factors that can be used include wind speed, wind direction, air temperature, relative humidity, and precipitation. On the other hand, the height of the TIBL can be considered as the height of the shallow convective boundary layer [50,51]. In the previous study, a TIBL forms if the wind blows from the sea and the boundary layer height is lower than 500 m, as indicated by the ceilometer [52]. However, as the minimum value of the BLH is about 600 m at 20 lidar sites, using the threshold value of 500 m to identify the TIBL is not reasonable. Overall, the TIBL is an important factor when considering the BLH over the coastal areas. Thus, extra efforts must be made to select or screen the cases of TIBL, such as obtaining meteorological data, determining the direction of sea breeze, and setting the threshold value, for example, to 500 m for different stations. Finally, all this work will be conducted in the near future with enough time.

Due to the limited number of lidar observation sites, the quality of the ECMWFs-derived BLHs is still under investigation. We will perform further evaluations for as long as more ground-based lidar observations are available. Special attention should be paid when using the ECMWFs BLH over coastal areas and mountain areas.

Author Contributions: Z.Z.: methodology, software, writing—original draft, and supervision; L.M.: methodology, editing, and supervision; C.L.: data curation, investigation. All authors have read and agreed to the published version of the manuscript.

Funding: This research was funded by the National Natural Science Foundation of China, grant number 41975019.

Institutional Review Board Statement: Not applicable.

Informed Consent Statement: Not applicable.

Data Availability Statement: Lidar data were provided by courtesy of AD-Net (http://www-lidar.nies.go.jp/AD-Net, accessed on 23 November 2022). The ECMWFs reanalysis data are accessible via http://apps.ecmwf.int/datasets/data/interim-full-moda/levtype=sfc/, accessed on 23 November 2022.

Acknowledgments: The authors gratefully acknowledge the anonymous reviewers for their excellent comments and efforts. The authors would like to thank AD-Net and the ECMWFs for providing the data used in our analysis.

Conflicts of Interest: The authors declare no conflict of interest.

References

1. Arya, S.P. *Introduction to Micrometeorology*, 2nd ed.; Academic Press: Cambridge, MA, USA, 2001; pp. 1–450.
2. Stull, R.B. *An Introduction to Boundary Layer Meteorology*; Kluwer Academic Publishers: Dordrecht, The Netherlands, 1988; pp. 1–27.
3. Wood, R. Stratocumulus clouds. *Mon. Weather Rev.* **2011**, *140*, 2373–2423. [CrossRef]
4. Seibert, P.; Beyrich, F.; Gryning, S.E.; Joffre, S.; Rasmussen, A.; Tercier, P. Review and intercomparison of operational methods for the determination of the mixing height. *Atmos. Environ.* **2000**, *34*, 1001–1027. [CrossRef]
5. Seidel, D.J.; Ao, C.O.; Li, K. Estimating climatological planetary boundary layer heights from radiosonde observations: Comparison of methods and uncertainty analysis. *J. Geophys. Res.-Atmos.* **2010**, *115*, D16113. [CrossRef]
6. Guo, J.; Deng, M.; Lee, S.S.; Wang, F.; Li, Z.; Zhai, P.; Liu, H.; Lv, W.; Yao, W.; Li, X. Delaying precipitation and lightning by air pollution over the Pearl River Delta. Part I: Observational analyses. *J. Geophys. Res.-Atmos.* **2016**, *121*, 6472–6488. [CrossRef]
7. Vogelezang, D.H.P.; Holtslag, A.A.M. Evaluation and model impacts of alternative boundary-layer height formulations. *Bound.-Lay. Meteorol.* **1996**, *81*, 245–269. [CrossRef]
8. Hennemuth, B.; Lammert, A. Determination of the atmospheric boundary layer height from radiosonde and lidar backscatter. *Bound.-Lay. Meteorol.* **2006**, *120*, 181–200. [CrossRef]
9. Sawyer, V.; Li, Z. Detection, variations and intercomparison of the planetary boundary layer depth from radiosonde, lidar and infrared spectrometer. *Atmos. Environ.* **2013**, *79*, 518–528. [CrossRef]
10. Beyrich, F. Mixing height estimation from sodar data-a critical discussion. *Atmos. Environ.* **1997**, *31*, 3941–3953. [CrossRef]
11. Eresmaa, N.; Karppinen, A.; Joffre, S.M.; Räsänen, J.; Talvitie, H. Mixing height determination by ceilometer. *Atmos. Chem. Phys.* **2006**, *6*, 1485–1493. [CrossRef]
12. Dai, C.; Wang, Q.; Kalogiros, J.A.; Lenschow, D.H.; Gao, Z.; Zhou, M. Determining boundary-layer height from aircraft measurements. *Bound.-Lay. Meteorol.* **2014**, *152*, 277–302. [CrossRef]
13. Chan, K.M.; Wood, R. The seasonal cycle of planetary boundary layer depth determined using cosmic radio occultation data. *J. Geophys. Res.-Atmos.* **2013**, *118*, 12422–12434. [CrossRef]
14. Liu, J.J.; Huang, J.P.; Chen, B.; Zhou, T.; Yan, H.R.; Jin, H.C.; Huang, Z.W.; Zhang, B.D. Comparisons of PBL heights derived from CALIPSO and ECMWF reanalysis data over China. *J. Quant. Radiat. Transf.* **2015**, *153*, 102–112. [CrossRef]

15. Zhang, W.; Guo, J.; Miao, Y.; Liu, H.; Zhang, Y.; Li, Z.; Zhai, P. Planetary boundary layer height from CALIOP compared to radiosonde over China. *Atmos. Chem. Phys.* **2016**, *16*, 9951–9963. [CrossRef]
16. Wang, L.; Schöck, M.; Chanan, G.A. Atmospheric turbulence profiling with SLODAR using multiple adaptive optics wavefront sensors. *Appl. Optics* **2008**, *47*, 1880–1892. [CrossRef] [PubMed]
17. Kovadlo, P.G.; Shikhovtsev, A.Y.; Kopylov, E.A.; Kiselev, A.V.; Russkikh, I.V. Study of the optical atmospheric distortions using wavefront sensor data. *Russ. Phys. J.* **2021**, *63*, 1952–1958. [CrossRef]
18. Wilson, R.W. SLODAR: Measurement of the height of optical turbulence using the Shaka-Hartman wavefront sensor. *Mon. ne. R. Astron. Soc.* **2002**, *337*, 103–108. [CrossRef]
19. Shikhovtsev, A.Y. A method of determining optical turbulence characteristics by the line of sight of an astronomical telescope. *Atmos. Ocean Opt.* **2022**, *35*, 303–309. [CrossRef]
20. Nishizawa, T.; Sugimoto, N.; Matsui, I.; Shimizu, A.; Hara, Y.; Itsushi, U.; Yasunaga, K.; Kudo, R.; Kim, S.W. Ground-based network observation using Mie-Raman lidars and multi-wavelength Raman lidars and algorithm to retrieve distributions of aerosol components. *J. Quant. Spectrosc. Radiat. Transf.* **2017**, *188*, 79–93. [CrossRef]
21. Nishizawa, T.; Sugimoto, N.; Matsui, I.; Shimizu, A.; Higurashi, A.; Jin, Y. The Asian dust and aerosol lidar observation network (AD-Net): Strategy and progress. In Proceedings of the 27th International Laser Radar Conference, New York, NY, USA, 5–10 July 2015. [CrossRef]
22. Sugimoto, N.; Nishizawa, T.; Liu, X.; Matsui, I.; Shimizu, A.; Zhang, Y.; Kim, Y.J.; Li, R.; Liu, J. Continuous Observations of Aerosol Profiles with a Two-Wavelength Mie-Scattering Lidar in Guangzhou in PRD2006. *J. Appl. Meteorol. Clim.* **2009**, *48*, 1822–1830. [CrossRef]
23. Sugimoto, N.; Nishizawa, T.; Shimizu, A.; Matsui, I.; Jin, Y.; Higurashi, A.; Uno, I.; Hara, Y.; Yumimoto, K.; Kudo, R. Continuous observations of atmospheric aerosols across East Asia. *SPIE Newsroom* **2015**. [CrossRef]
24. Shimizu, A.; Nishizawa, T.; Jin, Y.; Kim, S.W.; Wang, Z.F.; Batdorj, D.; Sugimoto, N. Evolution of a lidar network for tropospheric aerosol detecti on in East Asia. *Opt. Eng.* **2016**, *56*, 031219. [CrossRef]
25. Sugimoto, N.; Uno, I.; Nishikawa, M.; Shimizu, A.; Matsui, I.; Dong, X.; Chen, Y.; Quan, H. Record heavy Asian dust in Beijing in 2002: Observations and model analysis of recent events. *Geophys. Res. Lett.* **2003**, *30*, 1640. [CrossRef]
26. Shimizu, A.; Sugimoto, N.; Matsui, I.; Arao, K.; Uno, I.; Murayama, T.; Kagawa, N.; Aoki, K.; Uchiyama, A.; Yamazaki, A. Continuous observations of Asian dust and other aerosols by polarization lidars in China and Japan during ACE-Asia. *J. Geophys. Res.-Atmos.* **2004**, *109*, D19S17. [CrossRef]
27. Seidel, D.J.; Zhang, Y.; Beljaars, A.; Golaz, J.-C.; Jacobson, A.R.; Medeiros, B. Climatology of the planetary boundary layer over the continental United States and Europe. *J. Geophys. Res.-Atmos.* **2012**, *117*, D17106. [CrossRef]
28. Han, B.; Zhou, T.; Zhou, X.; Fang, S.; Huang, J.; He, Q.; Huang, Z.; Wang, M. A new algorithm of atmospheric boundary layer height determined from polarisation lidar. *Remote Sens.* **2022**, *14*, 5436. [CrossRef]
29. Zhang, M.; Tian, P.; Zeng, H.; Wang, L.; Liang, J.; Cao, X.; Zhang, L. A comparison of wintertime atmospheric boundary layer heights determined by tethered balloon soundings and lidar at the site of SACOL. *Remote Sens.* **2021**, *13*, 1781. [CrossRef]
30. Guo, J.; Miao, Y.; Zhang, Y.; Liu, H.; Li, Z.; Zhang, W.; He, J.; Lou, M.; Yan, Y.; Bian, L.; et al. The climatology of planetary boundary layer height in China derived from radiosonde and reanalysis data. *Atmos. Chem. Phys.* **2016**, *16*, 1330–13319. [CrossRef]
31. Zhang, H.; Zhang, X.; Li, Q.; Cai, X.; Fan, S.; Song, Y.; Hu, F.; Che, H.; Quan, J.; Kang, L.; et al. Research progress on estimation of atmospheric boundary layer height. *Acta Meteorol. Sin.* **2020**, *78*, 522–536. [CrossRef]
32. Palm, S.P.; Benedetti, A.; Spinhirne, J. Validation of ECMWF global forecast model parameters using GLAS atmospheric channel measurements. *Geophys. Res. Lett.* **2005**, *32*, 109–127. [CrossRef]
33. Shimizu, A.; Sugimoto, N.; Matsui, I. Detailed description of data processing system for lidar network in East Asia. In Proceedings of the 25th International Laser Radar Conference, St. Petersburg, Russia, 5–9 July 2010.
34. Korhonen, K.; Giannakaki, E.; Mielonen, T.; Pfüller, A.; Laakso, L.; Vakkari, V.; Baars, H.; Engelmann, R.; Beukes, J.P.; Van Zyl, P.G.; et al. Atmospheric boundary layer top height in South Africa: Measurements with lidar and radiosonde compared to three atmospheric models. *Atmos. Chem. Phys.* **2014**, *14*, 4263–4278. [CrossRef]
35. New, M.; Hulme, M.; Jones, P. Representing Twentieth-Century Space-Time Climate Variability. Part II: Development of 1901-96 Monthly Grids of Terrestrial Surface Climate. *J. Clim.* **2000**, *13*, 2217–2238. [CrossRef]
36. Wood, R.; Bretherton, C.S. Boundary layer depth, entrainment, and decoupling in the cloud-capped subtropical and tropical marine boundary layer. *J. Clim.* **2004**, *17*, 3576–3588. [CrossRef]
37. Edwards, J.M.; Beljaars, A.C.M.; Holtslag, A.A.M.; Lock, A.P. Representation of Boundary-Layer Processes in Numerical Weather Prediction and Climate Models. *Bound.-Lay. Meteorol.* **2020**, *177*, 511–539. [CrossRef]
38. Kursinski, E.R.; Hajj, G.A.; Schofield, J.T.; Linfield, R.P.; Hardy, K.R. Observing Earth's atmosphere with radio occultation measurements using the Global Positioning System. *J. Geophys. Res.-Atmos.* **1997**, *102*, 23429–23465. [CrossRef]
39. Hajj, G.A.; Ao, C.O.; Iijima, B.A.; Kuang, D.; Kursinski, E.R.; Mannucci, A.J.; Meehan, T.K.; Romans, L.J.; Juarez, M.D.; Yunck, T.P. Champ and SAC-C atmospheric occultation results and intercomparisons. *J. Geophys. Res.-Atmos.* **2003**, *109*, D06109. [CrossRef]
40. Tang, G.; Zhang, J.; Zhu, X.; Song, T.; Münkel, C.; Hu, B.; Schäfer, K.; Liu, Z.; Zhang, J.; Wang, L.; et al. Mixing layer height and its implications for air pollution over Beijing, China. *Atmos. Chem. Phys.* **2016**, *16*, 2459–2475. [CrossRef]
41. Zhou, T.; Xie, H.; Bi, J.; Huang, Z.; Huang, J.; Shi, J.; Zhang, B.; Zhang, W. Lidar measurements of dust aerosols during three field campaigns in 2010, 2011 and 2012 over Northwestern China. *Atmosphere* **2018**, *9*, 173. [CrossRef]

42. Zhang, Z.; Huang, J.; Chen, B.; Yi, Y.; Liu, J.; Bi, J.; Zhou, T.; Huang, Z.; Chen, S. Three-year continuous observation of pure and polluted dust aerosols over northwest China using the ground-based lidar and sun photometer data. *J. Geophys. Res.-Atmos.* **2019**, *124*, 1118–1131. [CrossRef]
43. Sugimoto, N.; Jin, Y.; Shimizu, A.; Nishizawa, T.; Yumimoto, K. Transport of mineral dust from africa and middle east to east asia observed with the lidar network (AD-Net). *SOLA* **2019**, *15*, 257–261. [CrossRef]
44. Qi, S.; Huang, Z.; Ma, X.; Huang, J.; Zhou, T.; Zhang, S.; Dong, Q.; Bi, J.; Shi, J. Classification of atmospheric aerosols and clouds by use of dual-polarization lidar measurements. *Optics Express.* **2021**, *29*, 23461–23476. [CrossRef]
45. Honda, N.; Coulibaly, S.; Funasaka, K.; Kido, M.; Oro, T.; Shimizu, A.; Matsumoto, T.; Watanabe, T. Comparison of the concentration of suspended particles and their chemical composition near the ground surface and dust extinction coefficient by Lidar. *Biol. Pharm. Bull.* **2022**, *45*, 709–719. [CrossRef] [PubMed]
46. Zhang, S.; Huang, Z.; Li, M.; Shen, X.; Wang, Y.; Dong, Q.; Bi, J.; Zhang, J.; Li, W.; Li, Z.; et al. Vertical structure of dust aerosols observed by a ground-based raman lidar with polarization capabilities in the center of the Taklimakan desert. *Remote Sens.* **2022**, *14*, 2461. [CrossRef]
47. Su, T.; Li, Z.; Kahn, R. A new method to retrieve the diurnal variability of planetary boundary layer height from lidar under different thermodynamic stability conditions. *Remote Sens. Environ.* **2020**, *237*, 111519. [CrossRef]
48. Odintsov, S.L.; Gladkikh, V.A.; Kamardin, A.P.; Nevzorova, I.V. The height of the region of intense turbulent heat transfer in a stably stratified boundary layer of the atmosphere. Part 2: Relationship with surface meteorological parameters. *Atmos Ocean Opt.* **2021**, *34*, 117–127. [CrossRef]
49. Pavlov, A.N.; Shmirko, K.A.; Stolyarchuk, S.Y. Features of the structure and dynamics of the planetary boundary layer in the Ocean-continent zone. Part II. Summer period. *Atmos Ocean Opt.* **2013**, *26*, 285–292. [CrossRef]
50. Prabha, T.V.; Venkatesan, R.; Mursch-Radlgruber, E.; Rengarajan, G.; Jayanthi, N. Thermal internal boundary layer characteristics at a tropical coastal site as observed by a mini-SODAR under varying synoptic conditions. *J. Earth Syst. Sci.* **2002**, *111*, 63–77. [CrossRef]
51. Sicard, M.; Pérez, C.; Rocadenbosch, F.; Baldasano, J.M.; García-Vizcaino, D. Mixed-layer depth determination in the barcelona coastal area from regular lidar measurements: Methods, results and limitations. *Boundary-Layer Meteorol.* **2006**, *119*, 135–157. [CrossRef]
52. Wei, J.; Tang, G.; Zhu, X.; Wang, L.; Liu, Z.; Cheng, M.; Münkel, C.; Li, X.; Wang, Y. Thermal internal boundary layer and its effect on air pollutants during summer in a coastal city in North China. *J. Environ. Sci.* **2018**, *70*, 37–44. [CrossRef]

remote sensing

Technical Note

A Novel Approach for the Global Detection and Nowcasting of Deep Convection and Thunderstorms

Richard Müller *, Axel Barleben, Stéphane Haussler and Matthias Jerg

Deutscher Wetterdienst, Frankfurter Str. 135, 63067 Offenbach, Germany; axel.barleben@dwd.de (A.B.); stephane.haussler@dwd.de (S.H.); matthias.jerg@dwd.de (M.J.)
* Correspondence: richard.mueller@dwd.de

Abstract: Thunderstorms are among the most common and most dangerous meteorological hazards in the world. They cause lightning and can lead to strong wind gusts, squall lines, hail and heavy precipitation combined with flooding, and therefore pose a threat to health and life, can cause enormous property damage and also endanger flight safety. Monitoring and forecast of thunderstorms are, therefore, important topics. In this work, a novel method for the detection and forecast of thunderstorms and strong convection is presented. The detection is based on the global GLD360 lightning data in combination with satellite information from the satellite series Meteosat, HIMAWARI and GOES, covering the complete geostationary ring. Three severity levels are defined depending on the occurrence of lightning and the brightness temperature difference of the water vapour channels and the infrared window channel (~10.8 μm). The detection of thunderstorms and strong convection is the basis for the nowcasting up to 2 h, which is performed with the optical flow method TV-L1. This method provides the needed atmospheric motion vectors for the extrapolation of the thunderstorm movement. Both, the validation results as well as the feedback of the customers show the great value of the new NowCastSat-Aviation (NCS-A) method. For example, the Critical Success Index (CSI) is, with 0.64, still quite high for the 60 min forecast of severe thunderstorms. The method is operated 24/7 by the German Weather Service (DWD), and is used to provide thunderstorm information to aviation customers and the central weather forecast unit of DWD.

Keywords: thunderstorms; cumulonimbus; convection; nowcasting; lightning

Citation: Müller, R.; Barleben, A.; Haussler, S.; Jerg, M. A Novel Approach for the Global Detection and Nowcasting of Deep Convection and Thunderstorms. *Remote Sens.* **2022**, *14*, 3372. https://doi.org/10.3390/rs14143372

Academic Editor: Gad Levy

Received: 31 May 2022
Accepted: 26 June 2022
Published: 13 July 2022

1. Introduction

Thunderstorms are among the most common and most dangerous meteorological hazards in the world. They are often associated with heavy rainfall, hail, wind gusts, squall lines and violent lightning. Phenomena that pose a threat to life, health, infrastructure and the environment. Thunderstorms are also referred to as Cumulonimbus clouds (Cbs) in meteorology. Any cloud that produces lightning is, by definition, a Cb.

While aircraft are well protected against direct lightning strokes as a result of the phenomena known as Faraday cage, turbulences and icing pose serious risks to aircraft and passengers. Therefore, for a safe transport, the location and expected severity of Cbs must be known. Aircraft are equipped with a board radar. However, these radars have only a short range, a limited viewing angle and the coverage is additionally hampered by "shadowing" by Cbs or optical thick clouds. Thus, relying only on the board radar could result in misleading judgement about the size and position of the Cbs and Cb clusters and cannot be exclusively used for the decision of the optimal and safest route in areas with Cbs. It is mainly suitable for spontaneous short-term evasive manoeuvres. Thus, gridded data of Cb information with a large geographical coverage are needed for forecast horizons from 0 to 3 h. The availability of such information in the cockpit improves the early Cb reconnaissance and, thus, the early selection of the safest flight routes—see Figure 1 for illustration.

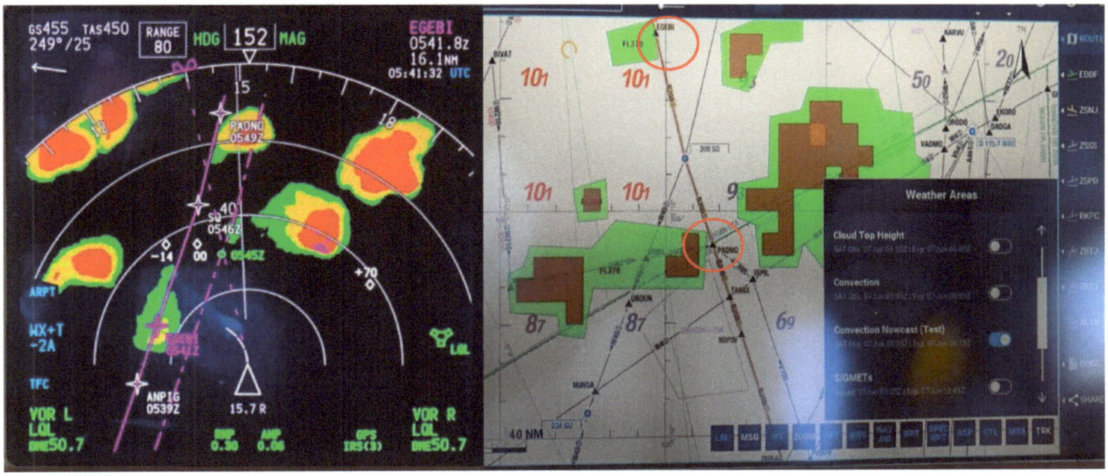

Figure 1. (**Left hand**): Photograph of the on-board radar display of a Lufthansa flight with heading 152 degree near way-point EGEBI (China) towards waypoint PADNO. The NCS-A information (**right hand**) provided in the EFB tablet computer in the cockpit supports situational awareness, e.g., that the passage of the southeast way-point PADNO could be possible and that there is likely no strong convection behind it (in the South, the lower part of the map).

Thunderstorm detection and nowcasting are also offered by operators of lightning detection networks—see [1–6]. However, these services are commercial and do not use satellite information for the estimation of severity levels and atmospheric motion vectors and are, therefore, based on a reduced set of information. The Nowcasting Satellite Application facility (NWC-SAF) [7] focuses on the development of software for satellite-based estimation of thunderstorms. However, the detection of Cbs based only on satellites is associated with a relatively high Flse Alarm Ratio (FAR) [8], which calls for the additional use of lightning data for Cb detection. Cbs are defined by the occurrence of lightning. Hence, using lightning data enables the detection of Cbs with a Probability of Detection (POD) about 100% and a FAR about 0%. It is obvious that this is a optimal basis for the nowcasting as it starts with the best possible POD and FAR values. Further, the NWC-SAF software is technically cumbersome and not optimised for a 24/7 operation at DWD. As a result of the request of key aviation customers the complete geostationary ring is covered by NCS-A, a feature not provided by NWC-SAF (Version 2018) . Finally, for many nowcasting applications a dense vector field is needed, e.g., [9]. DWD had quite good experience with modern Computer Vision techniques, e.g., [10,11]. They can be easily adapted to the different application fields as they provide a dense vector field based on a multi-scale approach. The parameters can be optimised for the respective application. Optical flow is used at DWD for turbulence [12], solar surface irradiance [9] and precipitation/radar nowcasting [13,14]. Thus, DWD wanted to use the established optical flow methods also for Cb nowcasting—an option not available within the NWC-SAF software. This motivates the development and operational implementation of a novel Cb nowcasting approach, referred to as NowCastSat-Aviation (NCS-A) version 1.0. NCS-A (Global Scan Service) is an operational 24/7 product from Deutscher Wetterdienst. NCS-A provides near real-time detection and predictions of convective cells across the global domain using the highest temporal and spatial resolution available from geostationary weather satellites. These includes METEOSAT, the European METEOrological SATellite [15], GOES, the US Geostationary Operational Environmental Satellite [16] and HIMAWARI [17], which means sunflower in English. The satellite information is combined with the global GLD360 lightning data from VAISALA [2–5]—see Section 2 for details. The novel aspects are the combination of modern Computer Vision methods with global lightning data, satellite information and numerical weather prediction data as well as the resulting definition of severity levels, which are described in more detail in the following paragraph. DWD also develops and 24/7 operates another Cb nowcasting system, which is called NowCastMIX-Aviation (NCM-A)—see James et al. [14]. NCM-A employs a multitude of data such as ground-based radar information, high-resolution lightning and model data. During the processing, the different information is combined by a fuzzy-logic technique to derive an optimal analysis of thunderstorm severity. However, due to the utilised sensors, the domain of NCM-A is limited and it cannot be applied to intercontinental long-haul flights. However, the viewing geometries of ground-based radar and satellites used by NCM-A and NCS-A, respectively, provide distinct insight into the characteristics of thunderstorms which complement each other. Therefore, using both nowcasting systems in combination, where possible, such as at air-traffic control centres, provides useful additional weather information for the air traffic controler [18]. Another global Cb nowcasting product is based on Convective Diagnosis Oceanic (CDO) algorithm. It is used to detect the area of storms that are most hazardous for aviation by a combination of geostationary satellite-based data and ground-based lightning data. A simple fuzzy-logic approach is used to combine the information from different input fields. The CDO input fields are the cloud top height, the Global Convective Diagnosis [19], the Overshooting Tops Detection algorithm [20] and the EarthNetworks global, ground-based lightning detection network. However, it is a commercial product, and central methods are not published in peer-reviewed journals or published at all. Furthermore, modern optical flow methods are to the knowledge of the authors not applied for the nowcasting. Therefore, the authors felt that the time was ripe to develop a new method using new techniques and approaches such as optical flow. The development was driven by a tight feedback loop with the users. In the process, some aspects of the CDO algorithm, e.g., the use of discretised severity levels, were adopted.

2. Materials and Methods: The Cb Detection and Nowcasting Method

In this section, the method for the detection and nowcasting of thunderstorms are described in more detail. For Cbs, three levels of severity are defined, which are discussed first. Then, the nowcasting method is described. The cloud top height of thunderstorms is of interest for the user as well. Thus, the method to derive CTH is discussed after that. The last issue covers the aspect of research to operations. The 24/7 implementation of the method as a basis for the end-user evaluation is documented here.

2.1. Detection and Definition of Severity Levels

The brightness temperature of the water vapour channels are used [8] for the satellite-based detection of Cbs. This concept is an adaptation of Schmetz et al. [21]. For the definition of the severity level the InfraRed (IR) window channel, ($\sim$10.8 μm) is used in addition. However, Cb detection based only on satellite data is associated with a relatively high FAR with sub-optimal POD [8]. In order to increase POD and reduce FAR, in particular for severe warnings, lightning data from VAISALA are used in addition [2–6]. The GLD360 data cover the globe and are based on Broadband VLF Radio Reception [22]. The use of receivers in the very low-frequency range (VLF; 3–30 kHz) enables the detection of radio pulses associated with lightning discharges over large distances (several thousand kilometers) [4,5]. The VLF frequency range enables an efficient routing of the signal through the ionosphere.

The severity of a thunderstorm provides important information for planning protective measures and was, therefore, an important user requirement. The definition of the severity levels are described in more detail below.

2.1.1. Light Convection

Light convection is defined as a region with neutral to in-stable layering of the atmosphere for satellite pixels with a brightness temperature difference of the water vapour channels ($BT_{6.2} - BT_{7.3}$) larger than -1. The regions with potential for thunderstorms are identified from numerical weather prediction model ICON [23] and are defined by a convective KO index of less than 2 [8]. Optically thick cold clouds can evolve into a thunderstorm and might be associated with strong convection. Thus, these clouds are also attributed to the light intensity level even when there is no lightning. Hence, the occurrence of lightning is not a precondition for the light convection level. The composition of the light convection is illustrated in Figure 2. The colour green is assigned to light convection.

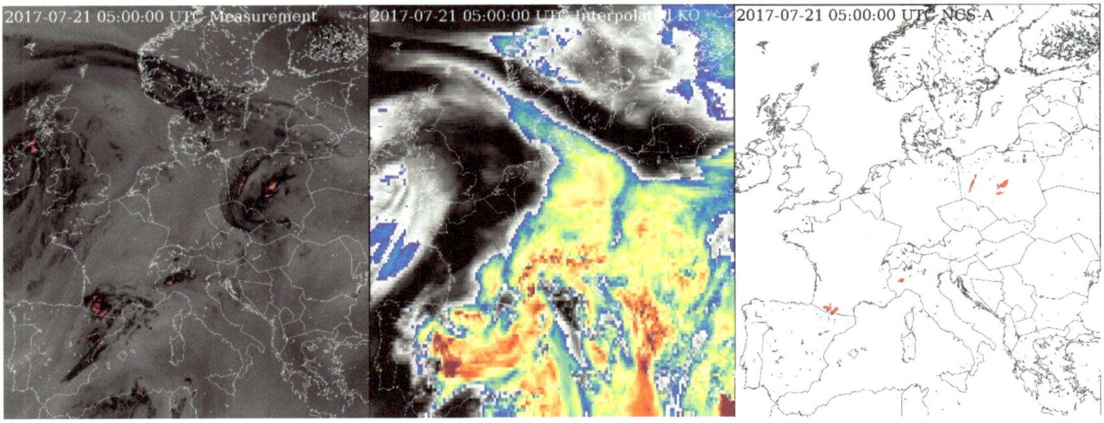

Figure 2. Illustration of the process for the definition of the light convection. The satellite data with BT larger than −1 (**left hand**) is folded with the NWP filter (**middle**) in order to derive the final product for light convection (**right hand**).

2.1.2. Moderate Convection

Furthermore, for moderate convection, the occurrence of lightning is not required. For this level, however, the clouds must be colder and thus closer to or in the tropopause. The brightness temperature of the water vapour channels and the IR window channel are used to define this level. If the differences exceed a specific threshold then deep convection is assumed to occur. The thresholds are defined as greater than 0.7 for the difference in the water water channels ($WV063 - WV073 > 0.7$) and greater than 2 for the BT difference of the water vapour channel and the window channel ($WV062 - IR10.8 > 2$). The latter condition can be used to identify overshooting tops, which are an indicator for strong updrafts associated with significant convection [20]. As for the light convection the NWP (KO) filter is used for this level. However, the likelihood for thunderstorms for OTs is quite high; hence, the NWP filtering is less important for this level than for the light level [8]. The colour yellow is assigned to moderate convection.

2.1.3. Severe Convection

The severity level for convection is defined as severe if lightning occurs, hence, the severe level is usually surrounded by the light or moderate level. The occurrence for lightning is a pre-requisite for the definition of severe convection within NCS-A. All lightning measurements occurring 15 min before the end of the latest satellite scan are taken into account and form the highest convection level "severe". The warning colour red is used for this level. It should be noted that the detection efficiency of the lightning measurements can vary somewhat and is particularly lower over large and remote ocean areas such as the South Pacific due to the lack of measurement sensors.

2.2. Cloud Top Height—CTH

The BT observed by the satellite corresponds to the temperature of the cloud top for optically thick clouds in the IR channel, since a black body can be assumed here [8]. As a consequence, Equation (1) is used to estimate the cloud top height

$$CTH = -(BT - T_{tropo})/LR + H_{tropo} \quad (1)$$

Here, BT is the brightness temperature of the IR window channel (~10.8 µm), T_{tropo} is the temperature of the tropopause and H_{tropo} is the height of the tropopause from the numerical weather prediction model ICON [23]. LR is the lapse rate and is set to 8 K/km.

This value represents a minor change from the value reported by Griffin et al. [24], which was determined by the expert knowledge of the DWD weather forecasters. Thus, the BT from IR channel at the top of the clouds is related to the cloud top height (flight level). The above mentioned "black body" assumption is not valid for semitransparent clouds and another method has to be applied. The respective method is referred to as water vapour H2O- intercept method applicable for upper-level semitransparent clouds [12].

2.3. Nowcasting

The nowcasting is performed with the optical flow method TV-L1 [10], which is provided as part of OpenCV [25,26]. The method allows the calculation of atmospheric motion vectors (AMVs) from two subsequent satellite images of the water vapour channels (ca. 6.2 µm) from the different geostationary satellites. The Cb image is then extrapolated in time based on the estimated atmospheric flow. Development and dissipation of cells is not captured by the method. The Cbs are only advected in time based on the AMVs. This limitation is an inherit feature and hold for all atmospheric motion vector methods. The TV-L1 parameters have been optimised for the nowcasting of Cbs—see Table 1. Forecast are calculated every 15 min and cover lead times up to to 2 h after the latest satellite scan. The temporal and spatial resolutions are 15 min and 0.1 degree (ca. 10 km) for satellite data and 1500 m for lightning.

Table 1. The NCS-A parameter settings used for TV-L1 .

Parameter	Value	Parameter	Value	Parameter	Value
Tau	0.15	Lambda	0.05	Theta	0.3
Epsilon	0.005	Outer Iterations	20	Inner Iterations	20
Gamma	0	Scales N	5	Scale Step	0.5
Warps	10	Median Filtering	1	-	-

2.4. 24/7 Implementation and Operation at DWD

2.4.1. Geotools

The processing of the satellite, lightning and NWP data is performed with a software package developed at DWD referred to as Geotools. The Geotools package is written in Python using Pytroll (https://pytroll.github.io/, accessed on 12 May 2022) for reading the satellite data and conducting basic image processing. Furthermore, Geotools are designed to merge the information from the various satellite sources and to apply the optical flow for the nowcasting, the geolocation and the polygonisation of the data. The latter is needed in order to reduce the necessary bandwidth for the transmission of the thunderstorm information to the flight desks. The software is optimised for speed in order to enable the processing of the complete geostationary ring and is also used for the polygonisation of the netcdf raster data.

2.4.2. 24/7 Processing

NowCastSat-Aviation NCS-A is hosted and operated 24/7 at DWD's High Performance Computer (HPC). For the 24/7 operation, Geotools are controlled by a shell script that processes the input and output data and calls the algorithms. EcFlow is used for the operational 24/7 call and monitoring of the job. EcFlow is a workflow package which has been developed to run a large number of programs in a controlled environment, providing restart and monitoring capabilities (via web page or email). It is used at DWD to run all operational suites on the HPC. EcFlow is developed and maintained by the European Center for Medium-Range Weather Forecasts (ECMWF)—see https://confluence.ecmwf.int/display/ECFLOW (accessed on 12 May 2022) for further details. The 24/7 processing uses multiple interfaces to the various data sources, the database, and the distribution systems that provide the resulting output data to external users. Figure 3 illustrates the end-to-end processing of NCS-A.

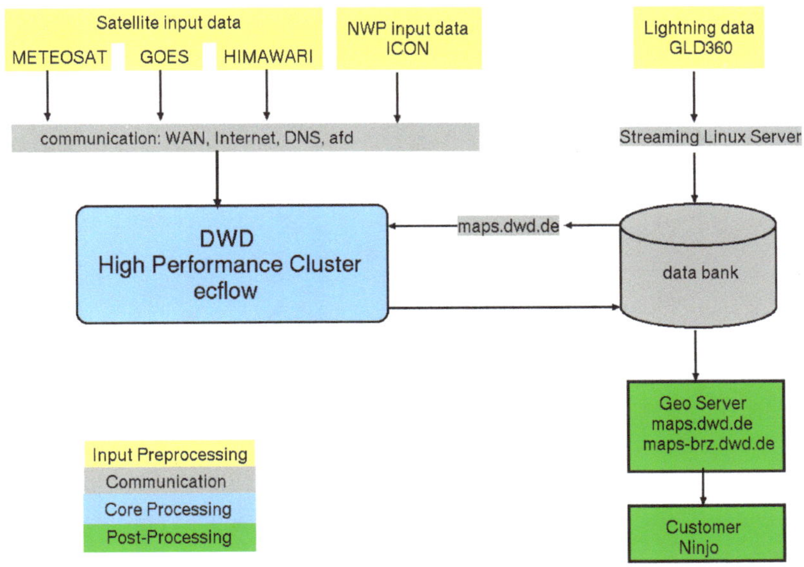

Figure 3. Illustration of the NCS-A end-to-end processing at DWD.

The Cb products are produced as raster data retrieval in cf conform netcdf Format. However, the data transfer rate to the cockpits of aeroplanes is limited. Thus, the raster data are polygonised to reduce the amount of data transferred to the cockpit. The Cb nowcasting product can be also visualised at DWD's meteorological workstation NinJo [27] and is, therefore, available to the routine forecasters. Figure 4 shows the visualisation of the NCS-A Cb severity product as displayed with DWD's geoweb service and the meteorological workstations.

2.5. Cockpit Implementation

For effective use of the data for route planning, the data are visualised in the cockpit with the aid of a tablet computer (Electronic Flight Bag EFB). Several customers have implemented the NCS-A data for visualisation in their specific visualisation tool. Figure 5 provides an overview about the existing implementations. The examples are eWAS from SITA (https://www.sita.aero/, accessed on 12 May 2022) and eRM as well as mPILOT from Lufthansa Systems (LHSys) (https://www.lhsystems.de/, accessed on 12 May 2022).

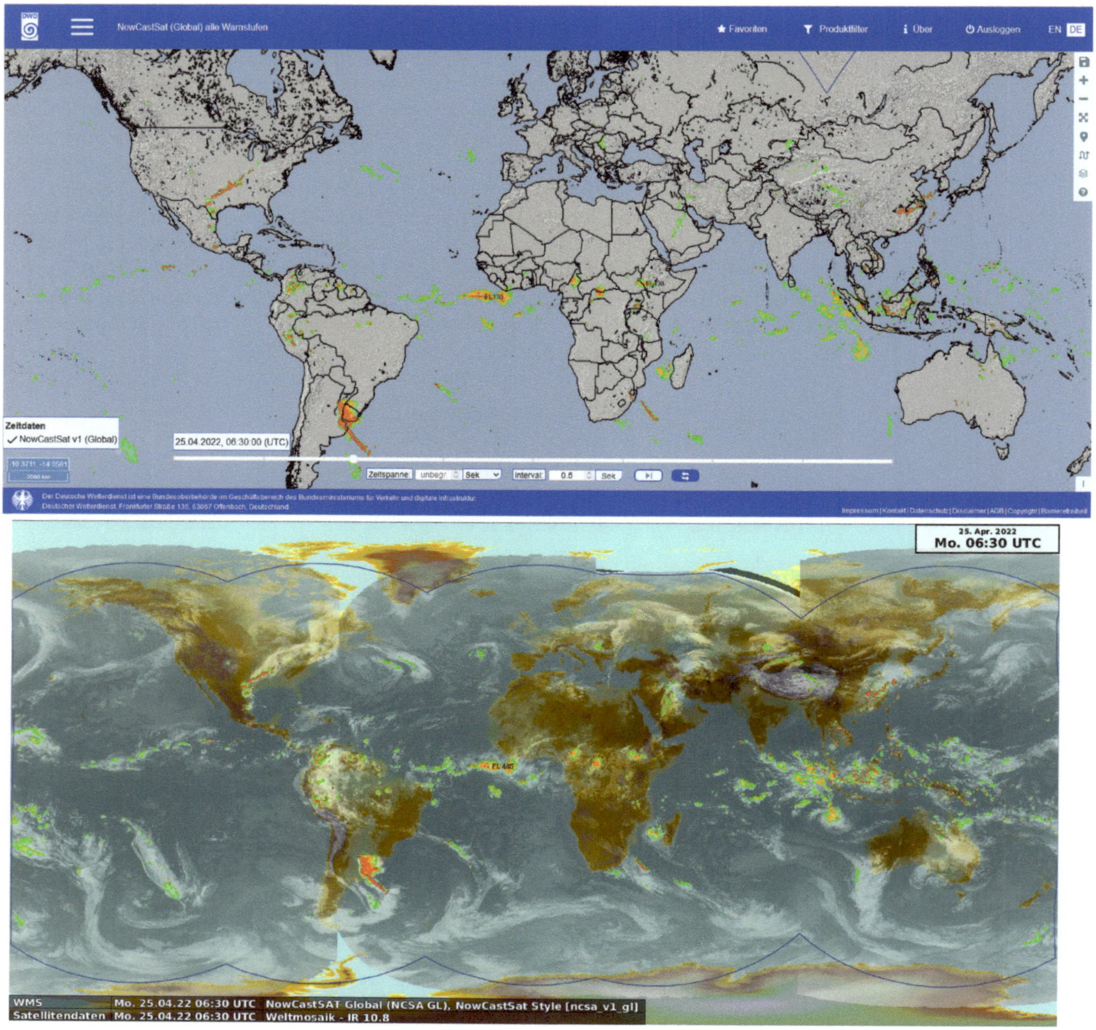

Figure 4. (**Top**): Visualisation of the NowCastSat-Aviation product in DWD's WebGIS Briefing System protoype. NCS-A covers the complete geostationary ring. The blue line indicates the satellite coverage. Light convection is shown in green, moderate in orange and severe in red. (**Bottom**): Visualisation of the NowCastSat-Aviation product in DWD's NinJo meteorological workstation. Both images show the convective situation on 25 April 2022 6:30 UTC. Note the tropical cyclone "Jasmine" in the Indian Ocean and a large mesoscale convective system (MCS) in the La Plata Basin, as well as other significant convective systems in the Altantic and Indo-Pacific. Both client systems are connected to DWD's Geoserver where the data are available per Web Mapping and Web Feature Service (WMS/WFS)—see Figure 3.

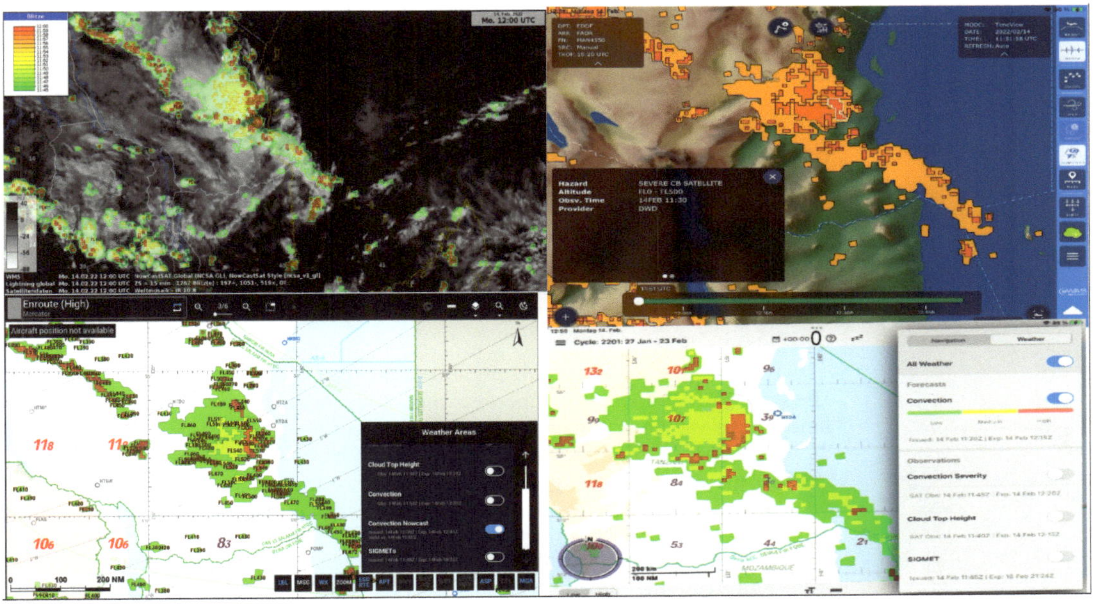

Figure 5. Technical implementation of NCS-A in various visualisation systems. (**Top left**): For forecasters in DWD workstation NinJo with observed satellite and lightning data underlayed. For pilots in different EFBs: (**top right**): eWAS/SITA, (**bottom left**): eRM/LHSys, (**bottom right**): mPILOT/LHSys. A mesoscale convective system is shown with severe convection (cloud top height in flight level 540) over Kenya on 14 February 2022 12 UTC.

3. Results

3.1. Validation of the Operational Nowcasting

Different methods and skill scores are available and applied for evaluation of forecasts. Therefore, the selection and decision of the method and skill scores should be quided by users of the respective forecast product. DWD is therefore in regular contact with aviation customers, e.g., pilots evaluating the products. The near-real-time Cb product reaches the pilot 15 to 20 min after the end of the satellite scan and is interpreted as a quasi-current convection analysis. The focus of the pilots is usually on the forecast periods of 30–60 min. Consequently, special consideration is given to this time interval.

Well established skill scores are used for the statistical analysis of the forecasts up to 2 h. In detail, the probability of detection (POD), the false alarm ratio (FAR) and the critical success index (CSI) [8,14,28]. These skill scores are based on 2 × 2 contingency tables [29] and are determined for the detection and nowcasting of the light and severe convection level of NCS-A compared to the measured lightning. Because the predictions are used in the form of polygons, verification on the basis of objects or pixels would be possible. Both methodological approaches have advantages [28]. Validation based on objects is not optimal to consider the severity or size of the Cbs for the evaluation of the scores. That is why the pixel- and object-based approaches are combined. Lightning can also occur outside the coldest cloud top because of the inclination of the discharge. In addition, horizontal displacement is also possible when the upper, coldest part of the cloud is blown away by strong winds. Finally parallax effects lead to a displacement between NCS-A polygons relative to the lightning events. To account for these spatial uncertainties, a distance of 0.3 degree between nowcasting (ncsa-light-polygon,ncsa-severe-polygon) and lightning measurements are accepted and counted as intersect. This distance of 0.3 degrees considers also the recommendation of the American Air Safety Authority FAA, according to which aircraft should maintain 20 miles lateral distance to Cbs. For each time step (00, 30, 60, 90,

120 min) it is analysed whether the NCS-A "light" and "severe" polygons intersect within the 0.3 degree search radius with a lightning polygon, which is formed from the discharges of the last 15 min of the respective time step. The NCS-A polygon is counted as correctly detected if there is an intersect, otherwise it is classed as a false detection.

In order to consider the size of the polygons, the pixels are counted in a second step. However, disadvantages of the pixel-based method are to be anticipated. Lightning cells and satellite objects cannot be expected to have the same size.

The months of August and September 2021 were used for the validation. Because an update of the NCS-A nowcasting takes place every 15 min, 5650 runs were verified against the lightning measurements for each investigated severity level. The validation is done for the light and subsequently for severe convection:

3.1.1. Light Convection

The resulting skill scores are presented in Figure 6 for the investigated forecast steps. The skill scores show the good performance of the method in particular up to 1 h forecast time. The scores correspond to comparable studies (e.g., best of James et al. [14], POD 0.75 FAR 0.32). The high POD at time step 0 is explained by the fact that due to the morphological filter for the polygons strong convection, i.e., lightning, are usually surrounded by green polygons (the polygons for light convection). Hence, allmost all lightning events are captured by the polygons for light convection for the time step 0. The optimum for POD, a value of 1, is not reached because of slight differences in the synchronisation of satellite scans and lightning data in combination with the polygonisation procedure. The FAR is relatively high for the time step zero. However, it has to be considered that lightning is not a must for the definition of the light convection and that satellite based detection of convection is limited by several handicaps, e.g., misclassification of stratiform cold clouds as Cb [8]. Further, not every cold convective cloud has permanent discharges at all times of the day. Nevertheless, respective convective systems pose still a danger to aviation, e.g., induced by the presence of downdrafts, heavy precipitation or hail, (clear air) turbulence and icing conditions.

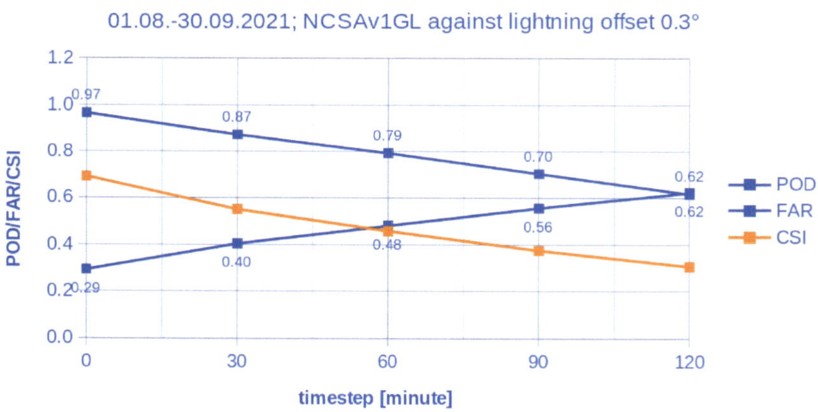

Figure 6. Skill Scores for light severity level of NCS-Av1GL as a function of lead time for analysis (timestep 0) and nowcasting after 30, 60, 90, 120 min against lightning measurement with an offset of 0.3 degree.

The POD, CSI and FAR values for the 1 h forecast are still good in comparison to the values for the analysis, demonstrating the good performance of the nowcasting method. This is also reflected in the comparison with the best values of James et al. [14] (POD 0.46 FAR 0.53). However, as mentioned in Müller et al. [8] comparison with values from the literature have to be interpreted carefully and can only provide a first hint. The optical flow method applied for the nowcasting does not account for decay or development of cells after the analysis time. Thus, the scores degrade with forecast lead time. Furthermore, uncertainties in the estimated optical flow (the velocity vectors) contribute to this degradation. Up to a time step of 90 min, the prediction performance can be considered sufficient and usable for operational applications. For general reliable nowcasting of more than 120 min improvements are necessary—see Section 4.

3.1.2. Severe Convection

Severe convection was validated analogously to light convection, but here the polygons of the severe convection were validated against the lightning measurements. The resulting skill scores are presented in Figure 7 for the investigated forecast steps. The definition of the severe level requires lightning for the detection of Cbs, hence the POD is about 100% and FAR about 0% at the time step 0, also referred to as analysis. These are much better starting values for the nowcasting (the extrapolation with optical flow) than for the light level. Thus, the nowcasting performs much better for the severe level. The POD, CSI and FAR values for the 1 h forecast are still very good. The CSI is with 0.64 significantly above the critical value of 0.5. This demonstrates the good performance of the nowcasting method. The optical flow method applied for the nowcasting does not account for decay of cells or development of cells after the analysis time. This is independent on the severity level and is a serious source for the degradation of the scores with forecast lead time. Up to a time step of 90 min, the prediction performance can be considered sufficient and usable for operational applications. For general reliable nowcasting of more than 120 min improvements are necessary—see Section 4.

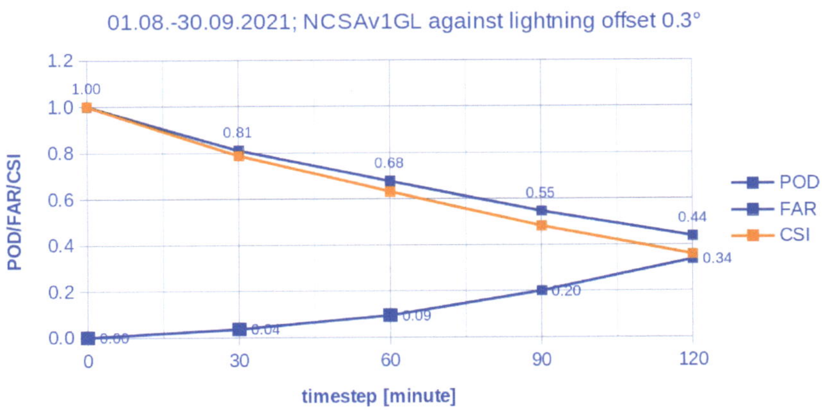

Figure 7. Skill Scores for severe severity level of NCS-Av1GL as a function of lead time for analysis (timestep 0) and nowcasting after 30, 60, 90, 120 min against lightning measurement with an offset of 0.3 degree.

3.2. User Evaluation

In addition to the validation study, which provides objective statistical values, the product was also evaluated by forecasters for specific cases. Forecaster of several European weather services have tested the convection product over Europe in summer 2021 within the annual testbed of the European Severe Storm Laboratory [30]: They summarised that "NowCastSAT's ability to detect the future location of cells was rated on a scale of 1 (bad in all cases) to 5 (spot-on). On average, it was rated 3.6 with grades ranging from 3 (good in about half of the cases) to 4 (good in most cases) and performance was best up to a forecast range of 60 min". Pilots (of the airlines Lufthansa and SWISS) came to similarly positive assessments in the form of evaluation reports during various trials in the last 1–2 years. Overall, these evaluations in combination with the validation results demonstrated the quality and usability of the novel NCS-A approach.

4. Discussion

The fact that the occurrence of lightning is not a pre-requisite for light and moderate convection level is also due to the reduced detection efficiency over oceans and rural areas with poor sensor coverage. Here, convection indicated by brightness temperature differences might show convection where lightning is not detected by the network.

Optical flow allows only advection of cells, decay or development of new cells are not captured. For example, cells developing after the satellite scan can not be detected by the method. Further all cells are extrapolated up to 2 h, but some cells might already decay before 2 h. DWD is currently developing a respective seamless prediction of convection based on the presented nowcasting approach and the NWP model ICON to overcome these limitations.

The ICOsahedral Nonhydrostatic D2 [23,31] is a nonhydrostatic model that enables improved forecasts of hazardous weather conditions with high-level moisture convection (super and multi-cell thunderstorms, squall lines, mesoscale convective complexes) due to improved physics in combination with its fine mesh size. The first validation results indicates the ability of the model to forecast Cbs with higher skill scores for lead times up to several hours. This would be ground breaking and a good basis for a seamless prediction of Cbs up to 8 h. The domain of ICON-D2 covers Germany and bordering countries with a spatial resolution of 2.2 km.

A first prototype based on simple blending is planned for the summer season 2022. For this, the data fusion will be performed with a simple blending. However, for an improved data fusion, the ANAKLIM++ method is envisioned in analogy to Urbich et al. [32]. First steps with ANAKLIM++ are planned for summer 2022. The novel seamless prediction product will be presented and discussed in detail in a forthcoming paper.

Another way to improve NCS-A is to use artificial intelligence. In Brodehl et al. [33], for example, strong indications are given that deep learning could improve satellite-based Cb nowcasting.

5. Conclusions

The novel NowCastSat-Aviation product was extensively validated and evaluated. The validation compiles an objective analysis based on the skill scores POD, FAR and CSI for 2 months. In addition, an evaluation was performed by external forecasters and pilots, which provide an important feedback concerning the practical usability. Overall, the results demonstrate that the product is quite useful for aviation and general weather forecasts up to 60–90 min forecasts. For example, the CSI for the severe level is, with 0.64, still quite high for the 60 min forecast. It should be emphasised that the product is available on the geostationary ring, i. H. the entire globe is covered in the range of $\pm$75 degrees north and south. This feature sets it apart from regional convection products and enables the use for global aviation applications, thereby meeting the requirements of key customers. The use of the global lightning data GLD360 of VAISALA improve the quality of the product significantly. Furthermore, together with the satellite information, they are a key to a reasonable definition of severity levels. Finally, the use of open source software from the Pytroll and OpenCV libraries is of great benefit to the flexibility of the method and its maintenance and allows to overcome significant technical drawbacks associated with the NWC-SAF software. Currently, the formation or decay of cells is not considered in the nowcasting of NCS-A. This weakness could be addressed by intelligent combination with lightning prediction from NWP or by using deep learning.

Author Contributions: A.B. carried out the operational implementation and is responsible for the further development of monitoring and 24–7 operation of the process. He performed the evaluation of the method and contributed to the development of the method. S.H. has played a major role in the development of NCS-A. Among other things, he developed the Geotools and optimised the optical flow for Cbs. R.M. contributed to the development of the method and wrote large parts of the paper. M.J. supervised the project. All authors contributed to the writing of the manuscript. All authors have read and agreed to the published version of the manuscript.

Funding: This research received no external funding.

Data Availability Statement: The data are available for customers after signing of corresponding contracts.

Acknowledgments: The authors would like to thank the following people for their valuable feedback and suggestions that improved NCS-A: L. Lewe, W. Greiner, O. Matz and A. Ritter from Lufthansa Airlines, M. Gerber and J. Vetsch from SWISS Airlines, A. Medlhammer and A. Erhegyi from Lufthansa Systems GmbH. V. Santos and A. Laj from SITA, W. Rumler from PACE Aerospace and M. Schneider from Eurocontrol MUAC.

Conflicts of Interest: The authors declare no conflict of interest.

Sample Availability: Samples of the compounds are available from the authors.

Abbreviations

The following abbreviations are used in this manuscript:

Cb	Comulinmbus Cloud, Thunderstorm
CSI	Critical Success Index
CTH	Cloud top Height
DWD	Deutscher Wetterdienst
FAR	False Alarm Ratio
GLD	Globale Lightning Detection
GOES	Geostationary Operational Environmental Satellite
h	hour(s)
HIMAWARI	Sunslower
MDPI	Multidisciplinary Digital Publishing Institute
METEOSAT	METEOrological SATellite
min	minute(s)
MSG	Meteosat Second Generation
NCS-A	Nowcast Satellite Aviation
POD	Probability Of Detection
TV-L1	Total Variation L1 norm
LD	Linear dichroism

References

1. Betz, H.; Schmidt, K.; Oettinger, W.; Montag, B. Cell-tracking with lightning data from LINET. *Adv. Geosci.* **2008**, *17*, 55–61. [CrossRef]
2. Pohjola, H.; Mäkelä, A. The comparison of GLD360 and EUCLID lightning location systems in Europe. *Atmos. Res.* **2013**, *123*, 117–128. [CrossRef]
3. VAISALA. *VAISALA GLD360 Global Dataset—Understanding the GLD360 Dataset*; Vaisala Corporation, Head Office: Helsinki, Finland, 2022.
4. Said, R.K.; Cohen, M.B.; Inan, U.S. Highly intense lightning over the oceans: Estimated peak currents from global GLD360 observations. *J. Geophys. Res. Atmos.* **2013**, *118*, 6905–6915. [CrossRef]
5. Said, R.; Murphy, M. GLD360 Upgrade: Performance Analysis and Applications. In Proceedings of the 24th International Lightning Detection Conference and 6th International Lightning Meteorology Conference, San Diego, CA, USA, 18–21 April 2016.
6. Pessi, A.T.; Businger, S.; Cummins, K.L.; Demetriades, N.W.S.; Murphy, M.; Pifer, B. Development of a Long-Range Lightning Detection Network for the Pacific: Construction, Calibration, and Performance. *J. Atmos. Ocean. Technol.* **2009**, *26*, 145–166. doi: 10.1175/2008JTECHA1132.1. [CrossRef]
7. Autones, F. *Algorithm Theoretical Basis Document for Rapid Development Thunderstorms*; Technical Report; NWC-SAF: Madrid, Spain, 2013.

8. Müller, R.; Haussler, S.; Jerg, M. The Role of NWP Filter for the Satellite Based Detection of Cumulonimbus Clouds. *Remote Sens.* **2018**, *10*, 386. doi: 10.3390/rs10030386. [CrossRef]
9. Urbich, I.; Benidx, J.; Müller, R. A Novel Approach for the Short-Term Forecast of the Effective Cloud Albedo. *Remote Sens.* **2018**, *10*, 955. [CrossRef]
10. Zach, C.; Pock, T.; Bischof, H. A duality based approach for realtime TV-L1 optical flow. In Proceedings of the Joint Pattern Recognition Symposium, Heidelberg, Germany, 12–14 September 2007; pp. 214–223.
11. Sánchez, J.P.; Meinhardt-Llopis, E.; Facciolo, G. TV-L1 optical flow estimation. *Image Process. Online* **2013**, *3*, 137–150. doi: doi: 10.5201/ipol.2013.26. [CrossRef]
12. Barleben, A.; Haussler, S.; Müller, R.; Jerg, M. A Novel Approach for Satellite-Based Turbulence Nowcasting for Aviation. *Remote Sens.* **2020**, *12*, 2255. doi: 10.3390/rs12142255. [CrossRef]
13. Ayzel, G.; Heistermann, M.; Winterrath, T. Optical flow models as an open benchmark for radar-based precipitation nowcasting (rainymotion v0.1). *Geosci. Model Dev.* **2019**, *12*, 1387–1402. doi: 10.5194/gmd-12-1387-2019. [CrossRef]
14. James, P.; Reichert, B.; Heizenreder, D. NowCastMIX: Automatic Integrated Warnings for Severe Convection on Nowcasting Time Scales at the German Weather Service, Weather and Forecasting. *Weather. Forecast.* **2018**, *33*, 1413–1433. [CrossRef]
15. Schmetz, J.; Pili, P.; Tjemkes, S.; Just, D.; Kerkmann, J.; Rota, S.; Ratier, A. An introduction to Meteosat Second Generation (MSG). *Bull. Am. Met. Soc.* **2002**, *83*, 977–992. [CrossRef]
16. GOES. *GOES-R Series Data Book*; CDRL PM-14; National Aeronautics and Space Administration; GOES-R Series Program Office Goddard Space Flight Center: Greenbelt, MD, USA, 2019.
17. Himawari. *Himawari-89—Himawari Standard Data User's Guide*; Version 1.3; Japan Meteorological Agency: Tokyo, Japan, 2017.
18. Schneider, M. (Maastricht Upper Area Control Centre (MUAC), Maastricht Airport, The Netherlands); Barleben, A. (DWD, Offenbach, Germany). User Feedback Loop with M. Schneider from MUAC Starting April 2019 and Finalized 31 March 2022. Personal communication, 2022.
19. Mosher, F. Detection of deep convection around the globe. In Proceedings of the 10th Conference on Aviation, Range and Aerospace Meteorology, Portland, OR, USA, 13–16 May 2002; American Meteorological Society: Boston, MA, USA, 2002; pp. 289–292.
20. Bedka, K.; Brunner, J.; Dworak, R.; Feltz, W.; Otkin, J.; Greenwald, T. Objective satellite-based detection of overshooting tops using infrared window channel brightness temperature gradients. *Appl. Meteorol. Climatol.* **2010**, *49*, 181–202. doi: 10.1175/2009JAMC2286.1. [CrossRef]
21. Schmetz, J.; Tjemkes, A.; Gube, M.; van der Berg, L. Monitoring deep convection and convective overshooting with Meteosat. *Adv. Space Res.* **1997**, *19*, 433–441. [CrossRef]
22. Cohen, M.B.; Inan, U.S.; Paschal, E.W. Sensitive Broadband ELF/VLF Radio Reception With the AWESOME Instrument. *IEEE Trans. Geosci. Remote Sens.* **2010**, *48*, 3–17. doi: 10.1109/TGRS.2009.2028334. [CrossRef]
23. Zängl, G.; Reinert, D.; Rípodas, P.; Baldauf, M. The ICON (ICOsahedral Non-hydrostatic) modelling framework of DWD and MPI-M: Description of the non-hydrostatic dynamical core. *Q. J. R. Meteorol. Soc.* **2015**, *141*, 563–579. doi: 10.1002/qj.2378. [CrossRef]
24. Griffin, S.M.; Bedka, K.M.; Velden, C.S. A Method for Calculating the Height of Overshooting Convective Cloud Tops Using Satellite-Based IR Imager and CloudSat Cloud Profiling Radar Observations. *J. Appl. Meteorol. Climatol.* **2016**, *55*, 479–491. doi: 10.1175/JAMC-D-15-0170.1. [CrossRef]
25. OpenCV. OpenCV Homepage. OpenCV Is an Open-Source, Computer-Vision Library for Extracting and Processing Meaningful Data from Images. Available online: http://opencv.org/ (accessed on 11 April 2022).
26. Bradski, G. The OpenCV Library. *Dr. Dobb's J. Softw. Tools* **2000**, *120*, 122–125.
27. Joe, P.; Koppert, H.J.; Heizenreder, D.; Erbshaeusser, B.; Raatz, W.; Reichert, B.; Rohn, M. Severe weather forecasting tools in the NinJo workstation. In Proceedings of the World Weather Research Program Symposium on Nowcasting and Very Short Range Forecasting, Toulouse, France, 5–9 September 2005.
28. Zinner, T.; Forster, C.; de Coning, E.; Betz, H.D. Validation of the Meteosat storm detection and nowcasting system Cb-TRAM with lightning network data—Europe and South Africa. *Atmos. Meas. Tech.* **2013**, *6*, 1567–1583. [CrossRef]
29. Wilks, D.S. *Statistical Methods in the Atmospheric Sciences*; International Geophysics Series; Elsevier: Amsterdam, The Netherlands, 2006; Volume 91.
30. Gatzen, C.; Pucik, T.; Groenemeijer, P. *Report on the Evaluation of DWD Nowcast and Warning Products at the ESSL Testbed 2021*; DWD internal 3053581 19-KAP; Technical Report DWD internal 3053581 19-KAP; European Severe Storms Laboratory: Wessling, Germany, 2021.
31. Reinert, D.; Prill, F.; Frank, H.; Denhard, M.; Baldauf, M.; Schraff, C.; Gebhardt, C.; Marsigli, C.; Zängl, G. *DWD Database Reference for the Global and Regional ICON and ICON-EPS Forecasting System*; Technical Report Version 2.1.7; DWD: Offenbach, Germany, 2021.
32. Urbich, I.; Bendix, J.; Müller, R. Development of a Seamless Forecast for Solar Radiation Using ANAKLIM++. *Remote Sens.* **2020**, *12*, 3672. [CrossRef]
33. Brodehl, S.; Müller, R.; Schömer, E.; Spichtinger, P.; Wand, M. End-to-End Prediction of Lightning Events from Geostationary Satellite Images. *Preprints* **2022**. doi: 10.20944/preprints202206.0238.v1. [CrossRef]

remote sensing

MDPI

Review

Machine-Learning for Mapping and Monitoring Shallow Coral Reef Habitats

Christopher Burns [1,*], Barbara Bollard [1] and Ajit Narayanan [2]

[1] Drone Lab, School of Engineering, Computer and Mathematical Sciences, Auckland University of Technology, WZ Building, Level 11, 6 St Paul Street, Auckland 1010, New Zealand; barbara.bollard@aut.ac.nz
[2] School of Engineering, Computer and Mathematical Sciences, Auckland University of Technology, WZ Building, Level 11, 6 St Paul Street, Auckland 1010, New Zealand; ajit.narayanan@aut.ac.nz
* Correspondence: cburns@aut.ac.nz

Abstract: Mapping and monitoring coral reef benthic composition using remotely sensed imagery provides a large-scale inference of spatial and temporal dynamics. These maps have become essential components in marine science and management, with their utility being dependent upon accuracy, scale, and repeatability. One of the primary factors that affects the utility of a coral reef benthic composition map is the choice of the machine-learning algorithm used to classify the coral reef benthic classes. Current machine-learning algorithms used to map coral reef benthic composition and detect changes over time achieve moderate to high overall accuracies yet have not demonstrated spatio-temporal generalisation. The inability to generalise limits their scalability to only those reefs where in situ reference data samples are present. This limitation is becoming more pronounced given the rapid increase in the availability of high temporal (daily) and high spatial resolution (<5 m) multispectral satellite imagery. Therefore, there is presently a need to identify algorithms capable of spatio-temporal generalisation in order to increase the scalability of coral reef benthic composition mapping and change detection. This review focuses on the most commonly used machine-learning algorithms applied to map coral reef benthic composition and detect benthic changes over time using multispectral satellite imagery. The review then introduces convolutional neural networks that have recently demonstrated an ability to spatially and temporally generalise in relation to coral reef benthic mapping; and recurrent neural networks that have demonstrated spatio-temporal generalisation in the field of land cover change detection. A clear conclusion of this review is that existing convolutional neural network and recurrent neural network frameworks hold the most potential in relation to increasing the spatio-temporal scalability of coral reef benthic composition mapping and change detection due to their ability to spatially and temporally generalise.

Keywords: remote sensing; machine-learning; deep-learning; coral reefs; mapping; change detection; spatio-temporal generalisation

Citation: Burns, C.; Bollard, B.; Narayanan, A. Machine-Learning for Mapping and Monitoring Shallow Coral Reef Habitats. *Remote Sens.* **2022**, *14*, 2666. https://doi.org/10.3390/rs14112666

Academic Editor: John Burns

Received: 7 April 2022
Accepted: 28 May 2022
Published: 2 June 2022

1. Introduction

Coral reef benthic composition maps have become an essential component in marine science and management [1]. In order for these maps to be most effective, they need to represent the underlying benthic classes as accurately as possible. The field of coral reef remote sensing science has established a number of key findings that have led to improvements in coral reef benthic composition map accuracies. These include: identifying the effects of spatial and spectral resolution [2–4]; the number of classes mapped [5–8]; the spatial and spectral similarities between benthic classes [4,8]; the use of either pixel or object-based classification methods [9]; and the in situ reference data collection method used [10]. The development of image pre-processing algorithms such as correction for absorption and scattering of light in the water column [11], sunglint removal [12], and atmospheric correction [13] have also contributed to improvements in the ability to accurately map coral reef benthic composition using remotely sensed imagery.

Another factor that affects the accuracy of a coral reef benthic composition map, which is in the control of a remote sensing scientist, is the choice of the machine-learning algorithm used to classify the coral reef benthic classes. To this end, these machine-learning algorithms use labelled samples in the form of either individual pixels or objects (grouped pixels) from within an image that are representative of specific classes in order to train an algorithm to subsequently classify all pixels or objects. In the field of coral reef benthic composition mapping, machine-learning algorithms are now more commonly used than manual delineation by expert interpretation [14], expert manual class assignment [5,15], and expert-derived ruleset development [9,16–20] This is because machine-learning algorithms are less subjective and more easily repeatable [21]. The most commonly used machine-learning algorithms for coral reef benthic composition mapping are the *k*-Nearest Neighbours (*k*-NN), Maximum Likelihood Classification (MLC), Minimum Distance to Means (MDM), Random Forest (RF), and Support Vector Machine (SVM) (Tables 1 and 2). A key advantage of these algorithms is the fact that they are able to achieve moderate to high overall accuracies with only small amounts of training data (i.e., <1000 training samples per class). To date, this has been a prerequisite for coral reef benthic mapping due to the logistical complexities inherent in acquiring in situ reference samples from a coral reef site, which are usually in the form of georeferenced benthic photographs [10].

The scarcity of coral reef benthic in situ reference data is perhaps the reason almost all related publications, to date, use training and testing data derived from the same specific reef, or reef area, within the extent that they are mapping. While this is suitable for mapping coral reef benthic composition at the target locations, where in situ reference data are present, the inherent bias between the training and testing samples will result in the machine-learning algorithm overfitting on this localised data. Therefore, it is unlikely that the trained algorithm can then generalise to mapping the coral reef benthic composition of a new reef (spatial generalisation), or even new imagery of the same reef (temporal generalisation) since remote sensing conditions may vary between images. The inability to spatially or temporally generalise therefore affects the spatio-temporal scalability of coral reef benthic mapping by limiting the target extents to only those reefs or reef areas that contain in-situ reference data.

In addition to mapping coral reef benthic composition at one point in time, ongoing monitoring is essential for conservation management. Coral reef benthic composition monitoring using remotely sensed imagery, to date, has relied primarily on post-classification comparison change detection (PCCCD), which classifies each individual image separately [14,22–24] before subsequently overlaying the classified images in order to identify where change has occurred. The accuracy of PCCCD methods is therefore dependent upon the accuracy of each individual classified map in the sequence. Since each map in the multi-temporal sequence is usually classified using similar machine-learning algorithms as those used in coral reef benthic composition mapping, the same primary limitation in terms of spatio-temporal scalability, the inability to generalise, is also present in PCCCD methods used for coral reef benthic change detection. The inability to generalise is becoming more pronounced given the rapid increase in the availability of high temporal (daily) and high spatial resolution (<5 m) multispectral satellite imagery [25].

Coincident with the increase in the availability of multispectral satellite imagery are advancements in deep learning frameworks [26,27]. Of particular interest to remote sensing scientists are convolutional neural networks (CNNs), which, in the field of land cover mapping, have demonstrated superior overall accuracies compared to traditional machine-learning classification algorithms such as object-based *k*-nearest neighbour (OBIA–KNN), object-based support vector machine (OBIA–SVM), and object-based random forest (OBIA–RF) [28,29], as well as spatio-temporal generalisation [30]. CNNs have only recently been used in the field of coral reef benthic mapping [31–33] but have already demonstrated high overall accuracies and spatio-temporal generalisation [32]. In relation to change detection,

Long Short-Term Memory networks (LSTMs) and recurrent convolutional neural networks (ReCNN) have been applied to binary (changed or unchanged) and multi-class land cover change detection and achieved superior overall accuracies compared to traditional land cover change detection methods, and they have demonstrated binary spatio-temporal generalisation [34,35]. To the best of our knowledge, however, LSTMs have not yet been applied to coral reef benthic change detection.

This review focuses on the most commonly used machine-learning algorithms applied to map coral reef benthic composition and detect changes over time using multispectral satellite imagery in order to identify their advantages as well as the primary limitations impacting their spatio-temporal scalability. Recent publications that have used CNNs for coral reef benthic mapping, and LSTMs in the field of land cover change detection are subsequently reviewed with a focus on their potential utility for increasing the spatio-temporal scalability of coral reef benthic mapping. Three keywords, 'coral reef habitat mapping', 'coral reef mapping using machine learning', and 'coral reef change detection' were searched in ScienceDirect and Google Scholar. For each keyword the first 100 results were screened in ScienceDirect and the first 10 pages in Google Scholar. No date range for publications was applied. We limited publications to only those that used multispectral satellite imagery and mapped shallow coral reef habitat classes. Multispectral satellite imagery was chosen since it is the most easily accessible imagery type for researchers, covers the largest area globally, and has the highest temporal frequency compared to airborne and/or UAV imagery. Publications focusing only on seagrass or coral reef geomorphological classes were excluded. Book chapters, theses, reports, and reviews were also excluded. From 600 screened results, 87 publications were deemed relevant based on the title and/or the abstract. Of these, 6 could not be accessed and 15 after being reviewed did not include enough detail to be included, as seen in Tables 1–3, leaving a total of 66 publications. A further five additional publications were manually searched since they were referenced a number of times within these 66 publications and therefore deemed important to include in this review. These publications were: [7,9,15,17,21]. One additional publication [33], was brought to our attention because it used a CNN for coral reef benthic mapping. In total, 72 publications based on coral reef mapping and change detection have been included in this review.

For CNN publications used in Sections 2.2 and 2.2.1, in addition to recurrent neural networks (RNNs) in Section 2.3.2, a keyword search methodology was not applied. CNNs were included in this review based on the fact they are starting to be used for coral reef benthic mapping. RNNs were included because they have demonstrated spatio-temporal generalisation in the field of land cover change detection, which current coral reef change detection methods have not demonstrated.

2. Machine-Learning Algorithms Applied to Coral Reef Benthic Mapping Using Multispectral Satellite Imagery

2.1. Pixel-Based Machine-Learning Classification Algorithms

Pixel-based machine-learning classification algorithms applied to mapping coral reef benthic composition, classify each individual pixel within an image as being one of a certain number of classes (i.e., coral, algae, and sand). In order for pixel-based algorithms to be feasible, two assumptions need to be met. First, each individual pixel needs to be represented by only one benthic class, meaning the spatial resolution of the pixel is higher than or similar to the target object. Secondly, pixels representing each class need to have similar spectral reflectance values to each other (i.e., all coral pixels need to have similar spectral reflectance signatures to each other, which are different to the spectral reflectance signatures of sand and other classes) [36]. For coral reef benthic composition mapping, the main pixel-based machine-learning algorithm used is MLC, based on the fact it is used by 47% of publications in Table 1 (In relation to calculating the number of publications that use each specific machine learning algorithm in Tables 1–3, only one algorithm per publication is used. When a publication compares multiple different machine learning algorithms, only the algorithm that achieves the highest overall accuracy is used in the calculation. This was done in order to ensure that publications that use one machine learning algorithm and those that are comparing multiple different ones are weighted equally in the calculation). MLC, which first assumes the data are normally distributed, takes the mean vector and covariance of each classes' spectral values into account before subsequently determining the membership of individual pixels to each class based on the highest statistical probability (maximum likelihood). Figure 1 illustrates equiprobability contours forming probability density regions around the mean of the class training samples that are used in order to determine the probability of pixel x belonging to each class. In this example pixel x would be classified as belonging to the coral class.

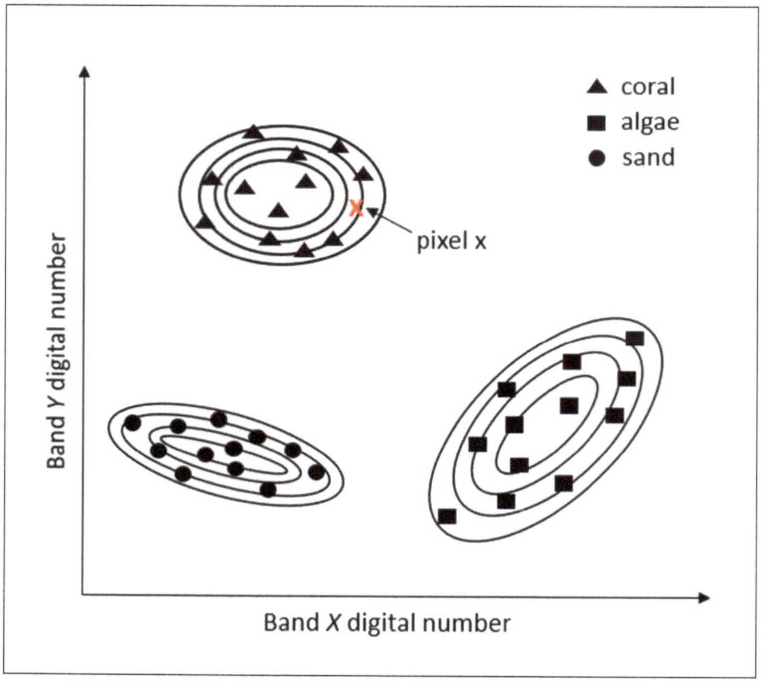

Figure 1. Basic conceptualisation of maximum likelihood classification illustrating 3 different classes with equiprobability contours.

Moderate to high overall accuracies can be achieved using pixel-based MLC to map coral reef benthic classes (Table 1) (All overall accuracies reported in this review, except for those in Section 2.3.2 have been rounded to the nearest whole number since there was not a consistent level of precision between all publications (i.e., some recorded overall accuracy to the nearest one decimal place while others to the nearest whole number)). Compared to the second most commonly used machine-learning algorithms in Table 1, which are RF, SVM, and MDM, each being used by 12% of publications, MLC has demonstrated higher overall accuracies when directly compared to MDM in Hossain et al. [37]. Wicaksono and Aryaguna [8], and Chegoonian et al. [38], however, found that RF and SVM, respectively, achieved higher overall accuracies when directly compared to MLC (Table 1).

Table 1. Pixel-based coral reef benthic mapping publications.

Authors	Sensor/Spatial Resolution	Pixel or Object-Based	Classification Algorithm(s)	Supervised or Unsupervised	Number of Benthic Habitat Classes	Accuracy (Overall Accuracy)
[39]	WorldView-3 (1.2 m).	Pixel-based.	SVM.	Supervised.	15	79%
[37]	QuickBird (2.4 m).	Pixel-based.	MLC, MD, *k*-NN, Parallelepiped classification (PP), and Fisher (F); then ensemble. Classification using Majority Voting (MV), Simple Averaging (SA), and Mode Combination (MC).	Supervised.	4	55% (MLC), 53% (MD), 54% (KNN), 41% (PP), 47% (F), 83% (MV), 71% (SA), and 68% (MC).
[8]	WorldView-2 (1.9 m).	Pixel-based.	MLC, RF.	Supervised.	2–26	74.01–22.15% (mean MLC), 95.97–76.83% (mean RF).
[40]	PlanetScope (3.7 m).	Pixel-based.	RF.	Supervised.	4	78% (Cemara Islands based on 500 trees), 61% (Gelang Island based on 500 trees); 79% (Cemara Islands based using Log and Entropy function), 61% (Geland Island using Square Root and Gini function).
[41]	Landsat-8 OLI (30 m).	Pixel-based.	Linear Discriminant Analysis (LDA).	Supervised.	4	80% (Palmyara Atoll), 79% (Kingman Reef), 69% (Howland Island), 71% (Baker Island Atoll), and 74% (Combined).
[42]	Planet Dove (4.7 m).	Pixel-based.	ISODATA classification.	Unsupervised.	8	63%
[38]	Landsat-8 (30 m).	Pixel-based.	MLC, SVM, ANN.	Supervised.	4	Lizard Island; 72% (ANN), 67% SVM, 67% MLC; Qeshm and Larak Islands; 58% (ANN), 68% (SVM), 66% (MLC).
[43]	IKONOS (4 m).	Pixel-based.	MLC.	Supervised.	6	82%
[10]	QuickBird-2 (2.4 m) local.	Pixel-based.	MDM.	Supervised.	21	Suva site: 69% (photo transect), 65% (spot check).
[44]	IKONOS (4 m).	Pixel-based.	MLC.	Supervised.	9	89% (Bawe) and 80% (Chumbe).
[45]	QuickBird (2.4 m).	Pixel-based.	MLC.	Supervised.	6	67% (no water column correction), 89% (with water column correction.

Table 1. *Cont.*

Authors	Sensor/Spatial Resolution	Pixel or Object-Based	Classification Algorithm(s)	Supervised or Unsupervised	Number of Benthic Habitat Classes	Accuracy (Overall Accuracy)
[46]	Landsat-7 ETM+ (30 m).	Pixel-based.	Ensemble of hybrid SVM Classifiers.	Supervised.	5	89%
[47]	QuickBird-2 (2.4 m), Landsat 5 TM (30 m).	Pixel-based.	MDM.	Supervised.	10–21	25–62%
[48]	Landsat 5 TM (30 m).	Pixel-based.	MLC.	Supervised.	7	76%
[7]	IKONOS (4 m).	Pixel-based.	MLC.	Supervised.	4, 8, 13	75% (4 classes), ~65% (8 classes), and 50% (13 classes).
[49]	Landsat TM (30 m).	Pixel-based.	MLC followed by contextual editing.	Supervised.	4, 8, and 13	~60%, ~40%, and ~25%, respectively.
[4]	Landsat TM (30 m).	Pixel-based.	MLC.	Supervised.	4, 6, 9	~57% (4 classes), ~53% (6 classes), and ~50% (9 classes).

For both Tables 1 and 2, in publications that compared multiple different sensors, only the sensors with the highest overall are in this table, since this is the most common metric reported in the majority of these publications. Furthermore, we aimed to make overall accuracies as consistent as possible between publications, which was difficult since many compared OA of different sensors, spatial resolution, classification algorithms etc., therefore, as a general rule we aimed to report the highest OA from each publication. This resulted in some including more information than others, which is a shortcoming of this review. Only the overall accuracies from the IKONOS sensor used in Mumby and Edwards. Ref. [7] are included since it was thought to be the most relevant to that particular publication even though it was not the highest overall accuracy achieved. While Andréfouët et al. [5] mapped multiple sites, the overall accuracies reported in this table are derived from their abstract, which is thought to be the average overall accuracy reported based on the number of classes mapped. When more than two accuracy metrics were used, only the overall accuracies were included.

In relation to spatio-temporal generalisation, only one publication in Table 1 demonstrated spatial generalisation [41] and none demonstrated temporal generalisation. Gapper et al. [41] used a pixel-based Linear Discriminant Analysis (LDA) classification algorithm applied to mapping two coral reef benthic habitat classes (coral and algae/sand) and two background classes (land/cloud and deep water) at four different coral reef sites using Landsat-8 multispectral imagery with a spatial resolution of 30 m. Their results showed the pixel-based LDA achieved an overall accuracy of 80% when trained and tested on the same reef, and 79%, 69%, and 71% when tested on three different reefs that the LDA had not been trained on; therefore, demonstrating spatial generalisation.

Although moderate to high overall accuracies can be achieved using pixel-based machine-learning algorithms, there are three inherent problems that affect pixel-based classification. The first is the problem of a mixed pixel, whereby a single pixel may contain multiple benthic classes, as illustrated in Figure 2a. The mixed pixel problem is a serious challenge for mapping coral reef benthic composition because of the spatial heterogeneity inherent in coral reef benthic habitats. The second problem is known as the 'salt and pepper effect' whereby single pixels are classified differently to neighbouring pixels surrounding them, since no information from the neighbouring pixels is considered during per-pixel classification [50]. The third problem is pixel redundancy resulting from the spatial resolution being much finer than the target objects as shown in Figure 2c, where a large number of pixels represent the same target feature [51,52]. Pixel redundancy is becoming more pronounced given the increasingly higher spatial resolutions of satellite imagery (<5 m).

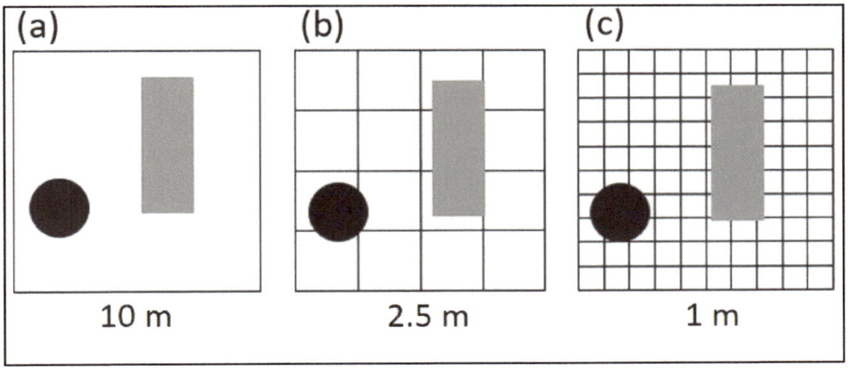

Figure 2. Illustration depicting the effect of spatial resolution on target objects. (**a**) At 10 m pixel resolution it is evident that pixel resolution is much lower (meaning the pixel is larger) than the objects resulting in a mixed class pixel. (**b**) At 2.5 m resolution the objects are closer to actual pixel resolution and are clearly separated meaning this 2.5 m resolution would not result in the mixed pixel problem and pixel redundancy would not be an issue. (**c**) At 1 m resolution objects are larger than pixels resulting in pixel redundancy. Illustration is based on Figure 1 in Blaschke [51].

Object-Based Image Analysis

An alternative approach to pixel-based classification is Object-based Image Analysis (OBIA), which is a segmentation approach that is not as severely affected by mixed-pixels, salt and pepper effects, and pixel redundancy. OBIA has also been referred to as Geographic Object Based Image Analysis (GEOBIA) [53] and Object-oriented Image Analysis [54]. The foundation for OBIA is image segmentation that dates back to the 1970s [55,56]. OBIA considers contextual information, which, in addition to spectral information, includes spatial dimensions of shape and compactness in order to generate relatively homogenous segments of grouped pixels that are semantically significant [51] (Figure 3b,c). Compared to individual pixels, these segmented objects have the advantage of containing much more spectral information such as the mean, median, minimum, and maximum spectral values per band in addition to mean ratios and variance [51].

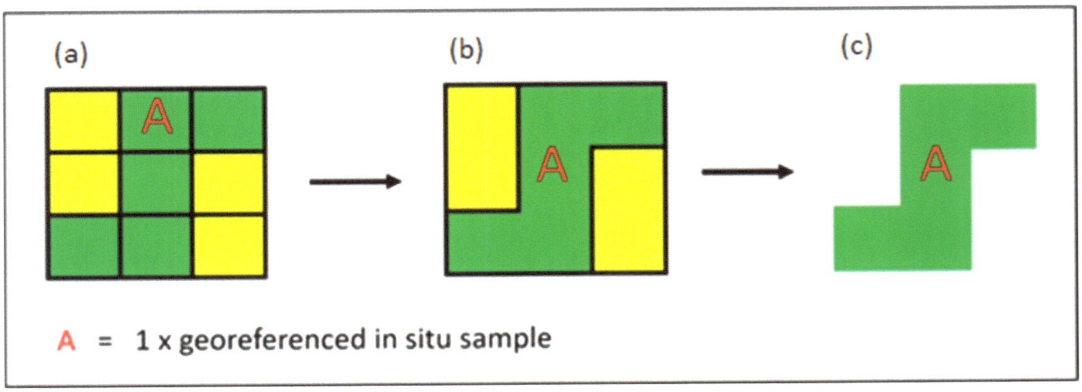

Figure 3. Example of object-based segmentation applied to an illustration depicting pixels. (**a**) Consists of 5 × green pixels and 4 × yellow pixels, (**b**) shows the result of object-based segmentation where all 5 × green pixels have been grouped into 1 × green object and 2 × yellow objects. The red A represents a georeferenced in situ reference sample which in (**a**) represents 1 × pixel, however, after OBIA segmentation in (**b**,**c**) it now represents 1 × green object consisting of 5 × grouped pixels.

Segmenting individual pixels into objects can be done using specialised software such as Trimble eCognition or ENVI [6,9,15–18,20,21,57–64]. Other approaches to generating segmented objects include unsupervised ISODATA clustering [65,66], unsupervised *k*-means cluster analysis [67], bag of features [68], seed pixel regional growing technique [69], texture analysis [70], image patches [33], and manual polygon digitisation [71,72].

It is important to note that OBIA using specialised software such as Trimble eCognition requires a user to determine a scale parameter. This scale parameter determines the output object size, which is difficult to identify since semantically significant regions are found at different scales [73]. While there have been techniques developed to objectively identify the optimal scale parameter, such as the estimation of scale parameter tool (ESP) developed by Dragut et al. [74], most coral reef benthic mapping publications that use OBIA determine the scale parameter based on subjective trial and error [6,21,59].

In the field of coral reef benthic composition mapping, OBIA has been applied on local [6,9,16,20,21,57,60–64,71] and regional scales [15,17,18,59]. In order to classify objects derived from OBIA segmentation, expert class assignment, whereby the map producer assigns classes to objects manually, has been shown to be an effective method for mapping coral reef benthic habitat classes on regional scales [15]. The limitations associated with expert class assignment, however, are the subjective nature of the map producer's classification ability, the limitation in the number of objects a map producer can manually label, and the difficulty in replicating this same classification with a new map producer. These issues therefore limit the spatio-temporal scalability of this approach. Phinn et al. [9] demonstrated how OBIA followed by expert-driven membership rulesets, which are used to assign classes to objects, can achieve overall accuracies of up to 78% for mapping 13 benthic habitat classes using Quickbird-2 multispectral satellite imagery with a spatial resolution of 2.4 m. OBIA followed by expert-driven membership rulesets is the most commonly used approach in Table 2, which is used by 23% of publications. Using expert-driven membership rulesets for classification, however, is also relatively subjective. For example, Phinn et al. [9] developed a total of 36 membership rules (based on variables such as brightness, standard deviation, blue/green band ratio, and others) with each rule containing an individual threshold that is iteratively determined by comparing the resultant segmented objects with expert knowledge of the reef, image interpretation, and references to in situ reference data in order to label reef scale classes (i.e., land, deep water, and shallow reef), geomorphic scale classes (i.e., outer reef flat and shallow lagoon), and benthic community scale classes (i.e., algae, seagrass, sand, and rock) [9].

Apart from expert class assignment or developing expert driven rulesets in the field of coral reef benthic composition mapping, one of the most commonly used machine-learning algorithms applied to classifying segmented objects derived from OBIA is the *k*-NN algorithm that is used by 12% of the publications in Table 2. The *k*-NN algorithm classifies segmented objects based on the class most represented by their *k* nearest neighbours [75,76]. *K* is a user-defined parameter that is the number of nearest neighbouring objects that are included in the majority voting process. When *k* is equal to one it is referred to as the Nearest Neighbour (NN) [21] and simply classifies an object based on the class of its nearest neighbour. The choice of *k* affects accuracy, as illustrated in Figure 4, whereby a value of $k = 1$ would result in *x* being classified as algae while a value of $k = 5$ would classify *x* as coral.

The accuracy of the *k*-NN algorithm is also affected (when not detected and separated) by the presence of class outliers [77]. Class outliers will be defined here as objects with classes other than their own surrounding them. These outliers can be the result of an insufficient number of training samples associated with the outlier class or skewed class distributions. Datasets used to train machine-learning algorithms for classifying coral reef benthic composition may be prone to containing class outliers since coral reef benthic classes are typically heterogenous in spatial distribution. Further, the logistical complexities of acquiring in situ reference samples from a coral reef location limits the quantity of samples that can be acquired.

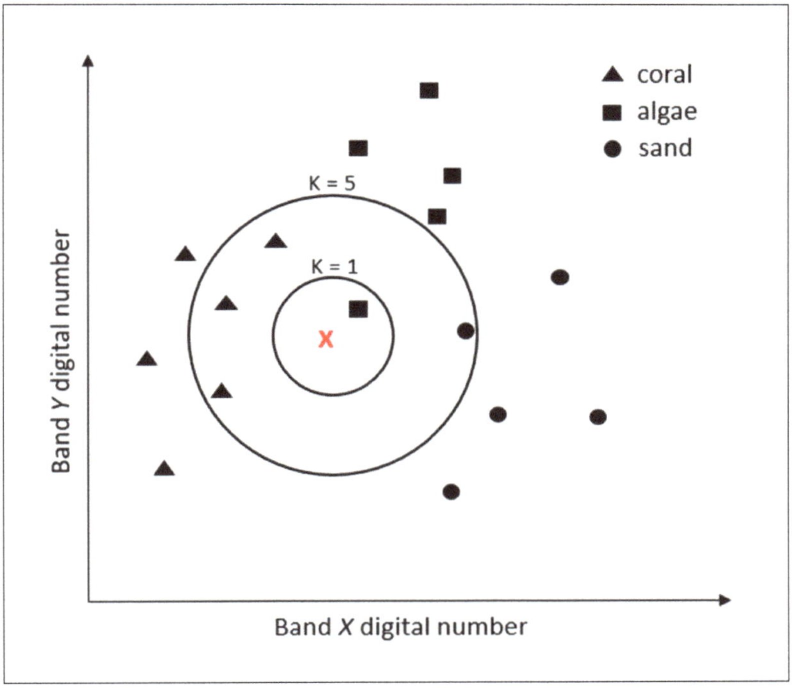

Figure 4. Basic conceptualisation of k nearest neighbour classification illustrating 3 classes. Circles show the class when $k = 1$ and when $k = 5$.

To address the fact that coral reef benthic habitat complexity varies across different coral reef geomorphologic zones [5], hierarchical OBIA segmentation approaches were developed by [9,15,21], which first segments a reef into different geomorphic zones and then re-segments each of these geomorphic zones individually in order to generate segments representing benthic habitat classes within each. Roelfsema et al. [21] compared a k-NN classification algorithm to the expert driven membership rulesets classification approach of Phinn et al. [9], applied to the same reef using the same in situ reference data and Quickbird-2 multispectral imagery. The results showed the object-based k-NN achieved an overall accuracy of 62% based on seven benthic classes, while the membership rulesets classification approach achieved an overall accuracy of 78% based on eleven benthic classes (Table 2). Although the overall accuracy of the object-based k-NN is lower, in addition to being less subjective, Roelfsema et al. [21] estimated that their hierarchical object-based k-NN approach is around twenty times faster to develop compared to Phinn et al. [9].

Another of the most commonly used object-based machine-learning classification algorithms applied to mapping coral reef benthic composition is the SVM algorithm, which is used in 17% of publications in Table 2. The modern formulation of an SVM, developed by Vapnik and Cortes [78], classify inputs by identifying decision boundaries that split input data points into two spaces that correlate to two different classes (Figure 5). This is done by mapping the input data to a new high-dimensional representation where the decision boundary (expressed as a hyperplane) is computed by maximising the distance between the decision boundary and the nearest data points from each class [79]. In order to find good decision boundaries, rather than explicitly computing the coordinates of the points in the new representation space, which is often computationally intractable, a kernel function is used. This maps two points in the initial representation space to the distance between these points in the target representation space, which is more efficient [79]

(Figure 5). Object-based SVM have been proven to achieve higher overall accuracies when directly compared to *k*-NN [63,68,72] (Table 2).

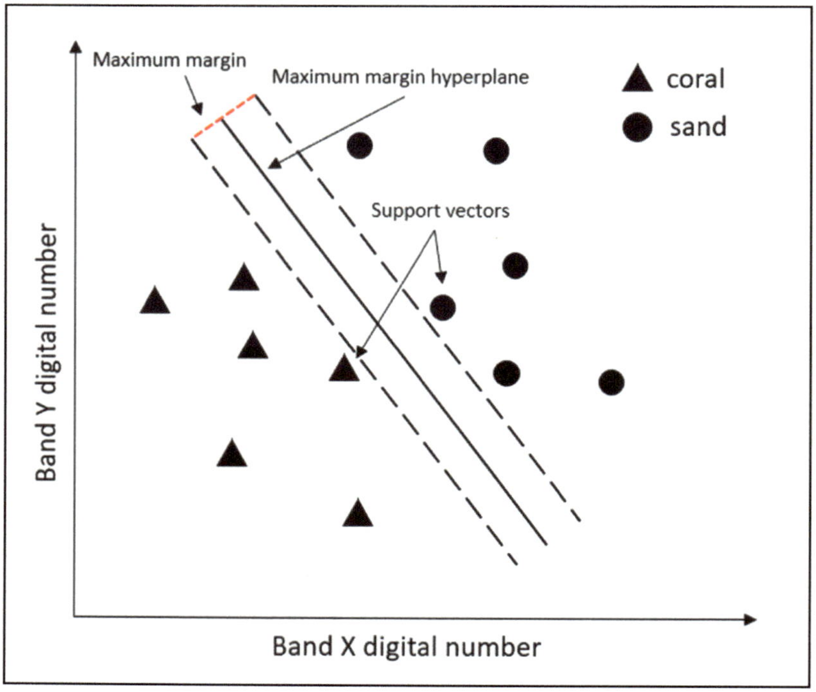

Figure 5. General conceptualisation of an SVM.

Table 2. Object-based coral reef benthic mapping publications.

Authors	Sensor/Spatial Resolution	Pixel or Object-Based	Classification Algorithm(s)	Supervised or Unsupervised	Number of Benthic Habitat Classes	Accuracy (Overall Accuracy)
[58]	Sentinel-2 (10 m).	Object-based.	OBIA–RF followed by expert-driven membership rulesets.	Supervised.	4	62%
[70]	Sentinel-2 (10 m).	Object-based.	Mean Texture Analysis followed by either RF or SVM.	Supervised.	4	71% (RF, highest), 73 (SVM, highest).
[59]	Worldview-2 (1.9 m), Planet Dove (5 m), Sentinel-2 (10 m), Landsat-8 (15 m).	Pixel-based and object-based.	Pixel-based RF and OBIA–RF followed by expert-driven membership rulesets.	Supervised.	8	78% (mean).

Table 2. *Cont.*

Authors	Sensor/Spatial Resolution	Pixel or Object-Based	Classification Algorithm(s)	Supervised or Unsupervised	Number of Benthic Habitat Classes	Accuracy (Overall Accuracy)
[17]	Landsat OLI (15 m).	Object-based.	OBIA followed by expert-driven membership rulesets.	Supervised.	6	50%
[31]	WorldView-2 (1.9 m), Sentinel-2 (10 m).	Object-based.	LAPDANN.	Supervised.	10	86% (trained/tested on same reef), 47% (trained/tested on data from Indian Ocean and Pacific Ocean simultaneously).
[32]	WorldView-2 (1.9 m), PlanetScope (3.7 m).	Object-based and pixel-based.	FCN–KNN, VGG16–FCN, DeepLab, SharpMask.	Supervised.	9	WorldView-2 imagery: 84% (FCN–KNN), 80% (VGG16–FCN), 81% (DeepLab), 80% (SharpMask); PlanetScope imagery: 73% (FCN–KNN), 73% (VGG16–FCN), 73% (DeepLab), 71% (SharpMask); Generalisation tests: 85% (FCN–KNN), 83% (VGG16–FCN), 78% (DeepLab), 82% (SharpMask).
[80]	WorldView-2 (1.9 m), Gaofen-2 (3.2 m).	Object-based.	CNN–SVM, CNN–RF, CNN, RF, SVM.	Supervised.	4	WorldView-2 data set 1: 92% (CNN–SVM), 91% (CNN–RF), 91% (CNN), 90% (RF), 89% (SVM); WorldView-2 data set 2: 86% (CNN_SVM), 85% (CNN–RF), 85% (CNN), 82% (RF), 84% (SVM) Gaofen-2 data set: 91% (CNN–SVM), 88% (CNN–RF), 89% (CNN), 87% (RF), 88% (SVM).
[33]	QuickBird (0.6 m) (benthic), GeoEye-1 (0.5 m) (seagrass).	Object-based.	CNN.	Supervised.	7 benthic, 4 seagrass.	90% (benthic), 91% (seagrass).
[67]	Sentinel-2 (10 m).	Object-based.	MD followed by post-classification filtering.	Supervised.	17 (incl. 5 non-coral reef benthic classes (i.e., mangroves, beach).	77%

Table 2. *Cont.*

Authors	Sensor/Spatial Resolution	Pixel or Object-Based	Classification Algorithm(s)	Supervised or Unsupervised	Number of Benthic Habitat Classes	Accuracy (Overall Accuracy)
[72]	WorldView-2 (1.9 m).	Object-based.	MLC, Neural Network (NN), SVM.	Supervised.	5	86% (MLC), 87%(NN), 93% (SVM).
[15]	WorldView-2 (1.9–2.4 m).	Object-based.	OBIA followed by manual class assignment.	Expert-derived.	Atlantic sites: 7 (aggregated Benthic cover type and geomorphology classes (i.e., Fore Reef Sediment with Algae), Non-Atlantic sites: 16.	81% (Atlantic sites), 90% (non-Atlantic sites).
[60]	WorldView-2 (1.9 m).	Object-based.	OBIA–RF, OBIA–Classification Tree Analysis (OBIA–CTA), OBIA–SVM.	Supervised.	14	89% (RF), 78% (CTA), 76% (SVM).
[61]	Planet Dove (3 m).	Object-based.	OBIA–KNN.	Supervised.	11	82%
[57]	GeoEye-1 (2 m).	Object-based.	OBIA and Jeffries–Matusita distance measure.	Supervised.	175	72%
[21]	QuickBird-2 (2.4 m).	Object-based.	OBIA–KNN.	Supervised.	7	62%
[68]	QuickBird (2.4 m).	Object-based.	Bag of Features (BOF) followed by either Bagging (BAG), KNN, or SVM then lastly a Weighted Majority Voting (WMV).	Supervised.	4	80% (BAG), 81% (KNN), 86% (SVM), 89% (WMV).
[71]	WorldView-2 (1.9 m).	Object-based.	SVM.	Supervised.	5	78%
[16]	Sentinel-2 (10 m).	Object-based.	OBIA with expert-driven membership rulesets.	Supervised.	6	49%
[19]	Landsat 8 (15 m).	Object-based.	OBIA with expert-driven membership rulesets.	Supervised.	5	33%

Table 2. *Cont.*

Authors	Sensor/Spatial Resolution	Pixel or Object-Based	Classification Algorithm(s)	Supervised or Unsupervised	Number of Benthic Habitat Classes	Accuracy (Overall Accuracy)
[69]	Landsat 7 ETM+ (30 m), Landsat 8 (30 m).	Object-based.	Seed pixel regional growing.	Supervised.	3 coral reef benthic and 2 non-benthic (i.e., land and human habitats).	75–99.7% based on 10 sites.
[20]	WorldView-2 (1.9 m).	Object-based.	OBIA with expert-driven membership rulesets.	Supervised.	4	76%
[62]	WorldView-2 (1.9 m).	Object-based.	OBIA-multinomial logistic discrete choice models.	Supervised.	8 benthic and 3 non-benthic (i.e., terrestrial vegetation).	85% (Vanua Vatu site).
[63]	Landsat 8 OLI (30 m).	Object-based.	OBIA–SVM, OBIA–RT, OBIA–DT, OBIA–KNN, OBIA–Bayesian.	Supervised.	7	73% (OBIA–SVM), 68% (OBIA–RT), 67% (OBIA–KNN), 66% (OBIA–Bayesian), and 56% (OBIA–DT).
[18]	QuickBird-2 (2.4 m), IKONOS (4 m).	Object-based.	OBIA with expert-driven membership rulesets.	Supervised.	14–17 (individual reefs), 20–30 (reef systems).	52–75%.
[9]	QuickBird-2 (2.4 m).	Pixel-based and object based.	OBIA with expert-driven membership rulesets; pixel-based MDM.	Supervised.	Heron Reef: 13 Ngderack Reef: 11 Navakavu Reef: 17.	78% (Heron Reef, object-based), 52% (Ngderack Reef, object-based), 65%, 57% (Navakavu Reef, object-based and pixel-based, respectively).
[64]	QuickBird-2 (2.4 m), IKONOS (4 m).	Object-based.	OBIA with expert-driven membership rulesets.	Supervised	22 benthic and 3 non-benthic (i.e., cloud).	67%.
[6]	QuickBird (0.6 m Pan-sharpened).	Pixel-based and object-based.	Pixel-based MLC and contextual editing; OBIA–NN.	Supervised.	5, 7, and 11.	59–77% (MLC), 61–76% (contextual editing), and 81–90% (OBIA–NN).
[66]	Landsat TM (30 m).	Object-based.	Unsupervised ISODATA Classification.	Unsupervised.	7	74%
[5]	IKONOS (4 m), Landsat 7 ETM+ (30 m).	Object-based (unsupervised segments and ground-truthed polygons).	Unsupervised segmentation followed by expert class assignment (applied to 2 reefs); MLC (applied to 7 reefs).	Unsupervised (2 reefs) and supervised (7 reefs).	4–5, 7–8, 9–11, >13.	77% (4–5 classes), 71% (7–8 classes), 65% (9–11 classes), and 53% (> 13 classes).

Table 2. *Cont.*

Authors	Sensor/Spatial Resolution	Pixel or Object-Based	Classification Algorithm(s)	Supervised or Unsupervised	Number of Benthic Habitat Classes	Accuracy (Overall Accuracy)
[65]	IKONOS (4 m).	Object-based.	MLC.	Supervised.	5	90% (Half Moon Bay), 89% (Tabyana Bay).

Unsupervised ISODATA clustering [65,66], unsupervised K-means cluster analysis [67], bag of features [68], seed pixel regional growing technique [69], image patches [33], manual polygon digitisation [71], and digitising regions of interest [57], have been grouped into the 'object-based' even though these polygons are not derived by utilizing a segmentation algorithm such as those derived from OBIA (eCognition for example). Additionally, the multi-scale feature extraction approach using bidimensional empirical mode decomposition (BEMD) [80] has also been grouped into 'object-based.' Publications that included both pixel-based and object-based classifications are included in Table 2 only, rather than Table 1 as well.

Second to the object-based SVM, and equal with the *k*-NN algorithm in terms of being used in Table 2, is the RF algorithm that is used by 12% of publications. RFs construct a wide variety of uncorrelated decision trees during training, with each individual tree consisting of a random sample (bootstrapped) of the training data, and then takes the aggregate (mode) of these individual trees as the output classification [81,82] (Figure 6). RFs are less prone to overfitting compared to individual decision trees [82] and they use the 'stochastic discrimination' method [83], which has been proven to reduce overfitting and a lack of generalisation properties [84–86]. These properties lead to RF algorithms gaining accuracy as they become bigger and more complex [81,83].

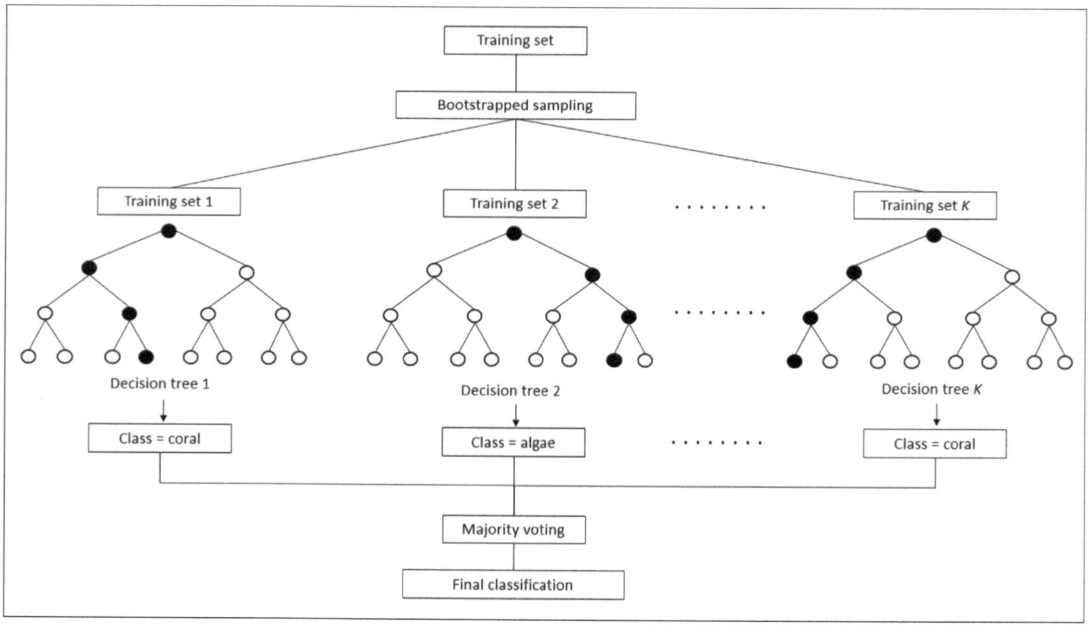

Figure 6. Basic conceptualization of random forest classification.

In Table 2 there are two publications that have directly compared object-based RF to object-based SVM algorithms. One found the object-based RF algorithms to be inferior in terms of overall accuracy compared to an object-based SVM [70], while the other found the object-based RF to be superior [40] (Table 2). In relation to spatio-temporal generalisation, Lyons et al. [59], as part of the Allen Coral Atlas project that aims to map shallow water coral reefs on a global scale, adopted a similar hierarchical mapping approach to Roelfsema et al. [21] in order to map fourteen geomorphic classes and eight benthic habitat classes on individual (~200 km^2 of reef) to regional scales (~200,000 km^2 of reef). A key difference, however, is that Lyons et al. [59] used an object-based RF algorithm rather than a *k*-NN algorithm, which is used by Roelfsema et al. [21]. When applied to mapping benthic habitat classes at different scales (individual reefs and regional scales), their frameworks mean that the overall accuracy is 78%. This overall accuracy, however, is based on validation data derived from within the regions where in situ training and validation data samples are present. Although this overall accuracy has been inferred to areas that contain no in situ training or validation data, for example, when scaling up from individual reefs to regional scales such as the entire GBR region, the accuracy for such areas is uncertain. Therefore, while their framework can be transferred to new areas provided that coincident in situ reference data (or expert derived polygons) is available, it is yet to be determined if it can spatially or temporally generalise to areas where no in situ reference training or validation data are present.

2.2. Convolutional Neural Networks

In relation to land cover mapping using remotely sensed imagery, convolutional neural networks (CNNs) have emerged as new algorithmic approaches for object detection [87], image classification [88], and image segmentation [89]. CNNs are a class of deep learning algorithms, which, put generally, take an input image (Figure 7) and convolve it through a series of successive layers in order to learn a hierarchical feature set that can subsequently be used for supervised classification at the final layer of the framework (Figure 8). In relation to remotely sensed imagery, the input layer is a tensor with shape dimensions equal to the image height multiplied by image width, and the number of channels (spectral bands) For instance, Figure 7 illustrates an example input from a multispectral satellite image with dimensions of 256 pixels (image height) × 256 pixels (width) × 4 channels (red, green, blue, and near-infrared bands). The input sample starts as 262,144 individual pixels, each consisting of four spectral bands. If there are 65,536 pixels representing 'coral' (white), this then leaves 196,608 pixels that are 'non-coral' (black), which are assigned a value of 0. This particular input sample will therefore consist of 65,536 individual pixels each with four spectral values. This input sample differs from the pixel-based samples in Figure 2, which each only contain one pixel with four spectral values, and also from the OBIA sample in Figure 3, since it does not contain grouped information (i.e., mean, minimum, maximum spectral values etc.) for the object itself, although CNNs do still extract contextual information from the input sample.

The fundamental building blocks of a CNN are the convolution operations that work by applying multiple filters each consisting of weights that either learn to find local patterns such as edges and textures or have been predefined to do so. Each filter is usually 3 × 3 in size and is applied across the width, height, and depth of the input image by computing the dot product between filter weights and the input pixel values. These values are then summed with their combined value outputting a summed activation. This summed activation is subsequently transformed by applying an activation function in order to allow the subsequent use of stochastic gradient descent with backpropagation. Backpropagation computes the gradient of the loss function in order to learn and update weights and biases in order to minimize the loss function, resulting in a more accurate identification of underlying features that best represent the input. If, however, the filters have been predefined to search for specific local patterns, such as edges in the case of an edge detection filter for example, then the gradient of the loss does not need to be learnt. For CNNs, the most commonly used activation function is the Rectified Linear Unit (ReLU), which is a calculation that returns input values if they are positive (>0.0) or returns a value of zero if the input value is 0.0 or less [90,91]. The bias is a learnable value that can be thought of as a threshold applied to the activation function in order for the activation function to return values that are most relevant to extracting the underlying features of the input layer. The output of a convolution layer is an activated feature map that characterises features detected by the filters.

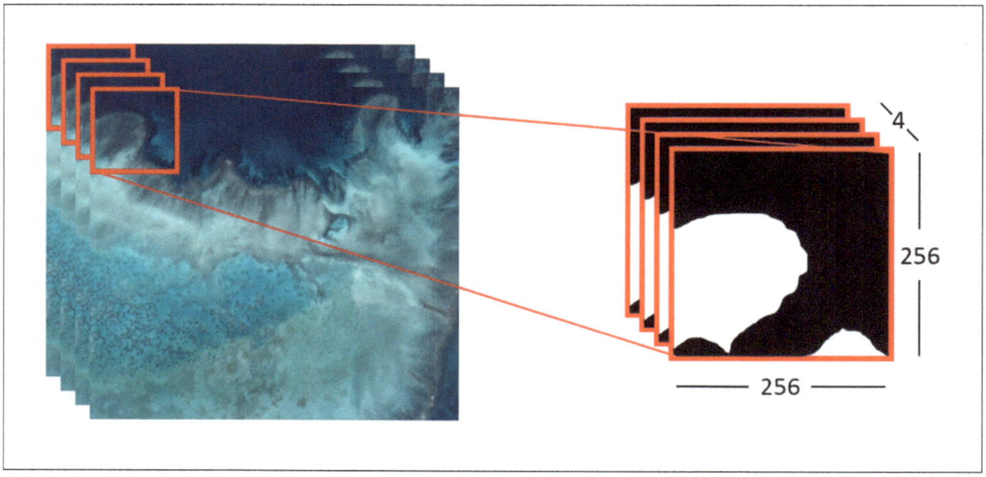

Figure 7. Example of a 256 × 256 × 4 training samples representing the class 'coral' (white) used as input to a CNN. The values 256 × 256 represent height and width in pixels while 4 represents the depth/number of bands in the multispectral image. The 256 × 256 × 4 sample is first extracted from the satellite image on the left. Next, all 'coral' pixels (white) are masked in order to assign the value of 0 to all non-coral pixels (black) from the training sample.

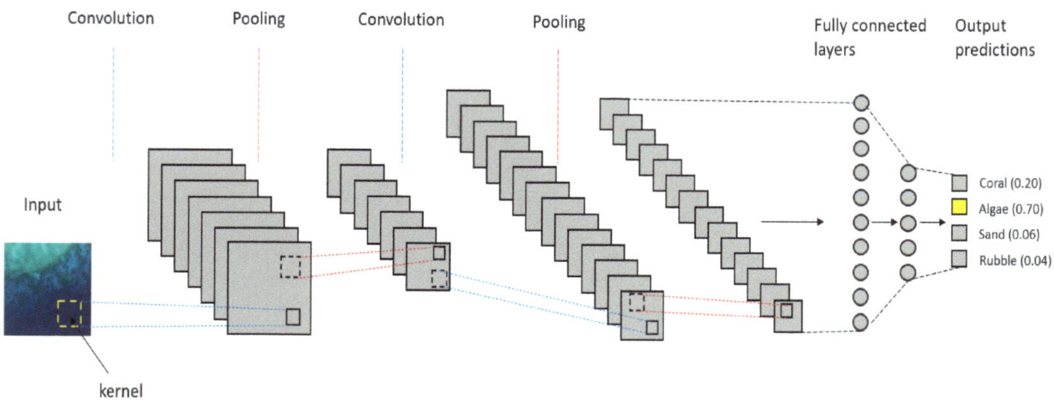

Figure 8. Illustration depicting a generic convolutional neural network consisting of convolutional, pooling, and fully connected layers.

In addition to the convolution operation, CNNs usually consist of pooling operations that are most commonly applied after a convolution operation. Pooling operations are used to reduce the dimensions (number of pixels) of the output feature map from the previous convolutional layer. For example, the max pooling operation uses a 2 × 2 filter applied across the height, width, and depth of the output feature map of the previous convolutional layer and outputs the maximum value of each channel. Unlike most convolution operations, max pooling uses a filter that is 2 × 2 in size rather than 3 × 3, and also uses a stride of two, therefore reducing the dimensionality of the feature map by a factor of two. Furthermore, max pooling filters do not learn weights like a convolutional filter. Max pooling, for example, does not have weights but instead takes the maximum value from each 2×2 filter covering the region of the feature map from the previous convolutional layer as its output. The resulting feature map after max pooling, therefore, only contains the most prominent features from the previous feature map.

A key difference between MLC, *k*-NN, SVM, and RF classification algorithms compared to CNNs is the incremental, layer by layer way in which CNNs extract increasingly complex representations from inputs and the fact that these representations are learned jointly. This means that, whenever the framework updates one of its internal features, all dependent features automatically adapt to this update without manual human intervention. This therefore allows CNNs to consist of tens to even hundreds of successive layers of representations [79]. In contrast, traditional machine-learning algorithms usually only transform the input data into one or two successive layers of representations such as decision trees in the case of RFs, or high-dimensional non-linear projections in the case of SVMs [79]. CNNs can therefore extract much more complex representations of the inputs that can then be used for supervised classification at the end of the network (Figure 8).

2.2.1. Fully Convolutional Neural Networks

While there are CNN frameworks capable of object detection and image classification, it is image segmentation frameworks that are most suited to mapping coral reef benthic composition using remotely sensed imagery. Image segmentation frameworks assign a class label based on a probability to each individual pixel. There have been a number of different deep learning-based approaches identified for image segmentation. For example, Cireşan et al. [92] developed a deep neural network (DNN) to segment neuronal membranes in electron microscopy images based on a sliding-window approach. For this, a patch (window) is placed around a pixel and a DNN is then used to classify the central pixel within each patch. Farabet et al. [93] developed a multiscale feature extraction framework for scene labelling (labelling each individual pixel based on the class it belongs to), and Pinheiro and Collobert [94] demonstrated the use of recurrent convolutional neural networks (RCNN) for scene labelling. Another approach was developed by Long et al. [95] using a fully convolutional neural network (FCN) framework for semantic segmentation; this surpassed previous approaches in terms of accuracy and learning speed. Building upon the FCN framework of Long et al. [95], Ronneberger et al. [96] developed an FCN they called U-Net (Figure 9), which was designed to work with very few training images and yields more precise segmentations.

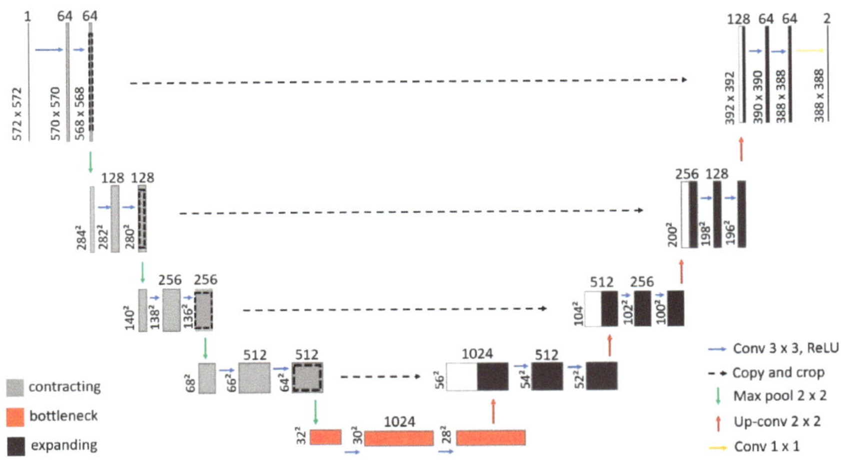

Figure 9. Illustration of U-Net based on the illustration of the U-Net framework by Ronneberger et al. [96].

Figure 9 illustrates the U-Net framework developed by Ronneberger et al. [96]. It starts with a contracting path that resembles a typical CNN framework, illustrated as contracting blocks in Figure 9, and consisting of convolutional and max pooling layers. Next, the contracting path transitions into a bottleneck consisting of two convolution layers followed by a ReLU and one up-convolution layer that initiates the expansive path. After each up-convolution the input is concatenated with a cropped version of the input from the corresponding block in the contracting path by using skip connections. The up-convolutions allow the framework to propagate contextual information to higher resolution layers, while the concatenation with corresponding contracting layers via skip connections allows for a more precise localisation [96]. At the final layer of the framework, the input passes through a 1×1 convolution with the number of output feature maps being equal to the number of classes that are being segmented. A pixel-wise softmax with a cross-entropy loss function is applied over the final feature map. Since its inception in 2015, U-Net has been used in a range of remote sensing publications [89,97–99].

2.2.2. Convolutional and Fully Convolutional Neural Networks Applied to Coral Reef Benthic Composition Mapping

Recently, a small number of publications have used CNN and FCN frameworks to map coral reef benthic composition using multispectral satellite imagery [31–33,80]. Mohamed et al. [33] used a simple CNN framework consisting of seven layers in order to classify seven coral reef benthic habitat classes located at Shiraho and four seagrass classes located at Fukido on Ishigaki, Japan, using pansharpened QuickBird (0.6 m) and GeoEye-1 (0.5 m) imagery, respectively. The inputs to their CNN framework were 1500 image patches that were each 2×2 pixels, which were placed around each correctly labelled in situ reference image location (an individual pixel) in horizontal and vertical directions. The results of their CNN frameworks showed an overall accuracy of 90% for mapping seven benthic classes at Shiraho and 91% for mapping four seagrass classes at Fukido.

In order to overcome the problem of spectral similarity between coral reef benthic classes, Wan and Ma [80] developed a CNN–SVM method that used spectral and multi-scale spatial information to map three different reef substrate classes (reef-depositional area, reef-clumping area, and submerged reef) and also 'seawater' using WorldView-2 (1.9 m) and Gaofen-2 (3.2 m) imagery covering Qilianyu Island, South China Sea. In addition to spectral and spatial information, Wan and Ma [80] use bidimensional empirical mode decomposition (BEMD) to extract multiscale information to incorporate into their CNN–SVM. The CNN is used to extract features of the four classes while the SVM is used in order to classify these features [80]. To determine the utility of their CNN–SVM framework, Wan and Ma [80] compared it to four other frameworks: SVM, RF, CNN, and CNN–RF, with the CNN–SVM achieving the highest overall accuracies (Table 2).

In relation to spatio-temporal generalisation, two publications have tested the utility of FCNs applied to coral reef benthic mapping [31,32]. Li et al. [32] developed an object-based fully convolutional neural network (FCN), which is similar to the U-Net framework. Unlike U-Net, however, for the contracting path Li et al. [32] use ResNet-50 adapted from [100], and for the expanding path a RefineNet [101,102] framework that uses residual convolution units (RCUs). Chained residual pooling (CRP) structures during the upsampling procedure allow additional context to be inferred. After initial segmentation using their FCN framework, they then used classified pixels in order to train a pixel-based k-NN algorithm based on the nearest ten points to classify coral, sediment, and seagrass. A conditional random field (CRF) on the k-NN algorithm was applied afterwards in order to reduce noise in the k-NN output and refine the boundaries.

Li et al. [32] refer to their framework as NeMO-Net and tested its capability by classifying five scenes, each with 256×256 patches covering Circa Island, Fiji, using WorldView-2 (1.9 m) imagery, and seven scenes, each with 256×256 patches using PlanetScope (3.7 m) imagery. The scenes were acquired over different dates and therefore significant spectral variations were present. NeMO-Net was compared to three other image FCN segmentation frameworks: VGG16–FCN, DeepLab, and SharpMask. The results showed that NeMO-Net with the *k*-NN applied to coral, sediments, and seagrass achieved the highest overall accuracy for both WorldView-2 and PlanetScope imagery (Table 2). In order to test if NeMO-Net is capable of spatio-temporal generalisation using WorldView-2 imagery, nine different patches each 256 × 256 covering a different island (Fulaga, Kabara, Mago, Matuka, Moala, Nayau, Totoya, Tuvuca, and Vanua Vatu) on different days were tested and compared to the three other FCN frameworks. The results showed that all four FCN frameworks are capable of spatio-temporal generalisation based on overall accuracies >77%, with NeMO-Net achieving the highest (85%).

In order to address the problem of limited training data for coral reef benthic mapping algorithms, Asanjan et al. [31] developed a framework they refer to as LAPDANN, which can use labeled samples from high-resolution WorldView-2 imagery (1.9 m) in order to accurately predict benthic classes in medium-resolution Sentinel-2 imagery (10 m); the classified Sentinel-2 imagery can then be downscaled to match the WorldView-2 spatial resolution of 1.9 m. LAPDANN consists of an improved version of a Domain Adaptation Neural Network (DANN) adapted from Ganin et al. [103] that is capable of learning from source domains and subsequently transferring learnt information to the target domain by implementing a three-part neural network consisting of: a generative network to extract domain invariant features, a U-Net framework to segment and classify input images, and a domain discriminative network to discern features from the source and target domains [31]. Next, a Laplacian Generative Adversarial Network (LAPGAN) framework proposed by Denton et al. [104] is adapted in order to downscale the segmented Sentinel-2 imagery to higher resolution (1.9 m) coral reef maps.

The results showed an overall accuracy of 86% for ten habitat classes consisting of a mix of benthic and terrestrial classes when training and testing on Peros Banhos Island in the Indian Ocean [31]. When testing the framework's ability to generalise to new geographical regions by training LAPDANN on data from the Indian Ocean and Pacific Oceans simultaneously (information on number of samples or which reefs in the Pacific Ocean are used for training are not included), the results showed an overall accuracy of 47%. Although this lower overall accuracy indicates an inability to generalise, Asanjan et al. [31] anticipate a higher overall accuracy provided more training data capturing variations in sensing conditions over multiple islands is used. They also tested the ability of LAPDANN to generalise by training on a subset of islands and then testing on an island that is new. The overall accuracy for this experiment, however, is not reported in Asanjan et al. [31].

2.3. Change Detection

2.3.1. Coral Reef Benthic Change Detection Methods

Mapping changes in coral reef benthic composition using multi-temporal satellite imagery is an inherently difficult task because atmospheric, water surface, water depth, and water clarity conditions may differ between multi-temporal satellite imagery. Compared to coral reef benthic composition mapping, there are a relatively small number of publications (Table 2). This is likely due to the lack of available in situ reference data sets that have been acquired consistently over such long timeframes. Benthic change detection publications can generally be separated into either pixel-based change detection (PBCD) or object-based change detection (OBCD).

PBCD methods, in general, first map benthic composition for each image separately by using either an unsupervised or supervised pixel-based machine-learning classification algorithm [22,23,105–111], or spectral unmixing [112]. The most used pixel-based machine-learning algorithm in PBCD is the Mahalanobis distance classification algorithm, which is used by 15% of publications in Table 3, followed by MLC and SVM, each used by 12%.

OBCD approaches on the other hand usually first generate segmented objects using either manual delineation [14,24,113] or specialized object-based software such as Trimble eCongition followed by the use of a machine-learning classification algorithm such as

MLC [105] or RF [114,115]. Aside from manual polygon delineation and class assignment, which is used by 12% of publications in Table 3, the RF algorithm is the most used object-based machine-learning algorithm in OBCD, used by 8% of publications.

Once coral reef benthic composition has been mapped within each multi-temporal image, in order to subsequently identify changes, the most common approach is to use post-classification comparison change detection (PCCCD), which overlays classified images in order to identify changes between either individual pixels (post-classification pixel-based change detection (PC-PBCD)) or class objects (post-classification object-based change detection (PC-OBCD)). Alternatively, pixel-based modelling [116], simulation [117], and statistical analysis [118,119] approaches have been used to identify benthic change. These approaches, however, are usually based on expert knowledge and may even require coincident classified benthic maps to validate the models' performance [116]. This makes them prone to the similar scalability limitations of the PBCD and OBCD approaches.

The accuracies of PC-PBCD and PC-OBCD, are dependent upon the accuracy of the initial benthic composition classification; therefore, the same primary limitation present in benthic composition mapping (the inability to spatially or temporally generalise) is also present in PC-PBCD and PC-OBCD. A further limitation with PC-PBCD or PC-OBCD is the fact that image classification occurs separately between images, thus leaving the change detection output maps prone to image misregistration errors. These errors are more pronounced in PC-PBCD compared to PC-OBCD, with PC-PBCD also being prone to the 'salt-and-pepper effect' [50,115]. However, PC- OBCD accuracy is dependent on how well the objects resulting from OBIA segmentation represent the underlying benthic classes. Therefore, the choice of an optimal scale parameter used to determine the size of objects resulting from the segmentation algorithm is very important [120].

An alternative approach to PC-OBCD, which reduces image misregistration errors, is multi-temporal image object analysis (MTOA), which simultaneously segments each image within a multi-temporal data set before using a machine-learning algorithm to predict benthic change type [115]. Zhou et al. [115] developed a multi-temporal OBCD (MT-OBCD) method that simultaneously segments images before using an RF algorithm to predict change types (i.e., '*reef sediment extension*', '*algae grow*'). To identify the optimal scale parameter, Zhou et al. [115] used the Estimate Scale of Parameter Tool that is based on the rate of local variance concept (ROC-LV). In this, the scale parameter is objectively determined by automatically increasing the local variance (LV) incrementally until the ROV-LV reaches a peak. At that point, it is considered to be the optimal scale parameter [74]. They applied this MT-OBCD method to Taiping Island, Zhongye Island, and two coral reef sites on the Barque Canada Reef in the South China Sea to predict four different coral reef benthic change types using QuickBird and World View-2 satellite imagery. They achieved an overall accuracy >90% for each site.

Table 3. Coral reef benthic change detection publications summary.

Authors	Pixel or Object-Based	Time-Series	Classification Method	Supervised or Unsupervised	Number of Classes Mapped	Change Detection Method
[121]	Pixel-based.	2015–2016	Radiometric normalization with pseudo invariant features (PIFs), multi-temporal depth invariant indices (DII), followed by SVM.	Supervised.	1 (bleached coral).	PCCCD.
[122]	Object-based.	2017–2019	Unsupervised ISODATA classification.	Unsupervised.	4	PCCCD.

Table 3. *Cont.*

Authors	Pixel or Object-Based	Time-Series	Classification Method	Supervised or Unsupervised	Number of Classes Mapped	Change Detection Method
[23]	Pixel-based.	2000–2014, 2002–2014, 2001–2015	Pixel-based-SVM.	Supervised.	2	PCCCD.
[112]	Pixel-based.	2009–2015	Spectral linear unmixing using IDL CONSTRAINED_MIN optimization algorithm followed by assigning class thresholds.	Supervised.	13	PCCCD.
[113]	Object-based.	2001–2015	Manual polygon delineation.	Supervised.	8 (habitat scenario trajectories).	PCCCD.
[114]	Object-based	2014–2016	Unsupervised IRMAD to detect areas of change, OBIA–RF to classify classes, overlaying images to perform supervised change detection.	Unsupervised and Supervised.	10 habitat classes and 5 classes of change type.	PCCCD.
[115]	Object-based (multiresolution segmentation) and pixel-based.	2013–2015	OBIA–RF change prediction, pixel-based-RF change prediction.	Supervised.	5 change types (i.e., reef sediments extension).	MT-OBCD.
[123]	Pixel-based.	1994–2014	Unsupervised Iterative self-organizing class analysis (ISOCLASS) followed by supervised reclassification based on visual interpretation.	Unsupervised and Supervised.	5	PCCCD.
[110]	Pixel-based.	2001–2014	SVM	Supervised.	11	PCCCD.
[22]	Pixel-based.	1987–2013	ISODATA clustering followed by unsupervised *k*-means classification; MLC.	Unsupervised and Supervised.	10 unsupervised, 5 supervised.	PCCCD.
[124]	Pixel-based.	2005–2008	MLC for mapping 5 classes then '*differences in reflectance values between two images within the coral classes were used to detect bleached corals.*'	Supervised.	5	PCCCD.
[24]	Object-based (manually delineated polygons).	1972–2007	Photo-interpretation based on manual polygon delineation.	Supervised.	3, 19, and 42 (based on level 1, 2, and 3 maps).	PCCCD.

Table 3. *Cont.*

Authors	Pixel or Object-Based	Time-Series	Classification Method	Supervised or Unsupervised	Number of Classes Mapped	Change Detection Method
[105]	Pixel-based and object-based.	2002–2004	Post-cyclone coral community structure maps: Photo-interpretation based on manual polygon delineation, pixel-based MLC, OBIA-MLC; Pre-cyclone community maps: *post-cyclone coral community structure classes were used to label pre-cyclone polygons based on consistent colour and texture visible on the images, and also accounting for proximity* [105].	Supervised.	20	PCCCD.
[111]	Pixel-based.	1991–2002	Parallelepiped classification.	Supervised.	6	PCCCD.
[14]	Object-based (manually delineated polygons).	1973–2007	Photo-interpretation based on manual polygon delineation.	Supervised.	15	PCCCD.
[106]	Pixel-based.	1984–2002	Mahalanobis distance classification.	Supervised.	4	PCCCD.
[116]	Object-based (timed automata model), Pixel-based (minimum distance classification).	2002–2004	A combined generic timed automata model of reef habitat trajectories and classified remotely sensed imagery based on MD classification.	Supervised.	36 (habitat classes).	PCCCD, Modelling (generic timed automata).
[119]	Pixel-based.	1990–2001	Unsupervised ISODATA classification followed by calculating the median coefficient of variation (COV). Images were then segmented by habitat to create habitat masks and also segmented by representative quadrants. The median COV for each habitat and quadrat were calculated before performing a Kruskall–Wallis nonparametric test to determine whether differences between the median COV values were significant at the 0.05 level.	Unsupervised and Supervised.	6 class habitat map, test for significant differences.	PCCCD, statistical analysis.

Table 3. *Cont.*

Authors	Pixel or Object-Based	Time-Series	Classification Method	Supervised or Unsupervised	Number of Classes Mapped	Change Detection Method
[125]	Pixel-based.	1987–2000	Multi-component change detection: image differencing to determine areas of significant change followed by MLC. Images were then '*combined*' to create a 'from-to change map.'	Supervised.	4 benthic classes each with 6 possible change types.	PCCCD.
[126]	Pixel-based.	1991–2003, 2000–2001	Unsupervised K-means clustering followed by PCA.	Unsupervised.	3	PCCCD.
[117]	Pixel-based.	1984–2000	Radiative transfer simulation and also an image normalisation method [127] followed by digital number comparison.	Supervised.	2 (radiative transfer simulation: bleached coral, healthy coral), 2 (normalisation method: slightly or non-bleached, severely bleached).	PCCCD (normalisation method), Modelling (radiative transfer simulation).
[109]	Pixel-based.	1984–2000	Mahalanobis Distance classifier.	Supervised.	4	PCCCD.
[108]	Pixel-based.	1981–2000	Mahalanobis Distance classifier.	Supervised.	4	PCCCD.
[128]	Pixel-based.	1998 (February)–1998 (August)	Image differencing based on mean (3 × 3) filtering, PCA, difference between local variation calculated as a standard deviation in a 3 × 3 neighbourhood.	Supervised.	1 (bleaching detection).	PCCCD.
[118]	Pixel-based.	1994–1996	Getis Statistic.	Supervised.	Test for significant difference.	PCCCD—Spatial autocorrelation.
[107]	Pixel-based.	1984–1999	Mahalanobis Distance classification.	Supervised.	4	PCCCD.

(Unsupervised ISODATA Clustering [22,122] and photointerpretation based on manually delineating polygons [14,24] has been grouped into 'object-based' even though these polygons are not derived by utilizing a segmentation algorithm such as those derived from OBIA. Accuracies are not reported since no consistent accuracy metric (i.e., overall accuracy) is used between all publications. Sensor/spatial resolution of imagery used is also not reported as it is in Tables 1 and 2 since it is not directly relevant in terms of identifying the change detection methodology and since accuracies are not reported (i.e., sensor/spatial resolution affects accuracies, therefore, should be included with accuracies). Pixel-based modelling [116], simulation [117] and statistical analysis [118,119] approaches have been grouped into PCCCD).

Due to the logistical complexities inherent to collecting in situ reference data from coral reef sites, there are very few consistent long-term in situ reference data sets available. Therefore, to map benthic changes over long time periods (>30 years), scientists have had to collate any available in situ reference data and imagery. This may not be consistent in terms of in situ reference data acquisition methodology, imagery type (i.e., aerial photographs and satellite imagery), image extent, and imagery acquisition dates. For example, in order to map fifty years of benthic habitat changes occurring on the outer reef flats of Grand Recif of Toliara in southwest Madagascar, Andréfouët et al. [129] used an imagery data set consisting of a combination of irregularly dated historical aerial photographs and satellite imagery from different satellite sensors (Quickbird, WorldView-2, Landsat-5, Landsat-7). In addition, there was a training and validation data set consisting of a combination of in situ reference data derived from benthic photo-transects as well as data derived from reference to hand drawn maps. Such mixed data sets make it difficult to train machine-learning algorithms to produce accurate benthic maps that can be used in PCCCD. In such instances, photointerpretation based on manual polygon delineation has been demonstrated as being a suitable alternative [14,129], although as previously mentioned, it is difficult to replicate since it is relatively subjective.

Because of the increase in high temporal and high spatial resolution satellite imagery now available to coral reef remote sensing scientists, it is now possible to ensure the consistent spatial and spectral resolution of imagery to be used for coral reef benthic change detection. For example, Planet Labs provide high spatial resolution (<5 m) multispectral satellite imagery covering the entire Earth's landmass with a daily temporal resolution. However, even with high temporal resolution satellite imagery, for a benthic change detection methodology to be relatively objective in nature compared to manual delineation, coincident in situ reference data are required. This requirement remains a primary limitation in relation to the spatio-temporal scalability of coral reef benthic change detection regardless of whether PC-PBCD, PC-OBCD, or MT-OBCD is used.

To address this limitation, Gapper et al. [23] investigated the use of an SVM algorithm applied to binary (coral and not coral) PC-PBCD and tested its spatial generalisability when applied to new reef sites that contained no site-specific in situ reference data. Four reef sites were used with an iterative classification process applied whereby each reef was first classified individually (training and test data derived from the same reef) using a site-specific SVM algorithm, then subsequently applying an SVM algorithm that has been trained on data derived from the three other reefs that they refer to as a controlled parameter cross-validation (CPCV) procedure. Site-specific overall accuracies ranged from 69% to 88% while the CPCV overall accuracies ranged from 65% to 81%. Although the CPCV overall accuracies were on average 10% lower than the overall accuracies of the site-specific SVM algorithms, and the fact this approach was tested on medium resolution (30 m) Landsat-7 and Landsat-8 imagery, which only allowed for binary (coral or not coral) PC-PBCD, Gapper et al. [23] has demonstrated the ability of an SVM to spatially generalise to new reefs, while maintaining moderate to high overall accuracies.

2.3.2. Recurrent Neural Networks Applied to Land Cover Change Detection Using Multispectral Satellite Imagery

Deep learning frameworks are also emerging as new algorithmic approaches for change detection using multispectral satellite imagery. So far though, they have not been applied to coral reef benthic change detection. When it comes to processing multi-temporal imagery, there is a subset of deep learning frameworks known as recurrent neural networks (RNNs) that are suited to dealing with sequential time series data [34,35]. A key difference between RNN frameworks compared to the PC-PBCD, PC-OBCD, and MT-OBCD methods currently used for coral reef benthic change detection is their ability to learn information relating to the difference between corresponding pixels within multi-temporal imagery (i.e., the difference between a pixel in the time-two image compared to the corresponding pixel in the time-one image) and store that learned information as a hidden state. This hidden state allows the RNN framework to detect changes in any new corresponding pixels added to the sequence (i.e., the corresponding pixel at time-three compared to the same pixel at time-two) without further training, since it still contains the stored memory of prior inputs.

While simple RNNs can suffer from the vanishing and exploding gradient problems, recurrent neural networks composed of Long Short-Term Memory units (LSTM), first introduced by Hochreiter and Schmidhuber [130], are less prone to this problem. Each LSTM unit contains a core memory cell that is regulated by three gates: an input gate, an output gate, and a forget gate (Figure 10). The core memory cell remembers information between multi-temporal data, while the input gate controls what new information is added into the cell, the forget gate controls what information is kept within the core memory cell or forgotten, and the output gate controls the information used to compute the output activation. For multi-temporal satellite imagery, recurrent connections between LSTM units at each time step allow the model to learn information relating to the difference between corresponding pixels (pixel from the time-two image and the corresponding pixel from the time-one image) that can subsequently be used for binary and/or multi-class change detection [34].

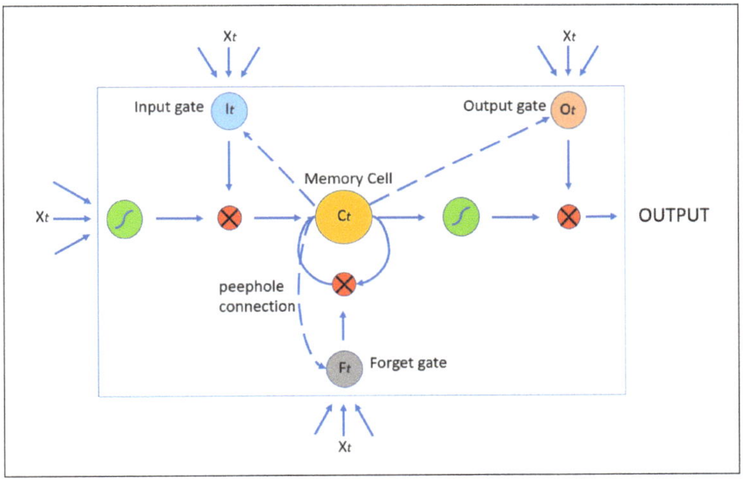

Figure 10. Basic conceptualisation of an LSTM unit where Xt represents the input vector at time t. It, Ot, Ft, and Ct represent the input gate, output gate, forget gate, and core memory cell, respectively. Blue arrows indicate directional information flow. Dotted blue arrows indicate peephole connections. This illustration is based on the illustration of an LSTM unit in Figure 5 depicted in Lyu et al. [34].

Lyu et al. [34], who were perhaps the first to use an RNN framework for the purpose of land cover change detection using multispectral satellite imagery, detected binary (changed and unchanged) and multi-class (i.e., city expansion, changed soil region, and changed water areas) changes at three different city sites in China using an LSTM framework they refer to as REFEREE (Figure 11). Looking at their REFEREE framework in Figure 11, it can be seen that first the input layer receives a 6-band pixel (pixel from a multispectral image with six spectral bands) that has been extracted from the T1 image (image at time-1). Next, the hidden layer receives this input and calculates its state information (which can be thought of as information about this particular pixel at this particular point in time) and stores this information. It is important to note that the hidden layer consists of LSTM units. The corresponding 6-band pixel from the T2 image (image at time-2) is then input to the hidden layer simultaneously, whereby change information between these two corresponding pixels can be learned by the current hidden layer. The label of '*changes*' or '*no changes*' can then be predicted in the decision layer.

Lyu et al. [34] compared their LSTM framework to four conventional change detection methods (change vector analysis (CVA), principal component analysis (PCA), iteratively reweighted multivariate alteration detection (IRMAD), and Slow Feature Analysis (SSFA)) for their binary change detection and three methods for their multi-class change detection (support vector machine (SVM), Decision Tree (DT), and a CNN). The results of their binary change detection experiments showed that their LSTM framework achieved the highest overall accuracies at three sites (the overall accuracies were all 98%), outperforming CVA (70–87%), PCA (74–90%), IRMAD (84–94%), and supervised slow feature analysis (SSFA) (95–98% (When overall accuracies are rounded to the nearest whole number of the LSTM of Lyu et al. [34], the highest is 98%, which is the same as the highest overall accuracy for SSFA, however, when rounded to the nearest one decimal place the LSTM is higher (98.4%) compared to the SSFA (97.6%)). Their multi-class change detection experiments resulted in their LSTM framework again achieving the highest overall accuracies (95–96%), outperforming CNN (92–93%), SVM (80–84%), and DT (70–71%). Lyu et al. [34] also demonstrated that their LSTM framework is capable of spatio-temporal generalisation when applied to binary change detection. This means it can transfer a learned change rule (information relating to the difference between corresponding pixels from multi-temporal imagery) to new multi-temporal imagery of a different city that the LSTM framework has not been trained on without requiring any extra learning processes. All imagery used in their transfer experiments had a similar spectral resolution (six bands) and the same spatial resolution (30 m). Their transfer experiments for binary change detection using different numbers of training samples ranging from 200–1000 resulted in overall accuracies ranging from 72–97%.

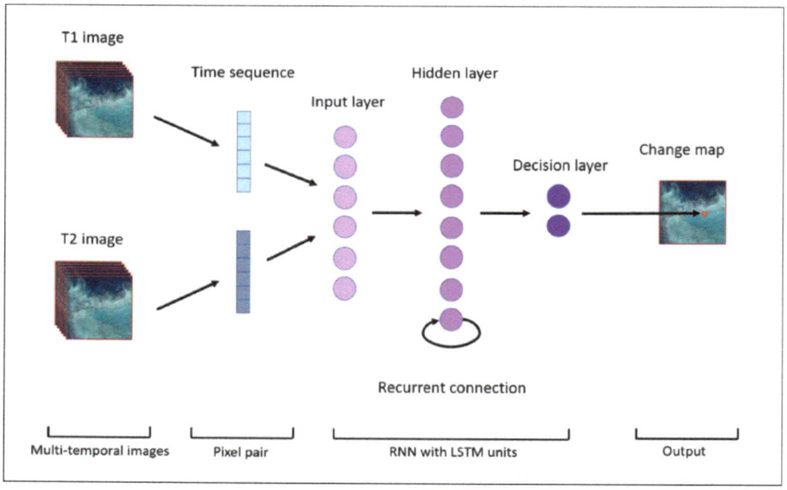

Figure 11. Illustration of the RNN-LSTM framework developed by Lyu et al. [34], which they refer to as REFEREE. The hidden layer consists of LSTM units as illustrated in Figure 10. This illustration is based on the illustration in Figure 4 depicted in Lyu et al. [34].

Another binary and multi-class land cover change detection framework was developed by Mou et al. [35] who used a recurrent convolutional neural network (ReCNN) that extracts joint spectral-spatial-temporal features from bi-temporal multispectral images. Their ReCNN combined the ability of a CNN to extract contextual features with an RNN's ability to model the temporal correlation between multi-temporal data. Their publication compared their ReCNNs' overall accuracy with the overall accuracy of six conventional land cover change detection methods (CVA, PCA, multivariate alteration detection (MAD), IRMAD, DT, and SVM). In addition, the LSTM framework referred to as REFEREE that was developed by Lyu et al. [34] was also compared, as well as three variations in the recurrent sub-network of their ReCNN framework. One had a fully connected RNN (ReCNN–FC), a second had LSTM units as the recurrent component (ReCNN–LSTM), and a third had Gated Recurrent Units (ReCNN–GRU). For both binary and multi-class change detection experiments, the overall accuracy of their ReCNN–LSTM achieved the highest overall accuracies ranging from 98–99%, which proved superior to the six conventional change detection methods (75–96%), as well as the REFEREE model of Lyu et al. [34] and the three variations in the recurrent sub-network of their ReCNN (95–99% (When the ReCNN–LSTM and ReCNN–GRU frameworks of Mou et al. [35] are rounded to the nearest whole number of 99%, they are the same. However, in Mou et al. [35] they round overall accuracies to the nearest two decimal points which shows the ReCNN–LSTM achieving the highest (98.70) compared to the ReCNN–GRU (98.64)). However, the spatio-temporal generalisability of their ReCNN framework was not tested.

3. Conclusions

3.1. Coral Reef Benthic Mapping

In relation to mapping coral reef benthic composition using multispectral satellite imagery, Tables 1 and 2 show a total of 47 papers; 17 are pixel-based and 30 are object-based. The most used pixel-based machine-learning classification algorithm is MLC, which is used by 47% of publications in Table 1, followed by RF, SVM, and MDM in second with each being used by 12%. In relation to object-based publications, aside from OBIA followed by expert driven membership rulesets (23%), SVM is the most used machine-learning classification algorithm, being used by 17%, followed by RF and *k*-NN in second place, each being used by 12%. Based on these two tables, we can conclude that object-based machine-learning classification algorithms are more commonly used compared to pixel-based. Furthermore, object-based machine-learning algorithms are not as prone to pixel-based issues mentioned in Section 2.1.

It is not possible to directly compare all machine-learning algorithms and change the detection methods covered in this review because of inconsistencies in classification schemes. These inconsistencies can be caused by a variety of differences in imagery (spectral and spatial resolutions), in situ reference data collection methodologies, the number and types of benthic classes mapped, and accuracy assessment protocols. It is clear that currently the main limitation in relation to spatio-temporal scalability is the requirement of coincident in situ reference data. It should be noted that this review focused on overall accuracy as the metric used to determine the accuracy of each machine-learning algorithm reviewed. However, the overall accuracy does not provide insight into class-specific accuracies. Therefore, further investigation is required to determine the utility of each algorithm in relation to class-specific accuracies.

In relation to spatio-temporal generalisation, only three publications in Tables 1 and 2 tested the generalisability of machine-learning algorithms [31,32,41], and one is still yet to verify the accuracy of their OBIA–RF s generalisability [59]. Based on the findings of these four publications, we believe there are two current approaches that have the potential to increase the spatio-temporal scalability of coral reef benthic mapping. The first is to use expert-derived training and validation samples [59,131]. This approach is necessary because existing pixel- and object-based machine-learning classification algorithms require coincident in situ reference data. Therefore, to increase their spatial scalability to a global scale, in situ reference data sets need to increase to a global scale as well. While [59], as part of the Allen Coral Atlas, have adopted this approach developed by Roelfsema et al. [131] on a global scale, they do not, however, report the accuracy of their object-based RF algorithm in areas outside the extent of their reference data; therefore, it remains currently unvalidated in those areas.

The second approach is to identify machine-learning algorithms capable of spatio-temporal generalisation. To this end, Gapper et al. [41] used pixel-based Linear Discriminant Analysis to map two benthic classes (coral and algae/sand). However, this was based on low resolution (30 m) Landsat-8 imagery, making it prone to pixel-based issues and was limited in terms of the number of benthic classes that could be mapped. Li et al. [32] demonstrated that na FCN framework combined with a *k*-NN classification algorithm can achieve high overall accuracies in relation to mapping three benthic classes (coral, seagrass, and sediment) using high resolution WorldView-2 (1.9 m) and PlanetScope (3.7 m) imagery and also demonstrated spatio-temporal generalisation. Furthermore, Asanjan et al. [31] developed the LAPDANN framework that can use labeled samples from high resolution WorldView-2 imagery (1.9 m) to accurately predict five benthic classes (reef crest—coralline algae ridge, fore-reef, back-reef pavement or sediment, back-reef coral framework, and seagrass meadows) in medium resolution Sentinel-2 imagery (10 m) with a high overall accuracy. This demonstrates an ability to generalise to different sensors and spatial resolutions. Based on these papers, the deep learning frameworks of Li et al. [32] and Asanjan et al. [31] appear to hold the most potential in relation to increasing the spatio-temporal scalability of coral reef benthic mapping.

3.1.1. Coral Reef Benthic Change Detection

In relation to change detection, Table 3 shows 26 papers that have mapped coral reef benthic change over time. Most (54%) of these use some form of pixel-based PCCCD. The most used pixel-based machine-learning algorithm used in PBCD is the Mahalanobis distance classification algorithm, used by 15% of publications in Table 3 followed by MLC and SVM, each used by 12%. Aside from manual polygon delineation and class assignment which is used by 12% of publications in Table 3, the RF algorithm is the most used object-based machine-learning algorithm in OBCD, used by 8% of publications. None of the 26 publications in Table 3 have demonstrated spatio-temporal generalisation. However, in this review we did identify the LSTM framework of Lyu et al. [34] that has proven superior in terms of overall accuracy compared to post-classification land-cover change detection methodologies. It has also demonstrated spatio-temporal generalisability. Coral reef benthic mapping and change detection may be a more difficult task to accomplish compared to land cover mapping and change detection because of the further complication of the water surface and water column light scattering and absorption affecting spectral reflectance. However, we believe it is reasonable to hypothesize the potential utility of the LSTM framework for binary and/or multi-class coral reef benthic change detection using multispectral satellite imagery.

3.1.2. Future Research

Given the increase in the number of publications using deep learning frameworks for the purpose of mapping and change detection using remotely sensed imagery, we acknowledge that other deep learning frameworks may also be candidates for spatio-temporal generalisation in relation to coral reef benthic mapping and change detection. However, based on examples in this review we believe there are four potential areas for further investigation relevant to increasing the spatio-temporal scalability of coral reef benthic mapping and change detection. These potential areas are to determine: (1) whether the FCN framework of Li et al. [32] (NeMO-Net) can generalise to different biogeographical regions that might exhibit different benthic compositions and spatial complexity; (2) whether the LAPDANN framework of Asanjan et al. [31] can spatially generalise when more training data-capturing variations in sensing conditions over multiple reefs is used; (3) whether the OBIA–RF of Lyons et al. [59], using the expert-derived training and validation data approach of Roelfsema et al. [131], can generalise to areas outside the extent of their reference data; and (4) the utility of the LSTM framework of Lyu et al. [34] for coral reef benthic change detection.

Author Contributions: Conceptualization, C.B., B.B. and A.N.; methodology, C.B., A.N. and B.B.; formal analysis, C.B.; data curation, C.B.; writing—original draft preparation, C.B.; writing—review and editing, C.B., A.N. and B.B. All authors have read and agreed to the published version of the manuscript.

Funding: This research received no external funding.

Acknowledgments: The authors would like to thank Akbar Ghobakhlou, Lennard Gillman, and Ashray Doshi for their advice related to this work. The authors would also like to thank the support of the AUT Doctoral Scholarship (Fees)—School of Engineering, Computer & Mathematical Sciences.

Conflicts of Interest: The authors declare no conflict of interest.

References

1. Hedley, J.D.; Roelfsema, C.M.; Chollett, I.; Harborne, A.R.; Heron, S.F.; Weeks, S.; Skirving, W.J.; Strong, A.E.; Eakin, C.M.; Christensen, T.R.L.; et al. Remote Sensing of Coral Reefs for Monitoring and Management: A Review. *Remote Sens.* **2016**, *8*, 118. [CrossRef]
2. Caras, T.; Hedley, J.; Karnieli, A. Implications of sensor design for coral reef detection: Upscaling ground hyperspectral imagery in spatial and spectral scales. *Int. J. Appl. Earth Obs. Geoinf.* **2017**, *63*, 68–77. [CrossRef]
3. Hochberg, E.J.; Atkinson, M.J.; Andréfouët, S. Spectral reflectance of coral reef bottom-types worldwide and implications for coral reef remote sensing. *Remote Sens. Environ.* **2003**, *85*, 159–173. [CrossRef]
4. Mumby, P.J.; Green, E.P.; Edwards, A.J.; Clark, C.D. Coral reef habitat mapping: How much detail can remote sensing provide? *Mar. Biol.* **1997**, *130*, 193–202. [CrossRef]
5. Andréfouët, S.; Kramer, P.; Torres-Pulliza, D.; Joyce, K.; Hochberg, E.J.; Garza-Pérez, R.; Mumby, P.J.; Riegl, B.; Yamano, H.; White, W.H.; et al. Multi-site evaluation of IKONOS data for classification of tropical coral reef environments. *Remote Sens. Environ.* **2003**, *88*, 128–143. [CrossRef]
6. Benfield, S.L.; Guzman, H.M.; Mair, J.M.; Young, J.A.T. Mapping the distribution of coral reefs and associated sublittoral habitats in Pacific Panama: A comparison of optical satellite sensors and classification methodologies. *Int. J. Remote Sens.* **2007**, *28*, 5047–5070. [CrossRef]
7. Mumby, P.J.; Edwards, A.J. Mapping marine environments with IKONOS imagery: Enhanced spatial resolution can deliver greater thematic accuracy. *Remote Sens. Environ.* **2002**, *82*, 248–257. [CrossRef]
8. Wicaksono, P.; Aryaguna, P.A. Analyses of inter-class spectral separability and classification accuracy of benthic habitat mapping using multispectral image. *Remote Sens. Appl. Soc. Environ.* **2020**, *19*, 100335. [CrossRef]
9. Phinn, S.R.; Roelfsema, C.M.; Mumby, P.J. Multi-scale, object-based image analysis for mapping geomorphic and ecological zones on coral reefs. *Int. J. Remote Sens.* **2012**, *33*, 3768–3797. [CrossRef]
10. Roelfsema, C. Integrating field data with high spatial resolution multispectral satellite imagery for calibration and validation of coral reef benthic community maps. *J. Appl. Remote Sens.* **2010**, *4*, 043527. [CrossRef]
11. Lyzenga, D.R. Remote sensing of bottom reflectance and water attenuation parameters in shallow water using aircraft and Landsat data. *Int. J. Remote Sens.* **1981**, *2*, 71–82. [CrossRef]
12. Hochberg, E.; Andrefouet, S.; Tyler, M. Sea surface correction of high spatial resolution ikonos images to improve bottom mapping in near-shore environments. *IEEE Trans. Geosci. Remote Sens.* **2003**, *41*, 1724–1729. [CrossRef]
13. Chavez, P.S. An improved dark-object subtraction technique for atmospheric scattering correction of multispectral data. *Remote Sens. Environ.* **1988**, *24*, 459–479. [CrossRef]
14. Scopélitis, J.; Andréfouët, S.; Phinn, S.; Chabanet, P.; Naim, O.; Tourrand, C.; Done, T. Changes of coral communities over 35 years: Integrating in situ and remote-sensing data on Saint-Leu Reef (la Réunion, Indian Ocean). *Estuar. Coast. Shelf Sci.* **2009**, *84*, 342–352. [CrossRef]
15. Purkis, S.J.; Gleason, A.C.R.; Purkis, C.R.; Dempsey, A.C.; Renaud, P.G.; Faisal, M.; Saul, S.; Kerr, J.M. High-resolution habitat and bathymetry maps for 65,000 sq. km of Earth's remotest coral reefs. *Coral Reefs* **2019**, *38*, 467–488. [CrossRef]
16. Hedley, J.D.; Roelfsema, C.; Brando, V.; Giardino, C.; Kutser, T.; Phinn, S.; Mumby, P.; Barrilero, O.; Laporte, J.; Koetz, B. Coral reef applications of Sentinel-2: Coverage, characteristics, bathymetry and benthic mapping with comparison to Landsat 8. *Remote Sens. Environ.* **2018**, *216*, 598–614. [CrossRef]
17. Roelfsema, C.M.; Kovacs, E.M.; Ortiz, J.C.; Callaghan, D.P.; Hock, K.; Mongin, M.; Johansen, K.; Mumby, P.J.; Wettle, M.; Ronan, M.; et al. Habitat maps to enhance monitoring and management of the Great Barrier Reef. *Coral Reefs* **2020**, *39*, 1039–1054. [CrossRef]
18. Roelfsema, C.; Phinn, S.; Jupiter, S.; Comley, J.; Albert, S. Mapping coral reefs at reef to reef-system scales, 10s–1000s km^2, using object-based image analysis. *Int. J. Remote Sens.* **2013**, *34*, 6367–6388. [CrossRef]
19. Roelfsema, C.; Kovacs, E.; Ortiz, J.C.; Wolff, N.; Callaghan, D.; Wettle, M.; Ronan, M.; Hamylton, S.; Mumby, P.; Phinn, S. Coral reef habitat mapping: A combination of object-based image analysis and ecological modelling. *Remote Sens. Environ.* **2018**, *208*, 27–41. [CrossRef]
20. Selgrath, J.C.; Roelfsema, C.; Gergel, S.E.; Vincent, A.C.J. Mapping for coral reef conservation: Comparing the value of participatory and remote sensing approaches. *Ecosphere* **2016**, *7*, e01325. [CrossRef]
21. Roelfsema, C.; Kovacs, E.; Roos, P.; Terzano, D.; Lyons, M.; Phinn, S. Use of a semi-automated object based analysis to map benthic composition, Heron Reef, Southern Great Barrier Reef. *Remote Sens. Lett.* **2018**, *9*, 324–333. [CrossRef]
22. El-Askary, H.; El-Mawla, S.H.A.; Li, J.; El-Hattab, M.; El-Raey, M. Change detection of coral reef habitat using Landsat-5 TM, Landsat 7 ETM+ and Landsat 8 OLI data in the Red Sea (Hurghada, Egypt). *Int. J. Remote Sens.* **2014**, *35*, 2327–2346. [CrossRef]

23. Gapper, J.J.; El-Askary, H.; Linstead, E.; Piechota, T. Coral Reef Change Detection in Remote Pacific Islands Using Support Vector Machine Classifiers. *Remote Sens.* **2019**, *11*, 1525. [CrossRef]
24. Scopélitis, J.; Andréfouët, S.; Phinn, S.; Done, T.; Chabanet, P. Coral colonisation of a shallow reef flat in response to rising sea level: Quantification from 35 years of remote sensing data at Heron Island, Australia. *Coral Reefs* **2011**, *30*, 951–965. [CrossRef]
25. Planet Labs. 2021. Available online: https://www.planet.com/products/monitoring/ (accessed on 10 November 2021).
26. Lecun, Y.; Bottou, L.; Bengio, Y.; Haffner, P. Gradient-based learning applied to document recognition. *Proc. IEEE* **1998**, *86*, 2278–2324. [CrossRef]
27. Krizhevsky, A.; Sutskever, I.; Hinton, G.E. ImageNet Classification with Deep Convolutional Neural Networks. In Proceedings of the Advances in Neural Information Processing Systems, Lake Tahoe, NV, USA, 3–6 December 2012; Volume 2.
28. Guirado, E.; Tabik, S.; Alcaraz-Segura, D.; Cabello, J.; Herrera, F. Deep-learning Versus OBIA for Scattered Shrub Detection with Google Earth Imagery: Ziziphus Lotus as Case Study. *Remote Sens.* **2017**, *9*, 1220. [CrossRef]
29. Liu, T.; Abd-Elrahman, A.; Morton, J.; Wilhelm, V.L. Comparing fully convolutional networks, random forest, support vector machine, and patch-based deep convolutional neural networks for object-based wetland mapping using images from small unmanned aircraft system. *GISci. Remote Sens.* **2018**, *55*, 243–264. [CrossRef]
30. Zhang, D.; Pan, Y.; Zhang, J.; Hu, T.; Zhao, J.; Li, N.; Chen, Q. A generalized approach based on convolutional neural networks for large area cropland mapping at very high resolution. *Remote Sens. Environ.* **2020**, *247*, 111912. [CrossRef]
31. Akbari Asanjan, A.; Das, K.; Li, A.; Chirayath, V.; Torres-Perez, J.; Sorooshian, S. Learning Instrument Invariant Characteristics for Generating High-Resolution Global Coral Reef Maps. In Proceedings of the ACM SIGKDD International Conference on Knowledge Discovery and Data Mining, Virtual Event, CA, USA, 23–27 August 2020.
32. Li, A.S.; Chirayath, V.; Segal-Rozenhaimer, M.; Torres-Perez, J.L.; Bergh, J.V.D. NASA NeMO-Net's Convolutional Neural Network: Mapping Marine Habitats with Spectrally Heterogeneous Remote Sensing Imagery. *IEEE J. Sel. Top. Appl. Earth Obs. Remote Sens.* **2020**, *13*, 5115–5133. [CrossRef]
33. Mohamed, H.; Nadaoka, K.; Nakamura, T. Semiautomated Mapping of Benthic Habitats and Seagrass Species Using a Convolutional Neural Network Framework in Shallow Water Environments. *Remote Sens.* **2020**, *12*, 4002. [CrossRef]
34. Lyu, H.; Lu, H.; Mou, L. Learning a Transferable Change Rule from a Recurrent Neural Network for Land Cover Change Detection. *Remote Sens.* **2016**, *8*, 506. [CrossRef]
35. Mou, L.; Bruzzone, L.; Zhu, X.X. Learning Spectral-Spatial-Temporal Features via a Recurrent Convolutional Neural Network for Change Detection in Multispectral Imagery. *IEEE Trans. Geosci. Remote Sens.* **2019**, *57*, 924–935. [CrossRef]
36. Goodman, J.A.; Purkis, S.; Phinn, S.R. *Coral Reef Remote Sensing. A Guide for Mapping, Monitoring and Management*; Springer: Berlin/Heidelberg, Germany, 2013; Volume 1.
37. Hossain, M.S.; Muslim, A.M.; Nadzri, M.I.; Teruhisa, K.; David, D.; Khalil, I.; Mohamad, Z. Can ensemble techniques improve coral reef habitat classification accuracy using multispectral data? *Geocarto Int.* **2020**, *35*, 1214–1232. [CrossRef]
38. Chegoonian, A.M.; Mokhtarzade, M.; Zoej, M.J.V. A comprehensive evaluation of classification algorithms for coral reef habitat mapping: Challenges related to quantity, quality, and impurity of training samples. *Int. J. Remote Sens.* **2017**, *38*, 4224–4243. [CrossRef]
39. da Silveira, C.B.L.; Strenzel, G.M.R.; Maida, M.; Gaspar, A.L.B.; Ferreira, B.P. Coral Reef Mapping with Remote Sensing and Machine Learning: A Nurture and Nature Analysis in Marine Protected Areas. *Remote Sens.* **2021**, *13*, 2907. [CrossRef]
40. Wicaksono, P.; Lazuardi, W. Random Forest Classification Scenarios for Benthic Habitat Mapping using Planetscope Image. In Proceedings of the International Geoscience and Remote Sensing Symposium (IGARSS), Honolulu, HI, USA, 25–30 July 2019. [CrossRef]
41. Gapper, J.J.; El-Askary, H.; Linstead, E.; Piechota, T. Evaluation of Spatial Generalization Characteristics of a Robust Classifier as Applied to Coral Reef Habitats in Remote Islands of the Pacific Ocean. *Remote Sens.* **2018**, *10*, 1774. [CrossRef]
42. Asner, G.P.; Martin, R.E.; Mascaro, J. Coral reef atoll assessment in the South China Sea using Planet Dove satellites. *Remote Sens. Ecol. Conserv.* **2017**, *3*, 57–65. [CrossRef]
43. Zapata-Ramírez, P.A.; Blanchon, P.; Olioso, A.; Hernandez-Nuñez, H.; Sobrino, J.A. Accuracy of IKONOS for mapping benthic coral-reef habitats: A case study from the Puerto Morelos Reef National Park, Mexico. *Int. J. Remote Sens.* **2012**, *34*, 3671–3687. [CrossRef]
44. Knudby, A.; LeDrew, E.; Brenning, A. Predictive mapping of reef fish species richness, diversity and biomass in Zanzibar using IKONOS imagery and machine-learning techniques. *Remote Sens. Environ.* **2010**, *114*, 1230–1241. [CrossRef]
45. Nurlidiasari, M.; Budiman, S. Mapping Coral Reef Habitat with and without Water Column Correction Using Quickbird Image. *Int. J. Remote Sens. Earth Sci. (IJReSES)* **2010**, *2*, 45–56. [CrossRef]
46. Henriques, A.P.; Neto, A.D.D.; Amaral, R.F. Classification of multispectral images in coral environments using a hybrid of classifier ensembles. *Neurocomputing* **2010**, *73*, 1256–1264. [CrossRef]
47. Roelfsema, C.; Phinn, S. Evaluating eight field and remote sensing approaches for mapping the benthos of three different coral reef environments in Fiji. *Proc. Remote Sens. Inland Coast. Ocean. Waters* **2008**, *7150*, 71500. [CrossRef]
48. Purkis, S.J.; Pasterkamp, R. Integrating in situ reef-top reflectance spectra with Landsat TM imagery to aid shallow-tropical benthic habitat mapping. *Coral Reefs* **2004**, *23*, 5–20. [CrossRef]
49. Mumby, P.J.; Clark, C.D.; Green, E.P.; Edwards, A.J. Benefits of water column correction and contextual editing for mapping coral reefs. *Int. J. Remote Sens.* **1998**, *19*, 203–210. [CrossRef]

50. Blaschke, T.; Lang, S.; Lorup, E.; Strobl, J.; Zeil, P. Object-Oriented Image Processing in an Integrated GIS/Remote Sensing Environment and Perspectives for Environmental Applications. *Environ. Inf. Plan. Politics Public* **2000**, *2*, 555–570.
51. Blaschke, T. Object Based Image Analysis for Remote Sensing. *ISPRS J. Photogramm. Remote Sens.* **2010**, *65*, 2–16. [CrossRef]
52. Brodrick, P.G.; Davies, A.B.; Asner, G.P. Uncovering Ecological Patterns with Convolutional Neural Networks. *Trends Ecol. Evol.* **2019**, *34*, 734–745. [CrossRef]
53. Hay, G.J.; Castilla, G. Geographic Object-Based Image Analysis (GEOBIA): A new name for a new discipline. In *Object-Based Image Analysis*; Lecture Notes in Geoinformation and Cartography; Springer: Berlin/Heidelberg, Germany, 2008; pp. 75–89. [CrossRef]
54. Blaschke, T.; Hay, G.J. Object-Oriented Image Analysis and Scale-Space: Theory and Methods for Modeling and Evaluating Multiscale Landscape Structure. *Int. Arch. Photogramm. Remote Sens.* **2001**, *34*, 22–29.
55. Haralick, R.M.; Shanmugam, K.; Dinstein, I.H. Textural Features for Image Classification. *IEEE Trans. Syst. Man Cybern.* **1973**, *SMC-3*, 610–621. [CrossRef]
56. Kettig, R.L.; Landgrebe, D.A. Classification of Multispectral Image Data by Extraction and Classification of Homogeneous Objects. *IEEE Trans. Geosci. Electron.* **1976**, *14*, 19–26. [CrossRef]
57. Ampou, E.E.; Ouillon, S.; Andréfouët, S. Challenges in rendering Coral Triangle habitat richness in remotely sensed habitat maps: The case of Bunaken Island (Indonesia). *Mar. Pollut. Bull.* **2018**, *131*, 72–82. [CrossRef]
58. Roelfsema, C.M.; Lyons, M.B.; Castro-Sanguino, C.; Kovacs, E.M.; Callaghan, D.; Wettle, M.; Markey, K.; Borrego-Acevedo, R.; Tudman, P.; Roe, M.; et al. How Much Shallow Coral Habitat Is There on the Great Barrier Reef? *Remote Sens.* **2021**, *13*, 4343. [CrossRef]
59. Lyons, M.B.; Roelfsema, C.; Kennedy, E.; Kovacs, E.M.; Borrego-Acevedo, R.; Markey, K.; Roe, M.; Yuwono, D.M.; Harris, D.L.; Phinn, S.R.; et al. Mapping the world's coral reefs using a global multiscale earth observation framework. *Remote Sens. Ecol. Conserv.* **2020**, *6*, 557–568. [CrossRef]
60. Wicaksono, P.; Aryaguna, P.A.; Lazuardi, W. Benthic Habitat Mapping Model and Cross Validation Using Machine-Learning Classification Algorithms. *Remote Sens.* **2019**, *11*, 1279. [CrossRef]
61. Li, J.; Schill, S.R.; Knapp, D.E.; Asner, G.P. Object-Based Mapping of Coral Reef Habitats Using Planet Dove Satellites. *Remote Sens.* **2019**, *11*, 1445. [CrossRef]
62. Saul, S.; Purkis, S. Semi-Automated Object-Based Classification of Coral Reef Habitat using Discrete Choice Models. *Remote Sens.* **2015**, *7*, 15894–15916. [CrossRef]
63. Wahidin, N.; Siregar, V.P.; Nababan, B.; Jaya, I.; Wouthuyzen, S. Object-based Image Analysis for Coral Reef Benthic Habitat Mapping with Several Classification Algorithms. *Procedia Environ. Sci.* **2015**, *24*, 222–227. [CrossRef]
64. Roelfsema, C.; Phinn, S.; Jupiter, S.; Comley, J.; Beger, M.; Paterson, E. The application of object based analysis of high spatial resolution imagery for mapping large coral reef systems in the West Pacific at geomorphic and benthic community spatial scales. In Proceedings of the International Geoscience and Remote Sensing Symposium (IGARSS), Honolulu, HI, USA, 15–30 July 2010. [CrossRef]
65. Maeder, J.; Narumalani, S.; Rundquist, D.C.; Perk, R.L.; Schalles, J.; Hutchins, K.; Keck, J. Classifying and Mapping General Coral-Reef Structure Using Ikonos Data. *Photogramm. Eng. Remote Sens.* **2002**, *68*, 1297–1306.
66. Call, K.A.; Hardy, J.T.; Wallin, D.O. Coral reef habitat discrimination using multivariate spectral analysis and satellite remote sensing. *Int. J. Remote Sens.* **2003**, *24*, 2627–2639. [CrossRef]
67. Mateos-Molina, D.; Antonopoulou, M.; Baldwin, R.; Bejarano, I.; Burt, J.A.; García-Charton, J.A.; Al-Ghais, S.M.; Walgamage, J.; Taylor, O.J. Applying an integrated approach to coastal marine habitat mapping in the north-western United Arab Emirates. *Mar. Environ. Res.* **2020**, *161*, 105095. [CrossRef]
68. Mohamed, H.; Nadaoka, K.; Nakamura, T. Assessment of Machine Learning Algorithms for Automatic Benthic Cover Monitoring and Mapping Using Towed Underwater Video Camera and High-Resolution Satellite Images. *Remote Sens.* **2018**, *10*, 773. [CrossRef]
69. Hossain, M.S.; Bujang, J.S.; Zakaria, M.H.; Hashim, M. Marine and human habitat mapping for the Coral Triangle Initiative region of Sabah using Landsat and Google Earth imagery. *Mar. Policy* **2016**, *72*, 176–191. [CrossRef]
70. Lazuardi, W.; Wicaksono, P.; Marfai, M.A. Remote sensing for coral reef and seagrass cover mapping to support coastal management of small islands. *IOP Conf. Ser. Earth Environ. Sci.* **2021**, *686*, 012031. [CrossRef]
71. Kabiri, K.; Rezai, H.; Moradi, M. Mapping of the corals around Hendorabi Island (Persian Gulf), using WorldView-2 standard imagery coupled with field observations. *Mar. Pollut. Bull.* **2018**, *129*, 266–274. [CrossRef] [PubMed]
72. Xu, H.; Liu, Z.; Zhu, J.; Lu, X.; Liu, Q. Classification of Coral Reef Benthos around Ganquan Island Using WorldView-2 Satellite Imagery. *J. Coast. Res.* **2019**, *93*, 466–474. [CrossRef]
73. Arbiol, R.; Zhang, Y.; Palà, V. *Advanced Classification Techniques: A Review*; Revista Catalana de Geografia: Enschede, The Netherlands, 2006; pp. 292–296.
74. Drăguţ, L.; Tiede, D.; Levick, S.R. ESP: A tool to estimate scale parameter for multiresolution image segmentation of remotely sensed data. *Int. J. Geogr. Inf. Sci.* **2010**, *24*, 859–871. [CrossRef]
75. Fix, E.; Hodges, J. *Discriminatory Analysis Nonparametric Discrimination: Small Sample Performance*; Project No. 21-49-004, Report No. 11, Contract No. AF 41(129)-31; USAF School of Aviation: Randolph Field, TX, USA, 1952.
76. Cover, T.; Hart, P. Nearest neighbor pattern classification. *IEEE Trans. Inf. Theory* **1967**, *13*, 21–27. [CrossRef]

77. Bhattacharya, G.; Ghosh, K.; Chowdhury, A.S. KNN Classification with an Outlier Informative Distance Measure. In *Proceedings of the Lecture Notes in Computer Science (Including Subseries Lecture Notes in Artificial Intelligence and Lecture Notes in Bioinformatics)*; Springer: Berlin/Heidelberg, Germany, 2017; Volume 10597.
78. Cortes, C.; Vapnik, V. Support-vector networks. *Mach. Learn.* **1995**, *20*, 273–297. [CrossRef]
79. Chollet, F. *Deep Learning with Python*; Manning Publications Co.: Shelter Island, NY, USA, 2017.
80. Wan, J.; Ma, Y. Multi-scale Spectral-Spatial Remote Sensing Classification of Coral Reef Habitats Using CNN–SVM. *J. Coast. Res.* **2020**, *102*, 11–20. [CrossRef]
81. Breiman, L. Random forests. *Mach. Learn.* **2001**, *45*, 5–32. [CrossRef]
82. Hastie, T.; Tibshirani, R.; Friedman, J. *Springer Series in Statistics the Elements of Statistical Learning*; Springer: New York, NY, USA, 2009; Volume 27.
83. Ho, T.K. The random subspace method for constructing decision forests. *IEEE Trans. Pattern Anal. Mach. Intell.* **1998**, *20*, 832–844. [CrossRef]
84. Kleinberg, E.M. Stochastic discrimination. *Ann. Math. Artif. Intell.* **1990**, *1*, 207–239. [CrossRef]
85. Kleinberg, E.M. An overtraining-resistant stochastic modeling method for pattern recognition. *Ann. Stat.* **1996**, *24*, 2319–2349. [CrossRef]
86. Kleinberg, E.M. On the algorithmic implementation of stochastic discrimination. *IEEE Trans. Pattern Anal. Mach. Intell.* **2000**, *22*, 473–490. [CrossRef]
87. Deng, Z.; Sun, H.; Zhou, S.; Zhao, J.; Lei, L.; Zou, H. Multi-scale object detection in remote sensing imagery with convolutional neural networks. *ISPRS J. Photogramm. Remote Sens.* **2018**, *145*, 3–22. [CrossRef]
88. Castelluccio, M.; Poggi, G.; Sansone, C.; Verdoliva, L. Land Use Classification in Remote Sensing Images by Convolutional Neural Networks. *arXiv* **2015**, arXiv:1508.00092.
89. Wagner, F.H.; Sanchez, A.; Tarabalka, Y.; Lotte, R.G.; Ferreira, M.P.; Aidar, M.P.; Gloor, E.; Phillips, O.L.; Aragao, L.E. Using the U-net convolutional network to map forest types and disturbance in the Atlantic rainforest with very high resolution images. *Remote Sens. Ecol. Conserv.* **2019**, *5*, 360–375. [CrossRef]
90. Nair, V.; Hinton, G.E. Rectified Linear Units Improve Restricted Boltzmann Machines. In Proceedings of the ICML 2010—Proceedings, 27th International Conference on Machine Learning, Haifa, Israel, 21–24 June 2010.
91. Glorot, X.; Bordes, A.; Bengio, Y. Deep Sparse Rectifier Neural Networks. *Proc. J. Mach. Learn. Res.* **2011**, *15*, 315–323.
92. Cireşan, D.C.; Giusti, A.; Gambardella, L.M.; Schmidhuber, J. Deep Neural Networks Segment Neuronal Membranes in Electron Microscopy Images. In Proceedings of the Advances in Neural Information Processing Systems, Lake Tahoe, NV, USA, 3–6 December 2012; Volume 4.
93. Farabet, C.; Couprie, C.; Najman, L.; LeCun, Y. Learning Hierarchical Features for Scene Labeling. *IEEE Trans. Pattern Anal. Mach. Intell.* **2013**, *35*, 1915–1929. [CrossRef]
94. Pinheiro, P.O.; Collobert, R. Recurrent Convolutional Neural Networks for Scene Labeling. In Proceedings of the 31st International Conference on Machine Learning, ICML 2014, Beijing, China, 21–26 June 2014; Volume 1.
95. Long, J.; Shelhamer, E.; Darrell, T. Fully Convolutional Networks for Semantic Segmentation. In Proceedings of the IEEE Computer Society Conference on Computer Vision and Pattern Recognition, Boston, MA, USA, 7–12 June 2015.
96. Ronneberger, O.; Fischer, P.; Brox, T. U-Net: Convolutional Networks for Biomedical Image Segmentation. In *Proceedings of the Lecture Notes in Computer Science (Including Subseries Lecture Notes in Artificial Intelligence and Lecture Notes in Bioinformatics)*; Springer: Berlin/Heidelberg, Germany, 2015; Volume 9351.
97. Flood, N.; Watson, F.; Collett, L. Using a U-net convolutional neural network to map woody vegetation extent from high resolution satellite imagery across Queensland, Australia. *Int. J. Appl. Earth Obs. Geoinf.* **2019**, *82*, 101897. [CrossRef]
98. Stoian, A.; Poulain, V.; Inglada, J.; Poughon, V.; Derksen, D. Land Cover Maps Production with High Resolution Satellite Image Time Series and Convolutional Neural Networks: Adaptations and Limits for Operational Systems. *Remote Sens.* **2019**, *11*, 1986. [CrossRef]
99. Clark, A.; McKechnie, J. Detecting Banana Plantations in the Wet Tropics, Australia, Using Aerial Photography and U-Net. *Appl. Sci.* **2020**, *10*, 2017. [CrossRef]
100. He, K.; Zhang, X.; Ren, S.; Sun, J. Deep Residual Learning for Image Recognition. In Proceedings of the IEEE Computer Society Conference on Computer Vision and Pattern Recognition, Honolulu, HI, USA, 21–26 July 2016; Volume 2016-December.
101. Lin, G.; Milan, A.; Shen, C.; Reid, I. RefineNet: Multi-Path Refinement Networks for High-Resolution Semantic Segmentation. In Proceedings of the Proceedings—30th IEEE Conference on Computer Vision and Pattern Recognition, CVPR 2017, Honolulu, HI, USA, 21–27 July 2017; Volume 2017-January.
102. Kemker, R.; Salvaggio, C.; Kanan, C. Algorithms for semantic segmentation of multispectral remote sensing imagery using deep learning. *ISPRS J. Photogramm. Remote Sens.* **2018**, *145*, 60–77. [CrossRef]
103. Ganin, Y.; Ustinova, E.; Ajakan, H.; Germain, P.; Larochelle, H.; Laviolette, F.; Marchand, M.; Lempitsky, V. Domain-Adversarial Training of Neural Networks. *J. Mach. Learn. Res.* **2016**, *17*, 1–35. [CrossRef]
104. Denton, E.; Chintala, S.; Szlam, A.; Fergus, R. Deep Generative Image Models Using a Laplacian Pyramid of Adversarial Networks. In Proceedings of the Advances in Neural Information Processing Systems, Montreal, QC, Canada, 7–12 December 2015; Volume 2015-January.

105. Scopélitis, J.; Andréfouët, S.; Phinn, S.; Arroyo, L.; Dalleau, M.; Cros, A.; Chabanet, P. The next step in shallow coral reef monitoring: Combining remote sensing and in situ approaches. *Mar. Pollut. Bull.* **2010**, *60*, 1956–1968. [CrossRef] [PubMed]
106. Palandro, D.A.; Andréfouët, S.; Hu, C.; Hallock, P.; Müller-Karger, F.E.; Dustan, P.; Callahan, M.K.; Kranenburg, C.; Beaver, C.R. Quantification of two decades of shallow-water coral reef habitat decline in the Florida Keys National Marine Sanctuary using Landsat data (1984–2002). *Remote Sens. Environ.* **2008**, *112*, 3388–3399. [CrossRef]
107. Palandro, D.; Andrefouet, S.; Muller-Karger, F.; Dustan, P. Coral reef change detection using Landsats 5 and 7: A case study using Carysfort Reef in the Florida Keys. In Proceedings of the International Geoscience and Remote Sensing Symposium (IGARSS), Sydney, NSW, Australia, 9–13 July 2001; Volume 2, pp. 625–627. [CrossRef]
108. Palandro, D.; Andréfouët, S.; Dustan, P.; Muller-Karger, F.E. Change detection in coral reef communities using Ikonos satellite sensor imagery and historic aerial photographs. *Int. J. Remote Sens.* **2003**, *24*, 873–878. [CrossRef]
109. Palandro, D.; Andréfouët, S.; Muller, P.H.; Dustan, P.; Hu, C.; Hallock, P. Detection of changes in coral reef communities using Landsat-5 TM and Landsat-7 ETM+ data. *Can. J. Remote Sens.* **2003**, *29*, 201–209. [CrossRef]
110. Iovan, C.; Ampou, E.; Andrefouet, S.; Ouillon, S.; Gaspar, P. Change Detection of Coral Reef Habitats from Multi-Temporal and Multi-Source Satellite Imagery in Bunaken, Indonesia. In Proceedings of the 2015 8th International Workshop on the Analysis of Multitemporal Remote Sensing Images, Multi-Temp, Annecy, France, 22–24 July 2015.
111. Nurlidiasari, M.; Budiman, S. Coral Reef Habitat Changing Assessment of Derawan Islands, East Kalimantan, Using Remote Sensing Data. *Int. J. Remote Sens. Earth Sci. (IJReSES)* **2010**, *2*, 32–44. [CrossRef]
112. Bajjouk, T.; Mouquet, P.; Ropert, M.; Quod, J.-P.; Hoarau, L.; Bigot, L.; Le Dantec, N.; Delacourt, C.; Populus, J. Detection of changes in shallow coral reefs status: Towards a spatial approach using hyperspectral and multispectral data. *Ecol. Indic.* **2019**, *96*, 174–191. [CrossRef]
113. Ampou, E.E.; Ouillon, S.; Iovan, C.; Andréfouët, S. Change detection of Bunaken Island coral reefs using 15 years of very high resolution satellite images: A kaleidoscope of habitat trajectories. *Mar. Pollut. Bull.* **2018**, *131*, 83–95. [CrossRef] [PubMed]
114. Ben-Romdhane, H.; Al-Musallami, M.; Marpu, P.R.; Ouarda, T.B.M.J.; Ghedira, H. Change detection using remote sensing in a reef environment of the UAE during the extreme event of El Niño 2015–2016. *Int. J. Remote Sens.* **2018**, *39*, 6358–6382. [CrossRef]
115. Zhou, Z.; Ma, L.; Fu, T.; Zhang, G.; Yao, M.; Li, M. Change Detection in Coral Reef Environment Using High-Resolution Images: Comparison of Object-Based and Pixel-Based Paradigms. *ISPRS Int. J. Geo-Inf.* **2018**, *7*, 441. [CrossRef]
116. Scopélitis, J.; Andréfouët, S.; Largouët, C. Modelling coral reef habitat trajectories: Evaluation of an integrated timed automata and remote sensing approach. *Ecol. Model.* **2007**, *205*, 59–80. [CrossRef]
117. Yamano, H.; Tamura, M. Detection limits of coral reef bleaching by satellite remote sensing: Simulation and data analysis. *Remote Sens. Environ.* **2004**, *90*, 86–103. [CrossRef]
118. LeDrew, E.F.; Wulder, M.; Holden, H. Change Detection of Satellite Imagery for Reconnaissance of Stressed Tropical Corals. In Proceedings of the International Geoscience and Remote Sensing Symposium (IGARSS), Honolulu, HI, USA, 24–28 July 2000; Volume 6.
119. Schuyler, Q.; Dustan, P.; Dobson, E. Remote Sensing of Coral Reef Community Change on a Remote Coral Atoll: Karang Kapota, Indonesia. In Proceedings of the 10th International Coral Reef Symposium (ICRS), Okinawa, Japan, 28 June–2 July 2004; Japanese Coral Reef Society: Tokyo, Japan, 2006; pp. 1763–1770.
120. Hussain, M.; Chen, D.; Cheng, A.; Wei, H.; Stanley, D. Change detection from remotely sensed images: From pixel-based to object-based approaches. *ISPRS J. Photogramm. Remote Sens.* **2013**, *80*, 91–106. [CrossRef]
121. Xu, J.; Zhao, J.; Wang, F.; Chen, Y.; Lee, Z. Detection of Coral Reef Bleaching Based on Sentinel-2 Multi-Temporal Imagery: Simulation and Case Study. *Front. Mar. Sci.* **2021**, *8*, 268. [CrossRef]
122. Wicaksono, P.; Wulandari, S.A.; Lazuardi, W.; Munir, M. Sentinel-2 images deliver possibilities for accurate and consistent multi-temporal benthic habitat maps in optically shallow water. *Remote Sens. Appl. Soc. Environ.* **2021**, *23*, 100572. [CrossRef]
123. Haya, L.O.M.Y.; Fujii, M. Mapping the change of coral reefs using remote sensing and in situ measurements: A case study in Pangkajene and Kepulauan Regency, Spermonde Archipelago, Indonesia. *J. Oceanogr.* **2017**, *73*, 623–645. [CrossRef]
124. Kabiri, K.; Pradhan, B.; Samimi-Namin, K.; Moradi, M. Detecting coral bleaching, using QuickBird multi-temporal data: A feasibility study at Kish Island, the Persian Gulf. *Estuar. Coast. Shelf Sci.* **2013**, *117*, 273–281. [CrossRef]
125. Vanderstraete, T.; Goossens, R.; Ghabour, T.K. The use of multi-temporal Landsat images for the change detection of the coastal zone near Hurghada, Egypt. *Int. J. Remote Sens.* **2006**, *27*, 3645–3655. [CrossRef]
126. Shapiro, A.C.; Rohmann, S.O. Summit-to-sea mapping and change detection using satellite imagery: Tools for conservation and management of coral reefs. *Proc. Rev. Biol. Trop.* **2005**, *53*, 185–193.
127. Yamano, H.; Tamura, M. Use of Landsat TM Data and Radiative Transfer Models for Monitoring Coral Reef Bleaching. In Proceedings of the International Geoscience and Remote Sensing Symposium (IGARSS), Sydney, NSW, Australia, 9–13 July 2001; Volume 5.
128. Ammenberg, P.; Lindell, T. Automated Change Detection of Bleached Coral Reefs. In Proceedings of the 7th International Conference, Remote Sensing for Marine and Coastal Environments, Miami, FL, USA, 20–22 May 2002.
129. Andréfouët, S.; Guillaume, M.M.M.; Delval, A.; Rasoamanendrika, F.M.A.; Blanchot, J.; Bruggemann, J.H. Fifty years of changes in reef flat habitats of the Grand Récif of Toliara (SW Madagascar) and the impact of gleaning. *Coral Reefs* **2013**, *32*, 757–768. [CrossRef]

130. Hochreiter, S.; Schmidhuber, J. Long short-term memory. *Neural Comput.* **1997**, *9*, 1735–1780. [CrossRef]
131. Roelfsema, C.M.; Lyons, M.; Murray, N.; Kovacs, E.M.; Kennedy, E.; Markey, K.; Borrego-Acevedo, R.; Ordoñez Alvarez, A.; Say, C.; Tudman, P.; et al. Workflow for the Generation of Expert-Derived Training and Validation Data: A View to Global Scale Habitat Mapping. *Front. Mar. Sci.* **2021**, *8*, 1–13. [CrossRef]

remote sensing

MDPI

Article

Performance and Uncertainty of Satellite-Derived Bathymetry Empirical Approaches in an Energetic Coastal Environment

Bertrand Lubac [1,*], Olivier Burvingt [1,2], Alexandre Nicolae Lerma [2] and Nadia Sénéchal [1]

1 U.M.R. 5805 EPOC, Avenue des Facultés, University of Bordeaux, 33405 Talence, France; olivier.burvingt@u-bordeaux.fr (O.B.); nadia.senechal@u-bordeaux.fr (N.S.)
2 Direction Nouvelle-Aquitaine, BRGM, 24 Av. Léonard de Vinci, 33600 Pessac, France; a.nicolaelerma@brgm.fr
* Correspondence: bertrand.lubac@u-bordeaux.fr

Abstract: Objectives of this study are to evaluate the performance of different satellite-derived bathymetry (SDB) empirical models developed for multispectral satellite mission applications and to propose an uncertainty model based on inferential statistics. The study site is the Arcachon Bay inlet (France). A dataset composed of 450,837 echosounder data points and 89 Sentinel-2 A/B and Landsat-8 images acquired from 2013 to 2020, is generated to test and validate SDB and uncertainty models for various contrasting optical conditions. Results show that water column optical properties are characterized by a high spatio-temporal variability controlled by hydrodynamics and seasonal conditions. The best performance and highest robustness are found for the cluster-based approach using a green band log-linear regression model. A total of 80 satellite images can be exploited to calibrate SDB models, providing average values of root mean square error and maximum bathymetry of 0.53 m and 7.3 m, respectively. The uncertainty model, developed to extrapolate information beyond the calibration dataset, is based on a multi-scene approach. The sensitivity of the model to the optical variability not explained by the calibration dataset is demonstrated but represents a risk of error of less than 5%. Finally, the uncertainty model applied to a diachronic analysis definitively demonstrates the interest in SDB maps for a better understanding of morphodynamic evolutions of large-scale and complex coastal systems.

Keywords: satellite-derived bathymetry; uncertainty; coastal; morphodynamics; multispectral; empirical model; cluster-based approach; Sentinel-2; Landsat-8

Citation: Lubac, B.; Burvingt, O.; Nicolae Lerma, A.; Sénéchal, N. Performance and Uncertainty of Satellite-Derived Bathymetry Empirical Approaches in an Energetic Coastal Environment. *Remote Sens.* **2022**, *14*, 2350. https://doi.org/10.3390/rs14102350

Academic Editor: Guillaume Ramillien

Received: 5 April 2022
Accepted: 10 May 2022
Published: 12 May 2022

1. Introduction

Coastal areas are major socio-ecological interfaces dramatically exposed to climate change and anthropogenic pressures [1,2]. Implementing effective climate change adaptation and disaster risk reduction policies requires correctly anticipating the response of coastal environments to long-term forcing factors and to the increase in intensity and frequency of hydrometeorological hazards. However, this remains a major challenge, particularly for morphodynamics models addressing multi-scale approaches [3]. For immediate needs, improving coastal impact model prediction involves systematic monitoring of coastal zones based on both space-based and in situ observation systems [4,5]. The availability of time-referenced bathymetry data with the appropriate temporal and spatial resolution is still greatly lacking [6] and appears to be a major key to the progress of morphodynamic models [7].

A major obstacle to the availability of accurate large-scale bathymetric data is the high cost (in terms of acquisition and/or post-treatment) associated with conventional observation systems, i.e., echo sounders on ships and LiDAR on aircraft. As a result, and due to constantly enhanced instrument features, radar [8] and optical [9] satellite imagery is expected to be a key complementary tool for generating bathymetry maps in the coming years. Two distinct approaches are used to retrieve Satellite-Derived Bathymetry

(SDB). The wave-based inversion exploits the relationship between water depth and surface gravity wave speeds and directions [10]. The radiative signal-based inversion exploits the relationship between water depth and water-leaving radiance for optically shallow waters [11]. Although the light propagation in the water column along the surface-bottom-surface path is limited by the diffuse attenuation coefficient of irradiance [12,13], this second approach is often the only alternative for environments controlled by wave and tidal current energy. In these mixed energy environments, the Doppler effect can negatively affect SDB accuracy of the wave-based inversion method [14].

Radiative signal-based inversion models were initially proposed in the late 1970s [15] and then reformulated in the 1990s [12,13]. Based on these formulations, many empirical [16] and physics-based algorithms [17,18] were developed over the past two decades. For physics-based algorithms, field data are not required, potentially ensuring reproducibility over space and time. However, their performances are strongly impacted by environmental noise and various sources of error performed on the water-leaving signal [19]. Empirical algorithms require a training dataset composed of sounding points in order to statistically calibrate the inversion model. These approaches are site- and time-dependent. Moreover, their accuracy depends on the spatial heterogeneity of substrate and water column optical properties. To overcome these limitations, empirical approaches are now associated with machine learning and multi-temporal techniques [20,21]. However, in most cases, regardless of approach, SDB suffers from a lack of precise quantification of uncertainty, limiting the massive scientific and operational exploitation of this product.

Uncertainty can be defined as a statistical parameter "characterizing the range of values within which the true value of a measurement is expected to lie as defined within a particular confidence level" [22]. Its accurate quantification remains a major issue for most ocean color products [23]. For SDB applications, uncertainty is depth-dependent and requires expressing it as a function of depth. However, in the scientific literature, metrics used to describe the SDB uncertainty are most often computed for all data without differentiating between the specific ranges of depth. Furthermore, the impact of extrapolating models beyond the domain of the calibration dataset on uncertainty is rarely addressed despite being a major issue due to often incomplete spatial data coverage [20]. Finally, the uncertainty associated with SDB is essential information for data interpretation, and, therefore, must be quantified when analyzing and interpreting bathymetric changes.

Objectives of this study are, therefore, to evaluate the performance and robustness of well-established SDB empirical models and to propose and validate an uncertainty model based on inferential statistics, and using a multi-scene approach for a mixed energy coastal environment.

2. Materials and Methods

2.1. Study Area

The Arcachon Bay is a semi-enclosed lagoon of approximately 180 km^2 located in SW France (44°40′ N, 1°10′ W) (Figure 1). The lagoon is connected to the Atlantic Ocean through a 10 km large tidal inlet, which disrupts the southerly longshore drift of the 110 km-long stretches of sandy coast. The inlet is a mixed energy environment [24], subject to meso- to macro-tidal conditions. The tidal range varies approximately from 1.5 m during neap tide conditions to 5 m during spring tide conditions [25]. The wave climate is energetic and strongly seasonally modulated with a monthly-averaged significant wave height Hs (peak wave period Tp), ranging from 1.1 m (8.5 s) in July with a dominant west-northwest direction to 2.4 m (13 s) in January with a dominant west direction [26]. Extreme wave conditions with the 100-year return Hs picking at 11.5 m [27] were observed during winter, where the significant wave height annually exceeds 6 m during storm events. The inlet was composed of 3 main subtidal morphological units, i.e., the spit platform, the Ebb-tidal delta, and the Flood-tidal delta, connected to a 9 m-deep northern channel and a 11 m-deep southern channel. Their morphology changes dynamically on timescales from months to years and decades [28,29].

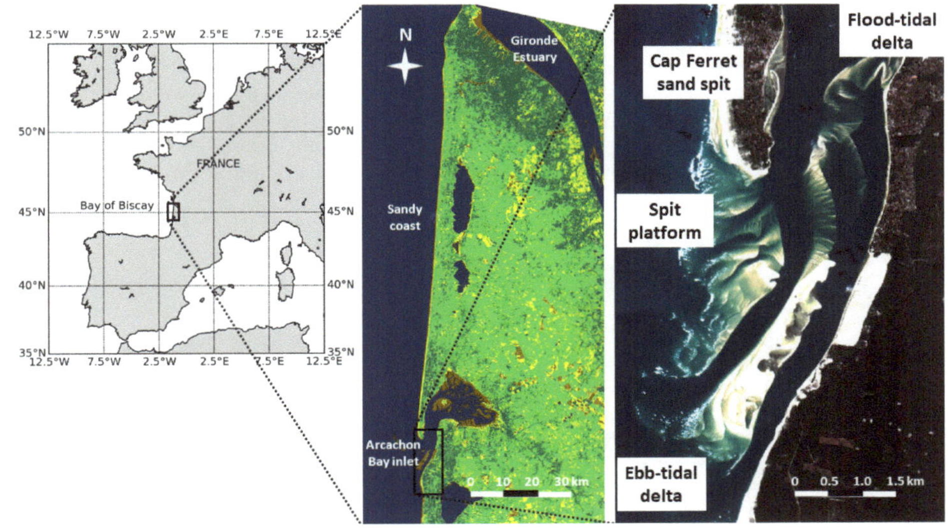

Figure 1. The Arcachon Bay inlet located south of the Gironde estuary at the end of the 110 km-long linear Gironde sandy coast (SW France). The three main morphological units of the study area are the spit platform connected to the Cap Ferret sand spit, the edd-tidal, and the flood-tidal deltas.

Optical properties of seawater along the Arcachon inlet change with hydrodynamics and seasonal conditions. They were mainly controlled by the suspended particulate and colored dissolved organic matters coming from the ocean, lagoon, or adjacent coasts. The particulate organic matter (POM) was mainly dominated by phytoplankton (89% on average) and a non-negligible contribution of anthropogenic (6%) and river (5%) sources [30]. The seasonal variability of total suspended particulate matter (SPM) was controlled by river discharges and biological and hydro-sedimentary processes [31]. Over the period 2010–2020, the mean annual values of the concentration of SPM, chlorophyll-a (Chl), and particulate organic carbon to nitrogen ratio (C/N ratio, proxy of the origin of POM) were 4.95 ± 4.78 mg/L, 1.86 ± 0.79 µg/L, and 8.21 ± 20.13 mol/mol, respectively, at the station Bouée 13 located north of the inlet (statistics extracted from the data set of the French Coastal Monitoring Network SOMLIT). Maximum values of Chl (2.42 ± 1.33 µg/L) were observed in spring due to phytoplankton blooms, while maximum values of SPM (7.90 ± 7.32 mg/L) were recorded in winter due to resuspension processes. These conditions generated mildly to moderately turbid seawaters. The satellite-derived vertically averaged diffuse attenuation coefficient for downwelling irradiance (K_d) at 545 nm ranges from 0.32 to 0.79 m^{-1}, with a mean value of 0.52 m^{-1} [11]. The bottom substrate of optically shallow waters, defined as seawater surfaces where the bottom significantly affects the remote sensing reflectance R_{rs}, was uniformly covered by medium quartz sand characterized by a cross-section ranging from 200 to 400 µm [28,32].

2.2. Ground Reference Bathymetry Data

Echo sound bathymetric data of the Arcachon Bay inlet were collected every year from April to October by the Direction Départementale des Territoires et de la Mer (DDTM) and the Syndicat Intercommunal du Bassin d'Arcachon (SIBA). These surveys were mainly carried out for the monitoring of navigation channels and the repositioning of navigation buoys; thus, bathymetric data generally covered only part of the study site. Due to the large lateral extent (almost 10 km from North to South), strong tidal currents and energetic waves, and shallow water areas (spit platform, inter-channels banks), the Arcachon inlet was a very challenging site to survey with conventional tools, which explained the lack of complete synchronous coverage. A comprehensive dataset of 450,837 echo sound data points collected between 2013 and 2020 was used in this study (Table 1). Echo sound data were acquired with an Odom Hydrotrac single-beam echosounder co-located with a Leica DGPS during low or high tide slack. The observed depths were then corrected for roll, pitch, yaw, GPS latency, and theoretical tide computed at the Arcachon-Eyrac reference tide gauge at the time of acquisition. Finally, the bathymetry (noted, Z_{situ}), which was associated with the corrected depth reduced within the appropriate vertical datum, was obtained after the correction of the local chart datum of the Lowest Astronomical Tide (LAT). Accuracy on Z_{situ} was assumed to be within 0.2 m.

Table 1. Date, location (ED: Ebb tidal Delta; FD: Flood tidal Delta; SP: Cap Ferret sand Spit Platform; CO: Coastal Ocean; CH: Channel), number of echo sound data points, and the minimum, maximum and median values of Z_{situ}, associated with the different bathymetric surveys.

Date	Location	Number of Points	Z_{min} (m)	Z_{max} (m)	Z_{median} (m)
15 October 2013	CH	2231	1.1	17.5	5.6
27 November 2013	ED; CH	5716	0.0	18.1	5.0
18 March 2014	ED; CH	155,077	0.0	19.7	5.4
16 April 2014	FD; CH; CO	40,620	0.0	25.5	6.4
28 May 2014	SP; CH	4727	0.0	21.1	4.0
23 September 2014	ED	3750	0.0	16.4	5.9
19 March 2015	ED	5266	0.0	17.3	5.5
15 April 2015	FD; SP; CO	102,422	0.2	24.4	8.1
25 September 2015	ED; CO	11,552	0.0	18.4	5.1
13 October 2015	ED	9406	0.0	20.3	3.7
22 March 2016	ED; CH	5960	0.0	19.0	6.1
15 May 2016	SP	2986	0.0	25.5	5.1
11 April 2017	ED; CH	3856	0.0	17.0	5.9
23 June 2017	FD; CH; CO	8804	0.0	22.0	6.6
20 September 2017	ED; SP	7283	0.0	18.3	4.5
15 November 2017	SP	8348	0.0	18.8	3.1
25 April 2018	ED	4446	0.0	16.3	5.6
29 May 2018	FD; CO	9276	0.0	25.2	7.9
08 October 2018	ED; CH; CO	7300	0.0	24.3	5.5
26 November 2018	SP	1988	0.0	18.0	4.6
20 March 2019	CO	2484	0.0	21.6	11.2
19 April 2019	ED; SP	8579	0.0	16.1	4.5
14 May 2019	FD; CH	4874	0.0	25.4	5.1
17 June 2019	FD; CH	3236	0.0	21.3	4.8
16 September 2019	ED; CH	6835	0.0	16.1	4.8
23 March 2020	ED	6669	0.0	16.1	5.0
01 July 2020	FD; CH; CO	4833	0.1	26.1	13.1
01 September 2020	SP	6817	0.0	17.4	2.5
19 October 2020	SP; CO	5496	0.0	25.9	5.1

2.3. Selection and Processing of Landsat-8 and Sentinel-2 Images

The Landsat-8 satellite mission was launched on 11 February 2013 by the National Aeronautics and Space Administration (NASA). Its payload comprised the Operational Land Imager (OLI) and the Thermal Infrared Sensor (TIRS). OLI provides enhanced instrument features for aquatic surface observation compared to former Landsat sensors [33]. OLI operates in the visible (VIS), near-infrared (NIR), and shortwave infrared (SWIR) for 9 spectral bands with a 12-bit radiometric resolution. Its spatial resolution was 30 m for multispectral images and 15 m for the panchromatic band. Landsat-8 Collection 1

Level 1 products were downloaded freely on https://earthexplorer.usgs.gov (accessed on 31 March 2022).

Sentinel-2A and 2B were twin polar-orbiting satellite missions launched in June 2015 and March 2017, respectively, by the European Space Agency (ESA). The multispectral instrument (MSI) mounted on the Sentinel spacecraft was very similar in design and requirements to OLI [34,35]. MSI acquired data in 13 spectral bands from VIS to SWIR with a spatial resolution ranging from 10 to 60 m and a radiometric resolution of 12 bit. Sentinel-2 Level 1C products associated with the T30TXQ tile were downloaded freely from https://scihub.copernicus.eu/ (accessed on 31 March 2022). OLI and MSI products were sufficiently consistent to be merged and to provide comprehensive observations for the monitoring of aquatic systems [36]. The combination of the products of these 3 missions yielded a global median average revisit interval of 2.9 days since the repeat cycle is 16-days for Landsat-8 and 5-days for Sentinel-2A/B [37].

SDB requires satellite images with high standards of quality, considerably reducing the number of usable images. The accuracy of SDB retrievals was strongly impacted by environmental noise caused by atmospheric and ocean surface effects [21]. Before applying atmospheric correction models, a pre-selection of satellite images was performed based on environmental criteria such as the absence of foam due to breaking waves, the absence of clouds that should not cover more than 10% of the images, and the absence of sun glint. An additional criterion of temporal proximity between bathymetry surveys and the date of image acquisition was also required, given that rapid bathymetry changes could be observed along the Arcachon inlet in response to energetic hydrodynamic conditions. For example, Capo et al. [11] demonstrated that, on average, sand bars along the Arcachon inlet migrated by about 10 m a month. Considering the spatial resolution of MSI products, only satellite images acquired within a 30-day time window of the bathymetry surveys were selected to avoid having larger bathymetric changes than the spatial resolution of images. Application of these 4 criteria to the 2013–2020 Landsat-8 and to the 2015–2020 Sentinel-2A/B archives allowed selection of 89 satellite images, among which 49 were Landsat-8 images and 40 were Sentinel-2A/B images (Figure 2a). Most of the images were acquired in October, September, April, and March (Figure 2b), when atmospheric and swell conditions (Figure 2c) were most suitable for acquiring high-quality images [38] and bathymetric field data, respectively.

In optically shallow waters, the remote sensing reflectance (R_{rs}) is a function of water depth, bottom albedo, and water column optical properties. Extracting R_{rs} from the top-of-atmosphere signal recorded by sensors requires accurate and consistent atmospheric corrections. In this study, the dark spectrum fitting atmospheric correction method (DSF), associated with the open access ACOLITE processor, was selected to process satellite images from Level-1C to Level-2A [39,40]. ACOLITE-DSF demonstrated high performance in deriving bathymetry from empirical SDB models [41]. This high performance compared to other atmospheric correction algorithms may be explained by less noisy ACOLITE products due to the assumption of spatially consistent aerosols [41,42]. Landsat-8 and Sentinel-2 A/B images were, therefore, processed to R_{rs} using the fixed aerosol optical thickness option for our 20 km $\times$ 20 km region of interest. Due to the large amount of bathymetry and satellite data, the sun glint correction option was not applied. All pixels contaminated by the sun glint effect were masked to avoid additional environmental noise for the assessment of the SDB empirical model performances. Finally, OLI and MSI data were projected on the same 30 m $\times$ 30 m grid to enable a joint statistical study to be carried out.

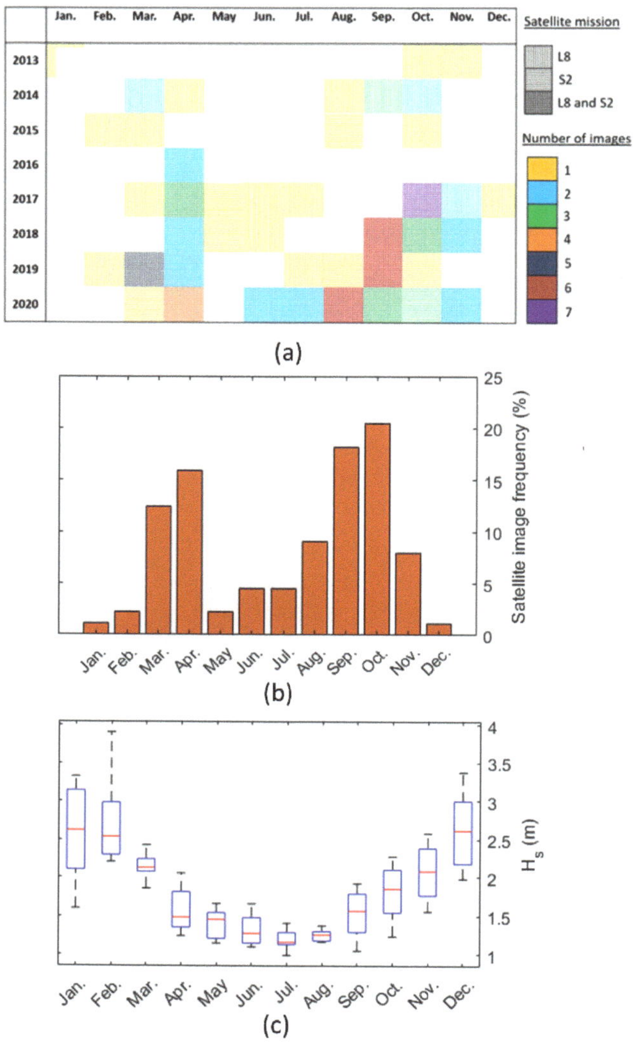

Figure 2. (**a**) Acquisition dates of the different Landsat-8 and Sentinel-2 A/B images belonging to the dataset; (**b**) distribution of these images per month; and (**c**) average monthly value of the significant wave height (H_s) computed over the period 2013–2020.

2.4. Inter-Comparison of SDB Empirical Model Performance

Three empirical approaches were selected for inter-comparison of performances. While many SDB empirical models were developed in the literature, the choice of these 3 approaches was motivated by the fact that they were representative of the 3 main well-established and emerging algorithm categories. The linear regression model (LRM) using log-transformed bands or log-transformed band ratios was originally developed by [43] and [44], and recently revisited by [41] for Sentinel-2A/B applications. The switching model (SM) [19] and the cluster-based regression model (CBR) [21] can be considered as an extension of LRM. SM was based on 2 LRM using the Red-Blue (LRM-RB) and Green-Blue (LRM-GB) log-transformed band ratios. This multi-conditional approach selects the most sensitive log-transformed band ratio for a predefined water depth range. LRM-RB was used for very shallow waters, while LRM-GB was used for deeper regions. CBR segments the region into different optical classes for which class-specific LRM were calibrated.

In practice, LRM is based on the following formulation:

$$Z_{\text{sat}} = m_1 X + m_0, \quad (1)$$

where Z_{sat} is the satellite-derived bathymetry (in meters) and X is the log-transformed band or log-transformed band ratio. The different parameterizations of X used in this study are log(B), log(G), log(R), log(B)/log(G), and log(R)/log(G), where B, G, and R are R_{rs} at the blue, green, and red bands, respectively. m_0 and m_1 are the coefficients of the linear regression established between the response variable Z_{situ} and the explanatory variable X, retrieved from a least-square approach.

The linear regression models calibrated for each satellite image have a range of application bounded by a lower limit (Z_{min}) and upper limit (Z_{max}) (Figure 3a,b). To compute Z_{min} and Z_{max}, the entire range of X values was divided into 20 intervals, bins, of equal sizes comprised between X_{min} and X_{max}. For each bin, the number of observations (N_{bin}) and the standard deviation of Z_{situ} values (σ_{bin}) was computed. When N_{bin} was lower than 30 or σ_{bin} higher than 1 m, observations of the bin were removed. The N_{bin} value of 30 was selected to ensure a statistical weight to the regression, while the choice of a 1 m-σ_{bin} filter was performed to remove observations associated with a strong environmental noise that generated large uncertainty on the Z_{situ} retrievals. The linear regression model was then calibrated from filtered observations. Z_{min} and Z_{max} were defined as the minimum and maximum values of Z_{sat} computed from Equation (1).

SM was based on a multi-conditional procedure. In step 1, Z_{max} and Z_{min} values associated with the different parameterizations of X are computed (Figure 3c). In step 2, the calibration dataset is filtered to remove all values of Z_{situ} higher than the lowest Z_{max} values (noted $Z_{\text{max-ref}}$). In step 3, the coefficient of determination (r^2) of the linear regression models calibrated for the different parameterizations of X are computed. In step 4, if the r^2 value of the parameterization of X associated with $Z_{\text{max-ref}}$ (noted X_{ref}) is higher than all other X parameterization r^2 values, the associated LRM is selected for the range of Z_{situ} values comprised between $Z_{\text{min-ref}}$ and $Z_{\text{max-ref}}$. The calibration dataset is then filtered to remove all values of Z_{situ} lower than $Z_{\text{max-ref}}$. X_{ref} is no longer considered in the procedure and the algorithm returns to step 2. In the opposite case, the LRM associated with X_{ref} is not considered and the algorithm returns to step 2.

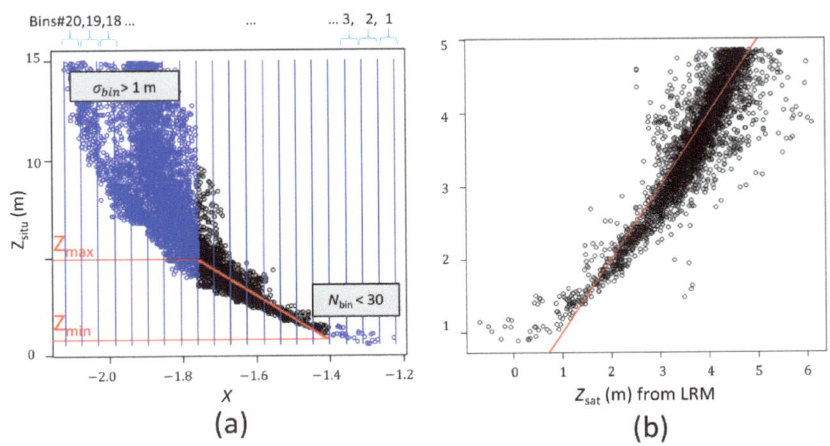

Figure 3. *Cont.*

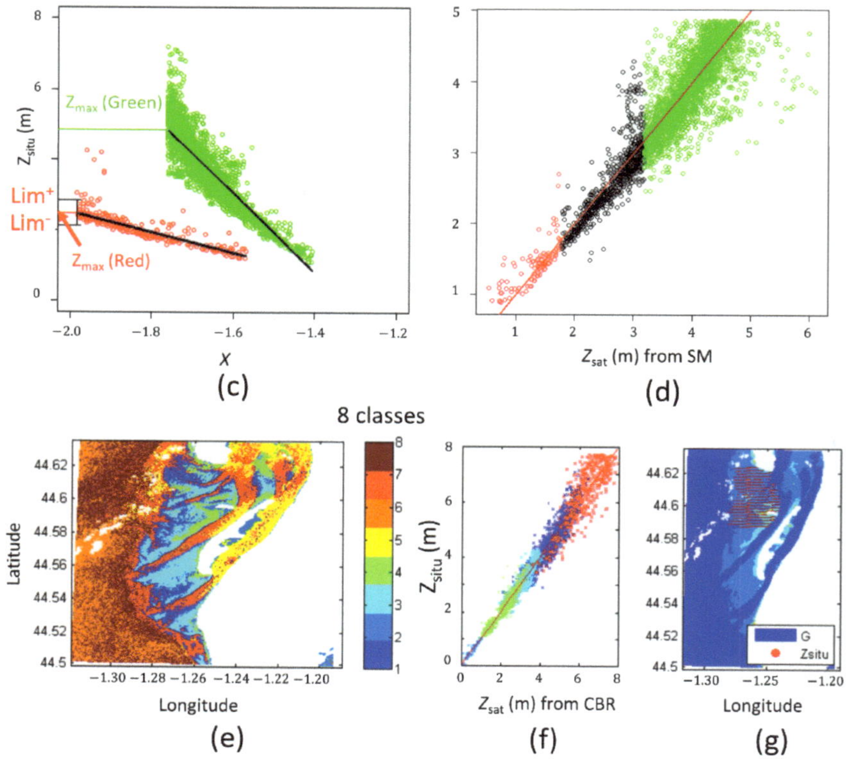

Figure 3. Illustration of the satellite-derived bathymetry linear regression (LRM), switching (SM), and cluster-based regression (CBR) methods for the 10 September 2020 Sentinel-2 cropped image. (**a**) Data filtration procedure for the green band. X is the log-transformed green band. Z_{situ} is the field-reduced depth (in m). The entire range of X values are divided into 20 bins of equal sizes comprised between X_{min} and X_{min}. N_{bin} and σ_{bin} are the number of observations and the standard deviation of Z_{situ} values per bins, respectively. When N_{bin} is lower than 30 or σ_{bin} higher than 1 m, observations of the bin are removed (blue points). The red line is associated with the linear regression model for the green band (LRM-G) computed from filtered observations (black points). Z_{min} and Z_{max} correspond to the minimum and maximum values of Z_{sat} computed from LRM-G. (**b**) Comparison between Z_{sat} derived from LRM-G and Z_{situ}. (**c**) Computation of the linear regression model for the red band (LRM-R), Z_{max} and switching points (lim^-, lim^+). (**d**) Comparison between Z_{sat} derived from SM using LRM-R and LRM-G and Z_{situ}. Black points are associated with the smoothing interval. (**e**) Example of classification of satellite image pixels into 8 optical classes using a k-means classifier. (**f**) Comparison between Z_{sat} derived from CBR and Z_{situ}. The color of points is associated with the color of optical classes. (**g**) Location of Z_{situ} data points used in this example for calibration of LRM, SM, and CBR.

If more than one LRM is selected, a procedure of weighting is applied between Z_{sat} values derived from the shallowest (LRM^-) and deepest (LRM^+) models to ensure a smooth transition (Figure 3d). The lower and upper bounds of the smoothing intervals are $lim^- = Z_{max} - \sigma_{bin}$ and $lim^+ = Z_{max} + \sigma_{bin}$, respectively, where Z_{max} and σ_{bin} are computed from the shallowest LRM. The equation of weighting is:

$$Z_{sat} = \alpha \, LRM^- + (1 - \alpha)LRM^+, \tag{2}$$

where

$$\alpha = (lim^{+} - \text{LRM}^{-})/(lim^{+} - lim^{-}). \tag{3}$$

This procedure for the switching method allows for the switching points (lim^{-}, lim^{+}) to automatically adapt when the water column optical properties change.

For the CBR method, we selected a k-means clustering algorithm for classifying each pixel into homogeneous optical classes, as in [21]. These unsupervised learning techniques showed high performance in differentiating between optically contrasted seawater classes [45,46]. The set of variables used for the classification was composed of B, G, R, and NIR was R_{rs} at the near-infrared band. Then, class-specific linear regression models were calibrated for each optical class using the LRM approach. To cover the optical variability of the entire study area, we selected a number of 8 optical classes. This number of classes was based on the assumption that optical conditions were controlled mainly by 3 independent environmental variables, each taking 2 outcomes, water depth (deep or shallow), sediment resuspension (yes or no), main origin of water column optical properties (ocean or lagoon), i.e., 8 possible combinations. The spatial distribution of optical classes varied from one satellite image to another, depending on hydrological conditions. One example of the distribution is presented in Figure 3f. It is important to note that pixels assigned to classes associated with optically deep waters or to classes that do not have enough Z_{situ} points for calibration (see the LRM procedure) will be automatically masked (Figure 3e,g).

Four statistical parameters were used to assess and compare the performance of the LRM, SM, and CBR methodologies applied to the 89 satellite images. The coefficient of determination (r^2), bias (Bias), mean relative absolute difference (MRAD), and root mean square error (RMSE) were computed from the matchup dataset (DS) composed of Z_{situ} and its estimator Z_{sat}. The number of observations (N^{DS}) varies with each pair of satellite images and field bathymetry dataset. Bias, MRAD, and RMSE are computed as follows:

$$\text{Bias} = \frac{1}{N^{DS}} \sum_{i=1}^{N^{DS}} (Z_{sat;i} - Z_{situ;i}), \tag{4}$$

$$\text{MRAD} = \frac{1}{N^{DS}} \sum_{i=1}^{N^{DS}} \frac{|Z_{sat;i} - Z_{situ;i}|}{Z_{situ;i}} \times 100\%, \tag{5}$$

$$\text{MSE} = \sqrt{\frac{1}{N^{DS}} \sum_{i=1}^{N^{DS}} (Z_{sat;i} - Z_{situ;i})^2}. \tag{6}$$

2.5. Assessment of SDB Uncertainty Using a Multi-Scene Approach

The calibration dataset often only partially covers the study area (Figure 3g). For pixels associated with areas that were not sampled and were, therefore, not included within the calibration dataset (here, referred as unsampled pixels), quantification of SDB uncertainty requires taking into account 2 terms (Figure 4). The first term was associated with the uncertainty of the SDB regression model. The second term was associated with the uncertainty generated by the extrapolation of the SDB model beyond the calibration domain. To estimate these 2 terms, an uncertainty model based on a multi-scene approach was proposed. We expressed the uncertainty as a positive value and assumed that error follows a normal distribution.

Combination of Landsat-8 and Sentinel-2 enables to have several image acquisitions in a 1 month-interval around the bathymetric survey. Assume we have N usable images for a given bathymetry survey. For each image, an SDB model was calibrated independently of other SDB models generating N bathymetry maps which can be associated with N random variables Z_{sat1}, Z_{sat2}, … Z_{satN}. Each variable was then separated into 2 parts comprising observations associated with the DS-matchup dataset, Z_{satj}^{DS}, and observations associated with the NS-unsampled pixels dataset, Z_{satj}^{NS}.

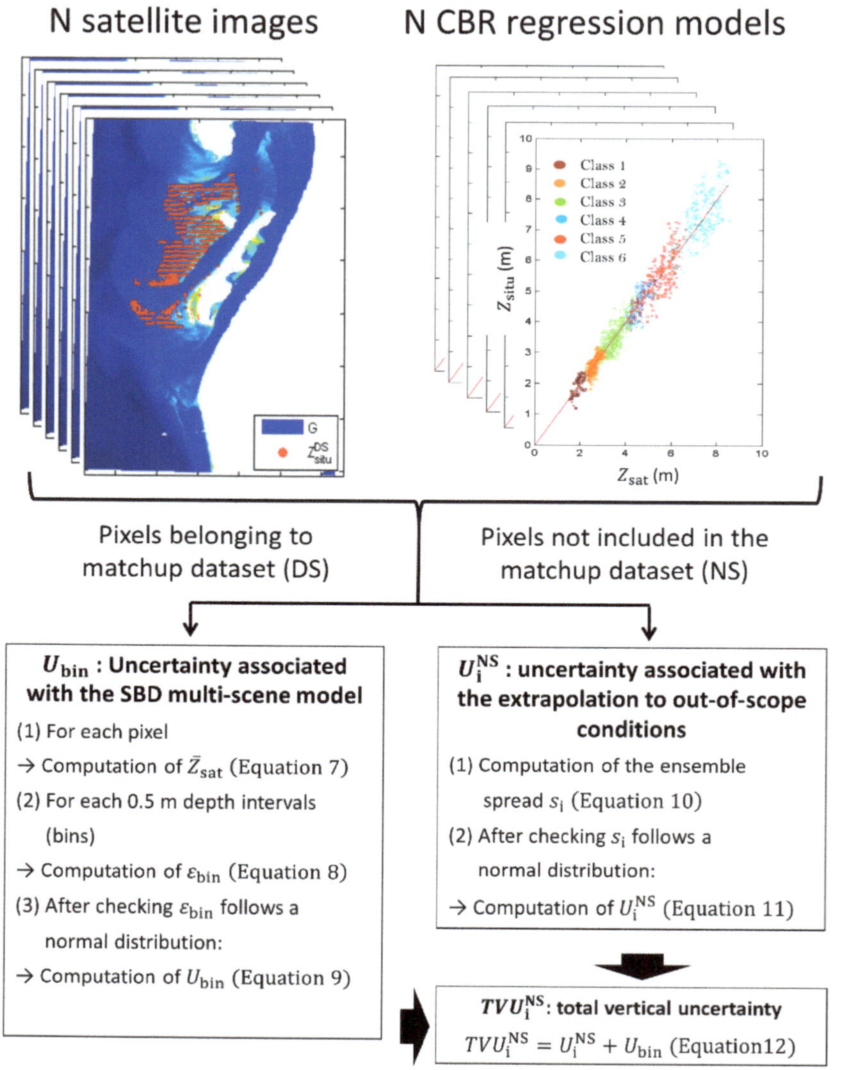

Figure 4. Flowchart describing the multi-scene algorithm used to quantify the 95% uncertainty on SDB.

The multi-scene approach allows to define a new variable $\overline{Z}_{sat}$ as:

$$\overline{Z}_{sat} = \frac{1}{N}\sum_{j=1}^{N} Z_{satj}^{DS}. \quad (7)$$

Quantification of uncertainty associated with the multi-scene model requires characterizing the statistical distribution of errors over DS for different levels of bathymetry. The entire range of $\overline{Z}_{sat}$ values were divided into bathymetry intervals of 0.5 m comprised between the minimum and maximum values. For each 0.5 meter-bin, the error term, ε_{bin}, was computed as:

$$\varepsilon_{bin} = \overline{Z}_{sat} - Z_{situ}. \quad (8)$$

For each bin, the normality of the ε_{bin} distribution was tested. If the normality was rejected, observations associated with the given bin were removed. For the other bins, the bias, $\bar{\varepsilon}_{bin}$, was computed as the averaged value of ε_{bin}, and the uncertainty at the 95% confidence level associated with the SDB multi-scene model was computed as:

$$U_{bin} = 1.96 \times \sigma_{\varepsilon_{bin}}, \quad (9)$$

where $\sigma_{\varepsilon_{bin}}$ is the standard deviation of ε_{bin}.

Quantification of the uncertainty associated with the extrapolation of the SDB model beyond the calibration domain was based on the model democracy theory or "one model one vote" [47]. This theory requires that variables $Z^{NS}_{sat1}, Z^{NS}_{sat2}, \ldots Z^{NS}_{satN}$ are reasonably independent, equally plausible and that the range of model predictions represents the uncertainty in the prediction. Independence is ensured by hydrodynamic conditions, which change from one image to another, and which do not allow inference on the spatial distribution of water column optical properties. The equal plausibility criterion is controlled by a statistical test on the probability density functions that are supposed to be the same. The last criterion is controlled by a test of normality for each pixels using the Shapiro–Wilk normality test. To avoid the problem of a small sample, N should be higher than 5 if possible.

For each observation of $Z^{NS}_{sat1}, Z^{NS}_{sat2}, \ldots Z^{NS}_{satN}$, associated with a given pixel i of the bathymetry maps, the spread of the ensemble is computed as:

$$s_i = \frac{1}{N-1} \sum_{j=1}^{N} \left(Z^{NS}_{satj;i} - \overline{Z}^{NS}_{satj;i} \right)^2. \quad (10)$$

The uncertainty at the 95% confidence level associated with the true value of $\overline{Z}^{NS}_{sat,i}$ and generated by the extrapolation of the SDB model beyond the calibration is then given by:

$$U^{NS}_i = t_{N-1;1-\alpha/2} \times \frac{s_i}{\sqrt{N}}, \quad (11)$$

where $t_{N-1;1-\alpha/2}$ is the critical value found from the t-distribution table. $N-1$ is the degree of freedom and α is the level of significance of the test. For a N and α value of 5 and 5%, respectively, $t_{4;0.975}$ is 2.776.

Finally, the total vertical uncertainty at the 95% confidence level associated with Z_{situ} is computed as:

$$TVU^{NS}_i = U^{NS}_i + U_{bin}. \quad (12)$$

Validation of the uncertainty model was carried out from a dataset composed of 6 satellite images acquired in an interval of 1 month (S2A, 6 October 2017;_S2A, 11 October 2017;_S2A and L8, 16 October 2017;_L8, 25 October 2017; S2A, 31 October 2017) around a bathymetric survey characterized by one of the greatest spatial coverages. DS was randomly separated into 2 parts comprising a calibration dataset (DS-C) and a validation dataset (DS-V), which represented 67% and 33% of the data, respectively. DS-C allows to compute U_{bin}, which was used to estimate TVU^{NS}_i for each observation of DS-V. DS-V allows to compute the error, ε^{DS-V}_i, associated with the SDB multi-scene model for each observation as:

$$\varepsilon^{DS-V}_i = \overline{Z}^{DS-V}_{sat,i} - Z^{DS-V}_{situ}. \quad (13)$$

The uncertainty model was finally evaluated by computing the percentage of absolute values of ε^{DS-V}_i lower than TVU^{NS}_i. To avoid any biais due to the random extraction of a specific dataset for DS-C and DS-V, an analysis of results was conducted on 100 random samplings.

3. Results

3.1. Hydrological Conditions and Spatio-Temporal Variability of R_{rs}

The dataset encompasses a wide range of hydrodynamic conditions. The number of images per tidal stage (TS) varies depending on flood (N = 29), ebb (N = 26), low tide (N = 19), and high tide (N = 15) conditions (Table 2). Tidal range (TR) and tidal level (TL) data were normally distributed with a mean value of 3.10 ± 0.84 m and 2.16 ± 0.92 m, respectively. Hs ranged from 0.25 to 2.20 m, with a mean value of 1.14 ± 0.50 m. Results showed that variables TL and TS (ANOVA *p*-value of 1.10–11), and Season and *Hs* (ANOVA *p*-value of 0.046) were not independent. However, there was a statistical relationship between TR and TS (ANOVA *p*-value < 0.001), which represents a sampling bias with respect to hydrodynamic conditions. Low tide and flood were mainly associated with high TR values, while high tide and ebb show on average low TR conditions. TR was also significantly correlated with TL, with a correlation coefficient value of −0.79 (*p*-value < 0.001).

Table 2. Descriptive statistics of environmental conditions (TS: Tidal Stage; TL: Tidal Level; TR: Tidal Range; *Hs*: significant wave height) associated with the 89 Sentinel-2 and Landsat-8 images.

Season		TS		TR (m)		TL (m)		*Hs* (m)	
Sp	20	HT	15	Mean	3.10	Mean	2.16	Mean	1.14
Su	24	LT	19	Sd	0.84	Sd	0.92	Sd	0.50
Fa	32	F	29	Min	1.50	Min	0.03	Min	0.25
Wi	13	E	26	Max	4.80	Max	3.60	Max	2.50

Sp: Spring; Su: Summer; Fa: Fall; Wi: Winter; HT: High Tide; LT: Low Tide; F: Flood; E: Ebb.

A principal component analysis (PCA) using the T-mode and S-mode orientations [48] was used to characterize the complexity of the spatio-temporal patterns of R_{rs} in the Arcachon inlet (Figure 5a). PCA was applied on G maps, as the green band was the most sensitive band for moderate values of SPM [49]. The S-mode PCA focuses on recurrent temporal patterns over space. The first principal component (PC1) explained 93.2% of the total variance. The spatial pattern associated with PC1 exhibited homogenous correlation values over the entire domain (Figure 5b). Statistical analyses performed between PC1 and environmental factors showed a significant seasonal influence with an ANOVA *p*-value lower than 0.001 (Table 3). The mean loading values for spring, winter, fall, and summer were 62.8, 40.0, −20.0, and −57.5, respectively. PC1 can be interpreted as describing a seasonal oscillation with high values of G in spring and winter probably associated with phytoplankton bloom and sediment resuspension processes, respectively, and low values in fall and summer.

The T-mode PCA focuses on recurrent spatial patterns over time. Results provide additional information to analyze more precisely spatial processes controlling the variability of G. The three first principal components accounted for 27.0%, 14.5%, and 6.9% of the variance. PC1 displayed an out-of-phase relationship between the G anomalies in the north and in the south (Figure 5c). A significant negative correlation was found between the PC1 and TL (*p*-value < 0.01) and between PC1 and TS (*p*-value < 0.01) (Table 3). This indicates that these anomalies of G were mainly controlled by tidal cycles. During low tide, positive anomalies of G were observed in the south associated with the downstream migration of turbid water dominated by lagoon optical properties, while negative anomalies observed during hide tide were correlated to the presence of less turbid oceanic seawater. PC2 displayed an out-of-phase relationship between the G anomalies on the western and eastern side of the channel (Figure 5d). TS significantly impacts the PC2 loadings with an ANOVA *p*-value lower than 0.05. The averaged loading value associated with the flood was strongly negative (−20.7), while the averaged loading value of the ebb was strongly positive (8.6). This indicates that PC2 describes a west-east oscillation of G controlled by the variation in the tidal current direction. During ebb, a strong west-east gradient of turbidity can be observed due to resuspension and advection of sediment on the eastern side of the channel. During the flood, the gradient was reversed, showing the highest turbidity values on the western side of the channel.

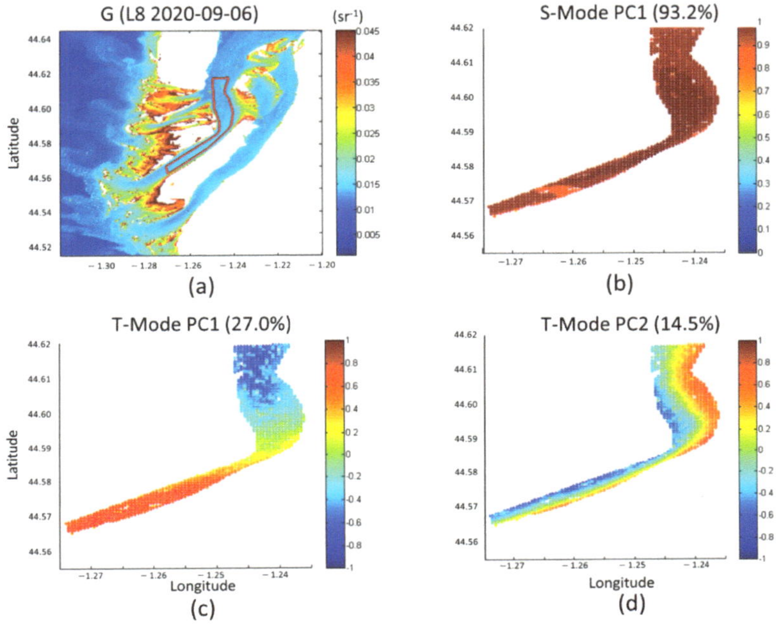

Figure 5. (**a**) Optically deep-water study area (red line) for analysis of the spatio-temporal variability of R_{rs} at the green band (noted G) using the 89 Landsat-8 and Sentinel-2 images; (**b**) the first principal component of the PCA using the S-mode orientation, which explains 93.2% of variance; (**c**) the first and; (**d**) the second principal components of PCA using the T-mode orientation, which explain 27.0% and 14.5% of the variance, respectively. Spatial patterns are presented as homogeneous correlation maps.

Table 3. *p*-values (significance code: "***" 0.001; "**" 0.01; "*" 0.05) of statistical analyses between the three first principal components using the S-mode and T-mode orientation and environmental variables (TS: Tidal Stage; TL: Tidal Level; TR: Tidal Range; *Hs*: significant Wave Height).

	S-Mode			T-Mode		
	PC1	PC2	PC3	PC1	PC2	PC3
Season	***	0.20	0.92	*	*	0.50
TS	0.34	*	0.12	**	*	***
TL	0.78	**	0.48	**	0.78	*
TR	0.58	*	0.33	*	0.27	0.82
Hs	0.56	*	0.33	0.13	0.61	0.85

3.2. Sensitivity of Linear Regression Models to Bathymetry Changes

The sensitivity of radiometric bands or ratios to bathymetry changes can be strongly impacted by the water column optical properties, which show very specific spatial patterns associated with the hydrodynamics forcing. To analyze this sensitivity, the mean and standard deviation of Z_{situ} associated with the 20 X bins were computed for each image and for the different parameterizations of X. For a given value of Z_{situ} mean, it was assumed that the higher the standard deviation, the lower the possibility of statistically differentiating this mean value from other means and, therefore, the lower the sensitivity.

Standard deviations associated with the G-band were lower than for the other X parameterizations, except for bathymetric values deeper than 9 m (Figure 6a). On the other hand, R-band showed the highest standard deviations except for bathymetric values that ranged from 2 m to 2.5 m. For G-band, the standard deviation was less than 1 m for all bathymetry lower than 4.4 m. This value decreased to 4.1 m, 3.8 m, 3.4 m, and 3.2 m, for B-band, B/G ratio, R-band, and R/G ratio, respectively. B-band and R-band showed a lower sensitivity than the G-band. This confirms a comprehensive sensitivity analysis carried out prior to this study, which demonstrates that the best performance for LRM and CBR was obtained by using G-band, while for SM, it was obtained by using the combination of R-band and G-band. In the following, the presentation of LRM, SM, and CBR results will be focused only on these *X* parameterizations.

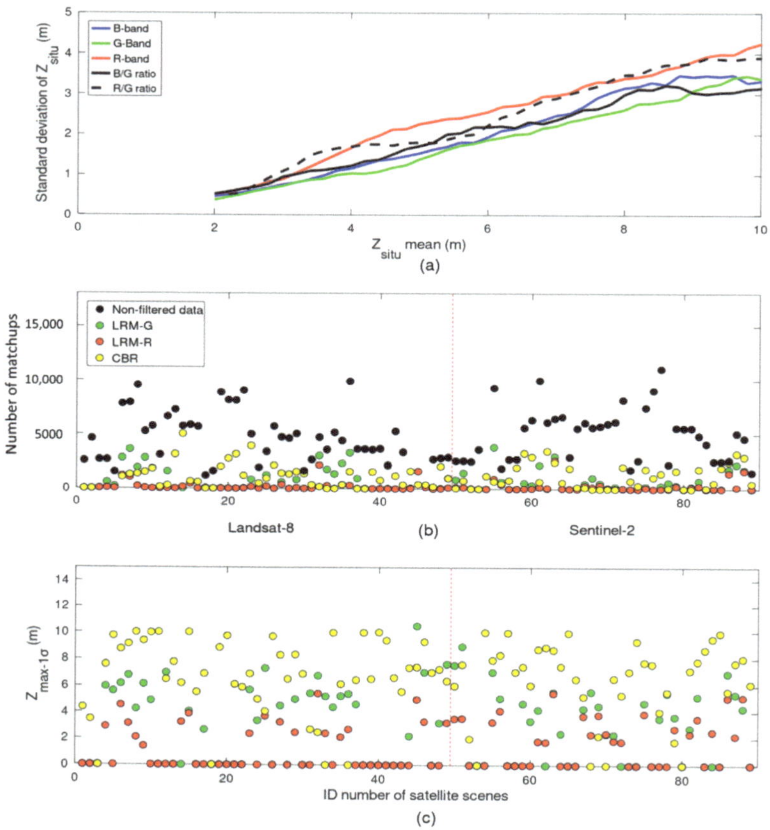

Figure 6. (**a**) Evolution of the standard deviation of Z_{situ} as a function of Z_{situ} mean computed over the 89 Landsat-8 and Sentinel-2 images for the different parameterizations of log-transformed bands and ratios; (**b**) number of matchups available for each of the 89 satellite images for non-filtered data and for 1σ-filtered data using LRM-G, LRM-R, and CBR approaches; (**c**) computation of the 1σ-filter maximum bathymetry for each of the 89 satellite images.

When the filter of 1σ was applied to the calibration dataset, the number of matchups significantly decreased from 915,326 points to 337,607 points, 130,583 points, and 19,737 points, for CBR, LRM using the log-transformed G-band (LRM-G), and LRM using the log-transformed R-band (LRM-R) methods, respectively (Figure 6b). A total of 70 satellite images had more than 100 matchups for CBR, whereas LRM-G and LRM-R had only 44 and 31 satellite images, respectively. It was interesting to note that the number of satellites images with more than 1000 and 10,000 matchups for CBR filter, was 43 and 3, respectively. Satellite images with few match-ups generated a 1σ-filtered maximum bathymetry ($Z_{max-1\sigma}$) value of 0. For CBR, 80 satellite scenes show a $Z_{max-1\sigma}$ value different of 0 (Figure 6c). The mean value of $Z_{max-1\sigma}$ was 7.3 ± 2.1 m. For LRM-G and LRM-R, the mean values were 5.1 ± 1.7 m and 3.2 ± 1.1 m, with a total number of images of 50 and 39, respectively.

Data filtering for CBR approach was applied on the different class-specific LRM-G. Each class was associated with specific optical conditions. The use of these classes allowed us to split the initial matchup dataset into class-specific matchup datasets for which the natural variability in the relationships between X and Z_{situ} was significantly reduced. This reduction of variability allowed us to reduce the number of matchups removed by the 1σ-filter.

3.3. Inter-Comparison of the Performance of Empirical SDB Approaches

The performance of the LRM-G approach was evaluated using the 50 satellite images for which the $Z_{max-1\sigma}$ value was different than 0. The median values for r^2, RMSE, Bias, and MRAD were 0.58, 0.54 m, 0.40 m, and 9.70%, respectively (Figure 7). The performance of SM using LRM-G and LRM-R was evaluated using a reduced dataset of 33 satellite images for which the $Z_{max-1\sigma}$ value was different than 0 for G-band and R-band. SM performance was slightly higher than LRM-G performance. The median values for r^2, RMSE, Bias, and MRAD were 0.64, 0.54 m, 0.39 m, and 8.99%, respectively. The CRB approach allowed us to evaluate the performance using a significantly larger dataset (N = 80). The median values for r^2, RMSE, Bias, and MRAD were 0.90, 0.53 m, 0.40 m, and 7.62%, which demonstrated higher performance than LRM-G and SA. When the analysis focused on a reduced dataset of the 33 best satellite scenes, performance became significantly higher. The median values for r^2, RMSE, Bias, and MRAD were 0.94, 0.46 m, 0.36 m, and 6.15%.

(a)
(b)
(c)

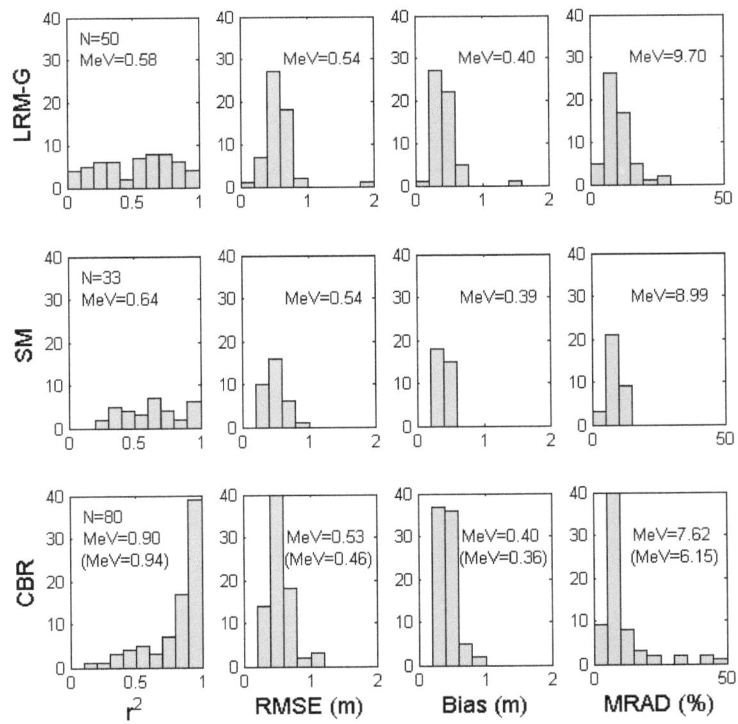

Figure 7. Distribution of performance of empirical SDB approaches for (**a**) the linear regression model using the log-transformed G-band (LRM-G); (**b**) the switching algorithm using LRM-G and LRM-R (SM);

and (**c**) the cluster-based regression model (CBR). Performance is quantified using the coefficient of determination (r^2), the root mean square error (RMSE), the bias (Bias), and the mean relative absolute difference (MRAD). N is the number of satellite images; MeV is the median value. The MeV value between parentheses is calculated using the 33 best satellite scenes.

3.4. *Validation of the SDB Uncertainty Model*

Validation of the multi-scene SDB uncertainty model was carried out only for the CBR approach that showed the best performance for regression models and allowed exploiting more satellite images and generating bathymetry maps over a larger depth range. After applying the 1σ-filter, the matchup dataset (DS), associated with the six satellite images acquired in October 2017, was composed of 3188 sounding points. The random point extraction procedures generated a calibration dataset (DS-C) and a validation dataset (DS-V) composed of 2136 and 1052 points, respectively (Figure 8a). The percentage of observations associated with an SDB error lower than the total vertical uncertainty, $P(\varepsilon_i^{DS-V} < TVU_i)$, was computed for each of the 100 random samplings (Figure 8b). The absolute frequency of $P(\varepsilon_i^{DS-V} < TVU_i)$ displayed a normal distribution with a mean value of 95.8% and a standard deviation of 0.5%. The minimum and maximum values were 94.6% and 97.2%, respectively. 90% of the $P(\varepsilon_i^{DS-V} < TVU_i)$ values were comprised between 95.1% and 96.7%. Only five samplings showed a value lower than 95%, indicating that the risk of failure of our multi-scene SDB uncertainty model was less than 5%.

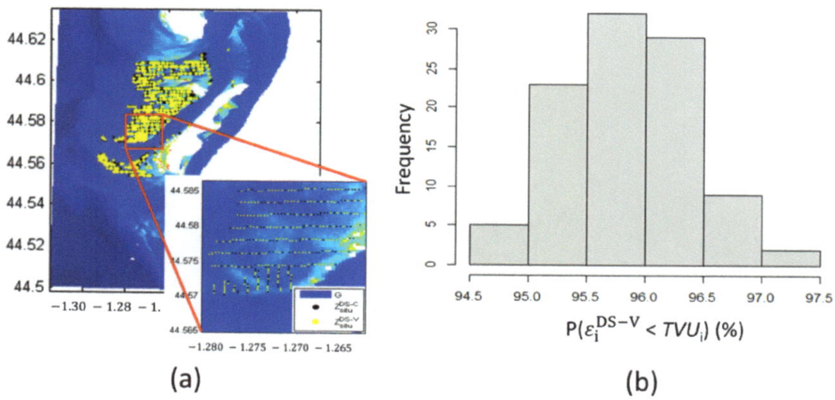

Figure 8. (**a**) Field bathymetry data point location of the matchup dataset DS associated with the 6 satellite images acquired in October 2017. Example of a random sampling where black points are associated with the calibration dataset (Z_{situ}^{DS-C}) and yellow points are associated with the validation dataset (Z_{situ}^{DS-V}). (**b**) Frequency distribution of the percentage of DS-V points showing an SDB error (ε_i^{DS-V}) lower than the total vertical uncertainty (TVU_i) generated from a set of 100 random samplings.

4. Discussion

4.1. *Impact of the Multi-Scene Approach on Uncertainty*

Three periods of 1 month each were identified to derive SDB and uncertainty maps from the multi-scene approach. The first period in October 2017 was covered by six satellite images. The second and third periods were centered on the months of September 2019 and September 2020, including, respectively, five and four satellite images of high quality. Field bathymetry datasets associated with these two last periods were composed of 2148 and 2802 sounding points, respectively. In comparison to the 2017 bathymetric survey, the spatial distributions of the bathymetric data points in 2019 and 2020 were smaller (Figure 9a,c,e). In 2020, field bathymetry data were mainly located on the spit platform, while the data collected in 2020, were gathered along the southern part of the inlet. Consequently, the bathymetry acquired in 2017, 2019, and 2020 showed different frequency distributions with median values of 3.8 m, 4.3 m, and 3.3 m, respectively.

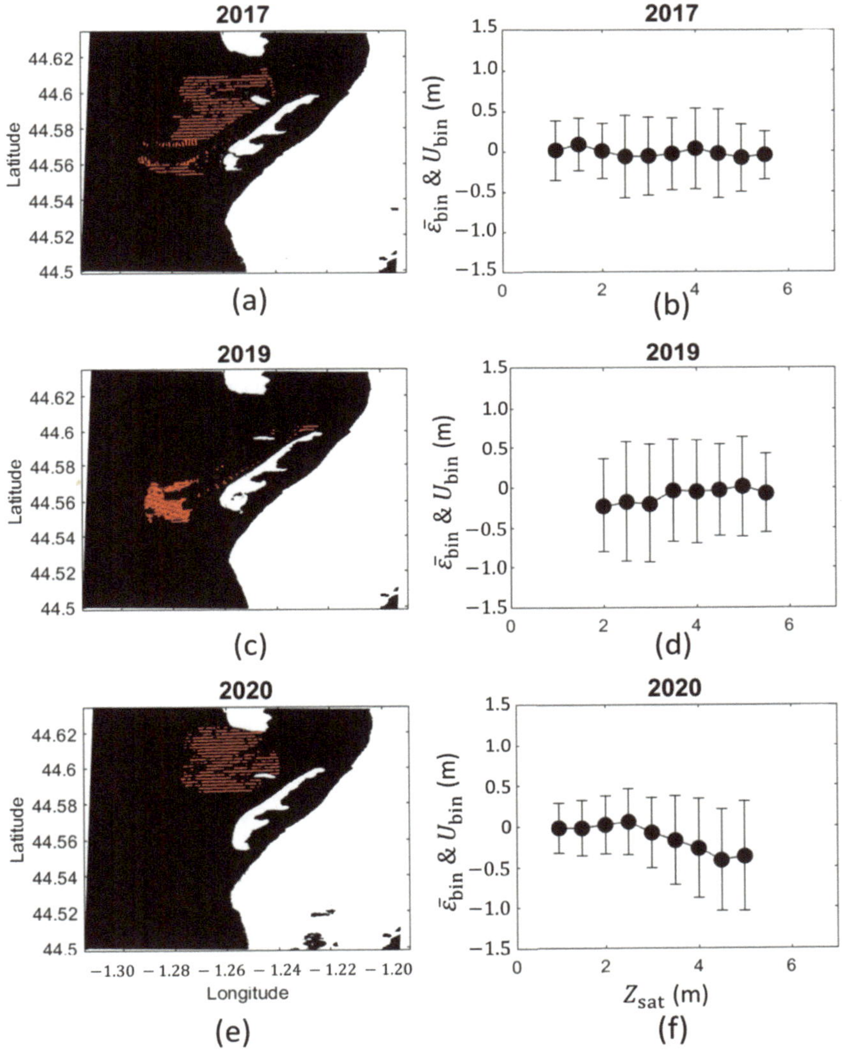

Figure 9. On the left panels, location of the field bathymetry data points is included within the matchup dataset; on the right panels, the 95% uncertainty (U_{bin}) and the bias ($\bar{\varepsilon}_{bin}$) computed per 0.5 m bathymetry intervals (bins) in (**a**,**b**) 2017, (**c**,**d**) 2019, and (**e**,**f**) 2020.

The bias ($\bar{\varepsilon}_{bin}$) and the 95% uncertainty associated with CBR (U_{bin}) were computed from DS for each 0.5 m bins (Figure 9b,d,f). $\bar{\varepsilon}_{bin}$ showed well-distributed values around zero, indicating the good performance of linear regressions. The average values of $\bar{\varepsilon}_{bin}$ for 2017, 2019, and 2020, were −0.02 m, −0.03 m, and −0.07 m, respectively. It can be noted that in 2020, $\bar{\varepsilon}_{bin}$ showed a significant deviation from zero for bathymetry higher than 3.5 m. This deviation was due to the low number of observations for these bathymetry ranges. The average values of U_{bin} computed from $\overline{Z}_{sat}$ were 0.67 m, 0.94 m, and 0.67 m in 2017, 2019, and 2020, respectively. When U_{bin} were computed from Z_{satj}, with j ranging from 1 to N, the general mean values of U_{bin} were 1.0 m, 1.54 m, and 0.96 m, respectively. These results highlighted that the multi-scene approach allowed us to significantly reduce the 95% uncertainty associated with the SDB regression models.

4.2. Impact of the Spatial Distribution of Sounding Point on Uncertainty

SDB and uncertainty maps, generated from the multi-scene approach, for the three 1-month time periods are presented in Figure 10. Most of the study site area was covered by the SDB models (Figure 10a–c), except for optically deep regions (ocean and tidal channels). The *TVU* map of 2017 shows values lower or equal to 1 m, except for the northern section of the inlet inner part where more turbid seawaters could generate higher uncertainties (Figure 10d). In 2019 and 2020, areas with uncertainties higher than 1 m (Figure 10e,f) showed larger cumulative surfaces than the ones in 2017. The mean values of *TVU* for each bin of Z_{sat} were displayed for the three years (Figure 10g–i). The average values were 0.89, 1.21, and 1.41 m, respectively. In 2017, all *TVU* averaged values were lower than 1.1 m, with a minimum observed for bathymetric values ranging from 1 to 1.5 m and a maximum value for the ones ranging from 4 to 4.5 m. In 2019, the maximum value of *TVU* was observed for bathymetric values comprised between 2 and 2.5 m, while the minimum value was observed for deeper waters. In 2020, the maximum *TVU* values were observed for deeper waters. These results may directly be associated with the specific spatial distribution of sounding points in 2019 and 2020, reducing the quality of the representativeness of the calibration dataset.

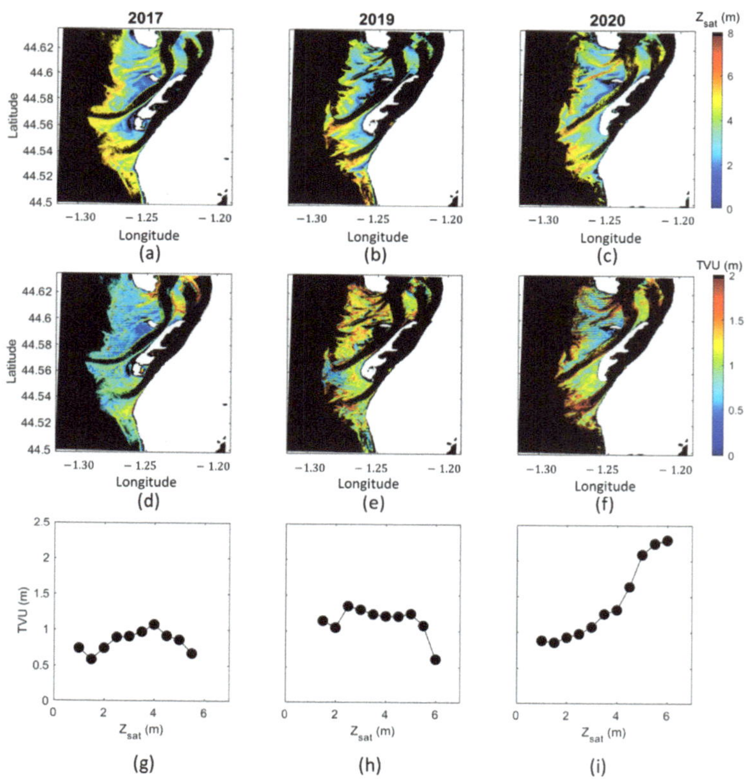

Figure 10. (**a–c**) SDB maps, (**d–f**) total vertical uncertainty (*TVU*) maps, and (**g–i**) *TVU* averaged per bins for 2017 (6 images), 2019 (5 images), and 2020 (4 images), computed from a multi-scene approach.

To better illustrate the impact of the spatial distribution of sounding points on the uncertainty model performance, three examples of the contrasted spatial distribution of sounding points for the 2017 dataset are discussed (Figure 11a). The first case (case 1) was associated with a spatially homogeneous point extraction from DS. One point out of three was extracted in order to generate DS-V. This selection ensured that DS-C was optically representative of DS-V. The second (case 2) and third (case 3) cases were associated with a geographic segmentation of DS-C and DS-V. For case 2, DS-C points were located on the northern part of the study area, while DS-C points were located on the southern part of the study area for case 3. These configurations may potentially affect the representativeness of the optical variability associated with DS-C, as the largest SDB errors, ε_{bin}, in DS were located in the northeast section of the study area. This specific distribution of ε_{bin} was due to a strong east-west optical gradient controlled by the tidal current direction.

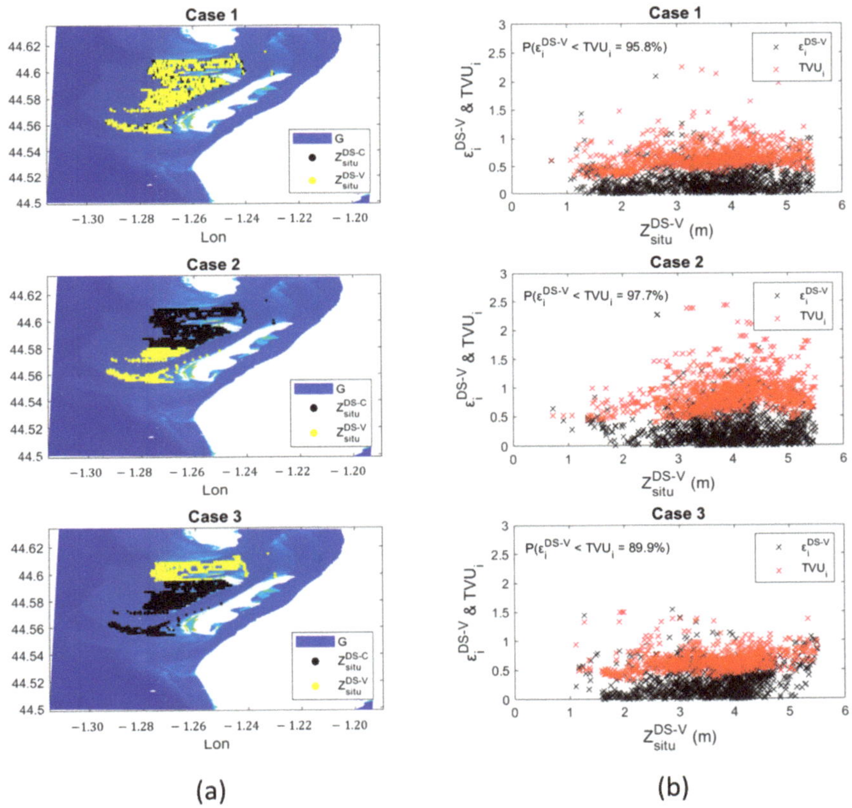

Figure 11. (**a**) Field bathymetry data point location associated with three contrasted spatial distribution of DS-C and DS-V (case 1: homogeneous distribution; case 2 and case 3: geographic segmentation) for the 2017 dataset. (**b**) Comparison between values of absolute error ε_i^{DS-V}(m), and total vertical uncertainty, TVU_i (m), computed for each observation of DS-V for case 1, case 2 and case 3.

ε_i^{DS-V} and TVU_i values were computed for each observation of DS-V for the three different cases (Figure 11b). Case 1 showed a $P(\varepsilon_i^{DS-V} < TVU_i)$ value of 95.8% was equal to the mean value of $P(\varepsilon_i^{DS-V} < TVU_i)$ computed over the 100 random samplings. For case 2, $P(\varepsilon_i^{DS-V} < TVU_i)$ showed a value of 97.7%, significantly higher than the maximum previously observed for the 100 random samplings. This case was a favorable configuration of the spatial distribution of sounding points. The risk of ε_i^{DS-V} being greater than TVU_i

was less than 5%. This result was due to lower optical variability in DS-V than in DS-C, which leads to overestimating U_{bin} and finally, TVU_i. Case 3 was the opposite configuration. The high optical variability associated with DS-V was not taken into account in DS-C and led to significantly underestimated U_{bin}. In this unfavorable scenario, 10.1% of observations had a $\varepsilon_\text{i}^{\text{DS-V}}$ value greater than TVU_i. However, this result should not be interpreted as a limitation of the uncertainty model but rather as a limitation of the classification approach selected in the CBR approach. Observations in DS-V associated with the strongest errors were classified among the classes that contained observations of DS-C despite a low spectral similarity. For future research, an alternative to the k-means clustering algorithm would be a fuzzy-c means clustering approach, which quantifies the performance of the pixels classification in relation to their optical properties [50,51]. Another alternative would be a supervised approach based, for example, on a random forests classifier [16]. This technique provides an accuracy map for the classification, which contains the indicator of the confidence degree of the classification.

4.3. Morphodynamics Application

SDB maps offer a unique opportunity to study morphological changes of large-scale tidal inlets, such as the Arcachon inlet, including the migration of sandbars, shoals, and tidal ebb and flood deltas. These maps can be used in semi-quantitative way by analyzing the longitudinal and latitudinal changes of specific isobaths. An example of the comparison between two isobaths (3 m and 4 m deep) extracted from SDB maps in 2017 and 2019 (Figure 12a–c) shows a southward migration of the ebb tidal delta and the formation of a large sandbar on the adjacent coast (−1.27°, 44.53°). The analysis of these morphological changes analyzed from SDB maps spanning years to decades supports the comprehensive understanding of the multi-annual dynamics of tidal inlet and the adjacent coast over relatively large spatial scales [52]. The supplementary use of uncertainty maps offers the opportunity to consider the analysis of vertical and volumetric morphological changes observed along the tidal inlet. The comparison of the 2017 and 2019 SDB maps (Figure 12d) showed that absolute vertical changes ranged from 0 to 4 m, while the sum of the uncertainties of 2017 and 2019 ranged from 0 to 3 m (Figure 12e). Despite high TVU values in 2019 due to low representativeness of the calibration dataset, several regions showed lower uncertainty values than bathymetric changes (Figure 12f). This result allows these changes to be assigned a high degree of confidence. For other regions, uncertainties were in the same order of magnitude as the vertical changes. However, they were meant to be significantly reduced in the future with the increase in the number and quality of the satellite images, along with enhancements of the bathymetry survey strategies and SDB algorithms.

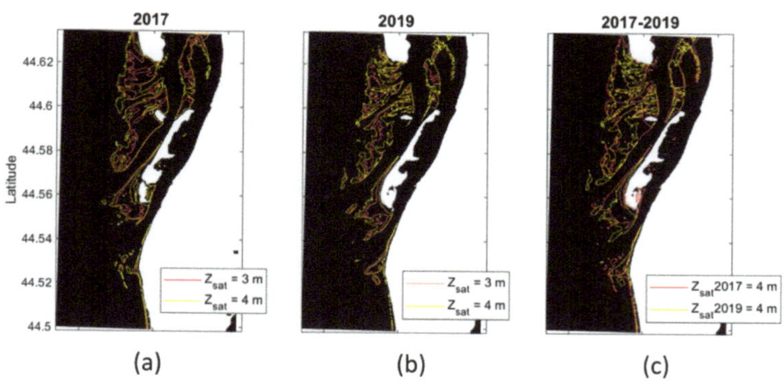

Figure 12. *Cont.*

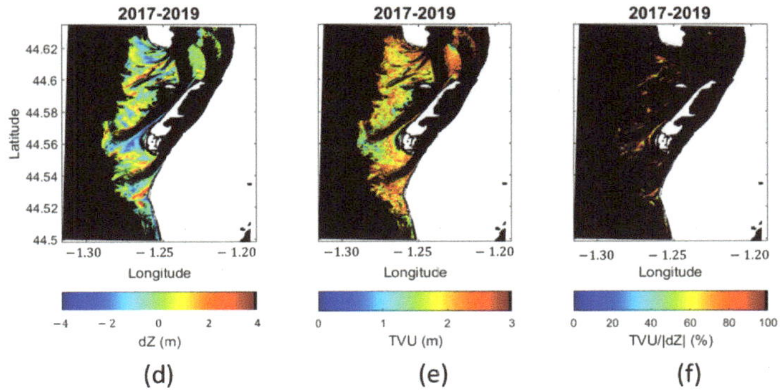

Figure 12. Maps of the 3-meter and 4-meter isobaths in (**a**) 2017 and (**b**) 2019. (**c**) comparison of the position of the 4-meter isobaths of 2017 and 2019. (**d**) Maps of the bathymetric vertical changes (*dZ*) between 2017 and 2019, (**e**) total vertical uncertainty (*TVU*), and (**f**) *TVU* to *dZ* absolute value ratio expressed in percentage.

5. Conclusions

In this study, the objectives are to evaluate the performance and robustness of well-established SDB empirical models and to propose and validate an uncertainty model for a mixed energy coastal environment. Uncertainty estimation provides new fundamental information for assessing the interpretability of SBDs and new perspectives for the use of these types of maps for the analysis of large-scale coastal systems.

The Arcachon lagoon inlet is characterized by a high spatial and temporal variability of water column optical properties mainly controlled by hydrodynamics and seasonal conditions. The CBR approach is the most appropriate approach to exploit the greatest number of images over the largest depth range and to reduce the natural optical variability of class-specific regression models, providing better performance and robustness than the other methods.

The total vertical uncertainty (*TVU*) associated with SDB empirical models depends on both the uncertainty generated by the regression model and the uncertainty due to the extrapolation of the SDB model beyond the calibration domain. The multi-scene approach developed in this study allows us to significantly reduce the uncertainty related to the regression models. The uncertainty associated with out-of-scope optical conditions is sensitive to the sampling of optical variability of the study area and consequently to the spatial distribution of sounding points. Furthermore, this component of *TVU* is impacted by the number of satellite images used for the multi-scene approach and by the associated hydrodynamic conditions. In the context of multisource bathymetric data merging approaches, these results demonstrate the need to adapt the sampling strategy of sounding points of bathymetric surveys to the optical conditions of the water column in order to generate a calibration dataset representative of the optical variability of the study site.

Author Contributions: Conceptualization, B.L. and O.B.; data curation, B.L. and A.N.L.; formal analysis, B.L. and O.B.; funding acquisition, A.N.L. and N.S.; investigation, B.L. and O.B.; methodology, B.L. and O.B.; project administration, A.N.L. and N.S.; supervision, B.L.; validation, B.L., O.B., A.N.L. and N.S.; writing—original draft, B.L. and O.B.; writing—review and editing, B.L., O.B., A.N.L. and N.S. All authors have read and agreed to the published version of the manuscript.

Funding: This research is part of the ARCADE project and was co-funded by the Region Nouvelle-Aquitaine, the Parc Naturel Marin du Bassin d'Arcachon, the Agence de l'Eau Adour-Garonne, the Syndicat Intercommunal du Bassin d'Arcachon, the Bureau de Recherches Géologiques et Minières and the Institut Français de Recherche pour l'Exploitation de la Mer.

Institutional Review Board Statement: Not applicable.

Informed Consent Statement: Not applicable.

Data Availability Statement: Landsat-8 Collection 1 Level 1 products were downloaded freely on https://earthexplorer.usgs.gov accessed on 16 March 2022. Sentinel-2 Level 1C products associated with the T30TXQ tile were downloaded freely from https://scihub.copernicus.eu/ accessed on 16 March 2022. Bathymetry data can be ordered from the Direction Départementale des Territoires et de la Mer 33 and the Syndicat Intercommunal du Bassin d'Arcachon (France).

Acknowledgments: The authors would like to thank the Direction Départementale des Territoires et de la Mer 33 and the Syndicat Intercommunal du Bassin d'Arcachon for kindly providing the bathymetric datasets. This research also significantly benefited from the high-quality of open-access satellite imagery provided by the European Space Agency (Sentinel-2 images) and the United States Geological Survey (Landsat-7 and Landsat-8 images).

Conflicts of Interest: The authors declare no conflict of interest. The funders had no role in the design of the study; in the collection, analyses, or interpretation of data; in the writing of the manuscript, or in the decision to publish the results.

References

1. Newton, A.; Weichselgartner, J. Hotspots of coastal vulnerability: A DPSIR analysis to find societal pathways and responses. *Estuar. Coast. Shelf Sci.* **2014**, *140*, 123–133. [CrossRef]
2. Nicholls, R.J.; Wong, P.P.; Burkett, V.R.; Woodroffe, C.D.; Hay, J.E. Climate change and coastal vulnerability assessment: Scenarios for integrated assessment. *Sustain. Sci.* **2008**, *3*, 89–102. [CrossRef]
3. Ranasinghe, R. Assessing climate change impacts on open sandy coasts: A review. *Earth-Sci. Rev.* **2016**, *160*, 320–332. [CrossRef]
4. Benveniste, J.; Cazenave, A.; Vignudelli, S.; Fenoglio-Marc, L.; Shah, R.; Almar, R.; Andersen, O.; Birol, F.; Bonnefond, P.; Bouffard, J.; et al. Requirements for a Coastal Hazards Observing System. *Front. Mar. Sci.* **2019**, *6*, 348:1–348:24. [CrossRef]
5. Lebbe, T.B.; Rey-Valette, H.; Chaumillon, E.; Camus, G.; Almar, R.; Cazenave, A.; Claudet, J.; Rocle, N.; Meur-Férec, C.; Viard, F.; et al. Designing coastal adaptation strategies to tackle sea level rise. *Front. Mar. Sci.* **2021**, *8*, 740602:1–740602:13. [CrossRef]
6. International Hydrographic Organization. International Hydrographic Publication C-55 Status of Hydrographic Surveying and Charting Worldwide. 2022. Available online: https://iho.int/uploads/user/pubs/cb/c-55/c55.pdf (accessed on 11 March 2022).
7. Jacob, B.; Stanev, E.V. Understanding the impact of bathymetric changes in the German bight on coastal hydrodynamics: One step toward realistic morphodynamic model. *Front. Mar. Sci.* **2021**, *8*, 640214:1–640214:18. [CrossRef]
8. Honegger, D.A.; Haller, M.C.; Holman, R.A. High-resolution bathymetry estimates via X-band marine radar: 1. beaches. *Coastal Eng.* **2019**, *149*, 39–48. [CrossRef]
9. Cesbron, G.; Melet, A.; Almar, R.; Lifermann, A.; Tullot, D.; Crosnier, L. Pan-european sateliite-derived coastal bathymetry—Review, user needs and future services. *Front. Mar. Sci.* **2021**, *8*, 740830:1–740830:15. [CrossRef]
10. Holman, R.; Plant, N.; Holland, T. cBathy: A robust algorithm for estimating nearshore bathymetry. *J. Geophys. Res.* **2013**, *118*, 2595–2609. [CrossRef]
11. Capo, S.; Lubac, B.; Marieu, V.; Robinet, A.; Bru, D.; Bonneton, P. Assessment of the decadal morphodynamic evolution of a mixed energy inlet using ocean color remote sensing. *Ocean Dyn.* **2014**, *64*, 1517–1530. [CrossRef]
12. Maritorena, S.; Morel, A.; Gentili, B. Diffuse reflectance of oceanic shallow waters: Influence of water depth and bottom albedo. *Limnol. Oceanogr.* **1994**, *39*, 1689–1703. [CrossRef]
13. Lee, Z.P.; Carder, K.L.; Mobley, C.D.; Steward, R.G.; Patch, J.F. Hyperspectral remote sensing for shallow waters: II deriving bottom depths and water properties by optimization. *Appl. Opt.* **1999**, *38*, 3831–3843. [CrossRef] [PubMed]
14. Honegger, D.A.; Haller, M.C.; Holman, R.A. High-resolution bathymetry estimates via X-band marine radar: 2. Effects of currents at tidal inlets. *Coastal Eng.* **2020**, *156*, 103626–103643. [CrossRef]
15. Lyzenga, D.R. Passive remote sensing techniques for mapping water depth and bottom features. *Appl. Opt.* **1978**, *17*, 379–383. [CrossRef]
16. Wei, C.; Zhao, Q.; Lu, Y.; Fu, D. Assessment of empirical algorithms for shallow water bathymetry using multi-spectral imagery of Pearl River delta coast, china. *Remote Sens.* **2021**, *13*, 3123. [CrossRef]
17. Dekker, A.G.; Phinn, S.R.; Anstee, J.; Bissett, P.; Brando, V.E.; Casey, B.; Fearns, P.; Hedley, J.; Klonowski, W.; Lee, Z.P.; et al. Intercomparison of shallow water bathymetry, hydro-optics, and benthos mapping techniques in australian and caribbean coastal environments. *Limnol. Oceanogr. Methods* **2011**, *9*, 396–425. [CrossRef]
18. Wei, J.; Wand, M.; Lee, Z.; Briceno, H.O.; Yu, X.; Jiang, L.; Garcia, R.; Wang, J.; Luis, K. Shallow water bathymetry with multi-spectral satellite ocean color sensors: Leveraging temporal variation in image data. *Remote Sens. Environ.* **2020**, *250*, 112035–112050. [CrossRef]
19. Botha, E.; Brando, V.; Dekker, A.; Botha, E.J.; Brando, V.E.; Dekker, A.G. Effects of per-pixel variability on uncertainties in bathymetric retrievals from high-resolution satellite images. *Remote Sens.* **2016**, *8*, 459. [CrossRef]
20. Caballero, I.; Stumpf, R. Towards routine mapping of shallow bathymetry in environments with variable turbidity: Contribution of Sentinel-2A/B satellites mission. *Remote Sens.* **2020**, *12*, 451. [CrossRef]
21. Geyman, E.C.; Maloof, A.C. A simple method for extracting water depth from multispectral satellite imagery in regions of variable bottom type. *Earth Space Sci.* **2019**, *6*, 527–537. [CrossRef]

22. International Hydrographic Organization. International Hydrographic Publication C-44 Standards for Hydrographic Surveys Edition 6.0.0. 2020. Available online: https://iho.int/uploads/user/pubs/standards/s-44/S-44_Edition_6.0.0_EN.pdf (accessed on 30 September 2020).
23. IOCCG. *Uncertainties in Ocean Colour Remote Sensing*; Mélin, F., Ed.; IOCCG Report series, No. 18; International Ocean-Colour Coordinating Group: Dartmouth, NS, Canada, 2019; ISBN 978-1-896246-68-0.
24. Hayes, M.O. Barrier island morphology as a function of tidal and wave regime. In *Barrier Island*; Leatherman, S.P., Ed.; Academic Press: New York, NY, USA, 1979; pp. 1–28, ISBN 0-12-440260-7.
25. Senechal, N.; Sottolichio, A.; Bertrand, F.; Goeldner-Gianella, L.; Garlan, T. Observations of waves' impact on currents in a mixed-energy tidal inlet: Arcachon on the southern French Atlantic coast. *J. Coast. Res.* **2013**, *65*, 2053–2058. [CrossRef]
26. Castelle, B.; Bujan, S.; Ferreira, S.; Dodet, G. Foredune morphological changes and beach recovery from the extreme 2013/2014 winter at a high-energy sandy coast. *Mar. Geol.* **2017**, *385*, 41–55. [CrossRef]
27. Nicolae Lerma, A.; Bulteau, T.; Lecacheux, S.; Idier, D. Spatial variability of extreme wave height along the Atlantic and channel French coast. *Ocean Eng.* **2015**, *97*, 175–185. [CrossRef]
28. Cayocca, F. Long-term morphological modeling of a tidal inlet: The Arcachon Basin, France. *Coast. Eng.* **2001**, *42*, 115–142. [CrossRef]
29. Nahon, A.; Idier, D.; Sénéchal, N.; Mallet, C.; Mugica, J. Imprints of wave climate and mean sea level variations in the dynamics of a coastal spit over the last 250 years: Cap Ferret, SW France. *Earth Surf. Process. Landf.* **2019**, *44*, 2112–2125. [CrossRef]
30. Liénart, C.; Savoye, N.; Bozec, Y.; Breton, E.; Conan, P.; David, V.; Feunteun, E.; Grangeré, K.; Kerhervé, P.; Lebreton, B.; et al. Dynamics of particulate organic matter composition in coastal systems: A spatio-temporal study at multi-systems scale. *Prog. Oceanogr.* **2017**, *156*, 221–239. [CrossRef]
31. Glé, C.; Del Amo, Y.; Sautour, B.; Laborde, P.; Chardy, P. Variability of nutrients and phytoplankton primary production in a shallow macrotidal coastal ecosystem (Arcachon Bay, France). *Estuar. Coast. Shelf Sci.* **2008**, *76*, 642–656. [CrossRef]
32. Pedreros, R.; Howa, H.L.; Michel, D. Application of grain size trend analysis for the determination of sediment transport pathways in intertidal areas. *Mar. Geol.* **1996**, *135*, 35–49. [CrossRef]
33. Pahlevan, N.; Lee, Z.; Hu, C.; Schott, J.R. Diurnal remote sensing of coastal/oceanic waters: A radiometric analysis for geostationary coastal and air pollution events. *Appl. Opt.* **2014**, *53*, 648–665. [CrossRef]
34. Drusch, M.; del Bello, U.; Carlier, S.; Colin, O.; Fernandez, V.; Gascon, F.; Hoersch, B.; Isoal, C.; Laberinti, P.; Martimort, P.; et al. Sentinel-2: ESA's Optical High-Resolution Mission for GMES Operational Services. *Remote Sens. Environ.* **2012**, *120*, 25–36. [CrossRef]
35. Pahlevan, N.; Scott, J.R.; Franz, B.A.; Zibordi, G.; Markham, B.; Bailey, S.; Schaaf, C.B.; Ondrusek, M.; Greb, S.; Strait, C.M. Landsat 8 remote sensing reflectance (Rrs) products: Evaluations, intercomparisons, and enhancements. *Remote Sens. Environ.* **2017**, *190*, 289–301. [CrossRef]
36. Pahlevan, N.; Chittimalli, S.K.; Balasubramaniam, S.V.; Velluci, V. Sentinel-2/Landsat-8 product consistency and implications for monitoring aquatic systems. *Remote Sens. Environ.* **2019**, *220*, 19–29. [CrossRef]
37. Li, J.; Chen, B. Global revisit interval analysis of Landsat-8-9 and Sentinel-2A -2B data for terrestrial monitoring. *Sensors* **2020**, *20*, 6631. [CrossRef]
38. Bru, D.; Lubac, B.; Normandin, C.; Robinet, A.; Leconte, M.; Hagolle, O.; Martiny, N.; Jamet, C. Atmospheric correction of multi-spectral littoral images using a PHOTONS/AERONET-based regional aerosol model. *Remote Sens.* **2017**, *9*, 814. [CrossRef]
39. Vanhellemont, Q.; Ruddick, K. Atmospheric correction of metre-scale optical satellite data for inland and coastal water applications. *Remote Sens. Environ.* **2018**, *216*, 586–597. [CrossRef]
40. Vanhellemont, Q. Sensitivity analysis of the dark spectrum fitting atmospheric correction for metre- and decametre-scale satellite imagery using autonomous hyperspectral radiometry. *Opt. Express* **2020**, *28*, 29948–29965. [CrossRef]
41. Caballero, I.; Stumpf, R.P. Retrieval of nearshore bathymetry from Sentinel-2A and 2B satellites in South Florida coastal waters. *Estuar. Coast. Shelf Sci.* **2019**, *12*, 451–474. [CrossRef]
42. Vanhellemont, Q.; Ruddick, K. Atmospheric correction of Sentinel-3/OLCI data for mapping of suspended particulate matter and chlorophyll-a concentration in Belgian turbid coastal waters. *Remote Sens. Environ.* **2021**, *256*, 112284. [CrossRef]
43. Lyzenga, D.R. Remote sensing of bottom reflectance and water attenuation parameters in shallow water using aircraft and Landsat data. *Int. J. Remote Sens.* **1981**, *2*, 71–82. [CrossRef]
44. Stumpf, R.P.; Holderied, K.; Sinclair, M. Determination of water depth with high-resolution satellite imagery over variable bottom types. *Limnol. Oceanogr.* **2003**, *48*, 547–556. [CrossRef]
45. Lubac, B.; Loisel, H. Variability and classification of remote sensing reflectance spectra in the eastern English Channel and southern North Sea. *Remote Sens. Environ.* **2007**, *110*, 45–58. [CrossRef]
46. Vantrepotte, V.; Loisel, H.; Dessailly, D.; Mériaux, X. Optical classification of contrasted coastal waters. *Remote Sens. Environ.* **2012**, *123*, 306–323. [CrossRef]
47. Knutti, R.; Hadorn, G.H.; Baumberger, D. Uncertainty Quantification Using Multiple Models—Prospects and Challenges. In *Computer Simulation Validation: Fundamental Concepts, Methodological Frameworks, and Philosophical Perspectives*; Beisbart, C., Saam, N.J., Eds.; Springer: Cham, Switzerland, 2019; pp. 835–855. [CrossRef]
48. Normandin, C.; Lubac, B.; Sottolichio, A.; Frappart, F.; Ygorra, B.; Marieu, V. Analysis of suspended sediment variability in a large highly turbid estuary using a 5-year-long remotely sensed data archive at high resolution. *J. Geophys. Res. Oceans* **2019**, *124*, 7661–7682. [CrossRef]

49. Novoa, S.; Doxaran, D.; Ody, A.; Vanhellemont, Q.; Lafon, V.; Lubac, B.; Gernez, P. Atmospheric corrections and multiconditional algorithm for multi-sensor remote sensing of suspended particulate matter in low-to-high turbidity levels coastal waters. *Remote Sens.* **2017**, *9*, 61. [CrossRef]
50. Moore, T.S.; Campbell, J.W.; Feng, H. A fuzzy logic classification scheme for selecting and blending satellite ocean color algorithms. *IEEE Trans. Geosci. Remote Sens.* **2001**, *39*, 1764–1776. [CrossRef]
51. Jackson, T.; Sathyendranath, S.; Mélin, F. An improved optical classification scheme for the Ocean Colour Essential Climate Variable and its applications. *Remote Sens. Environ.* **2017**, *203*, 152–161. [CrossRef]
52. Burvingt, O.; Nicolae Lerma, A.; Lubac, B.; Mallet, C.; Senechal, N. Geomorphological control of sandy beaches and dunes alongside a mixed-energy tidal inlet. *Mar. Geol.* 2022, *accepted*.

MDPI AG
Grosspeteranlage 5
4052 Basel
Switzerland
Tel.: +41 61 683 77 34

MDPI Books Editorial Office
E-mail: books@mdpi.com
www.mdpi.com/books

www.ingramcontent.com/pod-product-compliance
Lightning Source LLC
Chambersburg PA
CBHW041429210926
51850CB00013B/3385

* 9 7 8 3 7 2 5 8 1 4 7 7 0 *